W9-BNA-611

INTERMEDIATE ALGEBRA
Applications, Graphs, and Models

Fifth Edition

M. A. Munem
C. West
Macomb College

KENDALL/HUNT PUBLISHING COMPANY
4050 Westmark Drive   Dubuque, Iowa 52002

Cover photo "View of Statuary Hall" provided by Architect of the Capitol.

◆ Contents

Preface   vii

Review of Basic Concepts

R1 Notation and the Order of Operations   2

R2 Properties of Real Numbers   6

R3 Operations with Integers   12

R4 Operations with Rational Numbers   17

R5 Properties of Positive Exponents   23

R6 Formulas   27

1 Polynomials and Rational Expressions

1.1 Addition and Subtraction of Polynomials   38

1.2 Multiplication of Polynomials   42

1.3 Factoring Polynomials   47

1.4 Rational Expressions   54

1.5 Multiplication and Division of Rational Expressions   59

1.6 Addition and Subtraction of Rational Expressions   65

1.7 Complex Rational Expressions   70

1.8 Graphing Techniques   77

2 Linear Equations and Inequalities

2.1 Linear Equations in One Variable   86

2.2 Literal Equations and Formulas   92

2.3 Applied Problems   99

2.4 Equations Involving Rational Expressions   110

2.5 Linear Inequalities   117

2.6 Equations and Inequalities Involving Absolute Value   126

3 Linear Graphs and Functions

3.1 Graphing Linear Equations   136

3.2 Slope of a Line   142

3.3 Equations of Lines   155

3.4 Graph Linear Inequalities   164

3.5 Introduction to Functions   171

4 Systems of Equations and Inequalities

4.1 Graphical Solutions of Linear Systems in Two Variables   186

4.2 Algebraic Solutions of Linear Systems in Two Variables   193

4.3 Linear Systems in More than Two Variables   200

4.4 Systems as Models   206

4.5 Systems of Linear Inequalities   216

5 Exponents, Radicals, and Complex Numbers

5.1 Exponential Expressions and Equations   226

5.2 Radicals   233

5.3 Rational Exponents   241

5.4 Simplifying Radical Expressions   247

5.5 Operations with Radicals   252

5.6 Equations Involving Radicals and Rational Exponents   259

5.7 Complex Numbers   264

6 Quadratic Equations and Functions

6.1 The Factoring Method   276

6.2 Completing the Square   282

6.3 The Quadratic Formula   290

6.4 Equations Reducible to Quadratic Form   296

6.5 Quadratic Applications and Models   303

6.6 Nonlinear Inequalities   311

6.7 Quadratic Functions   318

6.8 Horizontal Parabolas and Circles   329

7 Exponential and Logarithmic Functions

7.1 Exponential Functions   340

7.2 Logarithmic Functions   348

7.3 Evaluating Logarithms   354

7.4 Graphs of Logarithmic Functions   360

7.5 Properties of Logarithms   368

7.6 Exponential and Logarithmic Equations   373

8 Topics in Algebra

8.1 Solutions of Linear Systems Using Matrices   386

8.2 Solutions of Linear Systems Using Determinants
     and Cramer's Rule   394

8.3 Linear Programming   403

8.4 Conic Sections: The Ellipse and Hyperbola   409

8.5 Systems of Nonlinear Equations and Inequalities   419

9 Sequences, Series and the Binomial Theorem

9.1 Sequences   430

9.2 Series and Summation   435

9.3 Arithmetic Sequences and Series   440

9.4 Geometric Sequences and Series   446

9.5 The Binomial Theorem   453

Answers to Odd Numbered Problems   A-1
Index   I-1

 **Preface**

This is a major revision of the previous edition. The overriding goal of this edition is to focus on *conceptual understanding* and *applicability* of intermediate algebra. In pursuit of that goal, we have taken into account market surveys and worked closely with a talented team of reviewers, users of previous editions, and colleagues who class tested the manuscript to ensure that the book is sufficiently *flexible*. The content of this edition represents our vision of how intermediate algebra can be taught. It is flexible enough that topics can easily be added or deleted, or the order changed.

Guiding Principles

In keeping with modern reform, this edition introduces topics from four perspectives, *geometric through graphs*, *numeric through data analysis* based on real-world situations, *analytic*, and *verbal*. We have also included many examples and problems that promote *visual thinking*. The text is filled with signals to the reader. Important terms are first introduced intuitively and identified in *italics*, then shown in **boldface** when they are defined and used in theorems. Labels in the margins highlight notes to students and help them avoid common errors. Tables appear throughout that summarize the procedures and supporting examples are discussed in the book. Special attention has been paid to point out the connection between the algebraic and geometric interpretations of important concepts.

Organization

In this edition, we continue to incorporate the best aspects of reform in a meaningful yet easy-to-use manner. Here are some of the variety of organizational changes.

1. **Review Material**: The review material from beginning algebra has been placed separately in Chapter R in the beginning of the book.
2. **Functions**: Functions are introduced in Chapter 3, and they are used throughout the book. Increased emphasis has been placed on the use of functions that describe interesting real-world situations.
3. **Systems of Equations**: In this edition, systems of linear equations are introduced in Chapter 4 to provide students with problem-solving tools. This will give the student the opportunity to translate problem situations into systems of linear equations. It also provides a useful reinforcement to translating problems into linear equations in one variable discussed in Chapter 2.
4. **Graphing**: Graphing is introduced in Chapter 1 of this edition and is integrated throughout the book. Sketching graphs by hand and equation recognition are fundamental parts of learning mathematics at this level. The text uses graphing tools to supplement and extend this process. *Interpreting graphs*, *exploring new ideas from graphs*, and *extracting information from graphs* are of primary importance in this edition.

5. **Applications and Models**: One of the goals of this edition is to link the student's own experience to real-world situations. For this reason, throughout the book, we have placed emphasis on applied problems drawn from a variety of disciplines. We also believe that we offer an approach to mathematical modeling that will capture the student's attention. Based on our classroom experiences, we have selected certain types of problems and organized their solutions through a sequence of questions that help students through the difficult thought processes of developing mathematical models.

6. **Revised Examples**: This edition makes algebraic skills attainable by providing numerous illustrative examples that are presented one step at a time. All examples, whether they are drill or applied, are revised and many new examples are added. They have a title so that students can see the purpose of each example. Each section includes solving applied problems from various disciplines, thus showing students the current application of these ideas. We focus on adding examples at the end of each section that are technology oriented.

7. **Revised Problem Sets**: All problem sets have been reorganized and rewritten. They have been checked and rechecked for accuracy by professors who class-tested this edition of the book. The exercises in each problem set have been grouped into three categories:

   (a) The first category is designed to provide the practice needed to *master the concepts* in the book. They are generally modeled after the examples and follow the order of presentation of the book.

   (b) The second category provides a broad range of applications and models that require students to *apply the concepts*. Special emphasis has been put on the design of these problems.

   (c) The third category challenges students to *develop and extend the concepts*. These problems encourage *critical thinking*, and provide an excellent opportunity to stimulate class discussions that foster *collaborative learning* among students.

   The problems in the last two categories offer the opportunity for writing assignments such as a report or a class project. Also new to this edition are many modeling data problems that ask students to find and interpret mathematical models from the real-life data.

8. **Use of Technology**: We use the generic term *grapher* to refer to *graphing calculators* and *computer software*. We provide examples and problems at the end of each section where technology explorations are used. This will enable students to visualize, discover, and explore the model from graphical and numerical viewpoints. We recommend the use of a grapher—not thoughtless key stroking—and stress that its use is not a substitute for understanding the concepts involved. Quite the contrary, we believe that one usually needs to understand the concepts well in order to verify and apply the results of the technology. Although the material in this book can be covered without the use of a grapher, it is likely that many students and instructors will want to make use of these devices. To assist these students and instructors, grapher's activities are clearly identified by the symbol **G** and can be omitted without the loss of continuity, if desired. Instructors and students are expected to use their own judgment to determine what specific calculator or software to use as well as where technology is useful.

9. **New Design:** The new design is more open and readable. Each section begins with a list of specific objectives. These objectives are restated in the margin at their point of use, to provide organizational breakdown of the section. The use of color is intended to highlight the important parts of graphs and important statements. All art has been redrawn for this edition.

10. **Review Problems and Chapter Tests**: As with previous editions, each chapter concludes with a review problem set. Also, a chapter test follows the comprehensive collection of chapter review problems. Chapter tests focus on the problems in the chapter so that students can determine if they are ready to take an actual class test.

Supplements

The fifth edition is accompanied by the following supplements package.
- **For the Instructor**: An instructor's test manual includes six different test forms for each chapter of the textbook. These tests have been written at graded levels of difficulty, with three of the tests for each chapter made up of multiple choice questions, and three of standard problem-solving questions. Answers to all test questions are provided.
- **For the Student**: A student solutions guide offers worked-out solutions to every other odd-numbered problem in the book. Chapter objectives are also included.

Acknowledgments

In preparing this edition, we have drawn from our own experience in teaching intermediate algebra as well as the feedback provided by our friends, colleagues, and students. We offer them all our grateful thanks. At the risk of omitting some, they include the following:

William Coppage, Wright State University
Linda Exley, DeKalb Community College—Clarkston
Merle Friel, Humboldt State University
Gene Hall, DeKalb Community College—Clarkston
Gus Pekara, Oklahoma City Community College
Stuart Thomas, Oregon State University
Roger Willig, Montgomery County Community College
Jim Wolfe, Portland Community College

Special thanks are due to our many colleagues at Macomb Community College who have taught from previous editions. We especially wish to thank the following colleagues who class-tested the manuscript of this edition and checked the solutions to the problems. They are Victoria Ackerman, Steve Fasbinder, and Marion Rappa.

M. A. Munem
C. West

Chapter R

REVIEW OF BASIC CONCEPTS

R1. Notation and the Order of Operations
R2. Properties of Real Numbers
R3. Operations with Integers
R4. Operations with Rational Numbers
R5. Properties of Positive Exponents
R6. Formulas

In our daily lives, we often have to evaluate algebraic expressions or formulas by substituting numbers for the letters involved. Suppose that, a spherical balloon is being inflated with helium. Its volume V is given by the formula

$$V = \frac{4}{3}\pi r^3$$

where r is the radius of the balloon. Find the volume (in cubic feet) of the balloon at the instant when its radius is 60 inches. Example 3 on page 29 solves this problem.

The study of Intermediate Algebra depends on successful completion of a Beginning Algebra course. This review material is intended to highlight particular topics that are important, and to establish a common language. Review is the key word in this introduction.

OBJECTIVES

1. Introduce Notation
2. Use Exponents
3. Use Order of Operations

R1 Notation and the Order of Operations

Introducing Notation

Algebra begins with a systematic study of the addition, subtraction, multiplication, and division operations which serves as a basis for all arithmetic calculations. In order to achieve generality, letters of the alphabet such as a, b, c, x and y are used to represent numbers. A letter that represents an arbitrary number is called a **variable.** The four arithmetic operations are described in symbols and words in Table 1, where a and b represent any two numbers.

TABLE 1

Operation	Operation in Symbols	Operations in Words
Addition	$a + b$	sum of a and b
Subtraction	$a - b$	difference of a and b
Multiplication	$ab, a \cdot b, a(b), (a)b$	product of a and b
Division	$a/b, a \div b, \dfrac{a}{b}, b \neq 0$	quotient of a and b

We can use the relationship between division and multiplication to clarify division involving zero. Suppose we try to divide 5 by 0. Let us say the answer is a number represented by the letter c. That is,

$$c = 5 \div 0$$

this means that $5 = c \cdot 0$. This is impossible, since $c \cdot 0 = 0$. In algebra, we say that $\frac{5}{0}$ is undefined. However, $0 \div 5 = 0$ since $0 = 0 \cdot 5$.
    In general,

Note that here $\dfrac{0}{0}$ is undefined.

$$\frac{a}{0} \text{ where } a \neq 0 \text{ is undefined, and } \frac{0}{b} = 0.$$

EXAMPLE 1   Translating Words into Symbols and Vice Versa

(a) Write the following statement in symbols.
                The sum of 4 and twice y is greater than 5.
(b) Express $4 + 3y > 10$ in words.

Solution   (a) $4 + 2y > 5$

(b) The sum of 4 and the product of 3 and y is greater than 10.

Table 2 shows six additional symbols that are used for comparing numbers and expressions.

TABLE 2

Comparison in Symbols	Interpretation in Words
$a = b$	a is equal to b
$a \neq b$	a is not equal to b
$a \leq b$	a is less than or equal to b
$a < b$	a is less than b
$a \geq b$	a is greater than or equal to b
$a > b$	a is greater than b
$a > 0$	a is positive
$a < 0$	a is negative
$a \geq 0$	a is nonnegative

Using Exponents

Algebraic notation is designed to clarify ideas and simplify calculations by allowing us to write expressions compactly and efficiently. For instance, the product $3 \cdot 3$ can be written as 3^2, read "3 squared." The product $3 \cdot 3 \cdot 3 \cdot 3$ can be written as 3^4, read "3 to the fourth power." The expression 3^4 is called an **exponential form**, whereas $3 \cdot 3 \cdot 3 \cdot 3$ is called an **expanded** or **factored form** of 3^4 and each number is called a **factor** of 3^4. Since $3 \cdot 3 \cdot 3 \cdot 3 = 81$, we say that the **value** of 3^4 is 81. The use of exponents provides a compact notation for products. Thus,

$$x \cdot x = x^2, \qquad x \cdot x \cdot x = x^3, \qquad \text{and} \qquad x \cdot x \cdot x \cdot x = x^4.$$

In general, if n is a positive integer, then

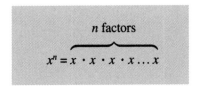

In using the **exponential notation** x^n, we refer to x as the **base** and n as the **exponent** or **power**. It should be noted that $x^1 = x$.

EXAMPLE 2   Using Exponential Notation
(a) Find the value of the expression 5^4.
(b) Write the expression $a \cdot b \cdot b \cdot b \cdot b \cdot c \cdot c$ in exponential notation.

Solution   (a) $5^4 = 5 \cdot 5 \cdot 5 \cdot 5 \cdot = 625$
(b) $a \cdot b \cdot b \cdot b \cdot b \cdot c \cdot c = ab^4c^2$

Using the Order of Operations

Consider the problem of evaluating the expression

$$2 + 3 \cdot 5$$

If we add 2 and 3 first, then multiply by 5, we get 25. If we take the product of 3 and 5 first, then add 2, we get 17. In order to avoid having two different values for the same expression, the following steps describe the order in which the operations should be performed.

The Order of Operations

1. **Grouping Symbols**   Perform operations inside grouping symbols. Grouping symbols include **parentheses ( )**, **brackets [ ]**, **braces { }**, and **fraction bars** —. Work from the innermost grouping symbols to the outermost.
2. **Powers**   Find the value of any powers indicated by exponents.
3. **Multiply (Divide)**   Perform all multiplications and divisions from left to right.
4. **Addition (Subtraction)**   Perform all additions and subtractions from left to right.

A fraction bar is a grouping symbol which groups the numerator *and the* denominator.

EXAMPLE 3   Using the Order of Operations

Find the value of each expression.

(a) $3 \cdot 2^3 + 10/5 - 3^2$

(b) $5[4^3 + 3(6^2 - 3 \cdot 2)]$

(c) $\dfrac{5^2 + 2 \cdot 5}{3 + 10 \div 5}$

Solution

(a)
$$\begin{aligned}
& 3 \cdot 2^3 + 10/5 - 3^2 && \text{Given} \\
&= 3 \cdot 8 + 10/5 - 9 && \text{Powers} \\
&= 24 + 2 - 9 = 17 && \text{Multiplication, division, addition, subtraction}
\end{aligned}$$

(b)
$$\begin{aligned}
& 5[4^3 + 3(6^2 - 3 \cdot 2)] && \text{Given} \\
&= 5[64 + 3(36 - 6)] && \text{Powers} \\
&= 5[64 + 3(30)] && \text{Innermost parentheses} \\
&= 5[64 + 90] && \text{Multiplication inside brackets} \\
&= 5[154] = 770 && \text{Add and multiply}
\end{aligned}$$

(c)
$$\begin{aligned}
& \frac{5^2 + 2 \cdot 5}{3 + 10 \div 5} && \text{Use of a fraction bar} \\
&= (25 + 10) \div (3 + 2) && \text{Simplify numerator and denominator separately} \\
&= 35 \div 5 = 7 && \text{Divide}
\end{aligned}$$

An expression formed from any combination of numbers and variables by using the four operations in Table 1 as well as raising to powers or taking roots is called an **algebraic expression**. If an algebraic expression consists of parts connected by plus or minus signs, each of the parts together with the sign preceding it is called a **term**. For example, $4x^2 + 3x - \frac{2}{x}$ is an algebraic expression with the following terms:

$$4x^2, \quad 3x, \quad \text{and} \quad -\frac{2}{x}$$

EXAMPLE 4   Evaluating an Expression for Specific Values

Evaluate $\dfrac{5x + y^2}{3}$ for $x = 12$ and $y = -3$.

Solution   Using parentheses to substitute 12 for x and -3 for y in the expression yields

$$\dfrac{5x + y^2}{3} = \dfrac{5(12) + (-3)^2}{3}$$

$= (60 + 9) / 3$

$= 69/3 = 23$ Fraction bar as the grouping symbol

EXAMPLE 5   Translating a Situation into Symbols

The cruise control of a car is set at y miles per hour (mi/hr), then the brakes are applied to drop the speed by 30 miles per hour. To pass a truck, the speed is doubled.
(a) Write an algebraic expression that describes the speed when passing the truck.
(b) Find the speed while passing the truck if the cruise control is initially set at:
 (i) 55 miles per hour
 (ii) 65 miles per hour.

Solution   (a) y The initial speed is y miles per hour.

 $y - 30$ The speed is dropped by 30 miles per hour.

 $2(y - 30)$ The speed is doubled while passing the truck.

(b) (i) If the initial speed is 55 mi/hr, then the speed passing the truck is
 $2(y - 30) = 2(55 - 30) = 2(25) = 50$ mi/hr.

(ii) If the initial speed is 65 mi/hr, then the speed passing the truck is
 $2(y - 30) = 2(65 - 30) = 2(35) = 70$ mi/hr.

 ## PROBLEM SET R1

In Problems 1–4, write each statement in symbols.

1. (a) The sum of 4 and 5 equals 9.
 (b) The sum of 5 and 4 equals 9.
 (c) The sum of 4 and 5 equals the sum of 5 and 4.
2. (a) The difference of 12 and 2 equals 10.
 (b) The difference of 2 and 12 does not equal 10.
 (c) The difference of 12 and 2 does not equal the difference of 2 and 12.
3. The product of x and 5 is greater than the quotient of x and 5.
4. The difference of twice x and 7 is less than or equal to 21.

In Problems 5–8, express each algebraic statement in words.

5. (a) $5x = 20$ 6. (a) $\dfrac{x}{3} = 7$

 (b) $5x - 4 = 20$ (b) $\dfrac{x}{3} + 2 = 7$

7. (a) $x + 2 < 10$ 8. (a) $3x \leq 15$
 (b) $10 > x + 2$ (b) $15 \geq 3x$

In Problems 9 and 10, write each expression in expanded form and find its value.

9. (a) 6^2 (b) 3^6 10. (a) 11^2 (b) 2^{11}

In Problems 11 and 12, write each expression in exponential form.

11. (a) $2 \cdot 2 \cdot x \cdot y \cdot y \cdot y$
 (b) $2 \cdot 2 \cdot 3 \cdot 3 \cdot y \cdot y \cdot z$
12. (a) $3 \cdot 3 \cdot 3 \cdot u \cdot u \cdot u$
 (b) $3 \cdot 3 \cdot 3 \cdot 4 \cdot u \cdot v \cdot v$

In Problems 13–20, apply the order of operations to find the value of each expression.

13. (a) $6 + 12 - 2 - 8$ 14. (a) $2 \cdot 3/3 \cdot 10$
 (b) $6 + (12 - (2 + 8))$ (b) $2 \cdot (3/3) \cdot 10$
15. (a) $3(2 + 3(7 - 4))$ 16. (a) $8(3 + (20 - 9)) + 6$
 (b) $3 \cdot 2 + 9(7 - 4)$ (b) $24 + 160 - 72 + 6$
17. $3^4 - 2^4$ 18. $(3 - 2)^4 + 7^3$
19. $(9 + 6)^2/5$ 20. $12 + 3 \cdot 4^2 - 5$

In Problems 21 and 22, find the value of each expression.

21. (a) $\dfrac{24}{6} + 2$ 22. (a) $\dfrac{10}{5} + 5$

 (b) $\dfrac{24}{6 + 2}$ (b) $\dfrac{10}{5 + 5}$

In Problems 23–28, evaluate each expression by substituting the given values for the variables.

23. $7 + 7x + 7y$ for $x = 2, y = 3$ 24. $\dfrac{x + y}{x - y}$ for $x = 5, y = 3$

25. (a) $10(x - y) - 4(x - y)$ for $x = 20, y = 10$
 (b) $6(x - y)$ for $x = 20, y = 10$

26. (a) $\dfrac{4x + 4y}{4x}$ for $x = 10, y = 10$

 (b) $1 + \dfrac{y}{x}$ for $x = 10, y = 10$

27. (a) $(x - y)^2$ for $x = 5, y = 2$
 (b) $x^2 - 2xy + y^2$ for $x = 5, y = 2$
28. (a) $(x + 1)^3$ for $x = 4$ (b) $x^3 + 3x^2 + 3x + 1$ for $x = 4$

In Problems 29–32, write each sentence in symbols.

29. (a) The product of 5 and the sum of 3 and 4 equals 35.
    (b) The sum of the product of 5 and 3 and the product of 5 and 4 equals 35.
30. (a) The product of 3 and the difference of 6 and 4 equals 6.
    (b) The difference of the product of 3 and 6 and the product of 3 and 4 equals 6.
31. (a) One number is 10 times 20 more than another.
    (b) One number is 20 more than 10 times another.
32. (a) One number is the square of the sum of two other numbers.
    (b) One number is the sum of the squares of two other numbers.

In Problems 33 and 34, express each algebraic statement in words.

33. (a) $3y - 1 \neq 5$
    (b) $3(y - 1) = 5$

34. (a) $\dfrac{y}{4} + 3 = 7$
    (b) $\dfrac{y + 3}{4} \neq 7$

In Problems 35–38, insert parentheses so that each statement is true.

35. (a) $4 + 5 \cdot 2 + 3 = 21$
    (b) $4 + 5 \cdot 2 + 3 = 45$
    (c) $4 + 5 \cdot 2 + 3 = 29$
36. (a) $12/3 + 1 + 2 = 7$
    (b) $12/3 + 1 + 2 = 5$
    (c) $12/3 + 1 + 2 = 2$
37. (a) $4 + 3 \cdot 8 - 1 + 6 = 31$
    (b) $4 + 3 \cdot 8 - 1 + 6 = 33$
    (c) $4 + 3 \cdot 8 - 1 + 6 = 7$
38. (a) $4 \cdot 6 + 1 - 5 + 4 \cdot 3 = 1$
    (b) $4 \cdot 6 + 1 - 5 + 4 \cdot 3 = 35$
    (c) $4 \cdot 6 + 1 - 5 + 4 \cdot 3 = 20$

In Problems 39–44, find the value of each expression, if possible. (Hint: Recall that division by zero is undefined.)

39. (a) $\dfrac{7(4 - 4)}{3(5 + 3)}$
    (b) $\dfrac{3(5 + 3)}{7(4 - 4)}$

40. (a) $\dfrac{9 + 3 \cdot 2}{24 - 8 \cdot 3}$
    (b) $\dfrac{24 - 8 \cdot 3}{9 + 3 \cdot 2}$

41. $\dfrac{25^2 + 18 \div 2 \cdot 3 + 8 \cdot 5}{2 \cdot 2}$

42. $\dfrac{3^2(3 + 2) + 3(3 + 2)^2}{5 \cdot 2 - 2(3 - 1)}$

43. $\dfrac{(5 + 7)^2 - (3 \cdot 4)^2}{5^2 + 7^2 - 3 \cdot 4^2}$

44. $\dfrac{5^2 + 7^2 - 3 \cdot 4^2}{(5 + 7)^2 - (3 \cdot 4)^2}$

In Problems 45 and 46, write the calculator key strokes necessary to find the value of each expression on your calculator.

45. (a) $3 + 4/5$
    (b) $(3 + 4)/5$
    (c) $4 + \dfrac{5(3 - 2)}{10}$

46. (a) $\dfrac{1}{3 + 4}$
    (b) $\dfrac{1}{3} + \dfrac{1}{4}$
    (c) $(4 + 17)(18 - 3)$

47. **House Value:** A house that originally costs $95,000 doubled in value over several years. Due to a recession, it then decreased in value by y. Write an expression for the value of the house after the recession. Find its value if $y = $5000 and $y = $7000.

48. **Sunset:** A week ago the sun set at 6:55 PM. Each day for the next five days the sun sets x minutes earlier. Write an expression for the time the sun sets on the sixth day. At what time would it set if $x = 4$ minutes?

49. **Depreciation:** A computer for an office costs w and will be used for five years at which time it will have a salvage value of $300. If the company uses straight line depreciation (the difference of cost and salvage value divided by useful life), write an expression for this depreciation amount. If the computer costs $4100, what is the depreciation amount?

OBJECTIVES

1. Introduce Sets
2. Create a Number Line
3. Categorize Numbers in Sets
4. Introduce the Properties of Real Numbers

R2 Properties of Real Numbers

In this section, we review sets of numbers and their properties.

Introducing the Notation of Sets

A **set** may be thought of as a collection of objects or numbers. An object in a set is called an **element** of the set. Capital letters, such as A, B, C, and D, are often used as labels for sets. Braces, { }, are used to enclose the elements of a set which are separated by commas. Thus, we write

$$A = \{1, 2, 3, 4, 5\}$$

to represent the set A whose elements are 1, 2, 3, 4, 5. The symbol $\in$ is used to mean that an element belongs to a set. The notation

$$3 \in A$$

is read "3 is an element of the set A". The symbol $\notin$ indicates that an element does not belong to a set.

The set $\{\varnothing\}$ does not represent the empty set since this set contains one element, the symbol $\varnothing$.

If a set contains no elements, we call it the **empty set** or the **null set** and is denoted by $\{\ \}$ or the symbol $\varnothing$. For example, the set of all numbers that are both odd and even is the empty set.

The set $A = \{1, 3, 5, 7, 11, 13\}$ contains a finite number of elements and is considered a *finite set*. The set of odd counting numbers

$$B = \{1, 3, 5, 7, \ldots\}$$

contains an infinite number of elements and is considered an *infinite set*. The three dots (ellipsis) indicate that the list of elements continues without ending.

Consider the sets $A = \{1, 2, 3\}$ and $B = \{1, 2, 3, 5\}$. Every element of set A is also an element of set B. We say, "A is a **subset** of B" and write

$$A \subseteq B$$

For example, the subsets of the set $\{1, 2, 3\}$ are $\varnothing$, $\{1\}$, $\{2\}$, $\{3\}$, $\{1, 2\}$, $\{1, 3\}$, $\{2, 3\}$ and $\{1, 2, 3\}$.

Sometimes it is easier to describe the elements of a set rather than listing them. In such cases, we may use a new notation called **set-builder notation** that includes conditions which define the set. Using this notation, we describe the set A that consists of all real numbers greater than 3 as

$$A = \{x \mid x > 3\}$$

which is read "the set of all real numbers x such that x is greater than 3".

Creating a Number Line

Sets of numbers and relations among such sets can be visualized by the use of a *number line* or *coordinate axis*. A **number line** is constructed by drawing a horizontal line, fixing a point called the **origin** and associating that point with the number 0. We choose an arbitrary *unit length* and mark off ticks to the right and left of the origin. The unit distance between consecutive tick marks is called the **scale**. The *positive numbers* extend to the right of the origin and *negative numbers* to the left (Figure 1).

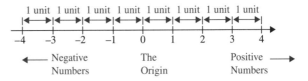

Figure 1

The point associated with a number on a number line is called its **graph**, and the number is called the **coordinate** of that point. For example, in Figure 2 the coordinates of points A, B, and C on the number line are the -3, 2, and 5, respectively.

Figure 2

In this textbook, we sometimes identify the number line by placing x at the end of the arrow. Thus, the coordinates of points A, B and C may also be referred to as $x = -3$, $x = 2$, $x = 5$ respectively.

Two numbers on a number line that are the same distance from the origin 0, but on opposite sides of 0 are called **negatives** or **opposites** of each other. For example, Figure 3 shows that the negative (or opposite) of 4 is −4, also the negative (or opposite) of −4 is written as −(−4) or 4. This example illustrates the fact that if a is a number, then

$$-(-a) = a.$$

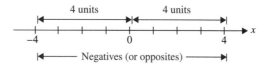

Figure 3

Categorizing Numbers in Sets

In algebra we encounter many different sets of numbers.

1. The set of **natural numbers**, N, is also called the **counting numbers** or **positive integers**,

$$N = \{1, 2, 3, 4, \ldots\}$$

This set is illustrated in Figure 4.

Figure 4

2. The set of **integers**, I, consists of the counting numbers, their negatives, and zero. Figure 5 shows the graph of the integers on a number line.

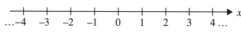

Figure 5

3. The **rational numbers**, Q, consist of all numbers that can be written in the form a/b, in which a and b are integers and $b \neq 0$. Note that because the denominator of the rational number a/b may equal 1, then every integer is considered a rational number. Some examples of rational numbers are

$$\frac{1}{2}, \quad \frac{4}{3}, \quad \frac{20}{1} = 20, \quad \frac{0}{3} = 0, \quad \text{and} \quad \frac{-7}{2}$$

In set-builder notation, the rational numbers are

$$Q = \left\{ \frac{a}{b} \,\middle|\, a, b \in I \ \text{and} \ b \neq 0 \right\}$$

Be careful when using a calculator to change a fraction to its equivalent decimal form. The repeating pattern may not be visible if it is more than 5 or 6 digits.

Every rational number can also be written in decimal form. To express a rational number as a decimal, divide the numerator by the denominator. When a rational number is written in decimal form, it can be shown that the decimal either terminates or is a repeating decimal. For example, the decimal forms of

$$\frac{2}{5} = 2 \div 5 = 0.4 \quad \text{and} \quad \frac{3}{4} = 3 \div 4 = 0.75$$

terminate, whereas the decimal forms of

$$\frac{2}{3} = 0.6666\ldots \quad \text{and} \quad \frac{1}{7} = 0.142857142857\ldots$$

are repeating and nonterminating. These decimals can be written with an over bar to indicate that portion of the decimals which repeats. Thus,

$$\frac{2}{3} = 0.6666\ldots = 0.\overline{6} \quad \text{and} \quad \frac{1}{7} = 0.142857142857\ldots = 0.\overline{142857}$$

EXAMPLE 1   Expressing Fractions as Decimals

Express each rational number as a decimal.

(a) $\dfrac{3}{8}$ (b) $\dfrac{-10}{3}$

Solution   (a) $\dfrac{3}{8} = 3 \div 8 = 0.375$

(b) $\dfrac{-10}{3} = -(10 \div 3) = -3.333\ldots = -3.\overline{3}$ ◈

Every terminating decimal can be changed to a fraction where the denominator is a power of 10. For example,

$$0.7 = \frac{7}{10^1} = \frac{7}{10} \quad \text{and} \quad 1.025 = \frac{1025}{10^3} = \frac{1025}{1000}.$$

We will see later in this book, that every decimal with a repeating pattern can also be written as a fraction.

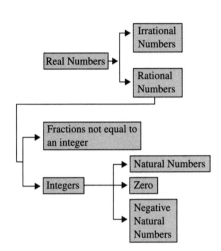

Real Numbers → Irrational Numbers, Rational Numbers

Fractions not equal to an integer

Integers → Natural Numbers, Zero, Negative Natural Numbers

Figure 6

4. The **irrational numbers**, J, have decimal expansions that neither terminate nor repeat digits. Examples of irrational numbers with their decimal forms are:

$$\sqrt{2} = 1.4142135\ldots \quad \text{and} \quad \pi = 3.1415927\ldots$$

To emphasize that a numerical value is only an approximation, we often use the symbol $\approx$ and write

$$\sqrt{2} \approx 1.41 \text{ and } \pi \approx 3.14.$$

Most calculators have a round off command. Check your manual to locate this command.

5. The **real numbers**, $\mathfrak{R}$, is the set of rational and irrational numbers. Thus, the real numbers are numbers which can be written as terminating or repeating decimals (rational numbers), or nonrepeating, nonterminating, decimals (irrational numbers). Figure 6 shows the relationship among these sets.

Introducing the Properties of Real Numbers

Now we consider basic properties of real numbers that serve as a foundation for algebra. The letters a, b, and c, in Table 1 represent any real number.

TABLE 1

Property	Symbolic Form	Example
Commutative Property for Addition	$a + b = b + a$	$3 + 4 = 4 + 3$
Commutative Property for Multiplication	$a \cdot b = b \cdot a$	$3 \cdot 4 = 4 \cdot 3$
Associative Property for Addition	$a + (b + c) = (a + b) + c$	$7 + (3 + 9) = (7 + 3) + 9$
Associative Property for Multiplication	$a \cdot (b \cdot c) = (a \cdot b) \cdot c$	$3 \cdot (5 \cdot 2) = (3 \cdot 5) \cdot 2$
Distributive Property	$a \cdot (b + c) = a \cdot b + a \cdot c$	$6 \cdot (5 + 7) = 6 \cdot 5 + 6 \cdot 7$
Identity for addition is 0	$a + 0 = 0 + a = a$	$7 + 0 = 0 + 7 = 7$
Identity for multiplication is 1	$a \cdot 1 = 1 \cdot a = a$	$8 \cdot 1 = 1 \cdot 8 = 8$
Additive inverse of a is $-a$	$a + (-a) = 0$	$4 + (-4) = 0$
Multiplicative inverse of a is 1/a	$a \cdot \dfrac{1}{a} = 1, a \neq 0$	$5 \cdot \dfrac{1}{5} = 1$

EXAMPLE 2   Applying Properties of Real Numbers

State the property that justifies each statement.
(a) $5 \cdot (1 \cdot 2) = (5 \cdot 1) \cdot 2$ (b) $5 \cdot (1 + 2) = (1 + 2) \cdot 5$
(c) $5 \cdot (1 + 2) = 5 \cdot 1 + 5 \cdot 2$ (d) $5 \cdot (1 + 2) = 5 \cdot (2 + 1)$

Solution   (a) This is an example of the associative property for multiplication.
(b) This is an example of the commutative property for multiplication.
(c) This is an example of the distributive property.
(d) This is an example of the commutative property for addition.

It is possible for the additive inverse to be positive. If $a = -5$, then $-a = -(-5) = 5$.

Many calculators have inverse keys. The additive inverse key is . It changes the sign of the displayed number. The multiplicative inverse key is 1/x or x^{-1}. It changes the displayed number to its multiplicative inverse by taking its reciprocal.

We now look at relationships on the set of real numbers. Table 2 lists the basic properties of equality.

TABLE 2

Property	Description
Reflexive	$a = a$
Symmetric	If $a = b$, then $b = a$
Transitive	If $a = b$ and $b = c$, then $a = c$
Substitution	If $a = b$, then a can be substituted for b in any statement involving b without affecting the truth of the statement.

EXAMPLE 3   Identifying Properties of Equality

State the property of equality that justifies each statement.
(a) If $\sqrt{9} = 3$, then $3 = \sqrt{9}$
(b) If $x + 1 = 4$ and $x = a$, then $a + 1 = 4$
(c) If $a = x$ and $x = 4$, then $a = 4$

Solution   (a) Symmetric property
(b) Substitution property
(c) Transitive property or substitution property

PROBLEM SET R2

In Problems 1 and 2, list all the possible subsets of each set.

1. (a) $\{0, 1\}$ (b) $\{a, b, c\}$
2. (a) $\{1/2\}$ (b) $\{1/2, 1.3/2, 2\}$

In Problems 3 and 4, use set-builder notation to describe each set.

3. (a) A is the set of all numbers less than or equal to zero.
   (b) B is the set of all negative numbers.
   (c) C is the set of all nonnegative numbers.

4. (a) A is the set of all numbers greater than zero.
   (b) B is the set of all positive numbers.
   (c) C is the set of all nonpositive numbers.

In Problems 5 and 6, describe each set using set notation.

5. (a) A is the set of integers within 5 units of the origin on a real number line.
   (b) B is the set of natural numbers within one unit of 10 on a number line.

(c) *C* is the set of all real numbers, except 5.

(d) *D* is the set of integers, except 5.

6. (a) *K* is the set of even integers.

(b) *L* is the set of odd integers.

(c) *M* is the set of numbers greater than 4 but not greater than 7.

(d) *N* is the set of all numbers between 0 and 1, excluding 0, but including 1.

In Problems 7 and 8, represent the elements of each set on a number line.

7. (a) $A = \left\{ \dfrac{22}{7}, \pi, \dfrac{223}{71} \right\}$

(b) $B = \{1.4141, \sqrt{2}, 1.4142\}$

8. (a) $C = \{-\sqrt{3}, -\sqrt{4}, -\sqrt{5}\}$

(b) $D = \left\{ \dfrac{1}{3}, 0.3, 0.\overline{3} \right\}$

In Problems 9–12, express each rational number as a decimal. If it repeats, use a bar over the repeating digits.

9. (a) $\dfrac{9}{10}$ 10. (a) $\dfrac{1}{3}$ 11. (a) $\dfrac{2}{9}$ 12. (a) $\dfrac{1}{11}$

(b) $\dfrac{9}{100}$ (b) $\dfrac{2}{3}$ (b) $\dfrac{5}{9}$ (b) $\dfrac{3}{11}$

(c) $\dfrac{9}{1000}$ (c) $\dfrac{4}{3}$ (c) $\dfrac{7}{9}$ (c) $\dfrac{12}{11}$

In Problems 13–16, identify all the subsets of the real numbers to which each of the following numbers belongs.

13. (a) -51 (b) 51

14. (a) $\dfrac{2}{10}$ (b) $\dfrac{10}{2}$

15. (a) $\sqrt{120}$ (b) $\sqrt{121}$

16. (a) $3.03030303\ldots$ (b) $3.030030003\ldots$

In Problems 17–20, use a calculator and round each number as indicated.

17. To four decimal places:

(a) π (b) $\dfrac{22}{7}$ (c) $\dfrac{512}{163}$

18. To five decimal places:

(a) $(\sqrt{2})^2$ (b) $(1.414)^2$ (c) $(1.4142)^2$

19. To four decimal places:

(a) $(1.732)^2$ (b) $(\sqrt{3})^2$ (c) $(1.733)^2$

20. To five decimal places:

(a) $(3.1622)^2$ (b) $(\sqrt{10})^2$ (c) $(3.1623)^2$

In Problems 21 and 22, find the negative (or opposite) of *a*, then graph *a* and its negative on a number line.

21. (a) $a = \dfrac{3}{5}$ 22. (a) $a = -0.7$

(b) $a = -\dfrac{2}{3}$ (b) $a = \dfrac{2}{3}$

In Problems 23–26, state the property or properties that justify each statement.

23. (a) $3(5 + 1) = 3(5) + 3(1)$

(b) $3(5 + 1) = (5 + 1)3$

(c) $3(5 + 1) = 3(1 + 5)$

24. (a) $2 + 3 \cdot 5 = 3 \cdot 5 + 2$

(b) $(2 + 3) \cdot 5 = 2 \cdot 5 + 3 \cdot 5$

(c) $2 + 3 \cdot 5 = 5 \cdot 3 + 2$

25. (a) $4 + (3 + 7) = (3 + 7) + 4$

(b) $4 + (3 + 7) = (4 + 3) + 7$

(c) $4 + (3 + 7) = (7 + 3) + 4$

26. (a) $7 \cdot (8 \cdot 9) = (7 \cdot 8) \cdot 9$

(b) $7 \cdot (8 \cdot 9) = 7 \cdot (9 \cdot 8)$

(c) $7 \cdot (8 \cdot 9) = (8 \cdot 9) \cdot 7$

In Problems 27–30, state the property of equality that justifies each statement.

27. If $4 = 2x - 1$, then $2x - 1 = 4$

28. If $4 = 2x - 1$ and $2x - 1 = y$, then $4 = y$

29. If $2x + 1 = 3$ and $x - 1 = 3$, then $2x + 1 = x - 1$.

30. If $x = 3$, then $2x + 1 = 2(3) + 1$

In Problems 31 and 32, plot each of the following on a number line:

31. $\dfrac{1}{2}, 0, -\sqrt{4}, \sqrt{2}, -\sqrt{10}$

32. $\dfrac{1}{4}, \dfrac{3}{8}, -\sqrt{8}, \pi, -\sqrt{15}$

In Problems 33 and 34, rewrite each expression using only the stated property.

33. (a) $(19 + 82) \cdot 7 = $ _____, distributive property

(b) $(19 + 82) \cdot 7 = $ _____, commutative property for addition

(c) $(19 + 82) \cdot 7 = $ _____, commutative property for multiplication

34. (a) $3x + 4x = $ _____, distributive property

(b) $5 \cdot \dfrac{1}{8} + 3 \cdot \dfrac{1}{8} = $ _____, distributive property

(c) $2\sqrt{7} + 5\sqrt{7} = $ _____, distributive property

In Problems 35 and 36, list the elements of each set that are (a) natural numbers *N*, (b) integers *I*, (c) rational numbers *Q*, (d) irrational numbers *J*, and (e) real numbers $\mathscr{R}$.

35. $\left\{ -\sqrt{9}, -\sqrt{5}, 0, \dfrac{25}{5}, \dfrac{25}{4} \right\}$

36. $\left\{ -\dfrac{100}{5}, -\dfrac{100}{12}, 0, 4.\overline{25}, 4.76776777677776\ldots \right\}$

37. **Tipping:** Calculate a 15% tip on a restaurant bill of $24.

(a) Multiply $(0.15)(24.00)$.

(b) Add a 10% tip to half of a 10% tip.

(c) Identify *a*, *b*, and *c* in $(a + b)c = ac + bc$ to help explain why the distributive property guarantees that a 15% tip equals a 10% tip plus a 5% tip.

$$(0.15)(24.00) = (0.10)(24) + (0.05)(24)$$

38. Give examples to show that the commutative and associative properties do not hold for subtraction.

39. Give examples to show that the commutative and associative properties do not hold for division.

40. Determine the number of subsets that exist for a set containing the following number of elements:

(a) 0   (b) 1   (c) 2   (d) 3   (e) 4

41. Which of the following indicates the number of subsets of a set containing *n* elements, 2^n or $2n$? (Hint: See Problem 40.)

42. (a) Explain why the associative property of multiplication allows us to write a triple product such as $12 \cdot 33 \cdot 42$ without parentheses.
    (b) Explain why the associative property of addition allows us to write a triple sum such as $12 + 30 + 45$ without parentheses.

In Problems 43–46, state the property that justifies each step.

43. $-2(-3) + 2(-3) = (-2 + 2)(-3)$ (a) _____
               $= 0(-3)$ (b) _____
               $= 0$ Definition of $0(-3)$

44. $5x + 7y + 15x = 5x + 15x + 7y$ (a) _____
               $= (5 + 15)x + 7y$ (b) _____
               $= 20x + 7y$ Definition of $5 + 15$

45. $(x + y) - y = x + (y - y)$ (a) _____
               $= x + (1 \cdot y - 1 \cdot y)$ (b) _____
               $= x + (1 - 1)y$ (c) _____
               $= x + 0 \cdot y$ (d) _____
               $= x + 0$ Definition of $0 \cdot y$
               $= x$ (e) _____

46. $x + (-x) = x(1) + x(-1)$ (a) _____
               $= x[1 + (-1)]$ (b) _____
               $= x \cdot 0$ (c) _____
               $= 0$ Definition of $x \cdot 0$

OBJECTIVES

1. Introduce Absolute Value
2. Add and Subtract Integers
3. Multiply and Divide Integers

R3 Operations with Integers

In this section we extend the rules for addition, subtraction, multiplication, and division of positive numbers to all integers.

Introducing Absolute Value

Although we can use the number line to add integers, it is often convenient to give a precise description of the addition process. For this purpose we consider the concept of absolute value. Figure 1 shows that points P and Q are 5 units from the origin.

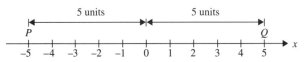

Figure 1

Geometrically, the *absolute value* of a number is the distance between the number and zero on the number line, regardless of direction. Since -5 and 5 each is 5 units from the origin, 0, they have the same absolute value, which we denote by two vertical bars. That is, $|-5|$ and $|5|$. The statements

$$|-5| = 5 \quad \text{and} \quad |5| = 5$$

are read as "the absolute value of -5 is 5" and "the absolute value of 5 is 5", respectively. In general, we define the absolute value of any real number x as follows:

DEFINITION OF ABSOLUTE VALUE

If x is a number, then $|x|$ is the **absolute value** of x, where

$$|x| = \begin{cases} x, & \text{if } x \geq 0, \quad x \text{ is positive or zero} \\ -x, & \text{if } x < 0, \quad x \text{ is negative} \end{cases}$$

For example according to this definition,

$$|3| = 3, \qquad |0| = 0, \qquad \text{and} \qquad |-2| = -(-2) = 2$$

EXAMPLE 1   Evaluating Absolute Value

Find the value of each expression.
(a) $|26|$ (b) $|-26|$ (c) $-|-26|$

Solution   We use the definition of absolute value.
(a) When $x > 0$, $|x| = x$, here $26 > 0$, so $|26| = 26$.
(b) When $x < 0$, $|x| = -x$, here $-26 < 0$, so $|-26| = -(-26) = 26$
(c) From part (b) we know that $|-26| = 26$. Thus, $-|-26| = -26$ ◈

Adding and Subtracting Integers

The justification for the addition rule is easily seen by drawing arrows on the number line to indicate a change in position. An arrowhead to the right indicates a positive move and an arrowhead to the left a negative move. The length of the arrow corresponds to the size of the move. For example, the sum $2 + 3$ can be interpreted on a number line as follows: start at the origin, 0, and move 2 units to the right. Then move 3 more units to the right to find the sum 5. Thus $2 + 3 = 5$ (Figure 2).

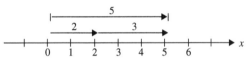

Figure 2

Similarly, to show the sum $3 + 2 = 5$, we start by moving 3 units to the right followed by moving 2 more units to the right to find the sum 5. We can also use the number line to add two negative numbers. For instance, the sum $(-2) + (-3)$ is interpreted as follows: start at the origin, move 2 units to the left, then move 3 more units to the left to find the sum -5 (Figure 3). That is $(-2) + (-3) = -5$.

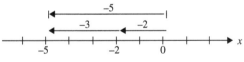

Figure 3

In general, when we add two numbers with like signs, we add their absolute values and keep their common sign.

For example, $(+2) + (+3) = +(|+2| + |+3|) = +5$ and
$$(-2) + (-3) = -(|-2| + |-3|) = -5$$

We can also use a number line to illustrate adding two numbers with different signs. We represent the sum $-3 + 8$ on a number line as follows: Start at the origin and move 3 units to the left, then move 8 units to the right to find the sum 5 (Figure 4). Thus $-3 + 8 = 5$.

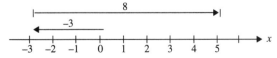

Figure 4

The following rules generalize the addition of integers:

1. If the signs of the numbers are the **same**, we add their absolute value and keep their common sign.
2. If the signs of the numbers are **different**, we find the **difference** of their absolute values and repeat the sign of the number with the largest absolute value.

EXAMPLE 2   Adding Integers

Find each sum.

(a) $(-15) + (-72)$ (b) $10 + (-230)$ (c) $-14 + 17 + 25$

Solution   (a) Since both numbers are negative, their sum is negative. That is, $(-15) + (-72) = -(|-15| + |-72|) = -87$

(b) The numbers have different signs and the negative number has the larger absolute value. That is, $10 + (-230) = -(|-230| - |10|) = -220$.

(c) Working from left to right, we add -14 and 17 first, then add 25 to obtain:
$-14 + 17 + 25 = 3 + 25 = 28$

Now we look at the operation of subtraction. The **difference,** $a - b$, is defined as the sum of a and the additive inverse of b. In symbols,

$$a - b \text{ means } a + (-b).$$

Thus, we can rewrite a difference as a sum by changing the sign of the number being subtracted; then apply the rules for addition. For example,

$$-7 - 3 = (-7) + (-3) = -10 \text{ and } 12 - (-5) = 12 + (+5) = 17$$

EXAMPLE 3   Subtracting Integers

Find each difference.

(a) $-99 - 12$ (b) $60 - (-50)$ (c) $(-37) - (-12)$

Solution   We rewrite each subtraction as an addition, and then add.

(a) $-99 - 12 = -99 + (-12) = -111$
(b) $60 - (-50) = 60 + 50 = 110$
(c) $-37 - (-12) = -37 + 12 = -25$

Multiplying and Dividing Integers

Multiplication is defined as repeated addition. For example, $4 \cdot 2$ means adding two four times; that is,

$$4 \cdot 2 = 2 + 2 + 2 + 2 = 8$$

Also, this product can be described as adding four twice; that is,

$$2 \cdot 4 = 4 + 4 = 8$$

The same approach applies to the product of a positive and a negative number. For instance, we interpret $4(-2)$ as

$$(-2) + (-2) + (-2) + (-2) = -8.$$

Figure 5 illustrates this product on a number line.

Figure 5

This example illustrates the general statement:

The product of a positive number and a negative number is negative.

Now consider the product of two negative numbers. To explore this multiplication, we will calculate $-2(-4 + 4)$ in two different ways. Adding the numbers inside the parentheses first yields

$$-2(-4 + 4) = -2(0) = 0$$

Applying the distributive property yields

$$-2(-4 + 4) = -2(-4) + (-2)4$$

We know $(-2)4 = -8$ and $-2(-4 + 4) = 0$, so by substitution we have

$$0 = -2(-4) + (-8)$$

The additive inverse of -8 is $+8$. Thus, the product -2 and -4 must be the positive number 8.

In general, we have the following rule for multiplying the numbers a and b, where a and b are positive numbers.

RULE FOR MULTIPLYING SIGNED NUMBERS

> 1. If the signs of the numbers are the **same**, the product is **positive**. That is, $a \cdot b = ab$ and $(-a)(-b) = ab$
> 2. If the signs of the numbers are **different**, the product is **negative**. That is, $a(-b) = -ab$ and $(-a)b = -ab$.

For example, $3 \cdot 5 = 15$, $(-3)(-5) = 15$ and $3(-5) = -15$.

When multiplying more than two integers, we note the following pattern:

$$(-1)(-2) = +2$$
$$(-1)(-2)(-3) = -6$$
$$(-1)(-2)(-3)(-4) = +24$$
$$(-1)(-2)(-3)(-4)(-5) = -120$$

The number of positive factors has no effect on the sign of the product.

As the pattern suggests, the product of integers is positive for an even number of negative factors and negative for an odd number of negative factors.

EXAMPLE 4   Multiplying More Than Two Integers

Determine whether each product is positive or negative. Then find each product.

(a) $3(-2)(-7)(6)(-2)$ (b) $-2(-2)(3)(-1)(-4)$

Solution   (a) Since three of the factors are negative, the product is negative. So that,
$$3(-2)(-7)(6)(-2) = -504$$
(b) Since four of the factors are negative, the product is positive. That is,
$$-2(-2)(3)(-1)(-4) = +48$$ ◈

Multiplication and division of numbers are related in the following way:

$$a \div b = c \text{ means } a = bc$$

The result of a division, c, is called a **quotient.** For example, $6 \div 2 = 3$ because $6 = 2(3)$ and 3 is called the quotient.

Two numbers are called the **multiplicative inverse** or **reciprocals** of each other, if their product is 1. For example 5 and $\frac{1}{5}$ are reciprocals, as are $\frac{-3}{7}$ and $\frac{-7}{3}$ because $5 \cdot \frac{1}{5} = 1$ and $\left(\frac{-3}{7}\right)\left(\frac{-7}{3}\right) = 1$. Using the concept of a multiplicative inverse, or reciprocal, division may be defined in terms of multiplication as follows:

DIVISION OF NUMBERS

If a and b are numbers, with $b \neq 0$, then
$$a \div b = a \cdot \frac{1}{b}$$

This definition says to divide a by b, multiply a by the multiplicative inverse of b. For example, $12 \div 3$ can be written as:

$$12 \div 3 = 12 \cdot \left(\frac{1}{3}\right) = 4$$

Quotients of nonzero numbers follow the same rules as multiplication. That is:
1. If two numbers have the **same signs**, their quotient is **positive**.
2. If two numbers have **different signs**, their quotient is **negative**.

EXAMPLE 5   Dividing Integers
Perform each division.
(a) $45 \div (-15)$ (b) $-45 \div 15$ (c) $-45 \div (-15)$

Solution   (a) $45 \div (-15) = 45 \cdot \frac{1}{(-15)}$ Multiply 45 by the reciprocal of -15
$= -3$

(b) $-45 \div 15 = (-45) \cdot \frac{1}{15}$ Multiply -45 by the reciprocal of 15
$= -3$

(c) $-45 \div (-15) = (-45) \cdot \frac{1}{(-15)}$ Multiply -45 by the reciprocal of -15
$= 3$

EXAMPLE 6   Applying the Order of Operations to Integers
Find the value of the expression.
$$(-3)^2 + 7(-6 + 4)/(-2)$$

Solution   The order of operations applies to all real numbers, so that
$(-3)^2 \neq -3^2$. The base of $(-3)^2$ is -3, whereas the base of -3^2 is 3.

$$(-3)^2 + 7(-6 + 4)/(-2) = (-3)^2 + 7(-2)/(-2) \quad \text{Parentheses}$$
$$= 9 + 7(-2)/(-2) \quad \text{Powers}$$
$$= 9 + (-14)/(-2) \quad \text{Multiplication}$$
$$= 9 + 7 = 16 \quad \text{Division, addition}$$

◆ PROBLEM SET R3

In Problems 1 and 2, find the value of each expression.

1. (a) $|-25|$ (b) $|25|$ (c) $-|-25|$
2. (a) $-|100|$ (b) $|-100|$ (c) $-|-100|$

In Problems 3 and 4, find each sum by using a number line.

3. (a) $-3 + 1$ (b) $-3 + 1 - 1$ (c) $-3 + 1 - 2$
4. (a) $2 + (-4) - 1$ (b) $4 + (-2) + 3$ (c) $2 + (-5) - 3$

In Problems 5 and 6, translate each phrase into symbols.

5. (a) 5 minus negative 3   (b) Negative 3 minus 1
6. (a) Negative 2 plus   (b) Negative 7 minus
       negative 4   negative 3

In Problems 7–10, find each sum.

7. (a) $15 + (-5)$ (b) $-15 + 5$
8. (a) $18 + (-18)$ (b) $7 - 12 + 5$
9. (a) $-8 + 25 + (-15)$ (b) $-8 + (-15) + 25$
10. (a) $-1.4 + 1.2 - 3$ (b) $-6.3 + (-3.6) + 2$

In Problems 11–14, find each difference.

11. (a) $-7 - 11$ (b) $-7 - (-11)$
12. (a) $-111 - (-111)$ (b) $-1212 - (1212)$
13. (a) $-503 - (-483)$ (b) $-483 - (-503)$
14. (a) $-1.6 - (-1.6)$ (b) $-4.3 - (-1.5)$

In Problems 15 and 16, find each product.

15. (a) $-100(-50)$ 16. (a) $24(18)$
     (b) $100(50)$ (b) $-24(18)$
     (c) $-100(50)$ (c) $-24(-18)$
     (d) $100(-50)$ (d) $24(-18)$

In Problems 17 and 18, determine whether each product is positive or negative. Then find the product.

17. (a) $-3(-5)(-15)$ 18. (a) $-2(7)(-10)(14)$
     (b) $-3(-5)(-15)(-20)$ (b) $-2(7)(-10)(-14)$

In Problems 19 and 20, find each quotient.

19. (a) $-22 \div (-15)$ 20. (a) $-288 \div 144$
     (b) $22 \div (-15)$ (b) $288 \div (-144)$
     (c) $-22 \div 15$ (c) $(-288) \div (-144)$

In Problems 21–30, find the value of each expression.

21. (a) $-2[9 + (-11)]$ (b) $-8[(-7) - (-3)]$
22. (a) $6[-8 - (-5)]$ (b) $7[-4 + (-3)]$
23. (a) $-5(-15 \div 3)$ (b) $-4[-35 \div (-7)]$
24. (a) $[28 - (-7)] \div (-2)$
     (b) $[-48 - (-12)] \div (-2)$
25. $-15 \div 3 + 2 \cdot (-5)$
26. $11(-16) \div (-4) - (-3)$
27. $0(-7)$
28. $0 \div (-2)$
29. $-3 \div 0$
30. $2(10) \div [(-3) + 3]$

In Problems 31–38, find the value of each expression.

31. (a) $(-5)^2$ (b) -5^2
32. (a) $(-1)^6$ (b) -1^6
33. (a) $(-4)^3$ (b) -4^3
34. (a) $\left(-\dfrac{1}{2}\right)^4$ (b) $-\left(\dfrac{1}{2}\right)^4$
35. $5(-2)^3(-3)^3 \div (-8)$
36. $5^2 \cdot (-3)2(-2)^3 \div 9$
37. $-2[(-3) - (-2)^3] + (4 - 6)$
38. $3 + [(-4) + 2]^3 \cdot (-8)$

In Problems 39–44, find the value of each expression given the indicated values of x and y.

39. For $x = -1$ and $y = -2$,
     (a) x^2y (b) xy^2
40. For $x = -3$ and $y = 5$,
     (a) $2xy$ (b) $-2xy$
41. For $x = -5$ and $y = -1$,
     (a) $(x + y)^2$ (b) $x^2 + 2xy + y^2$
42. For $x = 1$ and $y = -2$,
     (a) $(x - y)^2$ (b) $x^2 - 2xy + y^2$
43. For $x = -3$ and $y = -4$,
     (a) $\dfrac{x^2 + y^2}{x + y}$ (b) $\dfrac{x^2 - y^2}{x + y}$
44. For $x = -2$ and $y = -3$,
     (a) $\dfrac{x^3 - y^3}{x - y}$ (b) $\dfrac{x^3 + y^3}{x + y}$

45. **Temperature:** One night the thermometer outside read $18°F$ and in the morning it read $-2°F$.
     (a) Write an expression for the difference of night and morning temperatures.
     (b) Find the value of the expression.

46. **Finance:** On Monday you open a savings account with $270. You deposit $100 on Tuesday, withdraw $40 on Wednesday, and withdraw $65 on Thursday.
     (a) Write an expression that shows the activity in your account.
     (b) Find the value of the expression.

47. **Sports:** A football team starts with the ball on its own 20 yard line. In a series of three plays, the team loses 4 yards, gains 7 yards and then loses 5 yards.
     (a) Write an expression that shows the team's field position.
     (b) Find the value of the expression.

48. **Tides:** Suppose that on a typical day high tide was 6 feet and low tide was -2 feet.
     (a) Write an expression for the difference between the high and low tides.
     (b) Find the value of the expression.

OBJECTIVES

1. Reduce and Simplify Rational Numbers
2. Determine Ratios and Proportions
3. Multiply and Divide Rational Numbers
4. Add and Subtract Rational Numbers

R4 Operations with Rational Numbers

In this section we review operations with rational numbers. Recall that a rational number can be expressed as

$$\frac{a}{b}, \text{ where } a \text{ and } b \text{ are integers and } b \neq 0$$

with numerator a and denominator b. If $b = 0$, then $\frac{a}{b}$ is undefined.

Reducing and Simplifying Rational Numbers

Consider the rational number $\frac{15}{25}$. By writing the numerator and denominator of this number in factored form, and dividing out (canceling) the common factors, we have:

$$\frac{15}{25} = \frac{3 \cdot \cancel{5}}{5 \cdot \cancel{5}} = \frac{3}{5} \cdot 1 = \frac{3}{5}$$

Since the numerator and denominator of $\frac{3}{5}$ have no common factors other than 1 (or -1), we say that the rational number $\frac{3}{5}$ is **simplified** or **reduced to lowest terms**. In general, to *simplify* a rational number, or *reduce to lowest terms*, we use the following principle

FUNDAMENTAL PRINCIPLE
OF RATIONAL NUMBERS

> For any rational number $\dfrac{a}{b}$ and any number $k \neq 0$
>
> $$\frac{a \cdot k}{b \cdot k} = \frac{a}{b}$$

Thus, the fundamental principle says that multiplying or dividing the numerator and denominator of a rational number by the same nonzero number produces an equivalent rational number. Two rational numbers are said to be **equivalent** if they simplify to the same number. For example,

$$\frac{24}{28} = \frac{6 \cdot \cancel{4}}{7 \cdot \cancel{4}} = \frac{6}{7}$$

Thus, the rational numbers $\frac{24}{28}$ and $\frac{6}{7}$ are equivalent.

It should be noticed that for a rational number $\frac{4}{7}$

$$\frac{4}{7} = \frac{-4}{-7} = -\frac{-4}{7} = -\frac{4}{-7}$$

also,

$$-\frac{4}{7} = \frac{-4}{7} = \frac{4}{-7} = -\frac{-4}{-7}$$

EXAMPLE 1   Reducing Rational Numbers

Reduce each rational number to lowest terms, if possible.

(a) $\dfrac{16}{-48}$ (b) $\dfrac{-26}{-91}$ (c) $\dfrac{35}{26}$

Solution   By factoring the numerator and denominator of each rational number and dividing out common factors, we have

(a) $\dfrac{16}{-48} = \dfrac{1 \cdot 16}{-3 \cdot 16} = \dfrac{1}{-3} \dfrac{\cancel{16}}{\cancel{16}} = \dfrac{1}{-3} \cdot 1 = -\dfrac{1}{3}$

(b) $\dfrac{-26}{-91} = \dfrac{26}{91} = \dfrac{2 \cdot 13}{7 \cdot 13} = \dfrac{2}{7} \cdot \dfrac{\cancel{13}}{\cancel{13}} = \dfrac{2}{7} \cdot 1 = \dfrac{2}{7}$

(c) There are no common factors in 35 and 26, so $\frac{35}{26}$ is already simplified.   ◈

Determining Ratios and Proportions

A ratio $\frac{a}{b}$ is a quotient of two quantities a and b. The ratio $\frac{a}{b}$ can also be written as $a : b$. For example, if there are 2618 students and 154 teachers in a school, then the ratio of students to teachers is given by

$$\frac{2618}{154} = \frac{17 \cdot \cancel{154}}{1 \cdot \cancel{154}} = \frac{17}{1}$$

A very important use of ratio is *percent*. By definition, a **percent** is a ratio whose denominator is 100. *Percent means parts per 100 or parts of 100* and it is denoted by the symbol %. Rational numbers or decimals are often expressed as percents. For instance,

$$\frac{3}{100} \text{ or } 0.03 \times 100\% \text{ or } 3\%.$$

Thus, 3% of 60 is 0.03(60) = 1.8. Also, if we ask what percent of 400 is 36, then we write $\frac{36}{400} = \frac{9}{100} = 0.09 = 9\%$.

EXAMPLE 2 Using Percents
(a) What is 70% of 300?
(b) What percent of 2000 is 40?

Solution (a) $70\% \times 300 = 0.70 \times 300 = 210$
(b) $\frac{40}{2000} = 0.02 = 0.02 \times 100\% = 2\%$

As you can see, ratios are another name for rational numbers. Thus two ratios are *equivalent* if they simplify to the same number. When two ratios are equal, we call it a *proportion*. More formally, we say that:

> Two ratios $\dfrac{a}{b}$ and $\dfrac{c}{d}$ are **proportional** if $\dfrac{a}{b} = \dfrac{c}{d}$.

Notice that when the two ratios $\frac{2}{3}$ and $\frac{10}{15}$ are proportional the "cross products" are equal. That is, $\frac{2}{3} = \frac{10}{15}$ if and only if 2(15) = 3(10). In general we write:

> $\dfrac{a}{b} = \dfrac{c}{d}$ if and only if $ad = bc$.

EXAMPLE 3 Rewriting as a Ratio
Rewrite each statement as a ratio of equal values.
(a) 33 dimes to 22 nickels   (b) 12 meters to 30 centimeters

Solution (a) $\dfrac{\text{value of 33 dimes}}{\text{value of 22 nickels}} = \dfrac{33 \cdot 10}{22 \cdot 5} = \dfrac{3 \cdot \cancel{11} \cdot 2 \cdot 5}{2 \cdot \cancel{11} \cdot \cancel{5}} = \dfrac{3}{1}$

(b) $\dfrac{\text{12 meters as centimeters}}{\text{30 centimeters}} = \dfrac{12 \cdot 100}{30} = \dfrac{2 \cdot \cancel{6} \cdot \cancel{5} \cdot 20}{\cancel{5} \cdot \cancel{6}} = \dfrac{40}{1}$

Multiplying and Dividing Rational Numbers

The multiplication of two rational numbers is performed by finding the product of their numerators divided by the product of their denominators. That is, if a/b and c/d are rational numbers with $b \neq 0$ and $d \neq 0$, then

DEFINITION OF MULTIPLICATION

> $$\frac{a}{b} \cdot \frac{c}{d} = \frac{a \cdot c}{b \cdot d}$$

When two rational numbers containing common factors are to be multiplied, we may multiply the rational numbers then reduce the product to lowest terms. Another method is to factor the numerator and denominator, divide each by the common factors, then multiply.

EXAMPLE 4   Multiplying Rational Numbers

Find the product of

$$-\frac{22}{65} \cdot \frac{15}{11}$$

(a) Multiply the rational numbers, then reduce the product to lowest terms.

(b) Factor the numerator and denominator, divide by the common factors, then multiply.

Solution   (a) Multiply and then reduce

$$-\frac{22}{65} \cdot \frac{15}{11} = -\frac{22 \cdot 15}{65 \cdot 11} = -\frac{330}{715} = -\frac{6}{13} \cdot \frac{\cancel{55}}{\cancel{55}} = -\frac{6}{13}$$

(b) Factor, divide by common factors, then multiply

$$-\frac{22}{65} \cdot \frac{15}{11} = -\frac{2 \cdot \cancel{11}}{\cancel{5} \cdot 13} \cdot \frac{3 \cdot \cancel{5}}{\cancel{11}} = -\frac{6}{13} \qquad \diamondsuit$$

To divide two rational numbers, find the reciprocal of the divisor (second rational number) and multiply. If a/b and c/d are rational numbers with $b \neq 0$, $c \neq 0$ and $d \neq 0$ then

DEFINITION OF DIVISION

$$\frac{a}{b} \div \frac{c}{d} = \frac{a}{b} \cdot \frac{d}{c} = \frac{a \cdot d}{b \cdot c}$$

EXAMPLE 5   Dividing Rational Numbers

Perform each division and reduce the results to lowest terms.

(a) $\dfrac{5}{8} \div \dfrac{3}{4}$

(b) $\dfrac{39}{95} \div \dfrac{-26}{35}$

Solution   (a) $\dfrac{5}{8} \div \dfrac{3}{4} = \dfrac{5}{8} \cdot \dfrac{4}{3}$ Division as multiplication by the reciprocal

$$= \frac{5}{2 \cdot \cancel{4}} \cdot \frac{\cancel{4}}{3} \qquad \text{Factor and divide out common factor}$$

$$= \frac{5}{6}$$

(b) $\dfrac{39}{95} \div \dfrac{-26}{35} = \dfrac{39}{95} \cdot \dfrac{35}{-26} = -\dfrac{3 \cdot \cancel{13} \cdot \cancel{5} \cdot 7}{\cancel{5} \cdot 19 \cdot 2 \cdot \cancel{13}} = -\dfrac{21}{38} \qquad \diamondsuit$

Adding and Subtracting Rational Numbers

To add or subtract rational numbers having the same denominator, we add (or subtract) the numerators and repeat the denominator. In symbols:

$$\frac{a}{b} + \frac{c}{b} = \frac{a + c}{b} \quad \text{and} \quad \frac{a}{b} - \frac{c}{b} = \frac{a - c}{b}$$

EXAMPLE 6   Adding and Subtracting Rational Numbers with the Same Denominator

Combine the following rational numbers.

(a) $\dfrac{2}{7} + \dfrac{3}{7}$ (b) $\dfrac{7}{9} - \dfrac{5}{9}$ (c) $\dfrac{17}{20} + \dfrac{19}{20} - \dfrac{3}{20}$

Solution   (a) $\dfrac{2}{7} + \dfrac{3}{7} = \dfrac{2+3}{7} = \dfrac{5}{7}$

(b) $\dfrac{7}{9} - \dfrac{5}{9} = \dfrac{7-5}{9} = \dfrac{2}{9}$

(c) $\dfrac{17}{20} + \dfrac{19}{20} - \dfrac{3}{20} = \dfrac{17 + 19 - 3}{20} = \dfrac{33}{20}$ ◈

Some calculators can perform the arithmetic of rational numbers displaying answers in rational number form.

To add or subtract rational numbers having different denominators, we use *the fundamental principle of rational numbers* in reverse to change the form of each rational number so that they have a common denominator, then add (or subtract) as before.

DEFINITION OF ADDITION AND SUBTRACTION

$$\frac{a}{b} + \frac{c}{d} = \frac{ad}{bd} + \frac{cb}{db} = \frac{ad + bc}{bd} \quad \text{and}$$

$$\frac{a}{b} - \frac{c}{d} = \frac{ad}{bd} - \frac{cb}{db} = \frac{ad - bc}{bd}$$

In algebra, an improper rational number such as $\frac{29}{24}$ is preferred over the equivalent mixed number $1\frac{5}{24}$.

For example, to add 3/8 and 5/6, we have

$$\frac{3}{8} + \frac{5}{6} = \frac{3 \cdot 6}{8 \cdot 6} + \frac{5 \cdot 8}{8 \cdot 6} = \frac{3 \cdot 6 + 5 \cdot 8}{8 \cdot 6} = \frac{18 + 40}{48} = \frac{58}{48} = \frac{29}{24}$$

This process of using the product of the denominators as a common denominator often creates denominators larger than necessary. The problem generally will be less complicated if the *least common denominator*, LCD, is used. The **least common denominator** is the least common multiple of all the denominators. That is, it is the smallest positive integer that is a multiple of the denominators.

If the LCD is not obvious, then we find it as follows:

1. Factor each denominator into a product of *prime numbers*. (A **prime number** is a positive integer greater than 1 whose only factors are itself and 1.)
2. Form the product of all the different prime factors, raising each prime factor to the highest power that occurs in any of the denominators.
3. This product is the LCD.

For example, in the preceding example, 3/8 + 5/6, the prime factors of 8 are $2 \cdot 2 \cdot 2 = 2^3$ and the prime factors of 6 are $2 \cdot 3$. The LCD of the rational numbers is $2^3 \cdot 3 = 24$. We use the fundamental principle of rational numbers to build each rational number to an equivalent rational number with the denominator of 24 as follows:

$$\frac{3}{8} + \frac{5}{6} = \frac{3}{8} \cdot \frac{3}{3} + \frac{5}{6} \cdot \frac{4}{4} = \frac{9}{24} + \frac{20}{24} = \frac{29}{24}$$

EXAMPLE 7   Combining Rational Numbers with Different Denominators

Combine and simplify.

(a) $\dfrac{1}{12} + \dfrac{3}{40}$

(b) $\dfrac{11}{18} - \dfrac{7}{75}$

Solution   (a) To find the LCD, factor each denominator: $12 = 2^2 \cdot 3$, $40 = 2^3 \cdot 5$. The LCD is the product of every different factor to its highest power, $2^3 \cdot 3 \cdot 5$ is 120. Convert each rational number into an equal rational number with 120 as the denominator and then add.

$$\frac{1}{12} + \frac{3}{40} = \frac{1}{12} \cdot \frac{10}{10} + \frac{3}{40} \cdot \frac{3}{3} = \frac{10}{120} + \frac{9}{120} = \frac{19}{120}$$

(b) The LCD is $2 \cdot 3^2 \cdot 5^2 = 450$ since $18 = 2 \cdot 3^2$ and $75 = 3 \cdot 5^2$.

This can be easily checked with a calculator using decimal approximations.

$$\frac{11}{18} - \frac{7}{75} = \frac{11}{18} \cdot \frac{25}{25} - \frac{7}{75} \cdot \frac{6}{6} = \frac{275}{450} - \frac{42}{450} = \frac{233}{450}$$

PROBLEM SET R4

In Problems 1–6, reduce each rational number to lowest terms.

1. (a) $\dfrac{15}{45}$ (b) $\dfrac{270}{420}$

2. (a) $\dfrac{36}{90}$ (b) $\dfrac{48}{120}$

3. (a) $-\dfrac{65}{95}$ (b) $-\dfrac{150}{510}$

4. (a) $\dfrac{-75}{45}$ (b) $\dfrac{-210}{150}$

5. (a) $\dfrac{-12}{-18}$ (b) $\dfrac{-15}{-25}$

6. (a) $\dfrac{-270}{-450}$ (b) $\dfrac{-200}{-150}$

In Problems 7–10, rewrite each rational number as a percent.

7. (a) $\dfrac{4}{5}$ (b) $\dfrac{7}{20}$

8. (a) $\dfrac{3}{50}$ (b) $\dfrac{11}{25}$

9. (a) $\dfrac{3}{8}$ (b) $\dfrac{7}{8}$

10. (a) $\dfrac{2}{3}$ (b) $\dfrac{7}{16}$

In Problems 11–16, change each percent to a rational number.

11. (a) 7%   (b) 25%

12. (a) $9\dfrac{1}{2}\%$ (b) $\dfrac{3}{4}\%$

13. (a) $\dfrac{1}{2}\%$ (b) $\dfrac{5}{8}\%$

14. (a) 35%   (b) 75%

15. (a) 3.5%   (b) 10.6%

16. (a) 0.2%   (b) 6.4%

17. (a) What is 15% of 200?   (b) What is 4.2% of 200?

18. (a) What is 17% of 300?   (b) What is 8.5% of 300?

19. What percent of 400 is 80?

20. What percent of 156 is 26?

In Problems 21–24, rewrite each statement as a ratio.

21. (a) 15 inches to 25 inches   (b) 12 miles to 18 miles

22. (a) 40 feet to 65 feet   (b) 18 ounces to 24 ounces

23. (a) 75 seconds to 2 minutes   (b) 8 inches to 3 feet

24. (a) 4 feet to 4 yards   (b) 5 gallons to 12 quarts

In Problems 25–30, find each product by

(a) multiplying the rational numbers, then reducing the product to lowest terms, and

(b) dividing the common factors, then multiplying.

25. $\dfrac{5}{5} \cdot \dfrac{7}{5}$ 26. $-\dfrac{11}{7}\left(\dfrac{7}{11}\right)$

27. $\dfrac{-11}{24} \cdot \dfrac{24}{33}$ 28. $\dfrac{-40}{21}\left(\dfrac{-7}{20}\right)$

29. $\dfrac{11}{100} \cdot \dfrac{24}{132}$ 30. $\dfrac{-100}{90} \cdot \dfrac{45}{25}$

In Problems 31 and 32, divide the rational numbers and express the answers reduced to lowest terms.

31. (a) $\dfrac{21}{20} \div \dfrac{9}{15}$ 32. (a) $\dfrac{-5}{18} \div \dfrac{-25}{12}$

    (b) $\dfrac{110}{9} \div \dfrac{121}{9}$ (b) $\dfrac{-3}{20} \div \dfrac{9}{100}$

In Problems 33–36, combine the rational numbers and reduce the answers to lowest terms.

33. (a) $\dfrac{2}{5} + \dfrac{1}{5}$ (b) $\dfrac{7}{12} + \dfrac{5}{12}$

34. (a) $\dfrac{5}{11} - \dfrac{3}{11}$ (b) $\dfrac{7}{15} + \dfrac{-3}{15}$

35. (a) $\dfrac{-7}{13} - \dfrac{4}{13} + \dfrac{3}{13}$ (b) $-\dfrac{21}{25} - \dfrac{-16}{25} + \dfrac{7}{25}$

36. (a) $\dfrac{2}{5} - \dfrac{3}{8} + \dfrac{7}{20}$ (b) $\dfrac{5}{6} - \dfrac{2}{9} + \dfrac{4}{15}$

37. **Sports:** A football team won 11 of its 16 games played with no ties.
    (a) What is the ratio of wins to games played?
    (b) What percent of the games played did the team win?
    (c) What is the ratio of wins to losses?

38. **Gas Mileage:** A midsize car travels 252 miles on 9 gallons of gasoline. What is the ratio of miles to gallons?

39. **Marketing:** The price of a microwave oven went from $99 to $115. Find the ratio of increase in price to the original price.

40. **Marketing:** The price of a computer went from $1800 to $1100. Find the ratio of the decrease in price to the original price.

41. **Environment:** Environmentalists estimate that paper takes up about half of our landfill space. If newspapers are 2/5 of the paper in the landfills, what fraction of landfill space is taken up by newspapers?

42. **College Enrollment:** At a college, 3/5 of the students take a mathematics course. Of those, 1/3 take algebra. What fraction of the students take algebra?

43. **Land Distribution:** It is estimated that there are 2,300,000,000 acres of land in America. Native Americans own 1/50 of the land. How many acres of land are owned by Native Americans?

44. **Activities:** Suppose that a person sleeps 1/3 of the day and works 5/12 of the day. What fraction of the day is left for other activities?

OBJECTIVES

1. Introduction to the Properties of Exponents
2. Solve Applied Problems

Note the difference between $(-5)^4$ and -5^4.

R5 Properties of Positive Exponents

Recall from section R1 that the exponential expression x^n can be written as $x^n = x \cdot x \cdot x \cdot \ldots \cdot x$ where the right side of the equation is called the *expanded form*. For example,

  (i) $2^7 = 2 \cdot 2 \cdot 2 \cdot 2 \cdot 2 \cdot 2 \cdot 2 = 128$
  (ii) $(-2)^7 = (-2) \cdot (-2) \cdot (-2) \cdot (-2) \cdot (-2) \cdot (-2) \cdot (-2) = -128$
  (iii) $(-5)^4 = (-5) \cdot (-5) \cdot (-5) \cdot (-5) = 625$
  (iv) $-5^4 = -5 \cdot 5 \cdot 5 \cdot 5 = -625$

Introducing the Properties of Exponents

To multiply exponential expressions with the same base such as $a^2 \cdot a^3$, we can write the product as follows:

$$a^2 \cdot a^3 = (a \cdot a)(a \cdot a \cdot a) = a^5$$

Notice that the exponent of the product, a^5, is the sum of the exponents of the factors a^2 and a^3. That is,

$$a^2 \cdot a^3 = a^{2+3} = a^5$$

Expressions such as $x^4 y^2$ cannot be simplified because the bases are different.

In other words, when we multiply exponential expressions with the *same base*, repeat the base and add the exponents. In general, we have the following property.

PRODUCT PROPERTY OF EXPONENTS

If m and n are positive integers, then $a^n \cdot a^m = a^{n+m}$

Now we consider the meaning of $(a^2)^3$, observe that $(a^2)^3$ means a^2 raised to the third power. That is,

$$(a^2)^3 = a^2 \cdot a^2 \cdot a^2 = (a \cdot a) \cdot (a \cdot a) \cdot (a \cdot a) = a^{2 \cdot 3} = a^6$$

That is, to raise an exponential expression to a power, repeat the same base and multiply the exponents. In general, we have the following property

POWER PROPERTY FOR EXPONENTS

If m and n are positive integers, then $(a^n)^m = a^{n \cdot m}$

EXAMPLE 1   Illustrating the Properties of Exponents

Use the properties of exponents to simplify each expression.

(a) $x^{10} \cdot x^3$ (b) $(y^3)^4$ (c) $(-t^5)^6$

Solution   (a) $x^{10} \cdot x^3 = x^{10+3} = x^{13}$ Product Property
(b) $(y^3)^4 = y^{3 \cdot 4} = y^{12}$ Power Property
(c) $(-t^5)^6 = (-t^5)(-t^5)(-t^5)(-t^5)(-t^5)(-t^5) = t^{30}$ Expanded form   ◈

We can write the power of a **product** involving two bases such as $(a^2b^3)^4$ in expanded form as follows:

$$(a^2b^3)^4 = (a^2b^3)(a^2b^3)(a^2b^3)(a^2b^3)$$
$$= a^2a^2a^2a^2b^3b^3b^3b^3$$
$$= a^{2+2+2+2}b^{3+3+3+3}$$
$$= a^8b^{12}$$

If we multiply the exponents, the result is the same.

$$(a^2b^3)^4 = (a^2)^4(b^3)^4 = a^8b^{12}$$

In general, we have:

POWER OF A PRODUCT PROPERTY

If n is a positive integer, then $(ab)^n = a^nb^n$

We can also write the power of a **quotient** $\left(\frac{a}{b}\right)^3$ involving two different bases in expanded form as follows:

$$\left(\frac{a}{b}\right)^3 = \frac{a}{b} \cdot \frac{a}{b} \cdot \frac{a}{b} = \frac{a^3}{b^3}$$

Notice that if we raise the numerator and denominator to the third power, the result is the same.

$$\left(\frac{a}{b}\right)^3 = \frac{(a)^3}{(b)^3} = \frac{a^3}{b^3}$$

In general, we have the following property

POWER OF A QUOTIENT PROPERTY

If n is a positive integer, then $\left(\frac{a}{b}\right)^n = \frac{a^n}{b^n}, b \neq 0.$

EXAMPLE 2   Power Property of Exponents

Use the properties of exponents to simplify each expression.

(a) $(-x^3y)^4$ (b) $\left(\frac{x^2}{y}\right)^5$ (c) $\left(\frac{2x^5}{3^2y^2}\right)^3$

Solution   (a) $(-x^3y)^4 = (-x^3)^4y^4 = x^{12}y^4$ Power of a Product

(b) $\left(\frac{x^2}{y}\right)^5 = \frac{(x^2)^5}{y^5} = \frac{x^{10}}{y^5}$ Power of a Quotient

(c) $\left(\frac{2x^5}{3^2y^2}\right)^3 = \frac{(2x^5)^3}{(3^2y^2)^3}$ Power of a Quotient

$$= \frac{2^3(x^5)^3}{(3^2)^3(y^2)^3} = \frac{8x^{15}}{729y^6}$$ Power of a Product

To develop a property for dividing exponential expressions, consider the following quotient for $a \neq 0$:

$$\frac{a^6}{a^4} = \frac{a \cdot a \cdot a \cdot a \cdot a \cdot a}{a \cdot a \cdot a \cdot a} = \left(\frac{a}{a}\right) \cdot \left(\frac{a}{a}\right) \cdot \left(\frac{a}{a}\right) \cdot \left(\frac{a}{a}\right) \cdot a \cdot a$$

$$= 1 \cdot 1 \cdot 1 \cdot 1 \cdot a \cdot a = a^2$$

Here the exponent of the quotient, a^2, is the difference of the exponents of a^6 and a^4.

$$\frac{a^6}{a^4} = a^{6-4} = a^2$$

In general, we have the following property:

QUOTIENT PROPERTY FOR EXPONENTS

If m and n are positive integers where $m > n$, and $a \neq 0$, then
$$\frac{a^m}{a^n} = a^{m-n}$$

EXAMPLE 3   Illustrating the Quotient Property

(a) $\dfrac{x^5}{x^3}$ (b) $\dfrac{(-x)^5}{(-x)^3}$ (c) $\dfrac{-x^6}{(-x)^2}$ (d) $\dfrac{(2x)^5}{(3x)^2}$

Solution   (a) $\dfrac{x^5}{x^3} = x^{5-3} = x^2$

(b) $\dfrac{(-x)^5}{(-x)^3} = (-x)^{5-3} = (-x)^2 = (-x)(-x) = x^2$

(c) $\dfrac{-x^6}{(-x)^2} = \dfrac{-x \cdot x \cdot x \cdot x \cdot x \cdot x}{(-x)(-x)} = \dfrac{-x \cdot x \cdot x \cdot x \cdot x \cdot x}{x \cdot x} = -\dfrac{x^6}{x^2} = -x^{6-2} = -x^4$

In part d of example 3, we simplify first because of the different bases, $2x$ and $3x$, and then use the quotient property.

(d) $\dfrac{(2x)^5}{(3x)^2} = \dfrac{2^5 x^5}{3^2 x^2} = \dfrac{32}{9} x^{5-2} = \dfrac{32}{9} x^3$ ◇

The following is a summary of all these properties.

PROPERTIES OF EXPONENTS

If a and b are real numbers and m and n are positive integers, then

$a^m \cdot a^n = a^{m+n}$	**Product of Exponents**
$\dfrac{a^m}{a^n} = a^{m-n}, a \neq 0, m > n$	**Quotient of Exponents**
$(a^m)^n = a^{m \cdot n}$	**Power Property**
$(ab)^n = a^n b^n$	**Power of a Product**
$\left(\dfrac{a}{b}\right)^n = \dfrac{a^n}{b^n}, b \neq 0$	**Power of a Quotient**

Solving Applied Problems

Many applications require the use of very large numbers. For example, the speed of light is approximately

$$300,000,000 \text{ meters per second}$$

Scientific Notation expresses such numbers, using the form

$$s \times 10^n$$

where s is a number between 1 and 10, including 1 ($1 \leq s < 10$), and n is an integer. We write 300,000,000 in scientific notation as

$$3.0 \times 10^8$$

Notice in 3.0×10^8 that 3 is multiplied by $10^8 = 100,000,000$ (100 million), moving the decimal 8 places to the right.

Most calculators can express numbers in scientific notation by using the command **EE** or **EXP** (check your manual if you do not find these commands). To enter 3.0×10^8 on a calculator, enter 3.0, then **EE** or **EXP**, and the exponent 8. Any operations that the calculator can perform can be done in scientific notation. If the results of calculations have more digits than the calculator can display, the answer is automatically written in scientific notation. For example, we have seen that using the laws of exponents, the product $2^{15} \cdot 2^{25} = 2^{15+25} = 2^{40}$. On the calculator $2^{15} = 32,768$ and $2^{25} = 33,554,432$, but $2^{40} = 1.09951162778\text{E}12$. The result is written in scientific notation.

EXAMPLE 4   Using Scientific Notation

A light-year is the distance light travels in one year. Find this distance in meters (m). Express the answer in scientific notation.

Solution   The speed of light is approximately 300,000,000 meters per second. A light year is the distance that light travels in one year. So a light year is given by:

$$\frac{300,000,000 \text{ m}}{1 \text{ sec}} \cdot \frac{3600 \text{ sec}}{1 \text{ hr}} \cdot \frac{24 \text{ hr}}{1 \text{ day}} \cdot 365.25 \text{ days} = 9,467,280,000,000,000 \text{ m or}$$

$$9.46728 \times 10^{15} \text{ m}$$

Therefore, light travels 9.46728×10^{15} meters in one year.   ◈

EXAMPLE 5   Multiplying and Dividing Expressions in Scientific Notation

Use the properties of exponents to simplify each expression. Write each answer in scientific notation.

(a) $(4.0 \times 10^7)(2.0 \times 10^5)$ (b) $\dfrac{3.5 \times 10^7}{5 \times 10^4}$

Solution   (a) $(4.0 \times 10^7)(2.0 \times 10^5) = (4.0 \times 2.0)(10^7 \times 10^5)$
$$= 8.0 \times 10^{7+5}$$
$$= 8.0 \times 10^{12}$$

(b) $\dfrac{3.5 \times 10^7}{5 \times 10^4} = \left(\dfrac{3.5}{5}\right) \times \left(\dfrac{10^7}{10^4}\right) = 0.7 \times 10^{7-4} = 0.7 \times 10^3 = 7 \times 10^2$ ◈

◈ PROBLEM SET R5

In Problems 1–8, simplify each expression.

1. (a) $3^2 \cdot 3^3$
   (b) $x^2 \cdot x^3$
3. (a) $(-4)^5 \cdot (-4)^7$

2. (a) $2^4 \cdot 2^2$
   (b) $y^4 \cdot y^2$
   (b) $-x^5 \cdot (-x)^7$

4. (a) $(-10)^6 \cdot (-10)^4$ (b) $-y^6 \cdot (-y)^4$
5. $y^3 \cdot y^4 \cdot y^5$ 6. $(-y)^9 \cdot (-y)^3 \cdot (-y)$

7. (a) $\dfrac{13^5}{13^2}$ (b) $\dfrac{x^{50}}{x^{20}}$ 8. (a) $\dfrac{14^9}{14^6}$ (b) $\dfrac{x^{90}}{x^{60}}$

In Problems 9 and 10, simplify each expression.

9. (a) $\dfrac{-x^{25}}{x^{20}}$ (b) $\dfrac{(-x)^{25}}{(-x)^{20}}$ 10. (a) $\dfrac{-x^{13}}{x^{10}}$ (b) $\dfrac{(-x)^{13}}{(-x)^{10}}$

In Problems 11–14, simplify each expression by removing the parentheses.

11. (a) $(12^2)^3$ (b) $(x^{20})^{30}$
12. (a) $(13^2)^4$ (b) $(x^{20})^{40}$
13. (a) $(y^{13})^5$ (b) $(-y^5)^{13}$
14. (a) $(-y^{20})^3$ (b) $(-y^3)^{20}$

In Problems 15–20, simplify each expression.

15. (a) $-4\left(\dfrac{x^3y^2}{x^2y^3}\right)^2$ (b) $-\left(\dfrac{4x^3y^2}{x^2y^3}\right)^2$

16. (a) $\dfrac{3}{2}\left(\dfrac{a^5b^3}{a^2b^6}\right)^3$ (b) $-\left(\dfrac{3a^5b^3}{2a^2b^6}\right)^3$

17. $(x^2y^3)(xy^2)^2$ 18. $(-x^2y^3)(3xy^4)^2$

19. $\dfrac{x^2(y^3)^3}{(x^2y)^5}$ 20. $\dfrac{x^6y^4}{(xy^2)^2}$

In Problems 21 and 22, rewrite each number in scientific notation.

21. (a) 385,000   (b) 385 million   (c) 38,000
22. (a) 47,000   (b) 47 million   (c) 47 billion

In Problems 23–26, use the properties of exponents to simplify each expression. Write each answer in scientific notation.

23. (a) $(2 \times 10^5)(4.8 \times 10^9)$ (b) $\dfrac{2 \times 10^{15}}{4.8 \times 10^9}$

24. (a) $(3.4 \times 10^7)(2 \times 10^4)$ (b) $\dfrac{3.4 \times 10^7}{2 \times 10^4}$

25. $\dfrac{(9.75 \times 10^8) \cdot (1.5 \times 10^3)}{(7.5 \times 10^2) \cdot (1.1 \times 10^4)}$

26. $\dfrac{(1.86 \times 10^5) \cdot (2.4 \times 10^{19})}{(3.6 \times 10^7) \cdot (4.8 \times 10^{11})}$

In Problems 27–36, simplify each expression.

27. $\dfrac{a^2}{b^2} + \dfrac{b^2}{a^2}$ 28. $\dfrac{t^3}{u^3} + \dfrac{u^3}{t^3}$

29. $-\left(\dfrac{3a^5b^{13}}{2a^2b^6}\right)^3$ 30. $\left(\dfrac{-4t^{12}s^3}{8t^3s^2}\right)^5$

31. $\left(\dfrac{b^2}{3a^3}\right)^3\left(\dfrac{3a^6}{b}\right)^2$ 32. $\left(\dfrac{p^2q^{10}r^3}{q^3r^2}\right)^4$

33. $\dfrac{(a^5b^3)^2(a^2b^{10})^3}{(ab^2)(a^3b^7)^2}$ 34. $\dfrac{(a^5b^8)^{10}(a^5b^7)^2}{(4ab^2)(ab)^2}$

35. (a) $\dfrac{x^{2n+1}}{x^{2n-1}}$ (b) $\dfrac{x^{2n+1}}{x^{2n}}$

36. (a) $\dfrac{x^{2n+1}y^{2n}}{x^{2n-1}y^{2n-2}}$ (b) $\dfrac{x^{2n+1}y^{2n+1}}{x^{2n}y^{2n-1}}$

37. **Scientific Notation**: Rewrite each statement so that all numbers are expressed in scientific notation.
 (a) The diameter of the star Betelgeuse is approximately 358,400,000 kilometers.
 (b) Five thousand miles is about 300,000,000 inches.
 (c) Oil reserves in the United States are about 35,000,000,000 barrels.

In Problems 38–42, express each answer in scientific notation.

38. **Electronics:** A computer can do 86 million operations in 4.3 seconds. How long, in nanoseconds, does it take this computer to do one operation? (1 billion nanoseconds = 1 second)

39. **Astronomy:** In late 1993, shortly after it was repaired, the *Hubble Space Telescope* took a dramatic photo of M100, a spiral galaxy far from earth. Scientists estimate that the closest stars in M100 are 35 million light-years from us; whereas, the most distant stars are 80 million light-years from us. How many meters away from us are the closest and farthest stars?

40. **Physics:** The mass of the earth is 6×10^{21} tons. The mass of the sun is about 300,000 times the mass of the Earth. Use scientific notation to find the mass of the sun in tons.

41. **Chemistry:** What is the number of atoms in 15 grams of hydrogen if one gram of hydrogen contains 602,300,000,000,000,000,000,000 atoms?

42. **Astronomy:** The mass of the sun is approximately 1.97×10^{29} kilograms, and our galaxy (the Milky Way) is estimated to have a total mass of 1.5×10^{11} suns. The mass of the known universe is at least 10^{11} times the mass of the galaxy. Calculate the mass of the known universe.

OBJECTIVES

1. Introduce Dimensional Analysis
2. Apply Formulas from Geometry
3. Apply the Pythagorean Theorem
4. Apply Formulas from Business
5. Apply Formulas from Science

R6 Formulas

Many applications in mathematics and other sciences require the use of *formulas* for their solutions. A **formula** is an equation that expresses a relationship between several quantities. In this section, we shall use formulas from geometry, business and finance, physical sciences and medicine. As we work with formulas, it is often necessary to find a specific value when other quantities are known. To expedite this concept, we extend our work to include *dimensional analysis*.

Introducing Dimensional Analysis

In applications from mathematics and other sciences, we refer to the units of measure as **dimensions** and the study of these units is called **dimensional analysis**. In working with dimensional analysis, all the properties of algebra apply and we use *conversion factors* to allow us to change units of measure. For example, we write 3 feet = 1 yard as

$$\frac{3 \text{ feet}}{1 \text{ yard}}$$

and we read this conversion as *3 feet per yard.* Other examples of conversion factors include the following:

$$\frac{60 \text{ minutes}}{1 \text{ hour}} \text{ which reads "60 minutes per hour"}$$

$$\frac{24 \text{ hours}}{1 \text{ day}} \text{ which reads "24 hours per day".}$$

EXAMPLE1   Using Dimensional Analysis

According to the January 3, 1997 edition of *USA Today* a British pound was worth 1.7126 US dollars and $1 US was worth 115.93 Japanese yen. Find the conversion of 1 British pound to Japanese yen.

Solution

$$\frac{1.7126 \text{ US dollar}}{1 \text{ British pound}} \cdot \frac{115.93 \text{ yen}}{1 \text{ US dollar}} = \frac{198.54 \text{ yen}}{1 \text{ British pound}} \quad \diamondsuit$$

Applying Formulas from Geometry

There are several formulas in geometry that we use often. Table 1 shows formulas that give the area A and the perimeter P (or circumference C) of some plane figures.

TABLE 1

Name	Area (A)	Perimeter (P) Circumference (C)	Figures
Square	$A = s^2$	$P = 4s$	
Rectangle	$A = lw$	$P = 2l + 2w$	
Circle	$A = \pi r^2$	$C = 2\pi r$	
Triangle	$A = \frac{1}{2} bh$	$P = a + b + c$	

EXAMPLE 2   Finding the Cost to Carpet a Room

A rectangular room of a fitness center has dimensions 18 feet by 100 feet.
(a) Find the area of the room.
(b) Find the cost of the carpet for the room if the cost is $18.00 per square yard.

Solution   (a) Referring to Table 1, the area of the rectangle is given by:

$$A = lw = (18 \text{ ft}) (100 \text{ ft}) = 1800 \text{ ft}^2$$

Therefore, the area of the room is 1800 square feet.
(b) Because the carpet is priced by the square yard, we convert 1800 square feet to square yards.

$$\frac{3 \text{ feet}}{1 \text{ yard}} \cdot \frac{3 \text{ feet}}{1 \text{ yard}} = \frac{9 \text{ ft}^2}{1 \text{ yd}^2}$$

Throughout the book, some names and prefixes for units used in the solutions of applied problems and related arts are listed in Table 2.

TABLE 2

Name	Prefix
inch(es)	in
square inch	in^2
foot or feet	ft
square feet	ft^2
yard(s)	yd.
square yard	yd^2
centimeter(s)	cm
square centimeter	cm^2
meter(s)	m
cubic meter	m^3
feet per second	ft/sec
miles per hour	mi/hr

So 9 square feet equals 1 square yard.

$$1800 \text{ ft}^2 = 1800 \text{ ft}^2 \cdot \frac{1 \text{ yd}^2}{9 \text{ ft}^2}$$
$$= 200 \text{ yd}^2$$

Total cost, C, of the carpet is given by

$$C = \frac{\$18}{1 \text{ yd}^2}(200 \text{ yd}^2)$$
$$= \$3600$$

Therefore, the total cost of the carpet is $3600.
Common formulas for volumes and surface areas are given in Table 3.

TABLE 3

Name	Volume (V)	Surface Area (S)	Figure
Cube	$V = s^3$	$S = 6s^2$	
Rectangular Solid	$V = lwh$	$S = 2lh + 2hw + 2lw$	
Circular Cylinder	$V = \pi r^2 h$	$S = 2\pi r^2 + 2\pi rh$	
Sphere	$V = \frac{4}{3}\pi r^3$	$S = 4\pi r^2$	

EXAMPLE 3 Volume and Surface Area of a Balloon

A spherical balloon is being inflated with helium. At one instant, the radius of the balloon is 60 inches.
(a) Find the volume of the balloon in cubic feet.
(b) Find the surface area of the balloon in square feet. (Use $\pi \approx 3.14$)

Solution First we convert the radius to feet, so that

$$60 \text{ in} = (60 \text{ in}) \frac{1 \text{ ft}}{12 \text{ in}} = 5 \text{ ft.}$$

(a) Substituting 5 ft. for r in the volume formula, we have

$$V = \frac{4}{3}\pi r^3$$
$$= \frac{4}{3}(3.14)(5\text{ft})^3 = 523.33\text{ft}^3$$

Therefore, the volume of the balloon is 523.33 cubic feet.
(b) Now substituting 5 ft for r in the surface area formula, we have

$$S = 4\pi r^2$$
$$= 4(3.14)(5 \text{ ft})^2 = 314 \text{ ft}^2$$

Therefore, the surface area of the balloon is 314 square feet.

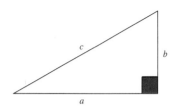

Figure 1

Applying the Pythagorean Theorem

The Pythagorean Theorem serves as a guideline for solving certain types of problems involving right triangles. The theorem relates the lengths of the sides of a right triangle. The side opposite the triangle's right angle is called the **hypotenuse**. The other two sides are called the **legs** of the triangle. Figure 1 shows a right triangle with legs a and b and hypotenuse c.

THE PYTHAGOREAN THEOREM

The sum of the squares of the lengths of the legs of a right triangle is equal to the square of the length of the hypotenuse. That is, $a^2 + b^2 = c^2$.

It should be noted that if $a^2 + b^2 = c^2$, then the triangle is a right triangle. This is commonly referred to as the *Converse of the Pythagorean Theorem.*

EXAMPLE 4 Using the Pythagorean Theorem

To prepare for an outdoor rock concert, the stage crew placed two large speaker towers on stage in such a way that the distance from the control booth to the first speaker tower is 25 feet, and the distance from the booth to the second tower is 60 feet. Assume the triangle formed is a right triangle with the distance between the towers as the hypotenuse (Figure 2). How far apart are the towers?

Solution Figure 2 shows the given lengths of the legs of the right triangle. Use the Pythagorean Theorem to find the length d of the hypotenuse as follows:

$$d^2 = (60 \text{ ft})^2 + (25 \text{ ft})^2$$
$$= 3600 \text{ ft}^2 + 625 \text{ ft}^2$$
$$= 4225 \text{ ft}^2$$

so $d = \sqrt{4225 \text{ ft}^2} = 65$ ft.

Therefore, the distance between the speaker towers is 65 feet. ◇

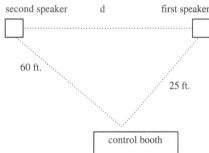

Figure 2

Applying Formulas from Business

Many types of investments and loans use formulas. For instance, the basic formula for computing simple interest, I, on an investment or loan is given by:

$$I = Prt$$

where P is the principal, r is the interest rate and t is time. The total amount, T, of money accumulated when P dollars earn at an interest rate of r for t years is given by

$$T = P + Prt$$

EXAMPLE 5 Computing Interest

A deposit of \$4,000 was left in an account that pays 7.5% simple interest.
(a) Find the amount of interest after 3 years.
(b) How much money is accumulated in this account after 3 years?

Solution   (a) Substituting \$4,000 for P, 0.075 per year for r and 3 years for t yields

$$I = Prt$$
$$= (\$4,000)\,\frac{0.075}{1\ \text{yr}}\,3\ \text{yrs}$$
$$= \$900$$

Therefore, the amount of interest earned is \$900.

(b) The total amount accumulated in the account after 3 years is given by:

$$T = P + I = P + Prt$$
$$= \$4000 + \$900 = \$4900$$

In the next example, we use the formula

$$A = P(1 + r)^t$$

to compute the amount A of accumulated principal and interest in a savings account if P is saved at an annual compounded interest rate r for t years. Compounded interest earns interest on both accumulated interest and principal.

EXAMPLE 6   Calculating Compound Interest in a Savings Account

Compute the amount of money in an Individual Retirement Account (IRA) if \$2000 is compounded annually at an interest rate of 6% for 30 years.

Solution   Substituting 2000 for P, 0.06 for r, and 30 for t yields

$$A = P(1 + r)^t$$
$$= 2000(1 + 0.06)^{30}$$
$$= 2000(1.06)^{30} = 11,486.98.$$

The total amount accumulated is approximately \$11,486.98.

Applying Formulas from Science

There are several formulas in science that we use quite often. They will be used periodically throughout this textbook. For example, if you travel at an average rate of 55 miles per hour for 4 hours, then the distance traveled is found by multiplying the rate by the time. Thus, the distance d is given by

$$d = \frac{55\ \text{miles}}{1\ \text{hour}}\,(4\ \text{hours}) = 220\ \text{miles}$$

This is stated as **distance equals rate times time**. In symbols, we use the formula

$$d = rt$$

EXAMPLE 7   Computing Distance

A motorist drove 3 hours on the Ohio Turnpike at 65 miles per hour and 2 hours on the Pennsylvania Turnpike at 55 miles per hour. What is the total distance traveled?

Solution   The distance traveled on the Ohio Turnpike is given by

$$d = rt = \frac{65\ \text{mi}}{1\ \text{hr}}\,(3\ \text{hrs}) = 195\ \text{mi}$$

The distance traveled on the Pennsylvania Turnpike is given by

$$d = rt = \frac{55\ \text{mi}}{1\ \text{hr}}\,(2\ \text{hrs}) = 110\ \text{mi}$$

The total distance traveled is $195 + 110 = 305$ miles.

The formulas $C = \frac{5}{9}(F - 32)$ and $F = \frac{9}{5}C + 32$ show the conversion from Fahrenheit to Celsius and from Celsius to Fahrenheit respectively, where C is the temperature reading on the Celsius scale and F is the temperature reading on the Fahrenheit scale.

EXAMPLE 8   Converting Temperatures

On a given day the temperature reading on the Canadian side of Niagara Falls is 25°C. What is the corresponding temperature on the American side?

Solution   In Canada the temperature reading is in Celsius while in the US the temperature is read in Fahrenheit. Using the formula, $F = \frac{9}{5}C + 32$

We have $F = \frac{9}{5}(25) + 32$

$$= 45 \quad + 32 = 77$$

Therefore, the temperature reading on the American side of Niagara Falls is 77°F.   ◈

◈ **PROBLEM SET R6**

In Problems 1–4, use the formula for the area of a rectangle ($A = lw$) and dimensional analysis to complete the following conversions. Note that 100 centimeters = 1 meter and 10 millimeters = 1 centimeter.

Item	Square Meters	Square Centimeters	Square Millimeters
1. Area of a sheet of paper		588	
2. Area of a desktop	1.5		
3. Area of a dollar bill		100	
4. Area of a postage stamp			480

In Problems 5 and 6, use dimensional analysis to complete each conversion (5280 feet = 1 mile).

5.   2.7 miles = _____ feet
6.   144 square inches = _____ square miles.

In Problems 7–18, find the area of the shaded region drawn in the given figure.

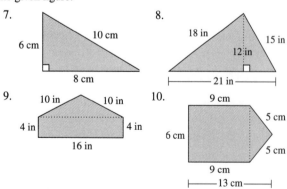

11.   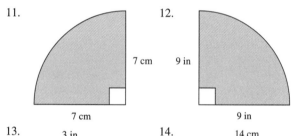   12.

13.   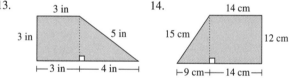   14.

15.   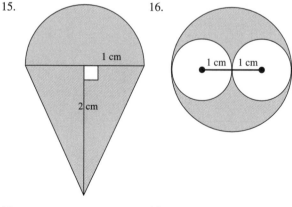   16.

17.   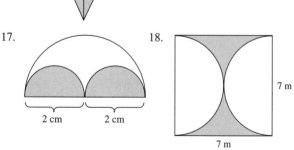   18.

In Problems 19–26, find (a) the volume and (b) the surface area of each solid figure.

19.

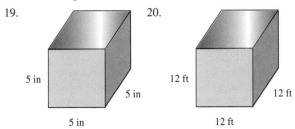

5 in
5 in
5 in

20.

12 ft
12 ft
12 ft

21.

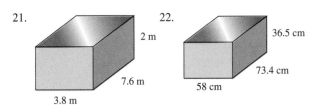

2 m
7.6 m
3.8 m

22.

36.5 cm
73.4 cm
58 cm

23.

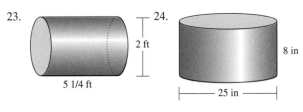

5 1/4 ft

24.

2 ft
8 in
25 in

25.

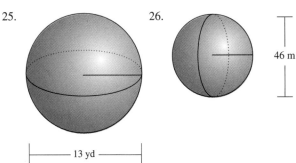

13 yd

26.

46 m

In Problems 27–32, use the Converse Pythagorean Theorem to determine if the triangle whose sides are a, b and c units is a right triangle.

27. $a = 5$, $b = 12$, $c = 13$
28. $a = 8$, $b = 15$, $c = 17$
29. $a = 12$, $b = 20$, $c = 24$
30. $a = 12$, $b = 30$, $c = 32$
31. $a = 12$, $b = 35$, $c = 37$
32. $a = 31$, $b = 19$, $c = 37$

In Problems 33–36, use the formula $I = Prt$ to complete the following table:

	Interest I (in dollars)	Principal P (in dollars)	Interest rate r	Time t (in years)
33.		2000	6%	3
34.		1345	4%	4
35.	312	1560		5
36.	1250.20	8930		2

In Problems 37–40, use the formula $d = rt$ to complete the following table:

	Distance d	Rate r	Time t
37.		550 miles/hour	4 hours
38.	2200 kilometers	110 kilometers/ hour	
39.	1000 feet		3.5 seconds
40.	685 miles	55 miles/hour	

In Problems 41–44, use the conversion formulas for temperature.

41. (a) Convert 25°C to Fahrenheit
    (b) Convert 25°F to Celsius
42. (a) Convert 53°C to Fahrenheit
    (b) Convert 53°F to Celsius
43. (a) Convert 17°C to Fahrenheit
    (b) Convert 17°F to Celsius
44. (a) Convert 2°C to Fahrenheit
    (b) Convert 2°F to Celsius

45. **Carpeting:** Suppose that we wish to carpet a rectangular room whose length is 14 feet and width 12 feet, with a carpet costing $12.95 per square yard and a pad costing $2.85 per square yard.
    (a) What is the cost of the carpet?
    (b) What is the cost of the pad?
    (c) What is the total cost of the carpet and the pad?

46. **Geometry:** A cement slab is 16 feet long, 12 feet wide and 6 inches thick.
    (a) How many cubic yards of concrete does the slab contain?
    (b) If "ready mix" concrete sells for $56 per cubic yard, what is the cost of the concrete?

47. **Geometry:** Figure 3 shows three tanks in the forms of a cube, a cylinder and a sphere with the given dimensions.

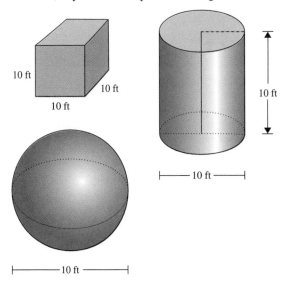

10 ft
10 ft
10 ft
10 ft
10 ft
10 ft

Figure 3

    (a) Which of the three tanks has the largest surface area?
    (b) Which of the three tanks has the largest volume?

48. **Geometry:** A tank in the shape of a right circular cylinder with a radius of 10 feet and a height of 14 feet is to be painted. According to the label on the can of paint, a gallon covers 400 square feet.

    10 ft

    14 ft

    (a) How many gallons of paint will be needed to paint the tank?
    (b) How many gallons of water will the tank hold? (1 cubic foot = 7.48 gallons)

49. **Geometry:** An orange grove, in the form of a square with sides 0.3 miles long is sold for $1,036,800.
    (a) What is the area of the grove in acres if one acre = 43,560 square feet and one mile = 5280 feet?
    (b) What is the cost per acre of this grove?

50. **Simple Interest:** If $6000 is borrowed from a bank at an interest rate of 6.5% for a period of 8 months, how much interest is owed?

51. **Travel Time:** Determine the time for the space shuttle to travel a distance of 6000 miles if its average speed is 27,200 kilometers per hour. (Use 1 mile = 1.6 kilometers.)

52. **Boston Marathon:** Find the average speed of the man holding the record time of 2 hours and 8 minutes in the Boston Marathon. The length of the course is 26 miles and 385 yards. (Use 1 mi = 1760 yds.)

53. **Doubling Principal:** Compute the amount of money in a savings account (use $T = P + Prt$).
    (a) A principal of $1000 is earning at a rate of 8% for 5 years.
    (b) A principal of $2000 is earning at a rate of 8% for 5 years.

54. **Doubling Interest Rate:** Compute the amount of money in a savings account (use $T = P + Prt$).
    (a) A principal of $5000 is earning at a rate of 4.75% for 6 years.
    (b) A principal of $5000 is earning at a rate of 9.5% for 6 years.

55. **Doubling Principal:** Compute the amount of money in a savings account (use $A = P(1 + r)^t$).
    (a) A principal of $1000 is earning at a rate of 8% compounded annually for 5 years.
    (b) A principal of $2000 is earning at a rate of 8% compounded annually for 5 years.

56. **Doubling Interest Rate:** Compute the amount of money in a savings account (use $A = P(1 + r)^t$).
    (a) A principal of $5000 is earning at a rate of 4.75% compounded annually for 6 years.
    (b) A principal of $5000 is earning at a rate of 9.5% compounded annually for 6 years.

CUMULATIVE REVIEW OF BASIC CONCEPTS

1. Write each statement in symbols.
   (a) Five added to the sum of -5 and 3.
   (b) The product of -2 and -6 decreased by 3.
   (c) The quotient of -30 and 5 decreased by 2.

2. Write an equivalent sentence in words.
   (a) $2(x + 1) < 5$ (b) $3x + 7 > 5x$
   (c) $\dfrac{x}{2} - 1 = 4$ (d) $5 - 8 \neq 7$

3. (a) Write 3^5 in expanded form.
   (b) Write the expression $3 \cdot 3 \cdot 3 \cdot x \cdot x \cdot x \cdot x$ in exponential form.

4. (a) Apply the order of operations to find the value of the expression
   $$(15 + 5 \cdot 3^2 - 7) \div (24 - 6 \cdot 3)$$
   (b) Find the value of the expression $x^3 - 3xy^2 + y^3$ if $x = 2$ and $y = 3$.

5. State the properties that justify each statement.
   (a) $4(5y) = (4 \cdot 5)y$ (b) $7(1) = 7$
   (c) $(3 + x) + 7 = (x + 3) + 7$ (d) $4 + (-4) = 0$
   (e) $7 + 0 = 7$ (f) $4(x - 2) = 4x - 8$
   (g) $(3 + 2) + x = (3 + x) + 2$ (h) $(2x - 3)4 = 8x - 12$

6. Consider the set of real numbers $\{-7, -1.35, 0, \frac{3}{4}, 1, \pi, \sqrt{2}, 5\}$.
   (a) Which of the numbers in this set are not integers?
   (b) Which numbers are rational numbers?
   (c) Which numbers are irrational numbers?

7. Express each rational number as a repeating decimal.
   (a) $\dfrac{3}{9}$ (b) $\dfrac{5}{18}$ (c) $\dfrac{2}{7}$ (d) $\dfrac{2}{11}$ (e) $\dfrac{-7}{13}$
   (f) Use a calculator to round off each number to four decimal places.

8. Find the value of each expression.
   (a) $|16| + |-4|$ (b) $|35| - |-20|$
   (c) $-16 + (-30)$ (d) $-15 - (-7)$
   (e) $-18 + 9 - 4$ (f) $(-8)(-7)$
   (g) $(-1)(-2)(-1)(-5)$ (h) $(-2)(-4)(-1)(-6)$
   (i) $(-63) \div (-9)$ (j) $(36) \div (-4)$
   (k) $(-5)^2 + (-1)^3$ (l) $5(4 - 6)^2 - 4(3 - 7)^2$
   (m) $\dfrac{3^2 + 4^2}{(3 - 4)^2}$ (n) $\dfrac{5^2 - 7^2}{(5 - 7)^2}$

9. Find the value of each expression.
   (a) $(2x + y)^2$ for $x = -1$ and $y = 3$.
   (b) $\dfrac{x^2 - y^2}{x + y}$ for $x = -2$ and $y = 1$.

10. (a) Reduce each rational number to lowest terms.
    $$\frac{525}{735}, \frac{-385}{-455}, \text{ and } \frac{-279}{310}$$
    (b) Rewrite each rational number as a percent.
    $$\frac{3}{5}, \frac{5}{16}, \text{ and } \frac{5}{8}$$
    (c) Change each percent to a rational number.
    $$4\%, \frac{3}{8}\%, \text{ and } 21\%$$
    (d) What is 12% of 340?
    (e) What is the ratio of 35 seconds to 2 minutes?

11. Perform each operation and simplify.
    (a) $\dfrac{5}{7} \cdot \dfrac{7}{10}$
    (b) $-\dfrac{11}{13} \cdot \dfrac{13}{11}$
    (c) $\dfrac{21}{10} \div \dfrac{7}{15}$
    (d) $\dfrac{-7}{18} \div \dfrac{-14}{12}$
    (e) $\dfrac{5}{12} + \dfrac{-3}{8}$
    (f) $\dfrac{3}{20} + \dfrac{7}{30}$
    (g) $\dfrac{9}{16} - \dfrac{-7}{12}$
    (h) $\dfrac{1}{30} - \dfrac{-9}{40}$
    (i) $\dfrac{1}{2} - \dfrac{1}{3} + \dfrac{1}{4}$
    (j) $\dfrac{1}{8} - \dfrac{1}{4} + \dfrac{1}{5}$

12. Simplify each expression.
    (a) $(-3)^4 \cdot (-3)^2$
    (b) $(x^3)^5$
    (c) $\dfrac{x^{10}}{x^4}$
    (d) $\dfrac{(-x)^{12}}{(-x)^5}$
    (e) $\left(\dfrac{x^2 y^3}{xy}\right)^3$
    (f) $\dfrac{x^7 y^4}{(xy^2)^2}$
    (g) $(x^3 y^2)^2 (x^4 y^5)^3$
    (h) $(x^2 y^4)^3 (xy^2)^2$

13. (a) Write the number 730,000,000 in scientific notation.
    (b) Write 2.35×10^5 in expanded form.
    (c) Simplify $\dfrac{(2.4 \times 10^7) \cdot (3.5 \times 10^4)}{(1.2 \times 10^5) \cdot (0.7 \times 10^2)}$.

14. Use the Converse of the Pythagorean Theorem to determine whether or not the triangle with sides a, b and c is a right triangle.
    (a) $a = 8, b = 15, c = 17$
    (b) $a = 5, b = 10, c = 13$

15. (a) Convert 11°C to Fahrenheit.
    (b) Convert 47°F to Celsius.

16. Use the formula $I = Prt$ to find I if:
    (a) $P = 1500, r = 3\%, t = 2$
    (b) $P = 1200, r = 3.7\%, t = 4$

17. A gambler begins his day in a casino with $100 and wins $70. After an hour, he doubled his money. The next hour he loses $123.
    (a) Write an expression with these numbers to describe the situation.
    (b) Find the value of the expression. That is, how much money does he have in the end?

18. Describe an everyday occurrence which exhibits the commutative property.

19. A dress that usually sells for $300 is marked down 25%. What is the sale price of the dress? Use the formula: sale price = regular price − markdown.

20. Suppose the regular price for a calculator is $90. If the calculator's price is marked down 20%, what is the markdown?

21. A carpet dealer marks up his merchandise 25% on dealer's cost. If the carpet costs the dealer $600, for how much will the dealer sell the carpet? Use the formula: retail price = cost + markup.

22. The regular price of a special computer monitor is $750. If it is reduced by $150, by what percent is the price reduced?

23. In an algebra class, 3/5 of the students are females. Of the female students, 1/3 earn a B in the course. What fraction of the students in the class earn a B in the course and are female?

24. Two cars cross an intersection. One is traveling due west at a rate of 60 miles per hour. The other is traveling due south at a rate of 50 miles per hour. How far are the two cars from each other after 3 hours? Round off the answer to two decimal places.

25. A man is standing on a dock pulling a rope 10 feet above water. The rope is 45 feet long and attached to his boat. When he pulls in 30 feet of rope, how far has he moved the boat? Round off to one decimal place.

26. A 16-foot ladder is leaning against a vertical wall. How far above the ground is the top of the ladder when the bottom of the ladder is 10 feet from the wall? Round off to one decimal place.

27. Figure 1 shows a design of a flower bed in the shape of a right triangle and two semicircles. Find the total area of the flower bed. Round off the answer to two decimal places.

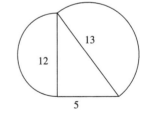

Figure 1

28. Compute the amount of money in an IRA account if $3000 is compounded annually at an interest rate of 7% for 25 years.

Chapter 1

POLYNOMIALS AND RATIONAL EXPRESSIONS

1.1 Addition and Subtraction of Polynomials
1.2 Multiplication of Polynomials
1.3 Factoring Polynomials
1.4 Rational Expressions
1.5 Multiplication and Division of Rational Expressions
1.6 Addition and Subtraction of Rational Expressions
1.7 Complex Rational Expressions
1.8 Graphing Techniques

A flower seed company has a rectangular piece of land which is used as a test plot. If x rows are plowed, the area (in square meters) is given by the polynomial

$$A = 15x^2 + 13x - 6$$

By factoring completely the right side of this formula, the length and width of the plowed land is expressed in terms of x. If we know how many rows are plowed, we can find the area, the length and the width of the land. Example 9 on page 52 explores this situation.

Flowerplot 245

In this chapter we perform the basic operations with polynomials and rational expressions. We introduce the Cartesian Coordinate system and graph equations of lines in the plane. In addition, we examine the advantages of a grapher in producing graphs.

OBJECTIVES

1. Define Polynomials
2. Add and Subtract Polynomials
3. Solve Applied Problems

1.1 Addition and Subtraction of Polynomials

In this section we extend the properties of real numbers to special algebraic expressions known as polynomials.

Defining Polynomials

Recall that *terms* of an algebraic expression are parts of the expression that are separated by plus or minus signs.

A **polynomial** is an algebraic expression in which all exponents of the variables are nonnegative integers and no variable appears in the denominator of a fraction. For example,

$$4x, \quad 3y - 5, \quad \text{and} \quad 3x^2 + 4x + 7$$

$\dfrac{3}{x^2}, \quad 4 - \dfrac{1}{y} \quad \text{and} \quad \sqrt{2t^{-2} + 1}$ are not polynomials.

are polynomials in one variable with one, two and three terms respectively.

Terms such as $5x^2$ and $-12x^2$ which differ only in their numerical coefficients are called **like terms** or **similar terms**. Terms such as $4x^2$ and $3x^3$ are unlike terms because the exponents of x are different. In particular, **monomials** are polynomials of one term, **binomials** are polynomials that have two unlike terms, and **trinomials** are polynomials that have three unlike terms. The term $-5xy$ has two variable factors, x and y, and one numerical factor, -5, which is called the *numerical coefficient*. The numerical coefficients of the terms in a polynomial are called the **coefficients** of the polynomial. For example, the coefficients of the polynomial

$$-2x^5 - 3x^4 + 5x^3 - 4x^2 + 8$$

When a term is "missing," its coefficient is 0, $0 \cdot x = 0$.

are $-2, -3, 5, -4, 0$, and 8.

The **degree of a nonzero term** in a polynomial is the sum of all the exponents of the variables in that term. The highest degree of all terms that appear with nonzero coefficients in a polynomial is called the **degree of the polynomial**. For instance, the degree of the monomial $5x^7$ is 7, and the degree of the binomial $5x^7 + 4x^8$ is 8 since the term $4x^8$ has the greatest degree, 8. A nonzero number such as 6 is called a **constant** polynomial with a zero degree. The number 0 is called the **zero polynomial** with no degree assigned to it.

We usually list terms in order of *descending powers*. For example, we write

$$4x^2 + 3x - 2$$

with powers of x decreasing from 2 to 1 to 0. The **leading coefficient** of a polynomial is the numerical factor of the highest degree term. The following chart contains specific illustrations of polynomials.

Polynomial	Type	Degree	Leading Coefficient	Coefficients of Each Term
5	monomial	0	5	5
$2x + 7$	binomial	1	2	2, 7
$-3x^2 - 5x - 1$	trinomial	2	-3	$-3, -5, -1$
$x^4 - x^2 + x + 1$	polynomial	4	1	$1, 0, -1, 1, 1$

EXAMPLE 1   Determining the Degree of a Polynomial

Find the degree of each polynomial and state the coefficient of each term.

(a) $4x^5 - 2x^3 + 1$ (b) $10 - 7x - 2x^2 + 3x^3$

Solution   (a) The degrees of the individual terms are 5, 3 and 0; the degree of the polynomial is the highest degree which is 5. The coefficient of each nonzero term in descending order is given by 4, -2, and 1 respectively.

(b) First, rewrite the polynomial in descending powers of x, $3x^3 - 2x^2 - 7x + 10$. The degree of the polynomial is 3, and the coefficient of each term is given by 3, -2, -7 and 10 respectively.

The next example shows how to evaluate a polynomial for a given value of x.

EXAMPLE 2   Evaluating a Polynomial

Find the value of the polynomial $40 - 2x - 9x^2$ for $x = 3$.

Solution   When $x = 3$, the value of the polynomial $40 - 2x - 9x^2$ is given by:

$40 - 2(3) - 9(3)^2$ Replace x with 3

$= 40 - 6 - 9(9)$ Order of operations

$= 40 - 6 - 81 = -47$ Simplify

Adding and Subtracting Polynomials

To find the sum of two or more polynomials, we use the associative and commutative properties of addition. Group like terms together and then combine these like terms by applying the distributive property. For example, we add $3x^2$ and $4x^2$ as follows:

$$3x^2 + 4x^2 = (3 + 4)x^2 = 7x^2$$

To find the difference of two polynomials, we change the signs of all the terms in the polynomial being subtracted and then add the resulting like terms. For example, we subtract $-6y^2$ from $-4y^2$ as follows:

$$-4y^2 - (-6y^2) = -4y^2 + 6y^2 = 2y^2$$

Horizontal and vertical formats for adding and subtracting polynomials are illustrated in the next two examples.

EXAMPLE 3   Using Horizontal Formats to Combine Polynomials

Use the horizontal format to perform each operation and simplify.

(a) $(4y^3 + 7y - 3) + (y^3 - y + 2)$ (b) $(4y^3 + 7y - 3) - (y^3 - y + 2)$

Solution   (a) Adding horizontally we have:

$(4y^3 + 7y - 3) + (y^3 - y + 2)$ Given

$= 4y^3 + 7y - 3 + y^3 - y + 2$ Remove parentheses

$= (4y^3 + y^3) + (7y - y) + (2 - 3)$ Group like terms

$= 5y^3 + 6y - 1$ Combine like terms

(b) We rewrite the subtraction problem as the addition of the additive inverse.

$(4y^3 + 7y - 3) - (y^3 - y + 2)$ Given

$= 4y^3 + 7y - 3 - y^3 + y - 2$ Interpret subtraction

$= (4y^3 - y^3) + (7y + y) + (-3 - 2)$ Group like terms

$= 3y^3 + 8y - 5$ Combine like terms

EXAMPLE 4   Using Vertical Format to Combine Polynomials

Use the vertical format to perform the indicated operations.

$$(x^2 - 3x + 4) + (2x^2 - x + 2) - (x^2 - 5x + 8)$$

Solution   We use a vertical format to add the first two polynomials by aligning like terms in the same columns.

$$\begin{array}{r} x^2 - 3x + 4 \\ 2x^2 - x + 2 \\ \hline 3x^2 - 4x + 6 \end{array}$$

Then we perform subtraction vertically.

$$\begin{array}{r} 3x^2 - 4x + 6 \\ -(x^2 - 5x + 8) \\ \hline 2x^2 + x - 2 \end{array} \quad \text{or} \quad \begin{array}{r} 3x^2 - 4x + 6 \\ - x^2 + 5x - 8 \\ \hline 2x^2 + x - 2 \end{array}$$

◈

Solving Applied Problems

There are many applications involving polynomials. One commonly used from physics is the polynomial given by the formula

$$h = -16t^2 + v_0 t + s_0$$

where h is the height of an object (in feet), v_0 is the initial velocity of the object (in feet per second) and s_0 is the initial height of the object (in feet).

EXAMPLE 5   Finding the Height of an Object

An object is thrown down from the top of a cliff 120 feet high, with an initial speed of 10 feet per second. Its height h after t seconds is given by the polynomial formula:

$$h = -16t^2 - 10t + 120$$

Find the height of the object after 2 seconds.

Solution   After 2 seconds, the height h of the object is given by:

$$h = -16(2)^2 - 10(2) + 120$$
$$= -64 - 20 + 120 = 36$$

Therefore, the height of the object after 2 seconds is 36 feet.   ◈

Another application involving polynomials is used in economics where polynomials can represent quantities. Suppose that C represents the total cost of producing x units of a particular commodity and R represents the total revenue from selling x units. From economics, the total *profit* is the *revenue*, R, minus the *cost*, C. That is, the total profit P obtained by producing and selling x units is given by

$$P = R - C$$

EXAMPLE 6   Finding a Profit

A manufacturer produces x framed posters at a total daily cost C (in dollars) of

$$C = 5x + 50$$

The total daily revenue R (in dollars) when x posters are sold is given by

$$R = (48 - 2x)x$$

(a) Find the polynomial that represents the daily profit P (in dollars).
(b) How much profit will the company make per day if it produces and sells 10 posters?

Solution　(a) We substitute $(48 - 2x)x$ for R and $5x + 50$ for C, so that
$$P = R - C = (48 - 2x)x - (5x + 50) = -2x^2 + 43x - 50$$
(b) To find the daily profit P, we replace x by 10,
$$P = -2x^2 + 43x - 50 = -2(10)^2 + 43(10) - 50$$
$$= -200 + 430 - 50 = 180$$

The daily profit from 10 posters is \$180.
To check this solution:
$$R = (48 - 2x)x = [48 - 2(10)]10 = \$280$$
$$C = 5x + 50 = 5(10) + 50 = \$100$$
$$P = R - C = 280 - 100 = \$180$$

PROBLEM SET 1.1

Mastering the Concepts

In Problems 1 and 2, find the degree of each polynomial and state the coefficient of each term.

1. (a) $3x^2 - 7x + 6$ 　　(b) $4 - 2x - 7x^2$
2. (a) $-u^2 - u + 5$ 　　(b) $81 - t^2 - 5t^3$

In Problems 3–6, find the value of each polynomial for the indicated value.

3. $x^3 - 4x$
   (a) $x = -2$ 　　(b) $x = 0$
   (c) $x = 3$ 　　(d) $x = 4$
4. $-x^4 - 2x^2$
   (a) $x = -1$ 　　(b) $x = 0$
   (c) $x = 2$ 　　(d) $x = 3$
5. $-3x^3 - 2x^2 - 3$
   (a) $x = -3$ 　　(b) $x = 0$
   (c) $x = 1$ 　　(d) $x = 2$
6. $4t^5 - t^2 + 6t + 3$
   (a) $t = -2$ 　　(b) $t = 0$
   (c) $t = 2$ 　　(d) $t = 3$

In Problems 7–24, perform each operation and simplify by using (a) the horizontal format and (b) the vertical format.

7. $(2x^2 - 3x + 2) + (5x^2 + 2x - 3)$
8. $(3x^2 + 2x + 1) - (5x^2 - 2x + 4)$
9. $(2t^2 + 7t + 8) - (t^2 - 3t + 5)$
10. $(6u^2 - 4u - 1) + (-u^2 - 2u + 2)$
11. $(4t^3 - 7t - 8) - (-2t^2 + 4t - 2)$
12. $(5y^3 - 3y^2 + 2y - 8) + (y - y^3 + 3y^2 - 1)$
13. $(x^6 - 2x^4 - 3x^2) - (x^5 - 2x^3 - 3x - 4)$
14. $(3x^4 - 4x^3 + 6x^2 + x - 1) + (4 - x + 2x^2 - 3x^3 - x^4)$
15. $(8x^2 + 3x - 7) + (-5x^2 + 2x + 1) + (2x^2 - 3x + 4)$
16. $(-2p^2 + 5p + 2) + (3p^2 - 2p + 3) + (p^2 - 3p - 7)$
17. $(t^3 - 2t^2 + 3t + 1) + (2t^3 + t^2 - 2t - 2) + (-t^3 + 3t^2 - 2t - 1)$
18. $(3y^4 - 4y^3 + y^2 - 2y + 3) + (7y^4 + 5y^3 + 2y^2 - y - 7) + (6y^3 - 6y + 5)$
19. $(2w^2 - 3w + 4) + (5w - 1 + w^2) - (w + 2w^2 - 6)$
20. $(-x^2 + 8x - 11) - (x^2 - 3x - 2) + (3x^2 - 10x + 3)$
21. $(-2x^2 + 5x + 2) + (3x^2 - 2x + 3) - (x^2 - 3x - 5)$
22. $(3x^3 - 2x^2 + 5x + 4) - (2x^3 + x^2 - 2x + 1) + (-2x^2 + 4x + 8)$
23. $(3y^4 - 4y^3 + y^2 - 2y + 4) + (7y^4 + 5y^3 + 2y^2 - y - 7) - (6y^3 - 6y + 5)$
24. $2(t^2 + t + 12) - 5(t^2 - 2t + 5) + 6(t^2 + 3t - 4)$

Applying the Concepts

25. **Free-Falling Object:** A stone is dropped from the top of a building. Its height h (in feet) t seconds after being dropped is given by the formula
$$h = -16t^2 + 64$$
Find the height of the stone at the specified times.
   (a) $t = 0$ 　(b) $t = \dfrac{1}{2}$ 　(c) $t = \dfrac{3}{4}$ 　(d) $t = 1$

26. **Horizontal-Moving Object:** A particle is moving along a horizontal straight line according to the formula
$$s = 4t^3 + 2t - 1$$
where s is the distance from a point 0 in meters and t is the time in seconds. Find the distance of the particle at the specified times.
   (a) $t = 1$ 　(b) $t = \dfrac{1}{2}$ 　(c) $t = \dfrac{1}{3}$

27. **Geometry:** The length of a rectangle is 5 feet more than its width. Let x be the width of the rectangle in feet.
   (a) Write a polynomial that represents the perimeter P of the rectangle and simplify the expression.
   (b) If the width of the rectangle is 4 feet, find its perimeter.

28. **Money:** A newspaper vending machine accepts dimes and quarters only. Let x represent the number of dimes in the machine at the end of the day. Assume that the machine has six fewer quarters than dimes.
   (a) Write an algebraic expression that represents the total value V of all coins.
   (b) Find the total value of the coins when the machine has 100 dimes.

29. **Profit:** A furniture company that produces hand-crafted desks estimates that the weekly total cost C of manufacturing x desks is given by

$$C = 6x^2 + 80x + 500$$

 (a) Find the polynomial for profit P if the total weekly revenue R obtained from selling x desks is given by $R = 500x$. (Use $P = R - C$)
 (b) How much profit per week will the company make if it manufactures and sells 25, 35 or 50 desks weekly?

30. **Biathlon Training:** While training for the biathlon this morning, you jogged for some time at a rate of 6 kilometers per hour. You then rode your bicycle for twice as much time at a rate of 10 kilometers per hour. Let t represent the time in hours that you jogged.
 (a) Write an algebraic expression for the total distance that you jogged and rode the bicycle. (Use the distance formula $d = rt$.)
 (b) Find the total distance after $\frac{1}{2}$ hour of training.

Developing and Extending the Concepts

31. Subtract $x^2y^2 - xy + 2$ from $2xy^2 - 3x^2y + 7$
32. Subtract $2xy^2 - 3x^2y$ from $4x^2y - 5xy - 4xy^2 + 6$
33. Subtract $1 - u + u^2 - u^3$ from $u^2 - 2u - 3$
34. Subtract $ab - a^2b + ab^2 - a^2b^2$ from
    $5ab + 2a^2b - 4ab^2 + 7a^2b^2$

In Problems 35 and 36, perform each operation and simplify.

35. $(7x^3y^2 - 3x^2y + 2) - (4 + x^2y - x^3y^2) - (2x^3y^2 + 5x^2y + 4)$
36. $(2xy - 3xz + 4yz) - (5xz - 3yz + 4xy)$
    $- (-2yz + 3xy - xz)$

37. Find the sum of the polynomial expressions and simplify.

$$\frac{1}{2}x^2 - 2x^3 + \frac{1}{3}x - 3, \ \frac{1}{2}x^3 - \frac{3}{2}x^2 + \frac{1}{2},$$

$$\frac{3}{4} + 2x - \frac{5}{2}x^2 + \frac{1}{6}x^3$$

38. Subtract the sum of $3x^3 - 2x^2 + 5$ and $3x^2 - x - 3$ from the difference of $2x^3 + 3x^2 + 7$ and $-4x^3 - x^2 + 13$.

39. (a) Show that the two polynomials $(2x - 3y)^2$ and $4x^2 - 12xy + 9y^2$ have the same value when $x = \frac{1}{2}$ and $y = \frac{1}{3}$.
 (b) Pick any other values of x and y and evaluate the two polynomials. What do you observe?

40. What must be subtracted from $3x + 2y + z$ to produce $2x - y - z$?

41. Remove grouping symbols and combine terms:

$$5x^2 - 3\{x - x[x + 4(x - 3)] - 5\}$$

42. Is the sum of two polynomials, each of degree 2, necessarily another polynomial of degree 2? Explain.

43. (a) Give an example of a polynomial of degree zero.
 (b) What is the degree of the product of a polynomial of degree zero and a polynomial of degree 2?

44. Given an example of two polynomials, each of degree 3, whose difference is:
 (a) a polynomial of degree 3.
 (b) a polynomial of degree 2.
 (c) a polynomial of degree 1.
 (d) a polynomial of degree 0.

45. Give an example of a polynomial that has no degree.

46. Two polynomials are equal if the coefficients of each like power terms are equal. Find the value of k so that

$$(3x^3 + 2x^2 + kx + 7) - (2x^3 + x^2 - 4x - 2) =$$
$$x^3 + x^2 + 7x + 5$$

OBJECTIVES

1. Multiply Monomials
2. Multiply Polynomials
3. Multiply Binomials
4. Use Special Products
5. Solve Applied Problems

1.2 Multiplication of Polynomials

To develop a method for multiplying polynomials, we start by multiplying monomials. Then we multiply polynomials by making repeated use of the distributive property. We also include special products in this section and certain techniques to perform multiplication of binomials quickly and efficiently.

Multiplying Monomials

To multiply two monomials, we use the properties of exponents along with the commutative and the associative properties as the next example shows.

EXAMPLE 1   Multiplying Monomials

Perform the multiplication and simplify.

$$(8x^3)(2x^2)$$

Solution $(8x^3)(2x^2) = (8 \cdot 2)(x^3 \cdot x^2)$
$$= 16x^{3+2} = 16x^5$$

Multiplying Polynomials

To multiply two polynomials with more than one term, we use the distributive property which plays a central role in the process and leads directly to the following rule

Multiply each term of the first polynomial by each term of the second and then combine like terms.

EXAMPLE 2   Multiplying a Monomial by a Polynomial

Find the product of $-2t^4$ and $-3t^3 - 7t + 4$ and then simplify.

Solution
$$-2t^4(-3t^3 - 7t + 4) = (-2t^4)(-3t^3) + (-2t^4)(-7t) + (-2t^4)(4)$$
$$= 6t^7 + 14t^5 - 8t^4$$ ◈

The multiplication of polynomials can be carried out by using a horizontal or a vertical format as the next example shows.

EXAMPLE 3   Multiplying Polynomials

Find the product of $(3t - 1)(2t^2 + 7t + 4)$.

Solution   First carry out the multiplication horizontally.

$(3t - 1)(2t^2 + 7t + 4)$ Use the distributive property
$= 3t(2t^2 + 7t + 4) - 1(2t^2 + 7t + 4)$
$= (3t)(2t^2) + (3t)(7t) + (3t)(4)$
$\quad - 1(2t^2) - 1(7t) - 1(4)$ Distributive property
$= 6t^3 + 21t^2 + 12t - 2t^2 - 7t - 4$ Multiply
$= 6t^3 + 19t^2 + 5t - 4$ Combine like terms

The same work is arranged vertically as follows:

$$
\begin{array}{r}
2t^2 + 7t + 4 \\
\times\ 3t - 1 \\
\hline
6t^3 + 21t^2 + 12t \qquad\qquad\quad \\
-2t^2 - 7t - 4 \\
\hline
6t^3 + 19t^2 + 5t - 4
\end{array}
$$

$\longleftarrow (2t^2 + 7t + 4)(3t)$
$\longleftarrow (2t^2 + 7t + 4)(-1)$ ◈

Multiplying Binomials

To multiply two binomials, once again we can apply the distributive property. We use the horizontal format and try to discover a method to enable us to perform the multiplication efficiently. For example, to multiply $(x + 2)(x + 3)$, we treat the binomial $x + 3$ as a single quantity and multiply as follows:

$$(x + 2)(x + 3) = x(x + 3) + 2(x + 3)$$
$$= x^2 + 3x + 2x + 6$$
$$= x^2 + 5x + 6$$

The result of the preceding multiplication is a trinomial whose terms are determined as follows:

First Term:

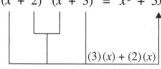

$$(x + 2)(x + 3) = x^2 + 5x + 6$$

Middle Term:

$$(x + 2)(x + 3) = x^2 + 5x + 6$$

$$(3)(x) + (2)(x)$$

Last Term:

$$(x + 2)(x + 3) = x^2 + 5x + 6$$

$$(2)(3)$$

A shortened process for multiplying the above binomials is called the **FOIL** method. It provides a speedy process for organizing the multiplication of the two binomials. **FOIL** helps the reader to remember the pattern: the product of the **First** terms, plus the product of the **Outside** terms, plus the product of the **Inside** terms, plus the product of the **Last** terms. The four operations are outlined in Figure 1.

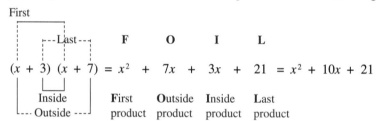

$$(x + 3)(x + 7) = x^2 + 7x + 3x + 21 = x^2 + 10x + 21$$

Figure 1

EXAMPLE 4   Multiplying Binomials by FOIL

Multiply the binomials and simplify the results.
(a) $(x + 2)(2x - 3)$
(b) $(7x + 2)(3x - 5)$

Solution   (a) We distribute $x + 2$ over the sum $2x - 3$ using the **FOIL** method.
   (i) $(2x)(x)$ is the product of the **First** terms.
   (ii) $(-3)(x)$ is the product of the **Outer** terms.
   (iii) $(2)(2x)$ is the product of the **Inner** terms.
   (iv) $(-3)(2)$ is the product of the **Last** terms.
   The Inner and Outer products can be added. Thus the product is

$$(x + 2)(2x - 3) = 2x^2 - 3x + 4x - 6 = 2x^2 + x - 6$$

(b) We distribute $7x + 2$ over $3x - 5$ using the **FOIL** method.
   (i) $(7x)(3x)$ is the product of the **First** terms.
   (ii) $(7x)(-5)$ is the product of the **Outer** terms.
   (iii) $(2)(3x)$ is the product of the **Inner** terms.
   (iv) $(2)(-5)$ is the product of the **Last** terms.
   The Inner and Outer products can be added. Thus the product is

$$(7x + 2)(3x - 5) = 21x^2 - 35x + 6x - 10$$
$$= 21x^2 - 29x - 10$$

Using Special Products

There are some special products that occur so frequently in algebra that it is worth learning special formulas dealing with them.

SPECIAL PRODUCTS OF POLYNOMIALS

Assume that a and b represent real numbers or algebraic expressions	
1. $(a + b)(a - b) = a^2 - b^2$	difference of squares
2. $(a + b)^2 = a^2 + 2ab + b^2$	squaring a binomial
3. $(a - b)^2 = a^2 - 2ab + b^2$	squaring a binomial
4. $(a + b)(a^2 - ab + b^2) = a^3 + b^3$	sum of cubes
5. $(a - b)(a^2 + ab + b^2) = a^3 - b^3$	difference of cubes

EXAMPLE 5   Multiplying by Using Special Products

Find each product.
(a) $(3x + 2)^2$ (b) $(1 - 7t)^2$
(c) $(2m - 3)(2m + 3)$ (d) $(u + 2)(u^2 - 2u + 4)$

Solution   (a) Using special product 2 with $a = 3x$ and $b = 2$, we have

$(x - 5)^2$ is not the same as $x^2 + 25$
$$(3x + 2)^2 = (3x)^2 + 2(3x)(2) + 2^2 = 9x^2 + 12x + 4$$

(b) Using special product 3 with $a = 1$ and $b = 7t$, we have
$$(1 - 7t)^2 = (1)^2 - 2(1)(7t) + (7t)^2 = 1 - 14t + 49t^2$$

(c) Using special product 1 with $a = 2m$ and $b = 3$, we have
$$(2m - 3)(2m + 3) = (2m)^2 - (3)^2 = 4m^2 - 9$$

(d) Using special product 4 with $a = u$ and $b = 2$, we have
$$(u + 2)(u^2 - 2u + 4) = u^3 + 2^3 = u^3 + 8$$

Solving Applied Problems

The product of polynomial expressions can be used to solve different types of
applied problems as the next example shows.

EXAMPLE 6   Finding the Area of a Sail

Figure 2 shows a sail in the shape of a triangle whose base is $2x + 3$ units and
whose height is $3x - 1$ units, where x is a variable.
(a) Find a polynomial expression to express the area of the sail in terms of x and
simplify the expression.
(b) Find the area of the sail if $x = 12$ feet.

Solution   (a) The formula for the area A of a triangle is given by

$$A = \frac{1}{2}bh$$

Substituting $2x + 3$ for b and $3x - 1$ for h, we have

$$A = \frac{1}{2}(2x + 3)(3x - 1)$$

By multiplying the polynomials, we have

$$A = \frac{1}{2}(6x^2 + 7x - 3)$$

$$A = 3x^2 + \frac{7}{2}x - \frac{3}{2}$$

(b) Replacing x by 12, we have

$$A = 3(12)^2 + \frac{7}{2}(12) - 1.5$$
$$= 3(144) + 7(6) - 1.5$$
$$= 432 + 42 - 1.5 = 472.5$$

so that the area is 472.5 square feet.

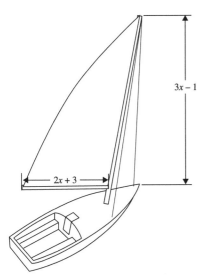

Figure 2

PROBLEM SET 1.2

Mastering the Concepts

In Problems 1–4, perform each multiplication and simplify.

1. (a) $(2x^2)(3x^4)$ (b) $(-5t^3)(6t^4)$
2. (a) $(7m^4)(-8m^3)$ (b) $(-3x^7)(6x^5)$
3. (a) $(3y)(5y^3)(-2y^2)$ (b) $(-10p^3)(2p)(-3p^4)$
4. (a) $(2b)(3b^5)(-2b^2)$ (b) $(5x^2)(6x)(-2x^4)$

In Problems 5–8, find each product and simplify.

5. (a) $-y(2y-3)$ (b) $t^2(3t+2)$
6. (a) $3x^3(2x-4)$ (b) $3u(2u^2-5)$
7. (a) $5y^2(3y^4+7y^3)$ (b) $-2x^2(2x^2+3x+5)$
8. (a) $4c^2(4c^3-2c^2+c)$ (b) $-3x^4(-2x+4x^2+5x^3-2)$

In Problems 9–22, use the vertical format to find each product.

9. $(3x+2)(3x^2+5x-4)$
10. $(2u-3)(u^2-3u+2)$
11. $(3y+5)(y^2-5y+7)$
12. $(4t^2+7)(t^3+5t-1)$
13. $(2u^2-3)(2u^2-2u+4)$
14. $(x^2+3)(x^2-2x+5)$
15. $(3x+2)(3x^2+5x-4)$
16. $(x-2y)(x^2-3xy-y^2)$
17. $(m^2+3)(m^3+2m^2-3m+4)$
18. $(4-x^2)(x^4-2x^2-3x+4)$
19. $(y^2-5y+3)(y^2+2y+3)$
20. $(u^2-2u+5)(u^2+3u+1)$
21. $(x^2+2xy-y^2)(x^3-3x^2y+y^3)$
22. $(2t^2-5t+s^2)(t^3-t^2s+s^2t-s^3)$

In Problems 23–30, perform the multiplication of the binomials and simplify the results.

23. (a) $(x+1)(x+2)$ (b) $(x-1)(x-2)$
24. (a) $(x+3)(x+4)$ (b) $(x-3)(x-4)$
25. (a) $(x+5)(x-3)$ (b) $(x-5)(x+3)$
26. (a) $(x-6)(x+2)$ (b) $(x+6)(x-2)$
27. (a) $(5x+4)(x-1)$ (b) $(5x-4y)(x+y)$
28. (a) $(3x+2)(2x-4)$ (b) $(3x-2y)(2x+4y)$
29. $(2x^2+3)(5x^2-1)$ 30. $(x^2+y^3)(x^3-y)$

In Problems 31–38, use a special product to perform the multiplication.

31. (a) $(w-7)(w+7)$ (b) $(7-w)(7+w)$
32. (a) $(2m-9)(2m+9)$ (b) $(9+2m)(9-2m)$
33. (a) $(x+1)^2$ (b) $(x-1)^2$
34. (a) $(x+4)^2$ (b) $(x-4)^2$
35. (a) $(4y^2+5z)^2$ (b) $(4y^2-5z)^2$
36. (a) $(9y^2+8z)^2$ (b) $(9y-8z)^2$
37. (a) $(x+y)(x^2-xy+y^2)$ (b) $(x-y)(x^2+xy+y^2)$
38. (a) $(x-2y)(x^2+2xy+4y^2)$
     (b) $(x+2y)(x^2-2xy+4y^2)$

Applying the Concepts

39. **Geometry:** The length of a rectangle is 5 feet more than its width. Let x (in feet) be the length of the rectangle.
    (a) Write an expression that represents the area of the rectangle.
    (a) Simplify the expression, then find the area of the rectangle if its length is 8 feet.

40. **Fliers Design:** Advertising fliers are to be printed on rectangular sheets of paper x centimeters long and $(2x-5)$ centimeters wide. Suppose the margins at the top and bottom are each 3 centimeters and the margins at the sides are each 2 centimeters (Figure 3).

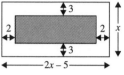

Figure 3

   (a) Write an expression that represents the area of the printed part of the flier.
   (b) Find the area of the printed part of the flier if $x = 15$ centimeters.

41. **Agriculture:** The total yield from a crop is the number of trees times the yield per tree. An orange grower has determined that if 120 trees are planted per acre, each tree will yield approximately 60 boxes of oranges over the growing season. If the number of trees is increased, the yield per tree will be reduced due to crowding. For each additional tree planted over 120, the yield is expected to decrease by 2 boxes per tree. Let x represent the number of trees planted over 120 per acre.
   (a) Write an expression that represents the total yield in terms of x.
   (b) Simplify the resulting expression and find the total yield if 125 trees are planted ($x = 5$).

42. **Package Design:** An open box is to be constructed by removing equal squares, each of length x centimeters, from the corners of a piece of cardboard that measures 48 centimeters wide and 32 centimeters long, and then folding up the sides (Figure 4).

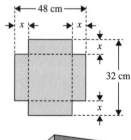
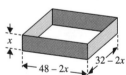

Figure 4

   (a) Write an expression that represents the volume of the box.
   (b) Simplify the resulting expression; then find the volume when $x = 6$ centimeters.

Developing and Extending the Concepts

In Problems 43–56, perform each operation and simplify.

43. $2x(1-2x)^2$ 44. $-4x^2(5x+10)^2$
45. $-(2x+7)(x+4)(x-3)$
46. $-(x-4)(3x-5)(x-3)$
47. $(x-4)(3x+4)^2$ 48. $(x+7)^2(3x-1)$
49. $3x(2x-1)^3$ 50. $-5x^2(10x-2)^3$
51. $(3x-2)^4$ 52. $(4-2x)^4$
53. $(x^2+3)(x^3+2x^2-3x+4)$
54. $(4-x^2)(x^4-2x^2-3x+4)$
55. $(x^2+2xy+y^2)(x^2-2xy+y^2)$
56. $(x^3-3x^2y+3xy^2-y^3)(x^3+3x^2y+3xy^2+y^3)$

OBJECTIVES

1. Factor Out Common Factors
2. Factor By Grouping
3. Factor Trinomials of the Form $x^2 + bx + c$
4. Factor Trinomials of the Form $ax^2 + bx + c$
5. Factor Special Products
6. Factor by Combining Methods
7. Solve Applied Problems

1.3 Factoring Polynomials

From the last section on multiplying polynomials we know that

$$(3x + 2)(5x - 1) = 15x^2 + 7x - 2$$

In this section, we *reverse* the process of multiplication and write a polynomial as a product of its factors. That is, we start with the expression $15x^2 + 7x - 2$ and obtain the factors $3x + 2$ and $5x - 1$. This process is called **factoring** and $3x + 2$ and $5x - 1$ are called **factors**. A polynomial that has no factors other than itself and 1 is called **prime**. When a polynomial is written as a product of prime factors, we say that it is **factored completely.**

Factoring Out Common Factors

The distributive property,

$$P(Q + R) = PQ + PR$$

provides a bridge between factors and products. If you write the distributive property in the "reverse" order,

$$PQ + PR = P(Q + R)$$

you obtain the general principle of common factoring. In the example above, P is a common factor of PQ and PR on the left side of the equation. The factors on the right side of the equation are P and $Q + R$.

The polynomial expression $x^2 + x$ can be written as $x \cdot x + x \cdot 1$. Thus, by applying the distributive property, we can factor out the common factor x. Therefore, the polynomial is factored completely as follows:

$$x^2 + x = x(x + 1)$$

When factoring a polynomial expression, we begin by determining the *greatest common factor* (GCF) of its terms. The GCF is the largest monomial that is a factor of each term of the polynomial. Once the GCF is found, we apply the distributive property and write the polynomial as a product, with the GCF as one of the factors. For example, to factor

$$28x^7 - 21x^4 + 35x^2$$

we see that $7x^2$ is the greatest common factor of the polynomial. By dividing each term of the polynomial by $7x^2$, the second factor is found.

$$28x^7 - 21x^4 + 35x^2 = 7x^2(4x^5 - 3x^2 + 5)$$

We can use the distributive property to check the factoring. Note that $4x^5 - 3x^2 + 5$ has no common factors.

EXAMPLE 1   Factoring Out Common Factors

Factor out common factors in each expression.
(a) $-x^2 + x$ (b) $(x + 2)5x + (x + 2)2$ (c) $2x^2 - 3y$

Solution   (a) We may factor out $-x$ and get $-x^2 + x = -x(x - 1)$ or we may factor out x and get $-x^2 + x = x(-x + 1)$. Both forms are acceptable and considered prime.
(b) The binomial $x + 2$ is the common factor and

$$(x + 2)5x + (x + 2)2 = (x + 2)(5x + 2)$$

(c) The expression $2x^2 - 3y$ has no common factors. It is prime.

Factoring by Grouping

The process of *factoring by grouping* terms is especially useful when a polynomial has four or more terms with no common monomial factor. In this case, we have to rearrange the terms in order to regroup them successfully. For example, to factor

$$3xm - 2x + 3ym - 2y$$

we begin by grouping the first two terms and the last two terms. Finding a common factor, we have

$$(3xm - 2x) + (3ym - 2y) = x(3m - 2) + y(3m - 2)$$

Notice that now there is a common factor, the binomial $(3m - 2)$, and the expression can be factored as

$$x(3m - 2) + y(3m - 2) = (3m - 2)(x + y)$$

The expression could also be rearranged as

$$(3xm + 3ym) - (2x + 2y) = 3m(x + y) - 2(x + y)$$
$$= (x + y)(3m - 2)$$

which is the same factorization.

EXAMPLE 2   Factoring by Grouping

Factor $ac - d - c + ad$ by grouping.

Solution $ac - d - c + ad = ac - c + ad - d$ Rearrange terms
$= c(a - 1) + d(a - 1)$ Take out common factor $a - 1$
$= (a - 1)(c + d)$ ◈

Factoring Trinomials of the Form $x^2 + bx + c$

We have seen that the product of two binomials may be a trinomial. This suggests that some trinomials may be factored as a product of two binomials. For example, consider the product of the binomials $x + 4$ and $x + 7$.

$$(x + 4)(x + 7) = x^2 + 7x + 4x + (4)(7)$$

$$= x^2 + \underline{11x} + 28$$

Product of the first terms x & x

Product of the last terms 4 & 7

Sum of inner and outer products $7x$ & $4x$

We notice that the product of the first terms of the binomials is the first term of the trinomial. The product of the last terms of the binomials is the last term of the trinomial. The sum of the products of the outer and inner terms of the binomial is the middle term of the trinomial. The relationship between the coefficients of the trinomial and the coefficients of the factors is illustrated in the following diagrams:

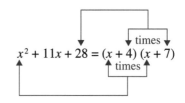

The coefficient of the middle term in the trinomial, namely 11, is obtained as follows:

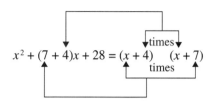

$$x^2 + (7 + 4)x + 28 = (x + 4)\,(x + 7)$$

These diagrams illustrate how to factor the trinomial $x^2 + 11x + 28$. We find two integers whose product is 28 and whose sum is 11. Thus the correct factorization is

$$x^2 + 11x + 28 = (x + 4)\,(x + 7).$$

The following strategy will shorten our work when factoring a trinomial.

TABLE 1

Trinomial	Signs of b and c	Signs of the Two Factors
$x^2 + bx + c$	signs of b and c are positive	positive
$x^2 - bx + c$	sign of b is negative and c is positive	negative
$x^2 - bx - c$	sign of c is negative	one sign is positive the other is negative

EXAMPLE 3   Factoring Trinomials

Factor each trinomial.

(a) $x^2 + 9x + 20$ (b) $x^2 + x - 20$

Solution   (a) To factor this trinomial, observe that all the coefficients are positive. So, we need two positive integers whose product is 20, and whose sum is 9. The integers are 4 and 5. Thus,

┌--- The product of 4 and 5 is 20

$$x^2 + 9x + 20 = (x + 4)\,(x + 5)$$

└--- The sum of 4 and 5 is 9

(b) Since the constant term of this trinomial is -20, we need two numbers with different signs so that their product is -20 and their sum is $+1$. The numbers are -4 and 5. Thus,

┌--- The product of -4 and 5 is -20

$$x^2 + x - 20 = (x + 5)\,(x - 4)$$

└--- The sum of -4 and 5 is 1

Factoring Trinomials of the Form $ax^2 + bx + c$

A trinomial whose leading coefficient is not 1 has many more possible factors than our previous examples. One possible method to factor such a trinomial is to list all the factors and identify the correct factorization. For example, the trinomial

$$6x^2 + 19x + 10$$

has two possible sets of factors of $6x^2$, $6x \cdot x$ and $3x \cdot 2x$, and two possible factors of $+10$, $1 \cdot 10$ and $2 \cdot 5$. All the possible binomials are shown in Table 2.

TABLE 2

Possible Factors	Sum of Product of Outside and Inside Terms	Middle Term of Trinomial
$(6x + 1)(x + 10)$	$60x + x$	$61x$
$(6x + 10)(x + 1)$	$6x + 10x$	$16x$
$(6x + 2)(x + 5)$	$30x + 2x$	$32x$
$(6x + 5)(x + 2)$	$12x + 5x$	$17x$
$(3x + 1)(2x + 10)$	$30x + 2x$	$32x$
$(3x + 10)(2x + 1)$	$3x + 20x$	$23x$
$(3x + 2)(2x + 5)$	$15x + 4x$	$19x$
$(3x + 5)(2x + 2)$	$6x + 10x$	$16x$

Only one pair of factors produces the correct middle term, $19x$. Thus,

$$(3x + 2)(2x + 5) = 6x^2 + 19x + 10$$

This method is appropriately called "factoring by trial and error." A shorter method is to consider the factors of $6 \cdot 10$, $a \cdot c$, whose sum is $+19$. If necessary we can construct a table (Table 3) to have an orderly search.

The factors of 60 whose sum is 19 are 4 and 15 and $4x + 15x = 19x$. Rewrite the trinomial as:

$$6x^2 + 19x + 10 = 6x^2 + 4x + 15x + 10$$

Now factor by grouping

$$= 2x(3x + 2) + 5(3x + 2)$$
$$= (3x + 2)(2x + 5)$$

TABLE 3

Factors of $+10 \cdot 6 = 60$	Sum = $+19$
1(60)	$1 + 60 = 61$
2(30)	$2 + 30 = 32$
3(20)	$3 + 20 = 23$
4(15)	$4 + 15 = 19$
5(12)	$5 + 12 = 17$
6(10)	$6 + 10 = 16$

EXAMPLE 4   Factoring Trinomials

Factor each trinomial:
(a) $6x^2 - 11x - 10$ (b) $6x^2 - 19x + 3$

Solution   (a) We must find two numbers whose product is $(-10) \cdot 6 = -60$ and whose sum is -11. Table 4 shows all the factors of -60.

The factors 4 and -15 add to -11. Rewrite the trinomial and factor by grouping. We have,

$$6x^2 - 11x - 10 = 6x^2 - 15x + 4x - 10$$
$$= 3x(2x - 5) + 2(2x - 5)$$
$$= (2x - 5)(3x + 2)$$

(b) We begin by writing

$$6x^2 - 19x + 3 = (\,?x - 3)(\,?x - 1)$$

The signs must be negative to produce the positive constant, 3, and the middle term -19. Since

$$6 = 6 \cdot 1 = 3 \cdot 2 = 2 \cdot 3 = 1 \cdot 6$$

there are four possible ways to fill the two question marks. We try them one at a time until we find that $(x - 3)(6x - 1)$ works. Therefore,

$$6x^2 - 19x + 3 = (x - 3)(6x - 1)$$

TABLE 4

Factors of $-10 \cdot 6 = -60$	Sum = -11
1(−60)	$1 - 60 = -59$
2(−30)	$2 - 30 = -28$
3(−20)	$3 - 20 = -17$
4(−15)	$4 - 15 = -11$
5(−12)	$5 - 12 = -7$
6(−10)	$6 - 10 = -4$

Factoring by Using Special Products

Success in factoring depends on our ability to recognize multiplication patterns in the polynomials to be factored. The following special products suggest such useful patterns (Table 5).

TABLE 5

Name	Pattern
1. perfect square	$a^2 + 2ab + b^2 = (a + b)^2$
2. perfect square	$a^2 - 2ab + b^2 = (a - b)^2$
3. difference of two squares	$a^2 - b^2 = (a - b)(a + b)$
4. sum of two cubes	$a^3 + b^3 = (a + b)(a^2 - ab + b^2)$
5. difference of two cubes	$a^3 - b^3 = (a - b)(a^2 + ab + b^2)$

EXAMPLE 5 Factoring by Using Special Products

Factor each binomial by using special products.
(a) $9x^2 - 25$ (b) $25x^2 - 64$ (c) $16x^2 + 49$

Solution Both $9x^2 - 25$ and $25x^2 - 64$ in (a) and (b) fit special product 3.
(a) $9x^2 - 25 = (3x)^2 - (5)^2 = (3x - 5)(3x + 5)$
(b) $25x^2 - 64 = (5x)^2 - (8)^2 = (5x - 8)(5x + 8)$
(c) $16x^2 + 49$ does not fit the pattern since it is the sum of two squares, not the difference. To justify this, try the following products:

$$(4x - 7)(4x + 7) = 16x^2 - 49$$
$$(4x - 7)(4x - 7) = (4x - 7)^2 = 16x^2 - 56x + 49$$
$$(4x + 7)(4x + 7) = (4x + 7)^2 = 16x^2 + 56x + 49$$

Notice that none of them is equal to $16x^2 + 49$; so $16x^2 + 49$ is not factorable. ◈

EXAMPLE 6 Factoring by Using Special Products

Factor each expression by using special products.
(a) $x^3 + 27$ (b) $125x^3 - 64y^6$

Solution (a) We apply special product 4 to get

$$x^3 + 27 = (x)^3 + (3)^3 = (x + 3)(x^2 - 3x + 9)$$

(b) By using special product 5, we obtain

$$125x^3 - 64y^6 = (5x)^3 - (4y^2)^3 = (5x - 4y^2)[(5x)^2 + (5x)(4y^2) + (4y^2)^2]$$
$$= (5x - 4y^2)(25x^2 + 20xy^2 + 16y^4)$$ ◈

Factoring by Combining Methods

Factoring some polynomials may require using more than one technique. The following strategy may be helpful when factoring polynomials.

TABLE 6

To Factor Any Polynomial

1. Use the distributive property to factor out the GCF, if there is one.

2. If the polynomial has four terms, try to factor by grouping.

3. If the polynomial has three terms, try to factor into two binomials.

4. If the polynomial has two terms, try the special products $3 - 5$.

EXAMPLE 7   Factoring Completely

Factor each expression completely.

(a) $3x^2 - 243$ (b) $-4z^3 + 24z^2 + 64z$ (c) $x^4 - 1$ (d) $5t^6 + 40$

Solution   (a) $3x^2 - 243 = 3(x^2 - 81)$ 3 is the greatest common factor
$\quad\quad = 3(x - 9)(x + 9)$ Use special product 3

(b) $-4z^3 + 24z^2 + 64z$
$\quad\quad = -4z(z^2 - 6z - 16)$ $-4z$ is the common factor
$\quad\quad = -4z(z - 8)(z + 2)$ Factor the trinomial

(c) $x^4 - 1 = (x^2 - 1)(x^2 + 1)$ Use special product 3
$\quad\quad = (x - 1)(x + 1)(x^2 + 1)$ Use special product 3 again

(d) $5t^6 + 40 = 5(t^6 + 8)$ 5 is the common factor
$\quad\quad = 5[(t^2)^3 + 2^3]$ Write in the form $a^3 + b^3$
$\quad\quad = 5(t^2 + 2)(t^4 - 2t^2 + 4)$ Use special product 4

Some polynomial expressions of degree higher than 3 can be factored in the same way as those of degrees 2 and 3 as the following example shows.

EXAMPLE 8   Factoring Higher-Degree Polynomials

Factor $x^4 + 2x^2 + 9$

Solution   It may be easier to see the structure of this polynomial with a substitution. We think of x^4 as $(x^2)^2$. If we write $u = x^2$, then $u^2 = x^4$, and

$x^4 + 2x^2 + 9 = u^2 + 2u + 9$ Substitution
$\quad\quad = u^2 + 6u - 4u + 9$ $u^2 + 6u + 9$ is a perfect square
$\quad\quad = (u + 3)^2 - 4u$
$\quad\quad = (x^2 + 3)^2 - 4x^2$ Use special product 3
$\quad\quad = (x^2 + 3 - 2x)(x^2 + 3 + 2x)$

Solving Applied Problems

Formulas from geometry and other sciences can be written in a factored form as the next example shows.

EXAMPLE 9   Area of a Field

A flower seed company uses a rectangular piece of land as a test plot. If x is the number of rows plowed in this plot, then the area (in square meters) plowed is given by the polynomial

$$A = 15x^2 + 13x - 6$$

Factor the right side of this formula completely. Find the area A, the length and width when 5 rows are plowed.

Solution   The right side of the formula is completely factored as follows:

$$A = (3x - 1)(5x + 6)$$

Since the area of a rectangle is length times width, $(A = lw)$, one of the factors represents the length and the other factor represents the width.

Replacing x by 5, we have:

$$A = [3(5) - 1][5(5) + 6]$$
$$= [15 - 1][25 + 6]$$
$$= (14)(31)$$
$$= 434 \text{ m}^2$$

So if 5 rows are plowed, the area of the field is 434 square meters. The length of the field is 31 meters and the width is 14 meters.

PROBLEM SET 1.3

Mastering the Concepts

In Problems 1–10, factor out the common factors in each expression.

1. (a) $x^2 - x$ (b) $4x^2 + 2x$
2. (a) $10x^2 - 5x$ (b) $4x^2 + 7xy$
3. (a) $a^2b - ab^2$ (b) $17x^3y^2 - 34x^2y$
4. (a) $12a^3b^2 + 36a^2b^3$ (b) $6ab^2 + 30a^2b$
5. (a) $12x^3y - 48x^2y^2$ (b) $4x^3 - 2x^2 + x$
6. (a) $4xy^2z + x^2y^2z - x^3y^3$
    (b) $x^3y^2 + x^2y^3 + 2xy^4$
7. (a) $9m^2n + 18mn^2 - 27mn$
    (b) $8xy^2 + 24x^2y^3 + 4xy^3$
8. (a) $3x^2y^2 + 6x^2z^2 - 9x^2$ (b) $2a^3b - 8a^2b^2 - 6ab^3$
9. (a) $3x(2a + b) + 5y(2a + b)$
    (b) $(2m + 3)x - (2m + 3)$
10. (a) $x(y - z) - (z - y)$ (b) $m(x - y) + (y - x)$

In Problems 11–16, factor each expression by grouping (if possible).

11. (a) $ax + ay + bx + by$ (b) $x^2a + x^2b + a + b$
12. (a) $ax^2 + b - bx^2 + a$ (b) $yz + 2y - z - 2$
13. (a) $ab^2 - b^2c - ad + cd$ (b) $2x^2 - yz^2 - x^2y + 2z^2$
14. (a) $x^3 + x^2 - 5x - 5$ (b) $x^2 - ax + bx - ab$
15. (a) $ax + bx + ay + by + a + b$
    (b) $2ax - b + 2bx - c + 2cx - a$
16. (a) $2ax + by - 2ay - bx$ (b) $x^2 + ax + bx + ab$

In Problems 17–32, factor each trinomial.

17. (a) $x^2 + 16x + 48$ (b) $x^2 - 16x + 48$
18. (a) $x^2 - 28x + 27$ (b) $x^2 + 28x + 27$
19. (a) $x^2 + 15x + 36$ (b) $x^2 + 20x + 36$
20. (a) $x^2 + 13x + 36$ (b) $x^2 + 12x + 36$
21. (a) $x^2 - 2x - 24$ (b) $x^2 + 2x - 24$
22. (a) $x^2 + x - 30$ (b) $x^2 - x - 30$
23. (a) $x^2 - 15x - 16$ (b) $x^2 - 6x - 16$
24. (a) $x^2 + 5x - 36$ (b) $x^2 + 16x - 36$
25. (a) $x^2 - 7x - 18$ (b) $x^2 - 17x + 30$
26. (a) $12 - x^2 - 4x$ (b) $40 - 3x - x^2$
27. (a) $2w^2 + 7w + 3$ (b) $2w^2 + 20w - 6$
28. (a) $5y^2 - 11y + 2$ (b) $4x^2 - 35xy - 9y^2$
29. (a) $6x^2 + 13x + 6$ (b) $6y^2 - y - 7$
30. (a) $12v^2 + 17v - 5$ (b) $6c^2 - 7cd - 3d^2$
31. (a) $12 - 2w^2 - 5w$ (b) $18x^2 + 101x + 90$
32. (a) $6rs + 5r^2 - 8s^2$ (b) $24x^2 - 67xy + 8y^2$

In Problems 33–40, factor each expression by special product (if possible).

33. (a) $x^2 - 100$ (b) $100 - x^2$
34. (a) $x^2 - 36$ (b) $36 - x^2$
35. (a) $16x^2 - 49$ (b) $16x^2 - 49y^4$
36. (a) $25x^2 - 64$ (b) $25x^2 - 64y^2$
37. (a) $16u^2 - 25v^2$ (b) $25m^2 - 49n^2$
38. (a) $(a - b)^2 - 100c^2$ (b) $144p^2 - (q - 3)^2$
39. (a) $x^2 + 30x + 225$ (b) $x^2 - 30xy + 225y^2$
40. (a) $x^2 - 22x + 121$ (b) $x^2 + 22xy + 121y^2$

In Problems 41–52, factor each expression completely.

41. (a) $x^3 + 1$ (b) $y^3 + 125$
42. (a) $27w^3 + z^3$ (b) $x^3y^3 - 64$
43. (a) $w^3 - 8y^3z^3$ (b) $64z^3 + 27b^3$
44. (a) $49x^2 - 16y^4$ (b) $16x^4y^4 - 1$
45. (a) $81x^4 - 1$ (b) $256x^4 - y^4$
46. (a) $625x^4 - 81y^4$ (b) $t^4 - 81y^4$
47. (a) $x^9 - 1$ (b) $y^9 + 512$
48. (a) $u^6 - 27$ (b) $(x + 2)^3 - y^3$
49. (a) $x^4 + x^2y^2 + y^4$ (b) $4x^4 + y^4$
50. (a) $9x^4 + 2x^2y^2 + y^4$ (b) $25x^4 + 4x^2y^2 + 4y^4$
51. (a) $5x^3 - 55x^2 + 140x$ (b) $x^2yz^2 - xyz^2 - 12yz^2$
52. (a) $16x^4 - x^2 + 6xy - 9y^2$ (b) $x^4 - x^2 + 4x - 4$

Applying the Concepts

53. **Geometry:** If the length of a rectangle is $4x + 3$ units and its area is $28x^2 + 13x - 6$ square units, use factoring to find an expression for its width in terms of x.

54. **Total Revenue:** The total revenue generated by selling $4x + 5$ units of a commodity is given by $36x^2 + 17x - 35$ dollars. Use factoring to find an expression for price per unit in terms of x. [Hint: total revenue = (price of units sold) × (number of units sold)].

55. **Geometry:** The surface area A of an open cylindrical can of radius $2x + 1$ units and height $3x - 2$ units (Figure 1), is given by the formula

    $$A = 16\pi x^2 + 2\pi x - 3\pi$$

    (a) Factor the right side of this formula completely.
    (b) What is the surface area of the can if we replace x by 2 inches?

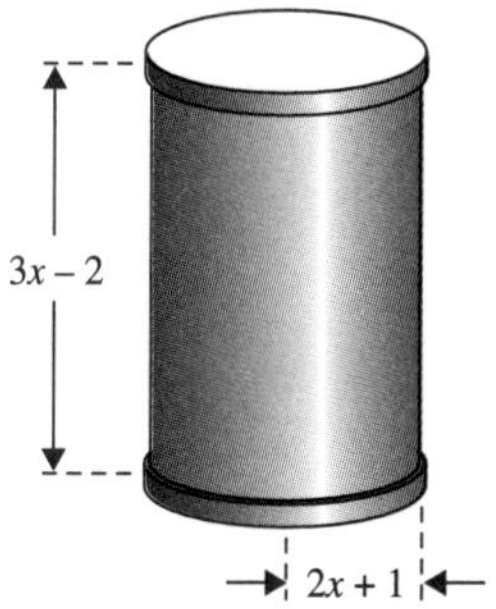

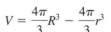

Figure 1

56. **Material for a Tank:** To construct a spherical cement tank whose inside radius is r units and outside radius is R units (Figure 2), the volume V of cement needed is given by the formula

    $$V = \frac{4\pi}{3}R^3 - \frac{4\pi}{3}r^3$$

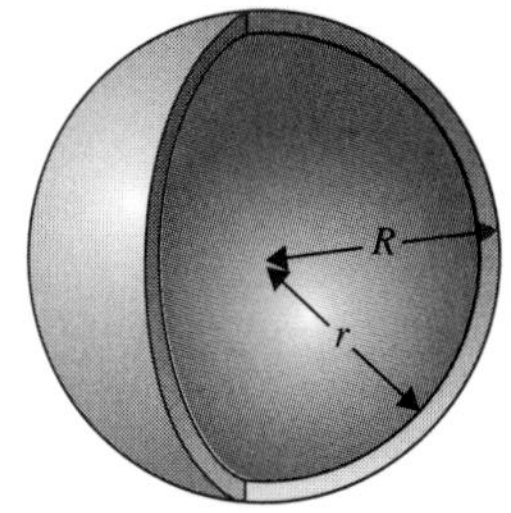

Figure 2

    (a) Factor the right side of this formula completely.
    (b) If $r = 9$ feet and $R = 9.25$ feet, how many cubic yards of cement are needed? (Use 9 cubic feet = 1 cubic yard)

Developing and Extending the Concepts

In Problems 57–62, factor each expression completely.

57. (a) $3x^5 - 48xy^4$ (b) $3t^4 - 24t$
58. (a) $54x^4 - 2xy^3$ (b) $36t - 4t(y + 3)^2$
59. (a) $64u^6 - 729$ (b) $3t^8 + 81t^2$
60. (a) $2u^7 - 128u$ (b) $7x^7y + 7xy^7$

61. (a) $2x^3 + x^2y - x^2 + 2xy + y^2 - y$
    (b) $2ax - b + 2bx - c + 2cx - a$
62. (a) $x^2 - y^2 + 2x - 2zy + 1 - z^2$
    (b) $x^2 + y^2 - z^2 - 9 + 2xy - 6z$
63. Find all values of b so that each trinomial is factorable.
    (a) $x^2 + bx - 8$ (b) $x^2 + bx + 24$

OBJECTIVES

1. Evaluate Rational Expressions
2. Reduce Rational Expressions
3. Build Rational Expressions to Higher Terms
4. Identify Equal Rational Expressions
5. Determine Signs of Rational Expressions
6. Solve Applied Problems

1.4 Rational Expressions

Recall from section R5 that a rational number can be expressed in the form:

$$\frac{a}{b}, \text{ where } a \text{ and } b \text{ are integers and } b \neq 0$$

In this section we extend our work to include expressions that can be written as a ratio of two polynomials. An algebraic expression of the form

$$\frac{P}{Q}, \text{ where } P \text{ and } Q \text{ are polynomials, with } Q \neq 0$$

is called a **rational expression**. Examples of rational expressions are:

$$\frac{2}{x}, \quad \frac{-3}{x - 2}, \quad \frac{x + 1}{x^2 + x - 3}, \quad \frac{c^2 + 4c - 21}{c^2 + c - 6}, \quad \text{and} \quad \frac{16x^2 - 9}{7}$$

Evaluating Rational Expressions

Rational expressions have specific values when real numbers are substituted for the variables (or letters). For instance, when $x = 4$, the value of the expression $\big(x/(x - 2)\big)$ is given by

$$\frac{x}{x - 2} = \frac{4}{4 - 2} = \frac{4}{2} = 2$$

In the same expression, when $x = 2$, we have

$$\frac{x}{x - 2} = \frac{2}{2 - 2} = \frac{2}{0}$$

which is undefined because the denominator is 0.

In general, a rational expression is **undefined** for those values of the variable which makes the denominator zero.

Unless told otherwise, we shall automatically assume that the variables in any rational expression are restricted to numerical values that will give a nonzero denominator. Throughout the book, we shall not always list these restrictions.

EXAMPLE 1   Determining when an Expression Is Undefined

Find the values of x for which the following expression is undefined.

$$\frac{(x - 1)^2}{(x - 1)(x - 2)}$$

Solution   The rational expression is undefined at those values of the variable which give the denominator a value of zero. These values are the solutions to the equation

$$(x - 1)(x - 2) = 0$$
$$x - 1 = 0 \text{ or } x - 2 = 0$$
$$x = 1 \text{ or } x = 2$$

For $x = 1$ we have

$$\frac{(x - 1)^2}{(x - 1)(x - 2)} = \frac{(1 - 1)^2}{(1 - 1)(1 - 2)} = \frac{0}{0(-1)} = \frac{0}{0}$$

and for $x = 2$, we have

$$\frac{(x - 1)^2}{(x - 1)(x - 2)} = \frac{(2 - 1)^2}{(2 - 1)(2 - 2)} = \frac{1}{1(0)} = \frac{1}{0}$$ ◈

Reducing Rational Expressions

Rational expressions, like rational numbers, can be written in **lowest terms, reduced to lowest terms** or **simplified** if the numerator and denominator have no common factors (other than 1 and -1). Thus, to *simplify* or *reduce* a rational expression, we factor both the numerator and the denominator into prime factors and then *cancel* common factors by using the fundamental principle of fractions.

FUNDAMENTAL PRINCIPLE OF FRACTIONS

> If $Q \neq 0$, and $K \neq 0$, then $\dfrac{PK}{QK} = \dfrac{P}{Q}$.

Cancellation is usually indicated by slanted lines drawn through the canceled factors. For instance,

$$\frac{x^2 + 3x}{x^2 + 2x - 3} = \frac{x\cancel{(x + 3)}}{(x - 1)\cancel{(x + 3)}} = \frac{x}{x - 1}$$

If one rational expression can be obtained from another either by canceling common factors or multiplying numerator and denominator by the same nonzero expression, then the two rational expressions are said to be **equivalent**. Thus, the canceling of common factors above shows that $\frac{x^2 + 3x}{x^2 + 2x - 3}$ and $\frac{x}{x - 1}$ are equivalent rational expressions.

EXAMPLE 2   Reducing Rational Expressions

Reduce each rational expression.

(a) $\dfrac{28x^3y}{21xy^2}$ (b) $\dfrac{y^2 - 3y}{y^2 - 9}$ (c) $\dfrac{6p^2 - 7p - 3}{4p^2 - 8p + 3}$

Solution   (a) $\dfrac{28x^3y}{21xy^2} = \dfrac{4x^2\cancel{(7xy)}}{3y\cancel{(7xy)}} = \dfrac{4x^2}{3y}$

(b) $\dfrac{y^2 - 3y}{y^2 - 9} = \dfrac{y\cancel{(y - 3)}}{(y + 3)\cancel{(y - 3)}} = \dfrac{y}{y + 3}$

(c) $\dfrac{6p^2 - 7p - 3}{4p^2 - 8p + 3} = \dfrac{(3p + 1)\cancel{(2p - 3)}}{(2p - 1)\cancel{(2p - 3)}} = \dfrac{3p + 1}{2p - 1}$

Building Rational Expressions to Higher Terms

The fundamental principle of fractions can also be used to build rational expressions to higher terms. This technique will be useful when we add and subtract rational expressions.

EXAMPLE 3   Building an Expression to Higher Terms

Replace each question mark with an appropriate expression so that the resulting expressions are equivalent.

(a) $\dfrac{3}{4} = \dfrac{?}{20t}$ (b) $\dfrac{2a}{3b} = \dfrac{?}{3ab - 3b^2}$ (c) $\dfrac{x + 3}{x - 2} = \dfrac{x^2 + 5x + 6}{?}$

Solution   (a) $\dfrac{3}{4} = \dfrac{(5t) \cdot 3}{(5t) \cdot 4} = \dfrac{15t}{20t}$ Replace ? by $15t$

(b) $\dfrac{2a}{3b} = \dfrac{(a - b)(2a)}{(a - b)(3b)} = \dfrac{2a^2 - 2ab}{3ab - 3b^2}$ Replace ? by $2a^2 - 2ab$

(c) $\dfrac{x + 3}{x - 2} = \dfrac{(x + 3)(x + 2)}{(x - 2)(x + 2)} = \dfrac{x^2 + 5x + 6}{x^2 - 4}$ Replace ? by $x^2 - 4$

Identifying Equal Rational Expressions

Recall from Section R4 that two rational numbers are equal (or equivalent) if they have the same value. We also checked to see whether they were equal by cross multiplication. Similarly, two rational expressions are equal if their cross products are equal. That is,

$$\frac{P}{Q} = \frac{R}{S} \text{ if and only if } PS = QR$$

For example, for all nonzero values of x and y,

$$\frac{3y}{xy} = \frac{3}{x}$$

since their cross products $(3y)x$ and $3(xy)$ are equal.

EXAMPLE 4   Identifying Equal Rational Expressions

Use cross multiplication to determine whether the rational expressions are equal.

(a) $\dfrac{14y}{7y + 21}$ and $\dfrac{2y}{y + 3}$ (b) $\dfrac{15x}{10x + 4}$ and $\dfrac{3x}{2x + 1}$

Solution   We compare the cross products
(a) $(14y)(y + 3) = 14y^2 + 42y$ and $(7y + 21)(2y) = 14y^2 + 42y$. Since the cross products are equal, the rational expressions are equal.
(b) $(15x)(2x + 1) = 30x^2 + 15x$ and $(10x + 4)(3x) = 30x^2 + 12x$. Since the cross products are not equal, the rational expressions are not equal.

Determining Signs of Rational Expressions

Every rational expression has three signs associated with it—the sign of the rational expression, the sign of the numerator, and the sign of the denominator. That is, if $\frac{P}{Q}$ is a rational expression, then

1. $-\dfrac{P}{Q} = \dfrac{-P}{Q} = \dfrac{P}{-Q}$

2. $\dfrac{P}{Q} = \dfrac{-P}{-Q} = -\dfrac{-P}{Q} = -\dfrac{P}{-Q}$

For example, we rewrite $\dfrac{-3}{4 - x}$ as $\dfrac{3}{x - 4}$ as follows:

$$\frac{-3}{4 - x} = \frac{(-1)(-3)}{(-1)(4 - x)} = \frac{3}{-4 + x} = \frac{3}{x - 4}$$

EXAMPLE 5 Changing the Signs of Expressions

Rewrite each rational expression with the denominator $x - y$

(a) $\dfrac{1}{y - x}$ (b) $\dfrac{-x}{y - x}$ (c) $-\dfrac{-1}{y - x}$

Solution (a) $\dfrac{1}{y - x} = \dfrac{(-1)(1)}{(-1)(y - x)} = \dfrac{-1}{-y + x} = \dfrac{-1}{x - y}$

(b) $\dfrac{-x}{y - x} = \dfrac{(-1)(-x)}{(-1)(y - x)} = \dfrac{x}{-y + x} = \dfrac{x}{x - y}$

(c) $-\dfrac{-1}{y - x} = +\dfrac{-1}{(-1)(y - x)} = \dfrac{-1}{-y + x} = \dfrac{-1}{x - y}$

Solving Applied Problems

Rational expressions can be used to describe real-world situations as the next example shows.

EXAMPLE 6 Solving a Revenue Problem

The total revenue R (in thousands of dollars) from the sale of a certain product is approximated by the formula

$$R = \frac{1000(t^2 - t - 2)}{t^2 + 2t - 8}, t > 2$$

where t is the number of years of production.

(a) Reduce the right side of this formula to lowest terms.

(b) Find the total revenue at the end of the fourth year.

Solution (a) $R = \dfrac{1000(t^2 - t - 2)}{t^2 + 2t - 8}$ Given

$= \dfrac{1000(t + 1)\,(t - 2)}{(t - 2)(t + 4)}$ Factor

$= \dfrac{1000(t + 1)}{(t + 4)}$ Reduce

(b) At the end of the fourth year, $t = 4$, so

$$R = \frac{1000(4 + 1)}{4 + 4} = \frac{5000}{8} = 625$$

Therefore, the total revenue is 625 thousand dollars ($625,000).

PROBLEM SET 1.4

Mastering the Concepts

In Problems 1 and 2, find the value of each rational expression at the specified numbers.

1. $\dfrac{4x}{x + 3}$

   (a) $x = 1$ (b) $x = -2$
   (c) $x = 3$ (d) $x = 0$

2. $\dfrac{x - 2}{2x - 5}$

   (a) $x = 2$ (b) $x = -2$
   (c) $x = \dfrac{1}{2}$ (d) $x = 0$

In Problems 3–6, determine the values of x for which each rational expression is undefined.

3. (a) $\dfrac{7x}{x - 3}$ (b) $\dfrac{x + 2}{x + 6}$

4. (a) $\dfrac{-11}{12 - 6x}$ (b) $\dfrac{x + 2}{(x + 2)(x - 8)}$

5. (a) $\dfrac{2x - 4}{(x - 6)(x + 7)}$ (b) $\dfrac{x - 1}{(x + 3)(x - 2)}$

6. (a) $\dfrac{x + 2}{x^2 + 2x - 3}$ (b) $\dfrac{1 - x}{12 - 4x - x^2}$

In Problems 7–20, reduce each rational expression to lowest terms.

7. (a) $\dfrac{12xy^2}{18x^2y}$ (b) $\dfrac{8x^2 - 4x}{4x}$

8. (a) $\dfrac{14x^2y^3}{21x^2y^4}$ (b) $\dfrac{6u^2 - 3u}{3u}$

9. (a) $\dfrac{x^2y - xy^2}{x^2 - xy}$ (b) $\dfrac{x + 4x^2y}{1 + 4xy}$

10. (a) $\dfrac{x^2 - 5x}{2x - 10}$ (b) $\dfrac{x^2 - 3x}{x^2y - 3xy}$

11. (a) $\dfrac{x^2 + x}{x^2 - x}$ (b) $\dfrac{x^2 - 9}{x^2 + 3x}$

12. (a) $\dfrac{x^2 - xy}{x^2y - xy^2}$ (b) $\dfrac{xy^2 + x^2y}{xy + y^2}$

13. (a) $\dfrac{y^2 + 6y + 8}{4y^2 + 16y}$ (b) $\dfrac{x^2 + 5x + 6}{3x^2 + 9x}$

14. (a) $\dfrac{y^2 + 4y}{y^2 + 6y + 8}$ (b) $\dfrac{4x^2 - 9}{6x^2 - 9x}$

15. (a) $\dfrac{x^2 + x - 12}{x^2 + 4x - 21}$ (b) $\dfrac{x^2 + 7x + 12}{x^2 + 5x + 6}$

16. (a) $\dfrac{3x^2 + 7x + 4}{3x^2 - 5x - 12}$ (b) $\dfrac{10x^2 + 29x - 21}{4x^2 + 12x - 7}$

17. (a) $\dfrac{x^2 + xy - 2y^2}{x^2 + 3xy + 2y^2}$ (b) $\dfrac{3x^2 - 12y^2}{x^2 + 4xy + 4y^2}$

18. (a) $\dfrac{2x^2 + 13x + 20}{2x^2 + 17x + 30}$ (b) $\dfrac{3 + 13m - 10m^2}{2m^2 + 5m - 12}$

19. (a) $\dfrac{x^3 - 4xy^2}{3x^3 - 2x^2y - 8xy^2}$ (b) $\dfrac{3x^2 - 6xy + 3y^2}{3x^2 - 3y^2}$

20. (a) $\dfrac{x^2 - 9}{a(x + 3) - b(x + 3)}$ (b) $\dfrac{4a^3 - 14a^2 + 12a}{24a + 4a^2 - 8a^3}$

In Problems 21–30, replace each question mark with an appropriate expression so that the resulting expressions are equivalent.

21. (a) $\dfrac{3}{4x} = \dfrac{?}{8x^2y}$ (b) $\dfrac{13x}{25} = \dfrac{?}{75y}$

22. (a) $\dfrac{7}{3x} = \dfrac{?}{6x^2y^2}$ (b) $\dfrac{5x}{4y} = \dfrac{20x^2y}{?}$

23. (a) $\dfrac{5m^3y}{7my^3} = \dfrac{?}{28m^5y^6}$ (b) $\dfrac{6}{11xy} = \dfrac{?}{33x^2y}$

24. (a) $\dfrac{3x}{4y} = \dfrac{?}{4xy + 4y^2}$ (b) $\dfrac{3}{x + y} = \dfrac{?}{5(x + y)^2}$

25. (a) $\dfrac{9(a - b)}{7a} = \dfrac{?}{7a^4(a - b)}$ (b) $\dfrac{mn}{m - n} = \dfrac{?}{m^3 - mn^2}$

26. (a) $\dfrac{a^3 - a}{a^2 - a} = \dfrac{?}{a + 1}$ (b) $\dfrac{1}{x - y} = \dfrac{?}{x^3y - xy^3}$

27. $\dfrac{2x - 3}{x^2 + 3x + 9} = \dfrac{?}{x^3 - 27}$ 28. $\dfrac{y - 3}{y^2 - y + 1} = \dfrac{?}{y^3 + 1}$

29. $\dfrac{x + 2}{x - 2} = \dfrac{?}{2x^2 - 7x + 6}$ 30. $\dfrac{3x + 2y}{5x + 3y} = \dfrac{?}{3y^2 - 4xy - 15x^2}$

In Problems 31–36, use cross multiplication to determine whether or not each pair of rational expressions is equal.

31. (a) $\dfrac{5}{4x}$ and $\dfrac{10}{8x}$ (b) $\dfrac{-7y}{6}$ and $\dfrac{21y}{-18}$

32. (a) $\dfrac{7a}{9b}$ and $\dfrac{8a}{10b}$ (b) $\dfrac{-14}{-9x}$ and $\dfrac{-12}{7x}$

33. $\dfrac{x + y}{a}$ and $\dfrac{7x + y}{7a}$ 34. $\dfrac{x + 2}{x^2 - 4}$ and $\dfrac{3}{3x - 12}$

35. $\dfrac{x + 2y}{5}$ and $\dfrac{2x^2 - 8y^2}{10x - 20y}$ 36. $\dfrac{x^2 - 9}{x - 3}$ and $\dfrac{2x + 6}{2}$

In Problems 37 and 38, rewrite each rational expression with the denominator $x - y$.

37. (a) $\dfrac{-7}{y - x}$ (b) $\dfrac{4}{-(x - y)}$

38. (a) $\dfrac{-xy}{y - x}$ (b) $\dfrac{3 - x}{y - x}$

Applying the Concepts

39. **Geometry:** The height h of a closed vertical cylindrical tank of radius r is given by the formula

$$h = \frac{S - 2\pi r^2}{2\pi r}$$

where S is the surface area of the tank. Determine the height of the tank if its surface area is 256π square feet and its radius is 6 feet.

40. **Investment:** In business the principal amount of investment can be determined by using the formula

$$P = \frac{A}{1 + rt}$$

where P is the principal amount of dollars invested at rate r, and A is the amount of money accumulated after t years. Find P when $A = \$580$, $r = 8\%$, and $t = 2$ years.

Developing and Extending the Concepts

In Problems 41–46, reduce each rational expression.

41. $\dfrac{3x^3 - 3xy^2}{3xy^2 + 3x^2y - 6x^3}$

42. $\dfrac{x^3 - 2x^2 + 5x - 10}{3x^5 + 15x^3 - x^2 - 5}$

43. $\dfrac{xz + xw - yz - yw}{xy + xz - y^2 - yz}$

44. $\dfrac{(2 + h)^3 - 8}{h}$

45. $\dfrac{x^{2n} - 9}{x^n + 3}$, where n is a positive integer.

46. $\dfrac{(1 + h)^2 + 2(1 + h) - 3}{h}$

OBJECTIVES

1. Multiply Rational Expressions
2. Divide Rational Expressions
3. Divide Polynomials
4. Solve Applied Problems

1.5 Multiplication and Division of Rational Expressions

Multiplication and division of rational expressions are based upon the corresponding properties of rational numbers covered in section R5.

Multiplying Rational Expressions

To multiply two rational expressions, we multiply their numerators and divide by the product of the denominators. That is, if (P/Q) and (R/S) are rational expressions, then

$$\frac{P}{Q} \cdot \frac{R}{S} = \frac{P \cdot R}{Q \cdot S}$$

For instance, $\dfrac{2x}{y^3} \cdot \dfrac{25x^4}{7y^2} = \dfrac{(2x)(25x^4)}{(y^3)(7y^2)} = \dfrac{50x^5}{7y^5}$

EXAMPLE 1   Multiplying Rational Expressions

Perform the multiplication.

$$\frac{4x}{5x - 1} \cdot \frac{5x + 1}{5x - 1}$$

Solution

$$\frac{4x}{5x - 1} \cdot \frac{5x + 1}{5x - 1} = \frac{(4x)(5x + 1)}{(5x - 1)(5x - 1)}$$

◆

The answers for multiplication and division problems may be left in factored form or in polynomial form.

This process of multiplying rational expressions can often be shortened by first factoring the expressions in the numerators and denominators, canceling common factors and completing the multiplication.

EXAMPLE 2   Multiplying Rational Expressions

Perform each multiplication by first factoring and then simplifying.

(a) $\dfrac{c}{c-1} \cdot \dfrac{c^2-1}{c^2}$ (b) $\dfrac{6t-6}{t^2+2t} \cdot \dfrac{t^2+4t+4}{2t^2+2t-4}$

Solution   (a) $\dfrac{c}{c-1} \cdot \dfrac{c^2-1}{c^2} = \dfrac{\cancel{c}(c-1)(c+1)}{(c-1)\cancel{c} \cdot c}$ Factor the numerator and denominator

$= \dfrac{c+1}{c}$ Cancel common factors

(b) $\dfrac{6t-6}{t^2+2t} \cdot \dfrac{t^2+4t+4}{2t^2+2t-4} = \dfrac{2 \cdot 3(t-1)(t+2)^2}{t(t+2)2(t+2)(t-1)}$ Factor the numerator and denominator

$= \dfrac{3}{t}$ Cancel common factors ◈

Dividing Rational Expressions

We divide two rational expressions in the same way as we divided rational numbers in section R5, by taking the reciprocal of the divisor, multiplying, and then simplifying the result. That is, if (P/Q) and (R/S) are rational expressions, then

$$\frac{P}{Q} \div \frac{R}{S} = \frac{P}{Q} \cdot \frac{S}{R} = \frac{P \cdot S}{Q \cdot R}$$

EXAMPLE 3   Dividing Rational Expressions

Perform each division and simplify the results.

(a) $\dfrac{3a^2}{5b} \div \dfrac{2a^2}{6b^3}$ (b) $\dfrac{a+b}{2} \div \dfrac{(a+b)^2}{6}$ (c) $\dfrac{x^2+5x+6}{x^2-4} \div \dfrac{x^2+4x+4}{x^2-4x+4}$

Solution   (a) We take the reciprocal of the divisor, factor, reduce and multiply.

$$\frac{3a^2}{5b} \div \frac{2a^2}{6b^3} = \frac{3a^2}{5b} \cdot \frac{6b^3}{2a^2} = \frac{3 \cdot \cancel{a^2} \cdot 2 \cdot 3 \cdot b^2 \cdot \cancel{b}}{5 \cdot \cancel{b} \cdot 2 \cdot \cancel{a^2}} = \frac{9b^2}{5}$$

(b) We take the reciprocal of the divisor, factor, reduce and multiply.

$$\frac{a+b}{2} \div \frac{(a+b)^2}{6} = \frac{a+b}{2} \cdot \frac{6}{(a+b)^2} = \frac{\cancel{a+b}}{2} \cdot \frac{2 \cdot 3}{(a+b)(a+b)} = \frac{3}{a+b}$$

(c) We take the reciprocal of the divisor, factor, reduce and multiply.

$$\frac{x^2+5x+6}{x^2-4} \div \frac{x^2+4x+4}{x^2-4x+4} = \frac{x^2+5x+6}{x^2-4} \cdot \frac{x^2-4x+4}{x^2+4x+4}$$

$$= \frac{(x+2)(x+3)}{(x-2)(x+2)} \cdot \frac{(x-2)(x-2)}{(x+2)(x+2)}$$

$$= \frac{(x-2)(x+3)}{(x+2)(x+2)} = \frac{(x-2)(x+3)}{(x+2)^2}$$ ◈

The next example demonstrates how we follow the order of operations to multiply and divide rational expressions.

EXAMPLE 4 Multiplying and Dividing Rational Expressions

Perform the indicated operations and simplify the result.

(a) $\dfrac{w^2 - w}{z^2 - z} \div \dfrac{w^2}{w - 1} \cdot \dfrac{z^2 w - zw}{w - 1}$ (b) $\dfrac{w^2 - w}{z^2 - z} \div \left[\dfrac{w^2}{w - 1} \cdot \dfrac{z^2 w - zw}{w - 1} \right]$

Solution (a) We follow the order of operations. First, we take the reciprocal of the second expression and multiply.

$$\frac{w^2 - w}{z^2 - z} \div \frac{w^2}{w - 1} \cdot \frac{z^2 w - zw}{w - 1} = \frac{w^2 - w}{z^2 - z} \cdot \frac{w - 1}{w^2} \cdot \frac{z^2 w - zw}{w - 1}$$

$$= \frac{\cancel{w}(w - 1)}{z(\cancel{z - 1})} \cdot \frac{\cancel{w - 1}}{\cancel{w} \cdot \cancel{w}} \cdot \frac{z\cancel{w}(\cancel{z - 1})}{\cancel{w - 1}} = w - 1$$

(b) Because of the parentheses, the divisor is the product of the two expressions inside the parentheses.

$$\frac{w^2 - w}{z^2 - z} \div \left[\frac{w^2}{w - 1} \cdot \frac{z^2 w - zw}{w - 1} \right] = \frac{w^2 - w}{z^2 - z} \div \left[\frac{w^2(z^2 w - zw)}{(w - 1)^2} \right] \quad \text{Multiply}$$

$$= \frac{\cancel{w}(w - 1)}{z(z - 1)} \cdot \frac{(w - 1)^2}{w^2 \cdot \cancel{w} \cdot z(z - 1)} = \frac{(w - 1)^3}{w^2 z^2 (z - 1)^2} \quad \begin{array}{l}\text{Take the reciprocal} \\ \text{of the divisor, mul-} \\ \text{tiply and reduce} \ \diamond\end{array}$$

Dividing Polynomials

Recall from arithmetic that if we divide 41 by 9, we obtain a *quotient* of 4 and a *remainder* of 5. This means that

$$41 = 4 \cdot 9 + 5$$

which can be interpreted as

$$41 \div 9 = 4 + \frac{5}{9}$$

In this process, we call 41 the **dividend** and 9 the **divisor**. Similarly, if we divide a polynomial A by another polynomial B, we obtain a *quotient* Q and a *remainder* R. As in arithmetic, this means that

$$A = Q \cdot B + R$$

and can be interpreted as

$$A \div B = Q + \frac{R}{B}$$

The actual division can be accomplished by algebraic *long division* which is similar to the process in arithmetic and is outlined in the following strategy:

LONG DIVISION OF POLYNOMIALS

1. **Arrange** both the divisor and the dividend in descending powers, noting any missing terms.
2. **Find** the first term of the quotient by dividing the first term of the dividend by the first term of the divisor.
3. **Multiply** the quotient obtained in step 2 by the entire divisor.
4. **Subtract** the product obtained in step 3 from the dividend, and bring down the next term.
5. **Repeat** the procedure in steps 2, 3, and 4 until the degree of the remainder is less than the degree of the divisor.

EXAMPLE 5   Using Long Division

Divide $6x^3 + 7x^2 + 4x + 5$ by $2x + 1$.

Solution

$$2x + 1 \overline{)6x^3 + 7x^2 + 4x + 5}$$ Arrange in descending powers

$$\begin{array}{r} 3x^2 \\ 2x + 1 \overline{)6x^3 + 7x^2 + 4x + 5} \end{array}$$ Divide $6x^3$ by $2x$

$$\begin{array}{r} 3x^2 \\ 2x + 1 \overline{)6x^3 + 7x^2 + 4x + 5} \\ 6x^3 + 3x^2 \end{array}$$ Multiply quotient, $3x^2$, by divisor, $2x + 1$

$$\begin{array}{r} 3x^2 \\ 2x + 1 \overline{)6x^3 + 7x^2 + 4x + 5} \\ -(6x^3 + 3x^2) \\ \hline 4x^2 + 4x \end{array}$$ Subtract the product from the dividend and bring down next term

$$\begin{array}{r} 3x^2 + 2x + 1 \\ 2x + 1 \overline{)6x^3 + 7x^2 + 4x + 5} \\ -(6x^3 + 3x^2) \\ \hline 4x^2 + 4x \\ -(4x^2 + 2x) \\ \hline 2x + 5 \\ -(2x + 1) \\ \hline 4 \end{array}$$ Repeat the process

$$\frac{6x^3 + 7x^2 + 4x + 5}{2x + 1} = \boxed{3x^2 + 2x + 1} + \frac{\boxed{4}}{\boxed{2x + 1}}$$

Remainder

Quotient   Divisor

To check,

(Quotient) (Divisor) + Remainder $= (3x^2 + 2x + 1)(2x + 1) + 4$

$= 6x^3 + 7x^2 + 4x + 5 = $ Dividend ⬦

The quotient in Example 5 involved division by a first degree binomial. The same process may be used for divisors of degrees greater than 1 as the next example shows.

EXAMPLE 6   Dividing Polynomials

Divide $-3w^3 + 2w^4 + 5w^2 + 2w + 7$ by $-w + w^2 + 1$.

Solution   First we arrange both polynomials in descending powers of w. The remaining steps are shown as follows:

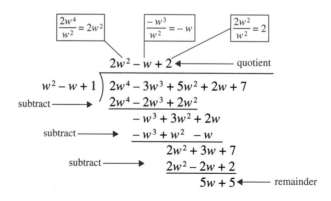

Solving Applied Problems

Multiplication and division of rational expressions can be used to find fuel consumption as the next example shows.

EXAMPLE 7   Fuel Consumption

On a recent trip, a motorist averaged y miles per hour and x miles per gallon.
(a) At this rate, find an expression for the fuel consumption in gallons per hour.
(b) Evaluate the expression from part (a) for $x = 25$ (miles per gallon) and $y = 55$ (miles per hour).

Solution   (a) The fuel consumption $= \dfrac{\dfrac{y \text{ mi.}}{\text{hr}}}{\dfrac{x \text{ mi.}}{\text{gal.}}} = \dfrac{y \text{ mi.}}{\text{hr.}} \div \dfrac{x \text{ mi.}}{\text{gal.}}$

$$= \frac{y \text{ mi.}}{\text{hr.}} \cdot \frac{\text{gal.}}{x \text{ mi.}}$$

$$= \frac{y \text{ gal.}}{x \text{ hr.}}$$

(b) Replacing x with 25 and y with 55 in part (a), we have

$$\frac{y}{x} = \frac{55}{25} = \frac{11}{5} = 2.2$$

The fuel consumption is 2.2 gallons per hour.

◆ PROBLEM SET 1.5

Mastering the Concepts

In Problems 1–10, perform each multiplication and reduce the result to lowest terms.

1. (a) $\dfrac{10}{9x} \cdot \dfrac{12x}{15y}$ (b) $\dfrac{3x}{7} \cdot \dfrac{14y}{27x^2}$

2. (a) $\dfrac{2x}{3y} \cdot \dfrac{12y}{5x}$ (b) $\dfrac{3x}{5yz} \cdot \dfrac{4y}{9x}$

3. $\dfrac{7x^3}{8y^4} \cdot \dfrac{16y}{21x^2}$

4. $\dfrac{5x^2y}{3a^2b} \cdot \dfrac{6ab}{10x^2}$

5. $\dfrac{-5xy^2}{3z} \cdot 6xz^2$

6. $-4a \cdot \dfrac{5b}{6bc}$

7. $\dfrac{4xy}{x-y} \cdot \dfrac{x-y}{8xy}$

8. $\dfrac{x+2}{3x} \cdot \dfrac{9x^2}{x+2}$

9. $\dfrac{x+4}{x^2} \cdot \dfrac{5x}{3x+12}$

10. $\dfrac{3x+3y}{xy} \cdot \dfrac{5x}{7x+7y}$

In Problems 11–20, perform each division and reduce the result to lowest terms.

11. (a) $\dfrac{11x}{5} \div \dfrac{22x}{25}$ (b) $\dfrac{9x}{4} \div \dfrac{36x}{15}$

12. (a) $\dfrac{8x}{9y} \div \dfrac{4x}{3y}$ (b) $\dfrac{9x}{4} \div \dfrac{15}{36x}$

13. $15xy \div \dfrac{x^2}{5xz}$

14. $\dfrac{5z^2}{3xy} \div 2z$

15. $\dfrac{a}{4bc} \div 16ab$

16. $\dfrac{6ab^2}{5} \div \dfrac{15a}{3b^2}$

17. $7xy \div \dfrac{-12xy^2}{3z}$

18. $-11x \div \dfrac{22x^2}{5z}$

19. $\dfrac{5x^2y}{x-2} \div \dfrac{15xy}{x-2}$

20. $\dfrac{3a+3}{a-1} \div \dfrac{7a+7}{9a-9}$

In Problems 21–40, perform each operation and simplify the result.

21. $\dfrac{m^2+mn}{mn} \cdot \dfrac{7n}{m^2-n^2}$

22. $\dfrac{m^2-4}{m+3} \cdot \dfrac{m^2-9}{m+2}$

23. $\dfrac{x^2-x}{x-1} \cdot \dfrac{x}{5x-5}$

24. $\dfrac{6x-12}{7x-21} \div \dfrac{2x-4}{x^2-9}$

25. $\dfrac{x^2-144}{x+4} \cdot \dfrac{x^2-16}{x+12}$

26. $\dfrac{3y^2-12}{3y^2-3} \cdot \dfrac{y-1}{2y+4}$

27. $\dfrac{2x-y}{x^2-y^2} \div \dfrac{4x^2-y^2}{x+y}$

28. $\dfrac{5x+10}{x^3} \cdot \dfrac{x+2}{x^4}$

29. $\dfrac{1-9a^2}{a+1} \div \dfrac{6a+2}{2a+2}$

30. $\dfrac{3x+6}{5x+5} \cdot \dfrac{10x+10}{x^2-6x-16}$

31. $\dfrac{3y^2-y-2}{3y^2+y-2} \cdot \dfrac{3y^2-5y+2}{3y^2+5y+2}$

32. $\dfrac{2x - y}{x^2 - y^2} \div \dfrac{4x^2 - y^2}{x + y}$

33. $\dfrac{2x^2 - 5x - 3}{3x^2 - 5x - 2} \div \dfrac{2x^2 + 11x + 5}{x^2 + 3x - 10}$

34. $\dfrac{2a - 6}{6a^2 - 15a} \cdot \dfrac{18a^3 - 45a^2}{4a^2 - 12a}$

35. $\dfrac{t^2 - 8t + 15}{t^2 + 2t - 35} \cdot \dfrac{t^2 + 9t + 14}{15 - 2t - t^2}$

36. $\dfrac{6y^2 + 7y - 3}{9y^2 - 25} \div \dfrac{12y^2 - y - 1}{12y^2 + 20y}$

37. $\dfrac{x^3 - 4x^2 + 3x}{x + 2} \div (x^2 - 3x)$

38. $\dfrac{y^2 - 3}{y^3 - 4y} \cdot \dfrac{y^2 - 4}{y^4 - 9}$

39. $\dfrac{a^3 + b^3}{2a^3 + 2a^2b} \cdot \dfrac{6a^2 - 6b^2}{3a^3 - 3a^2b + 3ab^2}$

40. $\dfrac{8x^3 - 27}{9x^2 - 3x + 1} \div \dfrac{4x^2 + 6x + 9}{27x^3 + 1}$

In Problems 41–48, perform the indicated operations and simplify.

41. $\left[\dfrac{27x^2}{16y^2} \cdot \dfrac{20y^3}{9x}\right] \div \dfrac{18x^3}{5y}$

42. $\left[\dfrac{10x^2}{17y^3} \cdot \dfrac{34y}{25x^3}\right] \div \dfrac{2y^3}{15x^4}$

43. $\left[\dfrac{x - 1}{x^2 - 4} \cdot \dfrac{2x + 4}{x^2 - 1}\right] \div \dfrac{2x + 2}{x - 2}$

44. $\left[\dfrac{7x^2 - 28}{3x^2 - 12x} \cdot \dfrac{9x^3 + 6x^2}{2x^2 - 4x}\right] \div \dfrac{x^3 + 2x^2}{2x^2 - 8x}$

45. $\left[\dfrac{x^2 - 1}{x + 2} \cdot \dfrac{x^2 - 4}{x - 3}\right] \div \dfrac{x^2 - 3x + 2}{x^2 - 9}$

46. $\left[\dfrac{y^2 - 3y + 2}{y^2 - 4} \cdot \dfrac{y + 2}{y - 1}\right] \div \left[\dfrac{y^2 - 4}{y - 2}\right]$

47. $\dfrac{x^2 - 1}{x + 1} \div \left[\dfrac{x^2 - 1}{x^2 + 2x + 1} \cdot \dfrac{x^2 + 1}{x^2 - 2x + 1}\right]$

48. $\dfrac{a^2 - ab}{ab + b^2} \div \left[\dfrac{a^2 - b^2}{a^2 + 2ab + b^2} \cdot \dfrac{a^2b + ab^2}{a^2 - 2ab + b^2}\right]$

In Problems 49–62, use the long division process to perform each division and check the results.

49. $(x^2 - 7x + 10) \div (x - 5)$
50. $(2x^2 + x - 6) \div (x + 2)$
51. $(2x^2 - 5x - 6) \div (2x - 1)$
52. $(6x^2 + 11x + 7) \div (3x + 1)$
53. $(8x^2 - 14x + 5) \div (2x - 3)$
54. $(12x^2 + 6x - 9) \div (4x - 2)$
55. $(x^3 + 3x^2 + 4x + 7) \div (x - 2)$
56. $(x^3 + 7x^2 + 3x + 5) \div (x - 1)$
57. $(x^4 + 2x^3 - 3x^2 + 5x + 1) \div (x + 1)$
58. $(x^4 - 3x^3 + 2x^2 - 6x + 3) \div (x - 3)$
59. $(x^4 + 5x^3 + 9x^2 + 5x - 4) \div (x^2 + 2x - 1)$
60. $(12x^2 - 9x^3 - 3x + 1) \div (1 - 3x^2)$
61. $(x^3 + 16x + 52 - 3x^2) \div (x^2 + 26 - 5x)$
62. $(3x^5 - 2x^4 - 10x^3 + 26x^2 - 25x - 2) \div (-5x + 3x^2 + 1)$

Applying the Concepts

63. **Fuel Economy:** On a recent trip a car was driven continuously for 17.5 hours. It traveled x miles and used y gallons of gasoline. The average rate of speed was $r = \frac{x}{17.5}$ miles per hour and the average rate of fuel consumption was $f = \frac{y}{17.5}$ gallons per hour.
    (a) Find an expression for the average mileage in miles per gallon.
    (b) Evaluate the expression from part (a) for $x = 900$ miles and $y = 45$ gallons.

64. **Pumping Rate:** A newly dug well pumps water into a reservoir at a rate of 55 gallons per minute. At this rate, find
    (a) the time (in hours) to pump x gallons into the reservoir
    (b) the time (in hours) to pump 400,000 gallons into the reservoir.

Developing and Extending the Concepts

In Problems 65–72, perform the indicated operations and simplify.

65. $\dfrac{8m^3 + n^3}{3m - 5n} \cdot \dfrac{9m^2 - 25n^2}{4m^2 - 2mn + n^2}$

66. $\dfrac{2x^2 - 5x - 3}{3x^2 - 5x - 2} \div \dfrac{2x^2 + 11x + 5}{x^2 + 3x - 10}$

67. $\dfrac{2x^3 + 16}{5x^3 - 135} \cdot \dfrac{x^2 - 9}{x^2 - 2x + 4}$

68. $\dfrac{6x^2 + 11x - 10}{3x^2 - 5x - 12} \div \dfrac{2x^2 + 9x + 10}{3x^2 + 10x + 8}$

69. $\dfrac{b - 1}{21 - 4a - a^2} \cdot \dfrac{b - 2}{b - b^3} \div \dfrac{2 - b}{a^2 + 6a - 7}$

70. $\dfrac{7}{5v^2(v + 3)} \div \left(\dfrac{v^2 - 5v + 6}{8v^2} \cdot \dfrac{21}{5v^2 - 45}\right)$

71. $\left(\dfrac{x^2 - 1}{x^2}\right) \cdot \dfrac{2}{x - 1} \div \dfrac{x^2 + 2x + 1}{x^3}$

72. $\dfrac{2y^2 + 5y - 3}{6y^2 - 5y - 6} \div \left(\dfrac{2y^2 + 9y - 5}{12y^2 - y - 6} \div \dfrac{2y^2 + y - 6}{4y^2 + 9y - 9}\right)$

73. Perform the indicated operations and simplify.
    (a) $\dfrac{1}{x} \div \dfrac{1}{x}$ (b) $\dfrac{1}{x} \div \dfrac{1}{x} \div \dfrac{1}{x}$ (c) $\dfrac{1}{x} \div \dfrac{1}{x} \div \dfrac{1}{x} \div \dfrac{1}{x}$
    (d) For what value(s) of x will the expressions in parts (b) and (c) be equal?
    (e) For what value(s) of x will the expression in part (c) be positive?
    (f) For what value(s) of x will the expression in part (c) be negative?

74. Use the pattern established in Problem 73 to find the value of the expression
    $$\dfrac{1}{x} \div \dfrac{1}{x} \div \dfrac{1}{x} \div \dfrac{1}{x} \div \dfrac{1}{x} \div \dfrac{1}{x}$$

75. Use long division to divide $(x^3 + y^3)$ by $(x + y)$, and then verify the special factoring formula
    $$x^3 + y^3 = (x + y)(x^2 - xy + y^2)$$

76. Use long division to divide $(x^5 - y^5)$ by $(x - y)$, and then verify the factoring formula
    $$x^5 - y^5 = (x - y)(x^4 + x^3y + x^2y^2 + xy^3 + y^4)$$

OBJECTIVES

1. Add and Subtract Rational Expressions with Common Denominators
2. Add and Subtract Rational Expressions with Different Denominators
3. Solve Applied Problems

1.6 Addition and Subtraction of Rational Expressions

Adding and Subtracting Rational Expressions with Common Denominators

To *add* or *subtract* rational expressions with common denominators, we add or subtract the numerators and place the result over the common denominator.

For example, we add $\dfrac{3}{7x}$ and $\dfrac{2}{7x}$ as follows:

$$\frac{3}{7x} + \frac{2}{7x} = \frac{3 + 2}{7x} = \frac{5}{7x}$$

This addition can also be justified by using the distributive property.

$$\frac{3}{7x} + \frac{2}{7x} = 3 \cdot \frac{1}{7x} + 2 \cdot \frac{1}{7x} = (3 + 2)\frac{1}{7x} = 5 \cdot \frac{1}{7x} = \frac{5}{7x}$$

Similarly, we subtract $\dfrac{2}{7x}$ from $\dfrac{3}{7x}$ as follows:

$$\frac{3}{7x} - \frac{2}{7x} = \frac{3 - 2}{7x} = \frac{1}{7x}$$

In general, we have:

ADDITION AND SUBTRACTION

> If $\dfrac{P}{Q}$ and $\dfrac{R}{Q}$ are rational expressions then
>
> $$\frac{P}{Q} + \frac{R}{Q} = \frac{P + R}{Q} \quad \text{and} \quad \frac{P}{Q} - \frac{R}{Q} = \frac{P - R}{Q}.$$

EXAMPLE 1   Adding and Subtracting with Common Denominators

Perform each operation and simplify the results.

(a) $\dfrac{2t}{t + 7} + \dfrac{14}{t + 7}$ (b) $\dfrac{25}{5 - y} - \dfrac{y^2}{5 - y}$

(c) $\dfrac{3x - 1}{x^2 + x - 12} - \dfrac{14 - 2x}{x^2 + x - 12}$

Solution   Since we have a common denominator, we add (subtract) the numerators and repeat the denominator. Thus,

(a) $\dfrac{2t}{t + 7} + \dfrac{14}{t + 7} = \dfrac{2t + 14}{t + 7} = \dfrac{2(\cancel{t + 7})}{\cancel{t + 7}} = 2$

(b) $\dfrac{25}{5 - y} - \dfrac{y^2}{5 - y} = \dfrac{25 - y^2}{5 - y}$ Subtract the numerators

$\dfrac{(\cancel{5 - y})(5 + y)}{\cancel{5 - y}} = 5 + y$ Factor and simplify

14 − 2x is the whole numerator because the fraction bar is an inclusion symbol.

(c) $\dfrac{3x - 1}{x^2 + x - 12} - \dfrac{14 - 2x}{x^2 + x - 12} = \dfrac{3x - 1 - (14 - 2x)}{x^2 + x - 12}$

$= \dfrac{3x - 1 + 2x - 14}{x^2 + x - 12}$

$= \dfrac{5x - 15}{x^2 + x - 12}$

$= \dfrac{5(\cancel{x - 3})}{(x + 4)(\cancel{x - 3})} = \dfrac{5}{x + 4}$ Factor and simplify

Adding and Subtracting Rational Expressions with Different Denominators

If the denominators of rational expressions are not the same, we use the fundamental principle of fractions so that they have a common denominator, and then we may use the above process. Even though any common denominator may be used, the problem will be easier if the least common denominator (LCD) is found. Recall from section R4 that the Least Common Denominator is the smallest denominator that is divisible by all the original denominators. Each expression in the sum or difference is then *built up* to an equivalent expression having the LCD as a denominator. The following strategy outlines how to identify the LCD.

FINDING THE LCD OF TWO
RATIONAL EXPRESSIONS

1. **Factor** each denominator completely into a product of prime factors.
2. **List** each different prime factor to the highest power which appears in any factored denominator.
3. The LCD is the **product** of these prime factors.

For example, to add the rational expressions

$$\frac{5}{a^2b} \text{ and } \frac{3}{ab^3},$$

we notice that the denominators are already factored. The different factors are a and b, the highest powers which appear are a^2 and b^3. So, the LCD is a^2b^3. To add these expressions, we build up the two expressions so that we have equivalent expressions with a^2b^3 as the common denominator. Thus,

$$\frac{5}{a^2b} + \frac{3}{ab^3} = \frac{5 \cdot b^2}{a^2b \cdot b^2} + \frac{3 \cdot a}{ab^3 \cdot a} = \frac{5b^2}{a^2b^3} + \frac{3a}{a^2b^3} = \frac{5b^2 + 3a}{a^2b^3}$$

EXAMPLE 2   Adding and Subtracting Rational Expressions

Perform each operation and simplify.

(a) $\dfrac{5}{x} + \dfrac{3}{y}$ (b) $\dfrac{7}{10x} + \dfrac{8}{15x} - 3$

(c) $\dfrac{5}{y^2 - y} - \dfrac{4}{y^2 - 1}$ (d) $\dfrac{1}{a - 3} + \dfrac{a - 2}{3 - a}$

Solution   (a) Here the LCD is xy, so that

$$\frac{5}{x} + \frac{3}{y} = \frac{5 \cdot y}{x \cdot y} + \frac{3 \cdot x}{y \cdot x}$$ Build each expression to the LCD xy

$$= \frac{5y + 3x}{xy}$$ Add the numerators

(b) $\dfrac{7}{10x} + \dfrac{8}{15x} - 3 = \dfrac{7 \cdot 3}{10x \cdot 3} + \dfrac{8 \cdot 2}{15x \cdot 2} - \dfrac{3 \cdot 30x}{30x}$ Build the expressions to the LCD, $30x$

$$= \frac{21 + 16 - 90x}{30x} = \frac{37 - 90x}{30x}$$ Add the numerators

(c) We factor each denominator, and then find the LCD. The following table is helpful in constructing the LCD of the two expressions.

Thus, the LCD is $y(y-1)(y+1)$ and we have

Denominators	Prime Factors		
$y^2 - y$	y	$y-1$	
$y^2 - 1$		$y-1$	$y+1$
LCD	y	$(y-1)$	$(y+1)$

$$\frac{5}{y^2-y} - \frac{4}{y^2-1} = \frac{5}{y(y-1)} - \frac{4}{(y-1)(y+1)} \qquad \text{Factor denominators}$$

$$= \frac{5(y+1)}{y(y-1)(y+1)} - \frac{4y}{(y-1)(y+1)y} \qquad \text{Add numerators, and simplify}$$

$$= \frac{5(y+1) - 4y}{y(y-1)(y+1)}$$

$$= \frac{5y + 5 - 4y}{y(y-1)(y+1)}$$

$$= \frac{y+5}{y(y-1)(y+1)}$$

(d) Recall that the signs in a rational expression can be changed without changing the value of the expression as follows:

$$\frac{a-2}{3-a} = \frac{-(a-2)}{-(3-a)} = \frac{2-a}{a-3}.$$

Thus, in changing the signs of the rational expression, we have two expressions with the same denominator.

$$\frac{1}{a-3} + \frac{a-2}{3-a} = \frac{1}{a-3} + \frac{2-a}{a-3} =$$

$$\frac{1+2-a}{a-3} = \frac{3-a}{a-3} = \frac{-(a-3)}{a-3} = -1 \qquad \diamond$$

EXAMPLE 3   Adding and Subtracting More than Two Expressions
Combine into a single rational expression and simplify.

$$\frac{x}{x+1} - \frac{x}{x-1} + \frac{2}{x^2-1}$$

Solution   The table to left shows how the LCD is obtained:
Therefore, the LCD is $(x+1)(x-1)$. We add

Denominators	Prime Factors	
$x+1$	$x+1$	
$x-1$		$x-1$
x^2-1	$x+1$	$x-1$
LCD	$(x+1)$	$(x-1)$

$$\frac{x}{x+1} - \frac{x}{x-1} + \frac{2}{x^2-1} = \frac{x}{x+1} - \frac{x}{x-1} + \frac{2}{(x-1)(x+1)}$$

$$= \frac{x(x-1)}{(x+1)(x-1)} - \frac{x(x+1)}{(x+1)(x-1)} + \frac{2}{(x+1)(x-1)}$$

$$= \frac{x(x-1) - x(x+1) + 2}{(x+1)(x-1)}$$

$$= \frac{x^2 - x - x^2 - x + 2}{(x+1)(x-1)}$$

$$= \frac{-2x + 2}{(x-1)(x+1)}$$

$$= \frac{-2(x-1)}{(x-1)(x+1)}$$

$$= \frac{-2}{x+1} \qquad \diamond$$

Solving Applied Problems

In applied problems, denominators can often include units of measure. In the formula

$$t = \frac{d}{r}$$

t is time, d is distance, and r is rate. Rate can be measured in many different ways, miles per hour, feet per second, or meters per minutes, and many other combinations. In order to add or subtract fractions with such denominators, the units of measure must be the same (i.e., a common denominator).

EXAMPLE 4 Finding Common Denominators Involving Dimensions.

Consider traveling x miles at 50 miles per hour and then $2x$ miles at 50 feet per second.

(a) Write a rational expression for total time, and then simplify it.

(b) Evaluate the expression for $x = 12$ miles.

Solution (a) First convert 50 feet per second to miles per hour so as to be consistent with the other units.

$$\frac{50 \text{ ft}}{1 \text{ sec}} = \left(\frac{50 \text{ ft}}{1 \text{ sec}}\right)\left(\frac{1 \text{ mi}}{5280 \text{ ft}}\right)\left(\frac{3600 \text{ sec}}{1 \text{ hr}}\right) \approx \frac{34 \text{ mi}}{1 \text{ hr}}$$

Using the formula, total time is

$$t = \frac{x}{50} + \frac{2x}{34}.$$

Since the units are the same, the table to the left helps us find the LCD.

Denominators	Prime Factors		
50	2	5^2	
34	2		17
LCD	2	5^2	17

$$t = \frac{x \cdot 17}{50 \cdot 17} + \frac{2x \cdot 25}{34 \cdot 25} = \frac{17x + 50x}{850} = \frac{67x}{850}$$

(b) for $x = 12$ miles, we have $\frac{67(12)}{850} \approx 0.95$ hour

The total time is about 0.95 hour which is about 57 minutes. ◈

PROBLEM SET 1.6

Mastering the Concepts

In Problems 1–42, perform each operation and simplify the results.

1. $\dfrac{5}{8x} + \dfrac{1}{8x}$

2. $\dfrac{7}{12t} + \dfrac{5}{12t}$

3. $\dfrac{6}{5x} - \dfrac{14}{5x}$

4. $\dfrac{3}{2y^2} - \dfrac{7}{2y^2}$

5. $\dfrac{5}{7x} + \dfrac{3}{7x}$

6. $\dfrac{2}{3t} + \dfrac{5}{3t}$

7. $\dfrac{7}{12t} - \dfrac{1}{12t}$

8. $\dfrac{15}{16y} - \dfrac{3}{16y}$

9. $\dfrac{t-1}{4k} + \dfrac{t+1}{4k}$

10. $\dfrac{3x-1}{5k} + \dfrac{2x+1}{5k}$

11. $\dfrac{2a}{a+3} + \dfrac{3}{a+3}$

12. $\dfrac{m}{m-3} - \dfrac{2}{m-3}$

13. $\dfrac{9}{y-5} + \dfrac{6}{y-5}$

14. $\dfrac{2u}{4u+3} + \dfrac{u}{4u+3}$

15. $\dfrac{2x}{x^2-4} - \dfrac{4}{x^2-4}$

16. $\dfrac{2t}{4t^2-9} - \dfrac{-3}{4t^2-9}$

17. $\dfrac{4}{16-x^2} + \dfrac{x}{16-x^2}$

18. $\dfrac{36}{6-v} + \dfrac{-v^2}{6-v}$

19. $\dfrac{14y-3}{2y^2+y-1} - \dfrac{2y+3}{2y^2+y-1}$

20. $\dfrac{2m+5}{m^2+m-2} + \dfrac{m+1}{m^2+m-2}$

21. $\dfrac{3x}{5y} + \dfrac{1}{7}$

22. $\dfrac{4}{x} - \dfrac{5}{3}$

23. $\dfrac{y}{4} + 3$

24. $y - \dfrac{1}{y}$

25. $\dfrac{a}{b^2} + \dfrac{a^3}{b^3} - \dfrac{1}{b}$

26. $\dfrac{1}{u} - \dfrac{t}{u} + \dfrac{t^2}{u^2}$

27. $\dfrac{5}{t^2} + \dfrac{3}{2t} + \dfrac{2}{3}$

28. $\dfrac{7}{4x^2} + \dfrac{5}{12x} + \dfrac{1}{6}$

29. $\dfrac{3}{x + 2} + \dfrac{5}{x - 1}$

30. $\dfrac{5}{x + 3} - \dfrac{5}{3x}$

31. $\dfrac{9}{y - 5} + \dfrac{6}{y - 3}$

32. $\dfrac{2a}{a - 3} + \dfrac{3}{a + 7}$

33. $\dfrac{x}{3x + 2} + \dfrac{1}{x - 4}$

34. $\dfrac{2u}{4u + 3} + \dfrac{u}{2u + 5}$

35. $\dfrac{3}{x^2 + x} + \dfrac{2}{x^2 - 1}$

36. $\dfrac{3v}{4v^2 - 1} + \dfrac{1}{2v + 1}$

37. $\dfrac{m - 5}{m^2 - 5m - 6} + \dfrac{m + 4}{m^2 - 6m}$

38. $\dfrac{3y}{y^2 + 7y + 10} - \dfrac{y}{y^2 + y - 20}$

39. $\dfrac{2}{a^2 - 4} + \dfrac{7}{a^2 - 4a - 12}$

40. $\dfrac{3x + 1}{2x^2 - 5x - 12} - \dfrac{x + 4}{6x^2 + 7x - 3}$

41. $\dfrac{c}{c^2 - 9} - \dfrac{c - 1}{c^2 - 5c + 6}$

42. $\dfrac{2t + 3}{3t^2 + 2t - 8} + \dfrac{3t + 4}{2t^2 + t - 6}$

In Problems 43–50, combine into a single rational expression and simplify.

43. $\dfrac{7}{4x^3y^2} + \dfrac{5}{12xy^3} - \dfrac{6}{6x^2y}$

44. $\dfrac{5}{x^2y} - \dfrac{3}{2xy^2} + \dfrac{8}{4x^3y^3}$

45. $\dfrac{2a}{a^2 - 1} + \dfrac{1}{a + 1} - \dfrac{1}{a - 1}$

46. $\dfrac{2}{u + 3} - \dfrac{1}{u - 3} + \dfrac{2u}{u^2 - 9}$

47. $\dfrac{x - 3}{x + 3} - \dfrac{x - 3}{3 - x} + \dfrac{x^2}{9 - x^2}$

48. $\dfrac{v}{v + 2} - \dfrac{v}{v - 2} + \dfrac{v^2}{v^2 - 4}$

49. $\dfrac{2t^2 - t}{3t^2 - 27} - \dfrac{t - 3}{3t - 9} + \dfrac{6t^2}{9 - t^2}$

50. $\dfrac{4}{a^2 - 4} + \dfrac{2}{a^2 - 4a + 4} + \dfrac{1}{a^2 + 4a + 4}$

Applying the Concepts

51. **Traveling Time:** Consider traveling x miles at 60 miles per hour and then y miles at 88 feet per second.
    (a) Write a rational expression for the total time, and then simplify it.
    (b) Evaluate the expression for $x = 700$ miles and $y = 1000$ miles.

52. **Work Rate:** You can do a job in x hours, while it takes your friend $x + 2$ hours to complete the same job. Write and simplify an expression for the part of the job you and your friend, working together, can complete in 1 hour.

53. **Filling a Pool:** An inlet pipe can fill a pool in x hours, while an outlet pipe can empty it in $x + 3$ hours. Write and simplify an expression for the part of the pool that is filled after 1 hour if both pipes are open.

54. **Work Rate:** A printing press can print a morning newspaper in x hours, a second press takes $x + 1$ hours, while a third press takes $x + 2$ hours to print it. In one hour the three presses working together can print $\frac{1}{x} + \frac{1}{x + 1} + \frac{1}{x + 2}$ of the job.
    (a) Combine and simplify this expression.
    (b) If $x = 6$ hours, find the value of this expression.

55. **Manufacturing:** A manufacturer estimates that the revenue (in dollars) from the sale of x units of a certain product is given by the expression
$$x^2 + 4x\left[4 - \dfrac{x}{16(x + 4)}\right].$$
    (a) Rewrite this expression as a single rational expression.
    (b) What is the revenue if 100 units were produced and sold?

56. **Electrical Circuit:** If three resistors of resistances R_1, R_2, and R_3 are connected in parallel, then the combined resistance R is given by the formula
$$\dfrac{1}{R} = \dfrac{1}{R_1} + \dfrac{1}{R_2} + \dfrac{1}{R_3}.$$
    (a) Combine and simplify the right side of the formula if R_1 is x ohms, R_2 is $x + 4$ ohms and R_3 is $2x + 3$ ohms.
    (b) Find the value of R (in ohms) if $x = 4$.

Developing and Extending the Concepts

In Problems 57 and 58, use a calculator to evaluate the expressions
(i) $\dfrac{1}{x} + \dfrac{1}{y}$ and (ii) $\dfrac{y + x}{xy}$ for the given values.

57. (a) $x = 1.5$, $y = 2.5$ (b) $x = -0.12558$, $y = 1.0285$
58. (a) $x = 0.25$, $y = -1.25$ (b) $x = -1.31$, $y = 1.25$

In Problems 59–62, combine into a single rational expression and simplify.

59. $\left(\dfrac{x}{y} + \dfrac{y}{x}\right)\left(\dfrac{x}{y} - \dfrac{y}{x}\right)$

60. $\left(\dfrac{1}{x + 2} - \dfrac{1}{x - 2}\right)\left(\dfrac{1}{x + 2} + \dfrac{1}{x - 2}\right)$

61. $\left(\dfrac{x^2 - 9}{x}\right)\left(\dfrac{1}{x + 3} + \dfrac{1}{x - 3}\right)$

62. $\left(\dfrac{x - 2}{x + 1} - \dfrac{x - 2}{x + 2}\right)\left(\dfrac{x^2 + 3x + 2}{x - 2}\right)$

63. In Calculus e^x can be approximated by the expression
$$1 + x + \dfrac{x^2}{2} + \dfrac{x^3}{6} + \dfrac{x^4}{24}.$$
    (a) Find the value of this expression at $x = 1$.
    (b) Simplify this expression first, then evaluate it at $x = 1$.
    (c) Use a calculator to approximate e^1. How does this approximation compare to the values in part (a) and (b)?

64. Identify the missing operation sign to make the expressions equal.
$$\dfrac{3x + 1}{x + 4} \;\boxed{?}\; \dfrac{2x + 5}{x + 4} = \dfrac{x - 4}{x + 4}$$

65. (a) Write $\dfrac{1}{x} + \dfrac{1}{y}$ as a single rational expression.

    (b) Evaluate the expression in part (a) for $x = 2$ and $y = 3$.

    (c) Evaluate $\dfrac{2}{x + y}$ for $x = 2$ and $y = 3$.

    (d) Are the two values in part (b) and (c) equal? Explain.

66. (a) Write $\dfrac{1}{x + 3} \cdot \dfrac{1}{y + 5}$ as a single rational expression.

    (b) Evaluate the expression in part (a) for $x = 2$ and $y = 3$.

    (c) Evaluate $\dfrac{1}{xy + 3y + 5x + 15}$ for $x = 2$ and $y = 3$.

    (d) Are the two values in part (b) and (c) equal? Explain.

OBJECTIVES

1. Simplify Complex Rational Expressions
2. Simplify Expressions with Negative Exponents
3. Solve Applied Problems

1.7 Complex Rational Expressions

So far, we have worked with rational expressions of the form (P/Q) where $Q \neq 0$. In this section, we look at a form of rational expressions with rational expressions in its numerator and/or denominator.

Simplifying Complex Rational Expressions

Rational expressions in which the numerators and/or the denominators are rational expressions are called **complex rational expressions**. Examples are:

$$\frac{\dfrac{5}{7}}{3}, \quad \frac{\dfrac{2}{x}}{\dfrac{11}{x}}, \quad \frac{3 + \dfrac{3}{y}}{9}, \quad \frac{\dfrac{5}{t}}{7 + \dfrac{2}{t}}, \quad \text{and} \quad \frac{p + \dfrac{3}{p}}{\dfrac{2}{2 - p} - \dfrac{p}{2 - p}}.$$

It is often necessary to simplify a complex rational expression. That is, to express it as a simple rational expression in lowest terms. There are two methods to accomplish this.

Method 1

Find the LCD of all the denominators within the complex rational expression. Then multiply the numerator and the denominator by the LCD and simplify the results. This is usually the most efficient method.

Method 2

Express the rational expression as a quotient of two rational expressions by adding or subtracting (if necessary) and then dividing and simplifying to get a single rational expression.

EXAMPLE 1   Simplifying a Rational Expression

Use methods 1 and 2 to simplify the expression $\dfrac{\dfrac{9}{xy^2}}{\dfrac{12}{x^2 y}}$

Solution   Method 1
The LCD of the numerator and the denominator is $x^2 y^2$.

$$\frac{\dfrac{9}{xy^2}}{\dfrac{12}{x^2 y}} = \frac{\dfrac{9}{xy^2} \cdot x^2 y^2}{\dfrac{12}{x^2 y} \cdot x^2 y^2} \qquad \text{Multiply by } \frac{x^2 y^2}{x^2 y^2}$$

$$= \frac{\overset{3}{\cancel{9x}}}{\underset{4}{\cancel{12y}}} = \frac{3x}{4y} \qquad \text{Simplify}$$

Method 2

We have a single rational expression in both the numerator and the denominator.

$$\frac{\dfrac{9}{xy^2}}{\dfrac{12}{x^2y}} = \frac{9}{xy^2} \div \frac{12}{x^2y} \qquad \text{Divide}$$

$$= \frac{\overset{3}{\cancel{9}}}{\underset{y}{\cancel{xy^2}}} \cdot \frac{\overset{x}{\cancel{x^2y}}}{\underset{4}{\cancel{12}}} = \frac{3x}{4y} \qquad \text{Multiply by the reciprocal of the denominator} \quad \diamond$$

EXAMPLE 2   Simplifying a Rational Expression

Use method 1 to simplify the expression $\dfrac{c - \dfrac{1}{d}}{1 - \dfrac{c}{d}}$

Solution   We multiply both the numerator and the denominator of the rational expression by d, the LCD.

$$\frac{c - \dfrac{1}{d}}{1 - \dfrac{c}{d}} = \frac{d}{d} \cdot \frac{c - \dfrac{1}{d}}{1 - \dfrac{c}{d}}$$

$$= \frac{d\left(c - \dfrac{1}{d}\right)}{d\left(1 - \dfrac{c}{d}\right)} \qquad \text{Multiply by } \dfrac{d}{d}$$

$$= \frac{d(c) - d\left(\dfrac{1}{d}\right)}{d(1) - d\left(\dfrac{c}{d}\right)} \qquad \text{Distributive property}$$

$$= \frac{cd - 1}{d - c} \qquad \text{Simplify} \qquad \diamond$$

EXAMPLE 3   Simplifying a Rational Expression

Use method 1 to simplify the expression $\dfrac{\dfrac{a}{b} - \dfrac{b}{a}}{\dfrac{1}{a} + \dfrac{1}{b}}$

Solution   The LCD of the four rational expressions in the numerator and denominator is ab.

$$\frac{\dfrac{a}{b} - \dfrac{b}{a}}{\dfrac{1}{a} + \dfrac{1}{b}} = \frac{\left(\dfrac{a}{b} - \dfrac{b}{a}\right)ab}{\left(\dfrac{1}{a} + \dfrac{1}{b}\right)ab} \qquad \text{Multiply by } ab$$

$$= \frac{\left(\dfrac{a}{b}\right)ab - \left(\dfrac{b}{a}\right)ab}{\left(\dfrac{1}{a}\right)ab + \left(\dfrac{1}{b}\right)ab} \qquad \text{Distributive Property}$$

$$= \frac{a^2 - b^2}{b + a}$$

$$= \frac{(a - b)\cancel{(a + b)}}{\cancel{b + a}} = a - b \qquad \text{Factor and simplify} \qquad \diamond$$

Simplifying Expressions with Negative Exponents

Recall from the properties of exponents that

$$\frac{a^m}{a^n} = a^{m-n}$$

If we allow m to equal n, then we have

$$\frac{a^m}{a^m} = a^{m-m} = a^0$$

But we know that

$$\frac{a^m}{a^m} = 1$$

Comparing the two expressions, we have the definition of the zero exponent.

$$a^0 = 1 \text{ for any real number } a \neq 0$$

For example, we have

$$7^0 = 1, \ (12)^0 = 1, \ (x + 3)^0 = 1, \text{ and } (x^2y^3)^0 = 1$$

Recall also from the properties of exponents that

$$a^m \cdot a^n = a^{m+n}$$

What happens if we allow one of the exponents to be negative and apply the above property? For example, if $m = 4$ and $n = -4$. Then

$$a^m \cdot a^n = a^4 \cdot a^{-4} = a^{4+(-4)} = a^0 = 1$$
$$\text{so } a^4 \cdot a^{-4} = 1$$

That is, a^{-4} must be the multiplicative inverse or reciprocal of a^4. In symbols we write,

$$a^{-4} = \frac{1}{a^4}$$

This reasoning leads to the following definition.

NEGATIVE EXPONENT

$$a^{-n} = \frac{1}{a^n} \text{ for } a \text{ nonzero and } n \text{ a positive integer.}$$

If n is a positive integer and $a \neq 0$, then

$$\frac{1}{a^{-n}} = \frac{1}{\dfrac{1}{a^n}} = 1 \div \frac{1}{a^n} = a^n$$

EXAMPLE 4   Using the Properties of Exponents

Rewrite each expression with no negative exponents.

(a) y^{-5} (b) $(-4)^{-3}$ (c) $\dfrac{1}{x^{-7}}$ (d) $\left(\dfrac{3}{4}\right)^{-2}$

Solution   Here we restrict all variables so that they represent nonzero real numbers.

(a) $y^{-5} = \dfrac{1}{y^5}$

(b) $(-4)^{-3} = \dfrac{1}{(-4)^3}$

$\qquad\qquad = \dfrac{1}{-64} = -\dfrac{1}{64}$

(c) $\dfrac{1}{x^{-7}} = x^7$

(d) $\left(\dfrac{3}{4}\right)^{-2} = \left(\dfrac{4}{3}\right)^2 = \dfrac{16}{9}$

EXAMPLE 5   Simplifying an Expression with Negative Exponents

Rewrite the expression with only positive exponents and simplify the result.

$$\frac{(x + h)^{-1} - x^{-1}}{x^{-1}}$$

Solution   First, we rewrite the expression with positive exponents.

$$\frac{(x + h)^{-1} - x^{-1}}{x^{-1}} = \frac{\dfrac{1}{x + h} - \dfrac{1}{x}}{\dfrac{1}{x}}$$

The LCD of both the numerator and the denominator of this expression is

$$x(x + h)$$

$$\frac{\dfrac{1}{x + h} - \dfrac{1}{x}}{\dfrac{1}{x}} = \frac{\left[\dfrac{1}{x + h} - \dfrac{1}{x}\right]}{\left[\dfrac{1}{x}\right]} \cdot \frac{x(x + h)}{x(x + h)} \qquad \text{Multiply by } \frac{x(x + h)}{x(x + h)}$$

$$= \frac{x(x + h)\dfrac{1}{x + h} - x(x + h)\left[\dfrac{1}{x}\right]}{x(x + h)\left[\dfrac{1}{x}\right]} \qquad \text{Use the distributive property}$$

$$= \frac{x - (x + h)}{x + h}$$

$$= \frac{-h}{x + h} \qquad\qquad\qquad\qquad \text{Simplify}$$

Solving Applied Problems

Sometimes, formulas contain rational expressions as the next example shows.

EXAMPLE 6   Solving an Electronics Problem

In electronics, a resistor is any device that offers resistance to the flow of electrical current. When a light bulb is on, it acts as a resistor. Suppose that two light bulbs are placed in a parallel circuit for a household. The total resistance R (in ohms) is given by the formula

$$R = \frac{1}{R_1^{-1} + R_2^{-1}}$$

where R_1 and R_2 represent the resistance of each light bulb.
(a) Simplify the complex rational expression.
(b) Find the total resistance R when $R_1 = 3$ ohms and $R_2 = 4$ ohms.

Solution   (a) The total resistance R can be written as follows:

$$R = \frac{1}{\dfrac{1}{R_1} + \dfrac{1}{R_2}}$$

$$= \frac{1(R_1 R_2)}{\left[\dfrac{1}{R_1} + \dfrac{1}{R_2}\right](R_1 R_2)} \qquad \text{Multiply by } \frac{R_1 R_2}{R_1 R_2}$$

$$= \frac{R_1 R_2}{R_1 + R_2} \qquad \text{Simplify}$$

(b) Substituting 3 for R_1 and 4 for R_2 we have

$$R = \frac{3(4)}{3 + 4} = \frac{12}{7} \text{ ohms} \qquad \qquad \diamondsuit$$

Negative exponents and complex rational expressions can be used to solve applied problems including *installment loans for cars or mortgages* as well as electronics problems. The formula

$$m = \frac{Pr}{12[1 - (1 + r/12)^{-12t}]}$$

is used to calculate the equal *monthly car* or *home mortgage payments, m,* required to pay off a borrowed *principal* amount of P dollars, at an annual *interest rate* of r, paid over a *term of t years*. The *total amount of interest, I,* charged (in dollars) is the difference between the total amount paid and the principal owed, P. This is given by the formula

$$I = 12mt - P$$

EXAMPLE 7   Determine a Car Payment

Suppose you want to buy a car and need to borrow $8,000 from the bank. The bank charges an annual interest rate of 11.5% on automobile loans with monthly payments over a term of 36 months (3 years). Calculate
(a) the monthly payment and
(b) the total interest charged.

Solution   (a) Substituting $P = 8,000$, $r = 0.115$, and $t = 3$ into the monthly car payment formula, we have

$$m = \frac{Pr}{12[1 - (1 + r/12)^{-12t}]} = \frac{8000 \cdot 0.115}{12[1 - (1 + 0.115/12)^{-12(3)}]}$$
$$= 263.81$$

Therefore, the monthly car payment is $263.81.

(b) Substitute $m = 263.81$ and $t = 3$ in the formula

$$I = 12mt - P \text{ we get}$$
$$I = 12(263.81)(3) - 8000 = 1{,}497.16$$

Therefore, the total interest charged is $1,497.16.

PROBLEM SET 1.7

Mastering the Concepts

In Problems 1–10, simplify each rational expression by rewriting the expression as a simple rational expression reduced to lowest terms. Use method 1.

1. (a) $\dfrac{\dfrac{2}{3x}}{\dfrac{x}{2y}}$ (b) $\dfrac{\dfrac{x}{y}}{\dfrac{z}{y}}$

2. (a) $\dfrac{\dfrac{3}{8x}}{\dfrac{5}{12x}}$ (b) $\dfrac{\dfrac{4}{15y}}{\dfrac{12}{5y}}$

3. (a) $\dfrac{\dfrac{-1}{5b}}{\dfrac{2a}{3b}}$ (b) $\dfrac{\dfrac{-x}{7}}{\dfrac{y}{14}}$

4. (a) $\dfrac{\dfrac{-xy}{t}}{\dfrac{yz}{x}}$ (b) $\dfrac{\dfrac{4}{x^2}}{\dfrac{y}{x}}$

5. (a) $\dfrac{\dfrac{-y}{5x}}{\dfrac{1}{5x^2}}$ (b) $\dfrac{\dfrac{3x}{11y}}{\dfrac{-1}{22y^2}}$

6. (a) $\dfrac{\dfrac{x^2y^2}{x}}{\dfrac{5xy^3}{16}}$ (b) $\dfrac{\dfrac{5ab^3}{16}}{\dfrac{a^2b^2}{8}}$

7. $\dfrac{\dfrac{1}{1+a} - 1}{3a}$

8. $\dfrac{2 - \dfrac{1}{y}}{y - 1}$

9. $\dfrac{3 + \dfrac{1}{x}}{2 - \dfrac{3}{x}}$

10. $\dfrac{5 + \dfrac{1}{x}}{1 - \dfrac{3}{x}}$

In Problems 11–18, simplify each rational expression by rewriting the expression as a simple expression reduced to lowest terms. Use method 2.

11. (a) $\dfrac{\dfrac{2}{3x}}{\dfrac{4}{5x}}$ (b) $\dfrac{\dfrac{17y}{6}}{\dfrac{34y}{9}}$

12. (a) $\dfrac{\dfrac{x}{y}}{x}$ (b) $\dfrac{x}{\dfrac{x}{y}}$

13. (a) $\dfrac{\dfrac{-a}{5b^2}}{\dfrac{6a^2}{2y}}$ (b) $\dfrac{\dfrac{1}{b^2}}{\dfrac{3}{b}}$

14. (a) $\dfrac{\dfrac{x^2}{y^2}}{\dfrac{y}{z}}$ (b) $\dfrac{\dfrac{-x^2}{y}}{\dfrac{-y^2}{z}}$

15. $\dfrac{x - \dfrac{y}{x}}{x - \dfrac{x}{y}}$

16. $\dfrac{\dfrac{x}{y} - 3}{\dfrac{x}{z} - 5}$

17. $\dfrac{\dfrac{1}{5} - \dfrac{1}{x}}{x - 5}$

18. $\dfrac{4 - \dfrac{x}{y}}{1 - \dfrac{4y}{x}}$

In Problems 19–36, simplify each rational expression by any appropriate method.

19. $\dfrac{\dfrac{2}{x} + \dfrac{7}{y}}{\dfrac{8}{x} - \dfrac{6}{y}}$

20. $\dfrac{\dfrac{x}{3} - \dfrac{y}{4}}{\dfrac{y}{5} - \dfrac{x}{6}}$

21. $\dfrac{1 + \dfrac{2}{m}}{1 - \dfrac{4}{m^2}}$

22. $\dfrac{\dfrac{9}{y^2} - 4}{\dfrac{3}{y} - 2}$

23. $\dfrac{\dfrac{c}{2d} - \dfrac{3d}{6c}}{\dfrac{6}{3c} - \dfrac{2}{d}}$

24. $\dfrac{\dfrac{2}{1+v}}{3 + \dfrac{1}{1+v}}$

25. $\dfrac{\dfrac{1}{1+t} - 2}{3 - \dfrac{1}{1+t}}$

26. $\dfrac{\dfrac{x^2 - y^2}{x + y}}{1 + \dfrac{x}{y}}$

27. $\dfrac{1 + x - \dfrac{1}{1 - x}}{1 - x - \dfrac{1}{1 + x}}$

28. $\dfrac{y + x + \dfrac{y^2}{x - y}}{x - y + \dfrac{y^2}{y + x}}$

29. $\dfrac{\dfrac{u}{v} + 1 + \dfrac{v}{u}}{\dfrac{u^3 - v^3}{uv}}$

30. $\dfrac{\dfrac{a}{b} - 1 + \dfrac{b}{a}}{\dfrac{a^3 + b^3}{a^2b + ab^2}}$

31. $\dfrac{\dfrac{3}{1 - x} + \dfrac{x}{x - 1}}{\dfrac{1}{x - 1}}$

32. $\dfrac{\dfrac{a}{b} - 2 + \dfrac{b}{a}}{\dfrac{a}{b} + 2 + \dfrac{b}{a}}$

33. $\dfrac{\dfrac{2}{t} + \dfrac{1}{1 + t}}{\dfrac{3}{t + 1} - \dfrac{1}{t}}$

34. $\dfrac{\dfrac{x^2 - y^2}{x + y}}{1 - \dfrac{x}{y}}$

35. $\dfrac{m + \dfrac{n}{n - m}}{n + \dfrac{m}{m - n}}$

36. $\dfrac{\dfrac{1}{a + b} - \dfrac{1}{a - b}}{\dfrac{4}{a^2 - b^2}}$

In Problems 37–50, rewrite each expression as a complex rational expression, and then simplify. Answers should have only positive exponents.

37. (a) $\dfrac{y^{-4}}{y^{-2}}$ (b) $\dfrac{x^4}{x^{-2}}$

38. (a) $\dfrac{1}{y^{-3}}$ (b) $\dfrac{\dfrac{1}{x^{-2}}}{\dfrac{3}{x^{-3}}}$

39. (a) $\dfrac{\dfrac{1}{m^{-4}}}{\dfrac{1}{m^{-2}}}$ (b) $\dfrac{\dfrac{1}{b^{-2}}}{\dfrac{1}{b^3}}$

40. (a) $\dfrac{\dfrac{1}{x^{-2}}}{\dfrac{1}{x^{-3}}}$ (b) $\dfrac{\dfrac{1}{10x^{-3}}}{\dfrac{2}{x^{-2}}}$

41. $\dfrac{8}{2^{-2} - 4^{-2}}$

42. $\dfrac{x^{-1} + y^{-1}}{xy}$

43. $\dfrac{a^{-1}b^{-2}}{a + b}$

44. $(a^{-1} + b^{-1})^{-1}$

45. $\dfrac{x^{-2} - y^{-2}}{x^{-1} - y^{-1}}$

46. $\dfrac{x^{-1} + y^{-1}}{x^{-2} - y^{-2}}$

47. $\dfrac{(x + 1)^{-1} + 1}{3(x + 1)^{-1}}$

48. $\dfrac{1 - 4(1 - x)^{-1}}{2 + 8(x - 1)^{-1}}$

49. $\dfrac{(3 + h)^{-2} - 3^{-2}}{h}$

50. $\dfrac{(1 + h)^{-3} - 1}{h}$

Applying the Concepts

51. **Dimensional Analysis:** Simplify each expression.

    (a) $\dfrac{\dfrac{70 \text{ miles}}{35 \text{ miles}}}{1 \text{ hour}}$ (b) $\dfrac{\dfrac{100 \text{ miles}}{20 \text{ miles}}}{2 \text{ hour}}$

    (c) $\dfrac{\dfrac{140 \text{ days}}{7 \text{ days}}}{1 \text{ week}}$ (d) $\dfrac{\dfrac{45 \text{ miles}}{1 \text{ hour}}}{\dfrac{150 \text{ gallons}}{1 \text{ hour}}}$

52. **Photography:** A camera lens has a characteristic measurement f, called its *focal length*. When an object is in focus, its distance from the lens is p and the distance from the lens to the film (or image) is q. The symbols p, q and f are related by the equation

    $$f = \dfrac{1}{\dfrac{1}{p} + \dfrac{1}{q}}$$

    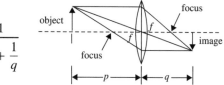

    (a) If $p = x + 10$ units and $q = x - 2$ units, simplify the complex rational expression.
    (b) If $p = 36$ centimeters and $q = 12$ centimeters, find f.

53. **Car Payments:** You are buying a car and need to borrow $16,000 from the credit union at an annual interest of 10.8%.
    (a) What is the monthly payment over a term of 4 years?
    (b) What is the total interest charged?

54. **Mortgage Payments:** A home is sold for $180,000 in which $\frac{5}{9}$ of its price is mortgaged at annual interest rate of 8.5%.
    (a) What is the monthly payment on the mortgage over a period of 30 years?
    (b) What is the total interest charged after 30 years of payments?

Developing and Extending the Concepts

In Problems 55 and 56, simplify each rational expression.

55. $\dfrac{\dfrac{x - 1}{x + 1} - \dfrac{x + 1}{x - 1}}{\dfrac{x - 1}{x + 1} + \dfrac{x + 1}{x - 1}}$

56. $\dfrac{\dfrac{2u - v}{2u + v} + \dfrac{2u + v}{2u - v}}{\dfrac{2u - v}{2u + v} - \dfrac{2u + v}{2u - v}}$

In Problems 57 and 58, rewrite each expression as a complex rational expression and simplify.

57. $\dfrac{2y^{-2} - y^{-1} - 1}{2y^{-2} - 3y^{-1} - 2}$

58. $\dfrac{x^{-3} + y^{-3}}{x^{-1} + y^{-1}} \cdot \dfrac{8}{x^{-2} - x^{-1}y^{-1} + y^{-2}}$

In Problems 59 and 60, simplify each expression.

59. (a) $1 + \dfrac{1}{2 + \dfrac{1}{2}}$ (b) $1 + \dfrac{1}{2 + \dfrac{1}{2 + \dfrac{1}{2}}}$

(c) $1 + \dfrac{1}{2 + \dfrac{1}{2 + \dfrac{1}{2 + \dfrac{1}{2}}}}$ (d) $1 + \dfrac{1}{x + \dfrac{1}{x + \dfrac{1}{x + \dfrac{1}{x}}}}$

60. (a) $1 + \dfrac{1}{1 + \dfrac{1}{2 + 2}}$ (b) $1 + \dfrac{1}{1 + \dfrac{1}{2 + \dfrac{2}{1 + 1}}}$

(c) $1 + \dfrac{1}{1 + \dfrac{1}{2 + \dfrac{2}{1 + \dfrac{1}{2}}}}$ (d) $1 + \dfrac{1}{1 + \dfrac{1}{x + \dfrac{x}{1 + \dfrac{1}{x}}}}$

OBJECTIVES

1. Plot Points in the Plane
2. Graph Linear Equations by Point Plotting
3. Use Graphing Technology

1.8 Graphing Techniques

In this section, we introduce the Cartesian coordinate system in a plane by means of two perpendicular lines. Then we introduce the basic concept of the graph of a linear equation by plotting points and by using a grapher.

Plotting Points in the Plane

The mathematical development that allows us to visualize data for certain activity is called the *Cartesian coordinate system*. We have seen that a number line associates points on the line with real numbers. Extending this concept, we can associate points on the plane with ordered pairs of real numbers. One reference system used to represent this association is called a **Cartesian** or **rectangular coordinate system**.

Figure 1 displays the *Cartesian, or rectangular, coordinate system*. It consists of two perpendicular number lines, one horizontal and one vertical, called the **coordinate axes**. These axes intersect at a point called the **origin**, which is the zero point on each number line. The horizontal line is often called the horizontal axis or the **x axis**. The positive portion of the x axis is to the right of the origin and the negative portion is to the left. The vertical line is called the vertical axis or the **y axis**, with the positive portion above the origin and the negative portion below it. A plane endowed with a Cartesian coordinate system is called a **Cartesian plane** or the **xy plane**.

The coordinate axes divide the plane into four regions called **quadrants**, which are described and displayed in Figure 2. All points in the xy plane lie on one of the four quadrants, or on the coordinate axes.

Any **ordered pair** of real numbers (a,b) can be represented as a point P in the coordinate system. The first number, a, called the **x coordinate** (or abscissa) of P, indicates the location of P to the right or left of the y axis. The second number, b, called the **y coordinate** (or ordinate) of P, indicates the location of P above or below the x axis. The coordinates of the origin are $(0,0)$. The point P with coordinates (a,b) is called the **graph** of the ordered pair (a,b). We plot, or locate, the position of (a,b) in the plane by placing a dot at the point P (Figure 1). The association between point P and ordered pair (a,b) seems so natural that in practice we write

$$P = (a,b)$$

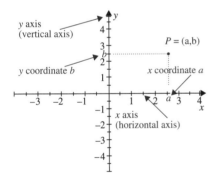

Figure 1

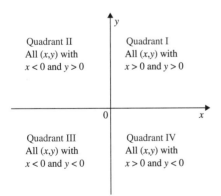

Figure 2

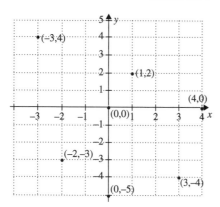

Figure 3

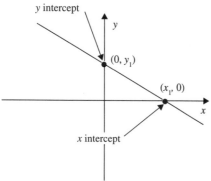

Figure 4

EXAMPLE 1 Plotting Points

Plot the points $(1,2)$, $(-3,4)$, $(-2,-3)$, $(4,0)$, $(0,0)$, $(0,-5)$, and $(3,-4)$, and specify their quadrant locations.

Solution Figure 3 shows that the points $(1,2)$, $(-3,4)$, $(-2,-3)$ and $(3,-4)$ lie in quadrants I, II, III, and IV respectively. The points $(4,0)$, $(0,0)$, and $(0,-5)$ lie on the coordinate axes. ◈

Graphing Linear Equations by Point Plotting

The **graph of an equation** in the two variables x and y is the set of all points (x,y) whose coordinates satisfy the equation. That is, the coordinates that make the equation true. One method of graphing equations is known as **a point plotting method**. In this method, we make a table of values by choosing several values for x and find the corresponding values for y. Then we plot these points and connect them with a smooth curve. The simplest curves in the plane are straight lines. Equations of lines in x and y are called **linear equations**; they have a degree of one. Some examples of linear equations are:

$$y = 3x,\ y = 2x - 6,\ \text{and } 3x - 2y = 6$$

We can graph a linear equation by plotting a few points and then drawing the line through them. In Figure 4, the graph crosses the x axis at $(x_1, 0)$ and the y axis at $(0, y_1)$. These ordered pairs are called the *intercepts* of the graph. The x value where the graph crosses the x axis is called the **x intercept**; the y value where the graph crosses the y axis is called the **y intercept**. These values often provide valuable information (Figure 4). To find the *x intercept*, set $y = 0$ and solve for x. To find the *y intercept*, set $x = 0$ and solve for y.

EXAMPLE 2 Graphing a Linear Equation

Graph the equation $y = 3x - 6$ and find the intercepts.

Solution First we list coordinates of some points in a table, where for each value of x, we obtain the value of y from $y = 3x - 6$.

Then we plot these points on the xy plane. The points with these coordinates appear to lie on a line. Figure 5 shows the graph of the line. The graph shows the x intercept is 2 and the y intercept is -6.

x	$3x - 6 = y$	(x, y)
-2	$3(-2) - 6 = -12$	$(-2,-12)$
-1	$3(-1) - 6 = -9$	$(-1,-9)$
0	$3(0) - 6 = -6$	$(0,-6)$
1	$3(1) - 6 = -3$	$(1,-3)$
2	$3(2) - 6 = 0$	$(2,0)$

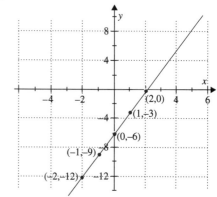

Figure 5

Using Graphing Technology

Equations can be graphed more quickly and efficiently using **graphers**. The term **grapher** refers to either a graphing calculator or a computer equipped with graphing software. The viewing window of a **grapher** is the portion of the xy plane shown on the screen. The viewing window can be defined manually by assigning a minimum x value (xMin), a maximum x value (xMax), position for tick marks (xScl), a minimum y value (yMin), a maximum y value (yMax) and position for tick marks (yScl). Most graphers have a default setting for a "friendly window." This is an assignment for xMin, xMax, xScl, yMin, yMax and yScl which will give easy to read values on the screen. Most viewing windows are not physically square; they do not have as many rows as columns of pixels (small dots that make up the picture). Usually graphers have a "squaring" feature, that is, a proportional window in which the graph will not be distorted. All graphers produce their pictures by the point plotting method. They start with the xMin value for x, find the corresponding value of y, and plot the point (x,y). The process is repeated at incremented values of x all the way across the screen. Graphers have the capability to TRACE . That is, to name the point that was plotted, and ZOOM , magnify the viewing window, on any part of a graph of an equation. This allows the *user* to approximate the x intercept and the y intercept of a graph.

All graphs generated by a grapher are designated by the symbol **G** . We will use the following procedure to graph an equation with a grapher.

> **Step 1.** Solve the equation for y in terms of x.
> **Step 2.** Select a viewing window.
> **Step 3.** Enter the equation and execute the graphing command.
> **Step 4.** Adjust the viewing window by zooming in or out until all the important parts of the graph are viewed.

G EXAMPLE 3   Using a Grapher

Generate the graph of the equation $2x - y = 4$ with a grapher and identify the x and y intercepts.

Solution   **Step 1.** Solve the equation for y in terms of x.
$$y = 2x - 4$$

Step 2. Select a window.
$$x\text{Min} = -8, x\text{Max} = 8, x\text{Scl} = 1$$
$$y\text{Min} = -8, y\text{Max} = 8, y\text{Scl} = 1$$

Step 3. Enter the equation. Execute the graphing command.
Step 4. Use ZOOM until the x and y intercepts appear on the screen, if necessary.

Figure 6 shows the graph of the equation. By using the TRACE feature, we see that the y intercept is -4 and the x intercept is 2. ◈

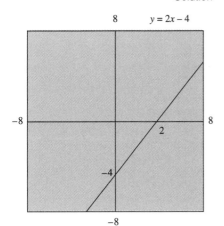

Figure 6

G EXAMPLE 4   Using a Grapher

Generate the graph of the equation $2x + y - 7 = 0$ and identify the x and y intercepts.

Solution   Solving for y in terms of x, we have
$$y = -2x + 7.$$
We know that the point at which the graph crosses the y axis must have an x coordinate of 0. Using the TRACE command, the point (0,7) is on the graph and the y intercept is 7. Continue tracing along the line until we reach the x intercept. This is the point where the line crosses the x axis and the y coordinate is 0. Depending upon the grapher, we may have to zoom to find the point on the line where the y coordinate is 0. The x intercept is 3.5. Figure 7 shows the line and the intercepts.   ◈

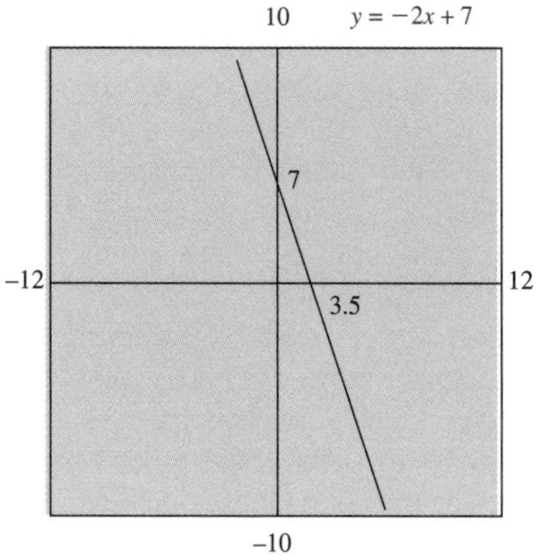

Figure 7

◈ **PROBLEM SET 1.8**

Mastering the Concepts

In Problems 1 and 2, plot the given set of points, and then specify their quadrant location.

1. (3,4), (−2,4), (5,−1), (0,7), (−7,0) and (0,−6).
2. (3,−2), (−2,3), (3,0), (0,−3), (−3,0) and (0,3).

In Problems 3–8, find the coordinates of the point A as described.

3. A is located three units to the left of the y axis and four units above the x axis.
4. A is located seven units to the right of the y axis and two units below the x axis.
5. A is located on the negative x axis, eight units from the origin.
6. A is located on the positive y axis, five units from the origin.
7. The coordinates of A are equal and A is located in the third quadrant five units below the x axis.
8. The coordinates of A are equal in absolute value and are opposite in sign. A is located in the fourth quadrant and 3 units below the x axis.

In Problems 9–26, graph each equation and locate the x and y intercepts.

9. $y = -2x + 2$
10. $y = 2x - 1$
11. $y = \dfrac{1}{2}x + 3$
12. $y = -\dfrac{1}{3}x + 2$
13. $3x + y = 2$
14. $-x + 2y = 4$
15. $3x - 2y = 6$
16. $2x - 3y = -6$
17. $2y - 6 = 0$
18. $3y + 4 = 0$
19. $y + 2 = 3(x - 1)$
20. $y - 1 = -2(x + 4)$
21. $\dfrac{x}{3} + \dfrac{y}{4} = 1$
22. $\dfrac{y}{2} - \dfrac{3x}{5} = 1$

23. $2x + 4 = 0$
24. $3x - 9 = 0$
25. $y = -x$
26. $y = 3x$

G In Problems 27–38, generate the graph of each equation by using the specified window setting on a grapher, and then use the TRACE and ZOOM features to identify the x and y intercepts.

27. $x + y = 4$
     xMin = −10, xMax = 10, xScl = 1
     yMin = −10, yMax = 10, yScl = 1
28. $x - y = 3$
     xMin = −10, xMax = 10, xScl = 1
     yMin = −10, yMax = 10, yScl = 1
29. $x - 2y = 20$
     xMin = −1, xMax = 26, xScl = 5
     yMin = −15, yMax = 5, yScl = 2
30. $2x - 6y + 3 = 0$
     xMin = −5, xMax = 5, xScl = 1
     yMin = −5, yMax = 5, yScl = 1
31. $5x - y + 6 = 0$
     xMin = −8, xMax = 8, xScl = 1
     yMin = −10, yMax = 10, yScl = 1
32. $3x + y + 4 = 0$
     xMin = −4, xMax = 4, xScl = 1
     yMin = −5, yMax = 5, yScl = 1
33. $2x + 5y - 10 = 0$
     xMin = −10, xMax = 10, xScl = 1
     yMin = −10, yMax = 10, yScl = 1
34. $y = -2x + 8$
     xMin = −10, xMax = 10, xScl = 1
     yMin = −10, yMax = 10, yScl = 1

35. $y = 15 - 2x$
     $x\text{Min} = -10, x\text{Max} = 10, x\text{Scl} = 1$
     $y\text{Min} = -20, y\text{Max} = 20, y\text{Scl} = 5$
36. $y = 2 + 3(x - 4)$
     $x\text{Min} = -10, x\text{Max} = 10, x\text{Scl} = 1$
     $y\text{Min} = -20, y\text{Max} = 20, y\text{Scl} = 5$
37. $y = \frac{x}{12} + \frac{1}{2}$
     $x\text{Min} = -8, x\text{Max} = 4, x\text{Scl} = 1$
     $y\text{Min} = -3, y\text{Max} = 3, y\text{Scl} = 1$
38. $y = 8(15 - x) + 3x$
     $x\text{Min} = -38, x\text{Max} = 55, x\text{Scl} = 10$
     $y\text{Min} = -40, y\text{Max} = 130, y\text{Scl} = 25$

Developing and Extending the Concepts

In Problems 39–44, determine the quadrant(s) in which each point is located.

39. $(x,3), x < 0$ 40. $(-4,y), y > 0$
41. $(x,y), x > 0, y < 0$ 42. $(-x,y), x < 0, y > 0$
43. $(x,y), xy > 0$ 44. $(x,y), xy < 0$

G In Problems 45–48, generate the graph of each equation by using the specified window setting on a grapher.

45. $y = |x|$
     $x\text{Min} = -10, x\text{Max} = 10, x\text{Scl} = 1$
     $y\text{Min} = -2, y\text{Max} = 10, y\text{Scl} = 1$
46. $y = |3 - 2x|$
     $x\text{Min} = -10, x\text{Max} = 10, x\text{Scl} = 1$
     $y\text{Min} = -10, y\text{Max} = 10, y\text{Scl} = 1$

47. $y = |2x + 14|$
     $x\text{Min} = -20, x\text{Max} = 20, x\text{Scl} = 5$
     $y\text{Min} = -10, y\text{Max} = 20, y\text{Scl} = 5$
48. $y = |2x - 10|$
     $x\text{Min} = -10, x\text{Max} = 10, x\text{Scl} = 1$
     $y\text{Min} = -10, y\text{Max} = 30, y\text{Scl} = 5$

G In Problems 49–54, generate the graph of each equation by using an appropriate window setting on a grapher. Then use the   TRACE   and   ZOOM   feature to find the x and y intercepts of each graph.

49. $5x + 7y = 35$ 50. $\dfrac{x}{5} - \dfrac{y}{3} = 1$

51. $\dfrac{1}{3}x + \dfrac{3}{4}y = 9$ 52. $2x + 3y = 30$

53. $x = 2y + 20$ 54. $\dfrac{x}{12} - \dfrac{y}{9} = 1$

◆ CHAPTER 1 REVIEW PROBLEM SET

1. Consider the polynomial expression
    $$5x^4 + x^3 + 2x^2 - x - 13$$

    (a) What is the degree of this polynomial?
    (b) What are the coefficients of the polynomial?
    (c) What is the value of the polynomial when $x = 2$?
2. Consider the polynomial expression
    $$-2x^4 + 5x^3 - 2x^2 + 7x + 11$$

    (a) What is the degree of this polynomial?
    (b) What are the coefficients of the polynomial?
    (c) What is the value of the polynomial when $x = -3$?

In Problems 3–12, perform each operation and simplify.

3. $(4x^3 + 3x^2) + (-5x^2 + 2x)$
4. $(3x^2 + 7x + 8) - (2x^2 + 3x + 2)$
5. $(-y^3 + 2y^2 + 3y - 7) - (-3y^3 + y^2 - 5y + 2)$
6. $(2x^3 + 4x^2 + 7x - 11) + (5x^2 - 3x + 5)$
7. $(a - b)(a^2 - 2ab + b^2)$
8. $(u^2 - u + 1)(2u^2 - 3u + 2)$
9. $(4y + 9x)(3y - 9x)$ 10. $(4u - 5v)(u - 3v)$
11. $(3x + 7)^2$ 12. $(m^2 - 3)(m^4 + 3m^2 + 9)$

In Problems 13–22, factor each expression completely.

13. $7x^2y - 21xy^3$ 14. $7x(y - x) + 14x^2(y - x)$
15. $5am^2 - bn + 5an - bm^2$ 16. $a^4 - b^4 - c^4 - 2b^2c^2$
17. $9u^3 - 81uv^2$ 18. $m^6p^6 - q^6$
19. $m^2 - 5m - 36$ 20. $6x^2 - 29x + 35$
21. $2a^3b - 10a^2b^2 - 28ab^3$ 22. $x^2yz - 6xy^2z - 16y^3z$

In Problems 23–26, perform each division.

23. $(5x^2 - 16x + 3) \div (x - 3)$
24. $(2y^2 + 17y + 21) \div (3 + 2y)$
25. $(x^3 - 6x^2 + 12x - 8) \div (x^2 - 4x + 4)$
26. $(x^4 - 3xy^3 - 2y^4) \div (x^2 - xy - y^2)$

In Problems 27–32, reduce each rational expression to lowest terms.

27. (a) $\dfrac{3m + 3n}{9m^2 - 9n^2}$ (b) $\dfrac{4y^2 - 9}{2y^2 - 3y}$

28. (a) $\dfrac{x^2 - 5x}{x^2 - 25}$ (b) $\dfrac{a^2 - 4b^2}{a^3 + 2a^2b}$

29. $\dfrac{x^2 - 7x + 12}{x^2 + 3x - 18}$ 30. $\dfrac{3y^2 - 17y - 28}{3y^2 + 10y + 8}$

31. $\dfrac{x^4 - 8x}{x^4 + 2x^3 + 4x^2}$ 32. $\dfrac{ax + ay - bx - by}{bx - by - ax + ay}$

In Problems 33 and 34, replace each question mark with an appropriate expression so that the resulting expressions are equivalent.

33. $\dfrac{3a}{2a - 2b} = \dfrac{?}{2a^2 - 2b^2}$ 34. $\dfrac{2t + 1}{t - 2} = \dfrac{?}{3t^2 - 4t - 4}$

In Problems 35–52, perform each operation and simplify.

35. (a) $\dfrac{x}{x + 3} + \dfrac{3}{x + 3}$ (b) $\dfrac{t + 1}{2t + 1} - \dfrac{t - 1}{2t - 1}$

36. (a) $\dfrac{3}{9 - t^2} - \dfrac{t}{9 - t^2}$ (b) $\dfrac{15t}{6t^2 - 5t - 6} + \dfrac{10}{6t^2 - 5t - 6}$

37. $\dfrac{3x}{4x^2 - 9} - \dfrac{x}{2x^2 + x - 6}$ 38. $\dfrac{7}{x^2 - y^2} + \dfrac{2}{x^2 - 2xy + y^2}$

39. $\dfrac{2}{2m^2 + m - 3} + \dfrac{1}{2m^2 + 5m + 3}$

40. $\dfrac{4x}{3x^2 + 5x - 2} - \dfrac{x}{2x^2 + 5x + 2}$

41. $\dfrac{4x^2 - 64}{2x^3 - 8x} \cdot \dfrac{x - 4}{x + 4}$

42. $\dfrac{4x^2 - 11x - 3}{6x^2 - 5x - 6} \cdot \dfrac{6x^2 - 13x + 6}{4x^2 + 13x + 3}$

43. $\dfrac{x^2 + 5x + 6}{x^2 + x - 2} \cdot \dfrac{x^2 + 3x - 4}{x^2 + 7x + 12}$

44. $\dfrac{x^2 - y^2}{x^2 + 2xy + y^2} \cdot \dfrac{3x + 3y}{6x}$ 45. $\dfrac{m^2 - m - 2}{m^2 - m - 6} \div \dfrac{m^2 - 2m}{2m + m^2}$

46. $\dfrac{y^2 - 5y - 6}{y^2 + y - 2} \div \dfrac{y^2 + y - 12}{y^2 + 3y - 4}$

47. $\dfrac{2x^2 - x - 6}{3x^2 - 11x - 4} \div \dfrac{2x^2 + 5x + 3}{3x^2 + 7x + 2}$

48. $\dfrac{8x^2 + 18x + 9}{6x^2 - 7x + 2} \div \dfrac{4x^2 + 7x + 3}{2x^2 + 9x - 5}$

49. $\dfrac{2 + 2(y - 1)^{-1}}{2(y - 1)^{-1}}$ 50. $\dfrac{x - x^{-1}}{x - 2 + x^{-1}}$

51. $\dfrac{\dfrac{1}{x} - \dfrac{1}{y}}{\dfrac{x^2 - y^2}{xy}}$ 52. $\dfrac{\dfrac{1}{2x - 3} - \dfrac{1}{2x + 3}}{\dfrac{x}{4x^2 - 9}}$

53. **Management:** Suppose that a company manufactures and sells x units of a product. Its profit P (in dollars) is given by the formula

$$P = -x^2 + 170x - 6600$$

Find the profit if the company makes and sells
(a) 1000 units   (b) 100 units   (c) 0 units

54. **Height of an Arrow:** An arrow is shot straight upward with an initial speed of 100 feet per second from a 200 foot high cliff. Its height, h, (in feet) after t seconds is given by the formula
$$h = -16t^2 + 100t + 200$$
What is the height of the arrow after
(a) 2 seconds   (b) 4 seconds   (c) 6 seconds?

55. **Geometry:** The lengths of the three sides of a triangle are consecutive odd numbers.
   (a) If x units is the length of the first side, find a formula for the perimeter of the triangle.
   (b) Find the perimeter of the triangle if $x = 41$ inches.

56. **Compound Interest:** Suppose that $1000 is invested in an account at an annual interest rate r for 2 years. The accumulated amount A in the account after 2 years is given by the formula
$$A = 1000(1 + r)^2$$
   (a) Write the right side of the formula in expanded form.
   (b) What is A if $r = 3\%$?

57. **Geometry:** Figure 1 shows a shaded area of a rectangular plot with two semicircles.
   (a) Find an expression for the total area in Figure 1.
   (b) Write the expression for the area in a factored form.
   (c) What is the total area if $r = 2$ meters? Round to one decimal place.

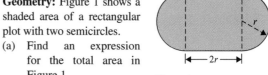

Figure 1

58. **Management:** A company's production P in thousands of units per year after t years from its starting date is given by the formula
$$P = \dfrac{3t^2 - 12t + 24}{t^2 + 2}$$
   (a) What is the company's production after 3 years from its starting date? Round to two decimal places.
   (b) after 5 years. Round to two decimal places.

59. **Physics:** The average speed V in miles per hour for a round trip made by an airplane is given by the formula

$$V = \frac{2}{\dfrac{1}{v_1} + \dfrac{1}{v_2}}$$

where v_1 (in miles per hour) is the average speed going to a destination and v_2 (in miles per hour) is the average speed returning.
   (a) What is the average speed for a round trip of an airplane with $v_1 = 450$ miles per hour and $v_2 = 500$ miles per hour? Round to the nearest whole number.
   (b) Simplify the complex fraction.

60. **Home Mortgage:** A home is purchased for $215,000, with a 15% down payment, and a 30 year mortgage on the balance at an annual interest rate of 7.25%. (see section 1.7 example 7)
   (a) What is the monthly payment on the mortgage?
   (b) When the mortgage is paid off, how much interest will have been paid?

G In Problems 61–64, use a grapher to graph each equation. Use the TRACE and ZOOM features to identify the x and y intercepts.

61. $5x - 2y - 10 = 0$ 62. $7x - y + 14 = 0$
63. $y = 3 - 2(x - 1)$ 64. $y = -4 + 3(x + 2)$

◆ CHAPTER 1 PRACTICE TEST

1. Find the degree of each polynomial and state the coefficient of each term. Also find the value of the polynomial for the indicated values.
   (a) $7x^2 - 3x + 5$: $x = -2$ and $x = 3$
   (b) $4x^3 + x - 1$; $x = 0$ and $x = -1$
2. Perform each operation and simplify.
   (a) $(3x^2 + 7x + 5) + (2x^2 - 5x - 2)$
   (b) $(6x^2 - 3x + 5) - (x^2 + 5x - 2)$
   (c) $(2x + 1)(2x^2 + x - 3)$
   (d) $(6x^3 + 21x^2 + 6x - 744) \div (3x - 12)$
3. Find each product.
   (a) $(7x + 1)^2$ (b) $(1 - 5x)^2$
   (c) $(3x - 2)(3x + 2)$ (d) $(x + 2)(x^2 - 2x + 4)$
4. Factor each expression.
   (a) $9m^2 - 100n^2$ (b) $a^2 - (b - c)^2$
   (c) $xy^2 - 4y^2 + 2x - 8$ (d) $x^2 + 16x - 17$
5. Simplify each expression.
   (a) $\dfrac{3x^2 - 5x - 2}{ax + 2b - bx - 2a}$ (b) $\dfrac{8x^2 - 14xy + 3y^2}{12x^2 + 5xy - 2y^2}$

6. Perform each operation and simplify.
   (a) $\dfrac{x^2 - 4}{x} \cdot \dfrac{2x}{3x - 6}$ (b) $\dfrac{x^2 + x - 2}{x + 2} \div \dfrac{x^2 + 2x - 3}{x + 3}$
   (c) $\dfrac{x}{x^2 - 1} + \dfrac{3}{4x - 4}$ (d) $\dfrac{b}{ab - b^2} - \dfrac{a}{b^2 + ab}$
7. Simplify each expression.
   (a) $\dfrac{1 + \dfrac{1}{x - 2}}{1 - \dfrac{1}{3x - 6}}$ (b) $\dfrac{x^{-1}}{x^{-1} + y^{-1}}$
8. Graph the equation $2x - 6y + 12 = 0$ and then find the intercepts of the graph.
9. A rectangular piece of cardboard measures 12 inches by 16 inches. An open box is formed by cutting squares of equal size from each of the four corners of the cardboard and folding the sides. Suppose that each square corner cut out has a side that measures x inches.
   (a) Find the area A of the base of the box in terms of x.
   (b) Simplify the expression, and then find the area of the base of the box if $x = 3$ inches.

Chapter 2

2.1 Linear Equations in One Variable
2.2 Literal Equations and Formulas
2.3 Applied Problems
2.4 Equations Involving Rational Expressions
2.5 Linear Inequalities
2.6 Equations and Inequalities Involving Absolute Value

LINEAR EQUATIONS AND INEQUALITIES

Throughout the world, people use the Celsius scale to determine temperature. In the United States, we use the Fahrenheit scale. The following formula

$$F = \frac{9}{5}C + 32$$

shows the connection between the readings C and F of the two scales. For example, to determine C when $F = -40$ or 32, we must solve the equations

$$-40 = \frac{9}{5}C + 32 \quad \text{and} \quad 32 = \frac{9}{5}C + 32$$

in Example 6 on page 95.

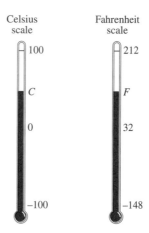

In this chapter we apply the algebraic skills developed in the preceding chapter to solve equations and inequalities. Both equations and inequalities are used to describe events from science and business as well as from social sciences and everyday life. Mathematics is quickly becoming the language not only of science but of much of human endeavor.

OBJECTIVES

1. Solve Linear Equations
2. Solve Linear Equations Involving Fractions
3. Classify Identities and Inconsistent Equations
4. Use Technology Exploration

2.1 Linear Equations in One Variable

An **equation** is a statement that two mathematical expressions are equal. Substituting a particular number for the variable produces an equation that is either true or false. If a true statement results from such a substitution, we say the number **satisfies** the equation and this number is called a **solution** or a **root** of the equation. For instance, 3 is a solution of the equation $4x + 7 = 19$, because the substitution 3 for x makes the equation a true statement. That is,

$$4(3) + 7 = 12 + 7 = 19$$

But -4 is not a solution, because

$$4(-4) + 7 = -16 + 7 = -9 \neq 19$$

Two or more equations are said to be **equivalent** if they have *exactly* the same solutions. For example, the following equations:

$$4x + 7 = 19$$
$$4x = 12$$
$$x = 3$$

are equivalent because they all have the same solution, namely, 3.

We can change an equation $P = Q$ into an equivalent one by performing any of the following operations.

Property	Let $P = Q$, then	Illustration Let $x = 3$
1. Addition (Subtraction) Property Add (subtract) the same quantity on both sides of the equation.	(a) $P + R = Q + R$ (b) $P - R = Q - R$	$x + 7 = 3 + 7$ $x - 7 = 3 - 7$
2. Multiplication (Division) Property Multiply (divide) both sides of the equation by the same nonzero quantity.	(a) $PR = QR, R \neq 0$ (b) $\dfrac{P}{R} = \dfrac{Q}{R}, R \neq 0$	$5x = (5)(3)$ $\dfrac{x}{6} = \dfrac{3}{6}$
3. Symmetric Property Interchange the two sides of the equation.	$Q = P$	$3 = x$

To **solve** an equation means to find all of its solutions. The usual method for solving an equation is to write a sequence of equations, starting with the given one, in which each equation is equivalent to the previous one. The last equation should express the solution directly.

Solving Linear Equations

Each of the following equations,

$$5x - 3 = 7, \qquad\qquad 5 + 4x = 33,$$
$$4 + 12m = 5m - 10, \qquad 5 + 8(x + 2) = 23 - 2(2x - 5)$$

and

$$\frac{x}{7} - \frac{x - 1}{2} = \frac{11}{4}$$

contain one variable, and each variable has degree one. They are examples of **linear** or **first degree equations**. In this section, we consider linear equations in one variable involving a polynomial of degree one. For instance, we solve the linear equation $5x - 3 = 7$ as follows:

$5x - 3 = 7$	Start with the given equation
$5x - 3 + 3 = 7 + 3$	Use the Addition Property
$5x = 10$	
$\dfrac{5x}{5} = \dfrac{10}{5}$	Use the Division Property
$x = 2$	An equivalent equation with an obvious solution

Check:

$$\text{If } x = 2, \text{ then}$$
$$5(2) - 3 = 10 - 3$$
$$= 7.$$

The solution is 2 because it makes the equation true.

It should be noted that this linear equation has only *one* solution; that is, no other value of x will make the original equation true.

In general, we may solve linear equations in one variable by using the following strategy.

STRATEGY FOR SOLVING
LINEAR EQUATIONS

1. COMBINE
  (a) Multiply each side of the equation by the least common denominator (LCD) if fractions exist in the equation.
  (b) Remove all parentheses or grouping symbols on each side of the equation.
  (c) Combine like terms on each side of the equation.

2. REARRANGE
  Apply the addition (or subtraction) property to collect all terms containing the variable for which we are solving on one side of the equation and all other terms on the other side.

3. ISOLATE
  Divide each side of the equation by the coefficient of the variable.

4. CHECK
  Substitute the solution into the original equation to verify that the equation is true.

EXAMPLE 1   Solving a Linear Equation
Solve each equation.
(a) $5 + 4x = 33$ (b) $4t + 7 = 3t + 5$

Solution   (a)
$$5 + 4x = 33 \qquad \text{Given}$$
$$5 + 4x - 5 = 33 - 5 \qquad \text{Rearrange}$$
$$4x = 28$$
$$\frac{4x}{4} = \frac{28}{4} \qquad \text{Isolate}$$
$$x = 7$$

Check: If $x = 7$, then $5 + 4(7) = 5 + 28 = 33$.
So that 7 is the solution and the only solution since this is a linear equation.

(b)
$$4t + 7 = 3t + 5 \qquad \text{Given}$$
$$4t + 7 - 7 = 3t + 5 - 7 \qquad \text{Subtract}$$
$$4t = 3t - 2 \qquad \text{Rearrange}$$
$$4t - 3t = -2 \qquad \text{Subtract}$$
$$t = -2$$

Check: If $t = -2$, then

Left Side	Right Side
$4(-2) + 7 =$	$3(-2) + 5 =$
$-8 + 7 = -1$	$-6 + 5 = -1$

Note that the "sides" of an equation are separated by the "=" sign.

EXAMPLE 2   Using the Distributive Property to Solve an Equation
Solve the equation

$$5 + 8(x + 2) = 23 - 2(2x - 5)$$

Solution   We begin by using the distributive property and combining like terms.
$$5 + 8(x + 2) = 23 - 2(2x - 5) \qquad \text{Given}$$
$$5 + 8x + 16 = 23 - 4x + 10 \qquad \text{Apply the distributive property}$$
$$8x + 21 = -4x + 33 \qquad \text{Combine like terms}$$
$$8x + 4x + 21 - 21 = -4x + 4x + 33 - 21 \qquad \text{Add } 4x - 21 \text{ to each side}$$
$$12x = 12 \qquad \text{Combine like terms}$$
$$x = 1 \qquad \text{Divide each side by 12}$$

Check: Let $x = 1$, then∆89

Left side	Right side
$5 + 8(1 + 2) =$	$23 - 2(2(1) - 5) =$
$5 + 8(3) =$	$23 - 2(-3) =$
$5 + 24 = 29$	$23 + 6 = 29$

EXAMPLE 3   Solving an Equation Involving Decimals
Solve the equation

$$1.3(2 - 5.2y) = 0.2 - (1.76y - 3.4)$$

Solution   Here we can use a calculator for some computations.
$$1.3(2 - 5.2y) = 0.2 - (1.76y - 3.4) \qquad \text{Given}$$
$$2.6 - 6.76y = 0.2 - 1.76y + 3.4 \qquad \text{Apply the distributive property}$$
$$-6.76y + 1.76y = 0.2 + 3.4 - 2.6 \qquad \text{Add } 1.76y \text{ and } -2.6 \text{ to each side}$$
$$-5y = 1.00 \qquad \text{Combine like terms}$$
$$y = -0.2 \qquad \text{Divide each side by } -5$$

If an equation involves decimals, the solution should be expressed in decimals.

A check will show that -0.2 is the solution.
We recommend that the reader check the answer.

Solving Linear Equations Involving Fractions

We indicated in step 1 of our strategy that when an equation contains fractions, we multiply each side of that equation by the least common denominator (LCD) to clear the fractions. This process produces an equivalent equation with integer coefficients which is easier to solve.

EXAMPLE 4 Solving Equations Involving Fractions

Solve each equation.

(a) $\dfrac{2}{3}x + \dfrac{1}{2} = \dfrac{5}{6}$

(b) $\dfrac{y-3}{4} - \dfrac{y-2}{3} = \dfrac{y-11}{6}$

Solution To find an equivalent equation without fractions, multiply each side of the equation by the LCD.

(a)

$$\dfrac{2}{3}x + \dfrac{1}{2} = \dfrac{5}{6} \qquad \text{Given}$$

$$6\left(\dfrac{2}{3}x + \dfrac{1}{2}\right) = 6\left(\dfrac{5}{6}\right) \qquad \text{Multiply each side by 6, the LCD}$$

$$6\left(\dfrac{2}{3}x\right) + 6\left(\dfrac{1}{2}\right) = 6\left(\dfrac{5}{6}\right) \qquad \text{Apply the distributive property}$$

$$4x + 3 = 5 \qquad \text{Multiply}$$

$$4x = 2 \qquad \text{Subtract 3 from each side}$$

$$x = \dfrac{2}{4} = \dfrac{1}{2} \qquad \text{Divide each side by 4}$$

Therefore, the solution is $\dfrac{1}{2}$. A check will show the answer is correct.

(b)

$$\dfrac{y-3}{4} - \dfrac{y-2}{3} = \dfrac{y-11}{6} \qquad \text{Given}$$

$$12\left(\dfrac{y-3}{4} - \dfrac{y-2}{3}\right) = 12\left(\dfrac{y-11}{6}\right) \qquad \text{Multiply each side by 12, the LCD}$$

$$12\left(\dfrac{y-3}{4}\right) - 12\left(\dfrac{y-2}{3}\right) = 12\left(\dfrac{y-11}{6}\right) \qquad \text{Apply the distributive property}$$

$$3(y-3) - 4(y-2) = 2(y-11) \qquad \text{Multiply}$$

$$3y - 9 - 4y + 8 = 2y - 22 \qquad \text{Apply the distributive property}$$

$$3y - 4y - 2y = -22 + 9 - 8 \qquad \text{Add } -2y, 9 \text{ and } -8 \text{ to each side}$$

$$-3y = -21 \qquad \text{Combine like terms}$$

$$y = \dfrac{-21}{-3} = 7 \qquad \text{Divide each side by } -3$$

Thus, 7 is the solution. A check will show the answer is correct.

Classifying Identities and Inconsistent Equations

Each of the equations considered in Examples 1–4 were true for only one value of the variable. These linear equations are called **conditional equations**. There are two other possibilities; identities and inconsistent equations. An equation that is true for all real numbers is called an **identity**. For instance,

$$7x + 2 = 10x + 2 - 3x$$

is an example of an *identity*.

An equation that has no solution is called an **inconsistent equation**. The equation

$$4(5x + 1) = 26x + 5 - 6x$$

is *inconsistent* as can be seen by simplifying each side.

$$20x + 4 = 20x + 5$$

There are no values of x which would make this equation true.

EXAMPLE 5 Classifying Equations

Determine which of the given equations is inconsistent and which is an identity.

(a) $3(4x + 5) = 4 + 12x$ (b) $2(x + 5) = 5x + 10 - 3x$

Solution

(a) $12x + 15 = 4 + 12x$ Apply the distributive property
              $15 = 4$ Subtract $12x$ from each side

Since $15 = 4$ is a false statement, the equation has no solution and it is called *inconsistent*.

(b) $2x + 10 = 5x + 10 - 3x$ Apply the distributive property
     $2x + 10 = 2x + 10$ Combine like terms

> Remember that 0 can be a solution to an equation. This is not the same as "no solution."

Notice that any number we substitute for x in the original equation leads to a true statement. Therefore, the solution consists of all real numbers and the equation is called an *identity*. ◈

Using Technology Exploration

We can use a grapher to solve all the linear equations we have considered so far. This is accomplished by first writing an equivalent equation for each linear equation with all the variables and the constants on the right side and zero on the left. That is, by writing each equation of the form $0 = mx + b$, and then graphing the line $y = mx + b$.

G EXAMPLE 6 Using a Grapher to Solve an Equation

Use a grapher to solve the equation

$$5 + 4x = 33.$$

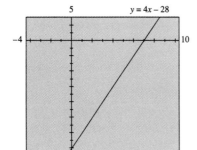

$y = 4x - 28$

Figure 1

Solution

We begin by writing an equivalent equation and setting it equal to zero.

$$0 = 4x - 28$$

Then we graph the equation

$$y = 4x - 28$$

with a grapher (Figure 1). The solution of the equation $0 = 4x - 28$ is the x intercept of the graph since $y = 0$ in the original equation. It appears from the graph in Figure 1 that there is an x intercept at $x = 7$. We can TRACE and ZOOM to the x intercept of the graph. Thus, 7 is the solution to the original equation $5 + 4x = 33$. This method can be used with any conditional equation. ◈

◈ PROBLEM SET 2.1

Mastering the Concepts

In Problems 1–50, solve each equation and check the results.

1. (a) $2x - 5 = 11$
   (b) $2x + 7 = 7$
2. (a) $3w - 8 = 13$
   (b) $4 - 3w = 16$
3. (a) $19 - 15x = -131$
   (b) $-21 = 15x - 131$
4. (a) $104 - 12t = 152$
   (b) $12t + 104 = 144$
5. (a) $18 - x = 5 + 3x$
   (b) $12 - 3m = 4m + 26$
6. (a) $2 - 5y = -3y + 6$
   (b) $2x - 14 = 7 - x$
7. (a) $2u + 1 = 5u + 18$
   (b) $2x - 3x = 7 - 15x$
8. (a) $7y - 15 = 3y - 10$
   (b) $7 + 8x - 12 = 3x - 8 + 5x$
9. $3y - 2y + 7 = 12 - 4y$
10. $2t - 9t + 3 = 6 - 5t$
11. $1 - 2(5 - 2y) = 26 - 3y$
12. $7t - 3(9 - 5t) = -5t$
13. $8(5x - 1) + 36 = -3(x - 5)$

14. $2(1 + 2y) = 3(2y - 4)$
15. $2(x + 5) - (3 - x) = 15$
16. $7(x - 3) = 4(x + 5) - 47$
17. $6(c - 10) + 3(2c - 7) = -45$
18. $5(x - 3) - (x - 1) = -14$
19. $3(x - 2) + 5(x + 1) = -45$
20. $4(x + 4) - 5(2 - x) = 3(6x - 2)$
21. $2[3x - (x - 3)] = 4(x - 3)$
22. $16 - 2[4 - 3(1 - x)] = 14 - 5x$
23. $\dfrac{t}{6} - \dfrac{t}{7} = 5$
24. $\dfrac{y}{2} - \dfrac{y}{3} = 4$
25. $\dfrac{4x}{3} - 1 = \dfrac{5x}{6}$
26. $\dfrac{x}{4} + \dfrac{2x}{3} = \dfrac{35}{12}$
27. $\dfrac{5x - 15}{7} - \dfrac{x}{3} = \dfrac{2}{3}$
28. $\dfrac{3y}{5} - \dfrac{13}{15} = \dfrac{y + 2}{12}$
29. $\dfrac{t + 5}{7} + \dfrac{t - 3}{4} = \dfrac{5}{14}$
30. $\dfrac{w - 2}{3} + \dfrac{w + 1}{4} = -3$
31. $\dfrac{4x + 1}{10} = \dfrac{5x + 2}{4} - \dfrac{5}{4}$
32. $\dfrac{4x - 1}{10} - \dfrac{5x + 2}{4} = -3$
33. $\dfrac{2y + 3}{3} + \dfrac{3y - 4}{6} = \dfrac{y - 2}{9}$
34. $\dfrac{x - 2}{2} + \dfrac{x + 3}{4} = \dfrac{6x + 7}{6}$
35. $\dfrac{x}{7} - \dfrac{x - 1}{2} = \dfrac{11}{4}$
36. $2z - \dfrac{9z + 1}{4} = \dfrac{2}{3}$
37. $\dfrac{2t - 1}{4} + \dfrac{3t + 4}{8} = \dfrac{1 - 4t}{12}$
38. $\dfrac{y - 2}{4} - \dfrac{y + 5}{6} = \dfrac{5y - 2}{9}$
39. $\dfrac{y + 9}{4} - \dfrac{6y - 9}{14} = 2$
40. $\dfrac{8t + 10}{5} - \dfrac{6t + 1}{4} = \dfrac{3}{20}$
41. $\dfrac{3x - 2}{3} + \dfrac{x - 3}{2} = \dfrac{5}{6}$
42. $\dfrac{3u - 6}{4} - \dfrac{u + 6}{6} + \dfrac{2u}{3} = 5$
43. $0.7x + 3 = 0.5x + 2$
44. $1.5y + 4 = 1.2y - 2$
45. $x - 0.1x - 4.5 = 0.75$
46. $0.6x - 2.5 = 0.03x + 8.9$
47. $0.02(y - 100) = 62 - 0.06y$
48. $0.05(t - 100) = 0.2t - 35$
49. $0.5x - 0.1x - 0.2x = 0.05$
50. $0.75x - 0.5x = 0.625x + 0.75$

In Problems 51–53, determine whether each equation is an identity, a conditional equation or an inconsistent equation.

51. (a) $6(x + 2) = 6x + 12$ (b) $4(x + 1) - 4 = 4x$
52. (a) $5(x + 3) = 5x + 3$ (b) $5(x + 1) - 5 = 0$
53. (a) $4(w - 2) + 2 = 3w - 7 + w$
    (b) $16 - 2[4 - 3(1 - x)] = 14 - 6x$

Applying the Concepts

54. **Work Rate Problem:** Two workers can dig trenches for a sprinkler system in t hours, where t satisfies the equation
$$\frac{t}{15} + \frac{t}{10} = 1$$
Find the required time t

55. **Coin Problem:** A vending machine accepts nickels, dimes and quarters. When the coin box is emptied, the total value of the coins is found to be $24.15. Suppose that the box contains n nickels, $n + 5$ dimes and $n/2$ quarters, where n satisfies the equation
$$0.05n + 0.10(n + 5) + 0.25(n/2) = 24.15$$
Solve the equation for n, and then find the number of coins of each kind in the box.

56. **Temperature Conversion:** A rule of thumb for converting a temperature from degrees Celsius, C, to degrees Fahrenheit, F, is given by the equation
$$F = 2C + 30 \qquad 0° \le C \le 100°$$
(a) If a bank thermometer registers 17° in the Celsius scale, what is the corresponding temperature in the Fahrenheit scale?
(b) If a thermometer registers 70° in the Fahrenheit scale, what is the corresponding temperature in the Celsius scale?

57. **Car Rental:** A car rental company charges for its midsize models $75 a week plus $0.15 for each mile driven for that week. The total charges C (in dollars) for a week is given by the equation
$$C = 75 + 0.15n$$
where n is the number of miles driven for a week.
(a) What is the total charge if a midsize rented car is driven 400 miles in a week?
(b) If the total charge for a week is $165, how many miles are driven for that week?

58. **Telephone Charges:** The cost of a long distance telephone call between two cities during business hours is $0.28 for the first minute and $0.15 for each additional minute or a fraction of a minute. The total cost C (in dollars) of long distance calls is given by the equation
$$C = 0.28 + 0.15(t - 1)$$
where t is the time for a call in minutes and $t > 1$.
(a) What is the total charge for a phone call that lasted 41 minutes?
(b) What is the length of a long distance phone call if the total cost is $3.28?

59. **Real Estate:** A real estate agent gets a commission of 7% of the sale price of a house. The net price N (in dollars) of a house after the commission is paid is given by the equation
$$N = p - 0.07p$$
where p (in dollars) is the sale price of the house.
(a) What is the net price of a house if it is sold for $140,000?
(b) If a couple wants to sell their house and have $148,800 after they pay the commission, what should be the selling price of the house?

Developing and Extending the Concepts

In Problems 60 and 61, solve each equation as follows:

    (a)   Multiply each side of the equation by a power of 10 that will eliminate the decimals.

    (b)   Without changing to integer form.

    (c)   Explain if you prefer method (a) or (b) and why.

60. $3y - 3(1.9y - 4.1) - 8 = 2(1.6y)$

61. $0.2(2x + 6.1) + 0.4(2.1x + 3) = 0.56$

G In Problems 62–65, use a grapher to solve each equation.

    (a)   Solve the equation first.

    (b)   Graph the given equation in y. The x intercept is the solution to the original equation.

    (c)   Simplify the right side of the given equation in y and graph this new expression.

    (d)   What do you see and why? If no mistake is made in simplifying, the given equation in y and your simplified equation in y have the same graphs.

62. $5x + 2(x + 1) = 23$
graph $y = 5x + 2(x + 1) - 23$
xMin $= -8$, xMax $= 8$, xScl $= 1$
yMin $= -30$, yMax $= 30$, yScl $= 1$

63. $7x - (3x - 4) = 12$
graph $y = 7x - (3x - 4) - 12$
xMin $= -8$, xMax $= 8$, xScl $= 1$
yMin $= -10$, yMax $= 10$, yScl $= 1$

64. $2(2x + 5) - 3x = 3$
graph $y = 2(2x + 5) - 3x - 3$
xMin $= -8$, xMax $= 8$, xScl $= 1$
yMin $= -10$, yMax $= 10$, yScl $= 1$

65. $5x - 4(x - 2) = 10$
graph $y = 5x - 4(x - 2) - 10$
xMin $= -8$, xMax $= 8$, xScl $= 1$
yMin $= -10$, yMax $= 10$, yScl $= 1$

66. Is the equation that expresses the associative property of addition

$$a + (b + c) = (a + b) + c$$

a conditional equation, an inconsistent equation, or an identity?

67. If the equation $a = b$ is true, what can you say about the equation $a + c = b + c$?

68. From geometry, two angles that are on opposite sides of the intersection of two lines are called *vertical angles*. It can be shown that vertical angles have the same measure.

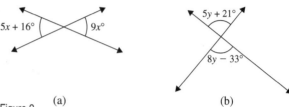

Figure 2   (a)   (b)

    (a)   Use Figure 2a to find x.

    (b)   Use Figure 2b to find y.

69. From geometry, the sum of the interior angles of a triangle is 180°. Use Figure 3 with the given measurements to find t.

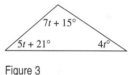

Figure 3

OBJECTIVES

1. Solve Literal Equations
2. Solve Formulas
3. Use Formulas to Solve Applied Problems
4. Use Technology Exploration

2.2 Literal Equations and Formulas

Thus far we have worked with equations that contain one variable, represented by a letter. To analyze a relationship between more than one variable or letter, we often use **literal equations**. In these equations, some of these letters represent variables, others represent constants. In mathematics, common examples of literal equations that express relationships between two or more quantities are called **formulas**. For example,

$$I = Prt, \quad F = \frac{9}{5}C + 32 \quad \text{and} \quad ax + by = c$$

are formulas for the simple interest, the conversion from Celsius to Fahrenheit, and the general form of a linear equation. These formulas describe real-life situations in mathematical terms.

Solving Literal Equations

Because literal equations and formulas are types of equations, we may use the equation solving strategy illustrated in section 2.1 to solve them. That is, we bring all terms containing the variable for which we wish to solve to one side of the equation, and all other terms to the opposite side. Then, we divide each side by the coefficient of that variable. As always, we must be careful not to divide by zero.

EXAMPLE 1   Solving a Literal Equation

Solve each equation.

(a) $4x - a = x + 8a$ for x

(b) $3x + 2y - 6 = 0$ for y

Solution   (a)

$4x - a = x + 8a$	Given
$4x - a - x = 8a$	Subtract x from each side
$4x - x = 8a + a$	Add a to each side
$3x = 9a$	Combine like terms
$x = 3a$	Divide by 3

Check: If $x = 3a$, then

Left side	Right side
$4(3a) - a =$	$(3a) + 8a =$
$12a - a = 11a$	$11a$

(b)

$3x + 2y - 6 = 0$	Given
$2y - 6 = -3x$	Subtract $3x$
$2y = -3x + 6$	Add 6
$y = \dfrac{-3x + 6}{2}$	Divide by 2
$y = \dfrac{-3x}{2} + 3$	

Check: Let $y = \dfrac{-3x}{2} + 3$, then

Left side	Right side
$3x + 2\left(\dfrac{-3x}{2} + 3\right) - 6 =$	0
$3x - 3x + 6 - 6 = 0$	0

EXAMPLE 2   Solving a Literal Equation Involving Fractions

Solve the equation $\dfrac{y - 3x}{b} = \dfrac{2x}{b} + y$ for y, $b \neq 0$

Solution

$\dfrac{y - 3x}{b} = \dfrac{2x}{b} + y$	Given
$b\left(\dfrac{y - 3x}{b}\right) = b\left(\dfrac{2x}{b} + y\right)$	Multiply each side by b, the LCD
$y - 3x = b\left(\dfrac{2x}{b}\right) + by$	Apply the distributive property
$y - 3x = 2x + by$	Multiply
$y - by - 3x = 2x$	Subtract by
$y - by = 2x + 3x$	Add $3x$
$y(1 - b) = 5x$	Combine like terms and factor y
$y = \dfrac{5x}{1 - b}$	Divide each side by $1 - b$, $b \neq 1$

A check will show the answer is correct.

Solving Formulas

Many applications from different sciences require the use of *formulas* for their solutions. For instance.

$$A = lw, \quad d = rt, \quad \text{and} \quad A = P(1 + rt)$$

are formulas for the area of a rectangle, the distance in terms of rate and time, and the accumulated amount of money at simple interest rate. To solve a formula for a specific variable or letter, we follow the same strategies as we do for solving equations.

EXAMPLE 3   Solving a Formula

Solve the formula $A = lw$ for w.

Solution

$A = lw$	Given
$lw = A$	Interchange the two sides
$w = \dfrac{A}{l}$	Divide each side by l

◈

EXAMPLE 4   Solving a Formula

The formula $A = P(1 + rt)$ gives the accumulated sum of money in dollars, when a principal of P dollars is invested at an interest rate r for t years. Solve the formula for t.

Solution

$A = P(1 + rt)$	Given
$P(1 + rt) = A$	Interchange the two sides
$P + Prt = A$	Apply the distributive property
$Prt = A - P$	Subtract P from each side
$t = \dfrac{A - P}{Pr}$	Divide each side by Pr

◈

Using Formulas to Solve Applied Problems

The process of describing a real-world situation in mathematical terms is call **mathematical modeling**. This process may involve constructing an equation or collecting data. Once a model is established, it can be used to analyze the situation and make predictions. Some mathematical models are very accurate because they can be described by formulas. Others are too complicated to be precisely modeled. For instance, if a volume is being sought in a real-world situation, we write a *formula* for the volume and replace quantities such as length, width and height by letters or values given in the situation. These quantities should be represented by the same units of measure such as feet, meters, or miles. Recall from section R5 that *dimensional analysis* is needed to convert from one system of measurement into another.

EXAMPLE 5   Modeling by Using a Formula for Volume

Each of the public tennis courts in a city measure 40 feet wide and 88 feet long. When these asphalt courts need to be resurfaced, sixteen 5 yard loads of asphalt are used for each court.
(a) Solve the volume formula $V = lwh$, for h (its depth), where l is the length of the court and w is the width.
(b) Determine the depth of asphalt for each court in inches.

Solution (a) First we solve the formula $V = lwh$ for h.

$$V = lwh \qquad \text{Given}$$
$$lwh = V \qquad \text{Interchange the sides}$$
$$h = \frac{V}{lw} \qquad \text{Divide each side by } lw$$

(b) Volume $= 16 \text{ loads} \cdot \dfrac{5 \text{ yd}^3}{1 \text{ load}} \cdot \dfrac{27 \text{ ft}^3}{1 \text{ yd}^3} = 2160 \text{ ft}^3$

With length = 88 ft and width = 40 ft, we have

$$h = \frac{V}{lw} = \frac{2160 \text{ ft}^3}{40 \text{ ft} \cdot 88 \text{ ft}} = \frac{2160 \text{ ft}^3}{3520 \text{ ft}^2} = 0.61 \text{ ft}$$

Therefore, the depth of the asphalt would be 0.61 foot rounded to two decimal places or $0.61 \text{ foot} \cdot \frac{12 \text{ inches}}{1 \text{ foot}} = 7.32$ inches, rounded to two decimal places. ◈

EXAMPLE 6 Modeling by Using a Formula from Chemistry

The formula $F = \frac{9}{5} C + 32$ expresses temperature F in degrees Fahrenheit in terms of the temperature C in degrees Celsius.

(a) Solve the formula for C.

(b) Complete the following table then use the table to predict when the temperature is the same in both scales.

F°	−40	32	212
C°			

Solution (a)

$$F = \frac{9}{5} C + 32 \qquad \text{Given}$$
$$F - 32 = \frac{9}{5} C \qquad \text{Subtract 32}$$
$$\frac{5}{9}(F - 32) = C \qquad \text{Multiply by } \frac{5}{9}$$
$$C = \frac{5}{9}(F - 32) \qquad \text{Interchange}$$

(b) If we replace F by −40, 32, and 212 in the above result, we obtain the corresponding values for C. Notice as the table shows, the temperature reading is the same, at −40 on both scales. ◈

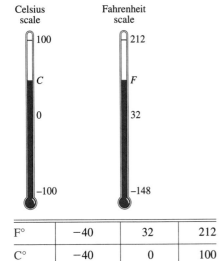

Celsius scale

| 100 |
| C |
| 0 |
| −100 |

Fahrenheit scale

| 212 |
| F |
| 32 |
| −148 |

F°	−40	32	212
C°	−40	0	100

Using Technology Exploration

Graphers can be used to graph straight lines. Quite often people get the impression that if they have a grapher, they don't need to use algebra. This is not the case. For example, to graph the equation

$$3x + 2y - 6 = 0$$

in Example 1b by a grapher, it is always required (by the grapher) to solve this equation for y in terms of x. Thus it provides a formula for calculating the y coordinate for a given x coordinate as the next example illustrates.

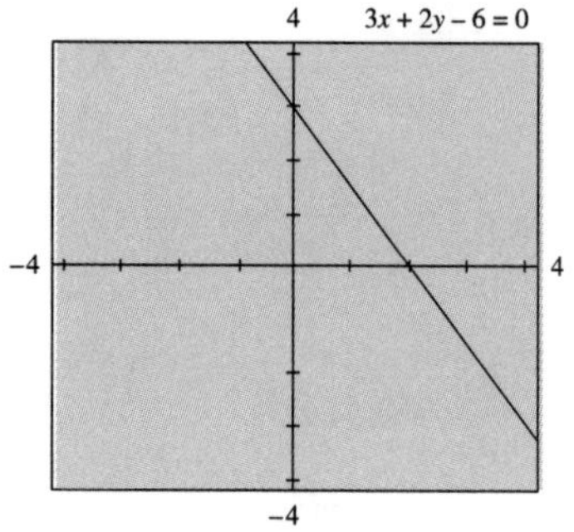

G EXAMPLE 7 Using a Grapher

Use a grapher to graph the equation

$$3x + 2y - 6 = 0$$

Solution   In Example 1b, we solved for y in terms of x and obtained

$$y = -\frac{3}{2}x + 3$$

The graph of this equation is shown in the viewing window in Figure 1. Notice that the x intercept of the graph is 2.

Figure 1

PROBLEM SET 2.2

Mastering the Concepts

In Problems 1–14, solve each equation for the indicated variable and check the results.

1. $6x + 7c = 37c$ for x
2. $4u - 19a = 5u$ for u
3. $ad + b = c$ for d
4. $ad - b = c$ for d
5. $12z - 4b = 6z - 7b$ for z
6. $13f + 6g = 8f - 9g$ for g
7. $4x - 3a - (10x + 7a) = 0$ for x
8. $27t - 4b - (15t - 6b) = 0$ for t
9. $5(4r - 3c) - 2(7r - 9c) = 0$ for r
10. $3(a - 2b) + 4(b + a) = 5$ for a
11. $5(mu - 2d) - 3(mu - 4d) = 9d$ for u
12. $9t + 7h - (11h - 13t) = 6t$ for t
13. $8(y - 2b) - 3(5y + 11b) = 0$ for y
14. $a(y - a) = -ab + 2b(y - b)$ for y

In Problems 15–24, solve each equation for y.

15. $\dfrac{x}{3} + \dfrac{y}{4} = 1$

16. $\dfrac{x}{7} - \dfrac{y}{4} = 1$

17. $\dfrac{x}{9} + \dfrac{y}{11} = 1$

18. $-\dfrac{x}{9} + \dfrac{y}{5} = 1$

19. $\dfrac{3x}{7} + \dfrac{2y}{9} = 1$

20. $\dfrac{7x}{11} - \dfrac{3y}{2} = 1$

21. $x = 0.01y - 0.03$

22. $8.7x = 3.9y + 9.6$

23. $\dfrac{x - 2y}{5} = \dfrac{3(y + 2)}{4} + 3$

24. $\dfrac{x + 3y}{5} = \dfrac{x - y}{2} + \dfrac{1}{2}$

In Problems 25–36, solve each formula for the indicated variable.

25. $A = P + Prt$, for P (Business)
26. $IR + Ir = E$, for r (Physics)
27. $\dfrac{T}{C} = \dfrac{x}{y}$ for C (Ecology)
28. $S(1 - r) = a$, for r (Mathematics)
29. $P = c - 30\%c$, for c (Business)
30. $BS = F + BV$, for B (Business)
31. $y = mx + b$, for x (Geometry)

32. $a = x - \dfrac{y_1}{y_2}$ for y_1 (Calculus)
33. $C = 2\pi r$, for r (Geometry)
34. $V = \pi r^2 h$, for h (Geometry)
35. $P = 2l + 2w$, for l (Geometry)
36. $A = 2\pi r^2 + 2\pi rh$, for h (Geometry)

In Problems 37 and 38, solve each formula for the indicated variable.

37. $V = \dfrac{Bh}{3}$ (volume of a pyramid)

    (a) For base B (b) For height h

38. $V = \dfrac{\pi r^2 h}{3}$ (volume of a cone)

    (a) For height h (b) For radius r

In Problems 39–42, solve each formula for the indicated variable.

39. $v = -32t + v_O$ (velocity in free-fall)

    (a) For initial velocity v_O (b) For time t

40. $A = \left(\dfrac{b_1 + b_2}{2}\right) h$ (area of a trapezoid)

    (a) For b_1 (b) For height h

41. $1 = \dfrac{x}{a} + \dfrac{y}{b}$ (intercept form of a line)

    (a) For a, the x intercept   (b) For b, the y intercept

42. $y - y_1 = m(x - x_1)$ (point-slope form of a line)

    (a) For slope m (b) For coordinate x_1

Applying the Concepts

43. **Circumference:** A runner jogged 5 miles in 35 laps around a circular track in a field house. What is the greatest distance measured straight across the track in feet? (Use $C = \pi d$, $\pi \approx 3.14$, and 5280 feet = 1 mile, round to one decimal place.)

44. **Area:** Suppose we wish to carpet the room shown in Figure 2. If the carpet, pad and labor total $18 per square yard, what is the cost?

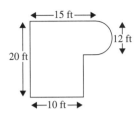

Figure 2

45. **Area:** You want to make a circular tablecloth with a ruffle along the bottom. The tablecloth is to hang 30 inches over the edge of a round table top (Figure 3).
    (a) Find the length of the ruffle. (Use $C = 2\pi r$)
    (b) If the ruffle costs $2.39 per yard, how much will it cost?

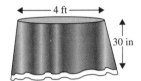

Figure 3

46. **Volume of a Teepee:** A tent is in the shape of a cone. Its volume is given by the formula
$$V = \frac{\pi r^2 h}{3}$$
where r is the radius of the circular base and h is the height of the tent (Figure 4).
    (a) Solve the formula for h.
    (b) What is the height of the tent if its volume is 1000 cubic meters and its radius is 8.7 meters.

Figure 4

47. **Hot Tub Volume:** A hot tub is in the shape of a right circular cylinder (Figure 5). It measures 6 feet in diameter and 4 feet in depth.
    (a) Find the volume V of the tub using $V = \pi r^2 h$ where r is the radius and h the depth.
    (b) Solve the formula for h.
    (c) Find the depth of a hot tub if its surface area is 266.9 square feet and its radius is 5 feet. (Use $S = \pi r^2 + 2\pi rh$)

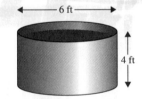

Figure 5

48. **Depth of Aquarium:** Suppose you have two fish aquariums. One measures 4 feet long and 18 inches wide with a depth of 2 feet (Figure 6). If you pour all the water into the other aquarium which is 7 feet long and 1.5 feet wide, what is the depth of water in the second aquarium (Use $V = lwh$)?

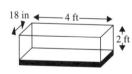

Figure 6

49. **G** **Tax Credit:** A certain state offers to give a tax credit for installing solar energy panels. The credit is modeled by the equation
$$T = 0.15(C - 1000)$$
where T is the tax credit and C is the cost of the panels and $C > 1000$.
    (a) Solve this equation for C in terms of T.
    (b) Use part (a) to determine the cost of panels if a tax credit of $252 is desired.
    (c) Use a grapher to graph $T = 0.15(C - 1000)$ in the viewing window: xMin $= 1000$, xMax $= 2520$, xScl $= 100$, yMin $= 0$, yMax $= 300$, yScl $= 50$.
    (d) Use the grapher to predict the tax credit for panels costing $1760 ($C = 1760$).

50. **Depreciation:** For tax purposes, companies use the depreciation formula
$$D = \frac{C - S}{n}$$
where D (in dollars) is the yearly depreciation of an item, C is its original cost, S is its salvage value and n is its useful life.
    (a) Solve the formula for S.
    (b) What is the salvage value of a copy machine if its original cost was $20,000 and its yearly depreciation is $2000 over a life of 10 years?

51. **Volume:** The volume V of a tank in the shape of a right circular cylinder of height h feet with hemispheres at each end (Figure 7) is given by the formula
$$V = \pi r^2 \left(h + \frac{4r}{3} \right).$$

    (a) Solve the formula for h.
    (b) The external liquid hydrogen tank of the space shuttle Endeavour is fabricated in this shape. Find the height h of the tank in meters if its total volume is 1174.86 cubic meters and the radius of each hemisphere is 4.2 meters. Round off the answer to two decimal places.

Figure 7

52. **Storage:** A cistern is to be fabricated in the shape of a right circular cylinder with a hemisphere at the bottom (Figure 8). The volume V of the cistern is given by the formula
$$V = \pi r^2 h + \frac{2}{3}\pi r^3$$

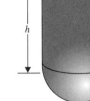

    (a) Solve this formula for h.
    (b) Find h if r is 3 feet and V is 810 cubic feet.

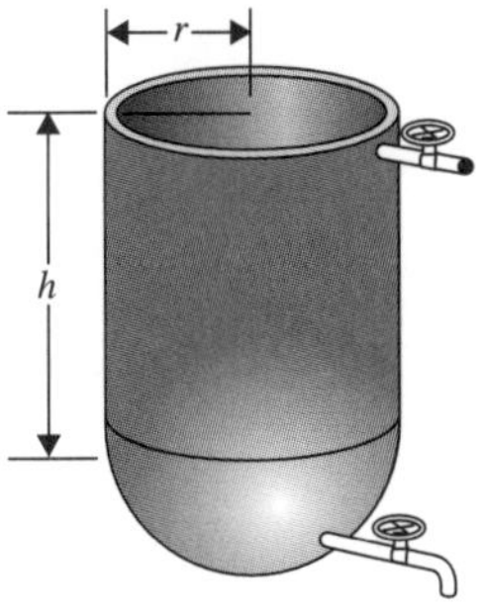

Figure 8

53. **Postal Charges:** In 1999 the U.S. Postal Service used the following charges to mail a first class item domestically: 33 cents for the first ounce and 22 cents for each additional ounce.
    (a) Write a formula for the total cost C (in dollars) for mailing a first class item whose weight is w (in ounces).
    (b) Solve the formula in part (a) for w in terms of C.
    (c) Use part (b) to determine the weight of a first class item if it costs $2.53 to mail it in the U.S.

G In Problems 54–56,
    (a) Solve each equation for y in terms of x.
    (b) Use a grapher to graph each equation in the given viewing window:
    $$x\text{Min} = -10,\ x\text{Max} = 10,\ x\text{Scl} = 1,$$
    $$y\text{Min} = -10,\ y\text{Max} = 10,\ y\text{Scl} = 1.$$
    (c) Use the ZOOM and TRACE features to find the x intercepts of each equation.

54. (a) $3x - 2y - 6 = 0$ (b) $2x - 3y + 6 = 0$
55. (a) $y + 7 + 2(x - 5) = 0$ (b) $y - 8 - 4(x - 2) = 0$
56. (a) $\dfrac{x}{3} + \dfrac{y}{4} = 1$ (b) $\dfrac{y}{2} - \dfrac{3x}{5} = 1$

Developing and Extending the Concepts

In Problems 57–60, solve each equation for the indicated variable.

57. $(ax + 7)(ax - 3) = ax(ax + 1)$, for x
58. $A = 2lw + 2lh + 2wh$ for l
59. $\dfrac{a^2 y + 8}{b} = \dfrac{a^2 y - 10}{3b}$ for y
60. $\dfrac{b - x}{a} = \dfrac{2a + b}{4} - \dfrac{3x}{5}$ for x

61. **Polygons:** The number of diagonals d of a polygon is one half the product of the number of sides n and the number of sides minus 3.
    (a) Write the formula for finding the number of diagonals d.
    (b) Find the number of diagonals for a six sided polygon.

62. **Polygons:** The measure m (in degrees) of each interior angle of a regular polygon having n sides is the product of 180 and two less than the number of sides divided by the number of sides.
    (a) Write the formula to find the measure of each interior angle m.
    (b) Solve the formula in part (a) for the number of sides n.

63. **Chain Length:** The approximate length L of a chain joining two sprockets (toothed wheels) of a bicycle of radii r and R (Figure 9) is given by the formula
    $$L = 2d + \frac{1}{2}(2\pi r) + \frac{1}{2}(2\pi R)$$
    where d is the distance between the center of the two sprockets. Consider a bicycle having a chain length of 64 inches and a distance between the center of the sprockets of 24 inches.
    (a) Find r, rounded to one decimal place, if $R = 3.5$ inches.
    (b) Find R, rounded to one decimal place, if $r = 1.5$ inches.

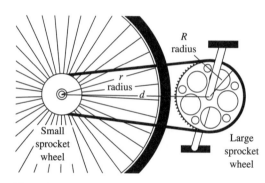

Figure 9

OBJECTIVES

1. Introduce a Problem Solving Format
2. Solve a Variety of Applied Problems
3. Use Technology Exploration

2.3 Applied Problems

In this section, we introduce a strategy for solving applied problems. These techniques were described in (the Hungarian mathematician) George Polya's book, "How to Solve It." The method focused on the problem solving process rather than just obtaining the correct answer.

Introducing a Problem Solving Format

One reason algebra is so important is that it can be used to solve problems in many disciplines, including business, engineering, economics, social sciences, and the physical sciences. Often the most challenging step in solving a word problem or application is translating the situation described in the problem into algebraic form. The complete result, including clearly stated assumptions and the appropriate formula, is a mathematical model. Polya developed the following three step process for solving applied problems.

PROBLEM SOLVING PROCESS

> Step 1. Make a plan
> Step 2. Write the solution
> Step 3. Look back

Step 1. Make a Plan

The following list is very useful in applying this plan.
(a) Identify the *given* (What information or assumptions are known).
(b) Identify the *goal* (What questions need to be answered).
(c) Draw a diagram whenever possible.
(d) Arrange and organize the given information in the form of a table or a chart whenever possible.
(e) Select one of the unknown quantities and label it with a symbol or a variable.
(f) Write algebraic relationships among the quantities involved, and then combine these relationships into a single equation.

Step 2. Write the Solution

(a) Solve the resulting equation in part (f) of step 1.
(b) Justify each step of the solution.

Step 3. Look Back

Interpret the solution in terms of the original problem to determine whether or not the answer is reasonable.

EXAMPLE 1   Solving a Number Problem

Find three consecutive odd integers whose sum is 63.

Solution   Make a Plan

The *given* is that there are three odd integers and they are consecutive. This means that these integers follow one another on the number line; that is, they are in order. Since they are consecutive odd integers they differ by 2. The *goal* is to find the three integers.

$$\text{Let } x = \text{ first odd integer, then}$$
$$x + 2 = \text{ second consecutive odd integer, and}$$
$$(x + 2) + 2 = x + 4 = \text{ third consecutive odd integer.}$$

The following diagram is a visual description of the *given*.

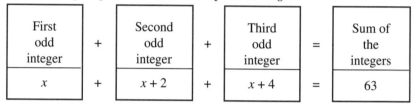

Translating the problem into symbols, we have the equation

$$x + (x + 2) + (x + 4) = 63$$

Write the Solution

Now we solve the equation

$x + (x + 2) + (x + 4) = 63$	Given
$3x + 6 = 63$	Remove parentheses
$3x = 57$	Subtract 6 from each side
$x = 19$	Divide by 3
$x + 2 = 19 + 2 = 21$	
$x + 4 = 19 + 4 = 23$	

Therefore, the three integers are 19, 21, and 23.

Look Back

These three integers, 19, 21, and 23, are consecutive and odd. Also, their sum is 63. So our answers, 19, 21, and 23, are reasonable.   ◈

As you try to work applied problems, you will notice the frequent occurrence of certain words in the statements of these problems. Table 1 presents some common English phrases used in word problems together with their symbolic translations, where x and y represent variables.

TABLE 1

English Phrases	Algebraic Expressions
The sum of x and y	$x + y$
The difference of x and y	$x - y$
The product of x and y	xy
The quotient of x and y	x/y
The square of the difference of x and y	$(x - y)^2$
Greater than x by y	$x + y$
y is less than x	$x - y$
Twice the sum of x and y	$2(x + y)$
The sum of twice x and y	$2x + y$
The sum of two consecutive integers	$x + (x + 1)$

Solving a Variety of Applied Problems

Our equation-solving skills enable us to tackle a wide range of applications.

EXAMPLE 2   Solving a Geometry Problem

The length of a rectangular soccer field is 20 meters more than twice the width. The field's perimeter is 310 meters. What are the dimensions of the field?

Solution   Make a Plan

The key words in this problem are *rectangular* and *perimeter*. The two unknowns are the length, l, and width, w, of the rectangular field. The perimeter P of a rectangle is found using the formula $P = 2l + 2w$. Here we make a sketch to help us visualize the situation involved (Figure 1). Let w = the width of the field in meters, $l = 2w + 20$ = the length of the field in meters. The following diagram is a visual description of the *given*.

w

$l = 2w + 20$

Figure 1

Two times the length of the rectangle	+	Two times the width of the rectangle	=	Perimeter of the rectangle
$2(2w + 20)$	+	$2w$	=	310

Translating the problem into symbols, we obtain the equation

$$2(2w + 20) + 2w = 310$$

Write the Solution

$2(2w + 20) + 2w = 310$	Given
$4w + 40 + 2w = 310$	Apply the distributive property
$6w + 40 = 310$	Combine like terms
$6w = 270$	Subtract 40
$w = 45$	Divide by 6
$2w + 20 = 110$	

The width of the soccer field is 45 meters and the length is 110 meters.

Look Back

We can check this result in the words of the problem. The perimeter is

$$P = 2l + 2w$$

$$= 2(110) + 2(45)$$

$$= 220 + 90$$

$$= 310.$$

Also, the length of 110 meters is 20 meters more than twice the width of 45 meters.   ◈

EXAMPLE 3   Cutting a Length of Wire

A wire is 150 centimeters long and is to be cut into three pieces. The second piece must be 6 centimeters shorter than three times the first, and the third piece must be 2 centimeters longer than two-thirds of the first. How long should the first piece be?

Solution   Make a Plan

The *given* is a length of wire which is 150 centimeters and is being cut into three pieces. Our *goal* is to determine the length of the first piece. We assign a variable to the first piece.

Let x = the length of the first piece in centimeters.
$3x - 6$ = the length of the second piece in centimeters.
$\frac{2}{3}x + 2$ = the length of the third piece in centimeters.

The following diagram is a visual description of the *given*.

length of the first piece	+	Length of the second piece	+	length of the third piece	=	Overall length
x	+	$3x - 6$	+	$\frac{2}{3}x + 2$	=	150

Translating the problem into symbols, we have the equation

$$x + (3x - 6) + \left(\frac{2}{3}x + 2\right) = 150$$

Write the Solution

$$x + 3x - 6 + \frac{2}{3}x + 2 = 150 \qquad \text{Remove parentheses}$$

$$\frac{14}{3}x - 4 = 150 \qquad \text{Combine like terms}$$

$$\frac{14}{3}x = 154 \qquad \text{Add 4 to each side}$$

$$\left(\frac{3}{14}\right)\left(\frac{14}{3}x\right) = \left(\frac{3}{14}\right)(154) \qquad \text{Multiply both sides by } \frac{3}{14}$$

$$x = 33$$

The length of the first piece is 33 centimeters.

Look Back

To check the results, we observe that the lengths of the three pieces are as follows:
The length of the first piece is 33 centimeters.
The length of second piece is $3(33) - 6 = 99 - 6 = 93$ centimeters.
The length of third piece is $\frac{2}{3}(33) + 2 = 22 + 2 = 24$ centimeters.
The sum of the length of the three pieces is $33 + 93 + 24 = 150$ centimeters. ◈

In many applied problems, rates, discounts, increases and decreases are often written as *percents* (%) which means *by the hundred*. Also, in most applications percents are expressed in decimals.

EXAMPLE 4   Solving an Investment Problem

A financial planner advised some customers to invest part of their money in a safe money market fund that paid $4\frac{1}{4}\%$ simple interest and the rest in a risky investment that earned $13\frac{3}{4}\%$ simple interest. If the customers invested \$24,000 that earned a total interest of \$1960 the first year, how much money did they invest in each fund?

Solution   Make a Plan

The formula for simple interest I is $I = Prt$ where P is the principal, r is the rate, and t is the time in years. The *given* is a total principal of \$24,000, two rates of interest earning a total of \$1960, and time is one year.

$$\text{Let } x = \text{amount (in dollars) invested at } 13\tfrac{3}{4}\%$$
$$24{,}000 - x = \text{amount (in dollars) invested at } 4\tfrac{1}{4}\%$$

We organize the information in the following table.

	Principal (dollars)	rate (% per year)	time (year)	interest (dollars)
Risky Investment	x	0.1375	1	$0.1375x$
Safe Investment	$24{,}000 - x$	0.0425	1	$0.0425(24{,}000 - x)$

We know that the sum of interest earned by both investments is a total of \$1960 for the year. The following diagram is a visual description of this *given*.

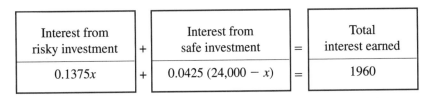

Translating the problem into symbols, we have the equation

$$0.1375x + 0.0425(24{,}000 - x) = 1960$$

Write the Solution

$0.1375x + 0.0425(24{,}000 - x) = 1960$	Given
$0.1375x + 1020 - 0.0425x = 1960$	Distributive property
$0.095x = 940$	
$x = 9894.74$	
$24{,}000 - x = 24{,}000 - 9894.74 = 14{,}105.26$	

Thus, \$9894.74 was invested in the risky investment and \$14,105.26 was invested in the safe investment.

Look Back

We check this solution by calculating the earnings from each investment.

$$\$9894.74 \text{ at } 13\tfrac{3}{4}\% \text{ yields } \$9894.74(0.1375) = \$1360.53$$
$$\$14{,}105.26 \text{ at } 4\tfrac{1}{4}\% \text{ yields } \$14{,}105.26(0.0425) = \$599.47$$

The total amount of interest is \$1960. The total amount invested is

$$\$14{,}105.26 + \$9894.74 = \$24{,}000.$$

EXAMPLE 5   Solving a Mixture Problem

A cereal processing plant introduces a recipe for a new breakfast cereal that combines rice crispies and wheat flakes. It is determined that a cup of rice crispies contains 2 grams of protein and a cup of wheat flakes contains 4.5 grams of protein. How much of each kind of cereal is included in one cup of the mixture that provides 3 grams of protein?

Solution   **Make a Plan**

We are combining into a single blend two kinds of cereal that have different amounts of protein. We identify the unknowns by using the fact that the total amount of protein in one cup is 3 grams.

Let x = the amount of rice crispies in one cup.
$1 - x$ = the amount of wheat flakes in one cup.

The table below summarizes the *given* information.

	Grams of protein in one cup of each cereal	Amount of cereal in one cup	Grams of protein
rice crispies	2	x	$2x$
wheat flakes	4.5	$1 - x$	$4.5(1 - x)$

The following diagram is a visual description of the *given*.

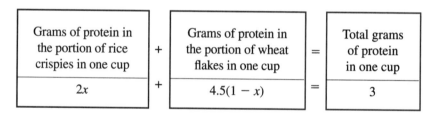

The equation that describes this model is:

$$2x + 4.5(1 - x) = 3$$

Write the Solution

$2x + 4.5(1 - x) = 3$ Given
$2x + 4.5 - 4.5x = 3$ Apply the distributive property
$\quad\quad -2.5x = -1.5$
$\quad\quad\quad x = \dfrac{-1.5}{-2.5} = \dfrac{3}{5} = 0.6$

and $1 - x = 1 - 0.6 = 0.4$

The final mixture must contain 0.6 cup of rice crispies and 0.4 cup of wheat flakes.

Look Back

The total amount of protein in one cup is:

$$2(0.6) + 4.5(0.4) = 3$$

These numbers check in the original problem.

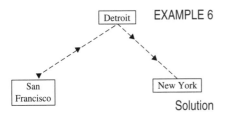

EXAMPLE 6   Solving a Rate Problem

On a recent trip across the country, a jetliner flew from San Francisco to Detroit at an average speed of 500 miles per hour. It then continued to New York at an average speed of 550 miles per hour. If the entire trip covered 3075 miles and the total flying time was 6 hours, what was the distance of each leg of the trip?

Solution   **Make a Plan**

Motion problems are based on the distance formula $d = rt$ where d is the distance, r is the average rate, and t is time. This trip consists of two legs: San Francisco to Detroit and Detroit to New York.

Let t = time (in hours) traveled on the first leg,
and $6 - t$ = time (in hours) traveled on the second leg

For each leg of the trip, we write expressions to represent rate, time and distance. This information is summarized in the following table.

	rate r (miles per hour)	time t (hours)	distance $d = rt$ (miles)
first leg of the trip	500	t	$500t$
second leg of the trip	550	$6 - t$	$550(6 - t)$

The following diagram is a visual description of the *given*.

Distance from San Francisco to Detroit	+	Distance from Detroit to New York	=	Total Distance
$500t$	+	$550(6 - t)$	=	3075

Since the total distance traveled is 3075 miles, we obtain the equation

$$500t + 550(6 - t) = 3075$$

Write the Solution

$$500t + 550(6 - t) = 3075 \qquad \text{Given}$$
$$500t + 3300 - 550t = 3075 \qquad \text{Apply the distributive property}$$
$$-50t = -225$$
$$t = 4.5$$

Substituting 4.5 for t, yields

$$500t = 500(4.5) = 2250 \text{ mi.}$$
$$550(6 - t) = 550(6 - 4.5) = 825 \text{ mi.}$$

Therefore, the first leg was 2250 miles long; the second was 825 miles.

Look Back

As a check, notice that the time spent traveling from San Francisco to Detroit is 4.5 hours. To find this distance, substitute the values of r and t in the equation $d = rt$, so that we have

$$d = 500(4.5) = 2250 \text{ miles}$$

The time spent traveling from Detroit to New York is $6 - 4.5 = 1.5$ hrs. Thus, the distance is

$$d = 550(1.5) = 825$$

Total distance traveled is $2250 + 825 = 3075$ miles.

Examples 4, 5, and 6 are *analogous*; they are solved using the same mathematical techniques. An important problem solving strategy is recognizing when a technique used to solve one problem can be used to solve another seemingly different, but analogous problem.

Using Technology Exploration

We can use a grapher to solve applied problems involving linear equations in one variable. We suggest the following guidelines for solving these applied problems with a grapher:
1. Use step 1: "make a plan".
2. Obtain a linear equation from step 1, and then write the equation in the form: $0 = $ expression and graph $y = $ expression.
3. Set the proper viewing window on your grapher to include all the important parts of the graph.

Let us solve the equation in example 2 with a grapher.

G EXAMPLE 7   Using a Grapher

Use a grapher to solve the equation $2(2w + 20) + 2w = 310$.

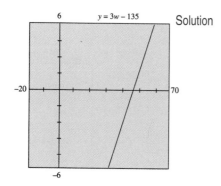

Figure 2

Solution   To graph this equation, we first rewrite it in the simpler form,

$3w + 20 = 155$ Divide each term by 2

$3w - 135 = 0$ Zero form

then graph $y = 3w - 135$. The viewing window in Figure 2 shows a portion of the graph where the line crosses the w axis at 45, which represents the solution of the equation. This can be confirmed by using the ZOOM and TRACE keys repeatedly. ◈

◈ **PROBLEM SET 2.3**

Mastering the Concepts

1. **Number Problem:** Find three consecutive even integers whose sum is 60.

2. **Number Problem:** (a) Find a number such that three more than twice the number is 57.
   (b) Find a number such that twice the sum of three and the number is 57.

3. **Number Problem:** Find three consecutive even integers such that twice the sum of the first and third is twelve more than twice the second.

4. **Age Problem:** A mother is five times as old as her son. If twice the age of the son is 3 years less than half of the age of the mother, how old are they?

5. **Number Problem:** Find three consecutive even integers such that twice the sum of the first and third is four more than three times the second.

6. **Home Address:** Four houses on one side of a street have addresses that are consecutive odd numbers. Find each address if the sum of these numbers is 23,672.

7. **Swimming Pool Dimensions:** The length of a rectangular swimming pool is 2 feet more than twice its width (Figure 3). The perimeter of the pool is. 94 feet. What are the length and width of the pool?

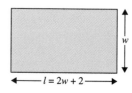

Figure 3

8. **Rectangular Dimensions:** A carpet layer wishes to determine the cost of carpeting a rectangular room whose perimeter is 52 feet. Suppose that the length of the room is 4 feet more than its width. Find
   (a) the length and width of the room;
   (b) the cost of the carpet if it costs $18.95 per square yard.

9. **Plot Dimensions:** A rectangular plot of farmland is bounded on one side by a river and the other three sides by a single-strand fence 797 meters long. What is the length

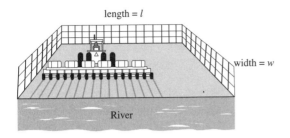

length = *l*

width = *w*

River

Figure 4

and the width of the plot if its length is 1 meter more than twice its width. No fence is used on the river side (Figure 4).

10. **Playground Dimensions:** A recreation department plans to build a rectangular playground enclosed by 1000 meters of fence. If the length of the playground is 4 meters less than three times the width, find the length and the width of the playground (Figure 5).

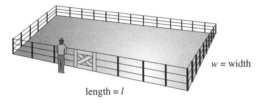

w = width

length = *l*

Figure 5

11. **Computer Monitor Dimensions:** The length of a rectangular computer monitor is 6 inches less than twice its width. The perimeter of the face of the monitor is 36 inches. Find the length and width of the monitor.

12. **Frame Dimensions:** The molding of a picture frame (the frame's perimeter) is 2.78 meters. The length of the finished frame is 0.1 meter more than twice the width (Figure 6). What are the dimensions of the frame?

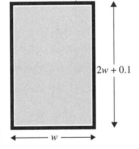

2w + 0.1

w

Figure 6

13. **Plumbing:** A plumber wishes to cut a 26 foot length of pipe into two pieces. The longer piece needs to be 7 feet less than twice the length of the shorter piece. Where should the plumber cut the pipe?

14. **Carpentry:** A carpenter needs two pieces of lumber which together are 21 feet long. The longer piece needs to be 1 foot longer than three times the length of the shorter piece. Find the length of each piece.

15. **Rope Cutting Problem:** A scout leader wishes to cut a rope for tents into three pieces whose lengths are consecutive odd integers. If the sum of the lengths of the first and the third piece is 58 feet, find the length of the middle piece.

16. **Cable Cutting Problem:** A cable installer has 105 meters of cable which is to be cut into three pieces. The second piece is 5 meters shorter than the first piece, and the length of the third piece is twice the length of the second. Find the length of each piece.

17. **Wire Cutting Problem:** A wire 180 centimeters long is to be cut into three pieces. The length of one piece is three times the length of the second. The length of the third piece is 12 centimeters shorter than twice the length of the second. Find the length of each piece.

18. **Fabric Cutting Problem:** A fabric store worker has a 57 yard bolt of cloth which is to be cut into three lengths. The length of the second piece is 5 yards shorter than twice the length of the first. The length of the third piece is 2 yards longer than twice the length of the first. Find the length of each piece.

19. **Investments:** Suppose an investor earned $2820 at the end of one year from a $32,000 investment in mutual funds. The investor bought two types of funds. One investment fund earned 9% annual interest, and the other earned 8.5% annual interest. How much was invested in each type of fund?

20. **Investments:** Suppose that $18,000 is invested in two types of bonds for one year. Part of the money is invested in a high-risk, high-growth fund that earned 24% annual interest. The rest is invested in a safe fund that earned 5% annual interest. If $2320 in interest is earned, how much was invested at each rate?

21. **Investments:** An inheritance is divided into two investments. One investment pays 7% annual interest whereas the second investment, which is twice as large as the first, pays 10% annual interest. If the combined annual income from both investments is $4050, how much money is the total inheritance?

22. **Assets of a Retirement Fund:** A portfolio manager purchased 10,000 shares in mutual stock funds and mutual bonds funds valued at $146,000 for a retirement fund. At the time of the purchase, the stock funds sold for $11 per share, while the bond funds sold for $17 per share. How many shares of each did the retirement fund purchase?

23. **Investments:** Suppose one investment in a money market fund earned 4.25% annual interest; the other investment in a stock fund earned 7.75% annual interest. Together the two investments totaled $3450. If a total of $230 for the year is earned, how much was invested in each fund?

24. **Auto Loan:** A customer purchased a car and financed $16,000. The customer borrowed part of the money from a bank charging 10% annual interest and the rest from a credit union at 8% annual interest. If the total interest for the year was $1390, how much was borrowed from the bank and how much was borrowed from the credit union?

25. **Juice Mixture:** A fruit juice company mixes pineapple juice that sells for $5.50 per gallon with 100 gallons of orange juice that sells for $3.00 per gallon. How much pineapple juice is used to make a pineapple-orange juice drink selling for $3.50 per gallon?

26. **Milk Mixture:** How many liters of whole milk containing 3.5% butterfat must be mixed with 3 liters of skim milk (containing no butter fat, 0% butterfat) to obtain a mixture containing 2% butterfat?

27. **Meat Mixture:** A meat market manager mixes hamburger having 30% fat content with hamburger that has 10% fat content in order to obtain 400 pounds of hamburger with 25% fat content. How much hamburger of each type should the manager use?

28. **Antifreeze Mixture:** A car radiator has a capacity of 8 quarts of coolant which is 20% antifreeze. A mechanic needs to drain part of the radiator fluid and replace it with 100% antifreeze to bring the coolant in the system to 60% antifreeze. How much of the original coolant should be drained and replaced by 100% antifreeze to achieve the desired mix?

29. **Gravel Driveway:** A farmer wishes to make a gravel driveway. The mixture needed is two parts pea gravel which costs $9.00 per ton and one part sand which costs $8.00 per ton. The gravel company gave a bid of $390. How much gravel and sand are they going to deliver?

30. **Seed Mixture:** A nursery has one kind of grass seed selling at 75 cents per pound and another kind of grass seed selling at $1.10 per pound. How many pounds of each kind should be mixed to produce 50 pounds of a mixture of seed that will sell for 90 cents per pound?

31. **Travel Distance:** A train travels through the mountains at an average speed of 35 miles per hour. It then continues on flat land at an average speed of 85 miles per hour. If the entire trip covers 1200 miles and each leg of the trip takes the same number of hours, how many miles is each leg?

32. **Travel Time:** Last weekend two college students were returning to school 200 miles away. During a blizzard, they were only able to average 25 miles per hour. For the rest of their trip, they averaged 55 miles per hour. If the entire trip took 4 hours, how long were they driving in the blizzard?

33. **Average Speed:** A cross country skier follows a trail to a nearby camp in 50 minutes. A snowmobile, averaging 16 miles per hour faster than the skier, starts from the same point on the trail and following the skier's track covers the same distance in 10 minutes. What was the average speed of the snowmobile?

34. **Travel Time:** On the first part of a 27.5 mile trip, the average speed was 48 miles per hour. Later due to traffic, the average speed was reduced to 35 miles per hour. If one spent five times as long on the first part as on the second part, what is the total time of the trip?

35. **Average Speed:** An Amtrak train leaves Las Vegas heading west towards Los Angeles. At the same time, 272 miles away a bus leaves Los Angeles traveling east along the same route towards Las Vegas. If the train averages 31 miles per hour faster than the bus and they pass each other in 2 hours, what are the average speeds of the train and the bus?

36. **Average Speed:** A helicopter travels 30 miles per hour faster than a speeding car. The car had a half hour head start, and both traveling the same route. If the helicopter overtakes the car in 1.5 hours, what is the speed of each?

Developing and Extending the Concepts

37. **Cash Register Change:** A fitness center usually needs three times as many $5 bills as $10 bills to transact its daily business. If the cash register contains $750 in fives and tens at the beginning of a business day, how many of each bill does the register contain?

38. **Coin Problem:** A vending machine accepts nickels, dimes and quarters. When the coins are emptied, the total value of the coins is found to be $24.15. Find the number of coins of each kind in the box, if there are twice as many nickels as quarters and 5 more dimes than nickels.

39. **Age Difference:** A man is 25 years older than his son. Fourteen years from now the man will be twice as old as his son will be. What are the present ages of the man and his son?

40. **Clothing Sale:** A clothing store has discounted the price of a sweater by 20%. If the sale price of the sweater is $49.95, what was the original price?

41. **Insurance Discount:** Some insurance companies reward students with safe driving records and good grades by discounting their annual premiums. If a student's annual premium is discounted by 15% to $874.65, what is the original annual premium?

42. **Computing Salaries:** An employee's new annual salary is $36,516, which includes a 5% pay raise and a 2.4% cost-of-living allowance. What was the employee's original salary?

43. **Tire Sale:** A tire was sold for $58.50 which included a 6% sales tax and an 11% federal tax. What was the price of the tire before the taxes?

44. **Consumer Cost:** A customer has $70 to spend in a grocery store. She spends $28.20 for canned goods. The remaining amount is needed to buy a total of 9 pounds of steak and chicken. If the cost per pound for steak and chicken is $6.95 and $2.80 respectively, how many pounds of each meat can she afford to buy?

45. **Tipping:** For groups of four or more diners, many restaurants automatically add a 15% gratuity as well as a tax of 8% to the original bill. If the bill (including gratuity and tax) for four people came to $193, how much was the tip?

46. **Credit Card Charge:** A customer carries an average daily balance on her credit card of $12.70 per day. The customer is offered a card from Bank A with no annual fee and 18% finance charge. Bank B offers a card with 15% finance charge and $39 annual fee. Which card is more economical for the customer? What average daily balance would give the same total yearly charges?

47. **Aerobics:** A runner burns 3 times as many calories running as walking the same distance. Suppose a runner burns a total of 765 calories running 3 kilometers and walking 8 kilometers. At this rate, how many calories are burned by running 1 kilometer?

48. **Operating Budget:** Suppose 60% of the salary budget for a college goes to faculty salaries, 25% to administrative salaries, 10% to support staff salaries, and the remaining $1,820,000 to maintenance salaries. How much money goes to each group?

49. **Work Force:** When a new factory opened recently, twice as many men as women applied for work. Five percent of all the applicants are hired. If 3% of the men who applied were hired, what percentage of the women who applied were hired?

50. **Stock Prices:** Two different stocks were sold for $3000 each. On one stock the gain was 10% of its original value; on the other the loss was 10% of its original value. Determine the total gain or loss from both stocks.

51. **Automobile Maintenance:** The mileage on a car's odometer reads 6253.6. The next service is scheduled to take place at about 10,000 miles. If the car is driven 170 miles per week, in how many weeks will it need to be serviced?

52. **Bank Charges:** A bank charges its customers a general fee of $6 per month plus an additional fee of 10 cents per check written from a checking account. Over the past nine months, a customer was charged $62 in service fees. How many checks did the customer write during that period?

53. **Geology:** While some climbers began a hike up Mt. St. Helens, the temperature at sea level was 20°C. The temperature dropped 5°C with every 1000 meters of elevation gained. When the climbers reached the top of their ascent, the temperature was 7.5°C. How high had they climbed?

54. **Car Rental:** For one of its compact cars, a car rental company charges customers $33 per day. In addition, the company charges 25 cents per mile for each mile driven over 400 miles. A customer rented a compact car for four days and paid $186.50. How many miles did the customer drive?

55. **Car Racing:** After 125 miles of The Indianapolis 500 car race, Car A has a lead of 1650 feet over Car B. If Car A averages 185 miles per hour and Car B averages 192 miles per hour, how long will it take Car B overtake Car A?

56. **Energy Saving:** By installing a $50 thermostat that automatically reduces the temperature setting while at work and at night, a condominium owner hopes to cut the annual heating bill by 15%. If the monthly heating bill is $85 before the new thermostat, how long will it take to recover the cost of the thermostat?

57. **Factory Outlet Sales:** A retail factory outlet store sells running shoes for $45 a pair. At this price, the profit on each pair of shoes is 25% of the cost to the retailer. What is the retailer's profit on each pair of shoes?

58. **Metal Alloys:** An archaeologist discovers a crown weighing 800 grams in the tomb of an ancient Egyptian king. Evidence indicates that the crown is made of a mixture of gold and silver. Gold weighs 19.3 grams per cubic centimeter; silver weighs 10.5 grams per cubic centimeter. It is found that the crown weighs 16.2 grams per cubic centimeter. How many grams of gold does the crown contain?

59. **Geometry:** The sum of the measures of the interior angles of a triangle is 180°. Is it possible that the three angles can have the following measurements? If yes, find the measures of the angles. If no, explain why.
    (a) Three consecutive integers. (Hint: Let the first angle measure x degrees.)
    (b) Three consecutive even integers. (Hint: Let the first angle measure $2b$ degrees.)
    (c) Three consecutive odd integers. (Hint: Let the first angle measure $2c + 1$ degrees.)

60. **Norman Doors and Windows:** The Norman style of architecture is characterized by massive construction and rectangular windows and doors surmounted by semi-circular arches (Figure 7). If such a door has an area of 29.4 square feet and a width of 3.3 feet, find its height h rounded to one decimal place.

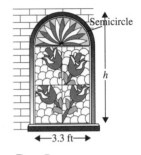

Figure 7

61. **Basketball Court:** The perimeter of a rectangular basketball court is 80 meters. Its length is 2 meters shorter than twice its width. Find the dimensions of the court.

62. **Fencing Cost:** The length of a rectangular yard is 10 meters less than 4 times its width. The total cost for a fence to enclose the entire yard is $2706. Find the dimensions of the yard if the fence costs $8.20 per meter.

63. **Package Design:** An open box is to be constructed by removing equal squares, each of length x inches, from the corners of a piece of cardboard and then folding up the sides (Figure 8). If the cardboard measures 12 inches by 16 inches, find the volume of the box if its length is twice its width.

Figure 8

64. **Biology:** A biologist wishes to determine the volume of blood in the circulatory system of an animal. She injects 6 milliliters of a 9% solution of a biologically inert chemical and waits until it is thoroughly mixed with the animal's blood. Then she withdraws a small sample of blood and determines that 0.03% of the sample consists of the biologically inert chemical. Find the original volume of blood in the circulatory system.

OBJECTIVES

1. Solve Equations Involving Rational Expressions
2. Solve Applied Problems
3. Use Technology Exploration

2.4 Equations Involving Rational Expressions

In Section 2.1 we solved equations containing fractions whose denominators did not contain variables. The technique was to multiply each side of the equation by the LCD of the fractions. In this section, we will proceed in a similar way to solve equations containing rational expressions with variables in the denominator.

Solving Equations Involving Rational Expressions

The strategy used in section 2.1 allows us to solve equations involving variables in the denominators such as

$$\frac{1}{2x} + \frac{2}{x} = \frac{5}{4} \quad \text{and} \quad \frac{1}{x-1} = \frac{2}{x+1}$$

In this case, we multiply each side of the equation by the same nonzero quantity, the LCD, thus transforming the equation into one that does not involve fractions. However, we have to avoid values of the variable that may cause the denominators of the original rational expression to be zero. For this reason, we *must* check the solution in each case. These restricted values cannot be solutions, since division by zero is undefined.

EXAMPLE 1   Solving an Equation Containing Rational Expressions

Solve the equation

$$\frac{4}{5t} + 2 = \frac{4}{t} + \frac{8}{5}$$

Solution

$$\frac{4}{5t} + 2 = \frac{4}{t} + \frac{8}{5}$$ Note that $t \neq 0$ for the expressions to be defined.

$$5t\left(\frac{4}{5t} + 2\right) = 5t\left(\frac{4}{t} + \frac{8}{5}\right)$$ Multiply each side by $5t$, the LCD

$$5t\left(\frac{4}{5t}\right) + 5t(2) = 5t\left(\frac{4}{t}\right) + 5t\left(\frac{8}{5}\right)$$ Apply the distributive property

$$4 + 10t = 20 + 8t$$ Simplify each side
$$4 + 2t = 20$$
$$2t = 16$$
$$t = 8$$

Check: Let $t = 8$, then

Left side	Right side
$\dfrac{4}{5(8)} + 2 =$	$\dfrac{4}{8} + \dfrac{8}{5} =$
$\dfrac{1}{10} + \dfrac{20}{10} =$	$\dfrac{1}{2} + \dfrac{8}{5} =$
$\dfrac{21}{10}$	$\dfrac{5}{10} + \dfrac{16}{10} = \dfrac{21}{10}$

Therefore, 8 is the solution.

EXAMPLE 2   Solving an Equation Containing Rational Expressions

Solve the equation

$$\frac{x}{2x-1} = \frac{2}{3}$$

Solution $\dfrac{x}{2x-1} = \dfrac{2}{3}$ Note $2x - 1 \neq 0$ so $x \neq \dfrac{1}{2}$

When two fractions are equal as in Example 2, *cross multiplication* can often be used to solve the equation. Thus, $\frac{x}{2x-1} = \frac{2}{3}$ has the same solution as $3x = 2(2x - 1)$ provided $x \neq \frac{1}{2}$.

$3(2x - 1)\dfrac{x}{2x-1} = 3(2x-1)\dfrac{2}{3}$ Multiply each side by $3(2x-1)$, the LCD

$\quad\quad 3x = 2(2x - 1)$ Apply the distributive property

$\quad\quad 3x = 4x - 2$

$\quad\quad -x = -2$

$\quad\quad\; x = 2$ Multiply both sides by -1

Check: Let $x = 2$, then

Left side: $\dfrac{2}{2(2)-1} = \dfrac{2}{4-1} = \dfrac{2}{3}$

Therefore, 2 is the solution.   ◈

EXAMPLE 3   Solving an Equation Involving Rational Expressions

Solve $\dfrac{3}{w-3} + 4 = \dfrac{w}{w-3}$

Solution $\dfrac{3}{w-3} + 4 = \dfrac{w}{w-3}$ $w \neq 3$

$(w-3)\left(\dfrac{3}{w-3} + 4\right) = (w-3)\left(\dfrac{w}{w-3}\right)$ Multiply by $w - 3$, the LCD

$(w-3)\left(\dfrac{3}{w-3}\right) + (w-3)4 = (w-3)\left(\dfrac{w}{w-3}\right)$ Apply the distributive property

$\quad\quad 3 + 4w - 12 = w$

$\quad\quad\quad\; 4w - 9 = w$

$\quad\quad\quad\quad\;\; 3w = 9$

$\quad\quad\quad\quad\;\;\; w = 3$

Check: Our solution process leads to 3 as the apparent solution. But 3 is a restricted value, as we noted at the beginning. If w is replaced by 3 in the original equation, the denominator on each side has a value of zero which is meaningless.

$$\frac{3}{3-3} + 4 = \frac{3}{0} + 4$$

Therefore, 3 is not a value which makes the equation true. In fact, it makes the rational expressions undefined. Hence, the original equation has *no solution* and 3 is called an **extraneous solution**.   ◈

Solving Applied Problems

Many applications involve equations with rational expressions. The problem-solving strategy is the same as we used in section 2.3.

EXAMPLE 4   Solving a Number Problem

One number is five times as large as a second number. The sum of their reciprocals is $\frac{2}{5}$. Find the two numbers.

Solution   **Make a Plan**

The key here is the word reciprocal. Let $x =$ the first number, then $5x =$ the second number.

Their reciprocals are $(1/x)$ and $1/(5x)$ respectively. Since zero has no reciprocal, $x \neq 0$. The following diagram is a visual description of the *given*.

Reciprocal of the first number		Reciprocal of the second number		$\frac{2}{5}$
$\dfrac{1}{x}$	$+$	$\dfrac{1}{5x}$	$=$	$\dfrac{2}{5}$

The equation which describes this situation is

$$\frac{1}{x} + \frac{1}{5x} = \frac{2}{5}$$

Write the Solution

$$\frac{1}{x} + \frac{1}{5x} = \frac{2}{5} \qquad x \neq 0$$

$$5x\left(\frac{1}{x} + \frac{1}{5x}\right) = 5x\left(\frac{2}{5}\right) \qquad \text{Multiply each side by } 5x, \text{ the LCD}$$

$$5 + 1 = 2x$$
$$6 = 2x$$
$$x = 3$$

and $5x = 15$

Look Back

We check the result by adding the reciprocals of the numbers 3 and 15

$$\frac{1}{3} + \frac{1}{15} = \frac{5}{15} + \frac{1}{15} = \frac{6}{15} = \frac{2}{5}$$

The solution checks in the original problem.   ◈

Problems concerning work done at a constant rate yield equations involving rational expressions. If a job can be completed in t hours, then $1/t$ of the job can be completed in 1 hour.

$$\frac{1 \text{ job}}{t \text{ hours}} = \frac{1}{t} \text{ job / hour}$$

The key to writing the equation is knowing that all the parts of the job add up to 1 complete job; that is, the whole is equal to the sum of its parts. In these problems, we will assume that work is done at a constant rate regardless of the number working.

EXAMPLE 5   Solving a Work Problem

Two cranes, operating together, can unload cargo ships at a rate of 1 ship every 4 hours. If one crane takes twice as long as the other to unload a ship, how long would it take each crane to unload a ship by itself?

Solution   **Make a Plan**

Let t = time (in hours) the faster crane unloads ship, then

   $2t$ = time (in hours) the slower crane unloads ship.

One method of solving this type of problem is to think in terms of how much of the job is done by one crane in 1 hour. Therefore, the amount of the job completed in 1 hour is $(1/t)$ for the faster crane and $1/(2t)$ for the slower one. We can organize the given information in the table below.

Cranes	Part of Job Completed in 1 Hour	Number of Hours	Part of Job Completed in 4 Hours
Faster Crane	$\dfrac{1}{t}$	4	$\dfrac{4}{t}$
Slower Crane	$\dfrac{1}{2t}$	4	$\dfrac{4}{2t} = \dfrac{2}{t}$

The following diagram is a visual description of the *given*.

Part of job completed by faster crane		Part of job completed by slower crane		One complete job
$\dfrac{4}{t}$	+	$\dfrac{4}{2t} = \dfrac{2}{t}$	=	1

Write the Solution

The equation is

$$\frac{4}{t} + \frac{2}{t} = 1 \qquad\qquad \text{Given}$$

$$\frac{6}{t} = 1 \qquad\qquad \text{Combine like terms}$$

$$t = 6$$

and $2t = 12$

The faster crane would take 6 hours to unload the ship by itself, and the slower crane alone would take 12 hours to unload the ship.

Look Back

If the faster crane takes 6 hours to unload the ship, in 4 hours it has done

$$\left(\frac{1 \text{ job}}{6 \text{ hours}}\right)(4 \text{ hours}) = \frac{2}{3} \text{ job}$$

and the slower crane has done $\left(\dfrac{1 \text{ job}}{12 \text{ hours}}\right)(4 \text{ hours}) = \dfrac{1}{3} \text{ job}$.

Together they have completed the job, $2/3 + 1/3 = 1$ and the faster crane operates twice as fast as the slower one.   ◈

EXAMPLE 6   Solving a Jet Stream Speed Problem

The *jet stream* is a narrow current of air encountered between altitudes of 25,000 and 35,000 feet. One day a flight from Los Angeles to New York (2460 miles) flying with the jet stream (with a tail wind) took the same amount of time as a flight from New York to Denver (1626 miles) flying against the jet stream (with a head wind). How fast was the jet stream? Assume that both planes fly 550 miles per hour in still air.

Solution   Make a Plan

A *tail wind* increases a plane's speed by applying force in the direction of travel, while a *head wind* decreases a plane's speed by applying force against the direction of travel.

We can use the formula $t = \frac{d}{r}$ to find the time for both flights. If we

let r = speed (in miles per hour) of the jet stream, then

$550 + r$ = speed (in miles per hour) of the plane with the jet stream,

and

$550 - r$ = speed (in miles per hour) of the plane against the jet stream.

The table below summarizes this information.

	Rate (miles per hour)	Distance (miles)	Time $= \frac{\text{Distance}}{\text{Rate}}$ (hours)
With the jet stream	$550 + r$	2460	$\dfrac{2460}{550 + r}$
Against the jet stream	$550 - r$	1626	$\dfrac{1626}{550 - r}$

The problem states that both flights take the same time. The following diagram is a visual description of the *given*.

$$\boxed{\begin{array}{c}\text{Time traveling}\\\text{with the jet}\\\text{stream}\end{array}} = \boxed{\begin{array}{c}\text{Time traveling}\\\text{against the jet}\\\text{stream}\end{array}}$$

Thus, we can form an equation as follows:

$$\frac{2460}{550 + r} = \frac{1626}{550 - r}$$

Write the Solution

$$\frac{2460}{550 + r} = \frac{1626}{550 - r} \qquad r \neq 550$$

$$2460(550 - r) = 1626(550 + r) \qquad \text{Cross multiply}$$

$$1{,}353{,}000 - 2460r = 894{,}300 + 1626r \qquad \text{Apply the distributive property}$$

$$458{,}700 = 4086r$$

$$r = 112.26$$

The jet stream was traveling about 112.26 miles per hour.

Look Back

Time traveling with the jet stream is $\dfrac{2460}{550 + 112.26} = \dfrac{2460}{662.26} = 3.71$ hours

Time traveling against the jet stream is

$$\frac{1626}{550 - 112.26} = \frac{1626}{437.74} = 3.71 \text{ hours}$$

Since the times are the same, the solution is acceptable.

Using Technology Exploration

By this time, you should be convinced that a grapher is a powerful tool in solving equations. Here we use a grapher to estimate the solutions of equations involving rational expressions. Let us select the equation in Example 1 as a sample.

G EXAMPLE 7   Using a Grapher

Use a grapher to solve the equation $\dfrac{4}{5t} + 2 = \dfrac{4}{t} + \dfrac{8}{5}$

Solution   First we write the equation in the equivalent form

$$\frac{4}{5t} + 2 - \frac{4}{t} - \frac{8}{5} = 0$$

Then we graph the associated equation

$$y = \frac{4}{5t} + 2 - \frac{4}{t} - \frac{8}{5}$$

to estimate the values of t at which the graph intercepts the horizontal axis. Figure 1 shows the graph of the above associated equation. The viewing window shows the portion of the graph which seems to cross the horizontal axis at 8. This can be confirmed by using the TRACE and ZOOM features. Therefore, 8 is the only solution.

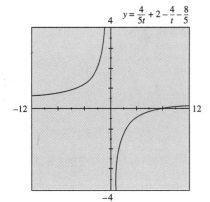

$y = \dfrac{4}{5t} + 2 - \dfrac{4}{t} - \dfrac{8}{5}$

Figure 1

PROBLEM SET 2.4

Mastering the Concepts

In Problems 1–30, note what values of the variable make the equation undefined. Then solve the equation and check the solutions.

1. $\dfrac{4}{3x} - \dfrac{3}{x} = \dfrac{10}{3}$

2. $\dfrac{1}{2y} + \dfrac{2}{y} = \dfrac{1}{8}$

3. $\dfrac{3}{t} - \dfrac{4}{5} = -\dfrac{1}{5}$

4. $\dfrac{5}{u} + 3 = \dfrac{4}{3}$

5. $\dfrac{5}{x} + \dfrac{3}{8} = \dfrac{7}{16}$

6. $\dfrac{3}{8u} - \dfrac{1}{5u} = \dfrac{7}{10}$

7. $\dfrac{2}{3x} + 1 = \dfrac{4}{3x} + \dfrac{1}{3x}$

8. $\dfrac{1}{y} + \dfrac{2}{y} = 3 - \dfrac{3}{y}$

9. $\dfrac{2}{3y} = \dfrac{5}{9}$

10. $\dfrac{3}{11x} = \dfrac{5}{12}$

11. $\dfrac{3}{x + 4} = \dfrac{3}{5}$

12. $\dfrac{6}{u + 2} = \dfrac{3}{4}$

13. $\dfrac{4x}{4x + 5} = \dfrac{3}{5}$

14. $\dfrac{4x + 5}{5x + 2} = \dfrac{3}{8}$

15. $\dfrac{3}{x + 7} = \dfrac{5}{x}$

16. $\dfrac{1}{2x} = \dfrac{1}{3x + 1}$

17. $\dfrac{8}{x - 3} = \dfrac{12}{x + 3}$

18. $\dfrac{3}{y - 2} = \dfrac{4}{y + 1}$

19. $\dfrac{y}{y - 2} - 1 = \dfrac{1}{y - 3}$

20. $\dfrac{2}{y - 3} - \dfrac{17}{2} = \dfrac{3}{4}$

21. $\dfrac{2t}{t - 3} + \dfrac{3}{t - 3} = -2$

22. $\dfrac{2}{x + 1} + \dfrac{5x}{x + 1} = 4$

23. $\dfrac{y}{y + 1} + 3 = \dfrac{4}{y + 1}$

24. $\dfrac{x}{x - 1} - 1 = \dfrac{3}{x - 1}$

25. $\dfrac{1}{x + 1} + \dfrac{2}{3(x + 1)} = \dfrac{1}{3}$

26. $\dfrac{2}{3x-2} - \dfrac{16x}{3(3x-2)} = -2$

27. $\dfrac{1}{x-1} - \dfrac{2}{1-x} = 0$ 28. $\dfrac{4}{4-x} + \dfrac{1}{x-4} = \dfrac{2}{x-4}$

29. $\dfrac{3x}{x+4} + \dfrac{4}{x+2} = 3$ 30. $\dfrac{2x}{x+3} + \dfrac{4}{x+4} = 2$

In Problems 31–36, solve each equation. Indicate whether the equation has an extraneous solution.

31. $\dfrac{x}{x-2} + \dfrac{2}{3} = \dfrac{2}{x-2}$ 32. $\dfrac{y}{y-6} - 3 = \dfrac{6}{y-6}$

33. $\dfrac{3x}{x-4} = 5 + \dfrac{12}{x-4}$ 34. $\dfrac{x}{x-3} = 2 + \dfrac{3}{x-3}$

35. $\dfrac{y+1}{y} = \dfrac{y-1}{y}$

36. $\dfrac{4}{t-5} + \dfrac{6}{t+5} = \dfrac{-40}{(t-5)(t+5)}$

In Problems 37–40, solve each equation for the indicated variable.

37. $\dfrac{ay}{b} = c + \dfrac{d}{b}$ for y 38. $\dfrac{3}{y} - \dfrac{4}{b} = -\dfrac{1}{b}$ for y

39. $\dfrac{b-x}{3} = \dfrac{2a-b}{4} - \dfrac{2x}{3}$ for x

40. $\dfrac{u+2c}{3} + \dfrac{u-3c}{2} = \dfrac{5}{12}$ for u

Applying the Concepts

41. **Number Problem:** One number is four times as large as another. Find the two numbers if the sum of their reciprocals is 5/28.

42. **Number Problem:** If 7/10 is added to four times the reciprocal of a number, the result is 3/4. Find the number.

43. **Sales Tax:** In the state of Michigan, the sales tax on a car that sold for $12,000 is $720. At this rate, how much is the sales tax on a car that sells for $18,500?

44. **Billboard Lighting:** The lighting for a billboard is generally provided by solar energy. Suppose that 5 energy panels generate 12 watts of power. How many panels are needed to generate 240 watts of power to light a certain billboard?

45. **Politics:** In a recent poll, 3000 people were asked their opinion about health care reform. If 37.5% of the people polled were men, how many women were polled?

46. **Journalism:** A national magazine pays freelance journalists by the word for published articles. Suppose a magazine pays $168 for an article that is 1050 words long. At this rate, how many additional words are needed in order for the journalist to earn $240 for an article?

47. **Roofing a House:** One roofing crew can finish the roof of a house in 10 hours. A second crew can roof the same house in 15 hours. If both crews were used, how long will it take to roof the house?

48. **Wallpapering an Office:** Working together, two workers can hang wallpaper for an office in 6 hours. If one worker can hang the wallpaper in 9 hours, how long would it take the other worker to do the office alone?

49. **Filling a Reservoir:** An inlet pump can fill a reservoir in 10 hours. An outlet pump can drain the same reservoir in 30 hours. How long would it take to fill the reservoir if both pumps were in use at the same time?

50. **Landscaping:** A landscaper can prepare a new lawn in 10 hours. If the landscaper hired a worker to assist, together they can do the job in 7 hours. How long would it take the assistant to complete this job alone?

51. **Snow Removal:** Suppose the snowplows in a city can clear a 12 inch snowfall in 10 hours, whereas the snow blowers take 6 hours. How long would it take them working together to clear the snow?

52. **Pollution Level:** At a garbage burning plant, one furnace reaches an air-quality alert level of pollution in 36 hours, whereas a second furnace reaches the same level in 45 hours. How long can both furnaces operate together without reaching that pollution alert level?

53. **Jet Stream Speed:** On a recent trip between two cities, it took the same time to fly 700 miles with the jet stream as it took to fly 550 miles against the jet stream. How fast is the jet stream if the plane flies 500 miles per hour in still air?

54. **Jet Stream Speed:** A commercial airliner has a speed of 450 miles per hour in still air. If the airliner flew a distance of 1020 miles with the jet stream in the same time it took to fly 780 miles against the jet stream, how fast is the jet stream?

55. **Average Speed:** A car traveled 126 miles in the same amount of time that a van traveled 99 miles. If the car went 15 miles per hour faster than the van, how fast did each vehicle travel?

56. **Current Speed:** A canoeist can paddle a canoe 6 kilometers per hour in still water. If the canoeist can paddle 1 kilometer upstream in the same amount of time that he can paddle 1.5 kilometers downstream, how fast is the current?

57. **Hovercraft Speed:** A hovercraft travels between Dover, England, and Calais, France, a distance of 38 kilometers. One day the wind was blowing so hard that the trip from Dover to Calais against the wind took twice as long as the return trip with the wind. If the wind was clocked at 10.7 kilometers per hour, how fast can the hovercraft travel with no wind?

58. **Car Rental:** At a special week sale, a car rental company advertised that a fleet of midsize cars can be rented for $225 and the same number of minivans may be rented for $325. If a minivan cost $10 more to rent than a midsize car, what is the weekly rental rate for each type of vehicle?

59. **Engineering:** Engineers define the gear ratio of two gears, A and B, to be the ratio of the number of teeth in gear A to the number of teeth in gear B (Figure 2). Suppose that the gear ratio of gear A to B is 3/2 and the sum of the teeth in both gears is 85. Find the number of teeth in each gear.

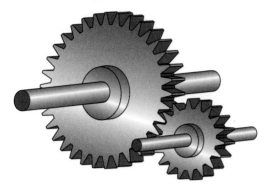

Figure 2

Developing and Extending the Concepts

In Problems 60 and 61 solve each formula for the indicated variable.

60. (a) $\dfrac{1}{p} + \dfrac{1}{q} = \dfrac{1}{f}$ for p 61. (a) $\dfrac{1}{r_1} + \dfrac{1}{r_2} + \dfrac{1}{r_3} = \dfrac{1}{R}$ for R

    (b) $R = \dfrac{2V - 2r}{I}$ for V (b) $\dfrac{E}{E_1} = \dfrac{R + R_1}{R_1}$ for E_1

62. Suppose that the triangles ABC and DEF are similar (Figure 3); that is, corresponding sides are proportional.
    (a) Write an equation that represents equal ratios.
    (b) Solve the resulting equation for x.
    (c) Find the lengths of the sides of the triangles.

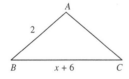 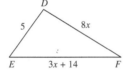

Figure 3

63. **Monthly Car Payments:** A credit union advertised that a monthly car payment is $29.50 per $1000 loaned. At this rate, what is the monthly payment for a $9000 car loan?

64. **Investment:** An investment of $1500 earns $90 at the end of each year. At this rate, how much additional money must be invested to earn $120 at the end of each year?

65. **Earthquake:** The secondary seismic waves (also known as s waves) from the epicenter of an earthquake reached a recording station 150 kilometers away at the same time the primary seismic wave (also known as p wave) reached another station 250 kilometers away from the epicenter.
    (a) How fast does each wave travel if the p wave travels 2 kilometers per second faster than the s wave?
    (b) Use the velocities of the two seismic waves from part(a) to find the distance to a recording station that recorded the s wave 2.5 seconds after the p wave.

66. **Ecology:** In a wildlife management study of a park, ecologists are to estimate the population of alligators. One year a researcher captured and tagged 300 alligators, and then released them. Later, another researcher captured 1000 alligators and observed that 27 of them were tagged. The capture-recapture method of estimating animal populations assumes that the ratio of tagged alligators to all alligators in the park is the same as the ratio of tagged alligators in subsequent samplings to the number in the sampling. Based on this method, approximately how many alligators are there in the park? (round to the nearest hundred)

OBJECTIVES

1. Introduce Interval Notation
2. Solve Linear Inequalities
3. Solve Compound Inequalities
4. Solve Applied Problems
5. Use Technology Exploration

2.5 Linear Inequalities

So far we have used linear equations to model certain types of applied problems. Certain relationships among quantities cannot always be described by linear equations. For instance, suppose that a mathematics department wishes to purchase a number of graphing calculators, where the cost of each calculator is $60 plus 8% sales tax. Also suppose that the delivery charge for the entire shipment is $50. To determine the number of calculators to be purchased in order to allocate "at most" $3200 for the entire order, we need to solve a linear inequality. In this section, we study such *linear inequalities*, which contain one or more variable terms separated by one or more of the following symbols:

    $>$ greater than $<$ less than

    $\geq$ greater than or equal to $\leq$ less than or equal to

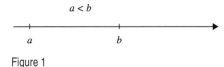

Figure 1

Recall that if point a lies to the left of point b on a number line (Figure 1), we say that b is *greater than* a (or equivalently, that a is *less than* b) and we write $b > a$ (or $a < b$). In other words, $b > a$ (or $a < b$) means that $b - a$ is positive. The symbols $<$, $>$, $\leq$ and $\geq$ are called **inequality signs** and the expressions on the left and right of these signs are called the **sides** or **members** of the inequality. The inequalities $b < a$ or $a > b$ are said to be **strict** because they do not allow the possibility of equality. However, the inequalities $b \leq a$ and $a \geq b$ which allow the possibilities of equality are said to be **nonstrict**. If the inequality $x \leq a$ is true when x is replaced by a or any number less than a, we say that these replacements **satisfy** the inequality. The set of all points on a real number line that satisfy an inequality is called the **graph** of the inequality.

Introducing Interval Notation

Certain sets of real numbers, called *intervals*, have an important role to play in the study of inequalities. The basic types of intervals are summarized in Table 1. (Note that a *parenthesis* is used to graph inequalities with symbols $<$ or $>$, and a *bracket* is used to graph inequalities with symbols $\leq$ or $\geq$.) In Table 1, we let a and b be real numbers such that $a < b$. The special symbols ∞ and $-\infty$, called **positive infinity** and **negative infinity**, are used to indicate that the interval extends indefinitely to the right or to the left.

TABLE 1

Interval Type	Inequality Notation	Interval Notation	Graph
Closed interval: Numbers between a and b, inclusive	$a \leq x \leq b$	$[a,b]$	
Open interval: Numbers between a and b	$a < x < b$	(a,b)	
Numbers greater than a	$x > a$	(a,∞)	
Numbers less than b	$x < b$	$(-\infty,b)$	
Half-open interval: Numbers greater than or equal to a and less than b	$a \leq x < b$	$[a,b)$	
Numbers greater than a and less than or equal to b	$a < x \leq b$	$(a,b]$	
Numbers greater than or equal to a	$x \geq a$	$[a,\infty)$	
Numbers less than or equal to b	$x \leq b$	$(-\infty,b]$	

EXAMPLE 1   Graphing Inequalities

Graph each inequality and express each using interval notation.

(a) $-3 \le x \le 5$ (b) $x \ge 4$ (c) $x < -3/2$

Solution   (a) The graph of this inequality is the set of all real numbers on the number line that are greater than or equal to -3 and less than or equal to 5 (Figure 2a). The interval notation is $[-3,5]$.

(b) The graph of $x \ge 4$ is the set of all real numbers on the number line that are greater than or equal to 4 (Figure 2b). The interval notation is $[4,\infty)$

(c) The graph of $x < -3/2$ is the set of all real numbers on the number line that are less than $-3/2$ (Figure 2c). The interval notation is $(-\infty,-3/2)$.

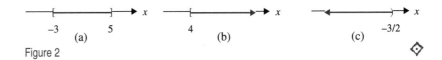

(a)   (b)   (c)

Figure 2

Solving Linear Inequalities

Inequalities such as

$$x - 3 < 5, \quad y + 2 \ge 7, \quad \text{and } 2x + 3 < 11 - 2x$$

are examples of linear inequalities in one variable. We **solve** an inequality for the variable x by finding all values of x for which the inequality is true. Such values are called **solutions**. The set of all solutions of an inequality is called its **solution set**. Two inequalities are said to be **equivalent** if they have exactly the same solution set. To solve a linear inequality, we proceed in much the same way as in solving linear equations. Use the following **properties** of **inequalities** to complete a solution.

Properties of Inequalities	Equivalent Inequality	Illustration
1. Addition (Subtraction) Property Add (subtract) any real number to each side of an inequality.	If $a < b$, then $a + c < b + c$ $a - c < b - c$	Since $5 < 13$, then $5 + 7 < 13 + 7$ $5 - 7 < 13 - 7$
2. Multiplication (Division) Property Multiplying (dividing) each side of an inequality by a *positive* number, yields an equivalent inequality.	If $a < b$, then $ac < bc, c > 0$ $\dfrac{a}{c} < \dfrac{b}{c}, c > 0$	$5 < 13$, then $5(7) < (13)(7)$ $\dfrac{5}{7} < \dfrac{13}{7}$
Multiplying (dividing) each side of an inequality by a *negative* number, yields an equivalent inequality with the order of the inequality reversed.	If $a < b$, then $ac > bc, c < 0$ $\dfrac{a}{c} > \dfrac{b}{c}, c < 0$	$5 < 13$, then $5(-7) > (13)(-7)$ $\dfrac{5}{-7} > \dfrac{13}{-7}$

Each of these properties is true if the symbols $<$ and $>$ are replaced by $\le$ and $\ge$, respectively. In addition a, b, and c can be either real numbers or algebraic expressions. Consider the effect of adding (subtracting) and multiplying (dividing) each side of the inequality $4 < 6$ by a positive number (see Table 2) or a negative number (see Table 3).

TABLE 2

	Add 2	Subtract 2	Multiply by 2	Divide by 2
Given	$4 < 6$	$4 < 6$	$4 < 6$	$4 < 6$
Result	$6 < 8$	$2 < 4$	$8 < 12$	$2 < 3$

Note that if $-3x > 0$, then $x < 0$, because the product of two negative numbers is positive. Thus, remember when you multiply or divide by a negative number, you must reverse the inequality symbol.

TABLE 3

	Add -2	Subtract -2	Multiply by -2	Divide by -2
Given	$4 < 6$	$4 < 6$	$4 < 6$	$4 < 6$
Result	$2 < 4$	$6 < 8$	$-8 > -12$	$-2 > -3$

Notice in Table 3, when we multiply or divide each side of the inequality $4 < 6$ by -2, the inequality symbol is reversed.

EXAMPLE 2 Solving a Linear Inequality

Solve the inequality $2x + 3 < 11 - 2x$ and graph the solution.

Solution

Remember that $x < 2$ is equivalent to $2 > x$. Traditionally, we rewrite inequalities so that the variable is on the left side.

$$2x + 3 < 11 - 2x \qquad \text{Given}$$
$$4x + 3 < 11 \qquad \text{Add } 2x \text{ to each side}$$
$$4x < 8 \qquad \text{Subtract 3 from each side}$$
$$x < 2 \qquad \text{Divide each side by 4}$$

The solution set of the inequality consists of all real numbers less than 2. In interval notation, $(-\infty, 2)$ is the solution set. The graph of this solution set is in Figure 3.

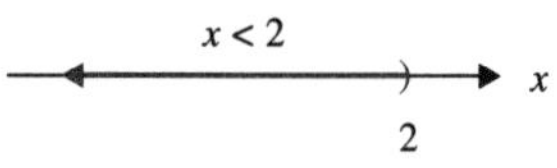

Figure 3

To check, choose a value in the solution set; the inequality is true. Also choose a value not in the solution set; the inequality is false.

Check: If $x = 1$, then

Left side	Right side
$2(1) + 3 = 5$	$11 - 2(1) = 9$

$5 < 9$ true

Check: If $x = 3$, then

Left side	Right side
$2(3) + 3 = 9$	$11 - 2(3) = 5$

$9 < 5$ false

EXAMPLE 3 Solving a Linear Inequality

Solve the inequality $4(x - 2) \geq 3(x - 2) - 4$ and graph the solution.

Solution

$$4(x - 2) \geq 3(x - 2) - 4 \qquad \text{Given}$$
$$4x - 8 \geq 3x - 6 - 4 \qquad \text{Apply the distributive property}$$
$$4x - 8 \geq 3x - 10 \qquad \text{Add like terms}$$
$$x - 8 \geq -10 \qquad \text{Add } -3x \text{ to each side}$$
$$x \geq -2 \qquad \text{Add 8 to each side}$$

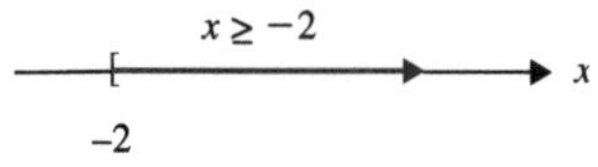

The solution set of the inequality consists of all real numbers greater than or equal to -2. In interval notation $[-2, \infty)$ is the solution set. The graph of this solution set is in Figure 4.

Figure 4

Check: If $x = -1$, then

Left side	Right side
$4(-1 - 2) = -12$	$3(-1 - 2) - 4 = -13$

$-12 \geq -13$ true

Check: If $x = -3$, then

Left side	Right side
$4(-3 - 2) = -20$	$3(-3 - 2) - 4 = -19$

$-20 \geq -19$ false

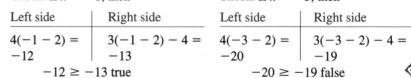

EXAMPLE 4   Solving a Linear Inequality

Solve the inequality $\dfrac{2-x}{2} \le \dfrac{x+1}{3}$ and graph the solution.

Solution

$$\dfrac{2-x}{2} \le \dfrac{x+1}{3}$$ Given

$$6\left(\dfrac{2-x}{2}\right) \le 6\left(\dfrac{x+1}{3}\right)$$ Multiply each side by 6, the LCD

$$3(2-x) \le 2(x+1)$$ Multiply

$$6 - 3x \le 2x + 2$$ Apply the distributive property

$$6 - 5x \le 2$$ Subtract $2x$ from each side

$$-5x \le -4$$ Subtract 6 from each side

$$x \ge \dfrac{4}{5}$$ Divide each side by -5 and reverse the inequality

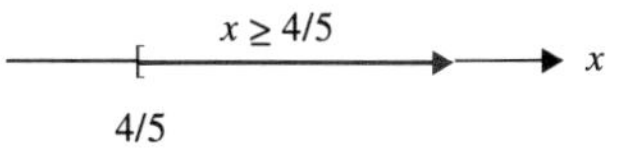

The solution set of the inequality consists of all real numbers greater than or equal to 4/5. In interval notation, [4/5,∞) is the solution set. The graph of this solution set is in Figure 5.

The check is left to the reader.   ◈

$x \ge 4/5$

x

4/5

Figure 5

Solving Compound Inequalities

Two inequalities connected with the words "and" or "or" are called **compound inequalities**. Examples of these inequalities are:

$$-1 \le x \quad \text{and } x \le 2, \quad x \le -5 \quad \text{or } x \ge 1$$

The solution set of a compound inequality with the word "and" is the set of real numbers *common* to the solution sets of each inequality. For example, the solution set of

$$-1 \le x \text{ and } x \le 2$$

consists of all real numbers x that are greater than or equal to -1 *and* less than or equal to 2. That is, all real numbers between -1 and 2 including the end points. This is also written as

$$-1 \le x \le 2.$$

Graphing these inequalities provides a visual idea of the solutions. Figure 6a shows the graph of the inequalities $-1 \le x$ and $x \le 2$.

Note that a compound inequality containing the word "or" may not be compressed as the "and" compound inequality is compressed.

The solution set of a compound inequality containing the word "or" is the set of real numbers in the solution set of either or both of the two inequalities. For example, the solution set of

$$x \le -5 \text{ or } x \ge 1$$

consists of all real numbers x that are less than or equal to -5, or all real numbers that are greater than or equal to 1 (Figure 6b).

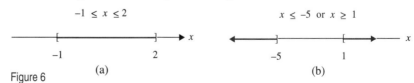

$-1 \le x \le 2$ $x \le -5$ or $x \ge 1$

x x

-1 2 -5 1

Figure 6   (a)   (b)

EXAMPLE 5 Solving Compound Inequalities

Solve each compound inequality. Express the solution set in interval notation and graph.

(a) $-3 < 2x + 1 \leq 5$ (b) $5x - 2 < -7$ or $5x - 2 > 8$

Solution (a) Our goal is to isolate the variable x in the middle.

$$
\begin{array}{lll}
-3 < 2x + \quad 1 \leq 5 & \text{Given} \\
-3 - 1 < 2x + 1 - 1 \leq 5 - 1 & \text{Subtract 1 from all three parts} \\
\quad -4 < 2x \qquad\qquad \leq 4 & \text{Combine like terms} \\
\quad -2 < x \qquad\qquad\ \leq 2 & \text{Divide each part by 2}
\end{array}
$$

Therefore, the solution set consists of all real numbers greater than -2 and less than or equal to 2. In interval notation the solution set is $(-2,2]$ and the graph is drawn in Figure 7a.

(b) We solve each inequality individually.

$$
\begin{array}{lll}
5x - 2 < -7 \ \text{ or }\ 5x - 2 > 8 & \text{Add 2 to each side of both inequalities} \\
5x \qquad < -5 \quad\ \ 5x \qquad > 10 & \text{Combine like terms} \\
\quad x \qquad < -1 \quad\ \ x \qquad > 2 & \text{Divide each side of both inequalities by 5}
\end{array}
$$

The solution set consists of all real numbers less than -1 or greater than 2. In interval notation the solution set is $(-\infty,-1)$ or $(2,\infty)$ and the graph is drawn in Figure 7b.

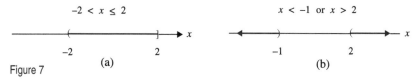

Figure 7   (a)   (b)

A check is left to the reader. ◈

TABLE 4

Word Statement	Algebraic Statement
x is at least 20	$x \geq 20$
x is at most 15	$x \leq 15$
x is no more than 20	$x \leq 20$
x is not less than 7	$x \geq 7$

Solving Applied Problems

The strategy for solving word problems in section 2.3 applies equally to word problems involving linear inequalities. Table 4 illustrates how inequality statements are translated into symbols. We use these symbols to solve the cost problem stated at the opening statement of this section.

EXAMPLE 6 Solving a Cost Problem

A mathematics department wishes to purchase a number of graphing calculators. Each calculator costs $60, plus 8% sales tax. In addition, a $50 delivery charge for the entire shipment must be paid. If the department has allocated at most $3200 for the purchase, how many calculators can be ordered?

Solution Make a Plan

The cost of each calculator is $60 + 0.08(\$60) = \$64.80$.

Let x = the number of graphing calculators, then
$$\$64.80x = \text{total cost of the graphing calculators.}$$

To solve this problem we write an inequality that requires the cost of the calculators, $64.80x$, plus the delivery charge, $50, to be less than or equal to $3200. The following diagram visualizes the *given*.

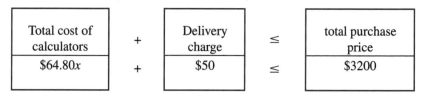

Write the Solution

We solve the inequality as follows:

$64.80x + 50 \leq 3200$	Given
$64.80x \leq 3150$	Subtract 50 from each side
$x \leq 48.6$	Isolate by dividing by 64.80

Enough money was allocated to order 48 or fewer calculators.

Look Back

Notice that we do not round up to 49, since

$$\$64.80(49) + \$50 = \$3225.20.$$

Such a purchase would be over the budget amount of $3200.

EXAMPLE 7   Temperature Range

The average daily range of temperature for a city during a late summer month varies from at least 72°F to no more than 98°F.
(a) Describe the temperature range for one day.
(b) If the average temperature decreases 2°F per day, describe the temperature range after one week.

Solution   ### Make a Plan

Let T represent the temperature during a summer month. We write a compound inequality stating that the temperature is to be between 72°F and 98°F inclusive. The temperature drops 2°F each day for a week. We decrease the temperatures by 7(2°F).

Write the Solution

(a) The average range of temperature is $72° \leq T \leq 98°$.
(b) Since the temperature decreases 2°F per day for a week, the temperature range after one week is

$$72° - 7(2°) \leq T \leq 98° - 7(2°)$$
$$72° - 14° \leq T \leq 98° - 14°$$
$$58° \leq T \leq 84°$$

Look Back

By decreasing the temperature 2°F each day for seven days, we see that 72°F drops to 58°F and 98°F to 84°F.

Using Technology Exploration

We can use a grapher to solve linear inequalities. The techniques are similar to those used in solving linear equations.
1. Rewrite the inequality in the zero form. That is, in the form $Q < 0$ or $Q > 0$ where Q is a linear expression.
2. Graph $y = Q$.
3. Examine the graph to determine the solutions of the inequalities that correspond to all values of x for which the values of y are below ($Q < 0$) or above the x axis ($Q > 0$).

We solve example 2 graphically.

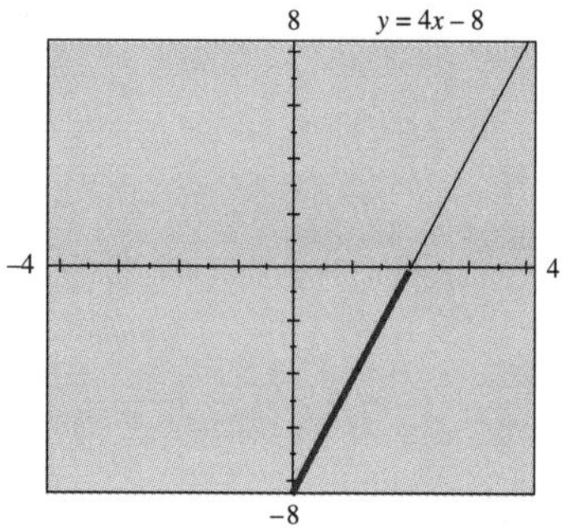

G EXAMPLE 8   Using a Grapher

Use a grapher to solve

$$2x + 3 < 11 - 2x$$

Solution   1. Rewrite the inequality in the zero form,

$$2x + 2x + 3 - 11 < 0$$
$$\text{or} \quad 4x - 8 < 0$$

2. Graph $y = 4x - 8$.
3. The solution set corresponds to all x values on the graph of the line for which the y values are below the x axis; that is , $y < 0$. The viewing window in Figure 8 shows the solution is $x < 2$ and the solution set is $(-\infty, 2)$.   ◈

Figure 8

PROBLEM SET 2.5

Mastering the Concepts

In Problems 1–8, graph each inequality and express each using interval notation.

1. (a) $-1 < x \le 1$
   (b) $-1 \le x < 1$
 (a) $x \ge -1$
   (b) $x < -1$
5. (a) $x \le 2$ or $x \ge 4$
   (b) $x \ge 2$ or $x \le 4$
   (a) $x \le -10$ or $x \ge -10$
   (b) $x \le -10$ and $x \ge -10$

2. (a) $-1 \le x \le 1$
   (b) $-1 < x < 1$
4. (a) $x < 1$
   (b) $x \le 1$
6. (a) $x \ge 4$ or $x \le 5$
   (b) $x \le 4$ or $x \ge 5$
8. (a) $x < 0$ or $x > 0$
   (b) $x < 0$ and $x > 0$

In Problems 9 and 10, state the property of inequalities that justifies each statement.

9. (a) If $x < 4$ then $2x < x + 4$
   (b) If $x < 4$ then $2x < 8$
   (c) If $y \le -2$ then $y + 2 \le 0$
   (d) If $y \le -2$ then $-y \ge 2$
10. (a) If $y \ge y - 1$ then $2y \ge 2y - 2$
    (b) If $y \ge y - 1$ then $-2y \le 2 - 2y$
    (c) If $x - 1 \le 4$ then $x \le 5$
    (d) If $a \le x$ then $a - 2 \le x - 2$

In Problems 11–40, solve each inequality and express the solution using interval notation. Also graph the solution on a number line.

11. $5 + 3x \ge 8$
13. $-3x + 4 < 14 + 2x$
15. $8 - 9x \le -x$
17. $2(3x - 2) < 7 + 4x - 1$
18. $6 - (7 - x) \ge 2(x - 3)$
19. $-(x - 1) > 2(x + 1/2)$
20. $4(3 - x) \ge 2(x - 1)$
21. $5x \ge -3(x - 2)$
22. $-3(x + 1) \ge -4(2x - 1)$
23. $-9(x - 3) - 8(4 - x) \ge -2x$
24. $-4(x - 1) \ge 2(x + 1) - 3$

12. $-5x + 2 > 12$
14. $x + 6 \le 4 - 3x$
16. $4 - x \ge 3x$

25. $3(x - 5) + 10 < 2(x + 4)$
26. $9(x + 2) < -6(4 - x) + 18$
27. $3(x + 2) - 2 \ge -(x + 5) + x$
28. $4(x + 4) > -2(x - 3) + 1$
29. $\frac{1}{9}(3x - 2) < \frac{1}{3}(1 - 4x)$
30. $\frac{1}{6}(2x - 7) < \frac{1}{2}(x + 1)$
31. $\frac{x}{3} + 2 < \frac{x}{4} - 2x$
32. $\frac{3x}{2} \ge -6 - \frac{x}{2}$
33. $\frac{x}{2} - \frac{x}{3} \le 4$
34. $\frac{x}{6} + 1 \ge \frac{x}{3}$
35. $\frac{4x - 2}{2} > \frac{3x + 6}{3}$
36. $\frac{x + 4}{2} < \frac{2x - 3}{3}$
37. $\frac{x + 4}{-2} \le \frac{x - 3}{5}$
38. $\frac{4x + 17}{7} > x + 2$
39. $\frac{5x + 1}{3} > \frac{3x + 5}{4}$
40. $\frac{2x + 4}{-5} \le \frac{3x - 3}{-3}$

In Problems 41–52, solve each compound inequality, express the solution in interval notation, and graph the solution on a number line.

41. $3 \le x + 1 \le 5$
43. $-9 < 3x < 6$
45. $-3 \le 4x - 1 \le 5$
46. $4x - 5 < -3$ or $4x - 5 > 3$
48. $5x - 4 > 1$ or $5x - 4 < -1$
49. $5 - 4x > 2$ or $5 - 4x < -2$
50. $4x - 7 \ge 3$ or $4x - 7 \le -3$
51. $3x - 1 < -7$ or $3x - 1 > 7$
52. $4x + 3 < -5$ or $4x + 3 > 5$

42. $-2 \le x - 3 \le 3$
44. $-4 \le 2x \le 8$
46. $1 < 8 - 3x < 12$

Applying the Concepts

53. **Sales Commission:** A sales representative for a tennis club earns a base salary of $500 per week plus a commission of $15 for every new membership sold. How many memberships must be sold in a week in order to earn at least $725 for that week?

54. **Weekly Income:** A newspaper carrier earns $7 per week plus $0.18 for each newspaper delivered to a home. How many newspapers should be delivered in order for the carrier to earn more than $30 per week?

55. **Telephone Charges:** Suppose the cost for an international telephone call is $3.50 for the first minute and $1.85 for each additional minute (or part of a minute). If you do not want your total cost to exceed $30, how long can you talk?

56. **Car Rental:** A vacationer has two choices of car rental agencies. The first agency charges a flat fee of $44.95 per day with unlimited miles. The second charges $19.95 per day plus $0.19 per mile.
  (a) Assuming a seven day rental, what range of miles driven would make the flat fee rental more economical?
  (b) What range of miles driven per day would make the second rental more economical?
  (c) What range of miles driven in three days would make the flat fee rental more economical?

57. **Test Averages:** A student scored 65, 80 and 74 on the first three tests during the term. What does the student need to score on the fourth test to ensure an average score that is above 75?

58. **Diet Program:** A clinic advertises that a person can reduce their weight by at least 1.5 pounds per week by exercising and special dieting. At the beginning of a diet program, a person weighs 195 pounds. What is the maximum number of weeks before this person's weight will be reduced to 170 pounds?

59. **Energy Consumption:** The average American home uses at least 90 but no more than 120 kilowatt hours of electricity per month.
  (a) Use an inequality to express the average range in kilowatt hours per day, assuming that one month is 30 days.
  (b) Use an inequality to express the average range in kilowatt hours per week and per year.

60. **Investment:** Suppose you invest $7000 in a mutual fund that pays simple interest for one year. If you earn at most $924 at the end of 18 months, what is the maximum annual interest rate for this investment?

61. **Geometry:** The length of a rectangle is 17 centimeters less than three times the width. Find, to the nearest positive integer, the width of the rectangle if the perimeter is between 222 and 238 centimeters.

62. **Travel Times:** A traveler flying between New York and Paris can choose among several planes with different average speeds. Suppose the slowest plane averages 515 miles per hour for the trip, and the fastest averages mach 1.30 (1.30 times the speed of sound which is 1100 feet per second).
  (a) Find the range in travel times for the 3645 mile trip.
  (b) Assume the jet stream (high-altitude winds) blows west to east at a velocity of 50 miles per hour. Find the range in travel time of this trip, traveling both with and against the jet stream.

Developing and Extending the Concepts

In Problems 63 and 64, solve each inequality, write it in interval notation and graph the solution set.

63. $2x - 3 \leq 3x + 1 \leq 4x - 5$

64. $2x - 1 \leq 3x + 7 \leq x + 9$

G In Problems 65 and 66, use the given grapher display and viewing window for each equation to find the solution set in interval notation of the associated inequalities.

65. $y = (6 - x) - (4x + 1)$
  (a) $6 - x \leq 4x + 1$
  (b) $6 - x > 4x + 1$

66. $y = (3x - 5) - (x + 1)$
  (a) $3x - 5 \leq x + 1$
  (b) $3x - 5 > x + 1$

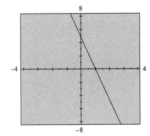 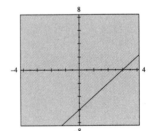

G In Problems 67 and 68, use a grapher to graph the associated equation for y in terms of x, and then use the graph of the equation to estimate the solution of the given inequality. Also write the solution in interval notation.

67. (a) $x + 7 \leq 4x - 8; y = (x + 7) - (4x - 8)$
   (b) $3x + 1 \geq 7x - 18; y = (3x + 1) - (7x - 18)$

68. (a) $\dfrac{2x}{9} - \dfrac{5}{6} > \dfrac{x}{12}; y = \left(\dfrac{2x}{9} - \dfrac{5}{6}\right) - \left(\dfrac{x}{12}\right)$
   (b) $\dfrac{x}{2} - \dfrac{3}{4} < \dfrac{7x}{4}; y = \left(\dfrac{x}{2} - \dfrac{3}{4}\right) - \left(\dfrac{7x}{4}\right)$

OBJECTIVES

1. Solve Equations Involving Absolute Value
2. Solve Inequalities Involving Absolute Value
3. Solve Applied Problems
4. Use Technology Exploration

2.6 Equations and Inequalities Involving Absolute Value

Recall from section R3 that

$$|x| = \begin{cases} x \text{ if } x \geq 0 \\ -x \text{ if } x < 0 \end{cases}$$

Geometrically, the absolute value of a number can be described on a number line as the distance between the number and zero. That is, $|a|$ represents the distance between a and 0 on a number line, regardless of whether the point a is to the left (Figure 1a) or to the right of 0 (Figure 1b).

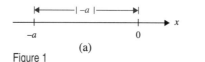
(a)

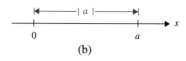
(b)

Figure 1

More generally,

$|a - b|$ is the number of units of distance between the point with coordinate a and the point with coordinate b.

This statement holds no matter which point is to the left of the other. That is, point a can be to the left (Figure 2a) or to the right (Figure 2b) of point b. For simplicity, we refer to the distance $|a - b|$ or $|b - a|$ as the distance between the numbers a and b, and we write, $|a - b| = |b - a|$.

(a)   (b)

Figure 2

For example, the distance between -4 and -7 is written as

$$|-4 - (-7)| = |-4 + 7| = |3| = 3 \quad \text{or} \quad |-7 - (-4)| = |-7 + 4| = |-3| = 3$$

Solving Equations Involving Absolute Value

An equation containing absolute value such as the following:
$$|x| = 3, |x + 2| = 5, \text{ and } |4x - 1| = 7$$

is called an **absolute value equation**. We can solve an absolute value equation such as $|x| = 3$ geometrically by interpreting $|x| = 3$ to mean those points on the number line which are exactly 3 units away from zero. There are two such points: -3 which is exactly three units to the left of zero, and $+3$ which is exactly three units to the right of zero (Figure 3). Therefore, the only solutions of the equation $|x| = 3$ are -3 and 3. In general, we solve an absolute value equation as follows:

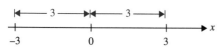

Figure 3

SOLVING AN ABSOLUTE
VALUE EQUATION

Let u be a variable expression, and a be a real number.
1. If $a > 0$, then $|u| = a$ is equivalent to $u = -a$ or $u = a$.
2. If $a = 0$, then $|u| = a$ is equivalent to $u = 0$.
3. If $a < 0$, then $|u| = a$ has no solution.

The strategy for algebraically solving an absolute value equation is to rewrite the equation as *two equivalent linear equations*, then solve the resulting linear equations.

EXAMPLE 1   Solving Absolute Value Equations

Solve each equation.
(a) $|x - 3| = 2$ (b) $|4x - 1| = 7$ (c) $|5x - 1| = -1$

Solution   (a) Using $u = x - 3$ and $a = 2$ in the equation $|u| = a$, we have,

$$x - 3 = -2 \text{ or } x - 3 = 2$$
$$x = 1 \quad \text{ or } \quad x = 5$$

The two solutions are 1 and 5. Check these solutions in the original equation.

(b) Using $u = 4x - 1$ and $a = 7$ in the equation $|u| = a$, we have,

$$4x - 1 = -7 \text{ or } 4x - 1 = 7$$
$$4x = -6 \qquad 4x = 8$$
$$x = \frac{-3}{2} \text{ or } \qquad x = 2$$

The solutions are $-\dfrac{3}{2}$ and 2. Check these solutions in the original equation.

(c) $|5x - 1| = -1$

This equation has *no solutions* because there is no value of x which would make $|5x - 1|$ negative.

Solving Inequalities Involving Absolute Value

As with equations involving absolute value, inequalities of the form

$$|x| < 3, \quad |x| > 3, \quad |x + 2| \geq 5, \quad |x - 3| < 7$$

can be solved by using geometric interpretation on a real number line. For example, $|x| < 3$ requires that the value of x is to be less than 3 units from zero. That is, x is between -3 and 3 which is written as $-3 < x < 3$. In interval notation the solution set is $(-3, 3)$ (Figure 4a). Similarly, $|x| > 3$ requires that the value of x is to be more than 3 units from zero. That is, $x < -3$ or $x > 3$. In interval notation the solution set is $(-\infty, -3)$ or $(3, \infty)$ (Figure 4b).

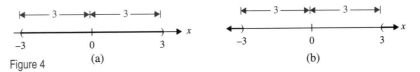

Figure 4   (a)   (b)

More generally, we have the following results

SOLVING AN ABSOLUTE
VALUE INEQUALITY

Statements comparable to these can be made for strict inequalities (those involving < or >).

Let u be a variable linear expression, and let a be a real number.
1. If $a \geq 0$, rewrite the **absolute value inequality**
   (a) $|u| \leq a$ as $-a \leq u \leq a$
   (b) $|u| \geq a$ as $u \leq -a$ or $u \geq a$
2. Solve the resulting linear inequalities.

EXAMPLE 2 Solving Absolute Value Inequalities

Solve each inequality and graph the solution set on a number line.

(a) $|x - 3| < 7$　　　　　　　　(b) $|7x - 2| \geq 9$

Solution (a) $|x - 3| < 7$　　　　Use $|u| < a$ where $u = x - 3$ and $a = 7$
　　　　　$-7 < x - 3 < 7$　　Equivalent inequalities
　　　　　$-4 < x < 10$　　　Add 3 to all three parts

Therefore, the solution set consists of all real numbers that are greater than -4 and less than 10; that is, between -4 and 10. In interval notation, the solution set is $(-4, 10)$ (Figure 5a).

(b) $|7x - 2| \geq 9$　　　　　　　Use $|u| \geq a$ for $u = 7x - 2$ and $a = 9$
　　$7x - 2 \leq -9$　or　$7x - 2 \geq 9$　Equivalent inequalities
　　　　$7x \leq -7$　　　　$7x \geq 11$　Add 2 to both sides of each inequality

　　　　$x \leq -1$　or　　$x \geq \dfrac{11}{7}$　Divide by 7

The solution set consists of all real numbers x such that $x \leq -1$ or $x \geq \dfrac{11}{7}$.

In interval notation, the solution set is $(-\infty, -1]$ or $[11/7, \infty)$ (Figure 5b).

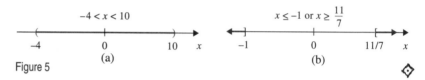

Figure 5　　　　　　　　(a)　　　　　　　　　(b)

EXAMPLE 3 Solving Absolute Value Inequalities

Solve each absolute value inequality.

(a) $|x - 3| < 0$　　　　　　　　(b) $|x - 3| \geq 0$

Solution (a) There are no solutions of $|x - 3| < 0$ because there are no values of x which would make the absolute value less than zero.

(b) Since the absolute value of every number is always greater than or equal to zero, the solution of $|x - 3| \geq 0$ is the set of all real numbers, $\mathcal{R}$.

Cut Point Method

All these inequalities can also be solved by an alternative method called the **cut point** method which is outlined in the following strategy.

Step 1. **Replace:** Replace the inequality symbol with $=$ and solve the related equation.

Step 2. **Separate:** Mark the numbers found in step 1 on the number line. These numbers divide the number line into intervals.

Step 3. **Test:** Select a test number from each interval. Substitute this test number in the original inequality to determine the intervals that make the inequality true.

Step 4. **State the Solution:** Test the numbers found in step 1 to confirm that the numbers are included in the solution set if the symbol is $\leq$ or $\geq$, and excluded if the symbol is $<$ or $>$. Then graph the solution set.

EXAMPLE 4   Solving Absolute Value Inequalities

Solve the inequality $|x - 5| < 3$

Solution   We carry out the cut point strategy.

Replace: Replace the symbol $<$ with $=$ and solve the related equation.

$$|x - 5| = 3$$

$$x - 5 = -3 \quad \text{or} \quad x - 5 = 3$$
$$x = -3 + 5 \quad \text{or} \quad x = 3 + 5$$
$$x = 2 \quad \text{or} \quad x = 8$$

The solutions of the related equations are

2 and 8

Separate: Mark the numbers 2 and 8 on a number line. These numbers divide the number line into 3 intervals denoted by A, B, and C.

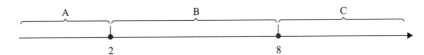

Test: Select test number 0 in A, 3 in B, and 9 in C.

0 in A	3 in B	9 in C
$\begin{aligned}|x - 5| &= |0 - 5| \\ &= |-5| = 5 < 3\end{aligned}$	$\begin{aligned}|x - 5| &= |3 - 5| \\ &= |-2| = 2 < 3\end{aligned}$	$\begin{aligned}|x - 5| &= |9 - 5| \\ &= |4| = 4 < 3\end{aligned}$
False	True	False

State the Solution: The solution set is the interval $(2,8)$ because all values of x that belong to $(2,8)$ satisfy the inequality (Figure 6).

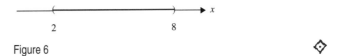

Figure 6

Solving Applied Problems

Equations and inequalities containing absolute values are extremely useful in solving applied problems, especially those involving data analysis. For instance, many specifications usually allow a certain amount of variation so that measured dimensions (such as thickness, distance, and so forth) differ from exact dimensions by acceptable errors, known as *tolerances*. The next example provides a way to represent such variations.

EXAMPLE 5   Solving a Manufacturing Problem

A manufacturer of helicopter blades determines that the measured thickness x of a blade must differ from the exact design thickness of 17.48 millimeters by no more than 0.12 millimeter (the tolerance is 0.12 millimeters). Write an absolute value inequality that expresses this relationship, and then solve the inequality to find the acceptable range of thicknesses.

Solution   Make a Plan

Let x = the measured thickness of the blade, then
$|x - 17.48|$ = the difference between the measured thickness and design thickness.

The tolerance of 0.12 millimeters can be expressed using the absolute value inequality: $|x - 17.48| \leq 0.12$

Write a Solution

$$\begin{array}{ll} |x - 17.48| \leq 0.12 & \text{Given} \\ -0.12 \leq x - 17.48 \leq 0.12 & \text{Equivalent inequality} \\ 17.36 \leq x \qquad\quad \leq 17.60 & \text{Add 17.48 to each part to isolate } x \end{array}$$

Therefore, the measured thickness of the blade must be greater than or equal to 17.36 millimeters and less than or equal to 17.60 millimeters.

Look Back

We can check this result by substituting the extreme values 17.36 and 17.60 for x in the inequality.

$$|17.36 - 17.48| = |-0.12| \leq 0.12$$
$$|17.60 - 17.48| = |0.12| \leq 0.12 \qquad\qquad ◈$$

Using Technology Exploration

To solve an equation involving absolute value such as

$$|ax + b| = c$$

by using a grapher, we first write the equation in the zero form $|ax + b| - c = 0$, then graph

$$y = |ax + b| - c.$$

The x values at which the graph intercepts the x axis is the solution for the equation. To solve inequalities such as

$$|ax + b| < c \text{ or } |ax + b| > c, (c > 0),$$

we graph $y = |ax + b| - c$ then select the interval(s) in which the x coordinates have corresponding y values either *below* the x axis where $y < 0$ or *above* the x axis where $y > 0$.

G EXAMPLE 6   Using a Grapher

Use a grapher to solve:
(a) $|x - 3| = 7$ (b) $|x - 3| < 7$ (c) $|x - 3| > 7$

Solution

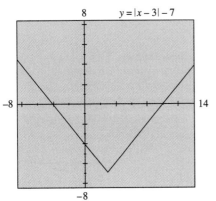

Figure 7

(a) To solve the equation $|x - 3| = 7$ by a grapher, we write the equation in the zero form $|x - 3| - 7 = 0$, and then graph $y = |x - 3| - 7$ using the viewing window xMin $= -8$, xMax $= 14$, xScl $= 2$, yMin $= -8$, yMax $= 8$, yScl $= 1$. The graph is a "V" shaped figure and it intercepts the x axis at -4 and 10 which happen to be the solutions of the equation (Figure 7).

(b) To solve the inequality $|x - 3| < 7$ graphically, we examine the graph of the equation $y = |x - 3| - 7$ in Figure 7. Observe that the values which make $y < 0$ are *below* the x axis and have x coordinates between -4 and 10. This means that the solution set of the inequality $|x - 3| < 7$ in interval notation is $(-4, 10)$.

(c) To solve the inequality $|x - 3| > 7$ graphically, we again examine the graph in Figure 7. The values that make $y > 0$ are *above* the x axis ($y > 0$) and have x coordinates which are greater than 10 or less than -4. This means that the inequality $|x - 3| > 7$ has a solution set in interval notation $(-\infty, -4)$ or $(10, \infty)$.   ◈

◈ PROBLEM SET 2.6

Mastering the Concepts

In Problems 1–20, solve each equation algebraically.

1. (a) $|x - 5| = 6$
    (b) $|5 - x| = 6$
2. (a) $|y - 3| = 4$
    (b) $|3 - y| = 4$
3. (a) $|x - 2| = 1$
    (b) $|x + 2| = 1$
4. (a) $|x - 4| = 2$
    (b) $|x + 4| = 2$
5. (a) $|y - 5| = 0$
    (b) $|y - 5| = -2$
6. (a) $|x - 7| = -1$
    (b) $|x + 7| = 0$
7. (a) $|3x| = 0$
    (b) $|-3x| = 0$
8. (a) $|-7x| = 14$
    (b) $|-7x| = -14$
9. (a) $|x| + 2 = 8$
    (b) $|x + 2| = 8$
10. (a) $|x| - 4 = 4$
    (b) $|x - 4| = 4$
11. $-4|5t| = 2$
12. $8|2x| = 4$
13. $|3t + 2| = 0$
14. $|5 - 4x| = 0$
15. $|3(x + 7)| = 12$
16. $|2(x - 10)| = 1$
17. $5 = \left|\dfrac{3x}{7} - \dfrac{2}{7}\right|$
18. $2 = \left|\dfrac{4x}{5} + \dfrac{1}{5}\right|$
19. $\left|\dfrac{5x}{2} + \dfrac{1}{2}\right| = \dfrac{5}{2}$
20. $\left|\dfrac{-5x}{2} - \dfrac{1}{2}\right| = \dfrac{5}{2}$

In Problems 21–44, solve each inequality algebraically. Write the solution set in interval notation and graph it on a number line.

21. (a) $|x - 4| \geq 5$
    (b) $|x - 4| < 5$
22. (a) $|x - 2| \geq 3$
    (b) $|x - 2| < 3$
23. (a) $|x + 14| \geq 25$
    (b) $|x + 14| < 25$
24. (a) $|x + 12| \geq 30$
    (b) $|x + 12| < 30$
25. $|3x| < 15$
26. $|6x| \geq 1$

27. (a) $|2x - 1| \leq 3$
    (b) $|2x - 1| \geq 3$
28. (a) $|8x + 3| \geq 3$
    (b) $|8x + 3| < 3$
29. (a) $|5x + 1| \geq -2$
    (b) $|5x + 1| \leq -2$
30. (a) $|2x - 12| \geq -3$
    (b) $|2x - 12| < 3$
31. $\left|\dfrac{x}{2}\right| > \dfrac{1}{2}$
32. $\left|\dfrac{1}{3}x\right| \leq 2$
33. $|2x - 1| - 2 < 3$
34. $|2x + 7| - 2 \geq 9$
35. $|3x + 1| + 2 \geq 7$
36. $|5x - 3| - 4 \leq 3$
37. $\left|\dfrac{2}{3}x - 1\right| \leq \dfrac{1}{3}$
38. $\left|\dfrac{x}{2} + 2\right| > \dfrac{5}{2}$
39. $\left|\dfrac{1}{3} - \dfrac{x}{6}\right| \geq \dfrac{1}{2}$
40. $\left|\dfrac{3x}{7} + \dfrac{2}{21}\right| < 1$
41. (a) $|x + 7| < 0$
    (b) $|x + 7| \leq 0$
42. (a) $2|x + 7| < 0$
    (b) $2|x + 7| \leq 0$
43. (a) $|x + 6| > 0$
    (b) $|x + 6| \geq 0$
44. (a) $3|x + 6| > 0$
    (b) $3|x + 6| \geq 0$

Applying the Concepts

45. **Body Temperature:** During and after surgery, research has determined that a patient's temperature T (in degrees Fahrenheit) must not differ from 98.6°F by no more than 2.3°F. Write an absolute value inequality to express this situation, and then solve the inequality.

46. **Manufacturing:** Manufacturers of compact discs (CDs) permit only very small tolerances. The measured diameter, x, of a CD may differ from the exact diameter of 117 mm by no more than 0.01 mm. Write an absolute value inequality to express this situation, and then solve the inequality.

47. **Electric Motor Design:** In the design of an electric motor, it was determined that the measured voltage needed to run the motor may not differ from the exact voltage of 220 by no more than 25 volts. Write an absolute value inequality that expresses this situation, and then solve the inequality to find the lower and upper limits for voltage for which the motor will run.

48. **Heating and Cooling:** A thermostat is set at 68°F so that the heating and cooling system will activate when the room temperature is more than 3.5°F from that setting. Write an absolute value inequality to express this relationship and solve the inequality.

49. **Car Production:** An automobile manufacturer predicted that the number of cars to be produced by the automobile industry for a given year must differ from the expected 4,000,000 demand for cars by no more than 500,000 cars. Use an absolute value inequality to describe this situation, and then solve this inequality.

50. **Stock Brokerage:** A stockbroker uses a computer program to be alerted on the price change of stocks. To trigger the alert, the current price x must differ from $16.25 by more than $0.75. Write an absolute value inequality to express this situation, and then solve the inequality.

51. **Medicine:** In administering medicine, a physician has determined that the dosage x cc for the patient must differ from the prescription of 2.5 cc. by no more than 0.2 cubic centimeters. Write an absolute value inequality that describes this situation. Then solve this inequality to determine the lower and upper limits of the dose of medicine to be given to the patient.

52. **Grading:** At the beginning of the semester, a mathematics professor informed the class that in order to receive a grade of C in this algebra course, a student's score x must differ from 71.5% by no more than 8%. Write an absolute value inequality that describes this situation, and then solve the inequality.

Developing and Extending the Concepts

In Problems 53 and 54, write an absolute value equation or inequality that represents each statement. Then solve the equation or inequality.

53. (a) The distance between x and 3 is 8.
    (b) The distance between x and 3 is almost 8.
54. (a) The distance between x and -5 is 4.
    (b) The distance between x and -5 is at least 4.

In Problems 55 and 56, write an absolute value inequality whose solution is shown in the graph.

55.

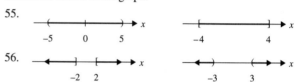

56.

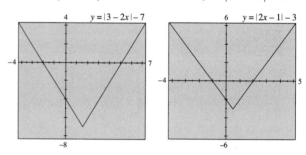

G In Problems 57 and 58, use the given grapher display to find the solution set, in interval notation, of each associated inequality.

57. $y = |3 - 2x| - 7$ 58. $y = |2x - 1| - 3$

(a) $|3 - 2x| \le 7$ (a) $|2x - 1| \ge 3$
(b) $|3 - 2x| > 7$ (b) $|2x - 1| < 3$

G In Problems 59 and 60, graph the equation, identify the x intercepts, and solve the inequality.

59. (a) $|2x + 3| \le 15; y = |2x + 3| - 15$
    (b) $|3x + 12| > 12; y = |3x + 12| - 12$
60. (a) $|x/2 + 4| \ge 5; y = |x/2 + 4| - 5$
    (b) $|4 - 2x| < 0; y = |4 - 2x|$

◆ CHAPTER 2 REVIEW PROBLEM SET

In Problems 1–22, solve each equation.

1. $3(x + 5) - 6x = 6$
2. $6x + 2(x - 3) + 5 = -25$
3. $5[3t - (2 - t)] = 3(2t - 1)$
4. $3[(2x - 5) - (x + 1)] = 3$
5. $0.13(x + 300) = 61 - 0.09x$
6. $0.003(t + 200) = 86 - 0.05t$
7. $0.09(15 - y) + 0.04\,y = 0.07(25)$
8. $0.05(300 - y) - 0.1y + 105 = 0$

9. $\dfrac{3x}{5} - \dfrac{2x}{3} = 1$
10. $\dfrac{t}{3} + \dfrac{8}{5} = \dfrac{8 - t}{5}$
11. $\dfrac{2x + 1}{4} - \dfrac{3x - 4}{5} = \dfrac{1}{2}$
12. $\dfrac{x - 3}{10} + \dfrac{x - 1}{6} = \dfrac{3}{5}$
13. $\dfrac{t - 4}{3} + \dfrac{2(t + 3)}{5} = 4$
14. $\dfrac{y - 3}{4} + \dfrac{y - 5}{6} = \dfrac{2y - 1}{3}$
15. $|3 - 2x| = 1$
16. $|7t + 4| = 3$

17. $\left|4x - 2\right| + 1 = 3$ 18. $\left|2y + 3\right| - 4 = 1$

19. $\left|4 - y\right| = \left|y - 2\right|$ 20. $\left|2x - 1\right| = \left|2x - 5\right|$

21. $\left|\dfrac{x - 3}{4}\right| + 4 = 4$ 22. $\left|\dfrac{5t - 3}{2}\right| + 2 = 6$

In Problems 23–30, solve each equation for the indicated variable.

23. $y = -2x + 5$, for x 24. $8x - 12b = 4(2x - 3b)$, for x

25. $A = \dfrac{1}{2}h(a + b)$, for b 26. $S = \dfrac{n}{2}(f + l)$, for f

27. $S = 2\pi r^2 + 2\pi rh$, for h

28. $A = 2hw + 2lw + lh$, for l.

29. $\dfrac{1}{a} + \dfrac{1}{b} = \dfrac{1}{c}$, for b 30. $y = \dfrac{3 - x}{x - 1}$, for x

In Problems 31–48, solve each inequality. Graph the solution set and give the solution in interval notation.

31. $8x + 14 \le 5x + 20$

32. $2(x - 3) \ge 3x + 4$

33. $3(x + 2) > 11 - 2(2 - x)$

34. $3(x - 4) + x < 2(6 + 2x)$

35. $2 \le x + 5 \le 8$

36. $2 \le 3x - 4 \le 8$

37. $2x - 1 < 5$ and $2x - 1 > 3$

38. $x < -1$ and $3x > x - 6$

39. $\left|2x - 3\right| \le 7$ 40. $\left|5 - 6x\right| \le 29$

41. $\left|x + 2\right| - 5 < -3$ 42. $\left|3x - 1\right| + 2 < -3$

43. $\left|\dfrac{x - 4}{2}\right| - 1 \ge 7$ 44. $\left|\dfrac{3x - 5}{6}\right| - 2 \ge 3$

45. $\left|\dfrac{x + 2}{2}\right| - 1 > 5$ 46. $\left|\dfrac{2x - 3}{5} + 7\right| + 6 < 8$

47. $\left|\dfrac{3x + 4}{5}\right| - \dfrac{2}{5} \le 1$ 48. $\left|2\left(\dfrac{3 - x}{5}\right)\right| - \dfrac{3}{5} < \dfrac{6}{5}$

49. For what values of x will each inequality be true?
    (a) $\left|3x + 2\right| \le 0$ (b) $\left|3x + 2\right| \ge 0$

50. (a) Criticize the following solution.

$$\frac{3}{x} \ge 5$$

$$\frac{3}{x} \cdot x \ge 5 \cdot x$$

$$3 \ge 5x$$

$$\frac{3}{5} \ge x$$

(b) Write the correct solution of $\dfrac{3}{x} \ge 5$.

51. Solve the inequality $-3\left|2x - 5\right| \ge -9$.

52. For what values of x will the expression

$$\frac{7}{\sqrt{4 - 2x}}$$

be a real number?

53. **Age Difference:** Milt is 15 years older than Linda. The sum of their ages is 85. Find Linda's age.

54. **Sales Tax:** The State of Michigan has a 6% sales tax on nonfood items. What is the original price of a computer if the total cost of the computer, including tax, is $1166?

55. **Geometry:** A rectangular pool has a perimeter of 68 feet. If the length of the pool is 4 feet more than twice its width, find the length and the width of the pool.

56. **Basketball:** In winning a basketball game, a professional basketball team made only 5 free throws (one free throw is worth one point). The remainder of their points came from two and three point baskets. If the number of baskets from the field (two or three point baskets) totaled 45, and if the team scored 110 points in the game, how many three point baskets did the team make in that game?

57. **Roofing:** Two ingredients are used to make a flat roof top filter. The first ingredient contains 35% coal tar, and the second contains 60% coal tar. How much of each ingredient should be mixed to make 60 tons of roof top filter that contains 50% coal tar?

58. **Investments:** A financial planner has $80,000 to invest. Part of the money is invested in a stock paying 9% annual simple interest for 2 years; the balance is invested in a mutual fund paying 7% annual simple interest for 3 years. If the total interest earned is $15,300, how much money is placed in each investment?

59. **Aircraft:** An aircraft manufacturer determines that one of its aircraft has twice as many seats as another. Find the number of seats for each aircraft if the total number of seats is 336.

60. **Biology:** The international whaling commission determined that the average weight of a blue whale is about 5 tons more than three times the average weight of a humpback whale. Determine the average weight of each whale if the total of their average weights is 117 tons.

61. **Painting:** A painter can paint a racquetball court in 8 hours. His assistant needs an additional 2 hours to paint the same court working by himself. How long will it take both of them working together to paint the court?

62. **Boating:** The speed of a motor boat in still water is 24 miles per hour. If the boat travels 27 miles upstream in the same time it takes to travel 45 miles downstream, what is the speed of the current?

63. **Telephone Fees:** A certain telephone service charges $1.57 for the first 3 minutes of a cellular call and $0.01 for each additional second. How long may one speak if the charge is not to exceed $4.72?

64. **Test Scores:** A math student has scores of 65, 68, 78 and 75 on her tests. Find the range of scores needed on her final examination so that she can earn a C in the course. Assume that the final examination counts as two tests and a grade of C is earned if the final course average is between 69.5 and 79.4.

G In Problems 65–70, use the TRACE feature of a grapher to solve each equation.

65. $5x + 12 = 2x - 3$

66. $2x + 5(x - 4) = 4x - 2(x - 10)$

67. $x + 3(2 + 4x) = 6(x + 1) + 5x$

68. $0.4x + 3.6 = 0.2x + 4.8$

69. $\dfrac{x + 1}{4} + \dfrac{3 + x}{5} = \dfrac{7}{4}$

70. $\dfrac{5x - 1}{3} - \dfrac{4x + 3}{5} = \dfrac{-1}{15}$

 ## CHAPTER 2 PRACTICE TEST

1. Solve each equation.
   (a) $3x + 2(45 - x) + 5 = 10$
   (b) $4(2y - 3) + 55 = 5(4 - 3y)$
   (c) $0.05(x + 2) + 0.10x = 2$

2. Solve each equation. Check for extraneous solutions.
   (a) $\dfrac{x}{8} + \dfrac{2x}{3} = \dfrac{19}{24}$
   (b) $\dfrac{3}{x - 3} + 4 = \dfrac{x}{x - 3}$

3. Solve each equation.
   (a) $|3y| = 24$ (b) $|2 - 3x| = 7$ (c) $|3x + 5| = 0$

4. (a) Solve for y, $3a + 6y = 8y - 9y$.
   (b) Solve the formula $A = Prt + P$ for P.

5. Graph each solution set on a number line.
   (a) $x \le -3$ (b) $-6 \le x \le 0$ (c) $x \le 5$ or $x > 8$

6. Solve each inequality. Express each answer in interval notation and show the solution on a number line.
   (a) $-4x \ge 12$
   (b) $15x < -45$
   (c) $-5 < 3x + 7 \le 10$
   (d) $\dfrac{2 - 3x}{2} \le \dfrac{2x + 1}{3}$
   (e) $|x| \le 3$
   (f) $|x - 4| \ge 1$

7. **Wholesale Pricing:** A carpet dealer marks up her merchandise by 21% over the wholesale cost. If a carpet is sold for $1,815, what is the wholesale price?

8. **Triangle:** The perimeter of a triangle is 17 inches. One side is twice as long as another, and the third is 5 inches long. What is the length of the shortest side?

9. **Current:** A sailboat travels 7 miles per hour in still water. The sailboat travels 20 miles upstream in the same time as it travels 50 miles downstream. What is the speed of the current?

10. **Numbers:** If 1 is added to five times a number, the result is greater than 9 subtracted from three times the number. Find all possible such numbers.

Chapter 3

LINEAR GRAPHS AND FUNCTIONS

3.1 Graph Linear Equations

3.2 Slope of a Line

3.3 Equations of Lines

3.4 Graph Linear Inequalities

3.5 Introduction to Functions

Highway engineers use the concept of slope to determine the *grade* of a road or the *change of elevation* of a hill. When they describe a 4% grade of a road, they mean that the road changes 4 feet vertically for every 100 feet of horizontal change. Using the 4% grade of a ramp, how much change of elevation will be permitted on the ramp if its horizontal change is 500 feet? Example 7 on page 150 provides a method to answer this question.

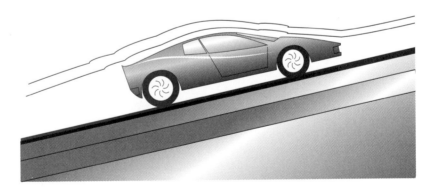

Graphs are often used in real world applications to provide a visual interpretation of changes in quantities. For instance, in meteorology, graphs might be used to illustrate the variation in air temperature during a day; in manufacturing, to express the relationship between the demand and the price of a product; in business, to show fluctuations in the stock market. In this chapter, we develop some basic geometric tools and apply these tools to graphing linear equations and inequalities in two variables. We also study the rate of increase or decrease in linear models of real world situations. In addition, the chapter includes the discussion of the important concept of a function which will be continued throughout the book.

OBJECTIVES
1. Graph by Point Plotting
2. Use Intercepts to Graph Equations
3. Graph Special Linear Equations
4. Solve Applied Problems
5. Use Technology Exploration

3.1 Graphing Linear Equations

Suppose that in 1990 a solar electric company declared a dividend of $3 per share. Each successive year they had a $0.25 increase in the dividend. Let the year 1990 correspond to $x = 0$ and successive years correspond to $x = 1, 2, 3, 4, \ldots$ and y dollars represents the dividend per share. This pattern is shown in the following table.

x	0	1	2	3	4	5	6
y	3.00	3.25	3.50	3.75	4.00	4.25	4.50

The graph in Figure 1 has been used to model this data. After examining the data, we notice that there is a pattern in the quantities involved. This pattern is expressed by the equation $y = 3 + 0.25x$ where x is a whole number and the equation is called a *linear equation in two variables*. In section 1.6, we considered linear equations in two variables. In particular, we plotted ordered pairs in the *rectangular coordinate system*, also called the *Cartesian coordinate system*. In this section we graph linear equations in more detail.

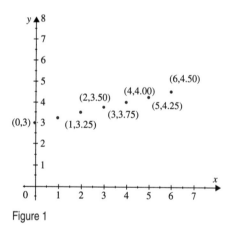

Figure 1

Graphing by Point Plotting

The Cartesian coordinate system enables us to explore the connection between algebra and geometry. This connection is illustrated by representing solutions of equations graphically. Recall that the *graph* of an equation in two variables x and y is the set of all points in the plane whose coordinates satisfy the equation. A *linear* (or *first degree*) *equation* in two variables is an equation that can be written in the *standard form*

$$Ax + By = C$$

where A, B, and C are real numbers and A and B are not both zero. The *graph of every linear equation in two variables is a straight line*. Point plotting is one technique for graphing linear equations. For instance, to graph the linear equation $y = 2x$, we are free to chose any value for x, then calculate the corresponding value for y. If we choose $x = -2$, the value for y is $y = 2(-2) = -4$. The associated ordered pair $(-2, -4)$ is said to satisfy the equation and is called a **solution** of the equation. Let us choose three values for x and find three ordered pair solutions. The following table shows three solutions.

The graph of this linear equation is a line and plotting these ordered pair solutions yields the graph of this straight line (Figure 2).

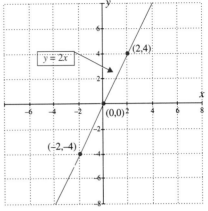

x	$2x = y$	Ordered Pair Solutions
-2	$2(-2) = -4$	$(-2, -4)$
0	$2(0) = 0$	$(0, 0)$
2	$2(2) = 4$	$(2, 4)$

Figure 2

EXAMPLE 1   Graphing by Point Plotting

Sketch the graph of the equation $3x + y = 6$.

Solution   By solving the equation for y in terms of x it is easier to construct a table and, we have

$$3x + y = 6 \quad \text{or} \quad y = -3x + 6$$

Now we make a table of values by choosing three values for x and finding the corresponding values for y. Then we plot these points and graph the line connecting them (Figure 3).

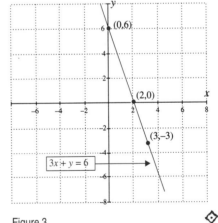

x	$-3x + 6 = y$	Ordered Pair Solutions
0	$-3(0) + 6 = 6$	$(0, 6)$
2	$-3(2) + 6 = 0$	$(2, 0)$
3	$-3(3) + 6 = -3$	$(3, -3)$

Figure 3

Using Intercepts to Graph Equations

Notice that the graph in Figure 3 crosses the x axis at the point $(2,0)$. Recall that this point is called the *x intercept point* and 2 is the *x intercept* of the graph. The graph also crosses the y axis at $(0,6)$, which is called the *y intercept point* and 6 is the *y intercept* of the graph. In section 1.6 we showed how to find the intercepts of a graph of a linear equation. In general, Table 1 on page 138 shows how to find the intercepts of any equation.

TABLE 1   Intercepts of the Graph of an Equation that Relates x and y

Terminology	Description	Illustration	Determination of Intercepts
x intercepts	The x coordinates of points where the graph intersects the x axis.		Set $y = 0$ and solve for x.
y intercepts	The y coordinates of points where the graph intersects the y axis.		Set $x = 0$ and solve for y.

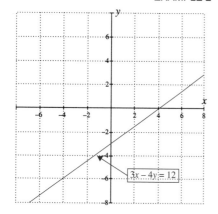

Figure 4

EXAMPLE 2   Using Intercepts to Graph an Equation

Find the intercepts and sketch the graph of the equation

$$3x - 4y = 12$$

Solution   The x intercept of the graph of the equation

$$3x - 4y = 12$$

is obtained by setting $y = 0$ and solving for x to get $3x = 12$ or $x = 4$. To find the y intercept, set $x = 0$ and solve for y to obtain $-4y = 12$ or $y = -3$. Thus the x and y intercepts of the graph are 4 and -3 respectively. Figure 4 shows the graph of the equation. ◈

We must be careful when graphing linear equations using only two points where the graph intercepts the coordinate axes. If one of the points is wrong, the graph will be wrong. As a precaution, we prefer to plot a third point as a *check point* to ensure the correctness of the graph.

Graphing Special Linear Equations

Equations such as $x = 2$ and $y = -3$ are both examples of linear equations in two variables. In both equations, one of the variables is missing. They are written in standard form as follows:

Equation	Standard Form
$x = 2$	$1x + 0y = 2$
$y = -3$	$0x + 1y = -3$

EXAMPLE 3   Graphing Special Linear Equations

Sketch the graph of each equation

(a) $y = -3$ (b) $x = 2$

Solution   (a) The equation $y = -3$ can be written in the form

$$y = 0x - 3$$

Thus, for any value of x selected, y will be -3. Figure 5a shows the graph of the equation $y = -3$.

(b) The equation $x = 2$ can be written as

$$x = 2 + 0y$$

Thus, for every value of y selected, x will have a value of 2. Figure 5b shows the graph of the equation.

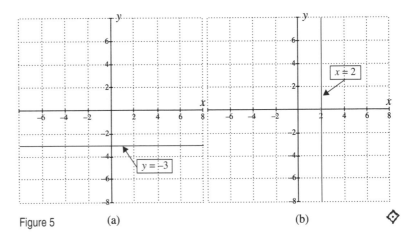

Figure 5   (a)   (b)

The x axis is a horizontal line with equation

$$y = 0.$$

The y axis is a vertical line with equation

$$x = 0.$$

Table 2 illustrates the characteristics that apply to all other equations of horizontal and vertical lines. Assume $h \neq 0$ and $k \neq 0$.

TABLE 2

Equation	Graph	x Intercept	y Intercept
$y = k$	horizontal line	none	k
$x = h$	vertical line	h	none

Solving Applied Problems

Graphs of linear equations are often used to show relationships between quantities represented by two variables. These variables can be designated by letters other than x and y.

EXAMPLE 4   Determining Cable Television Charges

A cable television company charges a $30 installation fee plus a $35 monthly service fee. The total charge C (in dollars) for its services is given by the linear model

$$C = 30 + 35t, \quad t \geq 0$$

where t is the number of months for the service.
(a) Sketch the graph of this equation using t as the horizontal axis and C as the vertical axis.
(b) Find and interpret the intercepts of the graph.
(c) Find the total cost for 9 months of service.
(d) If the total cost of service is $240, for how many months is the service used?

Solution   (a) Figure 6 shows the graph that models this situation.
(b) The C intercept of this graph is 30. This means that the total charges are $30 with zero monthly service. There is no t intercept.
(c) The total charges for 9 months of service is given by

$$C = 30 + 35(9) = 345.$$

Therefore, the total charges are $345.
(d) To find t when $C = $240, we solve the equation

$$240 = 30 + 35t$$
$$210 = 35t$$
$$6 = t.$$

Therefore, it takes 6 months of service to spend $240.

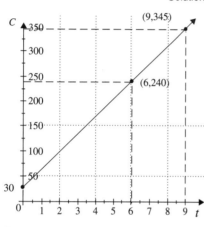

Figure 6

Using Technology Exploration

Graphers have TRACE features that show the horizontal and vertical coordinates of points on the graph.

G EXAMPLE 5   Finding a Temperature Conversion

The relationship between Fahrenheit temperature F and Celsius temperature C is given by the model

$$F = 1.8C + 32$$

(a) Use a grapher to sketch the graph of this model using C as the horizontal axis and F as the vertical axis.
(b) Use the TRACE feature of the grapher to determine the Fahrenheit temperature F when the Celsius temperature is
$$10°, 21.5°, 45.5°.$$

Solution   (a) Figure 7 is a display of the graph of the equation
$$F = 1.8C + 32.$$
(b) Using the TRACE feature of the grapher, we obtain the following temperatures

$$50°F, 70.7°F, \text{ and } 113.9°F$$

corresponding to

$$10°C, 21.5°C, \text{ and } 45.5°C$$

respectively.

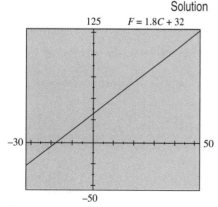

Figure 7

PROBLEM SET 3.1

Mastering the Concepts

In Problems 1–12, sketch the graph of each linear equation by point plotting.

1. (a) $y = x - 1$
    (b) $y = x + 1$
2. (a) $y = x + 2$
    (b) $y = x - 2$
3. (a) $3y = x$
    (b) $y = -3x$
4. (a) $4y = x$
    (b) $y = -4x$
5. $2x + 3y = 6$
6. $2y = 5x - 10$
7. $3x - 2y = 20$
8. $2x + 3y = 1$
9. $3x - 2y = 5$
10. $y = 5 - 3x$
11. $5x + 2y = 3$
12. $2x + 3y = 5$

In Problems 13–32, find the x and y intercepts of each equation and sketch the graph.

13. $3x - 4y = 24$
14. $2y = 3x + 6$
15. $2x = 3y + 1$
16. $-6x + 3y = 12$
17. $2x - 7y = 21$
18. $4x + 10y = 20$
19. $-6x + 3y = 12$
20. $2y - x = 5$
21. $10y = -6x + 14$
22. $3y + x = 4$
23. $2y - x = 0$
24. $y - 2x = 1$
25. $8y + 24 = 0$
26. $-2y = -7$
27. $2x - 3 = 0$
28. $4y + 5 = 0$
29. $2y + 7 = 0$
30. $4x - 8 = 0$
31. $3x - 1 = 0$
32. $2x + 3 = 0$

Applying the Concepts

33. **Car Rental:** A car rental company charges $80 per week plus 15 cents for each mile driven on its midsize cars. The total weekly cost C (in dollars) is given by the linear model

    $$C = 80 + 0.15m, \quad m \geq 0$$

    where m represents the weekly mileage driven.
    (a) Sketch the graph of this equation using m as the horizontal axis and C as the vertical axis.
    (b) Find and interpret the intercepts of the graph.
    (c) What is the cost C when the car has been driven 352 miles in a week?

34. **Photo Development:** A photo development center charges $4.00 for each roll of film developed plus 39 cents for each print. The total charge C (in dollars) is given by the linear model

    $$C = 4 + 0.39n, \quad n \geq 0$$

    where n represents the number of prints developed.
    (a) Sketch the graph of this equation using n as the horizontal axis and C as the vertical axis.
    (b) Find and interpret the intercepts of the graph.
    (c) What is the total cost for developing and printing 24 exposures?

35. **Cellular Phone Cost:** A cellular phone company charges $80 for each phone plus $30 per month under a special sale plan. The total cost C (in dollars) of operating the phone is given by the linear model

    $$C = 80 + 30t, \quad t \geq 0$$

    where t represents the number of months of operation of a cellular phone.
    (a) Sketch the graph of this equation using t as the horizontal axis and C as the vertical axis.
    (b) Find and interpret the intercepts of the graph.
    (c) How many months would it take to have a total cost of $290?

36. **Wages:** A fitness center offered its manager a $2,200 monthly salary plus 6% of the net monthly income of the center. The manager's monthly salary S (in dollars) is given by the linear model

    $$S = 2200 + 0.06n$$

    where n represents the monthly net income of the center (in dollars).
    (a) Sketch the graph of this equation using n as the horizontal axis and S as the vertical axis.
    (b) If the center's net income for a given month is $40,000, what will the manager's salary be for that month?
    (c) What is the net income for a month if the manager's salary for that month is $3,800?

Developing and Extending the Concepts

In Problems 37–40, sketch the graphs of each group of equations on the same coordinate system. Describe the changes you observe in the equations and in the graphs.

37. $y = x$, $y = 3x$ and $y = 5x$
38. $y = x + 1$, $y = 2x + 1$ and $y = 4x + 1$
39. $y = x - 1$, $y = x + 1$ and $y = x + 3$
40. $y = 2x - 2$, $y = 2x + 2$ and $y = 2x + 4$
41. For the equations $y = 3x + 4$ and $x = 3y + 4$
    (a) What is the shape of graphs that you expect and why?
    (b) Fill in the table for each equation.

x	$3x + 4 = y$	Ordered Pair Solutions	y	$3y + 4 = x$	Ordered Pair Solutions
-4			-4		
0			0		
2			2		

(c) Graph each equation on the same coordinate axes.
(d) Find an ordered pair which is a solution for both equations by solving the equation

$$\frac{x - 4}{3} = 3x + 4$$

42. For the equations $y = 5x + 2$ and $y = \dfrac{x - 2}{5}$

    (a) What is the shape of graphs that you expect and why?
    (b) Fill in the table for each equation.

x	$5x + 2 = y$	Ordered Pair Solutions	x	$\dfrac{x - 2}{5} = y$	Ordered Pair Solutions
-2			-8		
0			2		
2			12		

    (c) Graph each equation on the same coordinate axes.
    (d) Find an ordered pair which is a solution for both equations by solving the equation

$$\frac{x - 2}{5} = 5x + 2.$$

G In Problems 43–46, use a grapher to sketch the graph of each equation by using the specified setting on the viewing window. Use the TRACE and ZOOM feature to find the intercepts of each graph.

43. $y = -x/2 + 3$
    xMin $= -10$, xMax $= 10$, xScl $= 1$
    yMin $= -10$, yMax $= 10$, yScl $= 1$

44. $x - 2y = 10$
    xMin $= -1$, xMax $= 20$, xScl $= 5$
    yMin $= -15$, yMax $= 5$, yScl $= 5$

45. $y = -2x - 11$
    xMin $= -15$, xMax $= 10$, xScl $= 5$
    yMin $= -20$, yMax $= 10$, yScl $= 5$

46. $4y = -3x + 16$
    xMin $= -10$, xMax $= 10$, xScl $= 1$
    yMin $= -10$, yMax $= 10$, yScl $= 1$

G 47. Write each statement as an equation in two variables; solve the equations for y and graph. Adjust the viewing window to include the intercepts.
    (a) Twice the y value is 4 more than three times the x value.
    (b) Three times the y value is 5 less than twice the x value.

48. Table 3 shows data on the average prices paid, p, per gallon (in dollars) of unleaded gasoline in the southern part of the State of Michigan for each month, t, in 1996.

TABLE 3

Month t	Jan	Feb	Mar	Apr	May	Jun
Price p	1.15	1.10	1.10	1.20	1.20	1.25

Month t	Jul	Aug	Sep	Oct	Nov	Dec
Price p	1.28	1.18	1.18	1.15	1.15	1.10

    (a) Plot the data points and connect these points with a line segment. Use t as the horizontal axis.
    (b) Determine when the prices increased.
    (c) Determine when the prices decreased.
    (d) Determine when the prices remained constant.

49. One of the forms of depreciation for income tax purposes is the **straight line depreciation method** in which the value V dollars of a property is expressed in terms of t years by a linear equation. The graph in Figure 8 shows the value of a color printer purchased by a small business which is being depreciated over a period of 6 years. Use the graph in Figure 8.

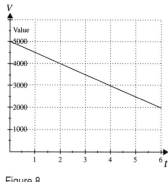

Figure 8

    (a) What is the original purchase price of the color printer?
    (b) What is the depreciated value of the color printer after 6 years?
    (c) What loss in value occurred after 1 year? 2 years? and 4 years?
    (d) Can you tell why the line is falling downward to the right?

G 50. Use a grapher to sketch the graph of each equation. Set your own viewing window. Use the TRACE feature to determine y when x is -1.1, 1.7 and 2.4.
    (a) $y = 3.5x - 1.3$ (b) $y = -2$ (c) $y = -1.4x + 5.4$

OBJECTIVES

1. Find the Slope of a Line
2. Determine Parallel and Perpendicular Lines
3. Solve Applied Problems
4. Use Technology Exploration

3.2 Slope of a Line

In section 3.1, we graphed equations of lines by point plotting and identifying their intercepts. A key geometric feature of a line is how sharply it rises or falls as we move from left to right, from point to point along the line. Visually this steepness is illustrated by comparing the vertical change to the horizontal change—an important characteristic called the *slope* of a line.

Finding the Slope of a Line

Informally, the idea of the slope is used when we refer to the steepness of a hill, the grade of a road or the pitch of a roof. In mathematics, the word slope has a precise meaning. Suppose that a nonvertical line contains two different points $P = (x_1, y_1)$ and $Q = (x_2, y_2)$. As we move along the line from P to Q, the **horizontal change** (also called the **run**) between P and Q is

$$\Delta x = x_2 - x_1$$

and the **vertical change** (also called the **rise**) is $\Delta y = y_2 - y_1$ (Figure 1), where the symbols Δx and Δy are read as "delta x" and "delta y" respectively. The ratio of the vertical change to the horizontal change is the *slope* of a line. More formally, we have the following definition.

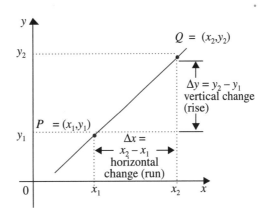

Figure 1

Definition: Slope of a Line

If $P = (x_1, y_1)$ and $Q = (x_2, y_2)$ are two different points of a nonvertical line, the **slope** m of the line containing P and Q is given by

$$m = \frac{\text{change in } y}{\text{change in } x} = \frac{\Delta y}{\Delta x} = \frac{y_2 - y_1}{x_2 - x_1}$$

It should be noted, as Figure 1 shows, that the slope of a line has the same value no matter which point we label (x_1, y_1) or (x_2, y_2).

$$\frac{y_2 - y_1}{x_2 - x_1} = \frac{y_1 - y_2}{x_1 - x_2}$$

Using the definition of the slope of a line containing any two points, we can find the slope of the line.

EXAMPLE 1   Finding the Slope of a Line

Use the coordinates of the given points on the graph of each line in Figure 2 to determine its slope.

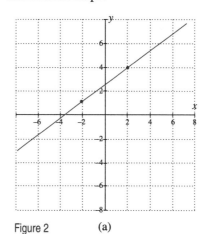

Figure 2   (a)   (b)

Solution   (a) Using Figure 2a, the coordinates of the given points are $(-2,1)$ and $(2,4)$, so we choose $x_1 = -2$, $y_1 = 1$, $x_2 = 2$ and $y_2 = 4$, and the slope m is given by

$$m = \frac{\Delta y}{\Delta x} = \frac{y_2 - y_1}{x_2 - x_1} = \frac{4 - 1}{2 - (-2)} = \frac{3}{4}$$

Notice that if we choose $x_1 = 2$, $y_1 = 4$, $x_2 = -2$ and $y_2 = 1$, the slope has the same value

$$m = \frac{\Delta y}{\Delta x} = \frac{y_2 - y_1}{x_2 - x_1}$$

$$= \frac{1 - 4}{-2 - 2} = \frac{-3}{-4} = \frac{3}{4}$$

Notice that the line rises from left to right.

(b) From Figure 2b, the coordinates of the given points are $(0,1)$ and $(4,-3)$, so we choose $x_1 = 0$, $y_1 = 1$, $x_2 = 4$ and $y_2 = -3$, and the slope m is given by

$$m = \frac{\Delta y}{\Delta x} = \frac{y_2 - y_1}{x_2 - x_1}$$

$$= \frac{-3 - 1}{4 - 0} = -1$$

Here the line falls from left to right.

The next example shows that a line is horizontal if $y_1 = y_2$ and it is vertical if $x_1 = x_2$.

EXAMPLE 2   Finding Slopes of Horizontal and Vertical Lines

Sketch the graph of each line containing the given points and find the slope.
(a) $(-1,4)$ and $(3,4)$ (b) $(2,-3)$ and $(2,2)$

Solution   Figure 3a shows the horizontal line containing the points $(-1,4)$ and $(3,4)$. Figure 3b shows the vertical line containing the points $(2,-3)$ and $(2,2)$.

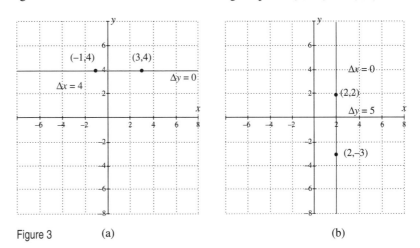

Figure 3   (a)   (b)

(a) The slope of the horizontal line in Figure 3a is given by

$$m = \frac{\Delta y}{\Delta x} = \frac{y_2 - y_1}{x_2 - x_1}$$

$$= \frac{4 - 4}{3 - (-1)} = \frac{0}{4} = 0$$

(b) The slope of the vertical line in Figure 3b is given by

$$m = \frac{\Delta y}{\Delta x} = \frac{y_2 - y_1}{x_2 - x_1}$$

$$= \frac{2 - (-3)}{2 - 2} = \frac{5}{0} \text{ which is undefined.}$$

It is very important to be able to visualize the difference between positive and negative slopes. As Figure 4a shows, the lines have positive slopes and they rise from left to right. Also, as Figure 4b shows, the lines have negative slopes and they fall from left to right.

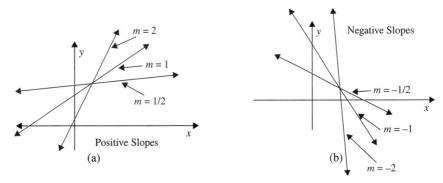

Figure 4

Table 1 on page 146 describes the connection between slopes and directions of lines.

TABLE 1   Slope and Direction of a Line

Slope Values (m)	Line Direction	Sample Graph
Positive	Rises from left to right	
Negative	Falls from left to right	
Zero	Horizontal	
Not defined	Vertical	

EXAMPLE 3   Graphing a Line with a Given Slope

Sketch the line that contains the point $(-2,3)$ and has the given slope.

(a) $m = \dfrac{3}{5}$ \qquad\qquad (b) $m = \dfrac{-5}{3}$

Solution   (a) The slope is

$$m = \frac{\Delta y}{\Delta x} = \frac{3}{5}.$$

If we start at the point $(-2,3)$ and move 5 units to the right and 3 units up, we find a second point $(3,6)$ on this line. We draw a line containing the two points (Figure 5a).

(b) The slope is

$$m = \frac{\Delta y}{\Delta x} = \frac{-5}{3} = -\frac{5}{3}.$$

Once again, if we start at the point $(-2,3)$ and move 3 units to the right and 5 units down, we find a second point $(1,-2)$ on this line. We draw a line containing the two points (Figure 5b).

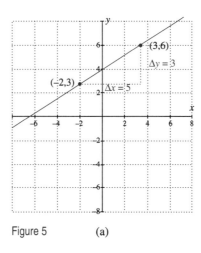
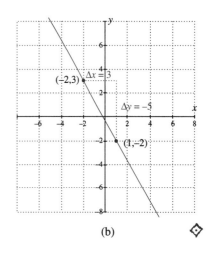

Figure 5   (a)   (b)

We can determine the slope of a line directly from its equation as the next example shows.

EXAMPLE 4   Finding a Slope of a Line Defined by an Equation

Find the slope of the line defined by the equation $2x - 3y = 12$.

Solution   The slope of this line can be found by using any two points on the line. For instance,

$$\text{if } x = 0 \text{ then } y = -4$$

and

$$\text{if } x = 9, \text{ then } y = 2 \text{ (Figure 6)}.$$

Using the points $(0,-4)$ and $(9,2)$ we find

$$m = \frac{\Delta y}{\Delta x} = \frac{y_2 - y_1}{x_2 - x_1}$$

$$= \frac{2 - (-4)}{9 - 0}$$

$$= \frac{6}{9} = \frac{2}{3}$$

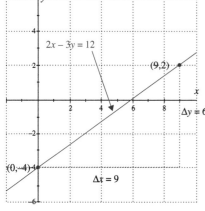

Figure 6

If we were to solve the equation in Example 4 for y we would have

$$y = \frac{2}{3}x - 4$$

We observe that the coefficient of x is 2/3 which is the same as the slope of the line. The constant, -4, of the equation is the y intercept.

In general, the equation of a line with slope m and y intercept b can be written in the form

$$y = mx + b$$

This equation is called the **slope-intercept form** of an equation of a line.

EXAMPLE 5   Using an Equation to Find the Slope

Express the equation $2x - 3y + 6 = 0$ in slope-intercept form and determine the slope and y intercept. Sketch the graph.

Solution   We solve the equation for y in terms of x.

$$2x - 3y + 6 = 0$$
$$-3y = -2x - 6$$
$$y = \frac{-2x}{-3} - \frac{6}{-3}$$
$$y = \boxed{\frac{2}{3}}\, x + \boxed{2}$$

$$\underbrace{\quad}_{\text{slope}} \qquad \underbrace{\quad}_{y \text{ intercept}}$$

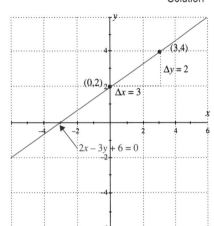

Figure 7

Thus the slope is 2/3 and the y intercept is 2. Since the slope is 2/3, we start at the point (0,2) and move 3 units to the right and 2 units up to find a second point (3,4) (Figure 7). ◈

Determining Parallel and Perpendicular Lines

The concept of slope provides a tool for determining whether lines in the Cartesian plane are parallel or perpendicular. To explore this idea, consider the relationship between the graphs of the lines, each with slope 2, and defined by the equations

$$y = 2x \text{ and } y = 2x + 4$$

Table 2 lists some solutions of both equations. By plotting these points and drawing the two lines on the same coordinate system, we see that the graph of $y = 2x + 4$ can be obtained by "shifting" the graph of $y = 2x$ four units upward (Figure 8a) . So the two lines are parallel.

TABLE 2

x	-2	-1	0	1	2
$y = 2x$	-4	-2	0	2	4
$y = 2x + 4$	0	2	4	6	8

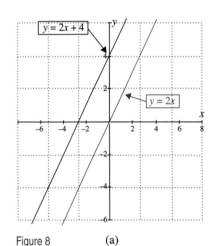
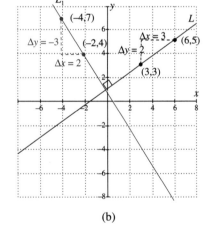

Figure 8   (a)   (b)

Now consider the graph of the line L whose equation is given by $y = \frac{2}{3}x + 1$ and whose slope is $m = \frac{2}{3}$ (Figure 8b). If we rotate this line 90° in the counterclockwise direction, we get a new line L_1 perpendicular to L.

By looking at the grid, we see that we can select two points P_1 and Q_1 on the new line L_1 whose coordinates are $(-4,7)$ and $(-2,4)$ respectively. Then the slope of the new line L_1 is given by

$$m_1 = \frac{4 - 7}{-2 - (-4)} = -\frac{3}{2}$$

Notice the slope of the new line L_1 compared to the slope of L; that is, the slope m_1 is the negative reciprocal of m, so that

$$m \cdot m_1 = \left(\frac{2}{3}\right)\left(-\frac{3}{2}\right) = -1$$

In general, we have the following properties:

PROPERTIES OF PARALLEL
AND PERPENDICULAR LINES

Suppose that L_1 and L_2 are two distinct nonvertical lines with slopes m_1 and m_2 respectively.
1. L_1 is parallel to L_2 if and only if $m_1 = m_2$.
2. L_1 is perpendicular to L_2 if and only if

$$m_1 = -\frac{1}{m_2} \quad \text{or} \quad m_1 \cdot m_2 = -1$$

All vertical lines are parallel to each other and are perpendicular to horizontal lines.

EXAMPLE 6   Determining if Two Lines Are Parallel or Perpendicular

Determine whether L_1 and L_2 are parallel or perpendicular. Then sketch the graphs of the lines.
(a) L_1 contains the points $(-4,1)$ and $(0,-5)$ and L_2 contains the points $(-2,5)$ and $(2,-1)$
(b) L_1: $-3x + 2y - 3 = 0$ and L_2: $2x + 3y - 10 = 0$

Solution   (a) The lines L_1 and L_2 that contain the points are graphed in Figure 9a. The slopes m_1 and m_2 of L_1 and L_2 are respectively given by

$$m_1 = \frac{\Delta y}{\Delta x} = \frac{y_2 - y_1}{x_2 - x_1} = \frac{-5 - 1}{0 - (-4)} = -\frac{6}{4} = -\frac{3}{2}$$

$$\text{and} \quad m_2 = \frac{\Delta y}{\Delta x} = \frac{-1 - 5}{2 - (-2)} = -\frac{6}{4} = -\frac{3}{2}$$

The two lines, L_1 and L_2, are parallel because their slopes are equal.
(b) We write both equations in slope-intercept form and compare their slopes.

$$\begin{array}{ll}
L_1: -3x + 2y - 3 = 0 & \qquad L_2: 2x + 3y - 10 = 0 \\
\quad 2y = 3x + 3 & \qquad \quad 3y = -2x + 10 \\
\quad y = \frac{3}{2}x + \frac{3}{2} & \qquad \quad y = -\frac{2}{3}x + \frac{10}{3}
\end{array}$$

Since $m_1 \cdot m_2 = (3/2)(-2/3) = -1$, the two lines are perpendicular (Figure 9b).

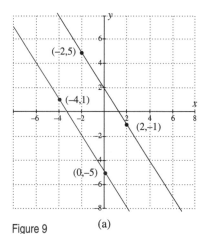

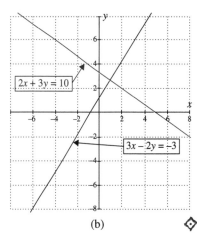

Figure 9   (a)   (b)

Solving Applied Problems

The concept of slope is used in many applications and models in which two quantities are changing at a fixed rate. For instance, numbers such as 2% and 4% are often used to refer to the **grade** of a road or the **change of elevation** on a hill or a mountain. (That is, the steepness of a road.) When we describe a road as having a 4% grade, we mean that the road changes 4 feet vertically for every 100 feet of horizontal change. Fitness centers also use the grade of a treadmill to measure its effect on the heart rate of its members. Another interpretation of slope is **the rate of change of one quantity per unit change of another**, such as fuel efficiency in miles per one gallon or wages in dollars per one hour. Also, the **pitch** of the roof of a house is an application of slope. The ratio of the rise to run determines the pitch or steepness of a roof.

EXAMPLE 7   Finding a Change of Elevation of a Ramp

A highway engineer has determined that an average grade of a ramp off an expressway is 4%. How much change of elevation will be permitted on a ramp with a 500-foot horizontal change?

Solution   Let y be the change of elevation in feet. Figure 10 shows a ramp with a convenient placement of the origin at the foot of the ramp.
Using the slope formula, we have,

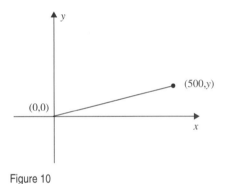

Figure 10

$$m = \frac{\Delta y}{\Delta x} = \frac{y_2 - y_1}{x_2 - x_1}$$

For 4% grade, we use $m = \frac{4}{100}$. Substituting for x and y in the slope formula, we have

$$\frac{4}{100} = \frac{y - 0}{500 - 0}$$

so that $100y = 4(500)$
or $y = 20$

Therefore, the change of elevation is 20 feet.

Using Technology Exploration

Graphers are often used to graph lines in the xy plane and help to explore the effect of the slope in linear equations. To sketch the graph of a linear equation such as $ax + by = c$ on a grapher, first we solve the equation for y in terms of x, then graph the resulting equation.

G EXAMPLE 8   Using a Grapher

(a) Use a grapher to sketch the graphs of the lines L_1 and L_2 and the graphs of L_3 and L_4 in the same "square" viewing window.
   (i) $L_1: y = -2x - 3$, $L_2: 2x + y - 4 = 0$
   (ii) $L_3: 5x - y - 4 = 0$, $L_4: x + 5y + 15 = 0$

(b) Use the slope concept to predict which two lines are parallel and which are perpendicular.

(c) Use the grapher display from part (a) to confirm your prediction in part (b).

Solution   (a) Figure 11a shows the graphs of the equations representing L_1 and L_2. Figure 11b shows the graphs of the equations representing L_3 and L_4.

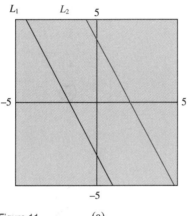

Figure 11   (a)   (b)

(b) We solve the equations of L_1 and L_2 for y to obtain

$$y = -2x - 3 \quad \text{and} \quad y = -2x + 4$$

The slopes are the same, -2, so the lines L_1 and L_2 are parallel.
By solving the equations for L_3 and L_4 for y, we obtain

$$y = 5x - 4 \quad \text{and} \quad y = -\frac{1}{5}x - 3$$

The slopes are 5 and $-\frac{1}{5}$. So $m_1 \cdot m_2 = -\frac{1}{5}(5) = -1$. Therefore the lines L_3 and L_4 are perpendicular.

(c) By examining the graphs visually, it appears that the lines L_1 and L_2 are parallel, while the lines L_3 and L_4 are perpendicular.   ◇

◆ PROBLEM SET 3.2

Mastering the Concepts

In Problems 1–4, use the coordinates of the indicated points on the graph of each line to find its slope.

1.

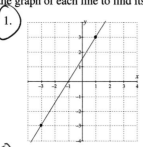

2.

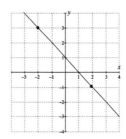

3.

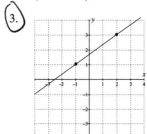

4.

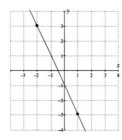

In Problems 5–22, sketch the graph of each line containing the given points, and then find the slope of each line. Indicate whether the line containing the pair of points rises, falls, is horizontal, or is vertical.

5. $(3,-2)$ and $(5,3)$ 6. $(3,4)$ and $(5,2)$
7. $(2,-1)$ and $(-3,4)$ 8. $(1,-4)$ and $(2,3)$
9. $(1,5)$ and $(-2,3)$ 10. $(0,0)$ and $(-2,5)$
11. $(0,0)$ and $(4,3)$ 12. $(-2,-3)$ and $(5,1)$
13. $(-1,4)$ and $(3,4)$ 14. $(-2,5)$ and $(3,5)$
15. $(3,2)$ and $(3,-4)$ 16. $(2,-1)$ and $(2,4)$
17. $(0,1)$ and $(2,0)$ 18. $(0,-2)$ and $(3,0)$
19. $(2.5,-3)$ and $(4.7,1)$ 20. $(3.4,1)$ and $(3.4,5)$
21. $(-2,-4)$ and $(2,4)$ 22. $(4.1,-1)$ and $(-4.1,3)$

In Problems 23–30, sketch the graph of the lines described in parts (a) and (b) on the same coordinate axis.

23. (a) $(0,2)$, $m = 2/3$ 24. (a) $(3,-2)$, $m = -2$
     (b) $(0,2)$, $m = -3/2$ (b) $(3,-2)$, $m = 1/2$
25. (a) $(-1,2)$, $m = 3$ 26. (a) $(4,1)$, $m = 4/5$
     (b) $(-1,2)$, $m = -1/3$ (b) $(4,1)$, $m = 5$
27. (a) $(-2,3)$, $m = 0$ (b) $(-2,3)$, m is undefined
28. (a) $(2,-0.5)$, $m = 0$
     (b) $(2,-0.5)$, m is undefined
29. (a) $(0,-5)$, $m = 5/8$ 30. (a) $(-5, 0)$, $m = 4/3$
     (b) $(0,-5)$, $m = -4$ (b) $(-2,6)$, $m = -3$

In Problems 31–36, express each equation in slope-intercept form, and then determine the slope and the y intercept. Also sketch the graph.

31. (a) $-3x - y - 6 = 0$ 32. (a) $2x + 3y - 12 = 0$
     (b) $x - y - 4 = 0$ (b) $x - 2y - 4 = 0$

33. (a) $x + y = 0$ 34. (a) $2x + y = 0$
     (b) $x - y = 0$ (b) $2x - y = 0$
35. (a) $y - 2 = 0$ 36. (a) $x - 3 = 0$
     (b) $x + 2 = 0$ (b) $y + 3 = 0$

In Problems 37–44, determine whether L_1 and L_2 are parallel, perpendicular or neither. Sketch a graph of the lines.

37. L_1 contains $(2,4)$ and $(3,8)$
     L_2 contains $(5,1)$ and $(4,-3)$
38. L_1 contains $(1,8)$ and $(-1,2)$
     L_2 contains $(0,-2)$ and $(2,4)$
39. L_1 contains $(2,4)$ and $(3,8)$
     L_2 contains $(8,-2)$ and $(-4,1)$
40. L_1 contains $(2,1)$ and $(6,3)$
     L_2 contains $(2,3)$ and $(3,1)$
41. $L_1: 3x - 2y = 7$ 42. $L_1: x - 2y - 2 = 0$
     $L_2: 6x - 4y = 5$ $L_2: 2x - 4y + 7 = 0$
43. $L_1: 2x - 3y = 10$ 44. $L_1: 4y - 3x + 5 = 0$
     $L_2: 3x - 2y = 6$ $L_2: 4x + 3y + 2 = 0$

Applying the Concepts

45. **Handicapped Access Ramp:** To design a handicapped access ramp, the law states that for every vertical rise of 1 foot, a 12 foot run is required (Figure 12).

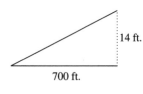

Figure 12

(a) What is the grade of the ramp?
(b) What is the horizontal distance from the foot of the ramp to a point 2.6 feet above the horizontal sidewalk in front of a building?

46. **Grade of a Ramp:** At an exit ramp off an expressway, cars ascend the ramp to an overpass. Suppose a point on the expressway is 14 feet below the ramp and 700 feet from the beginning of the exit ramp as in Figure 13.

(a) What is the grade of the ramp?
(b) If the horizontal distance from the beginning of the exit ramp to a point on the expressway directly below the overpass is 1640 feet, what is the vertical distance to the overpass above the expressway?

Figure 13

47. **Grade of a Treadmill:** A cardiologist changes the grade of a treadmill to measure the heart rate of a patient.
    (a) What is the grade of the treadmill if its vertical rise is 0.6 feet and the horizontal length of the treadmill is 8 feet?
    (b) What is the rise of a 10% grade of a 6-foot treadmill?

48. **Pitch of a Roof:** The *span of a roof* is the horizontal distance between the outside extensions of the roof, while the *rise* is the vertical distance from the top of the roof to the center of the span (Figure 14). The *pitch* is the ratio of the rise to the span of a roof. Determine

Figure 14

the rise of a house whose span is 28 feet with a pitch of
    (a) 1/4   (b) 1/2   (c) 4/7

49. **Rate of Change:** Suppose that the outside temperature rise on a certain day in January for a city in the Midwest is fairly linear from 6AM to 2PM. If the temperature at 6AM was $-10°F$, and by 2PM was up to $15°F$,
    (a) What was the hourly rate of change?
    (b) If the rate of increase of temperature is uniform during that day, estimate the temperature at noon time.

50. **Rate of Change:** On a certain day in 1973 the cost of one gallon of unleaded gasoline in a city was 27 cents. On the same date in 1997 the cost of a gallon of unleaded gasoline was $1.15 for the same city. Assume the increase in cost of unleaded gasoline is linear over these years.
    (a) What was the annual rate of change in the cost of one gallon of unleaded gasoline over this time period?
    (b) If this rate of increase would continue uniformly, predict the price of one gallon of unleaded gasoline on the same date in the year 2004.

Developing and Extending the Concepts

In Problems 51–53, if $P = (x_1, y_1)$ and $Q = (x_2, y_2)$ are two points in the xy plane, then the coordinates of the **midpoint** M of the line segment $\overline{PQ}$ are given by

$$M = \left(\frac{x_1 + x_2}{2}, \frac{y_1 + y_2}{2} \right)$$

The *midpoint* of a line segment is the point located halfway between the two endpoints of the line segment. For example, the coordinates of the midpoint of the line segment containing (3,6) and (5,8) are given by

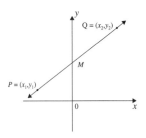

Figure 15

$$M = \left(\frac{3 + 5}{2}, \frac{6 + 8}{2} \right) = (4,7)$$

Find the coordinates of the midpoint M of the line segment $\overline{PQ}$.

51. (a) $P = (5,6)$ and $Q = (-7,8)$
    (b) $P = (-4,7)$ and $Q = (-3,0)$
52. (a) $P = (-2,3)$ and $Q = (4,-2)$
    (b) $P = (2,-5)$ and $Q = (-1,-3)$
53. $P = (8.32, 7.46)$ and $Q = (3.72, 6.38)$
54. **Map Coordinates:** If the map coordinates of two towns are (39.14, 61.78) and (49.07, 32.15), what are the coordinates of a point half way between them?
55. Many applications can be modeled by linear equations like

$$y = kx$$

which expresses a **direct variation** between the variables x and y. In this case we say that y is **directly proportional** to x, or y **varies directly as** x. The number k is called the **constant of variation.**
    (a) If y is directly proportional to x, and $y = 3$ when $x = 6$, find y when $x = 12$.
    (b) If you were to graph the linear equation in part (a), what role does the value of k play in the graph?
56. **Wages:** Linda's total weekly wage T (in dollars) varies directly as the number of hours h that she works. Suppose that her weekly pay for 40 hours of work is $340.
    (a) How much does she earn per hour?
    (b) Sketch the graph of the linear equation in part (a) and interpret the slope of the line.
57. **Relative Weight:** The weight W of an object on earth is directly proportional to the weight w of the same object on the moon. A 180-pound astronaut would weigh 30 pounds on the moon.
    (a) Write an equation that relates W to w.
    (b) Sketch the graph of the equation in part (a) and interpret the slope.

58. **Money Exchange:** The number of US dollars d is directly proportional to the number of Japanese yen y. On March 17, 1997, $100 US dollars bought 123.6 yen.
    (a) Find the constant of variation k that will convert US dollars to Japanese yen.
    (b) Sketch the graph of the equation in part (a) and interpret the slope.
    (c) At this rate of exchange, how many US dollars can be bought for 618 yen?

59. **Physics:** In physics, Hook's Law states that the distance d (in centimeters) that a spring will stretch is directly proportional to the mass m (in kilograms) of an object hanging from the spring.
    (a) Suppose that a 4 kilogram ball stretches a spring 50 centimeters. Find the constant of variation k.
    (b) Sketch the graph of the line in part (a) and interpret the slope.
    (c) Use the graph to predict how far the spring will stretch with a 6 kilogram ball.

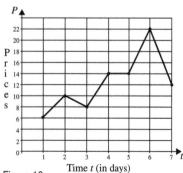

60. **Average Speed:** On a trip from Chicago, for 4 hours a motorist kept track of the distance traveled every half hour. She passed the given mile markers at the following times.

Distance in Miles	0	30	55	85	125	160	195	230	265
Time in Hours	0	0.5	1.0	1.5	2.0	2.5	3.0	3.5	4.0

    (a) Graph the data for the distance d in terms of time t where t represents the horizontal axis and d represents vertical axis.
    (b) Find and interpret the slope of each line segment in part (a).

61. **Stock Prices:** The graph in Figure 16 shows a stock's closing price for consecutive trading days.

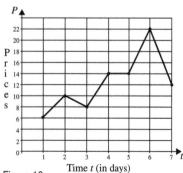

P
Prices
Time t (in days)
Figure 16

(a) What is the slope of each line segment?
(b) On what days did the price increase?
(c) On what days did the price remain constant?
(d) On what days did the price decrease?
(e) On what days did the price drop the fastest?

62. Use the concept of slope to show that the points $A = (-4,-1)$, $B = (0,2)$, $C = (2,2)$ and $D = (-2,-1)$ are the vertices of a parallelogram (a four sided figure with opposite sides parallel).

63. Use the concept of slope to determine whether the triangle with vertices A, B, and C is a right triangle.
    (a) $A = (-4,2)$, $B = (1,4)$, $C = (3,-1)$
    (b) $A = (8,5)$, $B = (1,-2)$, $C = (-3,2)$

64. Determine the value of k so that the given conditions are satisfied.
    (a) The lines $x - ky + 6 = 0$ and $2x + 4y - 5 = 0$ are parallel.
    (b) The lines $x - ky + 6 = 0$ and $2x + y = 4$ are perpendicular.

G 65. Given the equations of three lines:
    (i) $2x - 5y + 6 = 0$
    (ii) $2x = 5y + 3$
    (iii) $2x + 5y + 4 = 0$
    (a) Use the concept of slope to predict which lines are parallel.
    (b) Use a grapher to produce the graphs of these equations. (Let xMin $= -10$, xMax $= 10$, xScl $= 1$, yMin $= -8$, yMax $= 15$, yScl $= 1$.) Does the grapher display agree with your prediction?

G 66. Given the equations of three lines:
    (i) $x - 2y + 6 = 0$
    (ii) $2x + y + 3 = 0$
    (iii) $x + 3y - 5 = 0$
    (a) Use the concept of slope to predict which lines are perpendicular.
    (b) Use a grapher to produce the graphs of these equations. (Use a "square" window.) Does the grapher display agree with your prediction?

OBJECTIVES

1. Use the Slope-Intercept Form of a Line
2. Use the Point-Slope Form of a Line
3. Find Equations of Horizontal and Vertical Lines
4. Find Equations of Parallel and Perpendicular Lines
5. Solve Applied Problems
6. Use Technology Exploration

3.3 Equations of Lines

So far we have used linear equations to sketch graphs of lines. In this section, we reverse the process; we use information about the graphs of lines to find their equations. In the process, we examine different forms of linear equations so we can choose the form that is most appropriate for a given situation.

Using the Slope-Intercept Form of a Line

Example 5 of section 3.2 illustrated a procedure for graphing a line when we are given its slope and y intercept. If the slope m and the y intercept of a line are given, we can write the equation by substituting these values into the equation,

$$y = mx + b$$

which we called the *slope-intercept equation* of a line.

EXAMPLE 1   Finding an Equation of a Line in Slope-Intercept Form

Find an equation of each line with the given information.
(a) The line has a slope 4 and y intercept 3.
(b) The line has a slope 4 and contains the point $(1,7)$.

Solution   (a) Substituting $m = 4$ and $b = 3$ into the equation

$$y = mx + b$$

yields the equation $y = 4x + 3$

(b) To find the y intercept b, we substitute $m = 4$ and the coordinates of the point $(1,7)$ into the equation

$$\begin{aligned} y &= mx + b \\ \text{so that} \quad 7 &= 4(1) + b \\ \text{or} \quad b &= 3 \end{aligned}$$

Now we substitute 4 for m and 3 for b into the equation

$y = mx + b$ to obtain $y = 4x + 3$. ◈

Using the Point-Slope Form of a Line

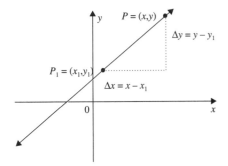

Figure 1

Suppose we are given a nonvertical line with slope m and a fixed point $P_1 = (x_1, y_1)$ (Figure 1). If $P = (x,y)$ is any other point on the line, then the slope m of the line is given by

$$m = \frac{y - y_1}{x - x_1}$$

Multiplying each side of the equation by $x - x_1$, we get the equation

$$m(x - x_1) = y - y_1$$
$$\text{or} \quad y - y_1 = m(x - x_1)$$

This form of a linear equation is called the *point-slope equation*.

POINT-SLOPE EQUATION

The **point-slope equation** of a line that contains the point (x_1, y_1) and has a slope of m, where m is defined, is given by

$$y - y_1 = m(x - x_1)$$

Thus, if we know the slope m of a line and the coordinates of one point on that line, we can find an equation.

EXAMPLE 2   Finding an Equation of a Line Given a Point and the Slope

Write the general equation of a line, $Ax + By = C$, containing the point $(-1, 3)$ with slope 3/2.

Solution   Using the point-slope equation, we have

$$y - y_1 = m(x - x_1) \qquad \text{Point-slope equation}$$

$$y - 3 = \frac{3}{2}(x - (-1)) \qquad \text{Substitute } -1 \text{ for } x_1, \text{ and } 3 \text{ for } y_1, \text{ and } m = 3/2$$

$$y - 3 = \frac{3}{2}(x + 1) \qquad \text{Simplify}$$

$$2(y - 3) = 3(x + 1) \qquad \text{Multiply each side by 2 the LCD}$$
$$2y - 6 = 3x + 3$$
$$-3x + 2y = 6 + 3$$
$$-3x + 2y = 9$$
$$\text{or } 3x - 2y = -9 \qquad \text{General equation} \qquad \diamondsuit$$

EXAMPLE 3   Finding the Equation of a Line Given Two Points

Find a slope-intercept equation of the line that contains the points $(-1, 5)$ and $(2, 20)$.

Solution   We first calculate the slope using $x_1 = -1$, $y_1 = 5$, $x_2 = 2$ and $y_2 = 20$.

$$m = \frac{\Delta y}{\Delta x} = \frac{y_2 - y_1}{x_2 - x_1} = \frac{20 - 5}{2 - (-1)} = \frac{15}{3} = 5$$

Substituting $m = 5$, $x_1 = -1$ and $y_1 = 5$ into $y - y_1 = m(x - x_1)$ yields

$$y - 5 = 5(x - (-1))$$
$$y - 5 = 5x + 5$$
$$y = 5x + 10 \qquad \diamondsuit$$

Finding Equations of Horizontal and Vertical Lines

In section 3.2 we learned that
1. A **horizontal line** has slope $m = 0$ with y intercept k and an equation $y = k$ for a constant k, $k \neq 0$.
2. A **vertical line** has undefined slope with x intercept h and an equation $x = h$ for a constant h, $h \neq 0$.

EXAMPLE 4   Finding an Equation of a Horizontal and Vertical Line

Find an equation of a line that contains $(3, 4)$ if the line is
(a) horizontal and (b) vertical (Figure 2).

Solution   (a) The equation of the horizontal line containing the point $(3, 4)$ is $y = 4$.
(b) The equation of the vertical line containing the point $(3, 4)$ is $x = 3$. $\diamondsuit$

Finding Equations of Parallel and Perpendicular Lines

Recall from section 3.2 that parallel lines have the same slopes while perpendicular lines have slopes whose product is -1.

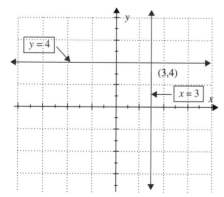

Figure 2

EXAMPLE 5   Using the Slopes of Parallel and Perpendicular Lines

Find a slope-intercept equation of the line containing the point (2,7) that is (a) parallel to the line $-5x + 4y = 8$, and (b) that is perpendicular to the line $-5x + 4y = 8$. Sketch the graph in each case.

Solution   First we express the equation $-5x + 4y = 8$ in slope-intercept form as follows:

$$4y = 5x + 8 \quad \text{or} \quad y = \frac{5}{4}x + 2$$

so the slope is $\frac{5}{4}$.

(a) The slope of a line parallel to the given line is also $\frac{5}{4}$. By letting $(x_1, y_1) = (2,7)$ and $m = \frac{5}{4}$, we have

$$y - y_1 = m(x - x_1)$$
$$y - 7 = \frac{5}{4}(x - 2)$$
$$y - 7 = \frac{5}{4}x - \frac{10}{4}$$
$$y = \frac{5}{4}x - \frac{5}{2} + 7$$
$$y = \frac{5}{4}x + \frac{9}{2} \quad \text{(Figure 3a)}$$

(b) The slope of a line perpendicular to the given line is the negative reciprocal, $-\frac{4}{5}$. Using the point-slope form with

$$(x_1, y_1) = (2,7) \text{ and } m = -\frac{4}{5},$$

we have

$$y - y_1 = m(x - x_1)$$
$$y - 7 = -\frac{4}{5}(x - 2)$$
$$y - 7 = -\frac{4}{5}x + \frac{8}{5}$$
$$y = -\frac{4}{5}x + \frac{8}{5} + 7$$
$$y = -\frac{4}{5}x + \frac{43}{5} \quad \text{(Figure 3b)}$$

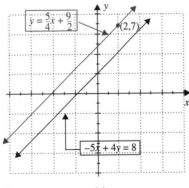

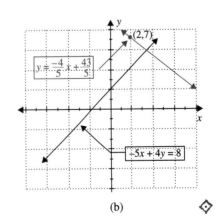

Figure 3   (a)   (b)

Solving Applied Problems

Many applied problems can be modeled by linear equations. Their graphs may be helpful in obtaining information about the relationship stated in the problems.

EXAMPLE 6   Solving a Straight Line Depreciation Problem

The graph in Figure 4 models the book value R of a van purchased by a small business firm.

(a) Express the book value R of the van as a linear equation in terms of a number of years t after the van was purchased.
(b) Find and interpret the R-intercept of the line.
(c) Find and interpret the slope of the line.
(d) Does the model make sense if the van is sold after 8 years? Explain.

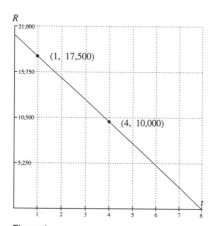

Figure 4

Solution   Using Figure 4, the point $(1, 17{,}500)$ represents the book value of the van after 1 year from its purchase, while the point $(4, 10{,}000)$ represents its book value after 4 years.

(a) To find the slope of the line, we substitute
$t_1 = 1$, $R_1 = 17{,}500$, $t_2 = 4$, $R_2 = 10{,}000$ into the slope formula to obtain

$$m = \frac{\Delta R}{\Delta t} = \frac{R_2 - R_1}{t_2 - t_1}$$

$$= \frac{17500 - 10000}{1 - 4}$$

$$= \frac{7500}{-3} = -2500$$

Substituting $m = -2500$ and $(t_1, R_1) = (1, 17500)$ into the point-slope form, we have

$$R - R_1 = m(t - t_1)$$
$$R - 17500 = -2500(t - 1)$$
$$R - 17500 = -2500t + 2500$$
$$R = -2500t + 20000$$

(b) The R intercept, \$20,000, represents the original purchase price of the van.
(c) The slope of the line, -2500, represents the rate of decrease in book value. This means that for each elapsed year, there is a reduction of \$2500 in the book value of the van. This is called **straight line depreciation**.
(d) When $R = 0$, then

$$0 = -2500t + 20{,}000$$
$$2500t = 20{,}000$$
$$t = 8$$

This means the van will be fully depreciated (its book value is zero) in 8 years. While its value might reach zero for accounting purposes, the van probably retains a **salvage value** for old vans.   ◈

EXAMPLE 7   Finding Temperature Conversions

The freezing temperature of water is 0° on the Celsius scale and 32° on the Fahrenheit scale. The boiling temperature of water is 100° Celsius and 212° Fahrenheit. Assume that the relationship between the temperature scales is linear.

(a) Determine an equation relating the temperature C in the Celsius scale to the temperature F on the Fahrenheit scale.
(b) Graph the equation of the line using F as the horizontal axis and C as the vertical axis.
(c) Find and interpret the intercepts of the line.
(d) Find and interpret the slope of the line.

Solution   (a) Using F for Fahrenheit and C for Celsius, the data points are

$$(F_1, C_1) = (32, 0) \text{ and } (F_2, C_2) = (212, 100).$$

Using the slope formula

$$m = \frac{C_2 - C_1}{F_2 - F_1}$$
$$= \frac{100 - 0}{212 - 32}$$
$$= \frac{5}{9}$$

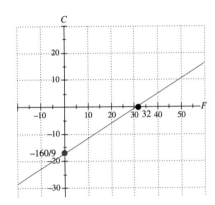

The equation of the line is

$$C - 0 = \frac{5}{9}(F - 32)$$

or

$$C = \frac{5}{9}(F - 32)$$

Figure 5

(b) Figure 5 shows the graph of the line.
(c) The F intercept is 32 and the C intercept is $-160/9$. This means
$$\text{when } F = 0°, C = (-160/9)°$$
and
$$\text{when } C = 0°, F = 32°.$$
(d) The slope of the line is $\frac{5}{9}$. This means that for a 5° change in Celsius, there is a 9° change in Fahrenheit measure.   ◈

Using Technology Exploration

We can use technology to "fit" a linear equation as closely as possible to a set of data. In this process, estimates are made of the values of one variable based on the values of another variable. The independent variable represents the basis for the prediction and is plotted on the horizontal or x axis. The dependent variable represents the prediction to be made and is plotted on the vertical or y axis. A valuable tool for recognizing and analyzing relationships is called the **scatter plot** or **scatter diagram**. The grapher gives us a powerful capability to plot all the data points as a scatter diagram and to find such linear models (called **linear regression models**) that "best fit" the data points.

G EXAMPLE 8 Establishing a Grading System

Table 1 shows a collection of data of midterm grades and final grades in a mathematics course,

TABLE 1

x	68	92	72	90	52	78	40	88
y	129	165	135	162	105	144	87	159

where x is the midterm grade based on 100 points and y is the final grade based on 200 points.

(a) Construct a scatter plot that exhibits the data points and indicates whether or not the relationship between x and y is linear.

(b) Assuming the relationship between x and y is approximately linear, the linear model $y = 1.5x + 27$ fits the data. Use a grapher to confirm the results.

(c) Find and interpret the slope and the y intercept of the line.

(d) Use this model to predict the final grade for a student whose midterm grade is 80.

Solution

(a) Figure 6 shows a plot of the data points in Table 1. The points seem to fall on a line. To confirm the relationship is linear, we show the rate of change is constant. For $x = 68$ to $x = 92$, we have

$$\frac{\Delta y}{\Delta x} = \frac{165 - 129}{92 - 68}$$

$$= \frac{36}{24}$$

$$= \frac{3}{2}$$

For $x = 72$ to $x = 90$, we have

$$\frac{\Delta y}{\Delta x} = \frac{135 - 162}{72 - 90}$$

$$= \frac{-27}{-18}$$

$$= \frac{3}{2}$$

Thus, we see that the rate of change between these points is 3/2.

(b) By direct substitution, the linear model $y = 1.5x + 27$ fits the data that relates the final grade y to the midterm grade x. The grapher also shows the line that fits the data points (Figure 6). This feature on a grapher is called a REGRESSION feature.

(c) The slope of the line is 3/2 as was shown in part (a). The y intercept is 27 from part (b). That is, this linear model predicts a score of 27 on the final grade for a score of 0 on a midterm.

(d) This model predicts that a midterm grade of 80 would result in a final grade of $1.5(80) + 27 = 147$. ◇

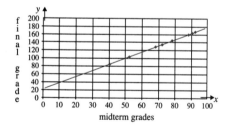

Figure 6

PROBLEM SET 3.3

Mastering the Concepts

In Problems 1–8, write an equation of the line in the slope-intercept form for the given graph.

1.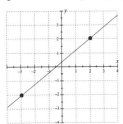
2.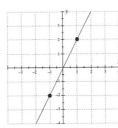
3.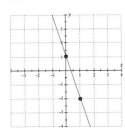
4.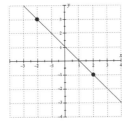
5.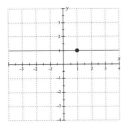
6.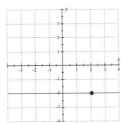
7.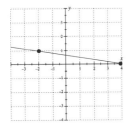
8.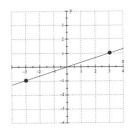

In Problems 9–16, write an equation of the line in slope-intercept form,

$$y = mx + b,$$

satisfying the following conditions:

9. $m = 2, b = -3$
10. $m = 3, b = 1$
11. $m = -2/3, b = 2$
12. $m = -1/2, b = 3$
13. $m = 3$ and contains $(0,2)$
14. $m = -4$ and contains $(0,-2)$
15. $m = -5/8$ and contains $(0,5)$
16. $m = -3/4$ and contains $(0,-5)$

In Problems 17–32, write an equation of the line satisfying the given conditions in (a) point-slope form, $y - y_1 = m(x - x_1)$, and (b) slope-intercept form, $y = mx + b$.

17. $m = 5$ and contains $(-1,2)$
18. $m = -3$ and contains $(7,3)$
19. $m = -3/7$ and contains $(5,-1)$
20. $m = -3/4$ and contains $(-4,6)$
21. $m = 3/8$ and contains $(0,0)$
22. $m = 4/5$ and contains $(-1,-2)$
23. $m = 0$ and contains $(3,-4)$
24. $m = 0$ and contains $(-4,2)$
25. $m = 5/3$ and contains $(-1,-1)$
26. $m = -5/11$ and contains $(1,3)$
27. contains $(2,3)$ and $(1,7)$
28. contains $(-2,3)$ and $(-4,2)$
29. contains $(4,-2)$ and $(4,-5)$
30. contains $(3,-5)$ and $(3,4)$
31. contains $(-5,-8)$ and $(-1,-9)$
32. contains $(-1,-4)$ and $(-3,5)$

In Problems 33–46, write an equation of the line satisfying the given conditions in general form, $Ax + By = C$.

33. $m = 2$ and contains $(-3,1)$
34. $m = -3/4$ and contains $(5,-2)$
35. $m = -2/3$ and contains $(4,0)$
36. $m = 1$ and contains $(4,7)$
37. $m = 4/5$ and contains $(3,-4)$
38. $m = 5/3$ and contains $(3,0)$
39. $m = -4/3$ and contains $(-8,2)$
40. $m = -5/6$ and contains $(-2,3)$
41. contains $(4,5)$ and $(5,-3)$
42. contains $(-1,2/3)$ and $(1/2,-1)$
43. contains $(-4,0.3)$ and $(3,-3.4)$
44. contains $(5,-1/2)$ and $(3/2,-3/4)$
45. m is undefined and contains $(-1,8)$
46. m is undefined and contains $(5,-4)$

In Problems 47–52, write an equation of the line: (a) in slope-intercept form, $y = mx + b$ and (b) in the general form

$$Ax + By = C,$$

satisfying the given conditions.

47. contains $(3,-1)$; parallel to $x + 2y + 7 = 0$
48. contains $(-2,1)$; parallel to $5x - 7y - 8 = 0$
49. contains $(1,5)$; perpendicular to $2x + 3y - 1 = 0$
50. contains $(-3,2)$; perpendicular to $x + 2y - 5 = 0$
51. has y intercept 3; parallel to $x - 2y = 5$
52. has y intercept 5; perpendicular to $2x - 3y = 6$

Applying the Concepts

53. **Telephone Charges:** A telephone company charges $1.80 for a 12 minute long distance domestic call, and $2.80 for a 20 minute call to the same number at the same time of day. Let the relationship between the cost C (in dollars) and the time t (in minutes) for a call be linear.
    (a) Write the information as two ordered pairs in the form (t,C).
    (b) Find an equation of the line that expresses C in terms of t and contains the data points.
    (c) Sketch the graph of the line using t as the horizontal axis and C as the vertical axis.

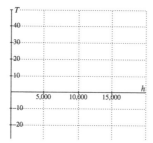

    (d) Use the graph to predict the cost of a 40 minute call.
    (e) Use the graph to predict the time in minutes for a call that costs $6.15.

54. **Temperature Changes:** An airline chart shows that the temperature T in degrees Fahrenheit at an altitude of 5,000 feet is 42°. At an altitude of 15,000 feet, the temperature is 5°. Assume the relationship between the air temperature T (in degrees Fahrenheit) and the altitude h (in feet above sea level) is linear for $0 \le h \le 20{,}000$.
    (a) Write the information as two ordered pairs in the form (h,T).
    (b) Find an equation of the line that expresses T in terms of h and contains the data points.
    (c) Sketch the graph of the line using h as the horizontal axis and T as the vertical axis.

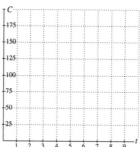

    (d) Use the graph to predict the air temperature at an altitude of 20,000 feet.
    (e) Use the graph to predict the altitude at which the air temperature is 0°.

55. **Cable Television Charges:** A cable television company charges a $30 installation fee and $35 per month for cable service.
    (a) Write a linear equation that determines the total cost C (in dollars) for t months of cable television service.
    (b) Sketch the graph of the equation in part (a) using t as the horizontal axis and C as the vertical axis.

(c) Use the graph to approximate the total cost for six months of service.
(d) Use the graph to approximate the number of months of cable television service if the total cost is $170.

56. **Pollution:** In 1999 tests show that water in a lake was polluted with 7 milligrams of mercury compounds per 1000 liters of water. To clean up the lake, it was determined that the pollution level must drop at the rate of 0.8 milligram of mercury compound per 1000 liters of water per year.
    (a) Write a linear equation that expresses the pollution level L (in milligrams per 1000 liters) in terms of t years.
    (b) Sketch the graph of the equation that represents this model if $t = 0$ corresponds to 1999 and $0 \le t \le 9$. Use t as the horizontal axis and L as the vertical axis.

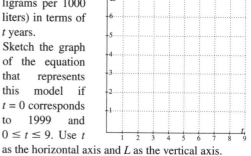

    (c) Use the graph in part (b) to predict when the lake will be free of mercury pollution.

57. **Straight-line Depreciation:** A business firm is depreciating the value of a photocopier originally purchased for $7500 over a period of 5 years. Suppose that the *trade-in* or *salvage value* of the photocopier at the end of 5 years is $1100.
    (a) Write a linear equation that expresses the value V (in dollars) of the photocopier in terms of time t (in years).
    (b) Sketch the graph of the equation that represents this model for $0 \le t \le 5$, using t as the horizontal axis and V as the vertical axis.

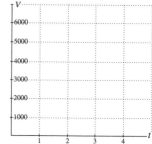

    (c) Use the graph of this model to determine the value of the copier after 2 years.

58. **Banking:** A bank charges its customers $4.00 per month as a maintenance fee for the first 12 checks and $0.30 for each additional check written on its checking account.
    (a) Write a linear equation that expresses the total monthly cost C (in dollars) in terms of the number of checks n written for more than 12 checks.
    (b) Sketch the graph of the line using the cost as the vertical axis and the number of checks as the horizontal axis.

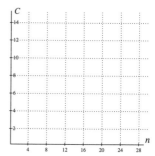

(c) Write a linear equation that expresses the total monthly cost C (in dollars) in terms of the number of checks n written for less than 12 checks.

(d) Sketch the graph of the line (c) using the cost as the vertical axis and the number of checks as the horizontal axis on the same axis as (b).

(e) What is the C intercept of the graph in part (d)? Interpret its meaning.

(f) Use the graph to predict the number of checks written for a month if the total monthly charges are $10.

59. **Aerobic Exercise:** Physiologists have determined that the maximum pulse rate for a person doing aerobic exercise is 220 minus the person's age. The exercise is considered effective when the person's pulse rate is between 60% and 80% of their maximum pulse rate. Assume that the relationship between the pulse rate P and the age t (in years) is linear.

(a) Write two linear equations which describe (i) the 60% pulse rate and (ii) the 80% pulse rate in terms of the age t.

(b) Use the models in part (a) to determine the interval of an effective pulse rate for a 40 year old person.

(c) Use the models in part (a) to determine a person's age whose pulse rate is between 102 and 136.

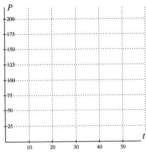

60. **Rental Occupancy:** A real estate company manages an apartment complex with 160 units. When the rent for each unit is $500 per month, all the apartments are occupied. However, when the rent for each unit is increased to $525 per month, 15 units become vacant. Assume that the relationship between the monthly rent R (in dollars) and the number n of occupied units is linear.

(a) Write an equation that expresses the monthly rent R in terms of the number n of occupied units for $50 \le n \le 160$.

(b) Use the model in part (a) to determine the monthly rent per unit if 120 units are occupied.

(c) Use the model in part (a) to determine how many units will be occupied if the monthly rent per unit is $600.

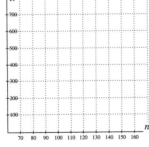

(d) Does the model make sense if the monthly rent per unit is $475? Why?

Developing and Extending the Concepts

61. (a) Suppose that a line L has nonzero x and y intercepts of a and b respectively. Show that an equation of L can be written in the **intercept form**

$$\frac{x}{a} + \frac{y}{b} = 1$$

(b) Use the result of part (a) to find an equation of the line in intercept form where (i) x intercept is 5 and y intercept is 6 (ii) x intercept is -2 and y intercept is 7.

62. Find a value of k so that each of the following conditions hold:

(a) The line $3x + ky + 2 = 0$ is parallel to the line $6x - 5y + 3 = 0$.

(b) The line $y = (2 - k)x + 2$ is perpendicular to the line $3x - y - 1 = 0$.

63. Let $P_1 = (-1, -3)$ and $P_2 = (5, 7)$ be two points.

(a) Find the midpoint M of the line segment $\overline{P_1 P_2}$.

(b) Find an equation in point slope form of the perpendicular bisector of the segment $\overline{P_1 P_2}$.

G 64. Table 2 shows the fine d (in dollars) imposed on a motorist by a certain state whose speed is v (in miles per hour), where 65 miles per hour is the speed limit.

TABLE 2

Speed v (mph)	70	75	80	85	90	95
Fine d (in dollars)	100	125	150	175	200	225

(a) Plot the data points (v horizontal axis and d vertical axis), and indicate whether or not the relationship between v and d is linear.

(b) Fit a regression line to the data points using the REGRESSION feature on a grapher, and graph the regression line.

(c) Find and interpret the slope and the d intercept of the line.

(d) Use this model to predict the fine imposed on a motorist whose speed is 100 miles per hour.

G 65. Table 3 shows the total cost C (in dollars) from selling n bagels by a fraternity at a college in one day.

TABLE 3

n bagels	0	5	10	50	100	200
C dollars	$50	$51.25	$52.50	$62.50	$75	$100

(a) Plot the data points from Table 3 and indicate whether or not the relationship between n and C is linear.

(b) Fit a regression line to the data points using the REGRESSION feature on a grapher, and graph the regression line.

(c) Find and interpret the slope of the line.

(d) Explain why it costs $50 if no bagels are sold.

G 66. For the following pair of points, find the slope-intercept form of the line. Use the linear **REGRESSION** feature on a grapher to confirm the equation of this line.

    (a) (65,25) and (-235,112)   (b) (210,100) and (450,748)

G 67. For the following ordered pairs, use the linear **REGRESSION** feature on a grapher to plot the scatter diagram. Do these points seem to lie on the same straight line? If so, find the slope-intercept form of the line. Use the linear **REGRESSION** feature on a grapher to confirm the equation of this line.

    (a) (9.8,2.9), (−4.6,1.1), and (1.4,1.85)

    (b) (4.5,3.7), (2.8,−4.4), and (−6.2,−3.1)

OBJECTIVES

1. Graph Linear Inequalities in Two Variables
2. Graph Compound Inequalities
3. Solve Applied Problems
4. Use Technology Exploration

3.4 Graphing Linear Inequalities

Graphing Linear Inequalities in Two Variables

Recall from section 3.1 that the graph of a linear equation in two variables is a straight line. Here we graph linear inequalities in two variables that can be written in the *standard form* $Ax + By < C$ where A, B and C are real numbers and A and B are not both zero. The symbol $<$ may be replaced by $\geq$, $\leq$, or $>$. Examples of linear inequalities are:

$$3x - y < 6, \quad 5x - y > 7, \quad 0x + y \leq 1, \quad \text{and} \quad x + 0y \geq 2$$

A **solution** of a linear inequality in two variables is an ordered pair of numbers (x,y) that makes the inequality true. All the ordered pairs that make the inequality true is called the **solution set**. The **graph** of an inequality in two variables represents its solution set. To graph a linear inequality in x and y, we first graph the **boundary line** of the related equation. This line divides the xy plane into two regions called **half planes**, one above the line and one below it (assuming this line is not vertical). To determine which half-plane satisfies the inequality, we test a point in either region.

For example, to solve the inequality $y \leq 3x + 6$, first we graph the boundary line $y = 3x + 6$ (Figure 1). This line divides the xy plane into an upper half-plane A and a lower half-plane C. The set of all points on the line is represented by B. The point $(0,0)$ is in the half-plane C. It makes the inequality true:

$$0 \leq 3(0) + 6.$$

The solution of the inequality $y \leq 3x + 6$ consists of all points in the half-plane C and all points on the boundary line B.

The following step procedure may be used to graph linear inequalities in two variables:

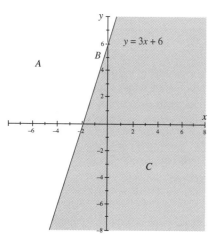

Figure 1

Graphing Linear Inequalities in Two Variables

Step 1. Sketch the graph of the boundary line obtained by replacing the inequality sign with an equal sign and graph this related equation. If the inequality has the symbol $<$ or $>$, draw a dashed line. A dashed line indicates that the boundary line is not part of the solution set. If the inequality has the symbols $\leq$ or $\geq$, draw a solid line to indicate that points on the line are solutions of the inequality.

Step 2. Determine which half-plane corresponds to the inequality. To do this, select any convenient **test point** (x,y) not on the boundary line. Substitute the coordinates of this test point into the original inequality.

  (i) If the point makes the original inequality true, shade the half-plane containing the test point.

  (ii) If the point makes the original inequality false, shade the other half-plane.

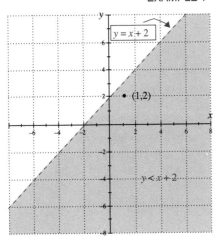

Figure 2

EXAMPLE 1   Graphing a Linear Inequality

Sketch the graph of the inequality $y < x + 2$

Solution   **Step 1.** The boundary line for this inequality is the graph of the related equation $y = x + 2$. We draw the graph of a dashed boundary line because the inequality contains the symbol $<$ (which is the strict inequality, less than but not equal to).

Step 2. To determine which half-plane is the solution, we select the test point $(1,2)$ not on the boundary line. Replacing x by 1 and y by 2 in the original inequality leads to the following statement

$$y < x + 2$$
$$2 < 1 + 2$$
$$2 < 3 \text{ which is true.}$$

Because $(1,2)$ satisfies the inequality, so does every point on the same side of the boundary line as $(1,2)$. Therefore, we shade the lower half-plane that contains the point $(1,2)$ (Figure 2). ◈

EXAMPLE 2   Graphing a Linear Inequality

Sketch the graph of the inequality $2x - y \leq 1$

Solution   **Step 1.** Because the inequality contains the symbol $\leq$ (meaning less than or equal to), we graph the related equation as a solid boundary line

$$2x - y = 1$$

Step 2. We test the inequality at the point $(1,-3)$ not on the boundary to determine which half-plane satisfies the inequality

$$2x - y \leq 1$$

When $x = 1$ and $y = -3$, we have

$$2(1) - (-3) \leq 1$$
$$5 \leq 1 \text{ False}$$

This point does not satisfy the inequality, so the correct half-plane to shade is the upper half-plane that does not contain the point $(1,-3)$ (Figure 3). ◈

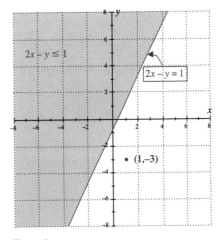

Figure 3

EXAMPLE 3   Graphing Special Linear Inequalities

Sketch the graph of each inequality

(a) $y \geq 3$ (b) $x < -1$

Solution (a) First we draw the boundary line as a solid horizontal line $y = 3$ since the symbol $\geq$ is used. Next we choose $(-2,1)$ as a test point and find that $1 \geq 3$ is false. Therefore, we shade the upper half-plane that does not contain the point $(-2,1)$ (Figure 4a).

(b) We begin by drawing the boundary line as the dashed line $x = -1$, since equality is not included. Then we choose $(0,3)$ as a test point and find that $0 < -1$ is false. Therefore, we shade the half-plane (to the left of the vertical boundary line) which does not contain the point $(0,3)$ (Figure 4b).

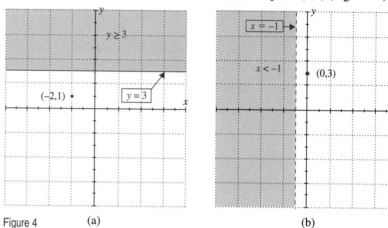

Figure 4   (a)   (b)

Graphing Compound Inequalities

In section 2.7 we solved compound inequalities containing one variable. Linear inequalities containing two variables can also be connected with the words "and" or "or" to form compound inequalities.

EXAMPLE 4   Graphing a Compound Inequality

Sketch the graph of the compound inequality

$$x \geq 2 \quad \text{and} \quad y \geq 4x + 3$$

Solution Figure 5a shows the graph of $x \geq 2$ to be the region to the right of and including the solid line $x = 2$. Figure 5b shows the region above and including the solid boundary line $y \geq 4x + 3$. Because the inequalities are connected with the word "and", the graph of the compound inequality is the area common to both or the intersection of the two graphs (Figure 5c).

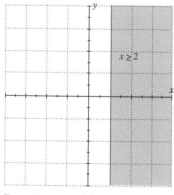 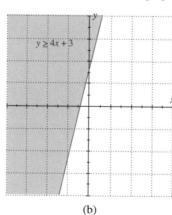 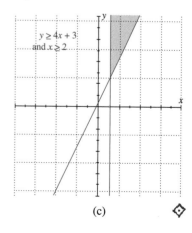

Figure 5   (a)   (b)   (c)

EXAMPLE 5   Graphing a Compound Inequality

Sketch the graph of the compound inequality

$$y \geq x - 2 \text{ or } y \leq 1$$

Solution   Figure 6a shows the graph of $y \geq x - 2$ as the region above and including the solid boundary line $y = x - 2$. Figure 6b is the graph of $y \leq 1$ as the region below and including the boundary line $y = 1$. This compound inequality uses the connector "or" so that the graph is all the points in one graph or the other as shown in Figure 6c.

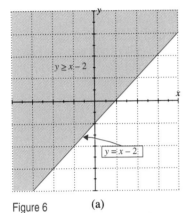 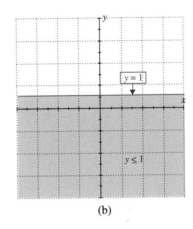 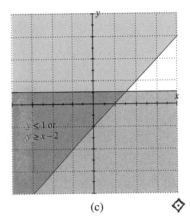

Figure 6   (a)   (b)   (c)

EXAMPLE 6   Identifying a Region by Using Linear Inequalities

Identify the region that satisfies all the conditions.

$$\begin{cases} x + 2y \leq 8 \\ x \geq 0 \\ y \geq 0 \end{cases}$$

Solution   The graph of the inequality $x + 2y \leq 8$ consists of all points in the plane on or below the solid line $x + 2y = 8$. The graph of $x \geq 0$ consists of all points on or to the right of the y axis, while the graph of $y \geq 0$ includes all points in the plane on or above the x axis. Therefore, the graph of these inequalities is the shaded region in the first quadrant including that part of the coordinate axes (Figure 7).

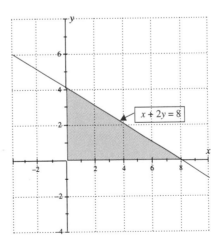

Figure 7

Solving Applied Problems

Linear inequalities in two variables are used in a variety of applications and models as the next example shows.

EXAMPLE 7   Solving a Nutrition Problem

A nutritionist determines that an apple contains 100 calories and a pear contains 120 calories. Suppose that a person eats an apple and a pear for a mid-morning snack several times a week. The person wishes to keep the total number of calories per week for the snack to be at most 1,180. If x and y represent the number of apples and pears consumed per week respectively, write a linear inequality to model this situation and graph it.

Solution   Since x and y represent the number of apples and pears consumed per week, the total number of calories consumed per week for the snack is modeled by the following linear inequality:

$$100x + 120y \leq 1180$$

with the restriction that $x \geq 0$ and $y \geq 0$. Figure 8 shows the shaded region which describes the solution set of this inequality.

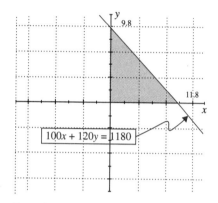

Figure 8

Using Technology Exploration

Linear inequalities can be graphed using a grapher with a **SHADE** feature.

G EXAMPLE 8   Using a Grapher to Graph a Linear Inequality

Use a grapher with a **SHADE** feature to graph the inequality $3x + 4y \leq 12$.

Solution   To use a grapher the inequality must first be solved for y in terms of x. That is,

$$3x + 4y \leq 12$$
$$4y \leq -3x + 12$$
$$y \leq -\frac{3}{4}x + 3$$

and graph the above inequality (Figure 9).

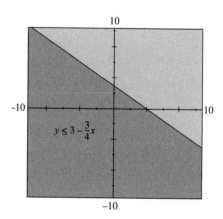

Figure 9

◈ PROBLEM SET 3.4

Mastering the Concepts

In Problems 1 and 2, match the shaded region with the appropriate inequality.

(i) $y > 2x - 3$ (ii) $x + y > 4$ (iii) $10y > -x + 5$
(iv) $y > x$ (v) $y < 3$ (vi) $x \geq 2$

1. (a)

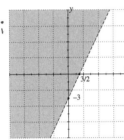

2. (a)

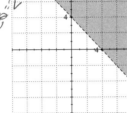

1. (b)

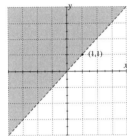

2. (b)

1. (c)

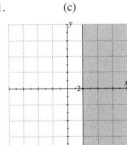

2. (c)

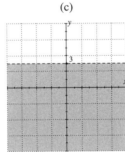

In Problems 3 and 4 write an inequality whose solution set is the given graph.

3. (a)

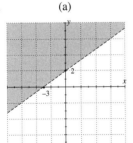

4. (a)

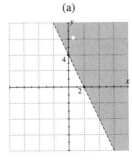

3. (b)

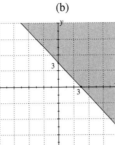

4. (b)

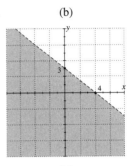

3. (c)

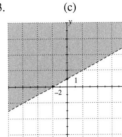

4. (c)

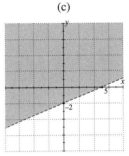

In Problems 5 and 6, the boundary line for the graph of the given inequality is shown. Shade the correct half-plane that indicates the solution of the inequality.

5. (a) $y \leq 2x - 3$ (b) $y \geq -3x$

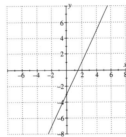
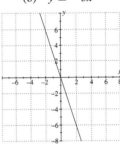

6. (a) $-2x + y < 6$ (b) $y \geq 2$

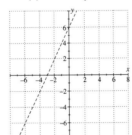
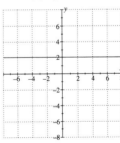

In Problems 7–20, sketch the graph of the given inequality.

7. $y \leq 2x + 5$ 8. $y \geq 3x + 4$
9. $y > 3x$ 10. $y \leq -4x$
11. $y > -2x + 3$ 12. $2y > 3x + 7$
13. $2x + y \leq 3$ 14. $5y \geq 7$

15. $3y \le -4$

16. $y \le -1$

17. $x \ge 2$

18. $2x - 6 < 0$

19. $x \le 2$

20. $x < -2$

In Problems 21 and 22, write a compound inequality whose solution is described in the given graph.

21.   (a)   22.   (a)

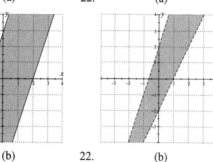

21.   (b)   22.   (b)

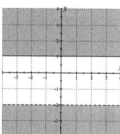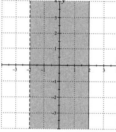

21.   (c)   22.   (c)

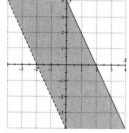

In Problems 23–32, sketch the graph of each compound inequality.

23. $y \ge -2$ and $x \le 1$

24. $y \le 2$ or $x \ge -1$

25. $y \le -3$ or $x \ge 2$

26. $y \ge -2$ and $x \le 1$

27. $2x + y < 3$ and $x + y < 1$

28. $-3x + y > 2$ and $x - 2y \le 4$

29. $2x + y < 4$ or $x - y < 2$

30. $y \ge 3$ or $y < -2x + 1$

31. $\begin{cases} 2x + y \le 10 \\ \quad x \ge 0 \\ \quad y \ge 0 \end{cases}$

32. $\begin{cases} 2x + 3y \le 18 \\ \quad x \ge 0 \\ \quad y \ge 0 \end{cases}$

Applying the Concepts

33. **Inventory Levels:** An electronic store sells two brands of VCRs, A and B. Customer demand indicates that it is necessary to stock at least twice as many units of brand A as of brand B. Due to limitation of space, there is room for no more than 100 units in the store. Write a linear inequality that describes all possibilities for stocking the two brands and produce its graph.

34. **Investments:** A retirement fund invests in two mutual funds, X and Y. The X fund pays 5% annual simple interest and the Y fund pays 6% annual simple interest. The total interest from both investments for one year must be at least $10,600. Write a linear inequality that describes this situation and produce its graph.

35. **Ticket Prices:** For an upcoming event in an auditorium, tickets are priced at $8 and $5. The total sales for that event must be at least $3000. Write a linear inequality that describes this situation and produce its graph.

36. **Fuel Economy:** A certain make of car gets 16 miles per gallon in city driving and 24 miles per gallon in highway driving. The car is driven at most 380 miles on a full tank of gasoline. Write a linear inequality that describes this situation and produce its graph.

Developing and Extending the Concepts

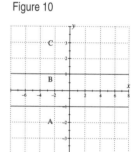

Figure 10

37. Figure 10 shows the graph of the equation $|x + y| = 2$. Use the graph to describe the region which indicates the solution of each inequality.

    (a) $|x + y| \le 2$

    (b) $|x + y| \ge 2$

38. Figure 11 shows the graph of the equation $|y| = 1$. Use the graph to describe the region which indicates the solution of each inequality.

    (a) $|y| \ge 1$

    (b) $|y| \le 1$

Figure 11

In Problems 39 and 40, sketch the graph of each absolute value inequality.

39. $|x - y| < 0$

40. $|y - 3x| \ge 0$

G In Problems 41–44, use a grapher with a SHADE feature to graph each inequality.

41. $y \le 6 - 3x$

42. $y \ge \dfrac{2}{3}x - 3$

43. $2x + y > 3$

44. $x - 2 < -y$

In Problems 45 and 46, write inequalities that describe the indicated quadrant.

45. (a) Quadrant II   (b) Quadrants I and IV
46. (a) Quadrant IV   (b) Quadrants III and IV

In Problems 47 and 48, sketch the graph of each inequality that fits the description.

47. A fast food restaurant has 20 Level I and 15 Level II employees. For a 7 hour day, each Level I employee is paid $50 and each Level II employee is paid $40. The daily payroll of the restaurant must never exceed $1,400.

48. A pen company manufactures two types of pens, A and B, which it sells for $2.75 and $1.50 respectively. Their total production is at most 15,000 pens per week. In order to remain profitable, their gross receipts must be at least $20,000.

OBJECTIVES

1. Define and Describe a Function
2. Evaluate Functions
3. Graph a Function
4. Find the Domain and Range of a Function
5. Solve Applied Problems
6. Use Technology Exploration

3.5 Introduction to Functions

Historically the term "function" was first used by Gottfried Leibniz (1646–1716) in the seventeenth century to denote the association of quantities with a curve. In the eighteenth century many of the concepts and the notation were further developed by the Swiss mathematician Leonhard Euler (1707–1783). This section discusses functions from both an algebraic and a graphic point of view. We will extend this discussion to different types of functions throughout this book.

Defining and Describing Functions

To understand the concept of functions, consider the following situations in which one quantity corresponds to another:

1. To each person there corresponds a birth date; that is, the month, day, and year born.
2. To each registered car there corresponds a unique license number.
3. To each circle there corresponds a unique area.

There are three characteristics common to all functions—an *input set*, an *output set*, and a *rule* that determines the association between the members of the two sets. The following table lists these characteristics for the above three examples.

TABLE 1

Set of Inputs	Set of Outputs	Rule (established by)
1. persons	birth dates	aging
2. cars	license numbers	licensing procedure
3. circles	areas	rule for area

These distinctions are central to the following definition.

DEFINITION OF A FUNCTION

A **function** is a rule of correspondence between two sets D and R so that for each *input* value in D there exists one and only one *output* value in R.

In this definition, all the input elements in D are called the **domain** of the function and all the output elements in R are called the **range** of the function. The rule of a function produces exactly one output in the range for each input in the domain. It is acceptable that different inputs may produce the same output.

EXAMPLE 1   Determining Whether a Correspondence Is a Function

Determine which of the following correspondences represent functions. Find the domain and range of the function.
(a) A child corresponds to his biological mother.
(b) A doctor corresponds to her patients.

Solution   Figure 1 shows the correspondences.

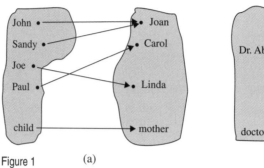 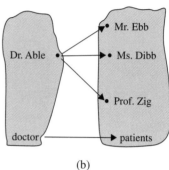

Figure 1   (a)   (b)

(a) This correspondence is a function because every child has one and only one biological mother. The domain is all the people born and the range is all the mothers.
(b) This correspondence is not a function because a doctor has more than one patient. Although this correspondence is not a function, its domain is the set of all doctors and its range is the set of all patients. ◈

Using example 1, we can make the following observation about a function:
1. Each input in the domain must be matched with one output in the range. In example 1a, everyone has only one mother.
2. Two or more inputs of the domain may be matched with the same output in the range. In example 1a, John and Sandy have the same mother, Joan; John and Sandy are called siblings.
3. No input of the domain is matched with two different outputs in the range. In example 1a, note that people do not have two biological mothers.

Evaluating Functions

Functions may be defined by an equation or a formula involving two variables, by a table of data, or by a graph. For instance, the equation

$$y = 7x - 3$$

represents y as a function of the variable x. We call x the domain element, the input and y the range element, the output. When the function is defined by an equation, calculations may be performed in order to determine which element of the range corresponds to each element of the domain. For example,

$$\text{for } x = -2, \, y = 7(-2) - 3 = -14 - 3 = -17$$
$$\text{for } x = 1, \, y = 7(1) - 3 = 7 - 3 = 4$$
$$\text{for } x = 3, \, y = 7(3) - 3 = 21 - 3 = 18$$

The inputs are values of x substituted into the equation to obtain the resulting output values of y. For a more concise notation, if we call the function f, we can use x to represent arbitrary inputs and $f(x)$ to represent the corresponding output. We read $f(x)$ as "f of x" or "the **value** of f at x". In this notation, the function is written as $f(x) = 7x - 3$ and the above calculations would be written as

$$f(-2) = 7(-2) - 3 = -17$$
$$f(1) = 7(1) - 3 = 4$$
$$f(3) = 7(3) - 3 = 18$$

Sometimes, instead of writing $f(x) = 7x - 3$, we might write $y = 7x - 3$ where it is understood that the value of y is calculated after choosing a value of x. That is, the value of y *depends* on the choice of the value of x. Because of this dependence, we refer to x as the **independent variable** and to y as the **dependent variable**.

The process of finding the value of $f(x)$ for a given value of x is called **evaluating a function**. Although we often use f as a convenient symbol to represent a function, other letters of the alphabet can be used to designate functions such as g and h. Also, letters other than x may be used to represent the independent variable. For example,

$$f(r) = 7r - 3, \quad g(t) = 7t - 3 \quad \text{and} \quad h(s) = 7s - 3$$

all define the same function. Note that a function f and a function value $f(x)$ are not the same: f is a rule of correspondence, but $f(x)$ is an output value of f.

EXAMPLE 2   Using Function Notation

Suppose that $f(x) = 5x + 7$, find the indicated value.
(a) $f(1)$ (b) $f(0)$ (c) $f(-4)$

Solution   To find the outputs we replace x by the inputs 1, 0, and -4 and simplify.
(a) $f(1) \quad = 5(1) + 7 = 5 + 7 = 12$
(b) $f(0) \quad = 5(0) + 7 = 0 + 7 = 7$
(c) $f(-4) = 5(-4) + 7 = -20 + 7 = -13$

In many applied problems the input can be a letter or an algebraic expression as the next example shows. ◈

EXAMPLE 3   Evaluating a Function

Let $f(x) = -3x + 4$. Find each value and simplify the result.

(a) $f(c)$ (b) $f(5b)$ (c) $f(t + 2)$ (d) $\dfrac{f(t + 2) - f(t)}{2}$

Solution   To find an output for a given input, we replace x in each case by the given letter or expression and simplify. Thus,
(a) $f(c) = -3c + 4$
(b) $f(5b) = -3(5b) + 4 = -15b + 4$
(c) $f(t + 2) = -3(t + 2) + 4 = -3t - 6 + 4 = -3t - 2$

$$\text{(d)} \quad \frac{f(t + 2) - f(t)}{2} = \frac{[-3(t + 2) + 4] - [-3t + 4]}{2}$$
$$= \frac{-3t - 6 + 4 + 3t - 4}{2}$$
$$= \frac{-6}{2} = -3$$

◈

Note that the value of this quotient is the *slope* of the line $y = -3x + 4$.

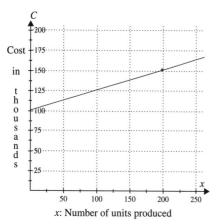

x: Number of units produced

Figure 2

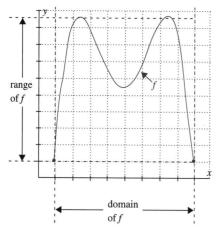

Figure 3

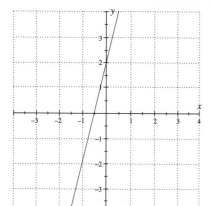

Figure 4

Graphing a Function

Functions can be described by tables and graphs. Such descriptions provide insight into many real world relationships. For example, the graph in Figure 2 shows the total cost (in thousands of dollars) for a manufacturing company that produces *x* units of its product. This is the graph of a function because each input (units produced) corresponds to exactly one output (cost). The horizontal scale represents the number of units produced, the domain elements, and the vertical scale, the range elements, represents the total cost *C* (in thousands of dollars).

From the graph we observe that $100,000 was spent initially ($x = 0$), and $150,000 was spent when 200 units were produced. Notice that as the number of units produced increases, the total cost also increases.

Quite often, businesses depend on equations that "closely fit" data-defined functions in order to predict future trends. For example, the linear function

$$f(x) = 0.25x + 100$$

approximates the data obtained from the graph of the function in Figure 2. To predict the cost (in thousands of dollars) spent to produce 230 units, we replace *x* by 230 and obtain

$$f(230) = 0.25(230) + 100 = 157.50$$

Therefore, we predict $157.50 (thousands of dollars) will be spent to produce 230 units.

Before we can analyze the graph of a function in the manner just illustrated in Figure 2, we must be able to sketch its graph. By definition, the **graph** of a function *f* is the set of all points $(x, f(x))$ for which *x* is in the domain of the function *f*. Graphically, *the domain of a function represents the set of all points on the horizontal axis (x axis) and the range represents the set of all points on the vertical axis (y axis)*. The domain and range of a function are found from its graph (Figure 3). We *graph* a function in the same way we graph an equation. We substitute representative numbers from its domain into the equation that defines the function. So we obtain points $(x, f(x))$ that lie on the graph, and then we plot these points and connect them with a line or a curve. Functions of the form

$$f(x) = mx + b$$

are called **linear functions** and their graphs are straight lines.

EXAMPLE 4   Graphing a Linear Function

Sketch the graph of the linear function

$$f(x) = 4x + 2$$

Solution   The graph of *f* is the same as the graph of the equation $y = 4x + 2$, which is obtained by finding the *x* and *y* intercepts, $-1/2$ and 2 respectively (Figure 4).   ◈

An examination of the graph of an equation can tell us whether or not the equation defines a function. Consider the graphs in Figure 5.

In Figure 5a, every choice of input, *x* value, will correspond to one and only one output, *y* value; so, the graph represents a function. In Figure 5b, a single input value (say $x = 2$) corresponds to two output values, 2 and -2, and so the graph cannot represent a function. In this case, the points $(2,2)$ and $(2,-2)$ are positioned on a vertical line above each other. This leads to the following test.

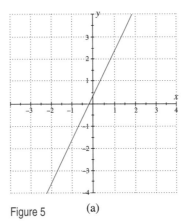
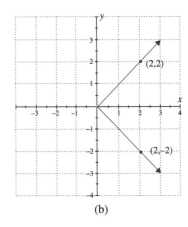

Figure 5   (a)   (b)

THE VERTICAL LINE TEST

> If a vertical line moving across a graph intersects the graph only one point at a time for each value in the domain, then the curve represents the graph of a function.

EXAMPLE 5   Using a Vertical Line Test

Use the vertical line test to determine which graph in Figure 6 represents a function.

Solution   Figure 6a represents the graph of a function because any vertical line intersects the graph only once. In Figure 6b, the vertical line crosses the graph in two points and so the graph does not represent the graph of a function.

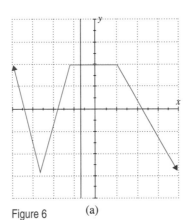
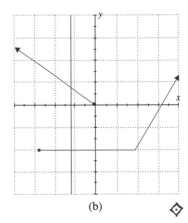

Figure 6   (a)   (b)   ◈

Finding the Domain and Range of a Function

By carefully examining the graph of a function, we may be able to determine its domain and range. It should be noted that when no endpoints are shown on the graph, we assume that the graph continues indefinitely. Also, unless otherwise stated, we assume that the domain of a function f is taken to be the set consisting of every real number for which the rule of f produces a real number.

EXAMPLE 6   Using Graphs to Determine the Domain and Range

Use the graph of the function f in Figure 7 to determine its domain and range.

Solution   The graph in Figure 7 appears to start with the end point on the left at $x = -2$ and the lowest point at $y = 0$. It extends indefinitely to the right and upward. Hence, the graph suggests that the domain of f consists of all real numbers such that $x \geq -2$ or $[-2, \infty)$ and the range consists of real numbers $y \geq 0$ or $[0, \infty)$.

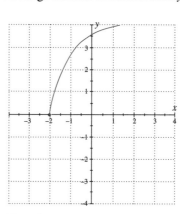

Figure 7

Solving Applied Problems

Functions can be used in many applications and mathematical models.

EXAMPLE 7   Determining the Cost of a Product

A company was established in 1994 and every year since its founding the cost per unit produced has increased at a constant rate of $0.35. In 1997 the cost per unit was $10.60, $t = 0$ represents 1994.

(a) Find a function that expresses the cost C (in dollars) per unit as a function of time t (in years).

(b) Sketch the graph of the function.

(c) Predict the cost per unit in the year 2008 assuming the rate of increase remains the same.

(d) In what year will the cost per unit be $10.95?

Solution   (a) Since the rate of increase is constant, the function is linear and of the form $C(t) = mt + b$. Here the slope is 0.35 and $C(3) = 10.60$. Thus we have,

$$C(3) = 10.60 = 0.35(3) + b$$
$$b = 10.60 - 1.05 \quad \text{and} \quad b = 9.55$$

The function is $C(t) = 0.35t + 9.55$

(b) Figure 8 displays the sketch of the graph.

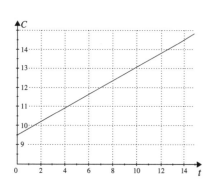

Figure 8

(c) In the year 2008, $t = 14$ and the cost per unit is

$$C(14) = 0.35(14) + 9.55 = \$14.45$$

(d) We are given $C(t) = 10.95 = 0.35t + 9.55$

$$1.40 = 0.35t$$
$$4 = t$$

So 4 years after 1994, that is 1998, the cost per unit will be $10.95.

EXAMPLE 8   Using Direct Variation

In photography, the time T (in seconds) to make an enlargement of a photo negative is directly proportional to the area A (in square inches) of the enlargement. Suppose it takes 30 seconds to make a 6 by 8 enlargement.

(a) Find a linear function T in terms of A to model this situation.

(b) Use part (a) to find the time T it takes to make a 12 by 16 enlargement.

Solution   (a) Since T is directly proportional to A, we write the equation

$$T = kA$$

where k is the constant of variation. For a given value of $A = 6(8) = 48$ square inches, there is exactly one value for $T = 30$ seconds, and so

$$30 = k(48)$$

$$k = \frac{30}{48} = \frac{5}{8}$$

The equation $T = \frac{5}{8}A$ describes a linear function.

(b) Substituting $12(16) = 192$ for A, we have

$$T = \frac{5}{8}(192) = 120$$

Therefore, it takes 120 seconds to make a 12 by 16 enlargement.

Using Technology Exploration

We can use a grapher as a tool to describe the domain and range of a function by using the TRACE and ZOOM features. The TRACE feature gives the x and y value for each of the points plotted. As a rule, the displayed portion of the graph with adjusted windows will give an indication of the domain and range of the function.

G EXAMPLE 9   Using a Grapher to Graph a Function

Use a grapher to graph the function $g(x) = |x| - 3$ in the viewing window $x\text{Min} = -8$, $x\text{Max} = 8$, $x\text{Scl} = 0$, $y\text{Min} = -8$, $y\text{Max} = 8$, $y\text{Scl} = 0$. Then visually estimate the domain and range of the function.

Solution   Figure 9a shows the graph of the function in the given window. In this window, the x coordinates on the graph vary from -8 to 8 and the y coordinates vary from -3 to 5. If we use the ⬭ ZOOM ⬭ feature of the grapher with a zoom factor of 5, then the graph of g appears in Figure 9b. In this case, the x coordinates vary from -40 to 40 and the y coordinates vary from -3 to 37.

Now we look for the inputs on the x axis that correspond to points on the graph by continuing to ⬭ ZOOM ⬭ out. We see, by using the ⬭ TRACE ⬭ feature and moving the cursor from left to right along the graph, that the inputs extend indefinitely to the right and left without changing the shape of the graph. Thus the displayed portion of the graph suggests that the domain of g consists of all the real numbers $\mathcal{R}$. To find the range, we examine the graph and look for the outputs on the y axis. Again we see that they extend as far down as -3 and up indefinitely. Thus the range of g appears to consist of all real numbers $y \geq -3$ or the interval $[-3, \infty)$. We confirm our results by examining the formula that defines the function. The domain, any real value of x, makes sense in this formula. The least value of $|x|$ is zero and three less is -3. Thus, the range elements are the real numbers greater than or equal to -3.

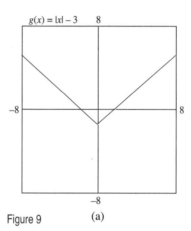

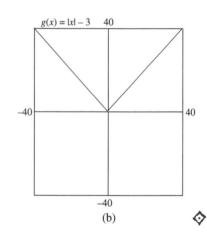

Figure 9   (a)   (b)   ◆

◆ **PROBLEM SET 3.5**

Mastering the Concepts

In Problems 1–4, determine which of the given correspondences represent a function. Find the domain and range of each function.

1.   (a)   (b)

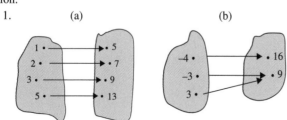

2.   (a)   (b)

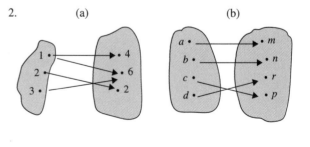

3. (a) Each US citizen corresponds to their social security number.
    (b) Each person's age corresponds to their height.
4. (a) Each person's height corresponds to their age.
    (b) Each VCR in a store corresponds to its price.

In Problems 5–14, find the indicated value of each given function.

5. $f(x) = 3x$
    (a) $f(-2)$
    (b) $f(0)$
    (c) $f(2)$
6. $g(x) = -3x$
    (a) $g(-3)$
    (b) $g(0)$
    (c) $g(3)$
7. $f(x) = x - 2$
    (a) $f(-2)$
    (b) $f(0)$
    (c) $f(2)$
8. $g(x) = 2 - x$
    (a) $g(-3)$
    (b) $g(0)$
    (c) $g(3)$
9. $f(x) = -4x + 3$
    (a) $f(0)$
    (b) $f(3/4)$
    (c) $f(-1/2)$
10. $g(x) = 5x + 2$
    (a) $g(-2/5)$
    (b) $g(0)$
    (c) $g(3/5)$
11. $h(x) = \dfrac{3x + 1}{5}$
    (a) $h(0)$
    (b) $h(-1/3)$
    (c) $h(2)$
12. $F(x) = \dfrac{-1 - 3x}{5}$
    (a) $F(0)$
    (b) $F(-1/3)$
    (c) $F(1/3)$
13. $f(x) = 6(2x - 3) + 4$
    (a) $f(0)$
    (b) $f(7/6)$
    (c) $f(3)$
14. $g(x) = 3(-2x + 1) - 5$
    (a) $g(0)$
    (b) $g(-1)$
    (c) $g(2)$

In Problems 15–18, find each value and simplify the results.

15. $f(x) = 7x + 3$
    (a) $f(t)$
    (b) $f(t + 2)$
    (c) $f(t) + 2$
    (d) $f(t + 2) - f(t)$
16. $g(x) = 3 - 7x$
    (a) $g(m)$
    (b) $g(m - 2)$
    (c) $g(m) - g(2)$
    (d) $g(m - 2) - g(2)$
17. $g(x) = 5 - 4x$
    (a) $g(2t)$
    (b) $g(t + h)$
    (c) $g(t) + h$
    (d) $g(t + h) - g(2t)$
18. $h(x) = 4x - 5$
    (a) $h(3p)$
    (b) $h(3p + 1)$
    (c) $h(3p) + 1$
    (d) $h(3p + 1) - h(3p)$

In Problems 19–24, find and simplify the value of the expression

$$\frac{f(t + h) - f(t)}{h}$$

This expression is called the *difference quotient* of f (see example 3d on page 173). Its value is equal to the slope of the line.

19. $f(x) = 3x + 11$
20. $f(x) = -7 + 4x$
21. $f(x) = 4 - 9x$
22. $f(x) = 5 - 13x$
23. $f(x) = -\dfrac{5}{2}x - 1$
24. $f(x) = \dfrac{3x}{7} + 4$

In Problems 25–32, sketch the graph of each function.

25. (a) $f(x) = x$
    (b) $g(x) = -x$

26. (a) $f(x) = \dfrac{x}{2}$
    (b) $g(x) = \dfrac{-x}{2}$
27. $f(x) = 3x - 4$
28. $g(x) = 4 - 3x$
29. $f(x) = \dfrac{-x}{4} + 1$
30. $g(x) = \dfrac{1}{2}x - 2$
31. $g(x) = -2$
32. $f(x) = 0$

33. Figure 10 shows the distance d (in miles) traveled in time t (in hours) over a period of 3.5 hours.

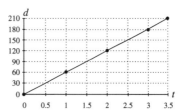

Figure 10

(a) Is this the graph of a function?
(b) What is the domain?
(c) What is the range?
(d) Estimate the distance traveled after 1.5 hours.
(e) Estimate the time it took to travel 180 miles.

34. Figure 11 shows the relationship between the number of units N sold of a product and the price p (in dollars) of that product.

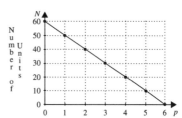

Figure 11

(a) Is this the graph of a function?
(b) What is the domain?
(c) What is the range?
(d) Estimate the number of units sold if the price is $4.50 per unit.
(e) Estimate the price per unit if 25 units were sold.

In Problems 35 and 36, use the vertical line test to determine whether or not each graph represents a function.

35.   (a)   (b)

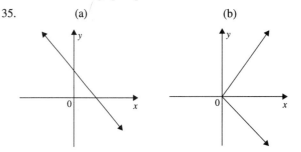

36.   (a)   (b)

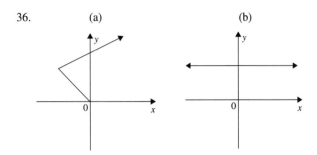

In Problems 37 and 38, use the graph of the given function to determine its domain and its range.

37.   (a)   (b)

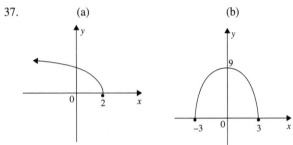

38.   (a)   (b)

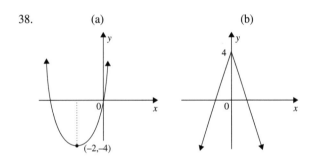

Applying the Concepts

39. **Car Rental:** The cost C (in dollars) for renting a midsize car for one day is shown in Figure 12 where m is the number of miles driven for that day.

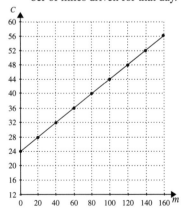

Figure 12

    (a) Find a function that models the cost C (in dollars) in terms of the number of miles m driven.

    (b) Use the model to estimate the cost of driving 130 miles for that day.

    (c) Use the model to determine the number of miles driven if the total cost for that day is $26.

    (d) Use the model to predict the total cost if 300 miles were driven for that day.

40. **Car Prices:** In 1990 the average price paid for a certain new car model was $16,800. The average price increased in a linear pattern to $24,290 in 1997.

    (a) Find a function f that models the average price P paid (in dollars) in terms of the time t (in years).

    (b) Use the data points to graph the function.

    (c) Use the graph to predict the average price of that same model in the year 2000.

    (d) Use the graph to predict the year in which the average price of this model is $26,430.

41. **Aerobics:** An aerobic dancer burns 350 calories per hour. The number of calories burned, C, is directly proportional to the number of hours danced, t.

    (a) Express C as a function of t to model the situation.

    (b) Sketch the graph of C.

    (c) Use the graph to determine the number of calories burned for an aerobic dancer who danced for 2.5 hours.

42. **Total Cost:** In 1998 the sales tax on a nonfood item in the state of Michigan was 6%.

    (a) Express the total cost C (in dollars) of a nonfood item as a function of the price p (in dollars) of the item.

    (b) Sketch the graph of $C = f(p)$ by using p as the horizontal axis and C as the vertical axis.

    (c) Use the graph to determine the total cost of an item that was purchased for $175.

    (d) Use the graph to predict the sale price before tax if its total cost is $200.

43. **Tuition Cost:** The yearly cost C (in dollars) of tuition and other required fees for attending a certain university can be modeled by the linear function

$$C(t) = 1500t + 7000, \ 1 \leq t \leq 10$$

where t is the number of years with $t = 1$ representing 1997.

    (a) Use this model to approximate the total yearly cost of attending this university in the year 2005.

    (b) Use this model to predict in what year will the total yearly cost of attending the university be $22,000.

    (c) Does this model make sense for predicting the total yearly cost after 30 years? Explain.

44. **Ideal Weight:** The ideal weight W (in pounds) of a man is estimated by multiplying his height h (in inches) by 5 and subtracting 190.

    (a) Express W as a linear function of h to model this situation.

    (b) Find the ideal weight of a man who is 6 feet tall.

    (c) Sketch the graph of this model.

    (d) What restrictions should be put on h to make this model reasonable?

45. **Ideal Shoe Size:** A manufacturer of modern scientifically designed walking shoes estimates the ideal shoe size S for a man is obtained by multiplying his foot length l (in inches) by 3 and subtracting 22.
    (a) Express S as a linear function of l to model this situation.
    (b) Find the shoe size of a man if the length of his foot is 10.5 inches.
    (c) Sketch the graph of this model.
    (d) What restrictions should be put on l to make this model reasonable?

46. **Ecology:** An ecologist investigating the effect of air pollution on plant life determines that the percentage, p (in percent) of diseased trees and shrubs at a distance of x (in kilometers) from an industrial city is given by the linear model
$$p(x) = 32 - 0.06x, 50 \le x \le 500$$
    (a) Find $p(100)$, $p(200)$, and $p(400)$.
    (b) Sketch the graph of p.
    (c) Use the graph to determine the percent p of diseased trees at a distance of 300 kilometers.

Developing and Extending the Concepts

In Problems 47–50, let $f(x) = 3x - 2$ and $g(x) = 2x + 7$. Write an expression for each function and simplify the results.

47. (a) $f(x) + g(x)$ (b) $f(x) - g(x)$
48. (a) $2f(x) + 3g(x)$ (b) $2f(x) - 3g(x)$
49. (a) $f(x) \cdot g(x)$ (b) $x \cdot f(x)$
50. (a) $\dfrac{f(x)}{g(x)}$ (b) $\dfrac{g(x)}{f(x)}$

51. Find an equation of a linear function such that $f(3) = 7$ and $f(4) = 9$.
52. Let $f(x) = 5x + 1$. Find
    (a) $f(s + t)$ (b) $f(s) + f(t)$
    (c) Is $f(s + t) = f(s) + f(t)$?   (d) $f(at)$
    (e) $a \cdot f(t)$ (f) Is $f(at) = a \cdot f(t)$?
53. Sketch the graph of a line that does not represent a function.
54. Suppose that $y = 2$ regardless of the value of x.
    (a) Is y a function of x? Explain.
    (b) Is x a function of y? Explain.
55. The following data was collected by measuring the weight y (in pounds) of a man on the moon and his weight x (in pounds) on Earth.

weight x (lbs.) on Earth	120	132	156	162
weight y (lbs.) on the moon	20	22	26	27

    (a) Plot the data points on a graph using x as the horizontal axis and y as the vertical axis.
    (b) Analyze the graph to determine a linear model for y as a function of x to fit the data.
    (c) Find and interpret the slope of the line.
    (d) Use the model to predict the weight of a person on Earth if his weight on the moon is 32 pounds.

56. Table 2 shows the average price of a movie ticket (in dollars) for a theater in a certain city.

TABLE 2

years (t)	1993	1994	1995	1996	1997
price p (dollars)	4.75	4.90	5.05	5.20	5.35

    (a) Plot the data points.
    (b) Use a grapher to find the (regression) line that approximates p as a function of t.
    (c) Use the graph to predict the price of a movie ticket in the year 2002.
    (d) Use the graph to predict when the average price of a ticket will be $8.95.

57. Let $f(x) = |2x - 1|$ and $g(x) = |1 - 2x|$
    (a) Find (i) $f\left(\dfrac{1}{2}\right)$ (ii) $g\left(\dfrac{1}{2}\right)$
              (iii) $f(0)$ (iv) $g(0)$
    (b) Use a grapher to sketch the graphs of f and g on the same viewing window. xMin $= -10$, xMax $= 10$, xScl $= 1$, yMin $= -10$, yMax $= 10$, yScl $= 1$.
    (c) Use the displayed portions of the graphs to estimate the domains and ranges of f and g.

58. Table 3 shows the rate at which the volume V (in cubic meters per hour) of water is entering a reservoir at a given time t (in hours). A negative rate indicates that the water is leaving the reservoir.

TABLE 3

Time t	1	2	3	4	5	6	7	8	9
Rate r	0	1	1	2	2	0	-1	-1	0

    (a) Plot the data points and connect them with line segments. Use the t axis as horizontal and the r axis as vertical
    (b) Use the graph to indicate where the volume of water is increasing, constant or decreasing.
    (c) Use the graph to indicate where the volume of water is increasing the fastest.

In Problems 59 and 60, use a grapher to graph all three functions on the same viewing window. Indicate how the graphs of $f(x) + c$ and $f(x + c)$ compare to the graph of $f(x)$. Use the displayed portion of the graphs to determine the range of each function.

59. (a) $f(x) = |x|$
    (b) $f(x) = |x| + 3$
    (c) $f(x) = |x + 3|$
60. (a) $f(x) = |x|$
    (b) $f(x) = |x| - 2$
    (c) $f(x) = |x - 2|$

CHAPTER 3 REVIEW PROBLEM SET

In Problems 1–6, sketch the graph of each linear equation by point plotting.

1. (a) $y = x - 3$
   (b) $y = x + 3$

2. (a) $y = \dfrac{3}{4}x$
   (b) $y = -\dfrac{3}{4}x$

3. (a) $y = 2x - 3$
   (b) $y = -2x + 3$

4. (a) $y = -3x + 1$
   (b) $y = -3x - 1$

5. $2x - 3y + 12 = 0$

6. $3x - 2y - 12 = 0$

In Problems 7–10, find the x and y intercepts of each equation and sketch the graph.

7. $2x - 5y = 10$

8. $-3x + 6y + 12 = 0$

9. $4x + 12 = 0$

10. $3y - 9 = 0$

In Problems 11–16, sketch the graph of each line containing the given points, and then find the slope of each line. Indicate whether this line rises, falls, is horizontal, or is vertical.

11. $(2,-2)$ and $(7,3)$

12. $(7,4)$ and $(11,12)$

13. $(-3,5)$ and $(1,3)$

14. $(-1,4)$ and $(2,1)$

15. $(3,0)$ and $(-5,0)$

16. $(3,4)$ and $(3,5)$

In Problems 17 and 18, given a point and the slope, sketch the graph of each pair of lines on the same coordinate system.

17. (a) $(1,2)$ and $m = \dfrac{3}{4}$
    (b) $(1,2)$ and $m = -\dfrac{4}{3}$

18. (a) $(4,-5)$ and $m = \dfrac{4}{7}$
    (b) $(4,-5)$ and $m = -\dfrac{7}{4}$

In Problems 19 and 20, express each equation in slope-intercept form, and then determine the slope and the y intercept. Also sketch the graph.

19. (a) $3x - 4y - 9 = 0$
    (b) $-3x + 4y + 9 = 0$

20. (a) $5x + 7y - 11 = 0$
    (b) $-7x + 5y - 11 = 0$

In Problems 21–24, determine whether L_1 and L_2 are parallel, perpendicular or neither. Sketch a graph of the lines.

21. L_1 contains $(1,3)$ and $(-1,-1)$
    L_2 contains $(2,1)$ and $(-3,-9)$

22. L_1 contains $(4,6)$ and $(-2,-3)$
    L_2 contains $(3,0)$ and $(6,-2)$

23. L_1: $3x + 2y - 6 = 0$
    L_2: $2x - 3y - 6 = 0$

24. L_1: $x - 2y - 5 = 0$
    L_2: $2x + y - 5 = 0$

In Problems 25–28, find an equation of each line L in slope-intercept form (if possible). Sketch the graph.

25. (a) $m = 3$ and contains $(1,1)$
    (b) $m = -2$ and contains $(-3,-2)$

26. (a) $m = -2$ and contains $(1,5)$
    (b) contains $(4,1)$ and $(2,3)$

27. (a) $m = 0$ and contains $(-3,4)$
    (b) m is undefined and contains $(5,3)$

28. (a) $m = 0$ and $b = -2$
    (b) m is undefined and contains $(1,7)$

In Problems 29–32, find an equation of the line L in point slope form and in general form. Sketch the graph.

29. L is parallel to the line $3x - 2y + 5 = 0$ and contains $(2,3)$.

30. L is parallel to the line $2x - 5y - 7 = 0$ and contains $(-3,4)$.

31. L is perpendicular to the line $2y + 3x + 6 = 0$ and contains $(-7,5)$.

32. L contains the point $(-1,2)$ and is perpendicular to the line segment containing $(3,4)$ and $(-5,6)$.

In Problems 33–38, sketch the graph of each inequality.

33. (a) $y \le -2$
    (b) $x \ge 3$

34. (a) $x \le 2$
    (b) $y \ge 3$

35. $y < -2x + 4$

36. $y > 3x - 7$

37. $y - 3x \ge 2$

38. $4x - 2y \ge 3$

In Problems 39 and 40, sketch the graph of each compound inequality.

39. (a) $y \ge -1$ and $x \le 2$
    (b) $x - 2y \le 6$ and $x + y \ge 2$

40. (a) $2x + y \le 6$ and $x + y \le 2$
    (b) $x \ge 2$ and $y \le -3$

In Problems 41–44, find the indicated value of each given function.

41. $f(x) = 2x + 3$
    (a) $f(3)$
    (b) $f(t + 1)$
    (c) $f(-2)$
    (d) $f(t + 1) - f(t)$

42. $g(x) = |x + 5|$
    (a) $g(-8)$
    (b) $g(-5) - g(0)$
    (c) $\sqrt{g(-14)}$
    (d) $g(2) - g(-2)$

43. $f(x) = \sqrt{4 - x^2}$
    (a) $f(0)$
    (b) $\dfrac{1}{f(0)}$
    (c) $f(-2)$
    (d) $f(-1)$

44. $h(x) = \dfrac{x}{|x|}$
    (a) $h(-2)$
    (b) $h(4)$
    (c) $h(-3)$
    (d) $h(5)$

In Problems 45 and 46, find and simplify the value of the expression

$$\frac{f(t + h) - f(t)}{h}$$

45. $f(x) = 3 - 5x$

46. $f(x) = \dfrac{2}{3}x + 5$

In Problems 47–50, sketch the graph of each function.

47. (a) $f(x) = 3x$
    (b) $g(x) = -3x$

48. (a) $f(x) = -\dfrac{x}{3}$
    (b) $g(x) = \dfrac{x}{3}$

49. $f(x) = 3 - 4x$

50. $g(x) = \dfrac{x}{4} + 3$

In Problems 51 and 52, determine whether or not the curve is the graph of a function. Use the graph of the functions to determine the domain and range of each function.

51.   (a)   (b)

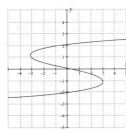

52.   (a)   (b)

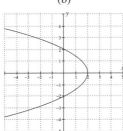

53.   **Gasoline Consumption:** The amount of gasoline A (in gallons) in a fuel tank of a sport utility vehicle can be related to the number of miles m driven by the model

$$A = 21 - 0.06m, 0 < m \leq 350$$

  (a)   Sketch the graph of this equation using m as the horizontal axis and A as the vertical axis.
  (b)   Find and interpret the intercepts of the graph.
  (c)   How much gasoline is left in the tank after driving 200 miles?
  (d)   If the gas gauge registers one-fourth of the tank (20 gallon capacity), how far has this vehicle been driven?

54.   **Diet Program:** The daily calorie allowance C to an ideal weight w (in pounds) of a young man is modeled by the equation

$$C = 18w - 100, 170 < w \leq 190$$

  (a)   Sketch the graph of this equation using w as the horizontal axis and C as the vertical axis.
  (b)   Use the graph to determine the calorie intake for a man whose ideal weight is 180 pounds.
  (c)   Compute the slope of this line and indicate the units.
  (d)   What is the significance of the slope in terms of this problem?

55.   **Printing Cost:** A printing company charges $20 for printing 300 business cards and $30 for printing 500 business cards. Assume the relationship between the cost C (in dollars) and the number n of cards is linear.
  (a)   Write an equation that expresses C in terms of n.
  (b)   Sketch the graph of this equation using n as the horizontal axis and C as the vertical axis.
  (c)   Use the graph to determine the cost of printing 750 business cards.

56.   **Appreciation:** A home is purchased for $225,000 in 1998 and is expected to double in value in the year 2010.
  (a)   Find a linear equation that expresses the cost C (in dollars) of the home after t years.
  (b)   Sketch the graph of the equation using t as the horizontal axis and C as the vertical axis.
  (c)   Use the equation to determine the cost of the home after 8 years.

57.   **Recycling:** Suppose that a recycling center recycled 500 tons of newspaper in the month of January and 530 tons of newspaper in the month of April. Suppose the increase was at a uniform rate. Let t denote the number of tons of newspaper recycled in the nth month (where $n = 1$ represents January). Assume the relationship between t and n is linear, where $1 \leq n \leq 12$.
  (a)   Find an equation that expresses t in terms of n.
  (b)   Find and interpret the slope of the line.
  (c)   In how many months does the amount of paper equal 570 tons?

58.   **Hockey Statistics:** Suppose that a professional hockey player scores 12 goals in the first 10 games of the season and assume this pace continues throughout the 82 games of the season.
  (a)   If the relationship is linear, express the number of goals N in terms of the number t of games played.
  (b)   What is the significance of the slope in this problem?
  (c)   Sketch the graph of this equation.
  (d)   Use this model to predict the approximate number of goals a player will score for the entire season.

G In Problems 59–62, use the fact that the *tangent line* to the graph of $y = x^2 + 1$ at the point (a,b) on the graph has a slope of $2a$.
  (a)   Find an equation of the tangent line to the graph of $y = x^2 + 1$ at each of the given points.
  (b)   Use a grapher to graph $y = x^2 + 1$ and each tangent line to see whether or not the line appears to be tangent to the curve at the given point.

59.   (1,2)   60.   (−1,2)
61.   (−0.5,1.25)   62.   (0.5,1.25)

◆ CHAPTER 3 PRACTICE TEST

1.   Use point plotting to sketch the graph of each linear equation.
  (a) $y = 4x$
  (b) $y = -2x + 3$
  (c) $x + 3y - 6 = 0$

2.   Find the x and y intercepts of each linear equation. Also sketch the graph.
  (a) $3x - 2y - 6 = 0$
  (b) $5x - 15 = 0$
  (c) $5y + 5 = 0$

3. Sketch the graph of each line L with the given information. Indicate whether the line rises or falls. Find the slope.
   (a) L contains $(-3,2)$ and $(-1,7)$
   (b) $m = -\frac{2}{3}$ and L contains $(2,5)$

4. Express the equation $5x - 3y - 15 = 0$ in slope-intercept form. Find the slope and the y intercept. Sketch the graph.

5. Determine whether the given lines are parallel or perpendicular or neither.

$$L_1 \text{ contains } (8,-6) \text{ and } (-4,3),$$
$$L_2 \text{ contains } (3,3) \text{ and } (9,11)$$

6. Find an equation of a line L in slope-intercept form and in general form. Also graph the line.
   (a) L contains $(0,5)$ and $(7,-2)$.
   (b) L contains $(-4,-1)$ and $m = -\frac{4}{3}$.
   (c) L contains $(1,-3)$ and is perpendicular to the line $7x - 3y - 4 = 0$.

7. Sketch the graph of each inequality.
   (a) $y \le -x + 3$ (b) $y \ge 2x + 5$ (c) $2x > 4$

8. Sketch the graph of the inequalities $x \le 3$ and $y \ge 3$ on the same coordinate system.

9. Let $g(x) = \sqrt{16 - x^2}$ find each value
   (a) $g(0)$ (b) $g(4)$ (c) $[g(2)]^2$ (d) $g(\sqrt{7})$

10. Find the domain and range of the graph in Figure 1.

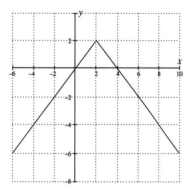

Figure 1

11. In business, the total cost y (in dollars) of producing x items of a product is given by an equation
$$y = mx + b$$
where b is the fixed cost and m is the variable cost. Suppose it cost \$4500 to start a business selling ice cream cones. Each cone costs \$0.45 to produce.
   (a) Write a specific cost equation for this business.
   (b) What is the cost y (in dollars) of producing 500 ice cream cones based on this equation?
   (c) How many ice cream cones can be produced if the total cost is \$5355?

12. The graph in Figure 2 shows the value of a photocopier over the first 5 years of ownership. Suppose that the trade in or salvage value of the photocopier is \$1000 after 5 years.

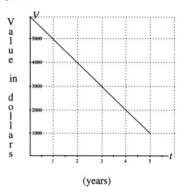

(years)

Figure 2

   (a) Write a linear equation that expresses the value V (in dollars) of the photocopier in terms of t (years).
   (b) What is the initial value of the photocopier?
   (c) What is the annual depreciation of the photocopier in each of the first 5 years?

13. Figure 3 shows a linear relationship between the number of computer video games N sold and the price P (in dollars) of each video game.

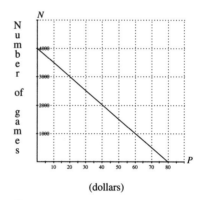

(dollars)

Figure 3

   (a) Express N as a function of P.
   (b) Estimate the number of video games sold if the price is \$20.
   (c) Estimate the price of each game if 3500 games are produced and sold.

G 14. Use a grapher to graph the functions
$$f(x) = |x - 1| \text{ and } g(x) = |1 - x|$$
on the same viewing window. Are the graphs of f and g the same?

Chapter 4

SYSTEMS OF EQUATIONS AND INEQUALITIES

4.1 Graphical Solutions of Linear Systems in Two Variables

4.2 Algebraic Solutions of Linear Systems in Two Variables

4.3 Linear Systems in More than Two Variables

4.4 Systems as Models

4.5 Systems of Linear Inequalities

In aviation the term *jet stream* is used to refer to a constant wind at high altitudes. Suppose a jetliner flies 2500 miles with the jet stream (with the wind) in 3.75 hours and makes the same trip against the jet stream (against the wind) in 4.4 hours. Assuming the jetliner travels at a constant speed and the wind blows at a constant rate, find the speed of the jetliner in still air and the speed of the wind. We use a system of linear equations to solve this problem. See Example 3 on page 209.

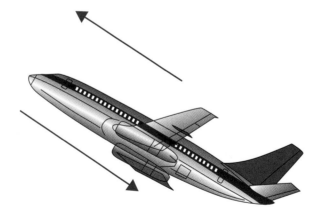

Up to now, we have solved equations and inequalities in one variable. In this chapter, we turn our attention to a set of two or more equations or inequalities containing two or more variables, called *systems*. Examples are:

$$\begin{cases} x + y = -7 \\ 3x - 4y = -35 \end{cases}, \quad \begin{cases} 3x - y + z = 5 \\ x + 2y - 2z = -3 \\ 2x + 3y + z = 4 \end{cases}$$

and

$$\begin{cases} x + 4y \le 9 \\ 5x - 2y < 7 \end{cases}$$

These systems are commonly used in geometry, science, business, and economics to describe relationships, limitations, and constraints. To *solve* a system of linear equations in two or more variables means to find values of the variables that make all equations in the system true.

OBJECTIVES

1. Solve a Linear System by Graphing
2. Explore the Geometry—Algebra Connection
3. Solve Applied Problems
4. Use Technology Exploration

4.1 Graphical Solutions of Linear Systems in Two Variables

Suppose we are to find the dimensions of a rectangular basketball court whose perimeter is 80 meters, and whose length is 2 meters shorter than twice its width. If we let x be the length of the court (in meters) and y be the width (in meters), we get the following system of linear equations:

$$\begin{cases} 2x + 2y = 80 \\ x \quad\quad = 2y - 2 \end{cases}$$

Such a collection of equations is called a **system of linear equations in two variables.** The **solution** of this system is a set of ordered pairs that makes both equations true. Now we ask the questions: Is there one such ordered pair? Is there more than one pair? Is there no pair at all? To answer these questions, we examine the graphs of both equations on the same coordinate system.

Solving a Linear System by Graphing

In this section we use *graphs* of linear equations to understand why some systems have one solution, others have an infinite number of solutions, and still others have no solution at all. The **graphical method** gives us a way of finding the numerical solution of a linear system of equations and also lends a visual meaning to the solution. In this case, the *point of intersection* of the two graphs of the linear equations in the system identify an ordered pair as the solution of the system.

A common error is to solve a system for only one of the two variables and not both variables.

To locate the point of intersection in a graph drawn by hand requires judgment and estimation usually resulting in limited accuracy. However, as we shall see, these graphs can be produced easily and with great accuracy by using a grapher.

EXAMPLE 1   Solving a System by Graphing

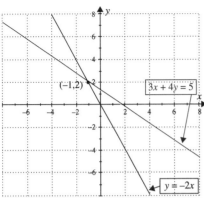

Figure 1

Solve the system $\begin{cases} 2x + y = 0 \\ 3x + 4y = 5 \end{cases}$ graphically.

Solution   First we solve each equation for y in terms of x:
From $2x + y = 0$, we have $y = -2x$ and from $3x + 4y = 5$, we have $4y = -3x + 5$
and $y = (-3/4)x + 5/4$.

Now we graph both equations on the same coordinate system. As Figure 1
indicates, the two lines intersect at the point $(-1,2)$, and this point is the solution of the system.

Check: Let $x = -1$ and $y = 2$, we have

first equation	second equation
$2x + y = 2(-1) + 2$	$3x + 4y = 3(-1) + 4(2)$
$\quad\quad = -2 + 2 = 0$	$\quad\quad = -3 + 8 = 5$

Thus $(-1,2)$ is the solution for the system because it makes both equations true. ◈

EXAMPLE 2   Solving a System by Graphing

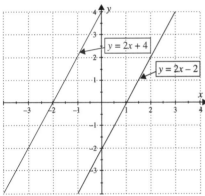

Figure 2

Solve the system $\begin{cases} 2x - y = 2 \\ -2x + y = 4 \end{cases}$ graphically.

Solution   Once again, we solve each equation for y: From $2x - y = 2$, we have
$y = 2x - 2$, and from $-2x + y = 4$, we have $y = 2x + 4$.

From the graph in Figure 2, the two lines appear to be parallel. If we examine both equations closely, we see that the slopes are the same. The y intercepts are -2 and 4. Thus, the lines are parallel and do not intersect; the system has no solution. ◈

EXAMPLE 3   Solving a System by Graphing

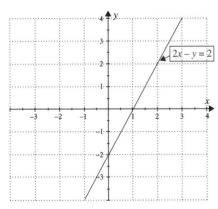

Figure 3

Solve the system $\begin{cases} 2x - y = 2 \\ -6x + 3y = -6 \end{cases}$ graphically.

Solution   We solve each equation for y in terms of x: From $2x - y = 2$, we have $y = 2x - 2$; from $-6x + 3y = -6$, we have $3y = 6x - 6$ or $y = 2x - 2$.

Figure 3 shows that both equations describe the same line. By examining the second equation, we see that it is just the first equation multiplied by -3. Thus, the two equations are equivalent and have the same ordered pair solutions. So, the system has an infinite number of solutions. ◈

Exploring the Algebra—Geometry Connection

In graphing a system of two linear equations in two variables, x and y, we expect the following three possible outcomes.
1. The equations describe lines which intersect at one point. Such a system has one solution (x,y), the point of intersection (see example 1). It is called a **consistent** system and the equations are said to be **independent**.
2. The equations describe parallel lines. The graphs have no point of intersection (see example 2). Such a system has no solutions and is called an **inconsistent** system.
3. The equations describe the same line. The graphs coincide (see example 3). Such a system has an infinite number of solutions and it is called a **consistent** system. The equations are said to be **dependent**.

The three possibilities show the algebra—geometry connection and are summarized in Table 1.

TABLE 1

Geometry	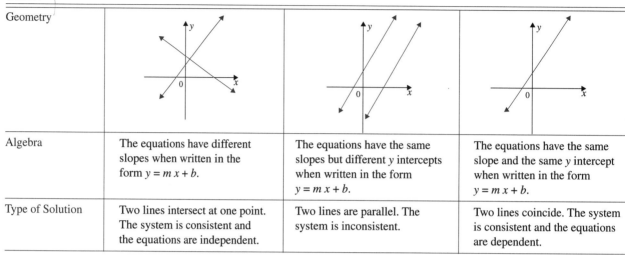		
Algebra	The equations have different slopes when written in the form $y = mx + b$.	The equations have the same slopes but different y intercepts when written in the form $y = mx + b$.	The equations have the same slope and the same y intercept when written in the form $y = mx + b$.
Type of Solution	Two lines intersect at one point. The system is consistent and the equations are independent.	Two lines are parallel. The system is inconsistent.	Two lines coincide. The system is consistent and the equations are dependent.

EXAMPLE 4   Solving Consistent and Inconsistent Systems

Write each equation in the given system in the slope-intercept form. Then sketch the graphs of the equations in each system on the same coordinate axes. Indicate whether the system is consistent or inconsistent; whether the equations are dependent or independent. Do not solve the system.

(a) $\begin{cases} x - 2y = 4 \\ -2x + 4y = -8 \end{cases}$ (b) $\begin{cases} 3x - 4y = -12 \\ \dfrac{1}{2}x = \dfrac{2}{3}y + 2 \end{cases}$ (c) $\begin{cases} 2x + 3y = 6 \\ -3x + 4y = 8 \end{cases}$

Solution   (a) $x - 2y = 4$ $\qquad\qquad -2x + 4y = -8$

$\qquad -2y = -x + 4$ $\qquad\qquad\quad 4y = 2x - 8$

$\qquad\quad y = \dfrac{1}{2}x - 2$ $\qquad\qquad\quad y = \dfrac{1}{2}x - 2$

These equations are equivalent because they have the same slope and the same y intercept. Figure 4a shows that the lines coincide. The system is consistent and the equations are dependent.

(b) $3x - 4y = -12$ $\qquad\qquad\qquad \dfrac{1}{2}x = \dfrac{2}{3}y + 2$

$\qquad -4y = -3x - 12$ $\qquad\qquad \dfrac{2}{3}y + 2 = \dfrac{1}{2}x$

$\qquad\quad y = \dfrac{3}{4}x + 3$ $\qquad\qquad\quad \dfrac{2}{3}y = \dfrac{1}{2}x - 2$

$\qquad\qquad\qquad\qquad\qquad\qquad y = \dfrac{3}{4}x - 3$

The equations have the same slope but different y intercepts. Figure 4b shows that the lines are parallel. The system is inconsistent.

(c) $2x + 3y = 6$ $\qquad\qquad\qquad -3x + 4y = 8$

$\qquad 3y = -2x + 6$ $\qquad\qquad\quad 4y = 3x + 8$

$\qquad\quad y = -\dfrac{2}{3}x + 2$ $\qquad\qquad\quad y = \dfrac{3}{4}x + 2$

The equations have different slopes. Figure 4c shows that the lines intersect at one point. The system is consistent and the equations are independent.

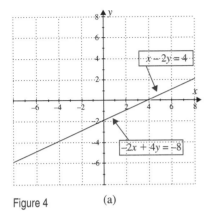

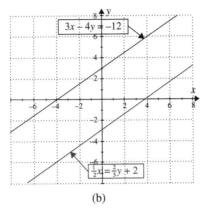

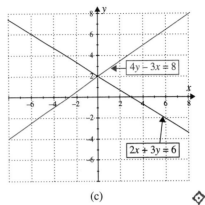

Figure 4   (a)   (b)   (c)

Solving Applied Problems

Some models from economics that arise in practical situations can be described by systems of linear equations. Suppose that p denotes the price per unit of a commodity, and q denotes the number of units of the commodity *demanded* in the market place at price p. The equation that relates q to p is called a **demand equation**. If this equation is linear, then the graph is a line (Figure 5a). Linear demand equations usually have negative slopes; as prices drop more people want to buy the commodity. On the other hand, let p still denote price, but let q denote the number of units *supplied* in the market place at price p. The equation that relates q to p is called a **supply equation**. If this equation is linear, then the graph is a line (Figure 5b). Linear supply equations usually have positive slopes; the more prices rise the more companies want to sell the commodity. If the demand and supply equations are graphed on the same coordinate system, the point where they intersect is called the **market equilibrium point** (Figure 5c). At this point, the quantity demanded will be equal to the quantity supplied. Under the usual interpretations of price, demand and supply, only the portions of the demand and supply graphs that fall in the first quadrant are economically meaningful.

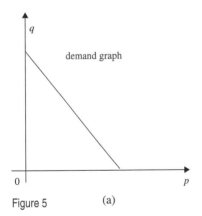

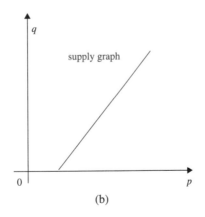

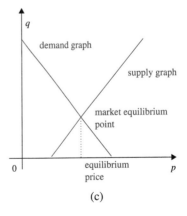

Figure 5   (a)   (b)   (c)

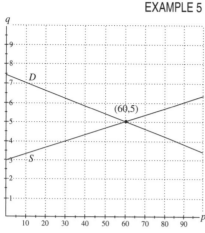

Figure 6

EXAMPLE 5   Modeling Supply and Demand

Suppose that the weekly demand D (in thousands) of a camera is related to the price of p dollars by the equation:

$$D(p) = -\frac{1}{25}p + 7.4$$

Also suppose that a manufacturer is willing to supply S (in thousands) of these cameras per week at a market price of p dollars per unit according to the supply equation:

$$S(p) = \frac{1}{30}p + 3$$

(a) Sketch the graphs of both equations on the same coordinate system. Use p as the horizontal axis and q the quantity supplied or demanded as the vertical axis.
(b) Find the market equilibrium point.
(c) Find the equilibrium price.

Solution   (a) First we graph both equations on the same coordinate system. Figure 6 shows the graphs of both equations.
(b) Using the graph in Figure 6, we see that the market equilibrium point is the point of intersection of the two graphs $(p,q) = (60,5)$.
(c) The equilibrium price is the p coordinate of the point of intersection, so that $p = 60$. This means that the demand is equal to the supply when the price of each camera is $60. ◇

Using Technology Exploration

Once we know the models that are described by linear systems of equations, we can produce those graphs easily with technological help. Unlike the handdrawn graphs, a grapher allows us to locate the coordinates of the point of intersection accurate to many decimal places. In this case, we graph the two equations in the same viewing window. Then we use the **TRACE** and **ZOOM** features or the **INTERSECT** feature to find the coordinates of the point of intersection of the graphs.

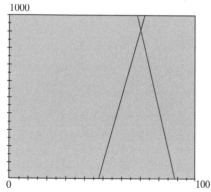

Figure 7

G EXAMPLE 6   Using a Grapher to Solve a System

The weekly demand D for tennis rackets is determined from the price p (in dollars) by the model: $D(p) = -50p + 4400$.

The weekly supply S of tennis rackets depends on the price p (in dollars) at which they are sold, according to the model: $S(p) = 40p - 1900$.

Use a grapher to determine the equilibrium price. That is, the price at which the supply equals the demand.

Solution   Figure 7 shows the graph of the system

$$\begin{cases} S(p) = 40p - 1900 \\ D(p) = -50p + 4400 \end{cases}$$

with a viewing window:

xMin = 0, xMax = 100, xScl = 10, yMin = 0, yMax = 1000, yScl = 100.

The point of intersection is (70,900) and the equilibrium price is $70.

PROBLEM SET 4.1

Mastering the Concepts

In Problems 1–12, solve each system graphically by sketching the graphs of the equations in the system on the same coordinate system. Then estimate the point of intersection (if any) of the two lines.

1. $\begin{cases} y = -x + 8 \\ y = 2x + 11 \end{cases}$

2. $\begin{cases} y = 2x \\ y = -3x - 5 \end{cases}$

3. $\begin{cases} x - 2y = 5 \\ 5x + y = 4 \end{cases}$

4. $\begin{cases} 2x = y + 3 \\ 4x - 2y = 5 \end{cases}$

5. $\begin{cases} y = 3x - 3 \\ 6x - 3y = 8 \end{cases}$

6. $\begin{cases} y = 2x - 3 \\ 4x - 2y = 6 \end{cases}$

7. $\begin{cases} x + 3y = 6 \\ 2x + 6y = 8 \end{cases}$

8. $\begin{cases} 2x - y = 6 \\ -4x + 2y = 10 \end{cases}$

9. $\begin{cases} 4x + 6y = 12 \\ 2x + 3y = -6 \end{cases}$

10. $\begin{cases} 2x + 5y = -9 \\ 3x - 4y = -2 \end{cases}$

11. $\begin{cases} 2x - 3y = -6 \\ -x + \dfrac{3}{2}y = 3 \end{cases}$

12. $\begin{cases} -2x + y = -8 \\ x - \dfrac{1}{2}y = 4 \end{cases}$

In Problems 13–18, the graphs of systems of linear equations are given. Use these graphs to determine the number of solutions of each system of equations. Do not solve the system.

13.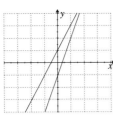

14.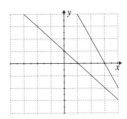

15.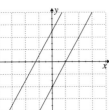

16.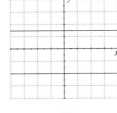

17.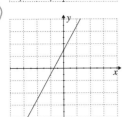

18.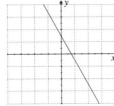

In Problems 19–24, write the two equations in the given system in slope-intercept form. Sketch the graphs of the equations in each system on the same coordinate system. Then indicate whether the system is consistent and dependent, consistent and independent, or inconsistent.

19. $\begin{cases} x + y = 6 \\ x - y = 4 \end{cases}$

20. $\begin{cases} x - y = -9 \\ x + y = -1 \end{cases}$

21. $\begin{cases} 2x + 3y = 5 \\ 4x + 6y = 7 \end{cases}$

22. $\begin{cases} -x + 2y = 1 \\ 3x - 6y = 5 \end{cases}$

23. $\begin{cases} 2x + 3y = 5 \\ -6x - 9y = -30 \end{cases}$

24. $\begin{cases} 2x - y = 0 \\ 4x - 2y = 0 \end{cases}$

In Problems 25–30, write the two equations in the given system in slope-intercept form. Indicate the number of solutions of the system. Do not solve the system.

25. $\begin{cases} 1.5x - 2y + 1 = 0 \\ 4y - 3x + 7 = 0 \end{cases}$

26. $\begin{cases} -2x = 3 \\ -4x + 8y = 7 \end{cases}$

27. $\begin{cases} x + y = 0.5 \\ x - y = -3 \end{cases}$

28. $\begin{cases} 3y = 2x - 6 \\ 3y = 2x + 4 \end{cases}$

29. $\begin{cases} x - y = 6 \\ -x + y = -6 \end{cases}$

30. $\begin{cases} 2x - 3 = 2y \\ y = x - 15 \end{cases}$

Applying the Concepts

31. **Modeling Supply and Demand:** In a particular city, the monthly supply and demand for a new type of tennis racket of price p (in dollars) and quantity q, are modeled by the equations:

$$\begin{cases} S(p) = 38p - 1710 \\ D(p) = -46p + 4002 \end{cases} \quad \$65 \le p \le \$75$$

(a) Sketch the graphs of the two equations on the same coordinate system. Let the horizontal axis be p and the vertical axis be q.

(b) Use the graphs of these models to estimate the equilibrium point. That is, at what price does the quantity supplied equal the quantity demanded?

32. **Modeling Supply and Demand:** In a particular area, the monthly supply and demand for a popular CD at price p (in dollars) is modeled by the equations:

$$\begin{cases} D(p) = -3p + 180 \\ S(p) = 4p + 110 \end{cases} \quad \$5 \le p \le \$15$$

(a) Sketch the graphs of the two equations on the same coordinate system.

(b) Use the graphs of these models to estimate the equilibrium point. That is, at what price does the supply equal the demand?

Developing and Extending the Concepts

33. Describe the graphs of the two equations in a given system that has (a) one solution, (b) no solution and (c) an infinite number of solutions.

34. How do the graphs of the two equations in a given system differ?
    (a) If the system is consistent and the equations are independent.
    (b) If the system is consistent and the equations are dependent.
    (c) If the system is inconsistent.

35. Describe how you might write a pair of equations in a given system that has (a) no solution or (b) an infinite number of solutions.

G 36. Figure 8 shows graphs of linear equations generated by a grapher in a standard viewing window. Which of the following ordered pairs could possibly be a solution for the system?

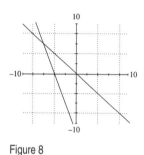

Figure 8

(a) (6,6)
(b) (−6,6)
(c) (−6,−6)
(d) (6,−6)

In Problems 37–40, determine the value of k so that the given system is classified as follows:

37. $\begin{cases} 2x - 3y = -6 \\ kx + 3y = 6 \end{cases}$ consistent and dependent

38. $\begin{cases} 2x + ky = 10 \\ y = 5 - x \end{cases}$ consistent and dependent

39. $\begin{cases} -x + 2y = 4 \\ x + ky = 0 \end{cases}$ inconsistent

40. $\begin{cases} x - ky = -5 \\ 2x + y = 2 \end{cases}$ inconsistent

In Problems 41 and 42, determine the restrictions on m and b so that the given system has the specified number of solutions.

41. $\begin{cases} y = m_1x + b_1 \\ y = m_2x + b_2 \end{cases}$ no solutions

42. $\begin{cases} y = m_1x + b_1 \\ y = m_2x + b_2 \end{cases}$ one solution

G In Problems 43–46, use TRACE and ZOOM features on a grapher to estimate the price p for which the supply and demand quantities are equal.

43. $\begin{cases} D(p) = 2100 - 100p \\ S(p) = -2400 + 400p \end{cases}$ $\$6 \le p \le \$12$

44. $\begin{cases} D(p) = 92 - 0.1p \\ S(p) = 45 + 0.05p \end{cases}$ $\$250 \le p \le \$400$

45. $\begin{cases} D(p) = 2514 - 14p \\ S(p) = -6880 + 90p \end{cases}$ $\$77 \le p \le \$110$

46. $\begin{cases} D(p) = 35 - 0.001p \\ S(p) = 12 + 0.005p \end{cases}$ $\$3500 \le p \le \$4300$

G In Problems 47 and 48, translate the given information into a system of linear equations. Use a grapher to estimate the solution of each system.

47. The sum of two numbers is −1 and their difference is 11. What are the two numbers?

48. A number is 4 less than a larger number. Two fifths of the larger number plus one third of the smaller is 6. What are the two numbers?

G 49. **Data Analysis:** The following table shows the total employment in a certain industry from the years 1997 to 1999.

Year	1997	1998	1999
Employment	898	838	758

(a) Use the REGRESSION feature of a grapher to find the regression line or "best fit" line for these data points. Let $x = 0$ correspond to 1997.

(b) From statistics, it can be shown that the solution to the system:

$$\begin{cases} 3m + 3b = 2494 \\ 5m + 3b = 2354 \end{cases}$$

gives the slope, m, and y intercept, b, of the "best fit" line. Solve this system graphically and compare m and b to the line in part (a).

50. **Van Rental:** Two van rental companies A and B offer the following options for renting their vans. Company A will rent a 24-foot van for $\$39.95$ plus $\$0.49$ per mile. Company B will rent the same van for $\$66$ plus $\$0.39$ per mile.
    (a) Suppose you are moving 200 miles round trip, which company is the cheapest to use?
    (b) Suppose you are moving 350 miles round trip, which company is the cheapest to use?
    (c) Write an equation for each company relating cost C (in dollars) as a function of miles, x.
    (d) Write and solve graphically an algebraic system to determine the number of miles traveled when the cost of renting from A is equal to the cost of renting from B. Hint: scale the x axis from 0 to 400 miles with increments of 20 miles and the C axis from $\$40$ to $\$250$ with increments of $\$20$.

1. Solve Linear Systems by Substitution
2. Solve Linear Systems by Elimination
3. Examine Types of Systems Algebraically
4. Solve Applied Problems Using Technology

4.2 Algebraic Solutions of Linear Systems in Two Variables

Although the graphing method provides an excellent means of approximating solutions to linear systems of equations in two variables, it does not always allow us to locate the exact solutions, especially if points of intersection are not located at integer values. For this reason, we use *algebraic methods* for solving these systems. We will see that such methods always give accurate results. In this section, we discuss two algebraic methods: the *substitution* and the *elimination methods*.

Solving Linear Systems by Substitution

The technique for solving a linear system of two equations in two variables by the **substitution method** is outlined in the following strategy.

SOLVING LINEAR SYSTEMS
BY SUBSTITUTION

1. Choose one of the equations; solve it for one of the variables in terms of the other variable.
2. Substitute the expression from step 1 for the variable in the other equation to obtain an equation in one variable.
3. Solve the equation resulting from step 2.
4. Substitute the value from step 3 back into either of the original equations to find the value of the second variable.

EXAMPLE 1   Solving a System by Substitution

Use the substitution method to solve the system.

$$\begin{cases} 3x + y = 7 \\ 2x - 3y = 1 \end{cases}$$

Solution

$$\begin{aligned} 3x + y &= 7 && \text{Choose first equation} \\ y &= -3x + 7 && \text{Solve equation 1 for } y \text{ in terms of } x \\ 2x - 3y &= 1 \\ 2x - 3(-3x + 7) &= 1 && \text{Substitute } -3x + 7 \text{ for } y \text{ in equation 2} \\ 2x + 9x - 21 &= 1 \\ 11x &= 22 \\ x &= 2 \\ 3(2) + y &= 7 && \text{Substitute 2 for } x \text{ in equation 1} \\ 6 + y &= 7 \\ y &= 1 \end{aligned}$$

Thus, (2,1) is the solution of the system.

Check:

Let $x = 2$ and $y = 1$

first equation	second equation
$3(2) + 1 = 7$	$2(2) - 3(1) = 1$

Therefore, (2,1) satisfies both equations.

EXAMPLE 2   Solving a System by Substitution
Use the substitution method to solve the system.

$$\begin{cases} \dfrac{x}{3} - \dfrac{y}{2} = \dfrac{8}{3} \\ -5x + 7y = -39 \end{cases}$$

Solution   We begin by rewriting the first equation without fractions.

$$6\left(\dfrac{x}{3} - \dfrac{y}{2}\right) = 6\left(\dfrac{8}{3}\right) \qquad \text{Multiply by 6, the LCD}$$

$$2x - 3y = 16$$

$$2x = 3y + 16 \qquad \text{Solve for } x$$

$$x = \dfrac{3}{2}y + 8$$

$$-5x + 7y = -39 \qquad \text{Second equation}$$

$$-5\left(\dfrac{3}{2}y + 8\right) + 7y = -39 \qquad \text{Substitute } \dfrac{3}{2}y + 8 \text{ for } x \text{ in equation 2}$$

$$\dfrac{-15}{2}y - 40 + 7y = -39$$

$$2\left[\dfrac{-15}{2}y - 40 + 7y\right] = 2(-39) \qquad \text{Multiply by 2, the LCD}$$

$$-15y - 80 + 14y = -78$$

$$-y = 2$$

$$y = -2$$

Substituting -2 for y in equation 2, we have

$$-5x + 7(-2) = -39$$

$$-5x - 14 = -39$$

$$-5x = -25$$

$$x = 5$$

Therefore,
$$\text{the solution is } (5, -2).$$
The reader is encouraged to check the solution.   ◈

Solving Linear Systems by Elimination

Another algebraic technique for solving linear systems of two equations in two variables is the **elimination method**. The strategy of this method is outlined in the following procedure.

SOLVING A SYSTEM BY ELIMINATION

1. Write each equation in the form $Ax + By = C$.
2. Multiply the terms of one or both equations by appropriate numbers so that the coefficients of one of the variables are opposites.
3. Add the resulting equations to produce a single equation in one variable, and solve that equation for the variable.
4. Substitute the value obtained in step 3 into either equation in the original system and solve this equation for the second variable.

In using the elimination method, our goal is to transform the original system into an "equivalent" system so that the coefficients of one of the variables are opposites. **Equivalent systems** of equations are those systems that have exactly the same solutions. The following three operations produce equivalent systems.

1. Interchange two equations in the system.
2. Multiply one equation of the system by a nonzero constant.
3. Multiply an equation in the system by a nonzero constant and add it to a given equation in the system.

EXAMPLE 3   Solving a System by Elimination

Use the elimination method to solve the system.

$$\begin{cases} 3x - 2y = 4 \\ 2x + y = 5 \end{cases}$$

Solution   Our objective is to transform our original system into an equivalent system so that the coefficients of one of the variables are opposites. We then add the equations to eliminate this variable. Notice that each equation in the system is of the form

$$Ax + By = C.$$

If we multiply each side of equation 2 by 2, the coefficients of the y terms will be opposites.

$$\begin{cases} 3x - 2y = 4 \xrightarrow{\hspace{1cm}\text{No change}\hspace{1cm}} \\ 2x + y = 5 \xrightarrow{\text{Multiply each side by 2}} \end{cases} \begin{cases} 3x - 2y = 4 \\ 4x + 2y = 10 \end{cases}$$

Now we add the two equations in the resulting system to obtain an equation in one variable.

$$\begin{cases} 3x - 2y = 4 \\ 4x + 2y = 10 \end{cases}$$
$$\overline{7x = 14}$$
$$x = 2$$

By substituting 2 for x into equation 1 in the original system, we have

$$3(2) - 2y = 4$$
$$6 - 2y = 4$$
$$-2y = -2$$
$$y = 1$$

The solution is (2,1).

Note that we could have substituted 1 for y in the equation

$$2x + y = 5$$

to obtain

$$2x + 1 = 5$$
$$2x = 4$$
$$x = 2$$

Thus, the same solution (2,1) is obtained.

EXAMPLE 4   Solving a System by Elimination

Use the elimination method to solve the system.

$$\begin{cases} 0.3x + 0.2y = 0.8 \\ 0.2x - 0.3y = 1.4 \end{cases}$$

Solution   Each equation in the system is of the form $Ax + By = C$. We multiply each side of both equations by 10 to clear the decimals.

$$\begin{cases} 0.3x + 0.2y = 0.8 \\ 0.2x - 0.3y = 1.4 \end{cases} \begin{matrix} \text{Multiply each side by 10} \\ \text{Multiply each side by 10} \end{matrix} \begin{cases} 3x + 2y = 8 \\ 2x - 3y = 14 \end{cases}$$

Next we make the coefficients of y opposites and eliminate y.

$$\begin{cases} 3x + 2y = 8 \\ 2x - 3y = 14 \end{cases} \begin{matrix} \text{Multiply each side by 3} \\ \text{Multiply each side by 2} \end{matrix} \begin{cases} 9x + 6y = 24 \\ 4x - 6y = 28 \end{cases}$$

Now we add the two equations to obtain an equation in one variable.

$$\begin{cases} 9x + 6y = 24 \\ 4x - 6y = 28 \end{cases}$$
$$\overline{13x = 52}$$
$$x = 4$$

By substituting 4 for x in equation 1 of the original system, we have

$$0.3(4) + 0.2y = 0.8$$
$$1.2 + 0.2y = 0.8$$
$$0.2y = -0.4$$
$$y = -2$$

Thus

the solution is $(4, -2)$.

Notice that we could have substituted 4 for x in equation 1 in the first "equivalent system" to obtain

$$3(4) + 2y = 8$$
$$12 + 2y = 8$$
$$2y = -4$$
$$y = -2$$

Thus the same solution $(4, -2)$ is obtained.   ◈

Examining Types of Systems Algebraically

So far each example we have solved in this section produced exactly one solution. The next example illustrates systems with either no solution or an infinite number of solutions. These possibilities should always be kept in mind when solving a linear system, regardless of the method used.

EXAMPLE 5   Solving Inconsistent and Dependent Systems

Use the elimination method to solve each system.

(a) $\begin{cases} 2x - 3y = 1 \\ -4x + 6y = -2 \end{cases}$ (b) $\begin{cases} -2x + 4y = -4 \\ 3x - 6y = 0 \end{cases}$

Solution   (a) Each equation in the system is of the form $Ax + By = C$. The coefficients of x and y will have opposite signs if we perform the following multiplication:

$$\begin{cases} 2x - 3y = 1 \quad \text{——Multiply each side by 2}\longrightarrow \\ -4x + 6y = -2 \text{——— No change ———} \longrightarrow \end{cases} \begin{cases} 4x - 6y = 2 \\ -4x + 6y = -2 \end{cases}$$

Adding, we get

$$\begin{cases} 4x - 6y = 2 \\ -4x + 6y = -2 \end{cases}$$
$$\overline{ 0 = 0}$$

We obtain the equation $0 = 0$ which is true for all values of x and y. This tells us that the system is *consistent* and the equations are *dependent*. Thus the system has an infinite number of solutions. If we write each equation in slope-intercept form, we have,

$$y = \frac{2}{3}x - \frac{1}{3}$$

We see that the graphs of these equations are identical (Figure 1a).

(b) Once again, each equation is of the form $Ax + By = C$. The coefficients of x and y will be opposites if we perform the following multiplication:

$$\begin{cases} -2x + 4y = -4\text{——Multiply each side by 3}\longrightarrow \\ 3x - 6y = 0\text{——Multiply each side by 2}\longrightarrow \end{cases} \begin{cases} -6x + 12y = -12 \\ 6x - 12y = 0 \end{cases}$$

Adding, we get

$$\begin{cases} -6x + 12y = -12 \\ 6x - 12y = 0 \end{cases}$$
$$\overline{ 0 = -12}$$

There are no values of x and y for which $0 = -12$ is true, so the system is *inconsistent* and has no solutions. Figure 1b shows the graph that verifies our conclusion.

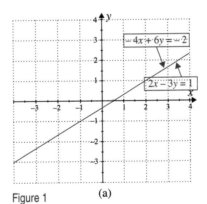

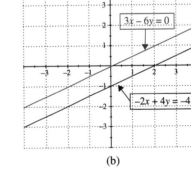

Figure 1   (a)   (b)

Solving Applied Problems Using Technology

Linear systems arise naturally as models for real-world situations. In business the point at which a company's costs equal its revenues is called the **break-even point**. If C represents the production cost (in dollars) of x units of a product, and R represents the revenue (in dollars) from the sale of x units of this product, then the break-even point is determined when $C = R$.

G EXAMPLE 6   Determining the Break-Even Point

A fitness equipment manufacturer produces a special kind of exercise bicycles. The cost C (in dollars) of producing x bicycles is given by

$$C = 220x + 180{,}000$$

The revenue R (in dollars) from selling x bicycles is given by

$$R = 310x$$

(a) Find the number of units produced and sold in order to break even.
(b) Sketch the graphs of the equations that represent C and R.
(c) Use the graph in part (b) to indicate when a loss occurs.
(d) Use the graph in part (b) to indicate when a profit occurs.

Solution   (a) The system of equations which models this situation is

$$\begin{cases} C = 220x + 180{,}000 \\ R = 310x \end{cases}$$

Since at the break-even point, $C = R$, we have the system

$$\begin{cases} C = 220x + 180{,}000 \text{—Multiply each side by } -1 \rightarrow \begin{cases} -C = -220x - 180{,}000 \\ C = 310x \text{————No change ————} \rightarrow \end{cases} C = 310x \end{cases}$$

Adding gives us

$$\begin{cases} -C = -220x - 180{,}000 \\ \ \ C = 310x \end{cases}$$
$$0 = 90x - 180{,}000$$
$$90x = 180{,}000$$
$$x = 2{,}000$$

Therefore, when 2000 units are produced and sold, the cost $620,000 of production equals the revenue $620,000.

(b) The graph of the system is in Figure 2. The viewing window is:

$$x\text{Min} = 0,\ x\text{Max} = 4000,\ x\text{Scl} = 500$$
$$y\text{Min} = 0,\ y\text{Max} = 800{,}000,\ y\text{Scl} = 200{,}000.$$

Using TRACE and ZOOM or INTERSECT we find

$$(2000, 620{,}000)$$

to be the point of intersection of the two lines which is the break-even point.

(c) A loss occurs when cost exceeds revenue; that is, when the cost line is above the revenue line. Therefore, when fewer than 2,000 units are produced and sold, the company suffers a loss.

(d) A profit occurs when revenue exceeds cost; that is, when the revenue line is above the cost line. Therefore, when more than 2,000 units are produced and sold, the company enjoys a profit. ◇

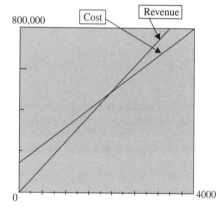

Figure 2

◆ PROBLEM SET 4.2

Mastering the Concepts

In Problems 1–18, use the substitution method to solve each system.

1. $\begin{cases} 2x - y = 5 \\ x + 3y = 13 \end{cases}$

2. $\begin{cases} 3x + y = 4 \\ 7x - 3y = 6 \end{cases}$

3. $\begin{cases} 5p - q = 13 \\ p + q = 1 \end{cases}$

4. $\begin{cases} 3r - 4t = -5 \\ 2r + 2t = 3 \end{cases}$

5. $\begin{cases} 2x - 3y = 4 \\ 3x + y = 1 \end{cases}$

6. $\begin{cases} 2x + 3y = 24 \\ 3x - 4y = -12 \end{cases}$

7. $\begin{cases} 3x - y = -2 \\ x + y = 6 \end{cases}$

8. $\begin{cases} 5x + 7y = 4 \\ -x + 6y = 14 \end{cases}$

9. $\begin{cases} 13a + 11b = 21 \\ 7a + 6b = -3 \end{cases}$

10. $\begin{cases} 2u - v = 3 \\ 3u + v = 22 \end{cases}$

11. $\begin{cases} 6x + y = -18 \\ -3x - \dfrac{1}{2}y = 9 \end{cases}$

12. $\begin{cases} 0.2m - 0.3n = 0.1 \\ 0.3m + 0.2n = 2.1 \end{cases}$

13. $\begin{cases} 0.5x - 1.2y = 0.3 \\ 0.7x + 1.5y = 3.6 \end{cases}$

14. $\begin{cases} \dfrac{3}{4}x + 3y = 30 \\ -5x - y = 28 \end{cases}$

15. $\begin{cases} 5p - q = 13 \\ p + 3q = -\dfrac{5}{3} \end{cases}$

16. $\begin{cases} 5r + s = 3 \\ -r - 2s = -\dfrac{12}{5} \end{cases}$

17. $\begin{cases} \dfrac{1}{2}x + \dfrac{1}{6}y = 1 \\ \dfrac{2}{3}x - \dfrac{1}{2}y = -3 \end{cases}$

18. $\begin{cases} \dfrac{3}{4}x - \dfrac{1}{12}y = -\dfrac{5}{2} \\ \dfrac{3}{8}x - \dfrac{1}{6}y = -\dfrac{1}{2} \end{cases}$

In Problems 19–40, use the method of elimination to solve each system.

19. $\begin{cases} 2x - 3y = -39 \\ 4x + 3y = 3 \end{cases}$

20. $\begin{cases} 2x + 3y = 18 \\ -2x + 5y = -34 \end{cases}$

21. $\begin{cases} 7x - 3y = -70 \\ 7x + 2y = -70 \end{cases}$

22. $\begin{cases} 5x - 2y = -12 \\ 3x - 2y = -4 \end{cases}$

23. $\begin{cases} 2x - y = 1 \\ x + y = 2 \end{cases}$

24. $\begin{cases} 2x + y = 10 \\ 3x - y = 5 \end{cases}$

25. $\begin{cases} 2r + 4s = 2 \\ -r + s = 8 \end{cases}$

26. $\begin{cases} 7p + q = 5 \\ -2p + q = 3 \end{cases}$

27. $\begin{cases} 7p + q = 3 \\ 5p + q = 6 \end{cases}$

28. $\begin{cases} 3y + 2z = 5 \\ 2y + 3z = 1 \end{cases}$

29. $\begin{cases} 4u - 3v = 1 \\ 3u + 4v = 6 \end{cases}$

30. $\begin{cases} 13y + 5z = 2 \\ 6y + 2z = 2 \end{cases}$

31. $\begin{cases} u + 3v = 9 \\ u - v = 1 \end{cases}$

32. $\begin{cases} 5u + v = 14 \\ 2u + v = 5 \end{cases}$

33. $\begin{cases} 3x + 2y = 4 \\ 5x + 3y = 7 \end{cases}$

34. $\begin{cases} 2x - 7y = 5 \\ x + y = -8 \end{cases}$

35. $\begin{cases} -3x + y = 3 \\ 4x + 2y = 10 \end{cases}$

36. $\begin{cases} 5x - 2y = 35 \\ x + 4y = 25 \end{cases}$

37. $\begin{cases} 0.1x + 0.3y = 0.4 \\ 0.2x + 0.9y = 0.3 \end{cases}$

38. $\begin{cases} 0.5x + 0.25y = 25 \\ 0.35x + 0.1y = 25 \end{cases}$

39. $\begin{cases} \dfrac{1}{2}x + \dfrac{1}{3}y = 13 \\ \dfrac{1}{5}x + \dfrac{1}{8}y = 5 \end{cases}$

40. $\begin{cases} \dfrac{1}{3}x - \dfrac{1}{4}y = 2 \\ \dfrac{1}{4}x - \dfrac{1}{2}y = 7 \end{cases}$

In Problems 41–46, use any algebraic method to solve each system. Also determine whether each system is consistent and independent, consistent and dependent or inconsistent. Confirm your results graphically.

41. $\begin{cases} 3x - 12y = 6 \\ 2x - 8y = 4 \end{cases}$

42. $\begin{cases} 2x + 6y = 8 \\ 3x + 9y = 12 \end{cases}$

43. $\begin{cases} 3x - 2y = 6 \\ -15x + 10y = 2 \end{cases}$

44. $\begin{cases} 4x - 9y = 12 \\ -8x + 18y = 10 \end{cases}$

45. $\begin{cases} \dfrac{1}{2}x + \dfrac{1}{3}y = 11 \\ \dfrac{1}{3}x - \dfrac{1}{5}y = 1 \end{cases}$

46. $\begin{cases} \dfrac{2}{3}x + \dfrac{3}{5}y = -17 \\ \dfrac{1}{2}x - \dfrac{1}{3}y = -1 \end{cases}$

Applying the Concepts

47. **Break-Even Analysis:** The cost C (in dollars) and the revenue R (in dollars) for producing and selling x units of a product are modeled by the equations:

$$\begin{cases} C = 8.5x + 75 \\ R = 10x \end{cases}$$

(a) Find the number of units that must be produced and sold in order to break even. That is, find the value of x for which $C = R$.

(b) Sketch the graphs of the equations that represent C and R.

(c) Use the graph in part (b) to indicate when a loss occurs.

(d) Use the graph in part (b) to indicate when a profit occurs.

48. **Break-Even Analysis:** Repeat problem 47 for the following cost and revenue models:

$$\begin{cases} C = 83x + 444 \\ R = 95x \end{cases}$$

49. **Health Insurance:** A family has a choice of two health insurance plans. The first option costs $400 per month and covers 80% of their health care expenses. The second option will cost $350 per month and will cover 70% of their health expenses. The total cost C_1 and C_2 (in dollars) of the two options after one year are modeled by the equations:

$$\begin{cases} \text{First option: } C_1 = 0.20x + 4800 \\ \text{Second Option: } C_2 = 0.30x + 4200 \end{cases}$$

where x (in dollars) is the total health care expenses for one year. At what level of health care expenses will both options cost the same amount of money for one year? That is, find the value of x for which C_1 and C_2 are equal.

50. **Car Rental:** A car rental agency offers two options for renting its cars. The first option charges $20 for the first day and $18 for each additional day. The second option charges $26 for the first day and $15 for each additional day. The total costs C_1 and C_2 (in dollars) of the two options are given by the models:

$$\begin{cases} C_1 = 18x + 20 \\ C_2 = 15x + 26 \end{cases}$$

where x is the number of additional days after the first. After how many additional days will both options cost the same amount of money? That is, find the value of x for which $C_1 = C_2$.

Developing and Extending the Concepts

In Problems 51–54, use the given substitution for u and v and the elimination method to solve each system.

51. $\begin{cases} 2(x+5)+3(y-2)=10 \\ (x+5)-3(y-2)=-4 \end{cases}$

Let $u=x+5, v=y-2$

52. $\begin{cases} 2(x-3)-(y+1)=-5 \\ 3(x-3)+2(y+1)=3 \end{cases}$

Let $u=x-3, v=y+1$

53. $\begin{cases} \dfrac{x+2}{2}+\dfrac{y+2}{3}=11 \\ \dfrac{x+2}{3}-\dfrac{y+2}{5}=1 \end{cases}$

Let $u=x+2, v=y+2$

54. $\begin{cases} \dfrac{-3(x-2)}{7}+\dfrac{2(y+4)}{5}=\dfrac{1}{35} \\ \dfrac{x-2}{5}+\dfrac{7(y+4)}{3}=\dfrac{1}{15} \end{cases}$

Let $u=x-2, v=y+4$

In Problems 55–58, use the substitution $u=1/x$ and $v=1/y$ and the method of elimination to solve each system.

55. $\begin{cases} \dfrac{2}{x}-\dfrac{1}{y}=9 \\ \dfrac{5}{x}-\dfrac{3}{y}=14 \end{cases}$

56. $\begin{cases} \dfrac{8}{x}+11=\dfrac{5}{y} \\ \dfrac{5}{x}-\dfrac{2}{y}=1 \end{cases}$

57. $\begin{cases} \dfrac{3}{x}+\dfrac{2}{y}=2 \\ \dfrac{1}{x}-\dfrac{1}{y}=9 \end{cases}$

58. $\begin{cases} \dfrac{4}{x}-\dfrac{3}{y}=1 \\ \dfrac{3}{x}-\dfrac{4}{y}=6 \end{cases}$

In Problems 59 and 60, determine the values of a, b or c so that the system:

59. (a) $\begin{cases} 4x-by=a \\ ax+2y=b \end{cases}$ has (3,2) as a solution

(b) $\begin{cases} 3x-cy=5 \\ 3x-4y=-2 \end{cases}$ is inconsistent

60. (a) $\begin{cases} 3x-2y=10 \\ cx+6y=5 \end{cases}$ is inconsistent

(b) $\begin{cases} x+cy=1 \\ 3x+18y=3 \end{cases}$ is consistent and dependent

G 61. **Data Analysis:** A system of equations can be used to find an equation of a line that fits a pair of points. That is, if $y=mx+b$ contains the point (3,5), then m and b must satisfy the equation $3m+b=5$. Find an equation of a line that fits the pair of points (3,5) and (7,8).

G 62. **Data Analysis:** As in problem 61, find the equation of a line that fits the pair of points $(-2,7)$ and $(4,-6)$.

OBJECTIVES

1. Solve Linear Systems With Three Variables By Elimination
2. Classify Linear Systems with Three Variables
3. Solve Applied Problems
4. Use Technology Exploration

4.3 Linear Systems in More than Two Variables

Many practical situations often lead to more than two equations with more than two variables. In this section, we examine linear systems with three variables such as:

$$\begin{cases} 2x-3y+2z=-1 \\ 4x+8y-3z=-16 \\ x+7y-3z=-8 \end{cases}$$

A **solution** of a linear system of three equations in three variables is an ordered triple (a,b,c) that satisfies all three equations in the system. For instance, the ordered triple $(-3,1,4)$ is a solution to the above system, since it satisfies each of the equations.

$$\begin{cases} 2(-3)-3(1)+2(4)=-1 \quad \text{true} \\ 4(-3)+8(1)-3(4)=-16 \quad \text{true} \\ (-3)+7(1)-3(4)=-8 \quad \text{true} \end{cases}$$

Solving Linear Systems with Three Variables by Elimination

Like the linear systems of two equations in two variables, linear systems with three variables are **equivalent** if they have the same solution. Graphical methods for illustrating the solutions of linear systems of three equations in three variables are impractical because such graphs may only be drawn in a three-dimensional coordinate system. Also, the substitution method is cumbersome when solving systems with three variables. The elimination method is the most suitable technique to solve this kind of a system. Its use is outlined in the following strategy.

SOLVING A LINEAR SYSTEM OF THREE
EQUATIONS IN THREE VARIABLES

1. Write each equation in the form $Ax + By + Cz = D$.
2. Select two equations in the system and eliminate one of the variables. The result is one equation in two variables.
3. Select one of the equations from step 2 and the third equation in the system. Eliminate the same variable from these equations. The result is another equation in the same two variables as step 2.
4. Solve this system of two equations in two variables.
5. Substitute the solution from step 4 into one of the original equations and solve for the third variable to complete the solution of the original system.
6. Check the solution in each of the original equations.

EXAMPLE 1   Solving a Linear System by Elimination

Use the elimination method to solve the system.

$$\begin{cases} x + 2y - z = -3 \\ 2x + 5y + 2z = 10 \\ -3x - 2y + 4z = 13 \end{cases}$$

Solution   We look at the coefficients of the variables and choose to eliminate x. First we use equations 1 and 2, and then use equations 1 and 3.

$$\begin{cases} x + 2y - z = -3 \text{ –Multiply each side by } -2 \to \\ 2x + 5y + 2z = 10 \text{ ——— No change ———} \to \end{cases} \begin{cases} -2x - 4y + 2z = 6 \\ 2x + 5y + 2z = 10 \end{cases}$$

Add these equations to eliminate x.   Add $y + 4z = 16$

$$\begin{cases} x + 2y - z = -3\text{— Multiply each side by } 3 \to \\ -3x - 2y + 4z = 13 \text{ ——— No change ———} \to \end{cases} \begin{cases} 3x + 6y - 3z = -9 \\ -3x - 2y + 4z = 13 \end{cases}$$

                                                             Add $4y + z = 4$

Now we have the new system.

$$\begin{cases} y + 4z = 16 \\ 4y + z = 4 \end{cases}$$

We continue to use the elimination method to solve this system. We eliminate y.

$$\begin{cases} y + 4z = 16\text{——Multiply each side by } -4 \longrightarrow \\ 4y + z = 4 \text{ ————— No change ————} \longrightarrow \end{cases} \begin{cases} -4y - 16z = -64 \\ 4y + z = 4 \end{cases}$$

                                                        Add $-15z = -60$

                                                              $z = 4$

Substitute 4 for z back into either equation in the system of two equations in two variables to obtain

$$y + 4(4) = 16 \qquad \text{or} \qquad 4y + 4 = 4$$
$$y = 0 \qquad\qquad\qquad\qquad y = 0$$

Now substitute 4 for z and 0 for y into any of the three original equations; we choose equation 1 to find x.

$$x + 2y - z = -3$$
$$x + 2(0) - 4 = -3$$
$$x = 1$$

Thus, the solution to the original system is (1,0,4). Once again we encourage the reader to check the answers. ◈

Classifying Linear Systems with Three Variables

As with the solutions of linear systems with two variables, linear systems of three variables produce three possible solutions. Although these systems do not lend themselves to graphical solutions, it is of interest to describe some possible solutions. The graph of a linear equation in three variables is a "plane" in three dimensional space. Thus, the solution of such a system is the intersection of three planes. These possibilities show the algebra—geometry connection and are summarized in Table 1.

TABLE 1

Geometry			
	The planes intersect at a point P. The system is *consistent* and the equations are *independent*.	All three planes do not have a common intersection. The system is *inconsistent*.	The planes intersect along a line. The system is *consistent* and the equations are *dependent*.
Algebra	The equations can be solved for x, y and z.	The variables cancel each other, and the statement is false.	The variables cancel each other, and the statement is true.
Type of Solution	The solution is the ordered triple (a,b,c).	No ordered triple is a solution.	An infinite number of ordered triple solutions.

EXAMPLE 2 Solving an Inconsistent System

Solve the system of equations by elimination.

$$\begin{cases} x + y - z = 9 \\ x + y - z = 7 \\ x - y + z = 3 \end{cases}$$

Solution First, we eliminate y using the first two equations.

In a given linear system of three equations in three variables, if one step produces an identity while another step produces a false equation, then the system is inconsistent.

$$\begin{cases} x + y - z = 9 \\ x + y - z = 7 \end{cases}$$ ——Multiply each side by -1 ⟶ $$\begin{cases} -x - y + z = -9 \\ x + y - z = 7 \end{cases}$$
——No change ⟶

Add $0 = -2$

Since $0 = -2$ is false, the system is inconsistent and has no solution. ◈

EXAMPLE 3 Solving a Consistent System

Solve the system of equations by elimination.

$$\begin{cases} x + 2y + 3z = 1 \\ 2x + 4y - z = 2 \\ 3x + 6y - 4z = 3 \end{cases}$$

Solution   First we eliminate z from the first two equations.

$$\begin{cases} x + 2y + 3z = 1 \text{────────No change ────────→} \\ 2x + 4y - z = 2 \text{────Multiply each side by 3────→} \end{cases} \begin{cases} x + 2y + 3z = 1 \\ 6x + 12y - 3z = 6 \end{cases}$$
$$\text{Add } 7x + 14y = 7$$

Next we eliminate z from the second and the third equations.

$$\begin{cases} 2x + 4y - z = 2 \text{—Multiply each side by } -4 \to \\ 3x + 6y - 4z = 3 \text{────────No change────────→} \end{cases} \begin{cases} -8x - 16y + 4z = -8 \\ 3x + 6y - 4z = 3 \end{cases}$$
$$\text{Add } -5x - 10y = -5$$

Now we solve the system.

$$\begin{cases} 7x + 14y = 7 \text{ ────────Divide each side by 7────────→} \\ -5x - 10y = -5 \text{────────Divide each side by 5────────→} \end{cases} \begin{cases} x + 2y = 1 \\ -x - 2y = -1 \end{cases}$$
$$\text{Add } 0 = 0$$

The identity $0 = 0$ indicates the system is consistent and the equations are dependent. ◇

Solving Applied Problems

In many real-world situations, linear systems of three equations in three variables serve as a mathematical model as the next example shows.

EXAMPLE 4   **Finding the Operating Cost of a Car at Different Speeds**

A car manufacturer determined that the cost C (in cents per mile) of operating a car is given by the function

$$C(x) = ax^2 + bx + c$$

where x is the speed of the car (in miles per hour) and a, b and c are constants.

(a) Write C in terms of x if the equation fits the data points (10,22), (20,20), and (50, 20). That is, find a, b, and c.

(b) Use the function in part (a) to determine the cost of operating the car at 70 miles per hour.

Solution   (a) Since the equation

$$C(x) = ax^2 + bx + c$$

fits the data points, we have

$$\begin{cases} 100a + 10b + c = 22 \\ 400a + 20b + c = 20 \\ 2500a + 50b + c = 20 \end{cases}$$

Solving this system for a, b and c, we get

$$a = 0.005,\ b = -0.35,\ \text{and } c = 25.$$

So that the equation relating speed to cost reads

$$C(x) = 0.005x^2 - 0.35x + 25$$

(b) Here we replace x by 70 to obtain

$$C(70) = 0.005(70)^2 - 0.35(70) + 25$$
$$C(70) = 25$$

Therefore, if the car is driven at 70 miles per hour, it will cost 25 cents per mile to operate. ◇

Using Technology Exploration

Although the graphical method is not practical when solving linear systems of equations, many graphers are equipped with a feature denoted by the symbol SIMULT which can solve these systems with relative ease. For instance, using the SIMULT feature, the solution of the system

$$\begin{cases} 2x - y + z = 3 \\ x + 4y - z = 6 \\ 3x + 2y + 3z = 16 \end{cases}$$

is (1,2,3).

⬥ PROBLEM SET 4.3

Mastering the Concepts

In Problems 1 and 2, determine whether or not the given ordered triple is a solution of the given system.

1. $\begin{cases} x + y + z = 6 \\ x + 2y + z = 9 \\ 2x + 2y - z = 6 \end{cases}$ (1,3,2)

2. $\begin{cases} -3x + 4y + z = -13 \\ 2x - 2y + 3z = -2 \\ 5x + 6y - 2z = 29 \end{cases}$ (2,−1,−3)

In Problems 3–16, use the method of elimination to solve each system.

3. $\begin{cases} x + y + 2z = 4 \\ x + y - 2z = 0 \\ x - y = 0 \end{cases}$

4. $\begin{cases} x + y + z = 2 \\ x + 2y - z = 4 \\ 2x - y + z = 0 \end{cases}$

5. $\begin{cases} x + y + z = 6 \\ x - y + 2z = 12 \\ 2x + y + z = 1 \end{cases}$

6. $\begin{cases} x + y + z = 6 \\ x - y + 2z = 7 \\ 2x + y + z = 14 \end{cases}$

7. $\begin{cases} x + y = 4 \\ 3x - y + 3z = 7 \\ 5x - 7y + 2z = -2 \end{cases}$

8. $\begin{cases} 7x + y + 3z = -6 \\ 4x - 5y + 6z = -27 \\ x + 15y - 9z = 61 \end{cases}$

9. $\begin{cases} 2p + q - 3r = 9 \\ p - 2q + 4r = 5 \\ 3p + q - 2r = 15 \end{cases}$

10. $\begin{cases} u + 3v - w = -2 \\ 7u - 5v + 4w = 11 \\ 2u + v + 3w = 21 \end{cases}$

11. $\begin{cases} 2r + 3s + t = 6 \\ r - 2s + 3t = 3 \\ 3r + s - t = 8 \end{cases}$

12. $\begin{cases} 8s + 3t - 18u = -76 \\ 10s + 6t - 6u = -50 \\ 4s + 9t + 12u = 10 \end{cases}$

13. $\begin{cases} a - 5b + 4c = 8 \\ 3a + b - 2c = 4 \\ 9a - 3b + 6c = 6 \end{cases}$

14. $\begin{cases} 2p + 3q - 2r = 3 \\ 8p + q + r = 2 \\ 2p + 2q + r = 1 \end{cases}$

15. $\begin{cases} 3x + y + 2z = 10 \\ -3x - 2y + 4z = -11 \\ 2y + z = 2 \end{cases}$

16. $\begin{cases} 2x - y + 2z = 2 \\ -2x + 3y + 4z = 10 \\ -2x + y - z = 0 \end{cases}$

In Problems 17–28, identify whether the given system is consistent and independent, consistent and dependent or inconsistent. Describe the solution geometrically; refer to Table 1.

17. $\begin{cases} p + q - r = 3 \\ p - 5q + r = 4 \\ -4p + 5q + r = 5 \end{cases}$

18. $\begin{cases} 2p + q - r = 4 \\ 4p - q + 2r = -2 \\ -2p + 2q - 3r = 6 \end{cases}$

19. $\begin{cases} x + y + z = 1 \\ 3x - 2y + 2z = 5 \\ 3x - 7y + z = 7 \end{cases}$

20. $\begin{cases} 3a - 5b - 2c = 9 \\ a - b + 2c = 3 \\ 2a - 3b = 1 \end{cases}$

21. $\begin{cases} 2p - q + 4r = -4 \\ 4p - 5q + 7r = 1 \\ -2p + 7q - 2r = 6 \end{cases}$

22. $\begin{cases} 3u + v - w = 8 \\ 2u - v + 2w = 3 \\ u + 2v - 3w = 5 \end{cases}$

23. $\begin{cases} a - 2b + c = -5 \\ -3a + 6b - 3c = 15 \\ 2a - 4b + 2c = -10 \end{cases}$

24. $\begin{cases} -2a - 4b + 6c = -8 \\ a + 2b - 3c = 4 \\ 4a + 8b - 12c = 16 \end{cases}$

25. $\begin{cases} \dfrac{p}{2} + \dfrac{q}{3} - \dfrac{r}{4} = -1 \\ \dfrac{p}{3} + \dfrac{r}{2} = 8 \\ \dfrac{2p}{3} + \dfrac{q}{3} - \dfrac{3r}{4} = -6 \end{cases}$

26. $\begin{cases} \dfrac{x}{6} + \dfrac{y}{3} + \dfrac{z}{2} = 1 \\ x - \dfrac{y}{2} + \dfrac{z}{2} = 1 \\ \dfrac{x}{6} - \dfrac{y}{2} + \dfrac{z}{3} = 0 \end{cases}$

27. $\begin{cases} 0.2x + 0.1y + 0.1z = 0.6 \\ 0.3x + 0.2y + 0.2z = 1.0 \\ -0.1x + 0.3y + 0.1z = 0 \end{cases}$

28. $\begin{cases} 0.5u + 1.5v - 0.5w = 2 \\ -1.5u - 2.5v + 0.5w = -4 \\ -0.5v + 1.5w = 7 \end{cases}$

Applying the Concepts

29. **Manufacturing:** A study of a manufacturing company shows that the profit P (in thousands of dollars) is a function of the number n (in hundreds) of full time employees as shown by the following equation:

$$P(n) = an^2 + bn + c, \ 0 \le n \le 100$$

Find the values of a, b and c such that this equation models the data (20,2000), (36,2512) and (70,200).

30. **Height of a Rocket:** A model rocket is launched and then it accelerates until the propellant burns out, after which it coasts upward to its highest point. The height h (in meters) above ground level t seconds after the burn out is modeled by the function

$$h(t) = at^2 + bt + c$$

Find the values of a, b and c such that this function models the data (1,110), (4,233), and (7,68).

31. **Rainfall:** A monthly rainfall data (x,y) for a city in a northwestern part of the United States fits the data (2,5), (4,3), and (8,1), where x represents a month, with $x = 1$ representing the month of January, and y represents the rainfall in inches. Find the values of a, b and c such that the function

$$f(x) = ax^2 + bx + c$$

models this data. Round off the answers to two decimal places.

32. **Electric Circuits:** In the electric circuits shown in Figure 1, I_1, I_2, and I_3 represent the amount of current (in amperes) flowing across the 1-ohm, 2-ohm, and 2-ohm resistors. The system of linear equations used to find the currents I_1, I_2, and I_3 is given by

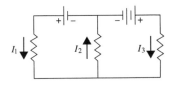

Figure 1

$$\begin{cases} I_1 + I_2 - I_3 = 0 \\ I_1 + 2I_3 = 12 \\ I_1 - 2I_2 = -4 \end{cases}$$

Solve the system for I_1, I_2, and I_3.

Developing and Extending the Concepts

33. Write a linear system of three equations in three variables that has $(3,2,-1)$ as a solution. How many possible systems can you have?

34. Use Table 1 to describe a way in which the graph of a linear system of three equations in three variables would indicate the system has:
    (a) one solution   (b) no solution
    (c) an infinite number of solutions.

In Problems 35–40, extend the substitution method to solve the given linear system of three equations in three variables.

35. $\begin{cases} x + y = 5 \\ x + z = 1 \\ y + z = 2 \end{cases}$

36. $\begin{cases} 2x + 3y = 28 \\ 3y + 4z = 46 \\ 4z + 5x = 53 \end{cases}$

37. $\begin{cases} x + y + 2z = 11 \\ x - y + z = 3 \\ 2x + y + 3z = 17 \end{cases}$

38. $\begin{cases} x - 3y = -11 \\ 2y - 5z = 26 \\ 7x - 3z = -2 \end{cases}$

39. $\begin{cases} 2p - q + r = 8 \\ p + 2q + 3r = 9 \\ 4p + q - 2r = 1 \end{cases}$

40. $\begin{cases} s + 3t - u = 4 \\ 3s - 2t + 4u = 11 \\ 2s + t + 3u = 13 \end{cases}$

In Problems 41 and 42, solve each system by using the substitutions $u = 1/x$, $v = 1/y$, and $w = 1/z$.

41. $\begin{cases} \dfrac{3}{x} - \dfrac{4}{y} + \dfrac{6}{z} = 1 \\ \dfrac{9}{x} + \dfrac{8}{y} - \dfrac{12}{z} = 3 \\ \dfrac{9}{x} - \dfrac{4}{y} + \dfrac{12}{z} = 4 \end{cases}$

42. $\begin{cases} \dfrac{3}{x} + \dfrac{1}{y} - \dfrac{1}{z} = 5 \\ \dfrac{4}{x} - \dfrac{1}{y} + \dfrac{2}{z} = 13 \\ \dfrac{2}{x} + \dfrac{2}{y} + \dfrac{3}{z} = 22 \end{cases}$

43. Determine the values of a, b and c so that

$$(a + b)x^2 + (c - 3a - 3b)x + 6a = 8x^2 - 17x + 30$$

44. The fraction $1/18$ can be written as

$$\frac{1}{18} = \frac{x}{2} + \frac{y}{3} + \frac{z}{9}$$

where x, y, and z are solutions to the system

$$\begin{cases} 4x + 3y + 2z = -7 \\ -4x + y + 3z = -24 \\ 3x - 2y - z = 8 \end{cases}$$

(a) Solve the system.
(b) Verify that the sum of the fractions is $1/18$.

G 45. Use a grapher to solve each system.

(a) $\begin{cases} 1.5a - 0.2b + 4.5c = 0.83 \\ 5a - 4b - 8c = 1.80 \\ 0.3a + 5b - 8c = -0.97 \end{cases}$

(b) $\begin{cases} 2p - 3.8q + 2.1r = 3.26 \\ 3.7p - 0.2q + 0.05r = 0.41 \\ 1.5p + q - 0.2r = 0.05 \end{cases}$

46. Suppose that the point of intersection of the graphs of the equations

$$\begin{cases} ax - y + 2z = 1 \\ 2x + by + 3z = 4 \\ x + y + cz = -9 \end{cases}$$

is $(3,-1,2)$. Find a, b, and c.

OBJECTIVES

1. Solve Applied Problems Involving Linear Systems with Two Variables
2. Solve Applied Problems Involving Linear Systems with Three Variables
3. Use Technology Exploration

4.4 Systems as Models

Quite often it is desirable or even necessary to use two or more equations involving more than one variable to model a real-world situation. In such a case, we construct a model by writing a system of equations that describes and answers questions about the situation. In section 2.3, we solved applied problems by using one equation with one variable. In keeping with our emphasis on modeling, we present in this section a number of applied problems using linear systems of equations. We will rely on the three-step format that we first employed in solving a single equation with one variable.

Problem Solving

Step 1. Make a plan.

Step 2. Write the solution.

Step 3. Look back.

Solving Applied Problems Involving Linear Systems with Two Variables

If an applied problem involves two unknowns, we use two variables and so we need a system of two equations.

EXAMPLE 1   Solving a Geometry Problem

The perimeter of the base of a rectangular box of pizza is 56 inches. The length l of one side of the box is 2 inches less than twice the width w. What are the length and width of the pizza box?

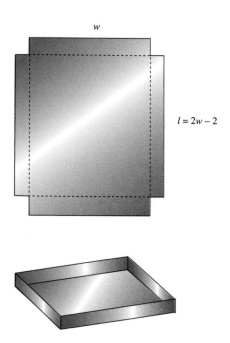

w

$l = 2w - 2$

Solution   Make a Plan

The two unknowns are the length and the width of the box. We use the formula for the perimeter P of a rectangle

$$P = 2l + 2w$$

where l is the length of the box in inches and w is the width of the box in inches. The following diagram is a visual description of the given.

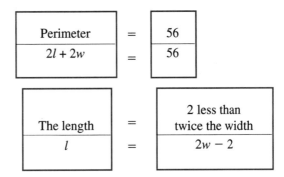

Translate the problem into symbols; then the length and width can be found by solving the system:

$$\begin{cases} 2l + 2w = 56 \\ l = 2w - 2 \end{cases}$$

Write the Solution

We use substitution to solve the system. From equation 2, we substitute the expression for l into equation 1.

$$2(2w - 2) + 2w = 56$$
$$4w - 4 + 2w = 56$$
$$6w - 4 = 56$$
$$6w = 60$$
$$w = 10$$
$$\text{and} \quad l = 2(10) - 2$$
$$l = 18$$

The width is 10 inches and the length is 18 inches.

Look Back

We check the results as follows:

The perimeter	length is two less than twice the width
$2l + 2w =$	$2(10) - 2 =$
$2(18) + 2(10) =$	$20 - 2 =$
$36 + 20 = 56$	18

EXAMPLE 2   Solving a Mixture Problem

A dietitian creates a nutritious breakfast by combining two types of grain. One type contains 10% protein and another contains 20% protein. How many grams of each type should be used to obtain a 100 gram mixture that contains 16% protein?

Solution   Make a Plan

The amount of protein is obtained by multiplying the quantity of grain by the percent of protein.

Let x = the number of grams of grain containing 10% protein.
y = the number of grams of grain containing 20% protein.

We organize the information in the following table.

	Amount of grain	Percent of protein	Total amount of protein
First Type	x	10%	$0.10x$
Second Type	y	20%	$0.20y$
Mixture	100	16%	0.16(100)

The quantity equation is

$$x + y = 100$$

The protein equation is

$$0.10x + 0.20y = 0.16(100)$$

So, we have the system of equations to solve.

$$\begin{cases} x + y = 100 \\ 0.1x + 0.2y = 16 \end{cases}$$

Write the Solution

$$\begin{cases} x + y = 100 \quad\text{—— No change ——}\longrightarrow \\ 0.1x + 0.2y = 16 \quad\text{——Multiply each side by } -10 \longrightarrow \end{cases} \begin{cases} x + y = 100 \\ -x - 2y = -160 \end{cases}$$

$$\text{Add } -y = -60$$
$$y = 60$$

Substitute this value into the original equation 1.

$$x + 60 = 100$$
$$x = 40$$

Therefore, the dietitian should mix 40 grams of the 10% grain with 60 grams of the 20% grain to get a mixture of 16% protein.

Look Back

Amount of Grain	Amount of Protein Mixed	Amount of protein required
40 + 60 = 100	0.10(40) + 0.20(60) = 4 + 12 = 16	0.16(100) = 16

EXAMPLE 3   Solving an Aviation Problem

A commercial jetliner flying with the wind takes 3.75 hours to fly the 2500 miles from Los Angeles to New York but 4.4 hours to fly the same distance from New York to Los Angeles against the wind. Assuming that the jetliner travels at a constant speed and the wind blows at a constant rate, find the speed of the jetliner in still air and the wind speed.

Solution   Make a Plan

Here we choose a variable to represent the speed of the jetliner in still air and another variable to represent the wind speed. Using these variables, we can express the speed of the jetliner with and against the wind. We also use the equation $d = rt$ to write expressions for the distance traveled by the jetliner.

$$\text{Let } j = \text{speed of the jetliner in still air and}$$
$$w = \text{speed of the wind.}$$

The following table visualizes these results.

	Rate (in miles per hour)	Time (in hours)	Distance (in miles)
With the wind (Los Angeles to New York)	$j + w$	3.75	$3.75(j + w)$
Against the wind (New York to Los Angeles)	$j - w$	4.40	$4.40(j - w)$

Now we use a linear system of two equations to determine how the expressions for the distance are related.

$$\begin{cases} 3.75(j + w) = 2500 \\ 4.40(j - w) = 2500 \end{cases}$$

Write the Solution

We solve the system as follows:

$$\begin{cases} 3.75(j + w) = 2500 \\ 4.40(j - w) = 2500 \end{cases} \begin{matrix} \text{Divide by 3.75} \\ \text{Divide by 4.40} \end{matrix} \rightarrow \begin{cases} j + w = 666.67 \\ j - w = 568.18 \end{cases}$$

$$\text{Add} \quad 2j = 1234.85$$
$$j = 617.43$$

To solve for w, substitute 617.43 in the equation $j + w = 666.67$ to obtain $w = 49.24$. Thus, the speed of the jetliner in still air is approximately 617.43 miles per hour and the wind speed is approximately 49.24 miles per hour rounded to two decimal places.

Look Back

To check the solution, we calculate the distance traveling with the wind and against the wind.

Distance with wind	Distance against wind
$3.75(617.43 + 49.24) =$ 2500.01	$4.40(617.43 - 49.24) =$ 2500.04

Solving Applied Problems Involving Linear Systems with Three Variables

With our focal theme on solving applied problems, we now consider linear systems with three variables that are used in modeling real-world situations. The method of solving such systems is an extension to the method of solving linear systems with two variables.

EXAMPLE 4   Solving an Investment Problem

A financial planner advised a customer to invest part of her $10,000 in a safe money market fund that paid 5% simple interest, part in bonds that paid 7% simple interest and the rest in a risky investment. If the risky investment earns 9% simple interest, the total earnings from all the investments for one year are $656. However, if the risky investment pays nothing for the year, then the total earnings from the first two investments are only $521. How much is invested in each fund?

Solution   Make a Plan

Here we have $10,000 to be divided into three investments:
money market, bonds and risky investments.
To find the amount invested at each rate, we let

x = the amount (in dollars) invested in the money market at 5%
y = the amount (in dollars) invested in bonds at 7%
z = the amount (in dollars) invested in a risky investment at 9%

Now, we use the formula

$$I = Prt$$

and analyze these investment in the following table.

Investment	Principal	Interest Rate	Interest
Money Market	x	5%	$0.05x$
Bonds	y	7%	$0.07y$
Risky	z	9%	$0.09z$

The three principals add to $10,000 and the sum of interests for one year is $656 and $521 depending on the return of the risky investment. The system of equations which represents this information is:

$$\begin{cases} x + y + z = 10000 \\ 0.05x + 0.07y + 0.09z = 656 \\ 0.05x + 0.07y = 521 \end{cases}$$

Write the Solution

We solve this system as follows:

$$\begin{cases} x + y + z = 10000 \\ 0.05x + 0.07y + 0.09z = 656 \\ 0.05x + 0.07y = 521 \end{cases} \begin{array}{l} \text{—— No change} \longrightarrow \\ \text{Multiply by 100} \rightarrow \\ \text{Multiply by 100} \rightarrow \end{array} \begin{cases} x + y + z = 10000 \\ 5x + 7y + 9z = 65600 \\ 5x + 7y = 52100 \end{cases}$$

We eliminate z from the first and second equations as follows:

$$\begin{cases} x + y + z = 10000 \\ 5x + 7y + 9z = 65600 \end{cases} \begin{array}{l} \text{Multiply each side by } -9 \rightarrow \\ \text{——No change——} \rightarrow \end{array} \begin{array}{r} -9x - 9y - 9z = -90000 \\ 5x + 7y + 9z = 65600 \\ \hline \text{Add } -4x - 2y = -24400 \end{array}$$

Now we have the following system of two equations in two variables:

$$\begin{cases} -4x - 2y = -24400 \\ 5x + 7y = 52100 \end{cases} \begin{array}{l} \text{Multiply each side by } 5 \rightarrow \\ \text{Multiply each side by } 4 \rightarrow \end{array} \begin{array}{r} -20x - 10y = -122000 \\ 20x + 28y = 208400 \\ \hline \text{Add } 18y = 86400 \\ y = 4800 \end{array}$$

Substitute this value for y in the equation

$$5x + 7y = 52100$$
$$5x + 7(4800) = 52100$$
$$5x = 18500$$
$$x = 3700$$

Finally, substitute these values into an original equation.

$$x + y + z = 10000$$
$$3700 + 4800 + z = 10000$$
$$z = 1500$$

Thus,

$$\$3700 \text{ was invested in mutual funds}$$
$$\$4800 \text{ in bonds}$$

and

$$\$1500 \text{ in the risky investment.}$$

Look Back

The sum of the investments is
$$\$10,000 = \$3700 + 4800 + 1500.$$
The money market fund earned
$$3700(0.05) = \$185.$$
The bonds earned
$$\$4800(0.07) = \$336.$$
The risky investment earned
$$\$1500(0.09) = \$135.$$
If the risky investment gives a 9% return, then the total return is:
$$\$185 + 336 + 135 = \$656.$$
The earnings without the risky investment is
$$\$185 + 336 = \$521.$$

G EXAMPLE 5   Solving a Manufacturing Problem

A manufacturer of portable radios produces three models of radios, model A, model B, and model C. The manufacturer determines that the total time allocated for production, assembly, and testing are 638 hours, 253 hours, and 159 hours respectively. The table lists the information for each model.

	Model A	Model B	Model C	Total
Production hours per radio	1.4	1.6	1.8	638
Assembly hours per radio	0.5	0.6	0.8	253
Testing hours per radio	0.3	0.4	0.5	159

Use a linear system to determine how many of each model can be produced using all the allocated time.

Solution   Make a Plan

$$\text{Let } a = \text{ the number of model } A \text{ radios produced}$$
$$b = \text{ the number of model } B \text{ radios produced}$$
$$c = \text{ the number of model } C \text{ radios produced}$$

Production requires $1.4a$ hours for model A, $1.6b$ for model B and $1.8c$ for model C for a total of 638 production hours, we have

$$1.4a + 1.6b + 1.8c = 638$$

Similarly, there are 253 hours available for assembly. We obtain,

$$0.5a + 0.6b + 0.8c = 253$$

The third equation considers that there are 159 hours available for testing, so we have

$$0.3a + 0.4b + 0.5c = 159$$

The following system describes this model:

$$\begin{cases} 1.4a + 1.6b + 1.8c = 638 \\ 0.5a + 0.6b + 0.8c = 253 \\ 0.3a + 0.4b + 0.5c = 159 \end{cases}$$

Using a grapher with   SIMULT   feature, we obtain the solution

$$a = 150, \ b = 110 \text{ and } c = 140$$

Write the Solution

Therefore, the manufacturer should produce 150 model A, 110 model B and 140 model C radios in order to use up the allotted time.

Look Back

By using the table, we have

$$\begin{cases} 1.4(150) + 1.6(110) + 1.8(140) = 638 \quad \text{true} \\ 0.5(150) + 0.6(110) + 0.8(140) = 253 \quad \text{true} \\ 0.3(150) + 0.4(110) + 0.5(140) = 159 \quad \text{true} \end{cases}$$ ◈

Using Technology Exploration

Many graphers are programmed to solve linear systems in two and three variables algebraically and graphically. However, we will only consider graphical solutions to linear systems with two variables.

EXAMPLE 6   Using a Grapher

Suppose that a vending machine contains $3.00 in nickels and dimes. If there are twice as many nickels as dimes, how many of each coin is in the machine?

Solution   Let n be the number of nickels and d the number of dimes. The value of the coins and the number of coins is represented by the system:

$$\begin{cases} 5n + 10d = 300 \\ n = 2d \end{cases}$$

We solve for n in terms of d to obtain the system:

$$\begin{cases} n = -2d + 60 \\ n = 2d \end{cases}$$

The viewing window for the grapher is:

$$x\text{Min} = 4,\ x\text{Max} = 18,\ x\text{Scl} = 1,$$
$$y\text{Min} = 0,\ y\text{Max} = 65,\ y\text{Scl} = 5$$

Figure 1 shows the intersection point (15,30) of the two lines on a grapher with d as the horizontal axis and n as the vertical axis. So there are 15 dimes and 30 nickels in the vending machine.

A grapher equipped with a SIMULT feature will verify that the algebraic solution of the system is 15 and 30.   ◈

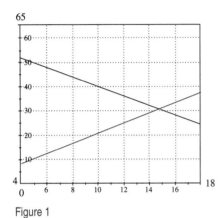

Figure 1

◈ PROBLEM SET 4.4

Mastering the Concepts

1. **Carpeting:** A carpet layer determines that the perimeter of a rectangular wedding hall is 900 feet.
   (a) Find the length and the width of the hall if its length is 50 feet more than its width.
   (b) If the carpet costs $12.75 per square yard installed, how much does it cost to carpet the hall?

2. **Dimensions of a Pool:** The perimeter of a rectangular swimming pool is 192 feet. Find the length and the width of the pool if its width is one-third of its length.

3. **Designing a Frame:** An artist wishes to frame a picture with 84 inches of framing material. How wide will the frame be if its width is 75% of its length?

4. **Fencing:** A rancher has 40 meters of fencing to enclose a rectangular pen next to a barn, with the barn wall forming the length of the pen (Figure 2). Find the length and width of the pen if its length is twice its width.

Figure 2

5. **Nutrition:** A dietitian arranges a special diet comprised of two types of foods, A and B. Each ounce of food A contains 10 units of calcium and 4 units of iron; each ounce of food B contains 6 units of calcium and 4 units of iron. How many ounces of foods A and B should be used to obtain a food mix that contains 92 units of calcium and 44 units of iron?

6. **Blending Oil:** An oil company mixes two types of oil, low-sulfur and high-sulfur. The low-sulfur oil has a sulfur content of 2%, while the high-sulfur oil has a sulfur content of 7%. How many barrels of each type should be used to obtain a mixture of 10,000 barrels with a sulfur content of 4%?

7. **Aviation:** An airliner flying with the wind takes 3.5 hours to fly 1,875 miles, and 4 hours to fly the same distance against the wind. Assuming that the airliner travels at a constant speed and the wind blows at a constant rate, find the speed of the airliner in still air and the wind speed. Round off the answers to two decimal places.

8. **Water Speed:** A cruising riverboat travels 36 miles downstream (with the current) in 2 hours. It travels the same distance up stream (against the current) in 3 hours. Assuming the riverboat travels at a constant speed in still water and the rate of the current is constant, find the speed of the riverboat in still water and the rate of the current.

9. **Cycling:** Two bicycle riders start at the same point and travel in opposite directions. One travels 5 kilometers per hour faster than the other. In 3 hours they are 119 kilometers apart. How fast is each traveling?

10. **Fitness:** A student walks 4 miles from home to school each day. She jogs back from school to home. She makes the trip walking to school in one hour and takes half an hour to return. How fast does she walk and jog?

11. **Investment Portfolio:** An inheritance of $40,000 is invested in two municipal bonds which pay 6% and 7% simple annual interest. If the annual interest from both bonds is $2,550, how much is invested at each rate?

12. **Investment Portfolio:** An investment club invested $35,000 in two accounts. During one year, the first account earned 5% annual simple interest and the second account earned 6% annual simple interest. The total interest from both investments for the year was $1880. How much was invested in each account?

13. **Investment Portfolio:** A total of $30,000 was invested in two mutual funds. One risky fund yields 10.5% annual simple interest; the other is a safe fund which earns 6%. At the end of the year, the combined interest from the two funds was $2340. How much was invested in each fund?

14. **Car Loan:** A new employee purchased a car and needed to borrow $16,000. He borrowed part of the money from the credit union at 9% annual simple interest and the remaining amount from his parents at 4% annual simple interest. If the total interest paid for one year was $1,190, how much was borrowed from the credit union?

15. **Coffee Mixtures:** A coffee shop mixes French roast coffee worth $9 per pound with kona coffee worth $4 per pound. How much of each type of coffee was used in order to obtain 20 pounds of a mixture worth $5 per pound?

16. **Window Cleaner Mixtures:** A window cleaner is made out of water and ammonia. Suppose that a window washer wishes to mix one brand of window cleaner that contains 5% ammonia with another brand that has 10% ammonia to form 100 liters of a mixture that is 7.5% ammonia. How many liters of each brand should be used (Figure 3)?

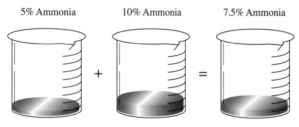

5% Ammonia   10% Ammonia   7.5% Ammonia

Figure 3

17. **Pharmaceutical Mixtures:** A pharmacist is to prepare a 50 fluid ounce solution that contains 54% glucose. The pharmacy has a 30% solution and a 90% solution in stock. How many ounces of each solution should be mixed to prepare the desired prescription?

18. **Antifreeze Mixtures:** An automobile radiator holds 16 liters of fluid. The radiator is filled with a mixture of 80% water and 20% antifreeze. How much of this mixture should be drained and replaced with pure antifreeze so that the final mixture will contain 40% antifreeze?

19. **Investment Portfolio:** A total of $20,000 is divided among three investments: a certificate of deposit paying 6% annual simple interest, stocks paying 7% annually, and municipal bonds paying 8% annually. The total return on the investments is $1390. The amount invested in the certificate of deposit plus the amount invested in municipal bonds is $11,000. Find the amount placed in each of the three investments.

20. **Geometry:** We know from geometry that the sum of the angles in any triangle is 180°. Suppose that the smallest angle is 78° less than the largest one. Also suppose that the middle angle is three times as large as the smallest angle. What is the degree measure of each angle (Figure 4)?

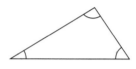

Figure 4

21. **Ticket Sales:** A concert was held in a sports arena that seats 20,000 people. Tickets were sold at $20, $15, and $10. There were twice as many $15 seats as the $20 seats. If the concert was sold out and grossed $260,000, how many of each kind were sold?

22. **Nutrition:** A hospital dietitian prepares a meal consisting of chicken, potatoes and peas. The dietitian determines that a patient meal should contain 415 calories, 50.5 grams of protein, and 553 grams of sodium. Table 1 lists the compositions of these mixes.

Use a linear system of three equations in three variables to determine how many servings of each should be used if all the allocation in Table 1 is used up.

TABLE 1

	3 ounce serving of chicken	one cup serving of potatoes	one cup serving of peas	total
Calories (cal)	140	160	125	415
Protein (gm)	27	4	8	50.5
Sodium (mg)	64	636	139	553

23. **Manufacturing:** A manufacturer of color printers produces three models of color printers: Model *A*, Model *B*, and Model *C*. The manufacturer determines that the total time allocated for production, assembly and testing is 100 hours, 100 hours, and 65 hours respectively. Table 2 lists the information for each model.

Use a linear system to determine how many of each model can be produced if all the allocated time is used up.

TABLE 2

One printer	Model *A*	Model *B*	Model *C*	Total
Production hours	2	1	3	100
Assembly hours	2	3	2	100
Testing hours	1	1	2	65

24. **Airport Shuttle:** An airport shuttle service has three sizes of vans. The biggest size van holds 12 passengers, the middle size holds 10 passengers, and the smallest size holds 6 passengers. On a certain day, the manager has 15 vans available to accommodate 152 passengers. However, to save on fuel costs, the manager decides to use 2 more of the available large size vans than the small size ones. How many vans of each kind should be used?

Developing and Extending the Concepts

25. **Number Problem:** The difference between two numbers is 12. The sum of the larger number and twice the smaller is 75. Find the numbers.

26. **Puzzle:** A boy and his girlfriend are arguing over the difference in their ages. The boy says that he is 3 years older than her. The girl claims that 4 years ago the sum of their ages was 23. If they are both right, how old is each now?

27. **Coin Problem:** A vending machine box contains 45 coins worth $8.70, all in quarters and dimes. How many quarters and how many dimes were there in the box?

28. **Plumber Wages:** A plumber charges a fixed charge plus an hourly rate for service on a house call. The plumber charged $70 to repair a water tank that required 2 hours of labor and $100 to repair a water tank that took 3.5 hours. Find the plumber's fixed charge and hourly rate.

29. **Height of Eiffel Tower:** The height of the Eiffel Tower (with the television mast added to the top) in Paris, France, is approximately 612 meters. An observer noticed that the difference in height between the tower and the television mast is approximately 262 meters. How tall is the tower (Figure 5)?

Figure 5

30. **Comparing Heights:** The sum of the heights of the tallest person and the shortest person in the world is approximately 3.30 meters. The difference in their heights is approximately 2.13 meters. How tall is each?

31. **Membership Fees:** A health club offers two membership plans. An active membership costs an initial fee of $175 plus $4 per visit. A social membership has no initial fee, but costs $7 per visit. If a person decides to join for one year, how often would that person have to visit for the active membership to be the less expensive plan?

32. **Tips:** Together two waiters worked a total of 14 hours. The more experienced waiter averaged $15 per hour in tips, whereas the newer waiter averaged $10 per hour. The waiters contributed 10% of their tips into a pool distributed to other workers who bus the tables. After paying into the pool, the waiters found that they had a total of $153 in tips. How many hours did each waiter work? How much did each waiter take home in tips?

33. **Shoe Sales:** A sports clothing store sells both running and tennis shoes. The tennis shoes sell for $46 per pair and the running shoes sell for $79 per pair. During a one-day sale, 260 pairs of shoes were sold for a total of $14,435. How many of each type of shoes were sold?

34. **Hours of Employment:** A student holds two part-time jobs totaling 20 hours per week. The first job pays $5.50 per hour and the second job pays $6.50 per hour. How many hours did the student work at each job if the gross weekly income is $117?

35. **Savings Accounts:** A couple engaged to be married decided they wanted to have $15,000 in savings by their wedding date. To reach their goal, the bride-to-be must double her savings and the groom-to-be must triple his savings. If they still have a total of $11,000 to set aside, how much does each one currently have in savings?

36. **Earnings:** One sales representative earned a base salary of $200 per week plus 6% commission on all sales. A second representative earned a base salary of $120 per week plus 8% commission on all sales. At the end of a special promotional campaign, the two representatives had earned $27,000 and $29,400, respectively. How long was the promotional campaign and how much did each representative sell?

37. **Car Sales:** Two friends compared notes about the cars they had bought recently. One friend paid 6% sales tax whereas the other paid 8.5% tax on the car. Together, their cars cost a total of $36,350 and they paid a total of $2546 in sales tax. How much did each car cost?

38. **Number Problem:** The sum of three numbers is 225. The first number is 50 more than the second. The third is 25 more than the sum of the second and the first. Find the numbers.

39. **Consumer:** An employee goes to breakfast at a nearby coffee shop. One day, he spent $2.20 for a glass of juice, one bagel, and 2 cups of coffee, excluding tax and tip. The second day he spent $2.35 for 2 bagels and 3 cups of coffee, excluding tax and tip. The third day he spent $2.75 for a glass of juice, 2 bagels, and a cup of coffee excluding tax and tip. What was the price of each item?

40. **Landscaping:** Figure 6 shows three circular flower beds which are mutually tangent to each other and with centers at A, B, and C and radii, r_1, r_2, and r_3 respectively. The distances between the centers of the beds are as follows:

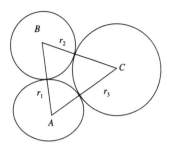

Figure 6

$\overline{AB} = 7$ feet, $\overline{BC} = 11$ feet, and $\overline{AC} = 10$ feet. Find the radii, r_1, r_2, and r_3, of the circular flower beds.

41. **Appliance Sales:** An appliance store sells dryers for $380, washing machines for $450, and microwave ovens for $140. On a certain day, the store sells 48 washers, dryers, and microwaves for total receipts of $12,630. If the number of microwaves sold for that day is twice the number of dryers, how many of each item was sold for that day?

42. **Work Rate:** Three painters A, B, and C can paint a court in a fitness center in 4 hours. When A and C work together, it takes 8 hours to paint the court. When A and B work together, it takes 6 hours to paint the court. How long would it take each of them to paint the court alone?

43. **Fertilizer Mixture:** A supplier of agriculture products has three types of fertilizer A, B, and C having nitrogen contents of 30%, 20%, and 15% respectively. The supplier plans to mix them, obtaining 600 pounds of fertilizer with a 25% nitrogen content. The mixture is to contain 100 pounds more of type C than of type B. How much of each type should be used?

G 44. **Coin Problem:** A vending machine which gives change must start each day with $12.50 in dimes and quarters. The number of dimes needs to be twenty more than the number of quarters. How many of each coin must be in the vending machine? Use a grapher to confirm your results.

G 45. **Coin Problem:** A cashier begins with 4 times as many nickels as quarters in the register. If the coins have a value of $3.60, how many of each coin is in the register? Use a grapher to confirm your results.

4.5 Systems of Linear Inequalities

OBJECTIVES

1. Graph Linear Systems of Inequalities
2. Solve Applied Problems
3. Use Technology Exploration

In Section 3.5, we have learned how to graph linear inequalities of the form
$$3x - 4y \le 12 \quad \text{or} \quad x - 2y \ge 6$$
Graphically, the solution set of each inequality is a line together with the region either below or above the line. Two inequalities considered together, such as
$$\begin{cases} 3x - 4y \le 12 \\ x - 2y \ge 6 \end{cases}$$
is called a **system of two inequalities** in two variables.

Graphing Linear Systems of Inequalities

Just as in systems of linear equations in two variables, the **graph** of a system of linear inequalities is the intersection of the graphs of each inequality in the system. That is, the set of all points that satisfy each inequality in the system. In this section, we limit our study of solutions of systems of inequalities to graphical techniques.

EXAMPLE 1   Solving a System of Inequalities

Sketch the graph of the solution set of the system of inequalities.
$$\begin{cases} 1 \le x \le 5 \\ 2 \le y \le 4 \end{cases}$$

Solution   This system is equivalent to the system

$$\begin{cases} x \geq 1 \\ x \leq 5 \\ y \geq 2 \\ y \leq 4 \end{cases}$$

Figure 1 shows the solution set of the system which consists of the region to the left and including the line $x = 5$, to the right and including the line $x = 1$, above and including the line $y = 2$ and below and including the line $y = 4$. ◇

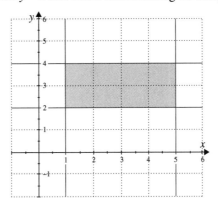

Figure 1

EXAMPLE 2   Solving a System of Inequalities

Sketch the graph of the solution set for the system of inequalities.

$$\begin{cases} x + y \geq 3 \\ 3x - y \geq 6 \end{cases}$$

Solution   In slope-intercept form we have the inequalities

$$x + y \geq 3 \qquad\qquad 3x - y \geq 6$$
$$ y \geq -x + 3 \qquad\qquad y \leq 3x - 6$$

Figure 2a shows the graph of $y \geq -x + 3$ whereas Figure 2b shows the graph of $y \leq 3x - 6$. Figure 2c shows the region common to both graphs. The solution set of the system consists of the points above and on $y = -x + 3$ and below and on $y = 3x - 6$, including the points on the line $y = 3x - 6$ from the point of intersection (2.25, 0.75).

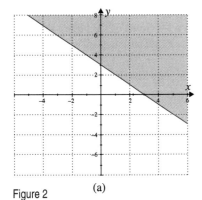

(a)

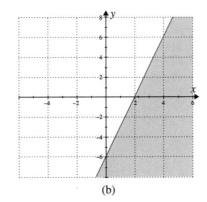

(b)

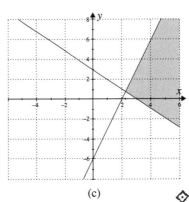

(c)   ◇

Figure 2

Systems of inequalities may include more than two inequalities as seen in the next example.

EXAMPLE 3   Solving a System of Inequalities

Sketch the graph of the solution set for the system of inequalities.

$$\begin{cases} 2x - y \geq 4 \\ 2x + 3y \leq 6 \\ x \geq 0 \\ y \geq 0 \end{cases}$$

Solution   Figure 3a shows the graph of $2x - y \geq 4$.
Figure 3b shows the graph of $2x + 3y \leq 6$.
Figure 3c shows the regions common to all graphs.
The solution set of the system consists of points

below the line $2x - y = 4$

and

below the line $2x + 3y = 6$.

The restriction

$$x \geq 0 \text{ and } y \geq 0$$

limits the solution to quadrant I, including the positive directions of the x and y axes and the origin.

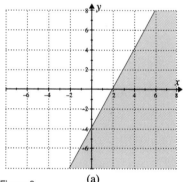
(a)

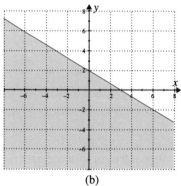
(b)

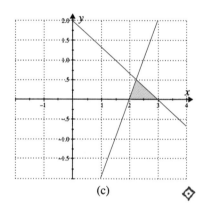
(c)

Figure 3

Solving Applied Problems

Many practical problems can be modeled by graphs of systems of inequalities.

EXAMPLE 4   Solving a Relief Problem

An airlift requires a total of at most 40 airplanes on a given day. Some of the planes are large and require a 4 person crew to fly. Some of the planes are small and require a 2 person crew. Both crews are selected from a pool of at most 120 crew members.
(a) Write a system of inequalities to describe this situation.
(b) Graph the solution set of this system.

Solution   Make a Plan

(a) First we find a system of inequalities with two variables to describe this situation. One inequality will describe the number of planes in the airlift, which is at most 40. The other will represent different arrangements of crew members to total at most 120 people.

Let x = number of large planes in the airlift
y = number of small planes in the airlift.

We form the system of inequalities as follows:

$$\begin{cases} x + y \le 40 \\ 4x + 2y \le 120 \\ x \ge 0, y \ge 0 \end{cases}$$

Write the Solution

(b) We graph both inequalities on the same coordinate system. The solution set of this system is the set of all ordered pairs that satisfy each inequality. That is, the intersection of the solution set of the individual inequalities. The inequalities are graphed as follows:

Shade the region below the solid line $x + y = 40$.
Shade the region below the line $4x + 2y = 120$ or $2x + y = 60$.
The restrictions on the number of planes and crew members limits us to quadrant I including the positive direction of the axes and the origin. Therefore, the solution of the system of inequalities, possible airlift schedules, is the set of all points within and on the boundaries of the shaded region in Figure 4.

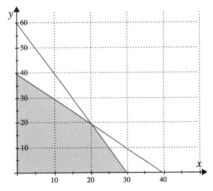

Figure 4

Look Back

All the points in the solution to this system of inequalities satisfy the conditions imposed in the problem. For example, (15,18) is a point in the shaded region. It would be possible to use 15 large airplanes and 18 small planes in the airlift. The number of planes being used $15 + 18 = 33$ is less than 40. There would be enough crew members: $4(15) + 2(18) = 96$ which is less than 120. Similarly, any point outside the shaded region would represent an impossible schedule. The point (21,24) is not in the shaded region; this would require more than 40 planes.   ◈

Using Technology Exploration

Graphers with SHADE features are helpful in graphing solution sets of systems of linear inequalities.

G EXAMPLE 5   Using a Grapher

To solve the system of linear inequalities with a grapher, first solve each inequality in the system for y in terms of x.

$$\begin{cases} x + y \geq 2 \\ -x + y \leq 2 \end{cases} \quad \text{becomes} \quad \begin{cases} y \geq -x + 2 \\ y \leq x + 2 \end{cases}$$

The solution is seen in the shaded part of the graph (Figure 5).   ◇

Figure 5

PROBLEM SET 4.5

Mastering the Concepts

In Problems 1–6, graph the solution set of each system of inequalities.

1. $\begin{cases} x \geq 2 \\ y \leq 3 \end{cases}$

2. $\begin{cases} x < -1 \\ y > 2 \end{cases}$

3. $\begin{cases} x \geq 0 \\ x + y < 4 \end{cases}$

4. $\begin{cases} y > 0 \\ x - y < 2 \end{cases}$

5. $\begin{cases} y \geq 0 \\ x - 2y \leq 6 \end{cases}$

6. $\begin{cases} x \geq -1 \\ x + 3y \geq 6 \end{cases}$

In Problems 7–10, write a system of linear inequalities in two variables whose solution set is represented by the shaded region.

7.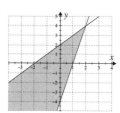

8.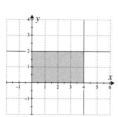

9.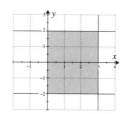

10.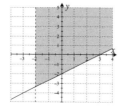

In Problems 11–14, sketch the graph of each system of linear inequalities.

11. $\begin{cases} x \geq 0, y \geq 0 \\ 3x + y \leq 6 \end{cases}$

12. $\begin{cases} x \geq 0, y \geq 0 \\ 2x - 3y \leq 12 \end{cases}$

13. $\begin{cases} x \geq 0, y \geq 0 \\ y \geq x, y \leq 4 \end{cases}$

14. $\begin{cases} x \geq 0, y \geq 0 \\ x \leq 4, y \leq 2 \end{cases}$

In Problems 15–30, sketch the graph of each system of linear inequalities.

15. $\begin{cases} x + y < 2 \\ 2x - y < -1 \end{cases}$

16. $\begin{cases} -x + y < 1 \\ -x + 3y \leq 5 \end{cases}$

17. $\begin{cases} x + y \leq 3 \\ x - y \geq 3 \end{cases}$

18. $\begin{cases} x + y > 5 \\ x - y < 9 \end{cases}$

19. $\begin{cases} x + 2y \leq 12 \\ x - y > 6 \end{cases}$

20. $\begin{cases} x + 2y > 12 \\ 3x - y < 1 \end{cases}$

21. $\begin{cases} x - y \geq 1 \\ x + y \leq 2 \end{cases}$

22. $\begin{cases} x - 2y \geq -2 \\ 2x + y \geq 2 \end{cases}$

23. $\begin{cases} 2x - 3y \leq -3 \\ 5x - 2y > 9 \end{cases}$

24. $\begin{cases} y \leq 2x + 4 \\ y \geq 3 - x \end{cases}$

25. $\begin{cases} x + y \leq 2 \\ -x + 3y \geq 4 \end{cases}$

26. $\begin{cases} 2x + y \geq 2 \\ x - 2y \leq 3 \end{cases}$

27. $\begin{cases} x \geq 0, y \geq 0 \\ 2x + y \leq 2 \\ x + 2y \leq 2 \end{cases}$

28. $\begin{cases} x \geq 0, y \geq 0 \\ x + y \geq 12 \\ 3x - y \geq 16 \end{cases}$

29. $\begin{cases} x \geq 0, y \geq 0 \\ x + y \geq 6 \\ 2x - 3y \geq 10 \end{cases}$

30. $\begin{cases} x \geq 0, y \geq 0 \\ x + y \leq 4 \\ 3x - y \leq 7 \end{cases}$

Applying the Concepts

31. **Investments:** An investment firm has at most $200,000 to invest in two stocks, A and B. Stock A sells at $56 a share and stock B at $86 a share; the total number of shares to be purchased of both stocks cannot exceed 3,000. Let x be the number of shares to be purchased of stock A and y be the number of shares to be purchased of stock B.
    (a) Write a system of inequalities that indicates the restrictions on x and y.
    (b) Graph the system showing the region of permissible values for x and y.

32. **Sales Promotion:** During a summer promotional campaign, a fitness center gave either one hour of free court time or a can of racquetball balls to new members. It cost the center $6 for each court time and $3 for each can of balls. The center could accommodate at most 179 new members and budgeted at most $900 for the incentives.
    (a) Write a system of inequalities that shows these restrictions.
    (b) Graph the system showing the possible combination of incentives.

33. **Comparing Salaries:** The sum of the weekly salaries of a server and a busser is at most $1200. The server makes at least $200 more per week than the busser.
    (a) Describe the distribution of salaries with a system of inequalities.
    (b) Sketch the graph of the system.
    (c) If the weekly salary of the server is $600, what is the most that the busser can make?

Developing and Extending the Concepts

In Problems 34–37, graph the system of inequalities.

34. $\begin{cases} |x + y| \le 3 \\ x \ge 0, y \le 0 \end{cases}$ 35. $\begin{cases} |x - y| < 4 \\ x \le 0, y \ge 0 \end{cases}$

36. $\begin{cases} |x| > |y| \\ x \le 0, y \ge 0 \end{cases}$ 37. $\begin{cases} |x| + |y| \le 3 \\ x \ge 0, y < 0 \end{cases}$

G In Problems 38–41, use the SHADE feature of a grapher to graph each system.

38. $\begin{cases} 2x + 4y < 3 \\ -2x - y > 4 \end{cases}$ 39. $\begin{cases} 3x - 5y \ge 15 \\ -2x + 6y \ge 12 \end{cases}$

40. $\begin{cases} 3x + y > 6 \\ x - y \ge 1 \end{cases}$ 41. $\begin{cases} 2x - y \le -2 \\ x - 2y \ge -2 \end{cases}$

◆ CHAPTER 4 REVIEW PROBLEM SET

In Problems 1–4, solve each system graphically by sketching the graphs of the equations in the system on the same coordinate system.
(a) Estimate the point of intersection (if any) of the two lines.
(b) Write each equation in slope-intercept form.
(c) Use the graph to indicate whether the system is consistent and independent, consistent and dependent or inconsistent.

1. $\begin{cases} x + y = 5 \\ x - y = 1 \end{cases}$ 2. $\begin{cases} x - 2y = 6 \\ 2x - 4y = 12 \end{cases}$

3. $\begin{cases} y = -2x + 6 \\ 4x + 2y = 8 \end{cases}$ 4. $\begin{cases} 3x - 4y = 1 \\ 2x - y = -1 \end{cases}$

In Problems 5–8, use the substitution method to solve each system.

5. $\begin{cases} x + 2y = 3 \\ 2x + y = 5 \end{cases}$ 6. $\begin{cases} 3x + 2y = 4 \\ 2x - 4y = 0 \end{cases}$

7. $\begin{cases} \dfrac{1}{2}x + \dfrac{1}{3}y = 8 \\ \dfrac{1}{4}x - \dfrac{1}{2}y = 0 \end{cases}$ 8. $\begin{cases} 0.04x + 0.2y = 8 \\ 0.06x - 0.1y = 4 \end{cases}$

In Problems 9–12, use the elimination method to solve each system (if possible). Determine whether the system is consistent and independent, consistent and dependent or inconsistent.

9. $\begin{cases} y = 4 - 2x \\ 2x - y = 4 \end{cases}$ 10. $\begin{cases} 2x + 3y = 5 \\ 3x + 5y = 7 \end{cases}$

11. $\begin{cases} x - y = 1 \\ 3x - y = -1 \end{cases}$ 12. $\begin{cases} 3x - 2y = 3 \\ 2x + 5y = -1 \end{cases}$

In Problems 13–24, use any method to solve each system (if possible).

13. $\begin{cases} x + y = \dfrac{9}{4} \\ 3x + 4y = 10 \end{cases}$ 14. $\begin{cases} x - y = 1 \\ x + \dfrac{5}{3}y = \dfrac{4}{3} \end{cases}$

15. $\begin{cases} x + y + 2z = 11 \\ x - y + z = 3 \\ 2x + y + 3z = 17 \end{cases}$ 16. $\begin{cases} x + 3y + z = 4 \\ 3x - 2y + 4z = 11 \\ 2x + y + z = 13 \end{cases}$

17. $\begin{cases} x - y + 2z = 1 \\ x - 4y + 3z = 1 \\ x + 2y - z = 1 \end{cases}$ 18. $\begin{cases} x - y - z = 5 \\ -2x + 2y + 2z = 2 \\ 4x + 2y + 2z = 8 \end{cases}$

19. $\begin{cases} x + y + z = 2 \\ x + y + z = 3 \\ x + y + z = 4 \end{cases}$ 20. $\begin{cases} 3x - y - z = 1 \\ 6x - 2y - 2z = 2 \\ 9x - 3y - 3z = 0 \end{cases}$

21. $\begin{cases} \dfrac{x}{2} + \dfrac{y}{3} - \dfrac{z}{4} = 3 \\ \dfrac{x}{3} + \dfrac{y}{2} = 5 \\ \dfrac{x}{2} + \dfrac{y}{3} - \dfrac{z}{4} = 3 \end{cases}$ 22. $\begin{cases} \dfrac{p}{6} + \dfrac{q}{3} + \dfrac{r}{2} = 5 \\ p - \dfrac{q}{2} + \dfrac{r}{2} = 6 \\ \dfrac{p}{6} - \dfrac{q}{2} + \dfrac{r}{3} = 0 \end{cases}$

23. $\begin{cases} \dfrac{1}{x} + \dfrac{3}{y} - \dfrac{4}{z} = 0 \\ \dfrac{1}{x} - \dfrac{1}{y} + \dfrac{2}{z} = 4 \\ \dfrac{1}{x} + \dfrac{2}{y} - \dfrac{1}{z} = 4 \end{cases}$ 24. $\begin{cases} \dfrac{2}{x} - \dfrac{1}{y} + \dfrac{5}{z} = 18 \\ \dfrac{2}{x} - \dfrac{3}{y} - \dfrac{4}{z} = -15 \\ \dfrac{1}{x} - \dfrac{4}{y} + \dfrac{3}{z} = 0 \end{cases}$

In Problems 25–32, sketch the graph of the solution set of each system of inequalities.

25. $\begin{cases} x \geq 2 \\ y \leq 2 \end{cases}$

26. $\begin{cases} x \geq 0 \\ y < 0 \end{cases}$

27. $\begin{cases} x \geq 1 \\ y \geq -1 \end{cases}$

28. $\begin{cases} x \geq y \\ y \geq x \end{cases}$

29. $\begin{cases} y - 3 \leq 0 \\ x + y > 2 \end{cases}$

30. $\begin{cases} y - 1 \leq 2 \\ x + 4 \geq 3 \end{cases}$

31. $\begin{cases} 2y - x < 4 \\ x + 2y \leq 6 \end{cases}$

32. $\begin{cases} 3y - x \geq -9 \\ x + 3y \geq -3 \end{cases}$

33. **Geometry:** The sum of the measure of two angles is 180°. If the difference between their angle measures is 32°, find the measure of each angle.

34. **Discount:** A publisher offered a college bookstore a 20% discount on hardback textbooks and a 10% discount on paperbacks. The regular price for the purchase of a shipment of books was $75,450, on which the total discount was $12,570. How much did the bookstore spend on each kind of book?

35. **Merchandising:** The total cost of a TV and a VCR is $890. If twice the cost of the VCR is $70 more than the cost of the TV, how much does the VCR cost?

36. **Furniture Sale:** A sofa and a matching chair were on sale for $1,990. If the sale price of the sofa was $485 less than twice the sale price of the chair, what was the sale price of the sofa?

37. **Stock Portfolio:** An investor owns stock in a company A selling at $44 per share, and stock in a company B selling at $72 per share. The investor owns 4 times as many shares of company A as of company B. If the investor's portfolio is worth $248,000, how many shares of each type does the investor own?

38. **Fencing:** A farmer has 720 feet of fencing to enclose a rectangular pasture. Because a river runs along the length of one side of the pasture, fencing will be needed on only three sides. Find the dimensions of the pasture if its length is 3 times its width.

39. **Air Travel:** Two airplanes leave the same airport at the same time and fly in opposite directions. Suppose that one airplane averages 120 miles per hour ground speed more than the other airplane. Find the ground speed of each airplane if they are 3000 miles apart in 3 hours.

40. **Coin Collection:** A coin collector has a collection of 174 coins consisting of dimes and nickels. If she thinks that each coin is worth 3 times its face value, then her collection is worth $40.35. How many of each type of coin does she have?

41. **Inventory:** A car dealer stocks minivans and sport utility vehicles and cannot hold more than 90 vehicles at one time. On the average, the manufacturer finances the dealership $15,000 on the purchase of a minivan and $18,000 on the purchase of a sport utility vehicle. How many vehicles of each type should the dealer stock if its total debt cannot exceed $1,446,000? Graph this region and interpret it.

42. **Investment:** A fund uses no more than $200,000 for two investments, one portion at 5% annual simple interest and the rest at 6% annual simple interest. Suppose that the total interest from both investments for one year is at least $10,600. Using inequalities, how much money should be put in each investment? Graph this region and interpret it.

G In Problems 43–46, use a TRACE feature of a grapher to find the approximate point of intersection graphically.

43. $\begin{cases} x + y = \dfrac{9}{4} \\ 3x + 4y = 10 \end{cases}$

44. $\begin{cases} 3x - 6y = 3.9 \\ 0.2x - 0.4y = 0.5 \end{cases}$

45. $\begin{cases} 0.2x + 0.3y = 0.7 \\ 0.4x - 0.5y = 0.3 \end{cases}$

46. $\begin{cases} 4.01x + 6.07y = 2.33 \\ 7.57x - 3.11y = 13.81 \end{cases}$

◆ CHAPTER 4 PRACTICE TEST

1. Solve each system graphically by sketching the graphs of the equations on the same coordinate system. Then estimate the point of intersection (if any) of the two lines. Write the two equations in each system in slope intercept form.

   (a) $\begin{cases} x - 2y = 1 \\ 2x + y = 7 \end{cases}$

   (b) $\begin{cases} 2x + y = 1 \\ 4x + 2y = 3 \end{cases}$

   (c) $\begin{cases} -x + 2y = 1 \\ 3x - 6y = -3 \end{cases}$

2. Indicate whether the systems in Problem 1 are consistent and dependent, consistent and independent or inconsistent.

3. Use any method to solve the systems. Determine whether or not the systems are consistent.

   (a) $\begin{cases} 2x - y = 3 \\ -3x + 2y = -7 \end{cases}$

   (b) $\begin{cases} \dfrac{3}{5}x - \dfrac{y}{5} = 1 \\ 2x + y = 20 \end{cases}$

   (c) $\begin{cases} x - y + z = 3 \\ x + y - z = 9 \\ x + y - z = 7 \end{cases}$

4. Determine whether or not $(-2,5,3)$ is a solution of the system.

$$\begin{cases} x + 2y - 3z = -1 \\ x + y - z = 0 \\ 3x - y + 3z = -2 \end{cases}$$

5. Graph the solution set of each system of inequalities.

(a) $\begin{cases} x \geq 1 \\ y < 2 \end{cases}$ (b) $\begin{cases} x - y \leq 3 \\ x + 2y \leq 4 \end{cases}$

6. Linda paid for her $4.25 lunch with 55 coins. If all of the coins were dimes and nickels, then how many were there of each type?

7. An investor lost twice as much in a mutual fund as she did in a bond fund. If her losses totaled $12,690, then how much did she lose in each market?

Chapter 5

EXPONENTS, RADICALS, AND COMPLEX NUMBERS

5.1 Exponential Expressions and Equations
5.2 Radicals
5.3 Rational Exponents
5.4 Simplify Radical Expressions
5.5 Operations with Radicals
5.6 Equations Involving Radicals and Rational Exponents
5.7 Complex Numbers

Did you ever wonder how a grandfather clock that keeps correct time is designed? In 1656 the first pendulum clock was built by Christian Huygens who discovered this model. The length of time T (in seconds) of one swing back and forth of a pendulum of length L (in feet) is given by the model

$$T = 2\pi \sqrt{\frac{L}{g}}$$

where g is the acceleration due to gravity ($g = 32$ feet per second squared). Find the time, T, when the length of the pendulum, L, is given by 1, 2, 2.25, 2.50, 2.75, 3.00, 3.25, 3.50, 3.75 and 4.00 feet. To answer this question see Example 8 on page 238.

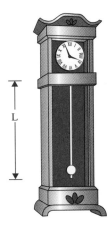

So far, we have encountered some examples of mathematical applications involving integer exponents. In this chapter, we reexamine some of these applications and models and raise some new questions concerning rational numbers as exponents. We shall see that rational exponents require a consideration of roots and radicals, and these in turn will lead us to investigate a new number system called the *complex numbers*. Functions and their graphs, including those produced by a grapher, will continue to play a role in the chapter.

OBJECTIVES

1. Extend the Properties of Positive Exponents
2. Solve Exponential Equations
3. Solve Applied Problems
4. Use Technology Exploration

5.1 Exponential Expressions and Equations

Extending the Properties of Positive Exponents

In section R5, we presented the properties of positive integral exponents. In this section, we will extend these properties to include all integer exponents, positive, negative and zero exponents. First, let us consider the definitions of zero and negative exponents.

ZERO AND NEGATIVE EXPONENTS

If a is a real number such that $a \neq 0$ and n is an integer, then

(1) $a^0 = 1$ (2) $a^{-n} = \dfrac{1}{a^n}$ (3) $\dfrac{1}{a^{-n}} = a^n$

The following results are direct consequences of the preceding definitions.

If $a \neq 0$ and $b \neq 0$, then

(1) $\left(\dfrac{a}{b}\right)^{-n} = \left(\dfrac{b}{a}\right)^n$ (2) $\dfrac{a^{-n}}{b^{-m}} = \dfrac{b^m}{a^n}$

EXAMPLE 1   Rewriting Expressions with Nonnegative Exponents

Rewrite each expression without negative exponents. Assume all variables are restricted to values for which all expressions represent real numbers.

(a) 7^{-1} (b) $\left(\dfrac{2}{3}\right)^{-2}$ (c) $(0.25)^{-1}$ (d) $\left(\dfrac{p}{q}\right)^{-3}$

(e) $(a + 2)^0$ (f) $\dfrac{1}{a^{-3}}$ (g) $\dfrac{y^{-2}}{x^{-2}}$ (h) $3^{-1} + 4^{-1}$

Solution   (a) $7^{-1} = \dfrac{1}{7}$ (b) $\left(\dfrac{2}{3}\right)^{-2} = \left(\dfrac{3}{2}\right)^2 = \dfrac{9}{4}$

(c) $(0.25)^{-1} = \left(\dfrac{1}{4}\right)^{-1} = \dfrac{4}{1} = 4$ (d) $\left(\dfrac{p}{q}\right)^{-3} = \left(\dfrac{q}{p}\right)^3 = \dfrac{q^3}{p^3}$

(e) $(a + 2)^0 = 1$ (f) $\dfrac{1}{a^{-3}} = a^3$

(g) $\dfrac{y^{-2}}{x^{-2}} = \dfrac{x^2}{y^2}$ (h) $3^{-1} + 4^{-1} = \dfrac{1}{3} + \dfrac{1}{4} = \dfrac{4 + 3}{12} = \dfrac{7}{12}$

Because of the nature of the definitions for zero and negative exponents, the properties of exponents still hold when the exponents are integers.

PROPERTIES OF INTEGER EXPONENTS

Let a and b be nonzero real numbers, and let m and n be integers. Then each of the following is true for all values of a and b for which both sides of the equation are defined:

(i) $a^m \cdot a^n = a^{m+n}$

(ii) $(a^m)^n = a^{mn}$

(iii) $(ab)^m = a^m b^m$

(iv) $\left(\dfrac{a}{b}\right)^m = \dfrac{a^m}{b^m}$

(v) $\dfrac{a^m}{a^n} = a^{m-n}$

EXAMPLE 2   Using Properties of Exponents

Rewrite each expression using only positive exponents and simplify the results.

(a) $x^6 \cdot x^{-2}$ (b) $(p^4)^{-3}$ (c) $(6y^3)^{-2}$ (d) $\left(\dfrac{3ab^0}{a^2(8b)^0}\right)^{-2}$ (e) $\left(\dfrac{x^{-4}(-3x)}{(-5x)^{-2}}\right)^{-1}$

Solution

(a) $x^6 \cdot x^{-2} = x^{6-2} = x^4$ Property (i) Product Property

(b) $(p^4)^{-3} = p^{-12} = \dfrac{1}{p^{12}}$ Property (ii) Power Property

(c) $(6y^3)^{-2} = 6^{-2} \cdot (y^3)^{-2}$ Property (iii)

$= \dfrac{1}{6^2} \cdot \dfrac{1}{(y^3)^2}$

$= \dfrac{1}{36} \cdot \dfrac{1}{y^6} = \dfrac{1}{36y^6}$

(d) $\left(\dfrac{3ab^0}{a^2(8b)^0}\right)^{-2} = \left(\dfrac{3a(1)}{a^2(1)}\right)^{-2}$ Definition of zero exponent

$= \left(\dfrac{3}{a}\right)^{-2}$ Property (v) Quotient Property

$= \left(\dfrac{a}{3}\right)^2 = \dfrac{a^2}{9}$

(e) $\left(\dfrac{x^{-4}(-3x)}{(-5x)^{-2}}\right)^{-1} = \dfrac{x^{-4(-1)}(-3x)^{-1}}{(-5x)^{-2(-1)}}$ Properties (iii) and (iv)

$= \dfrac{x^4}{(-5x)^2(-3x)}$

$= \dfrac{x^4}{25x^2(-3)x} = \dfrac{x^4}{-75x^3}$

$= -\dfrac{x}{75}$

Solving Exponential Equations

Up until now, we have focused on the properties of exponential expressions. Now we consider a strategy for solving exponential equations such as

$$2^x = 8, \qquad 4^{2x} = 16, \qquad \text{and} \qquad 3^{2x} = 9$$

For instance, to solve the exponential equation

$$5^{2x} = 25$$

first we rewrite the equation in the form $5^{2x} = 5^2$. Then observe that because the base of the exponentials is the same the exponents must be equal; that is, $2x = 2$ or $x = 1$. To solve this equation, we used the following property:

EXPONENTIAL PROPERTY

> Let a be a positive real number such that $a \neq 1$ and suppose that m and n are real numbers. Then
>
> $$a^m = a^n \text{ if and only if } m = n$$

EXAMPLE 3   Solving Exponential Equations

(a) $2^x = 16$ (b) $4^{-x} = 256$

Solution   (a) First we express 16 as a power of 2 so that both sides of the equation will have the same base. Because $16 = 2^4$, we can rewrite

$$2^x = 16 \text{ as } 2^x = 2^4$$
$$x = 4$$

The solution is 4. Check this in the original equation.

(b) $4^{-x} = 256$ Original equation
$ 4^{-x} = 4^4$ Property of exponential equation
$ -x = 4 \text{ or } x = -4$
The solution is -4.

Solving Applied Problems

In section R5, we saw that work in science and engineering often involves the use of very large as well as very small numbers. It is often convenient to represent such numbers in *scientific notation*. For example,

$$0.003 = 3 \times 10^{-3}$$
$$0.00007 = 7 \times 10^{-5}$$
$$0.00000059 = 5.9 \times 10^{-7}$$

EXAMPLE 4   Using Scientific Notation

A biologist determined that the radius r of a cell (in the form of a sphere) is 1.5×10^{-4} millimeter. Use the formula $V = (4/3)\pi r^3$ to find the volume of this cell. Round off to two decimal places.

Solution   We replace r by 1.5×10^{-4} in the formula to obtain

$$V = \frac{4}{3}\pi r^3$$
$$= \frac{4}{3}\pi(1.5 \times 10^{-4})^3$$
$$= 1.41 \times 10^{-11}$$

The volume of the cell is approximately 1.41×10^{-11} cubic millimeter.

Compound Interest: Bankers use the formula

$$S = P\left(1 + \frac{r}{n}\right)^{nt}$$

called the **compound interest formula** to model the **future value** S (in dollars) of the **present value** or **principal** P (in dollars) invested for a **term** of t years at an **annual interest rate** r **compounded** n **times** per year.

EXAMPLE 5   Hourly Wage Increase

Suppose that the starting hourly wage of a parttime employee in a fast food restaurant is \$5.50. After one year, the employee receives a 6% annual raise and may choose

(a) a 6% raise in 4 installments over the course of the year

or

(b) a 6% raise in one installment each year.

Find the hourly wage for each method after 3 years.

Solution   (a) For a 6% raise in 4 installments over the course of the year, we replace P by 5.50, n by 4, t by 3 and r by 0.06 in the formula.

$$S = P\left(1 + \frac{r}{n}\right)^{nt}$$

$$S = 5.50\left(1 + \frac{0.06}{4}\right)^{(4)(3)}$$

$$\approx 6.58$$

Therefore, the hourly wage of the employee after 3 years is approximately \$6.58.

(b) If a 6% raise is given in one installment per year, all the variables are the same except for $n = 1$. We have

$$S = P\left(1 + \frac{r}{n}\right)^{nt}$$

$$S = 5.50\left(1 + \frac{0.06}{1}\right)^{(1)(3)}$$

$$\approx 6.55$$

Therefore, receiving the same increase over 4 installments in one year, returns more money than over a single installment in one year.   ◇

Using Technology Exploration

A function of the form

$$f(x) = b^x$$

where $b > 0$, $b \neq 1$ and x is a real number is called an **exponential function with base** b. In the next example, we examine graphs of exponential functions and show their distinctive features.

G EXAMPLE 6   Graphing Exponential Functions
Let $f(x) = 2^x$ and $g(x) = \left(\frac{1}{2}\right)^x$

(a) Fill in the table for the given input values.

x	$f(x) = 2^x$	$g(x) = (1/2)^x$
-2	$f(-2) =$	$g(-2) =$
-1	$f(-1) =$	$g(-1) =$
0	$f(0) =$	$g(0) =$
1	$f(1) =$	$g(1) =$
2	$f(2) =$	$g(2) =$

(b) Plot the two functions on the same coordinate system to create curves that represent the graphs.
(c) Use a grapher to graph f and g in the same viewing window.
(d) Use the grapher display to indicate the domain and range of the functions f and g.

Solution   (a)

x	$f(x) = 2^x$	$g(x) = (1/2)^x$
-2	$f(-2) = 2^{-2} = 1/4$	$g(-2) = (1/2)^{-2} = 4$
-1	$f(-1) = 2^{-1} = 1/2$	$g(-1) = (1/2)^{-1} = 2$
0	$f(0) = 2^0 = 1$	$g(0) = (1/2)^0 = 1$
1	$f(1) = 2^1 = 2$	$g(1) = (1/2)^1 = \frac{1}{2}$
2	$f(2) = 2^2 = 4$	$g(2) = (1/2)^2 = \frac{1}{4}$

(b) Figure 1 shows the plotted points and the curve.
(c) Figure 2 shows the grapher viewing window for the functions f and g.

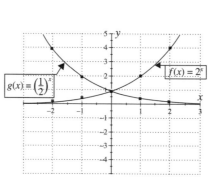

Figure 1

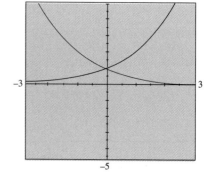

Figure 2

(d) The domain for both functions is the same: all real numbers. The range for both functions is the same: all positive real numbers. That is, $(0, \infty)$. ◇

EXAMPLE 7   Solving Exponential Equations Graphically
Solve the exponential equation $2^x = 16$ graphically.

Solution   Figure 3 shows the graphs of
$$y = 2^x \text{ and } y = 16$$
on a grapher. By using the ZOOM and TRACE features of the grapher, we see that the point of intersection of the two graphs is (4,16). The x coordinate, 4, of the point of intersection is the solution to the equation.   ◇

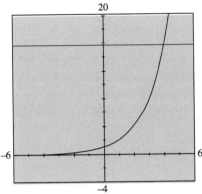

Figure 3

PROBLEM SET 5.1

Mastering the Concepts

In Problems 1–12, rewrite each expression in a simplified form using only positive exponents. (Assume all variables are restricted to values for which all expressions represent real numbers.)

1. (a) 4^0 (b) 8^0 (c) $(-3)^0$
2. (a) $(-2)^0$ (b) 7^0 (c) $(0.3)^0$
3. (a) $(2x)^0$ (b) $(7^{-1} + 3y)^0$ (c) $(0.17z)^0 + 6$
4. (a) $(-3y^2)^0$ (b) $(3 + x^2)^0$ (c) $2 + (3x - x)^0$
5. (a) 5^{-2} (b) 2^{-4} (c) $(-7)^{-2}$
6. (a) $(-3)^{-4}$ (b) $(-10)^{-1}$ (c) $(-5)^{-1}$
7. (a) $\left(\dfrac{4}{3}\right)^{-2}$ (b) $\left(\dfrac{1}{3}\right)^{-2}$ (c) $\left(\dfrac{1}{x}\right)^{-1}$
8. (a) $\left(\dfrac{1}{p}\right)^{-2}$ (b) $\left(\dfrac{2}{y}\right)^{-3}$ (c) $\dfrac{10^0}{10^{-3}}$
9. (a) $\dfrac{1}{5^{-2}}$ (b) $\dfrac{1}{x^{-4}}$ (c) $\dfrac{1}{y^{-3}}$
10. (a) $\dfrac{1}{4^{-3}}$ (b) $\dfrac{3}{t^{-2}}$ (c) $\dfrac{5}{y^{-4}}$
11. (a) $\dfrac{2^{-3}}{3^{-2}}$ (b) $\dfrac{x^{-4}}{y^{-2}}$ (c) $\dfrac{a^{-2}}{b^{-5}}$
12. (a) $\dfrac{5^{-2}}{3^{-2}}$ (b) $\dfrac{x^{-2}}{y^{-4}}$ (c) $\dfrac{a^{-3}}{b^{-4}}$

In Problems 13–26, use the properties of exponents to rewrite each expression so that it contains only positive exponents, and simplify the results. (Assume all variables are restricted to values for which all expressions represent real numbers.)

13. (a) $7^{-1} \cdot 7^3$ (b) $5^{-3} \cdot 5^{-2}$
14. (a) $9^{-3} \cdot 9$ (b) $3^{-4} \cdot 3^7$
15. (a) $x^{-3} \cdot x^{-2}$ (b) $y^{-7} \cdot y^{-2}$
16. (a) $(2^3)^{-2}$ (b) $(5^2)^{-1}$
17. (a) $[(-4)^2]^{-2}$ (b) $(p^{-2})^{-4}$
18. (a) $(3x)^{-4}$ (b) $(3y^{-4})^2$
19. (a) $(5^{-1}p^4)^{-2}$ (b) $(p^{-1}q^{-2})^5$

20. (a) $\left(\dfrac{5}{x}\right)^{-2}$ (b) $\left(\dfrac{3}{p^{-1}}\right)^{-1}$
21. (a) $\left(\dfrac{p^3}{q^2}\right)^{-2}$ (b) $\left(\dfrac{x^{-2}}{y^{-2}}\right)^{-1}$
22. (a) $\dfrac{8^{-3}}{8^{-5}}$ (b) $\dfrac{7^{-3}}{7^{-8}}$
23. $\left(\dfrac{b^{-1}}{3a^{-1}}\right)\left(\dfrac{3a}{b}\right)^{-1}$ 24. $\left(\dfrac{a^{-1}c^2}{a^{-2}c^{-1}}\right)^{-2}$
25. $\left(\dfrac{x^{-1}y^{-2}}{x^{-2}y^3}\right)^2$ 26. $\left(\dfrac{p^{-2}q^{-1}r^{-3}}{q^3r^{-1}}\right)^{-4}$

In Problems 27–32, solve each exponential equation.

27. (a) $3^x = 9$ (b) $2^x = 8$
28. (a) $4^x = 64$ (b) $5^x = 25$
29. (a) $2^{-3x} = 8$ (b) $2^{4x} = 16$
30. (a) $5^{-y} = 125$ (b) $6^{-x} = 36$
31. (a) $3^{-x} = 27$ (b) $4^{-x} = 16$
32. (a) $3^{2x} = 81$ (b) $5^{-2x} = 25$

Applying the Concepts

33. **Biology:** A certain virus is shaped like a sphere. If its radius is 0.000013 centimeter, find the volume of the virus. Express the answer in scientific notation and round off to 2 decimal places. ($V = \frac{4}{3}\pi r^3$)

34. **Computer Calculations:** A computer can do an arithmetic operation in 0.00000036 second. If the electricity in its circuits travels at the speed of light (186,000 miles per second), how far will the electricity travel in the time it takes the computer to complete 7 arithmetic operations? Express the answer in scientific notation and round off to two decimal places.

35. **Travel Time:** A satellite leaves the earth traveling at a uniform speed of 310,000 miles per hour. How long does it take for the satellite to reach the sun, if the average distance from the earth to the sun is 93,000,000 miles? Express the answer in scientific notation.

36. **Chemistry:** One gram of hydrogen contains 6.023×10^{23} atoms. What is the weight of one atom of hydrogen? Express the answer in scientific notation and round off to two decimal places.

37. **Hourly Wage Increase:** An employee's starting wage is $7.50 per hour. Find the hourly wage after 5 years if the total annual raise is 6%, and the raise is given in 2 installments over the course of the year.

38. **Compound Interest:** A deposit of $5000 is placed in a savings account. Find the balance in the account after 3.5 years if interest is compounded quarterly at a rate of 7.2%.

39. **Compound Interest:** Find the amount of money in an account if $2000 is deposited at 4% compounded monthly for 2 years and 3 months.

40. **Compound Interest:** Find the amount of money in a forgotten savings account if $5.00 is invested at 5% compounded quarterly for 50 years.

Developing and Extending the Concepts

G In Problems 41–44, use a calculator to simplify each expression. Round off to two decimal places and write each answer in scientific notation.

41. $\dfrac{(9.75 \times 10^{8}) \cdot (1.5 \times 10^{-3})}{(7.5 \times 10^{-2}) \cdot (1.1 \times 10^{-4})}$

42. $\dfrac{(5.7 \times 10^{4})^{2} \cdot (6.65 \times 10^{-5})^{3}}{(3.3 \times 10^{-7}) \cdot (3.8 \times 10^{-6})^{4}}$

43. $\dfrac{(1.12 \times 10^{-3}) \cdot (8.25 \times 10^{-5})}{(2.35 \times 10^{-6})^{2}}$

44. $\dfrac{(1.86 \times 10^{5}) \cdot (2.4 \times 10^{-9})}{(3.6 \times 10^{-7}) \cdot (4.8 \times 10^{-11})^{3}}$

In Problems 45 and 46, each of the following expressions is obtained in calculating compound interest. Write a story which would give rise to this expression and then use a calculator to evaluate it. Round off each answer to two decimal places.

45. (a) $1000\left(1 + \dfrac{0.08}{4}\right)^{4(3)}$ (b) $5000\left(1 + \dfrac{0.04}{12}\right)^{12(3)}$

46. (a) $10{,}000\left(1 + \dfrac{0.06}{12}\right)^{12(4)}$ (b) $4000\left(1 + \dfrac{0.05}{52}\right)^{52(3)}$

In Problems 47–50, evaluate each function at the indicated input values.

47. $f(x) = 3^{x}$
(a) $x = -2$ (b) $x = 0$ (c) $x = 2$

48. $g(x) = 3^{-x}$
(a) $x = -2$, (b) $x = 0$ (c) $x = 2$

49. $h(x) = 4^{x}$
(a) $x = -2$, (b) $x = 0$ (c) $x = 2$

50. $F(x) = 4^{-x}$
(a) $x = -2$, (b) $x = 0$ (c) $x = 2$

G In Problems 51–54, match each function with its graph. Then find the domain and range of the function from its graphical display. You may use a grapher to confirm these graphs.

51. $f(x) = 3^{x}$ 52. $G(x) = -3^{x}$
53. $h(x) = 3^{-x}$ 54. $G(x) = 3^{x} + 10$

(a) 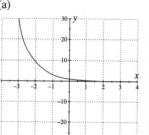   (b)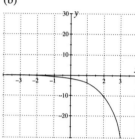

(c) 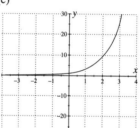   (d)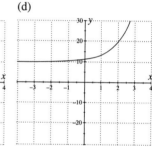

G In Problems 55–58, solve each equation graphically using the following methods. You may use a grapher.

(i) Graph $y = $ left side of the equation $-$ right side, then find the x intercept of the graph.

(ii) Graph $y_1 = $ left side of the equation and $y_2 = $ right side on the same coordinate system or viewing window. Then find the x coordinate of the point of intersection of the two graphs.

55. (a) $3^{x} = 9$ (b) $2^{x} = 8$
56. (a) $4^{x} = 64$ (b) $5^{x} = 25$
57. (a) $2^{-3x} = 8$ (b) $2^{4x} = 16$
58. (a) $5^{-x} = 25$ (b) $6^{-x} = 36$

In Problems 59–62, rewrite each expression without negative exponents and simplify. (Assume all variables are restricted to values for which all expressions represent real numbers.)

59. (a) $\dfrac{(a + b)^{-2}}{(a + b)^{-3}}$ (b) $x^{-2n} \cdot x^{2n+6}$

60. (a) $x^{3+n} \cdot x^{2-n}$ (b) $\dfrac{(x + 3y)^{-12}}{(x + 3y)^{10}}$

61. (a) $\left(\dfrac{3^{-1}x^{-2}y^{-5}}{9^{-1}x^{-1}y^{-3}}\right)^{-3}$ (b) $\left(\dfrac{u^{-4}v^3z^{-3}}{u^3(vz)^{-2}}\right)^{-4}$

62. (a) $\left[\left(\dfrac{a^{-1}}{b}\right)\cdot\left(\dfrac{b^{-1}}{a}\right)^{-1}\right]^{-2}$

   (b) $\left(\dfrac{a^{-4}b^2c^{-6}}{(ab)^{-2}(bc)^{-3}}\right)^{-1}$

At times, we rewrite expressions with negative exponents. In Problems 63 and 64, rewrite the expression using negative exponents.

63. (a) $\dfrac{1}{x^2y^4}$ (b) $\dfrac{5}{(2+a)^5}$

64. (a) $\dfrac{6}{x^7y}$ (b) $\dfrac{2}{(x+1)^7}$

OBJECTIVES

1. Find the nth Root
2. Find the Distance between two Points
3. Solve Applied Problems
4. Graph the Square Root Function
5. Use Technology Exploration

5.2 Radicals

Finding the nth Root

Recall that a *square root* of 25 is 5 because 5 is one of the two equal factors of 25. Similarly, a *cube root* of 8 is 2 because 2 is one of the three equal factors of 8. The following table shows the type of roots and equal factors.

Number	Equal Factors	Type of Root	Root
$25 = 5^2$	$5 \cdot 5$	square root	5
$25 = (-5)^2$	$(-5) \cdot (-5)$	square root	-5
$-8 = (-2)^3$	$(-2) \cdot (-2) \cdot (-2)$	cube root	-2
$16 = 2^4$	$2 \cdot 2 \cdot 2 \cdot 2$	fourth root	2

In general, we have the following definition

DEFINITION OF AN NTH ROOT

> Let a and b be real numbers, and n be a positive integer, then
> a is an **nth root** of b if $a^n = b$.

Some numbers have more than one nth root.
1. Any positive number has two square roots. For example, both 5 and -5 are square roots of 25 because
$$25 = 5^2 \quad \text{and} \quad 25 = (-5)^2$$
2. Negative numbers have no real square root, since no real number squared can be negative.
3. Zero has only one square root; namely, 0.
   For cube roots, the situation is different. Every number (positive, negative, or zero) has only one real cube root. Unlike square roots, it is possible to have a cube root of a negative number. To avoid ambiguity about which root of a number is considered, we generalize the above result.

NUMBER OF REAL NTH ROOTS OF A NUMBER a

If a is	n **is even root**	n **is odd root**
1. positive	two real nth roots	one real nth root
2. negative	no real roots	one real nth root

Radical notation is commonly used to indicate a root. We will use $\sqrt[n]{a}$ to represent the **nth root** of a. The symbol $\sqrt{}$ is called a **radical**, n is called the **index** of the radical and a is called the **radicand**.

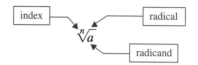

When the index is 2, it is usually omitted. That is, we use $\sqrt{}$ rather than $\sqrt[2]{}$ to represent square roots. If there are two nth roots, then $\sqrt[n]{a}$ denotes the positive root, called the **principal nth root**. We describe the nth root as follows:

NTH ROOT $\sqrt[n]{a}$

Radicand a	Index n is even	Index n is odd
positive	$\sqrt[n]{a}$ is positive nth root of a	$\sqrt[n]{a}$ is the nth root of a
negative	$\sqrt[n]{a}$ is not a real number	$\sqrt[n]{a}$ is the nth root of a

Integers such as 4, 9, 16, 25, 36, and 49 are called *perfect squares* because they have integer square roots. Similarly, integers such as 1, 8, 27, 64, 125, and 216 are called *perfect cubes* because they have integer cube roots.

EXAMPLE 1   Finding Roots of a Number

Find the principal roots of the given numbers.
(a) $\sqrt{49}$ (b) $-\sqrt{49}$ (c) $\sqrt{-9}$ (d) $\sqrt[3]{27}$ (e) $\sqrt[3]{-27}$

Solution

(a) $\sqrt{49} = 7$ because $7 \cdot 7 = 7^2 = 49$
(b) $-\sqrt{49} = -7$ because -7 is the opposite of $\sqrt{49}$
(c) $\sqrt{-9}$ is not a real number. Any real number squared is nonnegative
(d) $\sqrt[3]{27} = 3$ because $3 \cdot 3 \cdot 3 = 3^3 = 27$
(e) $\sqrt[3]{-27} = -3$ because $(-3)(-3)(-3) = (-3)^3 = -27$ ◈

Recall that the notation $\sqrt{x^2}$ indicates the positive square root of x^2 only. For example,

$$\sqrt{(-3)^2} = \sqrt{9} = 3$$

When variables are present in the radicand and it is unclear whether the variable represents a positive number or a negative number, absolute value bars are sometimes needed to ensure that the result is a positive number. With this in mind, we write

$$\sqrt{x^2} = |x|$$

Similar situations may occur when the index is any even positive integer. When the index is any *odd* positive integer, absolute value bars are not necessary. In general,

If a is a real number and n is a positive integer, $n \geq 2$, then:
1. $\sqrt[n]{a^n} = |a|$ if n is even   2. $\sqrt[n]{a^n} = a$ if n is odd.

EXAMPLE 2   Evaluating Radical Expressions

Evaluate each expression (if possible).

(a) $\sqrt{(-7)^2}$ (b) $\sqrt[3]{(-5)^3}$ (c) $\sqrt[3]{8x^3}$ (d) $\sqrt[4]{-2^4}$ (e) $\sqrt[4]{x^8}$

Solution   (a) $\sqrt{(-7)^2} = |-7| = 7$ since the index of the radical is even

(b) $\sqrt[3]{(-5)^3} = -5$ since the index of the radical is odd

(c) $\sqrt[3]{8x^3} = 2x$ since the index of the radical is odd

(d) $\sqrt[4]{-2^4} = \sqrt[4]{-16}$ is not a real number, because the radicand is negative

(e) $\sqrt[4]{x^8} = x^2$ since the expression under the radical is a perfect fourth power ◈

EXAMPLE 3   Using Absolute Value to Simplify Radicals

Evaluate each expression (if possible).

(a) $\sqrt{9x^2}$ (b) $\sqrt{(x-1)^2}$ (c) $\sqrt[3]{27y^3}$ (d) $\sqrt[5]{(x+3)^5}$

Solution   (a) $\sqrt{9x^2} = \sqrt{(3x)^2} = |3x| = 3|x|$

(b) $\sqrt{(x-1)^2} = |x-1|$

(c) $\sqrt[3]{27y^3} = \sqrt[3]{(3y)^3} = 3y$

(d) $\sqrt[5]{(x+3)^5} = x+3$ ◈

We can raise a number to the nth power and take the principal nth root of a number. For instance,

$$(\sqrt{9})^2 = 9 = 3^2 \text{ and } \sqrt{3^2} = \sqrt{9} = 3$$
$$(\sqrt[3]{27})^3 = 27 = 3^3 \text{ and } \sqrt[3]{3^3} = \sqrt[3]{27} = 3$$

In general,

> If a has a principal nth root, then $(\sqrt[n]{a})^n = a$ and $\sqrt[n]{a^n} = a$

For example,

$$(\sqrt{7})^2 = 7, \quad (\sqrt[3]{5})^3 = 5, \quad (\sqrt[5]{-2})^5 = -2, \quad \text{and } (\sqrt[4]{3})^4 = 3$$

Finding the Distance between Two Points

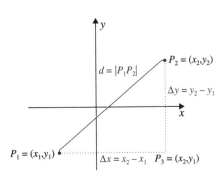

Figure 1

One of the most common applications of squares and square roots is the Pythagorean Theorem, reviewed in section R6. We use it to derive a formula for the distance between two points on a coordinate system. Let $P_1 = (x_1,y_1)$ and $P_2 = (x_2,y_2)$ be two points on an xy coordinate system. The distance formula permits us to find the length $\overline{P_1P_2}$ or the distance d between the two points. To accomplish this, we draw a right triangle with $\overline{P_1P_2}$ as the hypotenuse, one leg parallel to the x axis and the other leg parallel to the y axis. The two legs intersect at a point P_3 whose coordinates are (x_2,y_1) (Figure 1).

Now we find the change in x,

$$\Delta x = x_2 - x_1$$

and the change in y

$$\Delta y = y_2 - y_1$$

Because the triangle $P_1P_2P_3$ is a right triangle, the Pythagorean Theorem holds:

$$d^2 = |\overline{P_1P_2}|^2 = (\Delta x)^2 + (\Delta y)^2$$

so that

$$d = \sqrt{(x_2 - x_1)^2 + (y_2 - y_1)^2}$$

Once you understand the distance formula, you can use it in one of the two forms.

THE DISTANCE FORMULA

Note that $(x_2 - x_1)^2 = (x_1 - x_2)^2$ and $(y_2 - y_1)^2 = (y_1 - y_2)^2$

Let $P_1 = (x_1,y_1)$ and $P_2 = (x_2,y_2)$ be two points in the xy plane. The distance d between P_1 and P_2 is given by

$$d = \sqrt{(x_2 - x_1)^2 + (y_2 - y_1)^2} \text{ or } d = \sqrt{(x_1 - x_2)^2 + (y_1 - y_2)^2}$$

EXAMPLE 4   Finding the Distance between Two Points

Find the distance between the points $(-1,-2)$ and $(3,4)$.

Solution   Let $(x_1,y_1) = (-1,-2)$ and $(x_2,y_2) = (3,4)$ and apply the distance formula.

$$d = \sqrt{(x_2 - x_1)^2 + (y_2 - y_1)^2}$$

$\qquad = \sqrt{[3 - (-1)]^2 + [4 - (-2)]^2}$ Substitute coordinate points

$\qquad = \sqrt{4^2 + 6^2}$ Simplify

$\qquad = \sqrt{52}$

$\qquad = 2\sqrt{13} \approx 7.21$ Round to two decimal places

EXAMPLE 5   Applying the Distance Formula

Let $A = (1,2)$, $B = (3,1)$, and $C = (4,3)$.
(a) Plot A, B and C on the same coordinate system.
(b) Find the distances $|\overline{AB}|$, $|\overline{BC}|$, and $|\overline{AC}|$.
(c) Use the Converse of the Pythagorean Theorem to determine whether the points A, B and C form a right triangle.

Solution   (a) Figure 2 shows the plot of the points A, B, and C.
(b) Using the distance formula, we can find $|\overline{AB}|$, $|\overline{BC}|$, and $|\overline{AC}|$.

$$|\overline{AB}| = \sqrt{(3 - 1)^2 + (1 - 2)^2} = \sqrt{2^2 + (-1)^2} = \sqrt{5}$$
$$|\overline{BC}| = \sqrt{(4 - 3)^2 + (3 - 1)^2} = \sqrt{1^2 + 2^2} = \sqrt{5}$$
$$|\overline{AC}| = \sqrt{(4 - 1)^2 + (3 - 2)^2} = \sqrt{3^2 + 1^2} = \sqrt{10}$$

(c) Using the Converse of the Pythagorean Theorem (section R5), we have

$$|\overline{AB}|^2 + |\overline{BC}|^2 = (\sqrt{5})^2 + (\sqrt{5})^2$$
$$= 5 + 5$$
$$= 10 = |\overline{AC}|^2$$

Thus, the three points form a right triangle. The right angle is at B because side AC, opposite B, is the longest side and thus the hypotenuse.

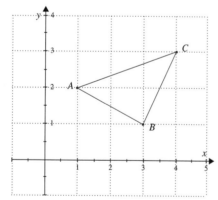

Figure 2

Solving Applied Problems

The square root concept can be used to solve many applied problems as the next example shows.

EXAMPLE 6   Solving an Applied Problem from Physics

A can of iced tea is h centimeters high. A hole is pierced in the side of the can y centimeters from the bottom (Figure 3). The stream of tea coming from that hole first hits the table d centimeters from the can, where d is given by the equation

$$d = \sqrt{y(h - y)} \qquad 0 < y < h$$

How far from the can will the tea hit the table if the can is 12 centimeters tall and the hole is pierced in the can 8 centimeters above the table? Round off to two decimal places.

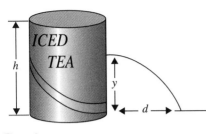

Figure 3

Solution   Replacing h by 12 and y by 8 in the above equation, we have

$$d = \sqrt{8(12 - 8)}$$
$$= \sqrt{32} \approx 5.66$$

Therefore, the stream of tea first hits the table 5.66 centimeters from the can. ◈

Graphing Square Root Function

The function $f(x) = \sqrt{x}$ represents a **square-root function**. Here the principal square root is a real number for $x \geq 0$. We will see that the output $f(x)$ is also nonnegative; that is, $f(x) \geq 0$.

EXAMPLE 7   Graphing a Square Root Function

(a) Let $f(x) = \sqrt{x}$, find $f(0), f(1), f(4),$ and $f(9)$.
(b) Plot the points $(0,f(0)), (1,f(1)), (4,f(4)),$ and $(9,f(9))$. Then connect these points with a smooth curve.
(c) Use the displayed graph of f to indicate the domain and range.

Solution   (a) $f(0) = \sqrt{0} = 0,$ $f(1) = \sqrt{1} = 1$
              $f(4) = \sqrt{4} = 2$ $f(9) = \sqrt{9} = 3$

(b) We record these points in a table format, plot them and connect them with a curve.
Figure 4 shows the graph of f.

(c) The graph of f suggests that the domain as well as the range of f are the sets of all nonnegative real numbers. The fact that the graph of f is in the first quadrant and the origin reinforces the nonnegative nature of the domain and range.

x	$f(x) = \sqrt{x}$	$(x,f(x))$
0	0	(0,0)
1	1	(1,1)
4	2	(4,2)
9	3	(9,3)

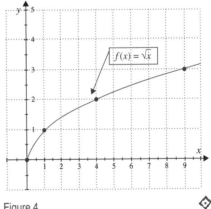

Figure 4   ◈

Using Technology Exploration

The use of a grapher is a perfect tool to help solve equations involving radicals.

The periodic motion of a swinging pendulum was first noticed by the Arab mathematician Ibn Yûnis (c. 1200 AD). In 1656 the first pendulum clock was built by Christian Huygens, who discovered this model. The length of time T (in seconds) of one swing back and forth of a pendulum of length L (in feet) is described by the model

$$T = 2\pi\sqrt{\frac{L}{g}}$$

where g is the acceleration due to gravity ($g = 32$ feet per second squared).

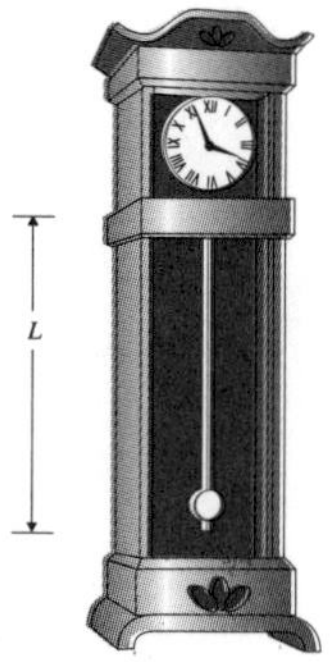

Figure 5

G EXAMPLE 8   Finding the Length of a Pendulum

In designing a grandfather clock that keeps correct time, it was determined that the pendulum took exactly 2 seconds to swing back and forth (Figure 5).

(a) Use the formula $T = 2\pi\sqrt{L/32}$ to complete the following table. (Round off to two decimal places.)

L(feet)	0	1	2	2.25	2.50	2.75	3.00	3.25	3.50	3.75	4.00
T(seconds)											

(b) Use a grapher to sketch the graph of $T = 2\pi\sqrt{L/32}$ and $T = 2$ in the same viewing window.

(c) Use the graphical display of part (b) to solve the equation $2\pi\sqrt{L/32} = 2$ by approximating the L coordinate of the point of intersection.

(d) Compare the results of the table output and the graph of the function to find the length L that will keep "correct time."

Solution   (a) The output values are given in the following table:

L(feet)	0	1	2	2.25	2.50	2.75	3.00	3.25	3.50	3.75	4.00
T(seconds)	0	1.11	1.57	1.67	1.76	1.84	1.92	2.00	2.08	2.15	2.22

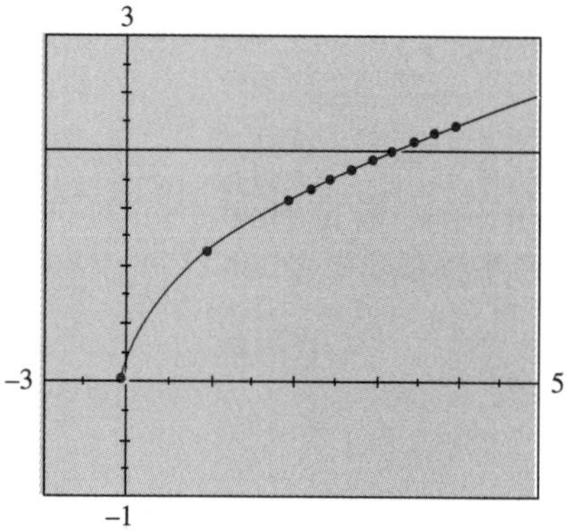

(b) Figure 6 shows the graphs of $T = 2\pi\sqrt{L/32}$ and $T = 2$ in the same viewing window.

(c) Using the TRACE and ZOOM features of the grapher, we see the point of intersection of the two graphs is (3.25, 2.00). Therefore, the length of the pendulum is approximately 3.25 feet.

(d) The table value and the graphical value seem to agree on the length of the pendulum to obtain the correct time. ◈

Figure 6

◈ PROBLEM SET 5.2
_____

Mastering the Concepts

In Problems 1 and 2, complete each statement.

1. (a) Since $5^2 = 25$, then _____ is a square root of 25.
   (b) Since $4^3 = 64$, then _____ is a cube root of 64.

2. (a) Since $3^4 = 81$, then _____ is a fourth root of 81.
   (b) Since $(2x)^5 = 32x^5$, then _____ is a fifth root of $32x^5$.

In Problems 3–12, find the principal roots of the given number. State the reason when it is not a real number.

3. (a) $\sqrt{16}$ (b) $\sqrt[3]{-8}$
4. (a) $\sqrt{144}$ (b) $-\sqrt{100}$
5. (a) $\sqrt{0.04}$ (b) $-\sqrt{0.36}$
6. (a) $\sqrt[3]{-0.008}$ (b) $-\sqrt{0.09}$
7. (a) $\sqrt[4]{81}$ (b) $\sqrt[4]{-81}$
8. (a) $\sqrt[3]{-64}$ (b) $\sqrt{-64}$
9. (a) $-\sqrt[4]{16/81}$ (b) $\sqrt{-9/25}$
10. (a) $\sqrt[5]{1/32}$ (b) $-\sqrt[5]{-1/32}$

11. (a) $-\sqrt[3]{1/8}$ (b) $-\sqrt[3]{-27/64}$
12. (a) $-\sqrt{-9/25}$ (b) $-\sqrt{81/144}$

In Problems 13–26, simplify each expression (if possible). Assume all variables represent positive real numbers.

13. (a) $-\sqrt{11^2}$ (b) $\sqrt{(-11)^2}$
14. (a) $-\sqrt{5^4}$ (b) $\sqrt{(-5)^4}$
15. (a) $\sqrt[3]{(-3)^3}$ (b) $-\sqrt[3]{(-3)^3}$
16. (a) $\sqrt[4]{-9}$ (b) $-\sqrt[4]{(-3)^4}$
17. (a) $\sqrt[3]{-8x^3}$ (b) $\sqrt{(-6y)^2}$
18. (a) $-\sqrt[3]{-8y^3}$ (b) $\sqrt{(-16)^2}$
19. (a) $\sqrt[3]{8x^9y^6}$ (b) $\sqrt{4x^2y^4}$
20. (a) $-\sqrt[3]{-125y^6}$ (b) $\sqrt[4]{16y^{12}}$
21. (a) $(\sqrt{5x})^2$ (b) $\left(\sqrt[3]{-3x}\right)^3$

22. (a) $-\sqrt[4]{-16y^4}$ (b) $\sqrt[3]{-27x^3y^9}$
23. (a) $(\sqrt{\sqrt{y^7}})^2$ (b) $(\sqrt[3]{-3x})^3$
24. (a) $(\sqrt[4]{4x})^4$ (b) $(\sqrt[5]{3y})^5$
25. (a) $(\sqrt[3]{-2x})^3$ (b) $(\sqrt[4]{-5t^2})^4$
26. (a) $(\sqrt[5]{2y^2})^5$ (b) $(\sqrt[6]{-4y})^6$

In Problems 27–32, simplify each expression. Assume that variables represent any real number.

27. (a) $\sqrt{4x^2}$ (b) $-\sqrt{25x^2}$
28. (a) $\sqrt{(-5x)^2}$ (b) $-\sqrt{36x^2}$
29. (a) $\sqrt{(-3x)^2}$ (b) $\sqrt[3]{y^6}$
30. (a) $\sqrt{100x^4}$ (b) $\sqrt{x^{12}}$
31. (a) $\sqrt[5]{x^5}$ (b) $\sqrt{(x-3)^2}$
32. (a) $-\sqrt{121x^{20}}$ (b) $-\sqrt{144y^{12}}$

In Problems 33–38, find the distance between the two points with the given coordinates. Round off to two decimal places when necessary.

33. (a) $(7,10)$ and $(1,2)$
   (b) $(-3,-4)$ and $(-5,-7)$
34. (a) $(-2,5)$ and $(3,-1)$
   (b) $(4,-3)$ and $(6,2)$
35. (a) $(-5,0)$ and $(-2,-4)$
   (b) $(-4,7)$ and $(0,-8)$
36. (a) $(6,2)$ and $(6,-2)$
   (b) $(-4,-3)$ and $(0,0)$
37. (a) $(2.3,4.3)$ and $(4.2,-5.7)$
   (b) $(3.5,-2.8)$ and $(-7.3,-9.8)$
38. (a) $(3.5,1.7)$ and $(-2.5,9.1)$
   (b) $(-8.4,5.2)$ and $(3.7,4.8)$

In Problems 39–44, evaluate each function at the indicated input values. Round off to two decimal places.

39. $f(x) = \sqrt{2x+3}$
   (a) $x=-1$ (b) $x=0$ (c) $x=2$
40. $f(x) = -\sqrt{3x+1}$
   (a) $x=0$ (b) $x=1$ (c) $x=3$
41. $g(x) = \sqrt[3]{x-8}$
   (a) $x=-1$ (b) $x=0$ (c) $x=3$
42. $g(x) = \sqrt[3]{2x+5}$
   (a) $x=-1$ (b) $x=3$ (c) $x=7$
43. $f(x) = -\sqrt{x^2+4}$
   (a) $x=-5$ (b) $x=0$ (c) $x=5$
44. $f(x) = \sqrt{x^3-5}$
   (a) $x=2$ (b) $x=3$ (c) $x=5$

Applying the Concepts

45. **Geometry:** Apply the distance formula and the converse of the Pythagorean Theorem to determine whether or not the triangle with vertices $A = (-3,1)$, $B = (3,10)$ and $C = (3,1)$ is a right triangle.

46. **Geometry:** Use the distance formula to determine whether or not the triangle with vertices $A = (5,-1)$, $B = (-6,5)$ and $C = (-2,4)$ is an isosceles triangle. (Two of the three sides are equal.)

47. **Biology:** A biologist estimates that the population P of a certain culture of bacteria is given by the model
$$P = 1000\sqrt[4]{t^5 + 10t^2 + 9}$$
where t is the time in hours since the culture was started. Find the population of the bacteria after (a) 2 hours, (b) 3 hours. Round off to the nearest integer.

48. **Electronics:** In electronics, the resonant frequency f (in hertz) for a tuned circuit is given by the formula
$$f = \frac{1}{2\pi\sqrt{LC}}$$
where L is the inductance and C is the capacitance. Find the frequency f for a tuned circuit in which $L = 3.57 \times 10^{-8}$ and $C = 1.21 \times 10^{-12}$ (round off to two decimal places).

49. **Physics:** The length of time T (in seconds) for a pendulum to complete one swing back and forth is given by the model
$$T = 2\pi\sqrt{\frac{L}{32}}$$
where L (in feet) is the length of the pendulum. Find the time of a complete swing back and forth for:
(a) a 2.4 foot pendulum to the nearest tenth of a second.
(b) The Hong Kong Bank of Canada pendulum in Vancouver, with a length of 90 feet.

50. **Physiology:** A manufacturer of scientifically designed walking shoes uses the equation $T = 2\pi\sqrt{0.068L}$ where T is the time (in seconds) for one swing of a leg of length L (in meters) (Figure 7). Find the time of the swing of a leg of length 0.81 meter.

Figure 7

51. **Satellite Orbit:** In order for a satellite to maintain an orbit of h kilometers, its velocity v (in kilometers/second) must be related to its height by the equation
$$v = \frac{626.4}{\sqrt{h+R}}$$
where $R = 6372$ kilometers is the radius of the earth. What is the speed of the satellite when it is in orbit 105 kilometers from the earth's surface?

52. **Physiology:** Physiologists define the threshold weight of a human being as the weight beyond which the risk of death increases significantly. For middle-aged males, their height h in feet is related to the threshold weight w in pounds by
$$h = 4.56\sqrt[3]{\frac{w}{100}} \qquad \text{where } 160 < w < 220$$
(a) Estimate h when $w = 175$.
(b) If the threshold weight of a man is 200 pounds, what is his height?

53. **Surface Area of a Tent:** A tent has the shape of a pyramid with a square base. Each side of the base measures 6 feet, and the center pole is 7 feet high (Figure 8). Suppose that one triangular section of the tent is burned due to a campfire. How much fabric is needed to replace the damaged triangular section? Use the formula $A = (1/2)\,hs$ and

$$l = \sqrt{\dfrac{s^2}{4} + h^2}$$

where s is the side of the base, h is the height of the pyramid, and l is the distance from the top of the pyramid to the base along the side of the tent.

Figure 8

54. **Building a Gazebo:** A homeowner wishes to build a gazebo at his waterfront home. The base of the gazebo is a hexagon (6 equal sides) with each side measuring 10 feet (Figure 9). The city ordinance requires that a foundation of concrete should be 4 inches thick. How many cubic yards of concrete are needed? Use

$$V = \dfrac{1}{2}\,nst\sqrt{r^2 - \dfrac{s^2}{4}}$$

Figure 9

where n = number of sides of the polygon, s = length of a side, t = thickness, and r = distance from center to a vertex.

Developing and Extending the Concepts

55. Let x be a real number. What restrictions are needed on n for each of the following equations to be true?
    (a) $\sqrt[n]{x^n} = x$ (b) $\sqrt[n]{x^n} = |x|$

56. Indicate the first step needed to simplify the expression

$$\sqrt{x^2 + 6x + 9}$$

Then simplify the expression.

In Problems 57–62, use the given function to find the output values.

57. $f(x) = \sqrt{1 - x}$
    (a) $f(-1)$ (b) $f(0)$ (c) $f(1)$
58. $k(x) = \sqrt{x} - \sqrt{x - 1}$
    (a) $k(-1)$ (b) $k(0)$ (c) $k(2)$
59. $h(x) = \sqrt{-x}$
    (a) $h(-1)$ (b) $h(0)$ (c) $h(1)$

60. $g(x) = \sqrt[3]{x}$
    (a) $g(-8)$ (b) $g(0)$ (c) $g(8)$
61. $F(x) = -\sqrt{1 - x}$
    (a) $F(-1)$ (b) $F(0)$ (c) $F(1)$
62. $j(x) = \sqrt{-x} - \sqrt[3]{x}$
    (a) $j(-1)$ (b) $j(0)$ (c) $j(1)$

G In Problems 63 and 64, determine the domain of each function algebraically. Then use a grapher to graph the function to support the algebraic results.

63. (a) $f(x) = \sqrt{x - 4}$ (b) $g(x) = \sqrt{4 - x}$
64. (a) $f(x) = \sqrt[3]{2x - 6}$ (b) $g(x) = \sqrt[4]{x + 1}$
65. Let $f(x) = \sqrt{x - 3}$
    (a) Complete the table.

x	3	4	7	9
$f(x)$				

    (b) Plot the points obtained in the table in part (a) and connect them with a smooth curve.

G (c) Use a grapher to graph f, and then use the graphical display to indicate the domain and range of f.

G (d) Find the solution of the equation $\sqrt{x - 3} = 2$ by identifying the x coordinate of the point of intersection of $y = \sqrt{x - 3}$ and $y = 2$.

66. Let $g(x) = \sqrt{x + 1}$
    (a) Complete the table.

x	-1	3	8	15
$g(x)$				

    (b) Plot the points obtained in the table in part (a) and connect them with a smooth curve.

G (c) Use a grapher to graph g, and then use the graphical display to indicate the domain and range of g.

G (d) Find the solution of the equation $\sqrt{x + 1} = 3$ by identifying the x coordinate of the point of intersection of $y = \sqrt{x + 1}$ and $y = 3$.

67. What can you say about a nonzero radicand if its cube root is:
    (a) doubled?   (b) tripled?
    (c) quadrupled?   (d) What is the pattern?
68. What can you say about a nonzero radicand if its square root is:
    (a) doubled?   (b) tripled?
    (c) quadrupled?   (d) What is the pattern?

OBJECTIVES

1. Define Expressions of the Form $a^{\frac{1}{n}}$
2. Define Expressions of the Form $a^{\frac{m}{n}}$
3. Convert Forms
4. Use Properties of Rational Exponents
5. Solve Applied Problems
6. Use Technology Exploration

5.3 Rational Exponents

Defining Expressions of the Form $a^{\frac{1}{n}}$

If all exponent properties hold even if some of the exponents are not integers, then we would have

$$(7^{1/2})^2 = 7^{2/2} = 7^1 = 7 \text{ and } (5^{1/3})^3 = 5^{3/3} = 5^1 = 5$$

That is, $7^{1/2}$ is a number whose square is 7, so $7^{1/2}$ is another name for a square root of 7. Similarly, $5^{1/3}$ is a number whose cube is 5, and so $5^{1/3}$ is another name for the cube root of 5. In general, we have the following definition

DEFINITION OF RATIONAL EXPONENTS

> Let a be a real number and n be an integer such that $n \geq 2$. If the principal nth root of a exists, we define
> $$a^{\frac{1}{n}} = \sqrt[n]{a}$$

The restriction on $a^{1/n}$ is the same as the restriction on $\sqrt[n]{a}$.

We need to be careful using the above definition when n is an even integer and a is negative because in such a case $a^{1/n}$ is not a real number.

EXAMPLE 1   Evaluating Expressions with Rational Exponents

Evaluate each expression (if possible).
(a) $4^{1/2}$ (b) $-4^{1/2}$ (c) $(-4)^{1/2}$ (d) $8^{1/3}$
(e) $-8^{1/3}$ (f) $(-8)^{1/3}$ (g) $0^{1/7}$

Solution   (a) $4^{1/2} = \sqrt{4} = 2$ (b) $-4^{1/2} = -\sqrt{4} = -2$
(c) $(-4)^{1/2} = \sqrt{-4}$ is not a real number   (d) $8^{1/3} = \sqrt[3]{8} = 2$
(e) $-8^{1/3} = -\sqrt[3]{8} = -2$ (f) $(-8)^{1/3} = \sqrt[3]{-8} = -2$
(g) $0^{1/7} = \sqrt[7]{0} = 0$

Defining Expressions of the Form $a^{m/n}$

Having defined expressions of the form $a^{1/n}$, we now are able to extend our definition to include expressions of the form $a^{m/n}$, where m/n is any rational number. Observe, for example, the two ways that $27^{2/3}$ can be evaluated on the assumption that the properties of exponents are to hold for rational numbers:

$$27^{2/3} = (27^{1/3})^2 = (\sqrt[3]{27})^2 = 3^2 \quad = 9$$

$$\text{or } 27^{2/3} = (27^2)^{1/3} = \sqrt[3]{27^2} \quad = \sqrt[3]{729} = 9$$

This suggests that we define $a^{m/n}$ to be $(\sqrt[n]{a})^m$ whenever the roots make sense. More formally, we have the following definition.

DEFINITION OF $a^{\frac{m}{n}}$ AND $a^{-\frac{m}{n}}$

> Suppose that m and n are integers with $n \geq 2$, and that the fraction m/n is reduced to lowest terms. Let a be a nonzero real number for which $\sqrt[n]{a}$ is defined. Then
> $$a^{\frac{m}{n}} = \left(a^{\frac{1}{n}}\right)^m = (\sqrt[n]{a})^m = \sqrt[n]{a^m} \text{ and } a^{-\frac{m}{n}} = \frac{1}{a^{\frac{m}{n}}} = \frac{1}{\sqrt[n]{a^m}}$$

EXAMPLE 2   Evaluating Expressions with Rational Exponents

Evaluate each expression.

(a) $4^{3/2}$ (b) $4^{-3/2}$ (c) $27^{4/3}$

(d) $(-8)^{-2/3}$ (e) $-8^{-2/3}$ (f) $(-4)^{3/2}$

Solution   (a) $4^{3/2} = (2^2)^{3/2} = 2^{2\cdot(3/2)} = 2^3 = 8$

(b) $4^{-3/2} = (2^2)^{-3/2} = 2^{2\cdot(-3/2)} = 2^{-3} = \dfrac{1}{8}$

(c) $27^{4/3} = (3^3)^{4/3} = 3^{3\cdot(4/3)} = 3^4 = 81$

(d) $(-8)^{-2/3} = [(-2)^3]^{-2/3} = (-2)^{3\cdot(-2/3)} = (-2)^{-2} = \dfrac{1}{4}$

(e) $-8^{-2/3} = -(2^3)^{-2/3} = -(2)^{3\cdot(-2/3)} = -2^{-2} = -\dfrac{1}{4}$

(f) $(-4)^{3/2} = [(-4)^{1/2}]^3$ is not a real number since $\sqrt{-4}$ is not a
                                             real number   ◈

An alternative method is to use radicals such as

(a) $4^{3/2} = (\sqrt{4})^3$ (d) $(-8)^{-2/3} = (\sqrt[3]{-8})^{-2}$

$\qquad = 2^3 = 8$ $\qquad = (-2)^{-2} = \dfrac{1}{4}$

Converting Forms

In the preceding definitions, we introduced the exponential notation $a^{1/n}$ and the radical notation $\sqrt[n]{a}$ to represent the nth root of a. We also used the notations $(\sqrt[n]{a})^m$ and $a^{m/n}$ to represent the powers of an nth root. In many situations, it is important to be able to convert expressions from radical forms to exponential forms and vice versa.

EXAMPLE 3   Converting to Radical Form

Convert each expression to radical form.

(a) $7^{1/2}$ (b) $4x^{3/5}$ (c) $y^{-2/3}$

Solution   (a) $7^{1/2} = \sqrt{7}$ (b) $4x^{3/5} = 4(\sqrt[5]{x})^3$ (c) $y^{-2/3} = \dfrac{1}{y^{2/3}}$

$\qquad\qquad\qquad\qquad\qquad = 4\sqrt[5]{x^3}$ $\qquad\qquad = \dfrac{1}{\sqrt[3]{y^2}}$ ◈

EXAMPLE 4   Converting to Exponential Form

Convert each expression to exponential form.

(a) $\sqrt{5}$ (b) $\sqrt[3]{x^2}$ (c) $\sqrt[5]{(3xy^2)^3}$

Solution   (a) $\sqrt{5} = 5^{1/2}$ (b) $\sqrt[3]{x^2} = x^{2/3}$ (c) $\sqrt[5]{(3xy^2)^3} = (3xy^2)^{3/5}$ ◈

Using the Properties of Rational Exponents

The properties of exponents listed in section 5.1 also apply to rational exponents, provided that all the quantities are defined. Therefore, we list these properties to indicate that they hold for all rational exponents.

PROPERTIES OF RATIONAL EXPONENTS

Let a and b be real numbers, and let p and q be rational numbers. Then, if all expressions are defined (as real numbers):

(i) $a^p \cdot a^q = a^{p+q}$

(ii) $(a^p)^q = a^{pq}$

(iii) $(ab)^p = a^p b^p$

(iv) $\left(\dfrac{a}{b}\right)^p = \dfrac{a^p}{b^p}$

(v) $\dfrac{a^p}{a^q} = a^{p-q}$

The following useful relationship is an immediate consequence of the exponent properties:

$$a^{m/n} = (a^{1/n})^m = (a^m)^{1/n}$$

EXAMPLE 5   Using the Properties of Exponents

Simplify each expression and write the result using only positive exponents.

(a) $7^{-1/2} \cdot 7^{5/2}$ (b) $(x^{3/5})^{-10}$ (c) $\dfrac{x^{2/3}}{x^{-1/5}}$ (d) $(125x^{-9})^{-2/3}$ (e) $\left(\dfrac{32}{x^{-5}}\right)^{-2/5}$

Solution   (a) $7^{-1/2} \cdot 7^{5/2} = 7^{-1/2+5/2}$ Property (i), Multiplication Property

$= 7^{4/2} = 7^2 = 49$

(b) $(x^{3/5})^{-10} = x^{(3/5)(-10)}$ Property (ii), Power Property

$= x^{-6} = \dfrac{1}{x^6}$

(c) $\dfrac{x^{2/3}}{x^{-1/5}} = x^{2/3 \, - \, (-1/5)}$ Property (v), Division Property

$= x^{10/15 \, + \, 3/15} = x^{13/15}$

(d) $(125x^{-9})^{-2/3} = (125^{-2/3})(x^{-9})^{-2/3}$ Property (iii)

$= (\sqrt[3]{125})^{-2}x^6$ Property (ii), Power Property

$= 5^{-2}x^6 = \dfrac{x^6}{25}$

(e) $\left(\dfrac{32}{x^{-5}}\right)^{-2/5} = \dfrac{(32)^{-2/5}}{(x^{-5})^{-2/5}}$ Property (iv)

$= \dfrac{(\sqrt[5]{2^5})^{-2}}{x^2} = \dfrac{2^{-2}}{x^2} = \dfrac{1}{4x^2}$

EXAMPLE 6   Using the Properties of Exponents

Suppose that x and y represent positive numbers. Simplify each expression and write the result using only positive exponents.

(a) $\dfrac{x^{1/5}y^{5/6}}{x^{4/5}y^{1/3}}$ (b) $x^{-2/3}(x^{4/3} + y^{1/5})$

Solution   (a) $\dfrac{x^{1/5}y^{5/6}}{x^{4/5}y^{1/3}} = x^{1/5-4/5} \cdot y^{5/6-1/3}$ Division Property

$= x^{-3/5} \cdot y^{3/6} = \dfrac{y^{1/2}}{x^{3/5}}$

(b) $x^{-2/3}(x^{4/3} + y^{1/5}) = x^{-2/3} \cdot x^{4/3} + x^{-2/3} \cdot y^{1/5}$ Distributive Property

$= x^{4/3-2/3} + x^{-2/3}y^{1/5}$ Product Property

$= x^{2/3} + \dfrac{y^{1/5}}{x^{2/3}}$

Solving Applied Problems

The compound interest model is also applicable if the money is invested at a specified compounding rate for a period of a fraction of a year such as 2 1/2 years, 3 2/5 years, and so on. The next example illustrates the use of the compound interest model for this case.

EXAMPLE 7　Earning Compound Interest

Suppose that $1000 is deposited in a bank that pays 3.5% annual interest compounded monthly. What is the value S (money accumulated) of the account after a term of 2/3 year?

Solution　Monthly compounding means that $n = 12$. Substituting 1000 for P, 0.035 for r, 12 for n, and 2/3 for t in the compound interest formula, we have

$$S = P\left(1 + \frac{r}{n}\right)^{nt}$$

$$= 1000\left(1 + \frac{0.035}{12}\right)^{12\left(\frac{2}{3}\right)}$$

$$= 1000\left(1 + \frac{0.035}{12}\right)^{8}$$

$$= 1023.57$$

Therefore, the accumulated amount of money after 2/3 year will be $1023.57. ◇

Using Technology Exploration

We now explore graphical results of the definition of $a^{m/n}$. Consider the function f defined by

$$f(x) = x^{2/3} = (\sqrt[3]{x})^2$$

We can input negative as well as positive values of x, since a cube root has an odd index. For example,

$$f(-8) = (\sqrt[3]{-8})^2$$
$$= (-2)^2 = 4$$
$$f(0) = (\sqrt[3]{0})^2$$
$$= (0)^2 = 0$$
$$f(3) = (\sqrt[3]{3})^2 \approx 2.08$$

G EXAMPLE 8　Graphing a Function
Let $f(x) = x^{2/3}$
(a) Find $f(-8), f(-6), f(-3), f(-1), f(0), f(1), f(3), f(6)$ and $f(8)$. Round to two decimal places when necessary.
(b) Plot the points in part (a) and connect them with a smooth curve.
(c) Indicate the domain and range of this function.
(d) Write $f(x) = x^{2/3} = (\sqrt[3]{x})^2$ and use a grapher to graph $y = (\sqrt[3]{x})^2$ and $y = 4$ on the same viewing window.
(e) Use part (d) to solve the equation $x^{2/3} = 4$.

Solution   (a) The following table shows the output values for the given input values.

x	−8	−6	−3	−1	0	1	3	6	8
$y = f(x) = x^{2/3}$	4	3.30	2.08	1	0	1	2.08	3.30	4

(b) Figure 1(a) shows the plotted points and the smooth curve connecting them.

(c) The graph suggests that the domain of f is the set of all real numbers and the range is the set of nonnegative real numbers or $[0,\infty)$.

(d) Figure 1(b) shows the graphs of $y = (\sqrt[3]{x})^2$ and $y = 4$ on the same viewing window.

Many graphers will incorrectly graph functions involving rational exponents of the form $x^{m/n}$, where $n \geq 2$, by mistakenly requiring $x \geq 0$ for odd root indexes. To avoid this, graph $y = x^{m/n}$ as $y = (x^m)^{1/n}$.

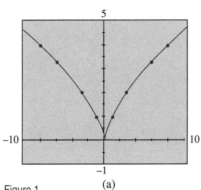
(a)

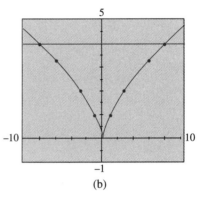
(b)

Figure 1

(e) Using the TRACE and ZOOM features of the grapher, we see that the two graphs intersect at the points $(-8,4)$ and $(8,4)$, which suggest the solutions of the equation are -8 and 8.   ◇

◇ PROBLEM SET 5.3

Mastering the Concepts

In Problems 1–14, find the exact value of each expression.

1. (a) $36^{1/2}$ (b) $49^{(-1/2)}$ (c) $27^{1/3}$

2. (a) $-16^{(-1/2)}$ (b) $-8^{(-1/3)}$ (c) $64^{(-1/3)}$

3. (a) $(-27)^{1/3}$ (b) $81^{1/4}$ (c) $-16^{1/4}$

4. (a) $(-64)^{1/3}$ (b) $49^{1/2}$ (c) $81^{(-1/4)}$

5. (a) $\left(\dfrac{8}{-27}\right)^{(-1/3)}$ (b) $\left(\dfrac{81}{25}\right)^{(-1/2)}$ (c) $\left(\dfrac{9}{16}\right)^{1/2}$

6. (a) $\left(\dfrac{1}{16}\right)^{(-1/4)}$ (b) $\left(\dfrac{1}{64}\right)^{(-1/3)}$ (c) $\left(\dfrac{16}{9}\right)^{(-1/2)}$

7. (a) $8^{2/3}$ (b) $-8^{4/3}$

8. (a) $(-8)^{4/3}$ (b) $(-27)^{2/3}$

9. (a) $(-32)^{3/5}$ (b) $128^{2/7}$

10. (a) $81^{3/4}$ (b) $(-64)^{4/3}$

11. (a) $(-16)^{3/2}$ (b) $(-36)^{3/2}$

12. (a) $27^{8/12}$ (b) $25^{7/14}$

13. (a) $(-8)^{(-4/3)}$ (b) $(-27)^{(-2/3)}$

14. (a) $9^{(-3/2)}$ (b) $(-32)^{(-4/5)}$

In Problems 15–20, convert to radical form. Do not simplify. Assume all variables represent positive real numbers. Use only positive exponents.

15. (a) $5^{1/3}$ (b) $6^{-1/4}$

16. (a) $11^{1/2}$ (b) $7^{-1/5}$

17. (a) $x^{2/7}$ (b) $y^{3/4}$

18. (a) $5y^{3/7}$ (b) $4t^{2/3}$

19. (a) $(8m)^{2/3}$ (b) $(4xy^2)^{2/5}$

20. (a) $(x + y)^{2/3}$ (b) $(x^2 + y^2)^{-1/2}$

In Problems 21–26, convert to rational exponents. Do not simplify. Assume all variables represent positive real numbers.

21. (a) $\sqrt{7}$ (b) $\sqrt{3}$

22. (a) $\sqrt{15}$ (b) $\sqrt{17}$

23. (a) $\sqrt[3]{m}$ (b) $\sqrt[5]{x^3}$

24. (a) $\sqrt[3]{x^2}$ (b) $\sqrt[4]{x^3}$

25. (a) $\sqrt[5]{(7x^2)^2}$ (b) $\sqrt[4]{(3xy)^2}$

26. (a) $\sqrt{(x+y)^3}$ (b) $\sqrt{x^2+y^2}$

In Problems 27–40, simplify each expression and write each answer using only positive exponents. Assume all variables represent positive real numbers.

27. (a) $2^{1/3} \cdot 2^{2/3}$ (b) $x^{-2/3} \cdot x^{5/3}$

28. (a) $5^{1/2} \cdot 5^{-3/2}$ (b) $y^{2/15} \cdot y^{-7/60}$

29. (a) $(5^{1/7})^{14}$ (b) $(x^{-7/4})^{18/7}$

30. (a) $(8^{-2/3})^{-6}$ (b) $(y^{-2})^{-15/2}$

31. (a) $(8p^9)^{4/3}$ (b) $\left(\dfrac{125}{y^3}\right)^{-1/3}$

32. (a) $(32u^{-5})^{-3/5}$ (b) $\left(\dfrac{x^{-5}}{32}\right)^{4/5}$

33. (a) $\dfrac{x^{2/3}}{x^{-1/3}}$ (b) $\left(\dfrac{x^{-3}}{y^2}\right)^{-1/6}$

34. (a) $\dfrac{x^{1/3}}{x^{-1/6}}$ (b) $\dfrac{y^{3/2}}{y^{-7/2}}$

35. $\left(\dfrac{x^{-10}y^8}{x^{-12}y^{-4}}\right)^{-1/2}$ 36. $\left(\dfrac{5^2 4^6}{5^{-4} 4^7}\right)^{-1/2}$

37. $\dfrac{(125x^{-5}y^7)^{-1/3}}{(64x^8y^2)^{-1/6}}$ 38. $\left(\dfrac{a^3 b^{3/2}}{a^{-3} b^{1/2}}\right)^{1/6}$

39. (a) $x^{1/2}(x^{3/2}+x^{-1/2})$ 40. (a) $x^{5/3}(x^{-2/3}+x^{1/3})$
(b) $x^{-1/2}(x+x^{1/2})$ (b) $x^{-2/3}(x^{2/3}+x^{-1/3})$

In Problems 41–44, evaluate each function at the indicated input values.

41. $f(x) = x^{\frac{2}{3}}$
(a) $x = -64$ (b) $x = 0$ (c) $x = 27$

42. $g(x) = x^{\frac{3}{2}}$
(a) $x = 0$ (b) $x = 4$ (c) $x = 9$

43. $h(x) = x^{-\frac{2}{3}}$
(a) $x = -8$ (b) $x = 1$ (c) $x = 27$

44. $F(x) = x^{-\frac{3}{5}}$
(a) $x = -32$ (b) $x = 0$ (c) $x = 32$

Applying the Concepts

45. **Shoe Design:** A man's shoe size s is approximately related to his height h in feet by the model
$$s = 0.769\, h^{3/2} \text{ where } 5 < h \le 9$$
(a) If a man is 6 feet tall, what is his shoe size? Round to the nearest half size.
(b) According to the Bible, Goliath was 6 cubits tall (approximately 9 feet). What shoe size would he have worn?

46. **Weight of a Whale:** The International Whaling Commission determined that the relationship between the weight W in tons and length L in feet of a sperm whale is given by the model
$$W = 0.000137 L^{3.18}$$
Determine the weight of a sperm whale which is 67.5 feet long. Round off to the nearest ton.

47. **Compound Interest Earned:** Suppose that $3000 is invested in a risky bond that earned 8% annual interest. Find the amount of money in the account
(a) after 2 1/2 years if the interest is compounded weekly.
(b) after 3 2/3 years if the interest is compounded monthly.

48. **Rate of Inflation:** Because of inflation, the value of a home often increases with time. State and local agencies use the formula
$$r = \left(\dfrac{S}{C}\right)^{1/t} - 1$$
to assess the annual rate of inflation r for tax purposes, where C (in dollars) is the cost of the home when new, S (in dollars) is the value of the home after t years. Suppose that a house cost $12,500 in 1962. In 1998 the same house is worth $75,000. What is the annual rate of inflation?

Developing and Extending the Concepts

In Problems 49–54, use the properties of rational exponents to simplify each expression. Express the answer using positive exponents. Assume all variables represent positive real numbers.

49. $\left(\dfrac{x^{2/3}}{y^{-3/4}}\right)^{12} \cdot (x^{-3/8}y^{1/4})^{-2}$ 50. $\left(\dfrac{81p^{-12}}{q^{16}}\right)\left(\dfrac{p^{-2/3}}{q^{1/3}}\right)^3$

51. $\dfrac{(3x^3y)^{1/3}(3xy^5)^{2/3}}{(9x^2y)^{1/2}}$ 52. $\dfrac{(5x^2y^3)^{3/4}(5x^2y^3)^{1/4}}{(5x^2y^3)^{-2}}$

53. (a) $(x^{1/2}-y^{-1/2})(x^{1/2}+y^{-1/2})$
(b) $(x^{3/2}-y^{3/2})(x^{3/2}+y^{3/2})$

54. (a) $(a^{-1/2}-3b^{-1/2})(a^{-1/2}+3b^{-1/2})$
(b) $(x^{3/2}-2y^{3/2})(x^{3/2}+2y^{3/2})$

In Problems 55 and 56, for each given function find the indicated values, if they are real. Round off to two decimal places if necessary.

55. $f(x) = x^{2/5}$
(a) $f(-32)$ (b) $f(-20)$
(c) $f(0)$ (d) $f(1)$
(e) $f(32)$

56. $g(x) = x^{3/2}$
(a) $g(0)$ (b) $g(1)$
(c) $g(7)$ (d) $g(9)$
(e) $g(16)$

57. Figure 2a shows a grapher display of $g(x) = x^{3/2}$ (black curve) and $h(x) = x^{3/2} - 8$ (red curve) in the same viewing window.
    (a) Indicate the domain and range of g and h.
    (b) Solve the equation $x^{3/2} = 8$ graphically by finding the x intercept of the graph of h.

58. Figure 2b shows the grapher display of $f(x) = x^{-3/2}$ (black curve) and $g(x) = x^{-2/3} - 1$ (red curve) in the same viewing window.
    (a) Indicate the domain and range of f and g.
    (b) Solve the equation $x^{-2/3} = 1$ graphically by finding the x intercept of the graph of g.

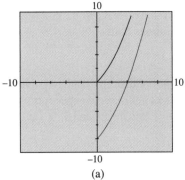

(a)

Figure 2

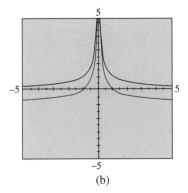

(b)

OBJECTIVES

1. Simplify Radicals
2. Solve Applied Problems
3. Use Technology Exploration

5.4 Simplifying Radical Expressions

Simplifying Radicals

In this section, we simplify radical expressions by using certain properties of radicals. For instance, we see that

$$\sqrt{4} \cdot \sqrt{9} = 2 \cdot 3 = 6 \text{ and that } \sqrt{4 \cdot 9} = \sqrt{36} = 6$$

Thus, we have $\sqrt{4} \cdot \sqrt{9} = \sqrt{4 \cdot 9}$.

Similarly,

$$\sqrt[3]{\frac{27}{8}} = \frac{3}{2} \text{ and that } \frac{\sqrt[3]{27}}{\sqrt[3]{8}} = \frac{3}{2}$$

so that

$$\sqrt[3]{\frac{27}{8}} = \frac{\sqrt[3]{27}}{\sqrt[3]{8}}$$

These examples are based on the following properties for multiplying and dividing radicals:

PROPERTIES OF RADICALS

Let a and b be real numbers and n be a positive integer with $n \geq 2$. Then, provided that $\sqrt[n]{a}$ and $\sqrt[n]{b}$ are real numbers,

(i) (Product Property) $\sqrt[n]{ab} = \sqrt[n]{a} \cdot \sqrt[n]{b}$

(ii) (Quotient Property) $\sqrt[n]{\frac{a}{b}} = \frac{\sqrt[n]{a}}{\sqrt[n]{b}}, b \neq 0$

To see that these properties are true, notice that property
(i) $\sqrt[n]{ab} = \sqrt[n]{a} \cdot \sqrt[n]{b}$ is the relationship $(ab)^{1/n} = a^{1/n} \cdot b^{1/n}$; and property
(ii) $\sqrt[n]{\dfrac{a}{b}} = \dfrac{\sqrt[n]{a}}{\sqrt[n]{b}}$ is the relationship $\left(\dfrac{a}{b}\right)^{1/n} = \dfrac{a^{1/n}}{b^{1/n}}$.

We can use these properties of radicals to simplify radical expressions by reading these properties from left to right or from right to left. A *simplest radical* is useful in comparing radicals without changing their value. A radical expression (or rational exponent expression) is said to be in **simplest form** or **simplified** if all of the following conditions hold.

Simplified Radical Form	**Simplified Rational Exponent Form**
1. There shall be no radicand containing polynomial factors to a power greater than or equal to the index of the radical.	1. All rational exponents shall be between 0 and 1.
2. No fraction appears within a radicand.	2. No fractions appear as the base of a rational exponent.
3. No radical appears in a denominator of a fraction.	3. No rational exponents appear in the denominator of a fraction.
4. The exponents of factors under the radical and the index of the radical have no common factors.	4. Rational exponents are reduced to lowest terms.

For the remainder of this chapter, we shall impose restrictions on variables so that each radical expression represents a real number.

EXAMPLE 1   Applying the Radical Properties
Simplify each radical expression by using the radical properties.

(a) $\sqrt{72}$ (b) $\sqrt[3]{-24}$ (c) $\sqrt{7/16}$ (d) $\sqrt[3]{5/27}$

Solution   (a) We factor 72 using perfect square factors such as $36 \cdot 2$, so we write
$$\sqrt{72} = \sqrt{36 \cdot 2} = \sqrt{36} \cdot \sqrt{2} = 6\sqrt{2} \qquad \text{Product Property}$$

$4 \cdot 9 \cdot 2$ is also a factorization of 72 using perfect square factors and
$$\sqrt{72} = \sqrt{4 \cdot 9 \cdot 2} =$$
$$\sqrt{4} \cdot \sqrt{9} \cdot \sqrt{2} =$$
$$2 \cdot 3 \cdot \sqrt{2} = 6\sqrt{2}$$

(b) Since the index is 3, we look for perfect cube factors and we write
$$\sqrt[3]{-24} = \sqrt[3]{(-8) \cdot 3} = \sqrt[3]{-8} \cdot \sqrt[3]{3} = -2\sqrt[3]{3} \quad \text{Product Property}$$

(c) $\sqrt{\dfrac{7}{16}} = \dfrac{\sqrt{7}}{\sqrt{16}} = \dfrac{\sqrt{7}}{4}$ Quotient Property

(d) $\sqrt[3]{\dfrac{5}{27}} = \dfrac{\sqrt[3]{5}}{\sqrt[3]{27}} = \dfrac{\sqrt[3]{5}}{3}$ Quotient Property ◈

EXAMPLE 2   Applying the Radical Properties
Use the properties of radicals to simplify each expression. Justify by using rational exponents. Assume that x is a positive real number.

(a) $\sqrt{4x^3}$ (b) $\sqrt[3]{24x^4y^7}$ (c) $\dfrac{\sqrt{12x^2}}{\sqrt{3}}$ (d) $\dfrac{\sqrt[3]{40xy^4}}{2\sqrt[3]{5x}}$

Solution

Radicals	Rational Exponents	Properties

(a) $\sqrt{4x^3} = \sqrt{4x^2 \cdot x}$ $\qquad$ $\sqrt{4x^3} = (4x^2 \cdot x)^{1/2}$ $\qquad$ Perfect square factors

$\qquad\quad = \sqrt{4x^2} \cdot \sqrt{x}$ $\qquad\qquad = (4x^2)^{1/2} \cdot x^{1/2}$ $\qquad$ Product property

$\qquad\quad = 2x \cdot \sqrt{x}$ $\qquad\qquad\quad = 2x \cdot x^{1/2}$

(b) $\sqrt[3]{24x^4y^7} = \sqrt[3]{(8x^3y^6)(3xy)}$ $\qquad$ $= (2^3x^3y^6 \cdot 3xy)^{1/3}$ $\qquad$ Perfect cube factors

$\qquad\qquad = \sqrt[3]{8x^3y^6} \cdot \sqrt[3]{3xy}$ $\qquad = (2^3x^3y^6)^{1/3}(3xy)^{1/3}$ $\qquad$ Product property

$\qquad\qquad = 2xy^2 \sqrt[3]{3xy}$ $\qquad\qquad = 2xy^2(3xy)^{1/3}$

(c) $\dfrac{\sqrt{12x^2}}{\sqrt{3}} = \sqrt{\dfrac{12x^2}{3}}$ $\qquad$ $\dfrac{(12x^2)^{1/2}}{3^{1/2}} = \left(\dfrac{12x^2}{3}\right)^{1/2}$ $\qquad$ Quotient property

$\qquad\qquad = \sqrt{4x^2} = 2x$ $\qquad\qquad = (4x^2)^{1/2} = 2x$

(d) $\dfrac{\sqrt[3]{40xy^4}}{2\sqrt[3]{5x}} = \dfrac{1}{2}\sqrt[3]{\dfrac{40xy^4}{5x}}$ $\qquad$ $\dfrac{(40xy^4)^{1/3}}{2(5x)^{1/3}} = \dfrac{1}{2}\left(\dfrac{40xy^4}{5x}\right)^{1/3}$ $\qquad$ Quotient property

$\qquad\qquad = \dfrac{1}{2}\sqrt[3]{8y^4}$ $\qquad\qquad\qquad = \dfrac{1}{2}(8y^4)^{1/3}$

$\qquad\qquad = \dfrac{1}{2}\sqrt[3]{(8y^3) \cdot y}$ $\qquad\qquad = \dfrac{1}{2}(2^3y^3)^{1/3} \cdot y^{1/3}$ $\qquad$ Perfect cube factors

$\qquad\qquad = \dfrac{1}{2} \cdot 2y\sqrt[3]{y}$ $\qquad\qquad\quad = \dfrac{1}{2}2y \cdot y^{1/3}$ $\qquad$ Product property

$\qquad\qquad = y\sqrt[3]{y}$ $\qquad\qquad\qquad = y \cdot y^{1/3}$ $\qquad\qquad$ ◈

In order to multiply or divide radical expressions with different indices, we begin by building their indices to a common index of individual radicals. This is accomplished by converting each expression to exponential form, then using the LCD of the rational exponents.

EXAMPLE 3   Simplifying Radical Expressions

Use rational exponents to rewrite each expression as a single expression in a simple form. Assume $x \geq 0$.

(a) $\sqrt[3]{\sqrt{64}}$ $\qquad\qquad$ (b) $\sqrt{3} \cdot \sqrt[3]{5}$ $\qquad\qquad$ (c) $\sqrt[4]{x^3} \cdot \sqrt[3]{x^2}$

Solution   (a) $\sqrt[3]{\sqrt{64}} = \sqrt[3]{8} = 2$

(b) $\sqrt{3} \cdot \sqrt[3]{5} = 3^{1/2} \cdot 5^{1/3}$ $\qquad$ Write radical as rational exponent

$\qquad\qquad = 3^{3/6} \cdot 5^{2/6}$ $\qquad$ LCD of 2 and 3 is 6

$\qquad\qquad = (3^3 \cdot 5^2)^{1/6}$ $\qquad$ Product property

$\qquad\qquad = (675)^{1/6} = \sqrt[6]{675}$ $\qquad$ Return to radical expression

(c) $\sqrt[4]{x^3} \cdot \sqrt[3]{x^2} = x^{3/4} \cdot x^{2/3}$ $\qquad$ Rewrite as rational exponents

$\qquad\qquad = x^{3/4+2/3}$ $\qquad$ Product rule for exponents

$\qquad\qquad = x^{9/12+8/12} = x^{17/12}$ $\qquad$ LCD of the denominator is 12

$\qquad\qquad = x^{12/12} \cdot x^{5/12}$ $\qquad$ Factor perfect 12th root

$\qquad\qquad = x\sqrt[12]{x^5}$ $\qquad$ Return to radical form $\qquad$ ◈

Solving Applied Problems

Several applications and models from different sciences involve radical expressions as the next example shows.

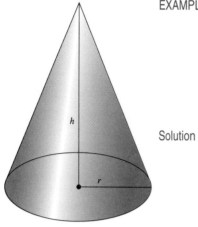

Figure 1

EXAMPLE 4 Finding the Surface Area of a Tent

The surface area A of a conical tent with radius r and height h (Figure 1) is given by the model

$$A = \pi r \sqrt{r^2 + h^2}$$

(a) If $h = 2r$, express A in terms of r in a simplified radical form.
(b) Find the surface area of a tent whose radius is 6.9 feet and whose height is 7.8 feet. Round off the answer to two decimal places.

Solution (a) If we replace h by $2r$ in the above model, we obtain

$$A = \pi r \sqrt{r^2 + h^2} = \pi r \sqrt{r^2 + 4r^2} =$$
$$= \pi r \sqrt{5r^2} = \pi r^2 \sqrt{5}$$

Note that r is positive since it is the radius of the tent.
(b) Replacing r by 6.9 and h by 7.8, we have

$$A = \pi(6.9)\sqrt{(6.9)^2 + (7.8)^2} \approx 225.74 \text{ square feet}$$ ◇

Using Technology Exploration

Technology can be used to verify the equality of radical expressions.

EXAMPLE 5 Using Technology to Compare Radical Forms

Let $f(x) = \sqrt{16x^3}$ and $g(x) = 4x\sqrt{x}$
(a) Find $f(1)$, $g(1)$, $f(2)$, $g(2)$, $f(3.7)$ and $g(3.7)$. Round off to two decimal places when necessary.
(b) What restrictions should be put on the inputs of f and g in order to obtain equal real number outputs?
(c) Use a grapher to graph f and g in the same viewing window.
(d) Use the grapher display to indicate the domain and range of f and g.
(e) Do the graphs show that $\sqrt{16x^3} = 4x\sqrt{x}$ on the restricted inputs of f and g?

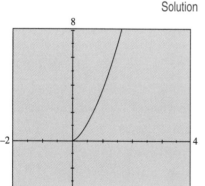

Figure 2

Solution (a)

$$f(x) = \sqrt{16x^3} \qquad\qquad g(x) = 4x\sqrt{x}$$
$$f(1) = \sqrt{16(1)^3} = \sqrt{16} = 4 \qquad g(1) = 4 \cdot 1 \cdot \sqrt{1} = 4$$
$$f(2) = \sqrt{16(2)^3} = 11.31 \qquad g(2) = 4 \cdot 2 \cdot \sqrt{2} = 11.31$$
$$f(3.7) = \sqrt{16(3.7)^3} = 28.47 \qquad g(3.7) = 4 \cdot 3.7 \cdot \sqrt{3.7} = 28.47$$

(b) Since the index of each radical is even, the inputs of f and g must be non-negative.
(c) Figure 2 shows the graphs of f and g in the same viewing window.
(d) The domain of f and g consists of the set of nonnegative real numbers or $[0,\infty)$. The range of f and g also consists of the nonnegative real numbers or $[0,\infty)$. Algebraically

$$\sqrt{16(x)^3} = \sqrt{16x^2 \cdot x} = \sqrt{16x^2} \cdot \sqrt{x} = 4x\sqrt{x}$$

The graphical display in Figure 2 supports the algebraic conclusion. ◇

 ## PROBLEM SET 5.4

Mastering the Concepts

In Problems 1–20, use the properties of radicals to simplify each expression. Write the answer in the simplest radical form. Assume that all variables represent positive real numbers.

1. (a) $\sqrt{27}$ (b) $\sqrt[3]{-54}$
2. (a) $\sqrt{162}$ (b) $\sqrt[3]{-250}$
3. (a) $\sqrt{5/4}$ (b) $\sqrt[3]{-5/8}$
4. (a) $\sqrt{5/16}$ (b) $\sqrt[3]{-11/125}$
5. (a) $\sqrt{48x}$ (b) $\sqrt[3]{24y^3}$

6. (a) $\sqrt{98p^3}$ (b) $\sqrt[3]{54t^4}$

7. (a) $\sqrt{32c^5}$ (b) $\sqrt[3]{-32p^6}$

8. (a) $\sqrt[3]{-16y^4}$ (b) $\sqrt[3]{250x^5}$

9. (a) $\sqrt{\dfrac{7}{4x^2}}$ (b) $\sqrt[5]{-32p^6}$

10. (a) $\sqrt{\dfrac{3}{25x^4}}$ (b) $\sqrt[5]{\dfrac{x}{y^{10}}}$

11. (a) $\sqrt{\dfrac{3}{25x^2}}$ (b) $\sqrt{\dfrac{5}{9x^4}}$

12. (a) $\sqrt{\dfrac{17}{y^4}}$ (b) $\sqrt[7]{\dfrac{3}{-a^{14}}}$

13. (a) $\dfrac{\sqrt{8a^3}}{\sqrt{2a}}$ (b) $\dfrac{\sqrt{14x^3}}{\sqrt{7x}}$

14. (a) $\dfrac{\sqrt{45x^3}}{\sqrt{3x}}$ (b) $\dfrac{\sqrt[3]{48x^4}}{\sqrt[3]{6x}}$

15. (a) $\dfrac{\sqrt[3]{10x^6}}{\sqrt[3]{2}}$ (b) $\dfrac{\sqrt{x^5}}{\sqrt{x}}$

16. (a) $\dfrac{\sqrt[3]{54x^7}}{\sqrt[3]{2x}}$ (b) $\dfrac{\sqrt[3]{128x^4}}{\sqrt[3]{2x}}$

17. $\sqrt[3]{125a^3b^6c^9}$ 18. $\sqrt[3]{-8a^6b^9c^3}$

19. $-\sqrt[3]{-27x^3y^9z^{12}}$ 20. $\sqrt[3]{81x^{12}y^4z^{16}}$

In Problems 21–28, use rational exponents to rewrite each expression as a single radical expression in simple form. Assume all variables represent positive real numbers.

21. (a) $\sqrt[3]{\sqrt{1024}}$ (b) $\sqrt{\sqrt[3]{x^6}}$

22. (a) $\sqrt{\sqrt[4]{p^8}}$ (b) $\sqrt[5]{\sqrt[3]{p^{15}}}$

23. (a) $\sqrt{3}\cdot\sqrt[3]{4}$ (b) $\sqrt[3]{4}\cdot\sqrt{7}$

24. (a) $\sqrt{5}\cdot\sqrt[3]{6}$ (b) $\sqrt[4]{7}\cdot\sqrt{3}$

25. (a) $\sqrt{x}\cdot\sqrt[3]{x}$ (b) $\sqrt[4]{x}\cdot\sqrt{x}$

26. (a) $\sqrt[3]{x}\cdot\sqrt[4]{x}$ (b) $\sqrt[3]{x}\cdot\sqrt[5]{x}$

27. (a) $\sqrt{5x}\cdot\sqrt[3]{5x}$ (b) $\sqrt[3]{4x}\cdot\sqrt{2y}$

28. (a) $\sqrt{3x}\cdot\sqrt[3]{3x}$ (b) $\sqrt[3]{5x}\cdot\sqrt{3y}$

Applying the Concepts

29. **Advertising:** A small company determines that if it spends x dollars on radio advertising, y dollars on TV commercials, and z dollars on newspaper advertising then its annual sale of p units of its product is modeled by the formula

$$p = \sqrt[5]{64x^3y^6z^5}$$

(a) Express p in a simplified radical form.

(b) Use the above model to determine the number of units sold if the company spends annually $300 on radio advertising, $400 on TV commercials, and $200 on newspaper advertising. Round off the answer to the nearest thousand.

30. **Geometry:** The volume V of a spherical hot air balloon is related to its surface area A by the formula

$$V = 0.094\sqrt{A^3}$$

(a) Use the formula $A = 4\pi r^2$, where r is the radius of the balloon to express V in terms of r in a simplified radical form.

(b) What is the volume of a balloon whose radius is 5.3 feet? Round off the answer to two decimal places.

Developing and Extending the Concepts

31. (a) Use a calculator to evaluate $\sqrt[4]{7}$. Round off to six decimal places.

(b) Use a calculator to evaluate $\sqrt{\sqrt{7}}$. Round off to six decimal places.

(c) Compare the results of parts (a) and (b).

32. What is the restriction on x in order for each equation to be true?

(a) $\sqrt{x}\cdot\sqrt{x} = x$ (b) $\sqrt{-x}\cdot\sqrt{x} = -x$

(c) $\sqrt[3]{x}\cdot\sqrt[3]{x}\cdot\sqrt[3]{x} = x$

33. (a) Use a calculator to evaluate each expression. Round off each result to three decimal places.

$$\sqrt{3},\ \sqrt{\sqrt{3}},\ \sqrt{\sqrt{\sqrt{3}}},$$

$$\sqrt{\sqrt{\sqrt{\sqrt{3}}}},\ \sqrt{\sqrt{\sqrt{\sqrt{\sqrt{3}}}}}$$

(b) If we continue to take the square root, what number does the value of the expression seem to be approaching?

34. What restrictions on a, b and c should be made in order for each of these expressions $\sqrt{a}\cdot\sqrt{b}\cdot\sqrt{c}$ and $\sqrt{abc}$ to be real numbers and $\sqrt{a}\cdot\sqrt{b}\cdot\sqrt{c} = \sqrt{abc}$?

G 35. Let $f(x) = \sqrt{4x^3}$ and $g(x) = 2x\sqrt{x}$

(a) Find $f(1)$, $g(1)$, $f(2)$, $g(2)$, $f(2.5)$, $g(2.5)$. Round off to two decimal places.

(b) What restrictions should be imposed on the inputs of f and g in order to obtain equal real number outputs?

(c) Use a grapher to graph f and g on the same viewing window.

(d) Use the grapher display to indicate the domain and range of f and g.

(e) Do the graphs show that $\sqrt{4x^3} = 2x\sqrt{x}$ for the restricted inputs of f and g?

36. Let $f(x) = \sqrt{x-1} \cdot \sqrt{x+1}$ and $g(x) = \sqrt{x^2-1}$
    (a) Find each value (if possible) of $f(3), g(3), f(1), g(1),$ $f(-2), g(-2), f(-3), g(-3)$. Round off to two decimal places.
    (b) Use a grapher to produce the graphs of f and g on the same viewing window. Are the graphs of f and g the same?
    (c) Use the grapher display in part (b) to indicate the domains of f and g.
    (d) According to the product rule $\sqrt{a} \cdot \sqrt{b} = \sqrt{ab}$, provided $\sqrt{a}$ and $\sqrt{b}$ are real numbers, $\sqrt{x-1} \cdot \sqrt{x+1} = \sqrt{x^2-1}$. Does this rule support the grapher display in part (b)?

In Problems 37–42, use the properties of radicals to simplify each expression. Assume that all variables represent positive real numbers.

37. $\sqrt[8]{x^{12}} \cdot \sqrt[8]{x^5 y^{-8}} \cdot \sqrt[8]{x^2 y^9}$

38. $\sqrt[3]{25y^2} \cdot \sqrt[3]{5y^4} \cdot \sqrt[3]{y^9}$

39. $\dfrac{\sqrt{324x^5 y} \cdot \sqrt{9x^2}}{\sqrt{25x^2 y}}$

40. $\dfrac{\sqrt{m^2 n} \cdot \sqrt{mn^4}}{\sqrt{mn^3}}$

41. $\dfrac{\sqrt[3]{p^2 q^3} \cdot \sqrt[3]{125p^3 q^2}}{\sqrt[3]{8p^3 q^4}}$

42. $\dfrac{\sqrt[3]{a^2 b^4} \cdot \sqrt[3]{a^4 b} \cdot \sqrt[3]{a^3 b^2}}{\sqrt[3]{ab^2} \cdot \sqrt[3]{a^2 b^7}}$

OBJECTIVES

1. Add and Subtract Radical Expressions
2. Multiply Radical Expressions
3. Divide Radical Expressions— Rationalize the Denominator
4. Solve Applied Problems
5. Use Technology Exploration

5.5 Operations with Radicals

In this section we develop skills for adding, subtracting, multiplying, and dividing radical expressions.

Adding and Subtracting Radical Expressions

Recall that the sum or difference of like terms of a polynomial can be simplified. To simplify these sums or differences, we use the distributive property. For example, $7x + 3x = (7+3)x = 10x$. Radicals with the same index and the same radicand are *like* radicals. For instance, $3\sqrt{6}$ and $-5\sqrt{6}$ are like radicals. We can add and subtract radicals in much the same way that we add or subtract like terms of a polynomial. For example,

$$3\sqrt{6} - 5\sqrt{6} = (3-5)\sqrt{6} = -2\sqrt{6} \text{ and}$$
$$7\sqrt{3} + 2\sqrt{3} = (7+2)\sqrt{3} = 9\sqrt{3}$$

In each case we applied the distributive property. The expression

$$\sqrt{7} + \sqrt{5}$$

$\sqrt{9} + \sqrt{16}$ is not the same as $\sqrt{9+16}$
$\sqrt{9} + \sqrt{16} = 3 + 4 = 7$ and
$\sqrt{9+16} = \sqrt{25} = 5$

is the sum of two unlike radicals which cannot be simplified and cannot be added. Although these radicals cannot be added, they can be approximated on a calculator, $\sqrt{7} + \sqrt{5} \approx 4.8818$ rounded to four decimals. Sometimes unlike radicals can be simplified as in this example,

$$\sqrt{9} + \sqrt{16} = 3 + 4 = 7$$

EXAMPLE 1   Combining Like Radicals

Combine like radicals. Assume $x \geq 0$.
(a) $12\sqrt{3} + 5\sqrt{3}$ (b) $7\sqrt[3]{2} - 3\sqrt[3]{2}$
(c) $3\sqrt{3x} - 6\sqrt{3x} + 5\sqrt{3x}$ (d) $3\sqrt{x} - 2\sqrt[3]{x} + 4\sqrt{x} + 7\sqrt[3]{x}$

Solution   (a) $12\sqrt{3} + 5\sqrt{3} = (12+5)\sqrt{3} = 17\sqrt{3}$
(b) $7\sqrt[3]{2} - 3\sqrt[3]{2} = (7-3)\sqrt[3]{2} = 4\sqrt[3]{2}$
(c) $3\sqrt{3x} - 6\sqrt{3x} + 5\sqrt{3x} = (3-6+5)\sqrt{3x} = 2\sqrt{3x}$
(d) $3\sqrt{x} - 2\sqrt[3]{x} + 4\sqrt{x} + 7\sqrt[3]{x} = 3\sqrt{x} + 4\sqrt{x} - 2\sqrt[3]{x} + 7\sqrt[3]{x}$
$$= 7\sqrt{x} + 5\sqrt[3]{x} \qquad \diamondsuit$$

Sometimes, two or more terms containing different radicals such as $5\sqrt{8} + 11\sqrt{18}$ can be combined when they are simplified algebraically as the next example shows.

EXAMPLE 2   Combining Terms Containing Different Radicals
Write radical expressions in simplest form, and then combine like radicals.

(a) $5\sqrt{8} + 11\sqrt{18}$ (b) $2\sqrt[3]{54} - 2\sqrt[3]{16}$ (c) $4\sqrt{12} + 5\sqrt{8} - \sqrt{50}$

Solution   (a) $5\sqrt{8} + 11\sqrt{18} = 5\sqrt{4 \cdot 2} + 11\sqrt{9 \cdot 2}$ Perfect square factors
$= 5\sqrt{4} \cdot \sqrt{2} + 11\sqrt{9} \cdot \sqrt{2}$ Product property
$= 5 \cdot 2\sqrt{2} + 11 \cdot 3\sqrt{2}$
$= 10\sqrt{2} + 33\sqrt{2} = 43\sqrt{2}$

(b) $2\sqrt[3]{54} - 2\sqrt[3]{16} = 2\sqrt[3]{27 \cdot 2} - 2\sqrt[3]{8 \cdot 2}$ Perfect cube factors
$= 2\sqrt[3]{27} \cdot \sqrt[3]{2} - 2\sqrt[3]{8} \cdot \sqrt[3]{2}$
$= 2 \cdot 3 \cdot \sqrt[3]{2} - 2 \cdot 2 \cdot \sqrt[3]{2}$
$= 6\sqrt[3]{2} - 4\sqrt[3]{2} = 2\sqrt[3]{2}$

(c) $4\sqrt{12} + 5\sqrt{8} - \sqrt{50} = 4\sqrt{4 \cdot 3} + 5\sqrt{4 \cdot 2} - \sqrt{25 \cdot 2}$
$= 4\sqrt{4} \cdot \sqrt{3} + 5\sqrt{4} \cdot \sqrt{2} - \sqrt{25} \cdot \sqrt{2}$
$= 4 \cdot 2\sqrt{3} + 5 \cdot 2\sqrt{2} - 5\sqrt{2}$
$= 8\sqrt{3} + 10\sqrt{2} - 5\sqrt{2}$
$= 8\sqrt{3} + 5\sqrt{2}$ These are unlike radicals and cannot be added ◈

Multiplying Radical Expressions

Multiplication of radical expressions is very similar to the multiplication procedures used to multiply polynomials. We will also make use of the product property,

$$\sqrt[n]{a} \cdot \sqrt[n]{b} = \sqrt[n]{ab}$$

EXAMPLE 3   Multiplying Radical Expressions
Perform each multiplication and simplify where possible. Assume $x \geq 0$.

(a) $\sqrt{5}(3\sqrt{7} + 2\sqrt{5})$ (b) $(\sqrt{3} - \sqrt{2})(2\sqrt{3} + \sqrt{2})$

(c) $(\sqrt{2} - 5\sqrt{3})^2$ (d) $(\sqrt{10x} - \sqrt{2})(\sqrt{10x} + \sqrt{2})$

Solution   (a) $\sqrt{5}(3\sqrt{7} + 2\sqrt{5}) = \sqrt{5} \cdot 3\sqrt{7} + \sqrt{5} \cdot 2\sqrt{5}$ Distribute
$= 3\sqrt{5 \cdot 7} + 2(\sqrt{5})^2$
$= 3\sqrt{35} + 2 \cdot 5$
$= 3\sqrt{35} + 10$

(b) $(\sqrt{3} - \sqrt{2})(2\sqrt{3} + \sqrt{2}) = \sqrt{3} \cdot 2\sqrt{3} + \sqrt{3} \cdot \sqrt{2} - \sqrt{2} \cdot 2\sqrt{3} - \sqrt{2} \cdot \sqrt{2}$
$= 2 \cdot 3 + \sqrt{6} - 2\sqrt{6} - 2$
$= 6 - \sqrt{6} - 2$
$= 4 - \sqrt{6}$

(c) $(\sqrt{2} - 5\sqrt{3})^2 = (\sqrt{2})^2 - 2 \cdot \sqrt{2} \cdot 5\sqrt{3} + (5\sqrt{3})^2$ Special product
$= 2 - 10\sqrt{6} + 75$
$= 77 - 10\sqrt{6}$

(d) $(\sqrt{10x} - \sqrt{2})(\sqrt{10x} + \sqrt{2}) = (\sqrt{10x})^2 - (\sqrt{2})^2$ Special product
$= 10x - 2$ ◈

Dividing Radical Expressions— Rationalizing the Denominator

We now consider several types of quotients involving radicals. The property $\sqrt[n]{a}/\sqrt[n]{b} = \sqrt[n]{a/b}$ is used when we divide two radical expressions. For example,

$$\frac{\sqrt{18}}{\sqrt{2}} = \sqrt{\frac{18}{2}} = \sqrt{9} = 3$$

$$\frac{\sqrt[3]{54y^4}}{\sqrt[3]{2y}} = \sqrt[3]{\frac{54y^4}{2y}} = \sqrt[3]{27y^3} = 3y$$

$$\text{and} \quad \frac{\sqrt{3}}{\sqrt{21}} = \sqrt{\frac{3}{21}} = \sqrt{\frac{1}{7}} = \frac{1}{\sqrt{7}}$$

Recall that to write the last expression in simplest radical form, no radical should appear in the denominator. To accomplish this, we multiply the numerator and denominator by $\sqrt{7}$. This is, in effect, multiplying by 1.

$$\frac{1}{\sqrt{7}} = \frac{1}{\sqrt{7}} \cdot \frac{\sqrt{7}}{\sqrt{7}} = \frac{\sqrt{7}}{7}$$

The process of removing radicals from the denominator so that the denominator contains only rational numbers is called **rationalizing the denominator**.

EXAMPLE 4   Rationalizing the Denominator

Rationalize the denominator and simplify.

(a) $\dfrac{6\sqrt{5}}{\sqrt{6}}$ (b) $\sqrt{\dfrac{27}{8}}$ (c) $\dfrac{2}{\sqrt[3]{5}}$

Solution   (a) $\dfrac{6\sqrt{5}}{\sqrt{6}} = \dfrac{6\sqrt{5} \cdot \sqrt{6}}{\sqrt{6} \cdot \sqrt{6}}$ Multiply by $\sqrt{6}/\sqrt{6}$ to create a perfect square in the denominator

$$= \frac{6\sqrt{30}}{6} = \sqrt{30}$$

(b) $\sqrt{\dfrac{27}{8}} = \dfrac{\sqrt{27}}{\sqrt{8}} = \dfrac{\sqrt{9 \cdot 3}}{\sqrt{4 \cdot 2}} = \dfrac{3\sqrt{3}}{2\sqrt{2}}$ Perfect square

$$= \frac{3\sqrt{3} \cdot \sqrt{2}}{2\sqrt{2} \cdot \sqrt{2}}$$ Multiply by $\sqrt{2}/\sqrt{2}$ to create a perfect square in the denominator

$$= \frac{3\sqrt{6}}{4}$$

(c) $\dfrac{2}{\sqrt[3]{5}} = \dfrac{2}{\sqrt[3]{5}} \cdot \dfrac{\sqrt[3]{25}}{\sqrt[3]{25}}$ Multiply by $\sqrt[3]{25}/\sqrt[3]{25}$ to create a perfect cube in the denominator

$$= \frac{2\sqrt[3]{25}}{\sqrt[3]{125}} = \frac{2\sqrt[3]{25}}{5}$$

To rationalize a denominator that contains binomial expressions involving radicals such as

$$\frac{7}{\sqrt{5} - \sqrt{3}}$$

we recall that

$$(a - b)(a + b) = a^2 - b^2$$

This suggests that we multiply both the numerator and denominator of the expression by the **conjugate** or **rationalizing factor** of the denominator. The expression $\sqrt{5} + \sqrt{3}$ is the conjugate or the rationalizing factor of $\sqrt{5} - \sqrt{3}$. Thus, we rationalize the denominator of the above expression as follows:

$$\frac{7}{\sqrt{5} - \sqrt{3}} = \frac{7(\sqrt{5} + \sqrt{3})}{(\sqrt{5} - \sqrt{3})(\sqrt{5} + \sqrt{3})}$$

$$= \frac{7(\sqrt{5} + \sqrt{3})}{(\sqrt{5})^2 - (\sqrt{3})^2} = \frac{7(\sqrt{5} + \sqrt{3})}{5 - 3}$$

$$= \frac{7\sqrt{5} + 7\sqrt{3}}{2}$$

By using a calculator and rounding to three decimals, we see that the decimal approximation of $7/(\sqrt{5} - \sqrt{3})$ is 13.888 and the decimal approximation of $(7\sqrt{5} + 7\sqrt{3})/2$ is 13.888 which confirms the equality of the expressions.

EXAMPLE 5   Rationalizing the Denominator

Rationalize the denominator of each expression.

(a) $\dfrac{5}{2 - \sqrt{3}}$

(b) $\dfrac{3\sqrt{2}}{\sqrt{5} + \sqrt{2}}$

Solution   (a) $\dfrac{5}{2 - \sqrt{3}} = \dfrac{5}{2 - \sqrt{3}} \cdot \dfrac{2 + \sqrt{3}}{2 + \sqrt{3}}$ Multiply by the conjugate

$$= \frac{5(2 + \sqrt{3})}{(2)^2 - (\sqrt{3})^2}$$ Special product

$$= \frac{10 + 5\sqrt{3}}{4 - 3} = 10 + 5\sqrt{3}$$

(b) $\dfrac{3\sqrt{2}}{\sqrt{5} + \sqrt{2}} = \dfrac{3\sqrt{2}}{\sqrt{5} + \sqrt{2}} \cdot \dfrac{\sqrt{5} - \sqrt{2}}{\sqrt{5} - \sqrt{2}}$ Multiply by the conjugate

$$= \frac{3\sqrt{2}(\sqrt{5} - \sqrt{2})}{(\sqrt{5})^2 - (\sqrt{2})^2}$$ Special product

$$= \frac{3\sqrt{2}(\sqrt{5} - \sqrt{2})}{5 - 2}$$

$$= \frac{3\sqrt{2}(\sqrt{5} - \sqrt{2})}{3}$$

$$= \sqrt{2}(\sqrt{5} - \sqrt{2}) = \sqrt{10} - 2$$ ◈

EXAMPLE 6   Reducing an Expression to Lowest Terms

Reduce the expression

$$\frac{(4 - \sqrt{12})}{2}$$

to lowest terms.

Solution $\dfrac{4 - \sqrt{12}}{2} = \dfrac{4 - \sqrt{4 \cdot 3}}{2}$ Simplify the numerator

$$= \frac{4 - 2\sqrt{3}}{2}$$

$$= \frac{2(2 - \sqrt{3})}{2} = 2 - \sqrt{3}$$ Factor the numerator and reduce

A decimal approximation confirms that the calculator value of each expression is 0.268.   ◈

Solving Applied Problems

Radical expressions can be used to model real world problems.

EXAMPLE 7 Modeling a Geometry Problem

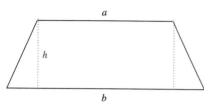

Figure 1

Figure 1 shows a piece of carpet in the form of a trapezoid whose area A is given by the formula

$$A = \frac{1}{2}[a + b]h$$

where a and b are the lengths of the parallel sides and h is the height.
(a) If $a = \sqrt{x}$ and $b = \sqrt{x} + 2$, find an expression for h in terms of x and A. Write it in a simplified radical form.
(b) Find h if $x = 9$ inches and $A = 24$ square inches.

Solution (a)

$$A = \frac{1}{2}[a + b]h$$

$$2A = (a + b)h \qquad \text{Multiply each side by 2}$$
$$(a + b)h = 2A \qquad \text{Use the symmetric property}$$
$$h = \frac{2A}{a + b} \qquad \text{Divide by } a + b$$

Replacing a by $\sqrt{x}$ and b by $\sqrt{x} + 2$, we have

$$h = \frac{2A}{\sqrt{x} + \sqrt{x} + 2} = \frac{2A}{2\sqrt{x} + 2} = \frac{2A}{2(\sqrt{x} + 1)} = \frac{A}{\sqrt{x} + 1}$$

$$= \frac{A}{\sqrt{x} + 1} \cdot \frac{\sqrt{x} - 1}{\sqrt{x} - 1} = \frac{A(\sqrt{x} - 1)}{x - 1}$$

(b) $h = \dfrac{A(\sqrt{x} - 1)}{x - 1} = \dfrac{24(\sqrt{9} - 1)}{9 - 1} = \dfrac{24(3 - 1)}{8} = \dfrac{24(2)}{8} = 6$

Therefore, the height is 6 inches. ◈

Using Technology Exploration

We can use graphing technology to study functions involving operations on radical expressions as the next example shows.

G EXAMPLE 8 Using a Grapher to Graph a Radical Expression
Let $f(x) = \sqrt{x} + \sqrt{x - 1}$.
(a) Find $f(1), f(4)$, and $f(6)$ rounded to two decimal places.
(b) Graph f in the viewing window

$$x\text{Min} = -2, x\text{Max} = 8, x\text{Scl} = 1,$$
$$y\text{Min} = -2, y\text{Max} = 8, y\text{Scl} = 1$$

(c) Use the graph to indicate the domain and range of f.

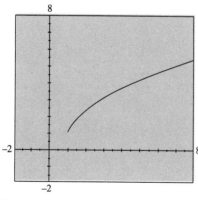

Figure 2

Solution (a) $f(x) = \sqrt{x} + \sqrt{x - 1}$
$$f(1) = \sqrt{1} + \sqrt{1 - 1} = 1.00$$
$$f(4) = \sqrt{4} + \sqrt{4 - 1} = 3.73$$
$$f(6) = \sqrt{6} + \sqrt{6 - 1} = 4.69$$
(b) Figure 2 shows the output of a grapher.
(c) The graph suggests that the domain of f is $x \geq 1$ or $[1, \infty)$ and the range is $y \geq 1$ or $[1, \infty)$. ◈

PROBLEM SET 5.5

Mastering the Concepts

In Problems 1–16, write each expression in simplest radical form and combine like terms. Assume all variables represent positive real numbers.

1. (a) $5\sqrt{7} + 3\sqrt{7}$ (b) $9\sqrt{11} + 8\sqrt{11} - 3\sqrt{11}$
2. (a) $8\sqrt{3} + 2\sqrt{3}$ (b) $8\sqrt[3]{4} - 3\sqrt[3]{4} + 2\sqrt[3]{4}$
3. (a) $7\sqrt{5} + 3\sqrt{5} - 2\sqrt{5}$
   (b) $8\sqrt[5]{2} - 4\sqrt[5]{2} + 3\sqrt[5]{2}$
4. (a) $\sqrt{x} - 4\sqrt{x} + 2\sqrt{x}$
   (b) $3\sqrt[3]{y} - \sqrt[3]{y} + 2\sqrt[3]{y}$
5. $\sqrt{x} + 3\sqrt{x} + 5\sqrt{y}$ 6. $3\sqrt{x} - 2\sqrt{y} + 7\sqrt{y}$
7. $\sqrt{18} + \sqrt{8}$ 8. $3\sqrt{18} + 4\sqrt{2}$
9. $2\sqrt{50} - 3\sqrt{128} + 4\sqrt{2}$
10. $5\sqrt{3} + 2\sqrt{12} - 2\sqrt{27}$
11. $4\sqrt{20} + 2\sqrt{45} + \sqrt{80}$
12. $\sqrt{72} - 2\sqrt{8} + \sqrt{2}$
13. $3\sqrt[3]{192} + 4\sqrt[3]{24} - 2\sqrt[3]{3}$
14. $\sqrt{75x} - \sqrt{3x} - \sqrt{12x}$
15. $\sqrt{27x} - 3\sqrt{12y} + 5\sqrt{3y}$
16. $4\sqrt{12} + 2\sqrt{27} - \sqrt{48}$

In Problems 17–28, find each product and simplify. Assume all variables represent positive real numbers.

17. (a) $\sqrt{2}(\sqrt{3} - \sqrt{2})$ (b) $\sqrt{3}(2 - 5\sqrt{3})$
18. (a) $\sqrt{3}(4 - \sqrt{2})$ (b) $\sqrt{6}(\sqrt{10} - \sqrt{15})$
19. (a) $\sqrt{3}(\sqrt{x} - \sqrt{6})$ (b) $\sqrt{n}(4 - 3\sqrt{n})$
20. (a) $\sqrt{a}(\sqrt{a} - 3)$ (b) $\sqrt{y}(5 - 2\sqrt{y})$
21. (a) $\sqrt{11}(2\sqrt{3} - 4\sqrt{11})$ (b) $(\sqrt{5} - 1)(2\sqrt{5} + 3)$
22. (a) $\sqrt{5}(4\sqrt{5} - 3\sqrt{2})$
    (b) $(3 - \sqrt{2})(2 + \sqrt{3})$
23. (a) $(4\sqrt{2} - \sqrt{3})(5\sqrt{2} + 2\sqrt{3})$
    (b) $(\sqrt{3} - 2)^2$
24. (a) $(3\sqrt{6} + 4\sqrt{2})(\sqrt{6} - \sqrt{2})$
    (b) $(\sqrt{5} - \sqrt{3})^2$
25. (a) $(2\sqrt{x} - 3)^2$ (b) $(\sqrt{5} - 3)(\sqrt{5} + 3)$
26. (a) $(1 + 3\sqrt{y})^2$ (b) $(\sqrt{7} - 2)(\sqrt{7} + 2)$
27. (a) $(3\sqrt{5} - \sqrt{3})(3\sqrt{5} + \sqrt{3})$
    (b) $(3\sqrt{x} - 11)(3\sqrt{x} + 11)$
28. (a) $(4\sqrt{x} - 5)(4\sqrt{x} + 5)$
    (b) $(2\sqrt{x} + \sqrt{7})(2\sqrt{x} - \sqrt{7})$

In Problems 29–44, rationalize each denominator and simplify. Assume all variables represent positive real numbers.

29. (a) $\dfrac{2}{\sqrt{3}}$ (b) $\dfrac{9}{\sqrt{21}}$
30. (a) $\dfrac{1}{\sqrt{7}}$ (b) $\dfrac{5}{\sqrt{6}}$
31. (a) $\dfrac{2\sqrt{3}}{\sqrt{6}}$ (b) $\sqrt{\dfrac{5}{8}}$
32. (a) $\dfrac{4\sqrt{3}}{\sqrt{7}}$ (b) $\sqrt{\dfrac{18}{5}}$
33. (a) $\dfrac{8}{7\sqrt{11}}$ (b) $\dfrac{10}{3\sqrt{5x}}$
34. (a) $\dfrac{4}{3\sqrt{5x}}$ (b) $\dfrac{7}{2\sqrt{3x}}$

35. $\dfrac{5}{\sqrt[3]{9}}$ 36. $\dfrac{5}{\sqrt[3]{7}}$
37. $\dfrac{3}{\sqrt{5} - 2}$ 38. $\dfrac{4}{\sqrt{n} + 3}$
39. $\dfrac{7}{\sqrt{10} + 3}$ 40. $\dfrac{\sqrt{6}}{\sqrt{6} - 2}$
41. $\dfrac{3}{\sqrt{5} + \sqrt{2}}$ 42. $\dfrac{\sqrt{6}}{\sqrt{6} - \sqrt{3}}$
43. $\dfrac{8}{2\sqrt{7} - \sqrt{5}}$ 44. $\dfrac{8}{6\sqrt{5} - 5\sqrt{3}}$

In Problems 45–50, simplify each expression, then use a decimal approximation to confirm the results.

45. $\dfrac{\sqrt{5} - 2}{\sqrt{7}}$ 46. $\dfrac{\sqrt{11} + 3}{\sqrt{8}}$
47. $\dfrac{\sqrt{7} + \sqrt{3}}{2\sqrt{3}}$ 48. $\dfrac{\sqrt{7} - \sqrt{5}}{3\sqrt{5}}$
49. (a) $\dfrac{12 + 9\sqrt{3}}{18}$ (b) $\dfrac{-6 + 4\sqrt{11}}{10}$
50. (a) $\dfrac{-15 + 9\sqrt{7}}{21}$ (b) $\dfrac{\sqrt{72} + 12\sqrt{2}}{18}$

Applying the Concepts

51. **Geometry:** The length of a rectangle is $12/\sqrt{6}$ inches and its width is $9\sqrt{3/2}$ inches. Find, in simplified radical form:
    (a) the area, $A = lw$
    (b) the perimeter, $P = 2l + 2w$.
    (c) Find a two decimal calculator approximation for parts (a) and (b).

52. **Geometry:** Figure 3 shows a triangle with sides of length $\sqrt{8}$ centimeters, $\sqrt{32}$ centimeters and $\sqrt{50}$ centimeters.
    (a) Find a simplified radical expression for the perimeter.
    (b) Find a two decimal calculator approximation for the perimeter.

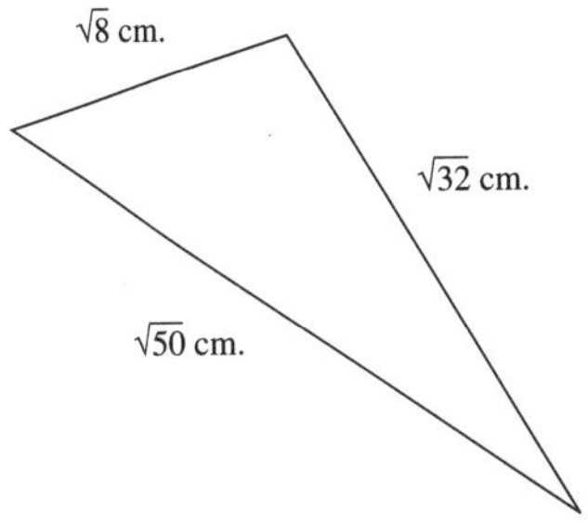

$\sqrt{8}$ cm.
$\sqrt{32}$ cm.
$\sqrt{50}$ cm.

Figure 3

Developing and Extending the Concepts

53. Find a two decimal calculator approximation for the expressions in (a) and (b).
    (a) $(\sqrt{63} + \sqrt{2})(\sqrt{63} - 3\sqrt{2})$
    (b) $57 - 2\sqrt{126}$
    (c) Show algebraically that
    $(\sqrt{63} + \sqrt{2})(\sqrt{63} - 3\sqrt{2}) = 57 - 2\sqrt{126}$.

54. Find a two decimal calculator approximation for the expressions in (a) and (b).

(a) $\dfrac{\sqrt{3}+1}{1-\sqrt{3}}$ (b) $-2-\sqrt{3}$

(c) Show algebraically that $\dfrac{\sqrt{3}+1}{1-\sqrt{3}} = -2-\sqrt{3}$.

55. Perform each operation and simplify.

(a) $5\sqrt{\dfrac{1}{12}} - 2\sqrt{\dfrac{1}{3}} + \dfrac{8}{\sqrt{3}}$

(b) $8\sqrt{\dfrac{25}{7}} - 3\sqrt{\dfrac{16}{7}} + 5\sqrt{\dfrac{4}{7}}$

(c) $\sqrt{\dfrac{5}{3}} - \dfrac{15}{\sqrt{15}} + \dfrac{7\sqrt{15}}{3}$

(d) Use a two decimal calculator approximation to support the results of (a), (b), and (c).

56. Rationalize the denominator of

$$\frac{1}{\sqrt{7}-\sqrt{5}+\sqrt{2}}$$ (Hint: rationalize twice)

In Problems 57–60, for each given function find:

(a) the indicated values, if they are real numbers. Round off to two decimal places.

(b) Match each function with the graphs in Figure 4.

(c) Indicate the domain and range of each function.

57. $f(x) = \sqrt{x-1} + \sqrt{x+1}$, find $f(1), f(2), f(3)$, and $f(5)$.

58. $g(x) = \sqrt{x-1} - \sqrt{2x}$, find $g(0), g(1), g(2)$, and $g(4)$.

59. $h(x) = \sqrt{1-x} + \sqrt{-x}$, find $h(-16), h(-3), h(-1)$, and $h(0)$.

60. $k(x) = \sqrt[3]{-x} + \sqrt[3]{1-x}$, find $k(-1), k(1), k(2)$, and $k(8)$.

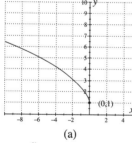

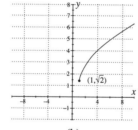

(a)

(b)

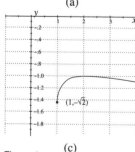

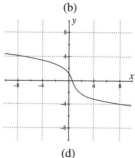

(c)

(d)

Figure 4

In Problems 61–64, perform each operation and simplify. Assume all variables represent positive real numbers.

61. (a) $\sqrt{m^3} - 2m\sqrt{m^5} + 3m\sqrt{m^7}$

(b) $\sqrt[5]{32xy^{13}} - 8\sqrt[5]{x^2y^6} + \sqrt[5]{x^{11}y^3}$

62. (a) $a\sqrt{a^3b} + a^2\sqrt{32ab} - \sqrt{162a^5b}$

(b) $\dfrac{3\sqrt[4]{m^9}}{m} - 5\sqrt[4]{m^5} + \sqrt[8]{m^{10}}$

63. (a) $\dfrac{\sqrt{18x^3}}{3y} - \dfrac{x\sqrt{32x}}{2y} + \dfrac{6x\sqrt{2x}}{y}$

(b) $x\sqrt[3]{27x} - 17\sqrt[3]{-x^4} + \dfrac{8\sqrt[3]{x^7}}{x}$

64. (a) $(\sqrt{x+3} - 2)(\sqrt{x+3} + 2)$

(b) $(\sqrt{3y-1} - 5)(\sqrt{3y-1} + 5)$

65. Let $f(x) = 2/(\sqrt{x+2} - \sqrt{x})$

(a) By rationalizing the denominator, show that $f(x)$ can also be written in the form $g(x) = \sqrt{x+2} + \sqrt{x}$

(b) Find $f(0), f(2), f(4), f(8)$, and $f(10)$ rounded to two decimal places.

(c) Use a grapher to graph f and g on the same viewing window. Are the graphs the same?

(d) Use the graphs to indicate the domain and range of f and g. Do the functions f and g have the same domain? Why or why not?

66. (a) Rationalize the denominator of $3/(\sqrt[3]{5} - \sqrt[3]{2})$ (Hint: recall the special product $(a-b)(a^2+ab+b^2) = a^3 - b^3$

(b) Use a three decimal calculator approximation to support the results of part (a).

In calculus it is sometimes desirable to *rationalize the numerator* of an expression. In this case, we multiply both the numerator and denominator of the expression by the conjugate of the numerator. For example, rationalize the numerator of $(\sqrt{6}-2)/2$.

$$\frac{\sqrt{6}-2}{2} = \frac{\sqrt{6}-2}{2} \cdot \frac{\sqrt{6}+2}{\sqrt{6}+2} \quad \text{Multiply by the conjugate}$$

$$= \frac{(\sqrt{6})^2 - (2)^2}{2(\sqrt{6}+2)} \quad \text{Special product}$$

$$= \frac{6-4}{2(\sqrt{6}+2)} = \frac{2}{2(\sqrt{6}+2)} = \frac{1}{\sqrt{6}+2}$$

A decimal approximation confirms that the calculator value of each expression is 0.225. In Problems 67 and 68, rationalize the numerator of the following expressions:

67. $\dfrac{\sqrt{10}-2}{2}$ 68. $\dfrac{\sqrt{x}+\sqrt{3}}{x-3}$

69. **Oceanography:** The cross section of a nautilus shell has a series of compartments similar to those in Figure 5a. As the sea creature grows, it vacates a chamber, builds a wall, and occupies the outer chamber. To build a wire model of this shell, calculate the total length of wire needed as an exact irrational number and as a two decimal place approximation. The frame must show the fourteen compartments as in Figure 5b with the given sides measuring 1 inch.

Figure 5   (a)   (b)

OBJECTIVES

1. Solve Equations Involving Radicals
2. Solve Equations by Extracting Square Roots
3. Solve Equations Involving Rational Exponents
4. Solve Applied Problems
5. Use Technology Exploration

5.6 Equations with Radicals and Rational Exponents

In this section we present techniques for solving equations involving radicals and rational exponents.

Solving Equations Involving Radicals

Equations of the form

$$\sqrt{x} = 3,\ \sqrt{5t + 1} = 4,\ \text{and}\ \sqrt[3]{5x + 2} = 3$$

are examples of **radical equations**. To solve an equation involving radicals, we use the following power rule.

POWER RULE

> If both sides of an equation are raised to the same power, then the solutions of the original equation are *included* in the solutions of the new equation. That is, all solutions of $P = Q$ are included in the solutions of $P^n = Q^n$, where n is a positive integer.

This rule does not guarantee that by raising both sides of an equation to a power, the new equation is equivalent to the original equation. That is, the two equations may not have the same solutions. For example, the solution of the equation $x = 2$ is 2. However, if we square each side of this equation, we obtain the new equation $x^2 = 4$, whose solutions are -2 and 2. Notice that the solution of the equation $x = 2$ is also a solution of $x^2 = 4$. The solutions of the original equation are contained in the solutions to the new equation. Because of this, *each solution of the new equation must be checked in the original equation when* n *is even, for a possible extra or false solution.* Such an extra solution is called an **extraneous solution**.

EXAMPLE 1   Solving a Radical Equation

Solve the equation $\sqrt{5x + 1} = 4$

Solution

If the index of the radical is odd, it is unnecessary to check the proposed solutions. Remember that it is always prudent to check all solutions.

Use the power rule and square each side of the equation to get

$$\left(\sqrt{5x + 1}\right)^2 = 4^2$$
$$5x + 1 = 16$$
$$5x = 15$$
$$x = 3$$

Check: If we replace x by 3, we have

$$\sqrt{5(3) + 1} = \sqrt{15 + 1} = \sqrt{16} = 4$$

Hence, 3 is the solution.   ◈

The solution of the radical equation in Example 1 is generalized in the following strategy for solving equations involving radicals.

STRATEGY FOR SOLVING EQUATIONS INVOLVING RADICALS

> **Step 1.** Write the equation so that one radical that contains variables is by itself on one side of the equation.
> **Step 2.** Raise each side of the equations to a power equal to the index of the radical.
> **Step 3.** Simplify each side of the new equation.
> **Step 4.** Repeat the above steps if the equation still contains a radical term and solve the new equation.
> **Step 5.** Check all proposed solutions in the original equation whenever a radical has an *even* index.

EXAMPLE 2   Solving Equations Involving Radicals

Solve each equation.

(a) $\sqrt{2x + 5} - 3 = 0$ (b) $\sqrt[3]{3t - 1} - 2 = 0$ (c) $\sqrt{4y^2 - 3} = 2y + 1$

Solution   (a) $\begin{aligned}\sqrt{2x + 5} - 3 &= 0 \\ \sqrt{2x + 5} &= 3 \\ (\sqrt{2x + 5})^2 &= 3^2 \\ 2x + 5 &= 9 \\ 2x &= 4 \\ x &= 2\end{aligned}$ Original equation
Isolate radical
Square each side
Simplify

Check: Substitute 2 for x in the original equation.
$\sqrt{2(2) + 5} - 3 = \sqrt{4 + 5} - 3 = \sqrt{9} - 3 = 0$ Therefore, 2 is the solution.

(b) $\begin{aligned}\sqrt[3]{3t - 1} - 2 &= 0 \\ \sqrt[3]{3t - 1} &= 2 \\ (\sqrt[3]{3t - 1})^3 &= 2^3 \\ 3t - 1 &= 8 \\ 3t &= 9 \\ t &= 3\end{aligned}$ Original equation
Isolate radical
Cube each side
Simplify

Since the index is odd, it is unnecessary to check for extraneous solutions. Therefore, 3 is the solution. It is always good practice to check solutions.

(c) $\begin{aligned}\sqrt{4y^2 - 3} &= 2y + 1 \\ (\sqrt{4y^2 - 3})^2 &= (2y + 1)^2 \\ 4y^2 - 3 &= 4y^2 + 4y + 1 \\ -4y &= 3 + 1 \\ -4y &= 4 \\ y &= -1\end{aligned}$ Original equation
Square each side
Simplify

Check: Substitute -1 for y in the original equation.

left side	right side
$\sqrt{4(-1)^2 - 3} =$	$2(-1) + 1 =$
$\sqrt{4 - 3} =$	$-2 + 1 =$
$\sqrt{1} = 1$	-1

The solution $y = -1$ is an extraneous solution because it does not make the original equation true. Therefore, this equation has no solution.   ◈

Solving Equations by Extracting Square Roots

At the beginning of this section, we indicated that the solutions of the equation $x^2 = 4$ are -2 and 2, because $(-2)^2 = 4$ and $2^2 = 4$. We can solve the equation $x^2 = 4$ by taking the square root of each side.

$$x^2 = 4$$
$$|x| = \sqrt{4}$$
$$x = \pm 2$$

This illustrates the following *square root property*.

SQUARE ROOT PROPERTY

If $au^2 = p$, which is equivalent to $u^2 = p/a$, $a \neq 0$, then

$$u = \sqrt{p/a} \text{ or } u = -\sqrt{p/a}$$

EXAMPLE 3   Solving Equations by the Square Root Property

Solve each equation.

(a) $2x^2 = 18$ (b) $(2x + 3)^2 = 36$

Solution   (a) $2x^2 = 18$ Original equation
                  $x^2 = 9$ Divide each side by 2
                  $x = \pm\sqrt{9}$ Extract square roots
                  $x = \pm 3$

Checking these solutions in the original equation verifies that -3 and 3 are solutions.

(b) $(2x + 3)^2 = 36$ Original equation
     $2x + 3 = \pm 6$ Extract square roots
     $2x + 3 = 6$ or $2x + 3 = -6$
     $2x = 3$ or $2x = -9$
     $x = 3/2$ or $x = -9/2$

The solutions are $3/2$ and $-9/2$. Check these solutions in the original equation.

Solving Equations Involving Rational Exponents

Equations involving rational exponents such as
$$x^{3/2} = 2, \; y^{2/3} = 4, \text{ and } (x + 1)^{1/5} = 1$$

can be solved by a procedure similar to that for solving equations involving radicals. To solve an equation of the form
$$x^{m/n} = a, \text{ for } x$$

we raise each side of the equation to the nth power or to the $\frac{n}{m}$th power. We must remember that an equation containing rational exponents (in lowest terms) with n even must be checked for extraneous solutions.

EXAMPLE 4   Solving Equations Involving Rational Exponents

Solve each equation.
(a) $x^{3/2} = 8$ (b) $(x - 3)^{2/3} = 4$

Solution   (a) $x^{3/2} = 8$ Original equation with n even
                  $(x^{3/2})^{2/3} = 8^{2/3}$ Raise each side to the power $\frac{2}{3}$
                      $x = 8^{2/3}$ Simplify
                      $x = 2^2 = 4$

This solution *must* be checked in the original equation because n, in the original equation, is even.
Check: Substituting 4 for x results in
$4^{3/2} = (2^2)^{3/2} = 2^3 = 8$ Therefore, 4 is the solution.

(b) $(x - 3)^{2/3} = 4$ Original equation

       $[(x - 3)^{2/3}]^3 = 4^3$ Raise each side to the third power

       $(x - 3)^2 = 64$
       $x - 3 = \pm\sqrt{64} = \pm 8$
       $x - 3 = 8$ or $x - 3 = -8$
       $x = 11$ $x = -5$

Check: $(11 - 3)^{2/3} = 8^{2/3} = 2^2 = 4$ and $(-5 - 3)^{2/3} = (-8)^{2/3} = (-2)^2 = 4$
Therefore, 11 and -5 are the solutions.

Solving Applied Problems

Many applied problems from different fields can be solved using equations involving radicals or rational exponents.

EXAMPLE 5   Applying the Pythagorean Theorem

A storage unit in the form of a right triangle is to be built with a hypotenuse of 25 feet. One leg is 7 feet long and the other is l feet long as in Figure 1.

Let $P = (c + 168,000)^{1/3}$ show the relationship between the cost (c in dollars) of building the shed and the perimeter, P in feet.
(a) Determine the length of the leg, l.
(b) Find the cost of building the shed.

Solution   (a) Using the Pythagorean Theorem,
$$25^2 = 7^2 + l^2$$
$$l^2 = 625 - 49$$
$$l = \sqrt{576} = 24$$

Therefore, the leg is 24 feet long.

(b) The perimeter of the triangle is $24 + 25 + 7 = 56$ feet, so we write
$$(c + 168,000)^{1/3} = 56$$
$$\left((c + 168,000)^{1/3}\right)^3 = 56^3 \qquad \text{Raise each side to the } 3^{rd} \text{ power}$$
$$c + 168,000 = 175,616$$
$$c = 7616$$

Therefore, the cost of the shed is $7616.   ◇

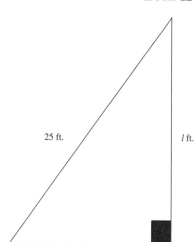

25 ft. l ft.

7 ft.

Figure 1

Using Technology Exploration

We can use a grapher to check the solutions of equations involving radicals and rational exponents.

G   EXAMPLE 6   Using Technology to Solve Equations
(a) Use a grapher to graph $y = \sqrt{5x + 1} - 4$. Find the x intercept of the graph to determine the solution of the equation $\sqrt{5x + 1} = 4$ graphically.
(b) Use a grapher to graph $y = \sqrt{5x + 1}$ and $y = 4$ in the same viewing window. Find the x coordinate of the point of intersection to solve the equation in part (a).

Solution   (a) Figure 2 shows a grapher display of the equation $y = \sqrt{5x+1} - 4$. Using the   TRACE   and   ZOOM   or the   ROOT   feature, the x intercept is 3.
(b) Figure 3 shows a grapher display of the two equations $y = \sqrt{5x + 1}$ and $y = 4$. Using   TRACE   and   ZOOM   or   INTERSECTION   feature, the point of intersection is (3,4).   ◇

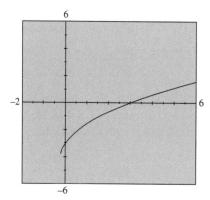

Figure 2

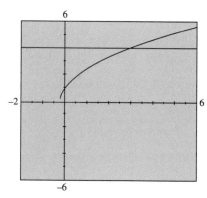

Figure 3

PROBLEM SET 5.6

Mastering the Concepts

In Problems 1–28, solve each equation. Check each solution.

1. (a) $\sqrt{x} - 3 = 0$ (b) $\sqrt{x-1} - 3 = 0$
2. (a) $\sqrt{2x} - 4 = 0$ (b) $\sqrt{2x-2} - 4 = 0$
3. (a) $\sqrt{x} = 5$ (b) $\sqrt{-x} = 2$
4. (a) $\sqrt{x} = 7$ (b) $\sqrt{-x} = 3$
5. $\sqrt{x} - 2 = 3$
6. $\sqrt{x+1} = 4$
7. $\sqrt{t+1} - 2 = 0$
8. $\sqrt{y-3} - 5 = 0$
9. $\sqrt{2y+5} - 4 = 0$
10. $\sqrt{6t-3} - 27 = 0$
11. $8 - \sqrt{y-1} = 6$
12. $2 + \sqrt{7m-5} = 6$
13. $\sqrt{x+1} + 4 = 0$
14. $\sqrt{5x-7} + 10 = 0$
15. $\sqrt{y+5} - \sqrt{y} = 0$
16. $\sqrt{5x-2} + 7 = 10$
17. $\sqrt{t+14} - \sqrt{5t} = 0$
18. $\sqrt{3x-5} - 3\sqrt{x} = 0$
19. $\sqrt{3y+1} = \sqrt{y+1}$
20. $\sqrt{2x+10} - 2\sqrt{x} = 0$
21. $\sqrt[3]{t+2} = -2$
22. $\sqrt[3]{3y-4} = 2$
23. $\sqrt[3]{3x-4} = \sqrt[3]{x+10}$
24. $\sqrt[3]{2-x} = 2\sqrt[3]{10-3x}$
25. $\sqrt{x^2+3x} = x + 1$
26. $\sqrt{4x^2-7} = 2x + 2$
27. $\sqrt{6y+2} - \sqrt{5y+3} = 0$
28. $\sqrt{30x+24} - 6\sqrt{x} = 0$

In Problems 29–38, solve each equation by the square root property.

29. (a) $x^2 = 64$ (b) $x^2 = 9$
30. (a) $x^2 = 4$ (b) $x^2 = 16$
31. (a) $4x^2 = 36$ (b) $2y^2 = 8$
32. (a) $2y^2 = 18$ (b) $3t^2 = 12$
33. $(x+2)^2 = 25$
34. $(x-1)^2 = 9$
35. $(2x+1)^2 = 36$
36. $(2y-5)^2 = 11$
37. $(4x-3)^2 - 16 = 0$
38. $(6x-5)^2 - 4 = 0$

In Problems 39–52, solve each equation.

39. $x^{2/3} = 4$
40. $y^{2/3} = 1$
41. $y^{1/5} = -3$
42. $y^{1/7} = -2$
43. $x^{2/3} = 16$
44. $x^{3/5} = 8$
45. $x^{4/7} = 16$
46. $x^{3/7} = -8$
47. $x^{-3/4} = 27$
48. $x^{1/6} = 2$
49. $(x-3)^{2/5} = 1$
50. $(x-1)^{5/2} = 32$
51. $(t-1)^{2/3} = 4$
52. $(5x-7)^{4/3} = 16$

Applying the Concepts

53. **Geometry:** The radius r of a sphere whose surface area is A is given by the formula
$$r = \sqrt{A/(4\pi)}$$
  (a) Find the surface area of a spherical ball whose radius is 4.9 inches. Round off the answer to two decimal places.
  (b) It is estimated that the radius of the moon is 1080 miles. Use the above formula to determine the surface area of the moon. Round off to the nearest square mile.

54. **Law Enforcement:** Law enforcement uses the model
$$v = \sqrt{10.5x}, \quad x \geq 0$$
to estimate the speed v of a car (in miles per hour) from the distance x (in feet) it skidded before it came to a stop on wet pavement. How fast was a car moving if it skidded 150 feet before it came to a stop? Round off to the nearest whole number.

55. **Utility Pole:** A wire is to be attached to support a 30 foot telephone pole. The wire must be anchored to the ground exactly 10 feet from the base of the pole and 15 feet above the base of the pole (Figure 4). If the length of the attached wire is represented by the expression $\sqrt[4]{x-2}$.
  (a) Find the length of the attached wire. Round off to one decimal place.
  (b) Formulate a radical equation and solve for x.

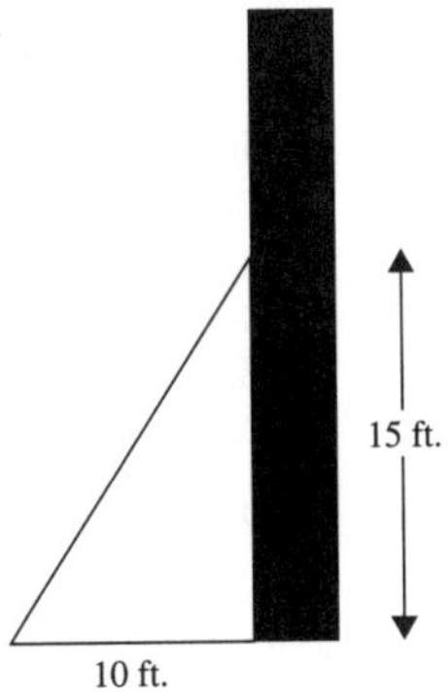

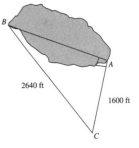

15 ft.

10 ft.

Figure 4

56. **Cable Problem:** A cable television company has a connector box located at point A on the shore of a lake. Nearby are two resorts B and C; B is located on the opposite shore from point A, and C is on land (Figure 5). The owners of the resorts would like to be serviced by the cable company. The distance from A to C is 1600 feet, the distance from B to C is 2640 feet and ABC forms a right triangle with angle A being a right angle.

B

A

2640 ft

1600 ft

C

Figure 5

  (a) Find the length of the underwater cable between A and B such that BC is the length of the hypotenuse.
  (b) It costs $1 per foot to run cable overland and $2 per foot to run underwater cable. Which is the cheapest route to wire resort B: run the cable underwater directly from A to B or to run the cable overland from A to C to B?

57. **Physics:** On earth a pendulum must be 39 inches long to be a "seconds" pendulum. That is, it takes two seconds for one complete swing back and forth of the pendulum. On the moon the pendulum must be 6.5 inches long to be a "seconds" pendulum. Find the value of g on earth and on the moon using the pendulum formula,
$$T = 2\pi\sqrt{\frac{L}{g}}$$
where T is the time (in seconds) for one complete swing back and forth, L is the pendulum length (in feet) and g is the gravitational constant.

58. **Speed of Sound:** The speed of sound, s, in air at 0° Celsius is 1087 feet per second. As the temperature T in degrees Celsius rises, the speed of sound will be modeled by the equation

$$s = 1087\sqrt{(T + 273)/273}$$

(a) Suppose you see a puff of smoke from a car that backfired 1500 feet away from you. If the temperature is 20° Celsius, how long does it take the sound to reach you?

(b) Suppose you hear thunder 2 seconds after you see lightning strike a tree which is 2250 feet from you. What is the temperature?

59. **Zoology:** Zoologists use the model

$$L = \sqrt[3]{w/400}$$

to relate the length L (in meters) of a lizard's body to its weight (in pounds). Determine the weight of a 1.75 meter monitor lizard. Round to the nearest pound.

60. **Population:** The population P (in thousands of people) of a small town is increasing according to the model

$$P = 15 + \sqrt{3t + 4}, t \geq 0$$

where t is time in years. When will the population reach twenty thousand people?

Developing and Extending the Concepts

In Problems 61–66, solve the following equations involving radicals. Check solutions for equations with even root indexes.

61. $\sqrt[4]{2x - 1} + 3 = 0$

62. $\sqrt[4]{t + 8} = \sqrt{3t}$

63. $\sqrt[4]{y^2 - 7y + 1} = \sqrt{y - 5}$

64. $\sqrt[4]{t^2 + 1} = \sqrt{t + 1}$

65. $\sqrt[3]{x^3 + 6x^2} = x + 2$

66. $\sqrt[3]{x^2 + 2x - 6} = \sqrt[3]{x^2}$

In Problems 67–70, solve and check each equation. Figure 6 shows these pair of equations graphed as functions. Use your results to match each pair of equations with one of the displays.

67. (a) $\sqrt{x} - 4 = 0$
    (b) $\sqrt{-x} - 4 = 0$

68. (a) $\sqrt{x - 1} - 5 = 0$
    (b) $\sqrt{1 - x} - 5 = 0$

69. (a) $\sqrt{2x + 1} + 1 = 0$
    (b) $\sqrt{2x + 1} - 1 = 0$

70. (a) $-5 - \sqrt{3x - 1} = 0$
    (b) $\sqrt{3x - 1} - 5 = 0$

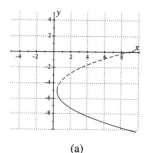

(a)

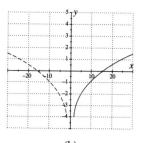

(b)

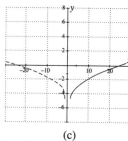

(c)

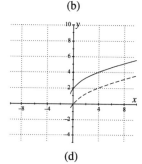

(d)

Figure 6

1. Simplify Numbers of the Form $\sqrt{-b}, b > 0$
2. Define Complex Numbers
3. Perform Operations with Complex Numbers
4. Find Powers of i
5. Solve Applied Problems
6. Use Technology Exploration

5.7 Complex Numbers

So far in this chapter, we emphasized that square roots of negative numbers, such as $\sqrt{-9}$, are not real numbers. In this section, we introduce a "new" number system that includes square roots of negative numbers. In 1545 the great Italian physician and mathematician Geronimo Cardano (1501–1576) discovered that solutions to cubic equations involved square roots of negative numbers. Gradually, other mathematicians began to accept the idea of calculating with square roots of negative numbers. In 1637 the French philosopher and mathematician René Descartes (1596–1650) introduced the terms "real" and "imaginary." Later, in 1748, the Swiss mathematician Leonhard Euler (1707–1783) used the symbol "i" for $\sqrt{-1}$. Finally, in 1832, the German mathematician Carl Friedrich Gauss (1777–1855) introduced the term "complex number." These numbers have been used in a wide variety of useful applications in electronics, engineering, and physics.

Simplifying Numbers of the Form $\sqrt{-b}, b > 0$

To launch our study of square roots of negative numbers we introduce the **imaginary unit**

$$i = \sqrt{-1}$$

For the time being, regard i or $\sqrt{-1}$ as an "invented number" with the property that

$$i^2 = -1$$

Using i, we can define the **principal square root** of a negative number as follows:

If $b > 0$, then $\sqrt{-b} = \sqrt{b(-1)} = \sqrt{b} \cdot \sqrt{-1} = \sqrt{b}\,i$

EXAMPLE 1   Simplifying Square Roots of Negative Numbers

Find the principal square root.

(a) $\sqrt{-4}$ (b) $\sqrt{-7}$ (c) $-\sqrt{-25}$ (d) $\sqrt{-16} + \sqrt{-9}$ (e) $\dfrac{\sqrt{-48}}{\sqrt{-3}}$

Solution   (a) $\sqrt{-4} = \sqrt{4(-1)} = \sqrt{4} \cdot \sqrt{-1} = 2i$
(b) $\sqrt{-7} = \sqrt{7(-1)} = \sqrt{7} \cdot \sqrt{-1} = \sqrt{7}\,i$
(c) $-\sqrt{-25} = -\sqrt{25(-1)} = -\sqrt{25} \cdot \sqrt{-1} = -5i$
(d) $\sqrt{-16} + \sqrt{-9} = \sqrt{16(-1)} + \sqrt{9(-1)}$
$$= \sqrt{16} \cdot \sqrt{-1} + \sqrt{9} \cdot \sqrt{-1} = 4i + 3i = 7i$$
(e) $\dfrac{\sqrt{-48}}{\sqrt{-3}} = \dfrac{\sqrt{48} \cdot \sqrt{-1}}{\sqrt{3} \cdot \sqrt{-1}} = \dfrac{\sqrt{48}\,i}{\sqrt{3}\,i} = \sqrt{\dfrac{48}{3}} = \sqrt{16} = 4$ ◇

To perform multiplication with square roots of negative numbers, we must first write the number in the i-form. That is, if a and b are positive real numbers, then

$$\sqrt{-a} \cdot \sqrt{-b} = \sqrt{a(-1)} \cdot \sqrt{b(-1)} = \sqrt{-1} \cdot \sqrt{a} \cdot \sqrt{-1} \cdot \sqrt{b}$$
$$= \sqrt{a} \cdot \sqrt{-1} \cdot \sqrt{b} \cdot \sqrt{-1} = \sqrt{a}\,i \cdot \sqrt{b}\,i = \sqrt{ab}\,i^2 = -\sqrt{ab}$$

EXAMPLE 2   Multiplying Square Roots of Negative Numbers

Find each product.
(a) $\sqrt{-4} \cdot \sqrt{-9}$ (b) $\sqrt{-3}(\sqrt{-27} - \sqrt{-4})$

Solution   (a) $\sqrt{-4} \cdot \sqrt{-9} = \sqrt{4}\,i \cdot \sqrt{9}\,i = 2i \cdot 3i = 6i^2 = -6$
(b) $\sqrt{-3}(\sqrt{-27} - \sqrt{-4}) = \sqrt{-3} \cdot \sqrt{-27} - \sqrt{-3} \cdot \sqrt{-4}$ Distributive Property
$$= \sqrt{3}\,i \cdot \sqrt{27}\,i - \sqrt{3}\,i \cdot \sqrt{4}\,i$$
$$= \sqrt{81}\,i^2 - 2\sqrt{3}\,i^2$$
$$= 9(-1) - 2(-1)\sqrt{3}$$
$$= -9 + 2\sqrt{3}$$ ◇

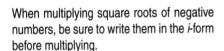

When multiplying square roots of negative numbers, be sure to write them in the i-form before multiplying.

Defining Complex Numbers

Numbers such as $2i$, $\sqrt{5}\,i$, and $-3i$ are called *pure imaginary numbers*. We form a new set of numbers by adding these imaginary numbers to real numbers.

DEFINITION OF A COMPLEX NUMBER

Any number, real or imaginary, of the form
$$a + bi$$
where a and b are real numbers and $i = \sqrt{-1}$, is called a **complex number.**

For example,

$$4 + 3i,\ 2 - 7i,\ 3 + \sqrt{5}\,i,\ \text{and}\ \frac{3}{4} - 2\sqrt{7}\,i$$

are complex numbers.

The form $a + bi$ is called the **standard form** for complex numbers. The number a is called the **real part** of the complex number. The number b is called the **imaginary part** of the complex number.

$$\underbrace{a}_{\text{real part}} + \underbrace{bi}_{\text{imaginary part}}$$

Every real number a is a complex number because the real number a can be written as

$$a = a + 0i$$

Similarly, if b is a real number, then

$$bi = 0 + bi$$

Two complex numbers $a + bi$ and $c + di$ are **equal** if and only if their real parts are equal and their imaginary parts are equal. That is,

$$a + bi = c + di \text{ if and only if } a = c \text{ and } b = d$$

EXAMPLE 3   Equality of Two Complex Numbers
Find x and y so that

$$5x - \sqrt{-36} = -15 + 3yi$$

Solution   First, we write the left side in standard form

$$5x - \sqrt{-36} = 5x - 6i$$

so, $5x - 6i = -15 + 3yi$
Thus, the real parts must be equal: $5x = -15$ and $x = -3$.
The imaginary parts must be equal: $-6 = 3y$ and $y = -2$.   ◈

Performing Operations with Complex Numbers

To add (or subtract) complex numbers, we add (or subtract) their real parts and then add (or subtract) their imaginary parts. The following rule is similar to the rule for combining like terms of a polynomial:

RULES FOR ADDING AND SUBTRACTING
COMPLEX NUMBERS

> For any real numbers a, b, c and d.
> 1. $(a + bi) + (c + di) = (a + c) + (b + d)i$
> 2. $(a + bi) - (c + di) = (a - c) + (b - d)i$

Notice that the sum and difference of complex numbers is again a complex number. Also, the commutative, associative and distributive properties for real numbers are also true for complex numbers.

EXAMPLE 4   Adding and Subtracting Complex Numbers
Perform each operation and simplify.
(a) $(5 + 6i) + (9 + 3i)$ (b) $(4 - 2i) - (-3 + i)$

Solution
(a) $(5 + 6i) + (9 + 3i) = (5 + 9) + (6i + 3i)$
$= (5 + 9) + (6 + 3)i$
$= 14 + 9i$

(b) $(4 - 2i) - (-3 + i) = (4 - [-3]) + (-2i - i)$
$= (4 + 3) + (-2 - 1)i$
$= 7 - 3i$ ◈

Complex numbers of the form $a + bi$ have the same form as a binomial, so we can multiply two complex numbers in the same way as we multiply binomials.

EXAMPLE 5   Multiplying Complex Numbers

Perform each multiplication.

(a) $4i(3 + 7i)$ (b) $(4 + 3i)(2 - 4i)$ (c) $(1 + 2i)(1 - 2i)$

Solution   (a) By applying the distributive property, we have

$$\begin{aligned}(4i)(3 + 7i) &= 4i(3) + 4i(7i)\\ &= 12i + 28i^2\\ &= 12i + 28(-1)\\ &= -28 + 12i\end{aligned}$$

(b) We multiply by using the FOIL method (see section 1.2)

$$\begin{aligned}(4 + 3i)(2 - 4i) &= 4 \cdot 2 - 4 \cdot 4i + 3i \cdot 2 - 3i \cdot 4i\\ &= 8 - 16i + 6i - 12i^2\\ &= 8 - 12(-1) - 16i + 6i \qquad i^2 = -1\\ &= 20 - 10i\end{aligned}$$

(c) This product has the form of the special product

$$\begin{aligned}(a - b)(a + b) &= a^2 - b^2\\ (1 + 2i)(1 - 2i) &= 1^2 - (2i)^2\\ &= 1 - 4i^2\\ &= 1 - 4(-1) = 1 + 4 = 5\end{aligned}$$

In example 5c, note that the product of two complex numbers can be a real number. This can happen with pairs of complex numbers of the form $a + bi$ and $a - bi$ called **complex conjugates**. In general, the product of a complex number and its conjugate is always a real number. That is,

$$\begin{aligned}(a + bi)(a - bi) &= a^2 - (bi)^2\\ &= a^2 - b^2i^2\\ &= a^2 - b^2(-1)\\ &= a^2 + b^2\end{aligned}$$

To perform the division

$$\frac{3i}{4 - 3i}$$

we look for a complex number in standard form that is equivalent to $3i/(4 - 3i)$. To do this, we need to eliminate i from the denominator, so we multiply the numerator and the denominator by the complex conjugate of the denominator, which is $4 + 3i$.

$$\begin{aligned}\frac{3i}{4 - 3i} &= \frac{3i(4 + 3i)}{(4 - 3i)(4 + 3i)}\\ &= \frac{12i + 9i^2}{16 - 9i^2}\\ &= \frac{-9 + 12i}{16 + 9}\\ &= \frac{-9 + 12i}{25}\\ &= \frac{-9}{25} + \frac{12}{25}i\end{aligned}$$

DIVIDING COMPLEX NUMBERS

There are no remainders when dividing complex numbers.

> To find the real and imaginary parts of a quotient of complex numbers, multiply the numerator and denominator by the complex conjugate of the denominator.

EXAMPLE 6   Dividing Complex Numbers

Perform each division and express the result in the form $a + bi$.

(a) $\dfrac{1}{4 + 3i}$

(b) $\dfrac{2 - 3i}{1 - 4i}$

Solution   (a) Multiplying the numerator and denominator by the conjugate of $4 + 3i$, we obtain

$$\frac{1}{4 + 3i} = \frac{4 - 3i}{(4 + 3i)(4 - 3i)}$$
$$= \frac{4 - 3i}{16 - 9i^2}$$
$$= \frac{4 - 3i}{16 + 9}$$
$$= \frac{4 - 3i}{25} = \frac{4}{25} - \frac{3}{25}i$$

(b) Multiplying the numerator and denominator by the conjugate of $1 - 4i$, we obtain

$$\frac{2 - 3i}{1 - 4i} = \frac{(2 - 3i)(1 + 4i)}{(1 - 4i)(1 + 4i)}$$
$$= \frac{2 + 8i - 3i - 12i^2}{1 - 16i^2}$$
$$= \frac{2 + 5i - 12(-1)}{1 - 16(-1)}$$
$$= \frac{14 + 5i}{17} = \frac{14}{17} + \frac{5}{17}i$$

Finding Powers of i

Positive integer exponents have the same meaning, in terms of repeated multiplication, for both complex numbers and real numbers. Integer powers of complex numbers are defined as they are for real numbers. In particular, for powers of i, we have

$$i^1 = i$$
$$i^2 = -1$$
$$i^3 = i^2 \cdot i = (-1)i = -i$$
$$i^4 = i^2 \cdot i^2 = (-1)(-1) = 1$$

If we continue the list, we repeat the same sequence of answers because we can replace every factor of i^4 by 1. Thus,

$$i^5 = i^4 \cdot i = 1i = i$$
$$i^6 = i^4 \cdot i^2 = 1(-1) = -1$$
$$i^7 = i^4 \cdot i^3 = 1(-i) = -i$$
$$i^8 = i^4 \cdot i^4 = 1 \cdot 1 = 1$$
$$i^9 = i^8 \cdot i = 1 \cdot i = i$$

and so on.

EXAMPLE 7   Simplifying Powers of i

Find each power of i.

(a) i^{18} (b) i^{27} (c) i^{105} (d) i^{-3}

Solution (a) $i^{18} = i^{16}i^2 = (i^4)^4 \cdot i^2 = 1^4(-1) = -1$

(b) $i^{27} = i^{24} \cdot i^3 = (i^4)^6 \cdot i^3 = 1^6(-i) = -i$

(c) $i^{105} = i^{104} \cdot i = (i^4)^{26} \cdot i = 1^{26} \cdot i = i$

(d) We note that $i^{-3} = 1/i^3$. By multiplying the numerator and denominator by i, we obtain a real number in the denominator, so that

$$i^{-3} = \frac{1}{i^3} = \frac{1 \cdot i}{i^3 \cdot i} = \frac{i}{i^4} = \frac{i}{1} = i$$

Solving Applied Problems

A famous puzzle was solved by Cardano using complex numbers.

EXAMPLE 8   Solving a Puzzle Problem

The puzzle is to find two numbers whose sum is 10 and whose product is 40.

Solution   Cardano suggested that the answers were the numbers 5 + RM15 and 5 − RM15 where RM15 meant "radix minus 15". He regarded these numbers as "fictitious" numbers which solved a "manifestly impossible" problem. Nowadays we use the notation for complex numbers to verify Cardano's conjecture.

$$(5 + \sqrt{-15}) + (5 - \sqrt{-15}) = (5 + \sqrt{15}i) + (5 - \sqrt{15}i) = 10, \text{ and}$$
$$(5 + \sqrt{-15})(5 - \sqrt{-15}) = (5 + \sqrt{15}i)(5 - \sqrt{15}i) = 5^2 - (\sqrt{15})^2 i^2$$
$$= 25 + 15 = 40$$

In 1797 the Norwegian surveyor C. Wessel used complex numbers to represent direction and distance on a two dimensional grid. The idea started from attempting to order complex numbers such as the following:

$$5, -3i, -3 + 3i, \text{ and } 4 + 3i$$

from smallest to largest. Every attempt had failed because there is no way of locating complex numbers on a number line. However, Gauss discovered that complex numbers can be graphed on a special coordinate system call the *complex plane*. The horizontal line (x axis) represents the real part of the complex number and is called the **real axis**. The vertical line (y axis) represents the imaginary part and is called the **imaginary axis**. The graph of the complex number

$$z = a + bi$$

is associated with the ordered pair (a,b) in the Cartesian coordinate system. Figure 1 shows the graph of the complex numbers 5, $-3i$, $-3 + 3i$, and $4 + 3i$.

Recall from section 2.6 the definition of absolute value as the distance from zero to the point associated to the number. We can use this same definition for the *absolute value* of a complex number, written $|a + bi|$. For example, $|-3 + 3i|$ represents the distance from the point $(-3,3)$ to $(0,0)$ on the complex plane (Figure 1). By the distance formula

$$|-3 + 3i| = \sqrt{(-3 - 0)^2 + (3 - 0)^2} = \sqrt{(-3)^2 + (3)^2} = \sqrt{18} = 3\sqrt{2}$$

Likewise, $|5| = |5 + 0i| = \sqrt{5^2 + 0^2} = \sqrt{25} = 5,$

$$|-3i| = |0 - 3i| = \sqrt{0^2 + (-3)^2} = \sqrt{9} = 3, \text{ and}$$
$$|4 + 3i| = \sqrt{4^2 + 3^2} = \sqrt{25} = 5$$

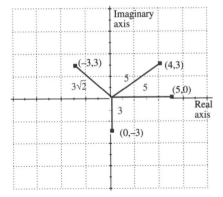

Figure 1

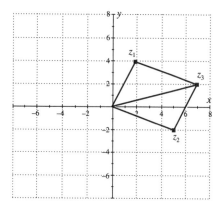

Figure 2

EXAMPLE 9   Adding Complex Numbers

Calculate $(2 + 4i) + (5 - 2i)$ and graph.

Solution $(2 + 4i) + (5 - 2i) = (2 + 5) + (4 - 2)i = 7 + 2i$

Let $z_1 = 2 + 4i$, $z_2 = 5 - 2i$ and $z_3 = 7 + 2i$. Figure 2 shows the points associated with these complex numbers. Notice that if we connect these points z_1, z_2 and z_3 with the origin, the figure formed is a parallelogram. The point diagonally opposite the origin in the parallelogram is the sum $z_1 + z_2$, that is $z_3 = z_1 + z_2$.   ◈

◈ PROBLEM SET 5.7

Mastering the Concepts

In Problems 1–10, find the principal square roots.

1. (a) $\sqrt{-16}$ (b) $\sqrt{-81}$
2. (a) $\sqrt{-25}$ (b) $\sqrt{-36}$
3. (a) $-\sqrt{-8}$ (b) $\sqrt{-72}$
4. (a) $-\sqrt{-12}$ (b) $-\sqrt{-18}$
5. (a) $-\sqrt{-9/16}$ (b) $-\sqrt{-25/4}$
6. (a) $-\sqrt{-3/4}$ (b) $-\sqrt{-5/16}$
7. (a) $\dfrac{\sqrt{-8}}{\sqrt{-2}}$ (b) $\dfrac{\sqrt{-32}}{\sqrt{-2}}$
8. (a) $\dfrac{\sqrt{-50}}{\sqrt{-2}}$ (b) $\dfrac{\sqrt{-72}}{\sqrt{-2}}$
9. $\sqrt{-8} + \sqrt{-72}$ 10. $\sqrt{-18} + \sqrt{-50}$

In Problems 11–14, find each product

11. $\sqrt{-4} \cdot \sqrt{-25}$ 12. $\sqrt{-8} \cdot \sqrt{-2}$
13. $\sqrt{-4}(\sqrt{-5} + \sqrt{-6})$ 14. $\sqrt{-49}(\sqrt{-3} - 2\sqrt{-2})$

In Problems 15 and 16, find x and y so that each statement is true.

15. $4x + 8i = 20 + 2yi$ 16. $(4x - 3) + 21i = 5 + (2y - 3)i$

In Problems 17–38, perform each operation and express the answer in the form $a + bi$.

17. (a) $(-3 + 6i) + (2 + 3i)$ (b) $(10 - 24i) + (3 + 7i)$
18. (a) $(-2 + 5i) + (-5 + i)$ (b) $(6 + 8i) + (6 - 8i)$
19. (a) $(-2 - 3i) - (-3 - 2i)$ (b) $(3 + 2i) - (3 - 2i)$
20. (a) $(7 + 24i) - (-3 - 4i)$ (b) $(10 - 8i) - (10 + 8i)$
21. $(5 - 7i) - (5 - 13i)$
22. $(6 - 8i) - (5 + 3i)$
23. $[(4 + 3i) - (3 + 6i)] + (2 - 3i)$
24. $(8 - 5i) + [(3 + 7i) - (2 - 5i)]$
25. $-7i(-2 + 4i)$ 26. $3i(2 + 3i)$
27. $(2 - 4i)(3 + 2i)$ 28. $(1 + 2i)(4 - i)$
29. $(2 + 3i)(3 - 5i)$ 30. $(-3 - 2i)(2 + 5i)$
31. $(3 - 7i)(2 + 3i)$ 32. $(1 - 6i)(2 + 5i)$
33. $(-7 - 2i)^2$ 34. $(1 + i)^2$
35. $(5 - 4i)(5 + 4i)$ 36. $(3 - 2i)(3 + 2i)$
37. $(5 - 12i)(5 + 12i)$ 38. $(2 - 7i)(2 + 7i)$

In Problems 39–44, perform each division and express the answer in the form $a + bi$.

39. (a) $\dfrac{1}{4 + 3i}$ (b) $\dfrac{1 - i}{-1 + i}$
40. (a) $\dfrac{1}{3 - 5i}$ (b) $\dfrac{3 + 4i}{-2 + 5i}$
41. (a) $\dfrac{7}{5 - 4i}$ (b) $\dfrac{7 + 2i}{1 - 5i}$
42. (a) $\dfrac{-3}{3 + 4i}$ (b) $\dfrac{4 - i^2}{2 - 7i}$
43. (a) $\dfrac{2 + i}{2 + 3i}$ (b) $\dfrac{5 - i^2}{11 + 2i}$
44. (a) $\dfrac{i}{-1 - i}$ (b) $\dfrac{1 + i}{2 - 3i}$

In Problems 45–48, find each power of i.

45. (a) i^{29} (b) $(-i)^{37}$ (c) i^{54}
46. (a) i^{49} (b) i^{65} (c) $(-i)^{76}$
47. (a) i^{108} (b) i^{-15} (c) i^{-7}
48. (a) i^{-18} (b) i^{-5} (c) i^{-14}

Applying the Concepts

49. **Alternating Current:** In 1893 Charles P. Steinmetz, an American electrical engineer, developed a theory of alternating currents based on complex numbers. He showed that *Ohm's Law* for alternating currents takes the form

$$E = IZ$$

where E is the *voltage*, I is the *current* and Z is the *impedance* given by the equation

$$Z = R + (X_L - X_C)i$$

where X_L is the *inductive reactance* and X_C is the *capacitative reactance*. Find E if $I = 1 - i$, $R = 3$, $X_L = 2$ and $X_C = 1$.

50. **Alternating Current:** Use the equation
$E = I[R + (X_L - X_C)i]$
to find I if $E = 3 + 2i$, $R = 4$, $X_L = 3$ and $X_C = 2$.

Developing and Extending the Concepts

In Problems 51–54, write each expression in the form $a + bi$.

51. $1 + i + (-i)^2 + i^3$ 52. $i - i^2 + (-i)^3 - i^4$
53. $5i^7 + 3i^2 - 2i$ 54. $2i^{84} - 3i^{100} + 2i^2$
55. Let $f(z) = z^2 - 2z + 2$, find and simplify $f(1 + i)$.
56. Let $f(z) = z^2 - 2z + 5$, find and simplify $f(1 + 2i)$.
57. Write the expression $(7 - 3i)^{-2}$ in the form $a + bi$.
58. Simplify the expression $i + i^2 + i^3 + i^4 + \ldots + i^{98} + i^{99} + i^{100}$.

59. (a) Draw on a coordinate system a parallelogram whose vertices are $A = (0,0)$, $B = (1,5)$, $C = (8,7)$, and $D = (7,2)$.
    (b) Find the length of the diagonal $\overline{AC}$.
    (c) Find the sum of the complex numbers represented by the ordered pairs B and D.
    (d) Find $|B + D|$.

CHAPTER 5 REVIEW PROBLEM SET

In Problems 1–8, rewrite each expression in a simplified form using only positive exponents. Assume all variables are restricted to values for which all expressions represent real numbers.

1. (a) $(-7)^0 + \dfrac{1}{2^{-1}}$ (b) $\dfrac{1 + 3^{-2}}{2^{-2}}$

   (c) $\dfrac{5^{-3}}{2^3} + \dfrac{2^{-3}}{5^3}$ (d) $\left(\dfrac{x^2}{y^3}\right)^{-2}$

2. (a) $(3)^0 + \dfrac{1}{7^{-2}}$ (b) $\dfrac{1}{2^{-3} + 3^{-2}}$

   (c) $\dfrac{2^{-1}}{3^2} + \dfrac{3^{-1}}{2^4}$ (d) $\left(\dfrac{x^{-3}}{y^2}\right)^2$

3. (a) $3^{-11} \cdot 3^{15}$ (b) $(2^{-1/3})^9$ (c) $\dfrac{b^{-6}}{b^{-11}}$

4. (a) $9^{-4/7} \cdot 9^{11/7}$ (b) $(3x^{-2})^{-2}$ (c) $(t^{-4})^{-3}$

5. (a) $x^{3/5} \cdot x^{-5/4}$ (b) $\dfrac{x^{-5}}{x^{-2}}$ (c) $(2x^{-3}y^{-4})^{-3}$

6. (a) $(3^{-1/2})^{-4}$ (b) $\dfrac{7^{-4}}{7^{-3}}$ (c) $(4^{-1}x^{-3})^2$

7. (a) $\dfrac{x^{-3}y^2}{x^{-2}y^{-3}}$ (b) $\dfrac{a^3b^{-4}}{a^{-4}b^2}$ (c) $\left(\dfrac{x^4}{2y^{-2}}\right)^{-2}$

8. (a) $\dfrac{m^{-4}n^{-3}}{m^{-7}n^{-2}}$ (b) $\dfrac{a^4b}{a^{-3}b^4}$ (c) $\left(\dfrac{x^3y^{-2}}{(xy)^{-1}}\right)^{-1}$

In Problems 9 and 10, solve each exponential equation.

9. (a) $2^{-x} = 4$ (b) $5^{2x} = 25$
10. (a) $3^{-2x} = 27$ (b) $7^{-x} = 49$

In Problems 11 and 12, find the principal root of the given number. State the reason when the number is not real.

11. (a) $\sqrt{81}$ (b) $\sqrt[3]{-64}$ (c) $\sqrt[4]{-16}$
12. (a) $\sqrt{25}$ (b) $\sqrt[3]{-216}$ (c) $\sqrt[4]{-64}$

In Problems 13 and 14, find the exact value of each expression.

13. (a) $16^{-3/4}$ (b) $-(-8)^{1/3}$ (c) $36^{-5/2}$
14. (a) $(-8)^{-2/3}$ (b) $-(-27)^{1/3}$ (c) $(-32)^{-3/5}$

In Problems 15 and 16, convert each expression to rational exponents.

15. (a) $\sqrt[3]{-7}$ (b) $\sqrt[5]{x^6}$
16. (a) $\sqrt[4]{3x^2 + 1}$ (b) $\sqrt[3]{(a + b)^2}$

In Problems 17 and 18, convert each expression to radical form.

17. (a) $7^{-2/3}$ (b) $(x + 1)^{2/5}$
18. (a) $5^{3/4}$ (b) $(x^2 + 7)^{-3/4}$

In Problems 19–30, use the properties of radicals to simplify each expression. Assume all variables represent positive real numbers.

19. (a) $\sqrt{125}$ (b) $\sqrt[3]{-125}$ (c) $\sqrt[4]{32t^5}$
20. (a) $\sqrt{8}$ (b) $\sqrt[3]{-45}$ (c) $\sqrt[3]{54x^7}$
21. (a) $\sqrt{4/25}$ (b) $\sqrt[3]{-7/27}$ (c) $\sqrt[3]{3/32}$
22. (a) $\sqrt[4]{5/81}$ (b) $\sqrt[3]{-11/8}$ (c) $\sqrt[3]{-3/128}$
23. (a) $\sqrt[5]{t^5/32}$ (b) $\sqrt[4]{16x^8/81}$ (c) $\sqrt[3]{-5/x^9}$
24. (a) $\sqrt{11/x^4}$ (b) $\sqrt[3]{-7/x^6}$ (c) $\sqrt[3]{3/x^{10}}$
25. (a) $\dfrac{\sqrt{125x^7}}{\sqrt{5x}}$ (b) $\sqrt{\sqrt{16x^8}}$
26. (a) $\dfrac{\sqrt[3]{-24m^{-11}}}{\sqrt[3]{3m^2}}$ (b) $\sqrt[3]{\sqrt[3]{\sqrt{x^{18}}}}$
27. (a) $\sqrt{8x^3y^4} \cdot \sqrt{2xy^2}$ (b) $\sqrt[3]{4x^2} \cdot \sqrt[3]{2x}$
28. (a) $\sqrt[3]{x^3x^2/x^8}$ (b) $\sqrt[7]{u^2u^7/u^{16}}$
29. (a) $\left(\dfrac{16x^4}{y^8}\right)^{-3/4}$ (b) $\left(\dfrac{x^5y^{10}}{32}\right)^{-3/5}$
30. (a) $\left(\dfrac{8m^{-6}}{27m^{-12}}\right)^{-1/3}$ (b) $\left(\dfrac{x^{-1}y^{-2/3}}{x^{-2}y^{-2/3}}\right)^{-3}$

In Problems 31–40, perform each operation and simplify. Assume all variables represent positive real numbers.

31. (a) $7\sqrt{2} - 3\sqrt{2} + 4\sqrt{2}$ (b) $\sqrt{128} + \sqrt{8}$
32. (a) $\sqrt{5} - 6\sqrt{5} + 2\sqrt{5}$ (b) $\sqrt{48} - \sqrt{12}$
33. (a) $\sqrt{63x} + 2\sqrt{112x} - \sqrt{252x}$
   (b) $\sqrt[3]{16y} - \sqrt[3]{54y} + \sqrt[3]{250y}$
34. (a) $\sqrt{3}(\sqrt{2} + \sqrt{5})$ (b) $\sqrt{5}(\sqrt{7} - \sqrt{2})$
35. (a) $(\sqrt{10} + \sqrt{2})^2$ (b) $(2\sqrt{x} - 3\sqrt{y})^2$
36. (a) $(2\sqrt{6} - 3\sqrt{2})^2$ (b) $(1 - \sqrt{2})^3$
37. (a) $(\sqrt{8} - \sqrt{2})(\sqrt{8} + \sqrt{2})$
   (b) $(\sqrt{a} - b)(\sqrt{a} + b)$
38. (a) $(\sqrt{3t} + \sqrt{5})(\sqrt{3t} - \sqrt{5})$
   (b) $(\sqrt{t} + \sqrt{5y})(\sqrt{t} - \sqrt{5y})$

39. (a) $(\sqrt{x} - \sqrt{y})(2\sqrt{x} + \sqrt{y})$
    (b) $(3\sqrt{x} - y)(\sqrt{x} + 2y)$
40. (a) $(\sqrt{8} + \sqrt{18})(\sqrt{6} + 1)$
    (b) $(\sqrt{x} + \sqrt{y})(2\sqrt{x} - 3\sqrt{y})$

In Problems 41–46, rationalize each denominator and simplify. Assume all variables represent positive real numbers. Use decimal approximations to confirm the results in Problems 41–44.

41. (a) $\dfrac{2}{\sqrt{7}}$ (b) $\dfrac{3}{5\sqrt{2}}$

42. (a) $\dfrac{-3}{\sqrt{5}}$ (b) $\dfrac{2}{7\sqrt{3}}$

43. (a) $\dfrac{1}{2 + \sqrt{3}}$ (b) $\dfrac{\sqrt{7}}{\sqrt{2} - \sqrt{3}}$

44. (a) $\dfrac{\sqrt{2}}{\sqrt{3} + \sqrt{7}}$ (b) $\dfrac{\sqrt{8} + 3}{3\sqrt{2} + 2}$

45. $\dfrac{\sqrt{x} + 1}{\sqrt{x} - 1}$ 46. $\dfrac{\sqrt{x} + \sqrt{3}}{\sqrt{x} - \sqrt{3}}$

In Problems 47 and 48, rationalize the numerator and simplify.

47. (a) $\dfrac{\sqrt{5} + 3}{2}$ (b) $\dfrac{\sqrt{7} + \sqrt{3}}{\sqrt{3}}$

48. (a) $\dfrac{\sqrt{3} - 1}{\sqrt{3} + 1}$ (b) $\dfrac{5 - \sqrt{3}}{\sqrt{5}}$

In Problems 49 and 50, find the distance between each pair of points. Round off each answer to two decimal places.

49. (a) (1,2) and (−2,1) (b) (−4,7) and (5,−3)
50. (a) (2,5) and (8,3) (b) (−2,3) and (−4,−9)

In Problems 51–54, solve each equation.

51. (a) $x^{1/6} = 2$ (b) $(x - 2)^{3/2} = 8$
52. (a) $y^{2/3} = 4$ (b) $(t - 3)^{1/5} = 1$
53. (a) $\sqrt{x - 2} = 3$ (b) $(x + 1)^2 = 4$
54. (a) $\sqrt[3]{x + 2} = -3$ (b) $(2x - 1)^2 - 9 = 0$

In Problems 55–60, write each expression in the form of $a + bi$, where a and b are real numbers.

55. (a) $4\sqrt{-9} - 2\sqrt{-25}$
    (b) $\sqrt{-16} \cdot \sqrt{-9}$ (c) $\dfrac{\sqrt{-16}}{\sqrt{-4}}$
56. (a) $\sqrt{-7} - 3\sqrt{-7}$
    (b) $\sqrt{-16} \cdot \sqrt{-25}$ (c) $\dfrac{\sqrt{-25}}{\sqrt{-16}}$
57. (a) $(2 - 3i) + (3 + 7i)$
    (b) $5i(1 - 7i)$ (c) $(2 - i)(2 + i)$
58. (a) $(2 - 3i)^2$
    (b) $(3 + 8i^3)(2 - 5i^5)$ (c) $\dfrac{2}{3 + 4i}$
59. (a) $\dfrac{3 + 2i}{4 + i}$ (b) $\dfrac{1 - i}{2 - 3i}$
60. (a) $\dfrac{2 + i}{-5 + i}$ (b) $\dfrac{3 + \sqrt{-16}}{2 - \sqrt{-9}}$

In Problems 61–66, evaluate each function at the indicated input values. Round off to two decimal places if necessary.

61. $f(x) = \left(\dfrac{1}{2}\right)^x$ (a) $x = -2$ (b) $x = 0$
    (c) $x = 3.1$ (d) $x = -2.4$

62. $g(x) = \left(\dfrac{2}{3}\right)^x$ (a) $x = -1$ (b) $x = 0$
    (c) $x = 2.3$ (d) $x = -3.1$

63. $f(x) = \sqrt{3x + 1}$ (a) $x = 2$ (b) $x = 0$
    (c) $x = 2.4$ (d) $x = -\dfrac{1}{3}$

64. $g(x) = \sqrt[3]{4x + 9}$ (a) $x = -1$ (b) $x = 0$
    (c) $x = 2.5$ (d) $x = -\dfrac{9}{4}$

65. $f(x) = x^{-2/3}$ (a) $x = -8$ (b) $x = 0$
    (c) $x = 3.1$ (d) $x = -4.3$

66. $g(x) = \sqrt{2 - x} + \sqrt{-x}$
    (a) $x = -2$ (b) $x = 0$
    (c) $x = -4$ (d) $x = -3.4$

67. **Compound Interest:** A deposit of $1000 is placed in a savings account. Find the balance in the account after 4.5 years if interest is compounded monthly at a rate of 4.7%.

68. **Navigation:** The horizon distance d (in miles) that can be viewed on the surface of the ocean at a viewing height h (in feet) above the surface of the ocean can be approximated by the model

$$d(h) = 1.4\sqrt{h}$$

(a) Find the horizon distance from a balloon at an altitude of 1200 feet to view a ship on the ocean.
(b) Is it possible to view a sailboat on the ocean at a distance of 20 miles from a blimp at an altitude of 205 feet?

69. **Geometry:** The radius r of a circle inscribed in a triangle of sides a, b and c (Figure 1) is given by the formula

$$r = \sqrt{\dfrac{(s - a)(s - b)(s - c)}{s}}, \ s = \dfrac{1}{2}(a + b + c)$$

Find the radius of a circular flower bed that can be inscribed in a triangular plot with sides 18 feet, 22 feet and 26 feet. Round off the answer to two decimal places.

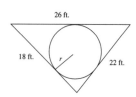

Figure 1

70. **Investment:** The annual rate of return r on an investment of p dollars that is worth s dollars after t years is modeled by the formula

$$r = \left(\dfrac{s}{p}\right)^{1/t} - 1$$

Find the annual rate of return of a stock that was purchased for $2.55 and sold 6 years later for $37.75. Round off to one decimal place.

G In Problems 71–76, (a) Use a grapher to graph each function. (b) Use the graph to indicated the domain and range of each function. (b) Find the x intercept of each graph to determine the solution of the associated equation.

71. $f(x) = \sqrt{3x + 1} - 4$; $\sqrt{3x + 1} = 4$

72. $f(x) = \sqrt{x + 3} - \sqrt{x - 2} - 1$;
$\sqrt{x + 3} - \sqrt{x - 2} = 1$

73. $f(x) = \sqrt{x + 4} - \sqrt{x - 1} - 1$;
$\sqrt{x + 4} = \sqrt{x - 1} + 1$

74. $f(x) = 3\sqrt{2x - 5} - \sqrt{x + 23}$;
$3\sqrt{2x - 5} = \sqrt{x + 23}$

75. $f(x) = \sqrt{x - 2} - x + 47$; $\sqrt{x - 2} + 47 = x$

76. $f(x) = \sqrt[3]{3x - 1} - 2$; $\sqrt[3]{3x - 1} = 2$

CHAPTER 5 PRACTICE TEST

1. Rewrite each expression in a simplified form using only positive exponents.
   (a) $(-10)^0$
   (b) $\left(\dfrac{-3}{7}\right)^{-2}$
   (c) $\dfrac{1}{3^{-4}}$
   (d) $x^{-11} x^7$
   (e) $\left(\dfrac{x^{-2}y^2}{x^{-3}y^{-3}}\right)^{-1}$

2. Solve each exponential equation.
   (a) $2^x = 16$ (b) $3^{-4x} = 81$ (c) $5^{2y} = 125$

3. Find the principal root of the given numbers. When it is not a real number, state the reason.
   (a) $\sqrt[4]{16}$ (b) $\sqrt{-25}$ (c) $-\sqrt{25}$ (d) $\sqrt[3]{-27}$

4. Simplify each expression (if possible). Assume all variables represent positive real numbers.
   (a) $-\sqrt{3^2}$ (b) $(\sqrt[3]{5x})^3$ (c) $\sqrt{9x^2}$ (d) $(\sqrt[4]{-3y})^4$

5. Find the distance between (1,2) and (7,10).

6. Let $f(x) = \sqrt[3]{x - 8}$. Find (a) $f(0)$ (b) $f(-19)$

7. Find the exact value of each expression.
   (a) $4^{-1/2}$ (b) $16^{3/4}$ (c) $(-27)^{2/3}$ (d) $(-8)^{-4/3}$

8. (a) Convert $\sqrt[7]{x^2}$ to an expression with a rational exponent.
   (b) Convert $(x)^{3/11}$ to a radical form.

9. Simplify each expression. Write each answer using only positive exponents.
   (a) $5^{1/2} \cdot 5^{-3/2}$ (b) $(3^{1/7})^{14}$
   (c) $\dfrac{4^{1/3}}{4^{-1/6}}$

10. Let $f(x) = x^{2/3}$. Find (a) $f(-8)$ (b) $f(0)$ (c) $f(27)$.

11. Use the properties of radicals to simplify each expression. Assume all variables represent positive real numbers.
    (a) $\sqrt[3]{54}$
    (b) $\sqrt{3/(25x^2)}$
    (c) $\sqrt[3]{-81x^4}$
    (d) $\dfrac{\sqrt{45x^3}}{\sqrt{3x}}$

12. Rewrite $\sqrt{3} \cdot \sqrt[3]{4}$ as a single radical expression.

13. Perform each operation and simplify.
    (a) $\sqrt{48} + \sqrt{12}$ (b) $\sqrt{3}(\sqrt{15} + \sqrt{6})$
    (c) $(\sqrt{7} + 2\sqrt{3})^2$ (d) $(\sqrt{7} - \sqrt{2})(\sqrt{7} + \sqrt{2})$

14. Rationalize the denominator and simplify.
    (a) $\dfrac{7}{\sqrt{21}}$
    (b) $\dfrac{\sqrt{3} + 1}{\sqrt{3} + 2}$

15. Let $f(x) = \sqrt{1 - x} + \sqrt{-x}$.
    Find (a) $f(0)$ (b) $f(-1)$ (c) $f(-4)$.

16. Solve each equation.
    (a) $x^{2/3} = 4$ (b) $x^{3/2} = 8$
    (c) $\sqrt{x - 1} = 3$ (d) $4x^2 = 25$

17. Write in the form of $a + bi$.
    (a) $3 + \sqrt{-9}$ (b) $(2 + 3i) + (-7 - i)$
    (c) $(2 - 5i)(1 + 2i)$ (d) $\dfrac{1 + i}{3 - 2i}$ (e) $\dfrac{1}{i^{-4}}$

18. An employee's starting wage is $5.50 per hour. Find the hourly wage after 3 years if the total annual raise is 5% and the raise is given in 2 installments over the course of the year.

Chapter 6

QUADRATIC EQUATIONS AND FUNCTIONS

6.1 The Factoring Method
6.2 Completing the Square
6.3 The Quadratic Formula
6.4 Equations Reducible to Quadratic Form
6.5 Quadratic Applications and Models
6.6 Nonlinear Inequalities
6.7 Quadratic Functions
6.8 Horizontal Parabolas and Circles

Projectile motion refers to the movement of an object projected into the air at or near the earth's surface. Galileo and his colleagues were the first people to describe projectile motion. Problems involving projectile motion occur in sports such as baseball and football. Suppose that a projectile is shot vertically upward from the top of a building 80 feet high with an initial speed of 64 feet per second. Its height h (in feet) above ground after t seconds is modeled by the equation

$$h = -16t^2 + 64t + 80$$

After how many seconds will the projectile hit the ground? Example 6 on page 279 shows the solution of this problem.

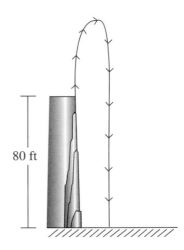

80 ft

An important part of the study of mathematics is learning to analyze real world situations and solve problems. Often, these situations obey basic algebraic models described by *quadratic functions* and *quadratic equations* of the form

$$f(x) = ax^2 + bx + c, a \neq 0 \text{ and } ax^2 + bx + c = 0, a \neq 0$$

In this chapter, we discuss various methods for solving quadratic equations and inequalities. Quadratic functions and properties of their graphs as well as graphs of circles conclude this chapter.

OBJECTIVES

1. Solve Quadratic Equations by Factoring
2. Solve Polynomial Equations by Factoring
3. Interpret Graphs of Quadratic Functions
4. Solve Applied Problems
5. Use Technology Exploration

6.1 The Factoring Method

An equation that can be written in the form

$$ax^2 + bx + c = 0$$

where a, b, and c are real numbers with $a \neq 0$ is called a **standard form** of a **quadratic equation** in one variable. In this section, we use factoring to solve quadratic equations and other related nonlinear equations.

Solving Quadratic Equations by Factoring

The method presented in this section for solving quadratic and other nonlinear equations is called the **factoring method**. The method is based on the following *zero factor property*.

ZERO FACTOR PROPERTY

> If a and b are real numbers, or algebraic expressions, and if
> $ab = 0$, then $a = 0$ or $b = 0$.

In other words, if the product of two real numbers (or algebraic expressions) is zero, then at least one number (or expression) must be zero.

For instance, to solve the equation

$$(x - 2)(x + 5) = 0$$

we use the zero factor property and conclude that either $x - 2$, or $x + 5$ must be zero. That is,

$$x - 2 = 0 \quad \text{or} \quad x + 5 = 0$$
$$x = 2 \qquad \qquad x = -5$$

Thus, the equation $(x - 2)(x + 5) = 0$ has exactly two solutions, 2 and -5. To check, replace x by 2 and then replace x by -5 in the original equation to obtain

$$(2 - 2)(2 + 5) = (0)(7) = 0 \text{ or } (-5 - 2)(-5 + 5) = (-7)(0) = 0$$

The general strategy for solving a quadratic equation by factoring is outlined as follows.

SOLVING QUADRATIC EQUATIONS
BY FACTORING

> 1. Write the equation in standard form; i.e., set equal to zero.
> 2. Factor the nonzero side of the equation completely.
> 3. Apply the zero factor property and set each factor equal to zero.
> 4. Solve the resulting linear equations.
> 5. Check each solution in the original equation.

EXAMPLE 1   Solving Quadratic Equations by Factoring

Solve each equation.

(a) $2x^2 - x = 1$ (b) $5x^2 = 15x$

Solution   (a)

$$2x^2 - x = 1 \qquad \text{Original equation}$$
$$2x^2 - x - 1 = 0 \qquad \text{Write in standard form}$$
$$(2x + 1)(x - 1) = 0 \qquad \text{Factor the quadratic expression}$$
$$2x + 1 = 0 \ \text{ or } \ x - 1 = 0 \qquad \text{Apply the zero factor property}$$
$$2x = -1 \qquad x = 1 \qquad \text{Solve}$$
$$x = -1/2$$

The solutions are $-1/2$ and 1.

Check: Substituting in the original equation:

Let $x = -1/2$ Let $x = 1$

$$2(-1/2)^2 - (-1/2) = 2(1/4) + 1/2 \qquad\qquad 2(1)^2 - 1 = 2 - 1$$
$$= 1/2 + 1/2 \qquad\qquad\qquad\qquad\qquad = 1$$
$$= 1$$

> To apply the zero factor property, the equation must be written in standard form. That is, one side of the equation must be zero.

(b)

$$5x^2 = 15x \qquad \text{Original equation}$$
$$5x^2 - 15x = 0 \qquad \text{Write in standard form}$$
$$5x(x - 3) = 0 \qquad \text{Factor the quadratic expression}$$
$$5x = 0 \ \text{ or } \ x - 3 = 0 \qquad \text{Apply the zero factor property}$$
$$x = 0 \qquad x = 3 \qquad \text{Solve}$$

The solutions are 0 and 3 and should be checked by substitution.

EXAMPLE 2   Solving a Nonstandard Quadratic Equation

Solve the equation $(t + 6)(t - 2) = -7$

Solution   Although this equation appears to be in factored form, one side is not equal to zero. We begin by multiplying the factors on the left side.

$$(t + 6)(t - 2) = -7 \qquad \text{Original equation}$$
$$t^2 + 4t - 12 = -7 \qquad \text{Multiply factors}$$
$$t^2 + 4t - 5 = 0 \qquad \text{Write in standard form}$$
$$(t + 5)(t - 1) = 0 \qquad \text{Factor quadratic expression}$$
$$t + 5 = 0 \ \text{ or } \ t - 1 = 0 \qquad \text{Zero factor property}$$
$$t = -5 \qquad t = 1$$

Therefore, the solutions are -5 and 1. Check these solutions in the original equation.

EXAMPLE 3   Solving a Quadratic Equation with a Repeated Solution

Solve the equation $2(x^2 + 3) + 5 = -2(x^2 + 6x) + 2$

Solution

$$2(x^2 + 3) + 5 = -2(x^2 + 6x) + 2 \qquad \text{Original equation}$$
$$2x^2 + 6 + 5 = -2x^2 - 12x + 2 \qquad \text{Apply the distributive property}$$
$$4x^2 + 12x + 9 = 0 \qquad \text{Write in standard form}$$
$$(2x + 3)(2x + 3) = 0 \qquad \text{Factor left side}$$
$$2x + 3 = 0 \ \text{ or } \ 2x + 3 = 0 \qquad \text{Apply zero factor property}$$
$$2x = -3 \qquad 2x = -3$$
$$x = \frac{-3}{2} \qquad x = \frac{-3}{2}$$

Note that even though the left side of the equation has two factors, the factors are the same. Thus, the two solutions are equal, and the only solution is $-3/2$. Check this solution in the original equation.

Solving Polynomials Equations by Factoring

The zero factor property can be extended to three or more factors as the next example shows.

EXAMPLE 4   Solving a Polynomial Equation

Solve the equation

$$p^3 + 4p^2 - 9p - 36 = 0$$

Solution   We factor the left side by grouping

$p^3 + 4p^2 - 9p - 36 = 0$	Original equation
$p^2(p + 4) - 9(p + 4) = 0$	Group to remove common factors
$(p + 4)(p^2 - 9) = 0$	Common factor $p + 4$
$(p + 4)(p - 3)(p + 3) = 0$	Factor completely
$p + 4 = 0$ or $p - 3 = 0$ or $p + 3 = 0$	Apply zero factor property
$p = -4 \qquad p = 3 \qquad p = -3$	

The solutions are

$$-4, 3, \text{ and } -3.$$

Check these solutions in the original equation.   ◇

Notice that the third degree equation in Example 4 has three factors and three solutions. This is not an accident. A polynomial equation can have at most as many real solutions as its degree.

Interpreting Graphs of Quadratic Functions

A function f of the form

$$f(x) = ax^2 + bx + c, a \neq 0$$

is referred to as a **quadratic function**. In the next example, we will see how solutions to a quadratic equation determine the x intercepts of the graph of the corresponding function f. Equivalently, we graph the function f and look for values of x for which $f(x) = 0$. The values of x for which

$$f(x) = 0$$

are the x intercepts and are called the **zeros** of f.

EXAMPLE 5   Analyzing a Graph of a Quadratic Function

Figure 1a shows the graph of the function

$$f(x) = x^2 - 7x + 10$$

(a) Find $f(1), f(2), f(3.5), f(4),$ and $f(5)$.

(b) Plot the ordered pairs obtained from part (a). Determine if these points lie on the graph of the function f.

(c) Solve the equation
$$x^2 - 7x + 10 = 0.$$

(d) What is the relationship between the solutions of the equation in part (c) and the x intercepts of the graph of f.

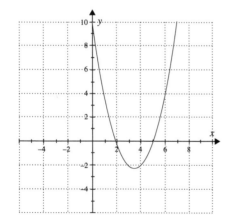

Figure 1a

Solution

(a) $f(x) = x^2 - 7x + 10$
$$f(1) = 1^2 - 7(1) + 10 = 4$$
$$f(2) = 2^2 - 7(2) + 10 = 0$$
$$f(3.5) = 3.5^2 - 7(3.5) + 10$$
$$= -2.25$$
$$f(4) = 4^2 - 7(4) + 10 = -2$$
$$f(5) = 5^2 - 7(5) + 10 = 0$$

(b) The ordered pairs are
$$(1,4), (2,0), (3.5,-2.25), (4,-2), \text{ and } (5,0).$$
By plotting these points, we see that all of them lie on the graph of f (Figure 1b).

(c) $x^2 - 7x + 10 = 0$
$$(x - 2)(x - 5) = 0$$
$$x - 2 = 0 \quad \text{or} \quad x - 5 = 0$$
$$x = 2 \quad \text{or} \qquad x = 5$$

(d) Observe that the solutions, 2 and 5, to the quadratic equation $x^2 - 7x + 10 = 0$ are also the x intercepts of the graph of $f(x) = x^2 - 7x + 10$. ◈

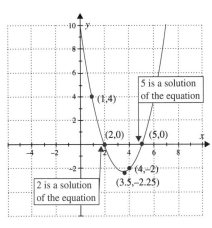

Figure 1b

Solving Applied Problems

Quadratic functions serve as models for many applications encountered in many fields. As always, finding out information about the function—the output values at particular input values: the domain, the range, and the intercepts—helps us answer questions about a particular model.

EXAMPLE 6   Analyzing a Path of a Projectile

A projectile is shot vertically upward from the top of a building 80 feet high with an initial speed of 64 feet per second. Its height h (in feet) above the ground after t seconds is given by the model

$$h(t) = -16t^2 + 64t + 80$$

After how many seconds will the projectile hit the ground?

Solution

We must find a value for t that results in a value of zero for h. That is, we set $h = 0$ and solve the resulting equation.

$$-16t^2 + 64t + 80 = 0 \qquad \text{Height, } h = 0$$
$$t^2 - 4t - 5 = 0 \qquad \text{Divide by } -16$$
$$(t - 5)(t + 1) = 0$$
$$t - 5 = 0 \quad \text{or} \quad t + 1 = 0$$
$$t = 5 \qquad\qquad t = -1$$

Since the time cannot be negative, 5 is the only meaningful solution for this model. This means the projectile will hit the ground about 5 seconds after it has been shot. Figure 2 shows the height of the projectile at each point in time. Notice the t intercept of the graph is 5 which is the solution of the equation. ◈

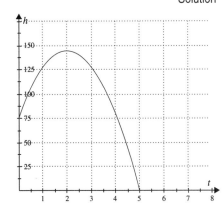

Figure 2

Using Technology Exploration

We can use technology to verify our solutions of a quadratic equation. Recall that the solution of an equation corresponds to either the x intercepts of the graph or to the x coordinates of the points of intersection of the graphs of y_1 = right side of the equation, and y_2 = left side of the equation on the same viewing window.

G EXAMPLE 7   Using a Grapher to Solve an Equation

Use a grapher to support the algebraic solution of the equation $2x^2 - x = 1$ in example 1a on page 277.

Solution   There are two ways for solving this equation graphically. The first is to find the x intercepts (Figure 3a) of the graph of

$$y = 2x^2 - x - 1$$

The second is to rewrite the equation as two equations

$$y_1 = 2x^2 - x \text{ and } y_2 = 1$$

then find the x coordinates (Figure 3b) of the points of intersection of the two graphs. Using the ZOOM and TRACE features, these coordinates are $-1/2$ and 1.

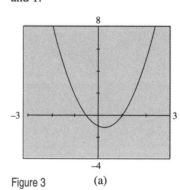

Figure 3   (a)   (b)

◆ # PROBLEM SET 6.1

Mastering the Concepts

In Problems 1–26, solve each equation by factoring.

1. (a) $x^2 + 7x + 6 = 0$ 2. (a) $x^2 + 25x + 24 = 0$
   (b) $x^2 - 12x + 11 = 0$ (b) $x^2 - 13x + 12 = 0$
3. (a) $c^2 - 6c + 8 = 0$ 4. (a) $15x^2 - 19x + 6 = 0$
   (b) $y^2 - 37y + 36 = 0$ (b) $t^2 - 6t + 5 = 0$
5. (a) $y^2 - 6y + 9 = 0$ 6. (a) $x^2 - 24x + 144 = 0$
   (b) $9t^2 - 12t + 4 = 0$ (b) $49z^2 + 14z + 1 = 0$
7. (a) $b^2 - 7b - 8 = 0$ 8. (a) $t^2 - t - 20 = 0$
   (b) $u^2 + 5u - 66 = 0$ (b) $y^2 + 9y - 10 = 0$
9. (a) $x^2 - 2x = 35$ 10. (a) $y^2 + y = 12$
   (b) $x^2 + 2x = 35$ (b) $y^2 - y = 12$
11. (a) $6x^2 + 11x = 10$ 12. (a) $6z^2 + 17z = 3$
    (b) $6x^2 - 11x = 10$ (b) $10t^2 - 11t = -3$
13. (a) $6u^2 + 7u = 20$ 14. (a) $4t^2 + 20 = 21t$
    (b) $10y^2 - 31y = 14$ (b) $4x^2 - 7 = 27x$
15. (a) $x^2 - 7x = 0$ 16. (a) $x^2 + 8x = 0$
    (b) $2m^2 - 9m = 0$ (b) $y^2 + 7y = 0$
17. $y(2y - 19) = 33$ 18. $x(5x + 2) = 3$
19. $x(x - 2) = 9 - 2x$ 20. $(x + 7)(x - 4) = 26$
21. $(2x + 1)(3x - 2) = 10$
22. $(2x - 1)(x + 2) = -3$
23. $4x^2 - (x - 1)^2 = 0$
24. $9y^2 - (y + 1)^2 = 0$
25. $6y^2 + 7y - 3 = -12y^2 - 54y + 4$
26. $16x^2 - 3x + 1 = -9x^2 + 17x - 3$

In Problems 27–34, solve each polynomial equation by factoring.

27. $(p - 1)(p + 2)(2p - 3) = 0$
28. $(3u + 1)(u + 3)(2u + 5) = 0$
29. $x(x - 2)(x + 3)(2x + 7) = 0$
30. $3x(2x + 1)(3x - 4)(4x - 5) = 0$
31. $x^3 + 2x^2 - x - 2 = 0$
32. $x^3 - 11x^2 - 4x + 44 = 0$
33. $2x^4 + 25x^3 + 12x^2 = 0$
34. $6x^4 + 5x^3 - 25x^2 = 0$
35. Figure 4 shows the graph of the function $f(x) = x^2 - 4x$.
    (a) Find $f(-4)$, $f(-2)$, $f(0)$, and $f(4)$.
    (b) Plot the ordered pairs obtained in part (a). Determine if these points lie on the graph of the function.
    (c) Solve the equation $x^2 - 4x = 0$ algebraically.
    (d) What is the relationship between the solutions of the equation in part (c) and the x intercepts of the graph of f?

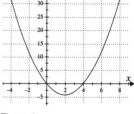

Figure 4

36. Figure 5 shows the graph of $f(x) = 4x - x^2$. Repeat the instructions (a), (b), (c), and (d) from problem 35 for the function f.

37. Figure 6 shows the graph of $f(x) = x^2 - 2x - 15$.
    (a) Find $f(-5)$, $f(-3)$, $f(-2)$, $f(0)$, and $f(5)$.
    (b) Plot the ordered pairs obtained from part (a). Determine if these points lie on the graph of the function.
    (c) Solve the equation $x^2 - 2x - 15 = 0$ algebraically.
    (d) What is the relationship between the solutions of the equation in part (c) and the x intercepts of the graph of f?

38. Figure 7 shows the graph of $f(x) = 15 - 2x - x^2$. Repeat the instructions (a), (b), (c), and (d) from problem 37.

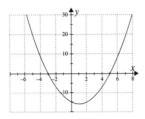

Figure 5

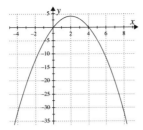

Figure 6

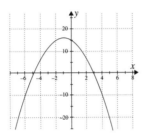

Figure 7

Applying the Concepts

39. **Free Falling Object:** The position of an object that is propelled at a speed of 64 feet per second straight up from the ground is given by the model
$$h(t) = 64t - 16t^2, t \geq 0$$
where h (in feet) is the height after t seconds. After how many seconds will the object return to the ground?

40. **Free Falling Object:** A ball is thrown vertically upward from the top of a building 48 feet high. After t seconds, the height h (in feet) is given by the model
$$h(t) = 48 + 32t - 16t^2, \quad t \geq 0$$
How long will it take the ball to reach the ground?

41. **Walking Time:** A night watchman makes his rounds by walking back and forth from his guard station to a nearby building. The distance (in feet) from his guard station after t minutes is given by the model
$$h(t) = -16t^2 + 80t, \quad t \geq 0$$
    (a) What restrictions on t should be made for this model to make sense?
    (b) How far away is the building from the guard station?
    (c) How long does it take the watchman to complete one round trip?

42. **Compound Interest:** A principal of $2,000 is deposited in an account at an annual interest rate r compounded annually. The amount A (in dollars) accumulated after 2 years in the account is given by the formula
$$A(r) = 2000(1 + r)^2$$
    (a) Calculate and explain the meaning of $A(0.03)$, $A(0.05)$ and $A(0.07)$. Round off to two decimal places.
    (b) If $A = \$2,050$ after 2 years, find the value of r.

Developing and Extending the Concepts

In Problems 43–46, solve each equation.

43. $(2x - 3)(x + 1) - 4(2x - 3) = 0$
44. $(1 - 2x)(1 + 2x) = (1 - 2x)^2$
45. $y(y + 1) - (y + 1) = y^2$
46. $(t + 1)(t + 2) - 6t = (2t - 1)(t - 1)$

In Problems 47–50, solve each equation for the indicated unknown.

47. $x^2 - 2ax - 15a^2 = 0$, for x
48. $12y^2 - 10my = 12m^2$, for y
49. $(at - bt)^2 = t(b - a)$, for t
50. $6m^2 + mb = 2b^2$, for m

In Problems 51–54, match the function with its graph. Then use the x intercepts of the graph to solve the associated equation.

51. $f(x) = x^2 - 3x - 4$ 52. $f(x) = x^2 - 9x + 18$
    $x^2 - 3x - 4 = 0$ $x^2 - 9x + 18 = 0$
53. $f(x) = 10x - x^2 - 25$ 54. $f(x) = 6 + x - x^2$
    $10x - x^2 - 25 = 0$ $6 + x - x^2 = 0$

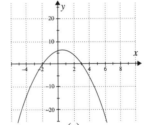

(a)

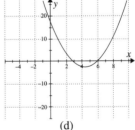

(b)

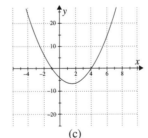

(c)

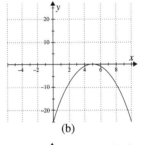

(d)

55. What relationship exists between the solutions of a quadratic equation and the graph of its associated function?

56. Use factoring to determine the x intercepts of the graph of the function $f(x) = 2x^3 + x^2 - 6x$. Use a grapher to graph the function and indicate the intercepts on the graph.

[G] In Problems 57–60, use a grapher to graph each function. Use the x intercepts of the graph to solve the associated equation.

57. $f(x) = x^2 - 3x - 10$
$x^2 - 3x - 10 = 0$

58. $f(x) = x^2 + 4x + 4$
$x^2 + 4x + 4 = 0$

59. $f(x) = x^2 - 5x - 24$
$x^2 - 5x - 24 = 0$

60. $f(x) = x^2 - 11x + 10$
$x^2 - 11x + 10 = 0$

OBJECTIVES

1. Solve Equations with Complex Solutions
2. Create a Perfect Square Trinomial
3. Solve Quadratic Equations by Completing the Square
4. Interpret Graphs of Quadratic Functions
5. Solve Applied Problems
6. Use Technology Exploration

6.2 Completing the Square

Solving Equations with Complex Solutions

Recall from section 5.6 how easily the square root property can be used to solve equations of the form

$$au^2 - b = 0, \quad a \neq 0$$

The solutions that were considered had been real numbers. But now that we have studied complex numbers, it makes sense to look for other types of solutions. For instance, although the quadratic equation $x^2 + 4 = 0$ has no real number solutions, it does have two solutions that are complex numbers. Namely, $-2i$ and $2i$. To check these solutions, we substitute $-2i$ and $2i$ for x. That is,

$$(-2i)^2 + 4 = -4 + 4 = 0 \text{ and } (2i)^2 + 4 = -4 + 4 = 0$$

In order to find complex solutions of a quadratic equation, we extend the use of the square root property to complex numbers. For example, to solve the equation $x^2 + 4 = 0$, we write

$$x^2 + 4 = 0$$
$$x^2 = -4$$
$$x = \pm\sqrt{-4}$$
$$x = \pm 2i$$

The solutions are $2i$ and $-2i$.

EXAMPLE 1   Obtaining Complex Solutions

Solve the equation $2(x - 1)^2 + 18 = 0$

Solution
$$\begin{array}{ll} 2(x - 1)^2 + 18 = 0 & \text{Original equation} \\ 2(x - 1)^2 = -18 & \text{Subtract 18 from each side} \\ (x - 1)^2 = -9 & \text{Divide each side by 2} \\ x - 1 = \pm\sqrt{-9} & \text{Square root property} \\ x - 1 = \pm 3i & \text{Complex roots} \\ x = 1 \pm 3i & \end{array}$$

The solutions are $1 + 3i$ and $1 - 3i$.

Check:

replace $1 + 3i$ for x	replace $1 - 3i$ for x
$2(1 + 3i - 1)^2 + 18$	$2(1 - 3i - 1)^2 + 18$
$= 2(3i)^2 + 18$	$= 2(-3i)^2 + 18$
$= 2(9i^2) + 18$	$= 2(9i^2) + 18$
$= 2(-9) + 18 = 0$	$= 2(-9) + 18 = 0$

Creating a Perfect Square Trinomial

Suppose that we are given the quadratic equation in standard form

$$x^2 + 8x - 4 = 0$$

Notice that the left side of this equation is not factorable. Because of this, we will try to convert it into the form

$$(x + k)^2 = h$$

The process of converting a quadratic equation into the latter form is called **completing the square**. The method is based on the familiar special product for the square of a binomial

$$x^2 + 2kx + k^2 = (x + k)^2$$

Thus, if we start with an expression of the form $x^2 + 2kx$, we see from the preceding equation that we can "complete the square" by adding k^2. We obtain the term k^2 by squaring one-half the coefficient of the first-degree term. That is, we complete the square by adding

$$\left[\frac{1}{2}(2k) \right]^2 = k^2$$

so that

$$x^2 + 2kx + \left[\frac{1}{2}(2k) \right]^2 = x^2 + 2kx + k^2 = (x + k)^2$$

coefficient of the x term

For example, given $x^2 + 8x$, we can complete the square as follows:

$$x^2 + 8x + \left[\frac{1}{2}(8) \right]^2 = x^2 + 8x + 16 = (x + 4)^2$$

coefficient of the x term

EXAMPLE 2 Creating a Perfect Square Trinomial

What term should be added to the given expression so it becomes a perfect square trinomial?

(a) $y^2 - 18y$ (b) $t^2 + 13t$

Solution In each case we complete the square by adding the square of one-half the coefficient of the first-degree term.

(a) $y^2 - 18y + \left[\frac{1}{2}(-18) \right]^2$ The term which needs to be added is 81

Therefore, the perfect square trinomial and its factored form are

$$y^2 - 18y + 81 = (y - 9)^2$$

(b) $t^2 + 13t + \left[\frac{1}{2}(13) \right]^2$ The term that needs to be added is 169/4

Therefore, the perfect square trinomial and its factored form are

$$t^2 + 13t + \frac{169}{4} = \left(t + \frac{13}{2} \right)^2$$

Solving Quadratic Equations by Completing the Square

Now we can use the square root property and the procedure of completing the square to solve quadratic equations. For instance, to solve the equation

$$x^2 + 8x - 4 = 0$$

we proceed as follows:

$$x^2 + 8x = 4 \qquad \text{Add 4 to both sides}$$

$$x^2 + 8x + \left[\frac{1}{2}(8)\right]^2 = 4 + \left[\frac{1}{2}(8)\right]^2 \qquad \text{Complete the square on the left side and add the same quantity to the right side}$$

$$x^2 + 8x + 16 = 4 + 16$$
$$(x + 4)^2 = 20$$

By using the square root property, we have $x + 4 = \pm\sqrt{20}$, so that

$x + 4 = \sqrt{20}$	$x + 4 = -\sqrt{20}$
$x + 4 = 2\sqrt{5}$	$x + 4 = -2\sqrt{5}$
$x = -4 + 2\sqrt{5}$	$x = -4 - 2\sqrt{5}$

Therefore, the solutions are

$$-4 + 2\sqrt{5} \text{ and } -4 - 2\sqrt{5}.$$

The above example of completing the square involves a quadratic equation in which the coefficient of the second degree term is 1. However, we can extend this procedure to any quadratic equation. When $a \neq 1$, we divide by a, creating a coefficient of 1 for the x^2 term; and then proceed to solve the equation by completing the square.

The following strategy summarizes the process of solving a quadratic equation

$$ax^2 + bx + c = 0$$

by completing the square.

1. Write the equation in the form $ax^2 + bx = -c$
2. If $a \neq 1$, divide both sides of the equation by a.
3. Complete the square by adding the square of half of

$$\frac{b}{a}, \quad \left[\frac{1}{2}\left(\frac{b}{a}\right)\right]^2,$$

   to both sides.
4. Express the trinomial as the square of a binomial.
5. Solve the resulting equation by using the square root property.

EXAMPLE 3   Solving Quadratic Equations by Completing the Square

Solve the equations by completing the square. Give both the exact solution and its two decimal place approximation.

(a) $x^2 + 4x - 2 = 0$ (b) $2x^2 + 3x - 6 = 0$

Solution   (a)

$$x^2 + 4x - 2 = 0 \qquad \text{Original equation}$$
$$x^2 + 4x = 2 \qquad \text{Add 2 to both sides}$$
$$x^2 + 4x + 4 = 2 + 4 \qquad \text{Add } [(1/2)(4)]^2 = 4 \text{ to both sides}$$
$$(x + 2)^2 = 6 \qquad \text{Write left side as the square of a binomial}$$
$$x + 2 = \pm\sqrt{6} \qquad \text{Use the square root property}$$
$$x = -2 \pm \sqrt{6}$$

The exact solutions are

$$-2 + \sqrt{6} \text{ and } -2 - \sqrt{6}.$$

The two decimal place approximations of the solutions are 0.45 and -4.45 respectively. Check these solutions in the original equation.

(b)

$$2x^2 + 3x - 6 = 0 \qquad \text{Original equation}$$
$$2x^2 + 3x = 6 \qquad \text{Add 6 to both sides}$$
$$x^2 + \frac{3}{2}x = \frac{6}{2} \qquad \text{Divide both sides by 2}$$
$$x^2 + \frac{3}{2}x + \frac{9}{16} = \frac{6}{2} + \frac{9}{16} \qquad \text{Add } \left[\left(\frac{1}{2}\right)\left(\frac{3}{2}\right)\right]^2 = \frac{9}{16} \text{ to both sides}$$
$$\left(x + \frac{3}{4}\right)^2 = \frac{57}{16} \qquad \text{Write left side as the square of a binomial}$$
$$x + \frac{3}{4} = \frac{\pm\sqrt{57}}{4} \qquad \text{Use the square root property}$$
$$x = \frac{-3 \pm \sqrt{57}}{4}$$

The exact solutions are

$$\frac{-3 + \sqrt{57}}{4} \text{ and } \frac{-3 - \sqrt{57}}{4}.$$

The two decimal place approximations of the solutions are

$$1.14 \text{ and } -2.64$$

respectively. Check these solutions in the original equation.

EXAMPLE 4   Solving an Equation with Complex Solutions

Solve the equation $8x^2 - 16x + 24 = 0$

Solution

$$8x^2 - 16x + 24 = 0 \qquad \text{Original equation}$$
$$8x^2 - 16x = -24 \qquad \text{Add } -24 \text{ to both sides}$$
$$x^2 - 2x = -3 \qquad \text{Divide both sides by 8}$$
$$x^2 - 2x + (-1)^2 = -3 + (-1)^2 \qquad \text{Complete the square}$$
$$(x - 1)^2 = -2$$
$$x - 1 = \pm\sqrt{-2} \qquad \text{Use the square root property}$$
$$x - 1 = \pm\sqrt{2}\, i$$
$$x = 1 \pm \sqrt{2}\, i$$

The solutions are $1 + \sqrt{2}\, i$ and $1 - \sqrt{2}\, i$. The first of these solutions is checked as follows.

$$8(1 + \sqrt{2}\, i)^2 - 16(1 + \sqrt{2}\, i) + 24$$
$$= 8(1 + 2\sqrt{2}\, i + 2i^2) - 16 - 16\sqrt{2}\, i + 24$$
$$= 8(1 + 2\sqrt{2}\, i - 2) - 16 - 16\sqrt{2}\, i + 24$$
$$= 8 + 16\sqrt{2}\, i - 16 - 16 - 16\sqrt{2}\, i + 24 = 0$$

Interpreting Graphs of Quadratic Functions

We now adopt the method of completing the square as an aid in analyzing graphs of quadratic functions of the form

$$f(x) = ax^2 + bx + c, a \neq 0$$

as the next example shows.

EXAMPLE 5   Analyzing a Graph of a Quadratic Function

Figure 1 shows the graph of the function

$$f(x) = x^2 + 10x + 23$$

(a) Find $f(-6)$, $f(-5)$, and $f(-4)$.
(b) Plot the resulting ordered pairs in part (a), and determine if they lie on the graph of f.
(c) Rewrite the equation for f in the form $f(x) = a(x - h)^2 + k$ by completing the square.
(d) Use the graph to indicate the range of f.

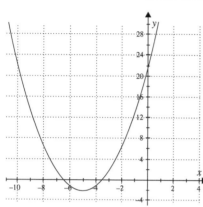

Figure 1

Solution   (a) The output values for the given input values for

$$f(x) = x^2 + 10x + 23$$

are as follows.

$$f(-6) = (-6)^2 + 10(-6) + 23 = 36 - 60 + 23 = -1$$
$$f(-5) = (-5)^2 + 10(-5) + 23 = 25 - 50 + 23 = -2$$
$$f(-4) = (-4)^2 + 10(-4) + 23 = 16 - 40 + 23 = -1$$

(b) By plotting the points $(-6,-1)$, $(-5,-2)$, and $(-4,-1)$, we see that all of these points lie on the graph of f.

(c) By completing the square, we can rewrite $f(x)$ as follows:

$$\begin{aligned}
f(x) &= x^2 + 10x + 23 \\
&= (x^2 + 10x) + 23 \\
&= (x^2 + 10x + 5^2) + 23 - 5^2 \qquad \text{Adding and subtracting 25} \\
&= (x + 5)^2 - 2 \qquad\qquad\quad \text{Factoring and simplifying}
\end{aligned}$$

(d) From part (a) we see that $(-5,-2)$ lies on the graph of f. Figure 1 visually suggests that this point is the lowest point on the graph. To be more certain, we could check the other output values in part (a) to be sure that $(-5,-2)$ is the lowest point. If this is the case, then the graph in Figure 1 suggests that the range of f consists of real numbers greater than or equal to -2, or the set of real numbers in the interval $[-2,\infty)$. Notice in part (c) that h is -5 and k is -2. This is not a coincidence as the next section shows. ◈

Solving Applied Problems

Perhaps the most important application of completing the square is to find the highest or lowest points on the graph of a quadratic function of the form

$$f(x) = a(x - h)^2 + k$$

It turns out that the coordinates of the *highest* or *lowest* point on the graph of f is given by $(h, f(h)) = (h, k)$.

EXAMPLE 6   Solving a Fuel Consumption Problem

The model

$$m = -\frac{1}{40}(v^2 - 70v) - \frac{5}{8}, \; 10 \le v \le 40$$

describes the fuel economy m (in miles per gallon) that a particular car can be driven at a speed v (in miles per hour).

(a) Use completing the square to rewrite the above equation in the form

$$m = -\frac{1}{40}(v - h)^2 + k$$

(b) Indicate where the highest point on the graph occurs, and visually check on the graph at what speed, v, will the miles per gallon, m, be the greatest.

Solution   (a) By completing the square, we have

$$m = -\frac{1}{40}(v^2 - 70v + 35^2) + \frac{1}{40}(35)^2 - \frac{5}{8}$$

$$= -\frac{1}{40}(v - 35)^2 + \frac{245}{8} - \frac{5}{8}$$

$$= -\frac{1}{40}(v - 35)^2 + 30$$

(b) The graph in Figure 2 visually suggests that $(35,30)$ is the highest point on the graph. This can be checked algebraically by substituting values of v in the given model and observing the mileage outputs. Figure 2 shows that at 35 miles per hour the fuel economy is 30 miles per gallon. ◇

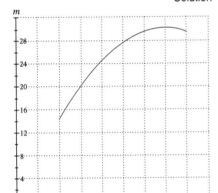

Figure 2

Using Technology Exploration

Once again, we can use technology to determine the coordinates of the highest and lowest points on the graph of a function as well as the range of a function.

G EXAMPLE 7   Using a Grapher to Find the Lowest Point

Use a grapher to support the conclusion of example 5d for the function $f(x) = x^2 + 10x + 23$

Solution   Using the ZOOM and TRACE features, the graph in Figure 3 visually suggests that the point $(-5,-2)$ is the lowest point on the graph of f. It also suggests that the range of f is $[-2,\infty)$. ◇

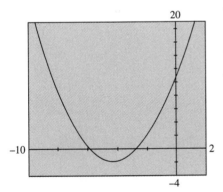

Figure 3

◇ ## PROBLEM SET 6.2

Mastering the Concepts

In Problems 1–6, obtain the complex solutions of each equation. Check the solutions.

1. (a) $x^2 + 64 = 0$
   (b) $4x^2 + 36 = 0$
2. (a) $9x^2 + 4 = 0$
   (b) $2y^2 + 18 = 0$
3. (a) $(2m + 1)^2 + 36 = 0$
   (b) $(y + 2)^2 + 25 = 0$
4. (a) $(x - 1)^2 + 9 = 0$
   (b) $(y + 3)^2 + 16 = 0$
5. $(3u - 6)^2 + 36 = 0$
6. $(3x - 2)^2 + 49 = 0$

In Problems 7–10, what term should be added to the given expression so it becomes a perfect square trinomial?

7. (a) $x^2 + 6x$
   (b) $t^2 - 8t$
8. (a) $y^2 - 10y$
   (b) $z^2 + 12z$
9. (a) $x^2 + 7x$
   (b) $t^2 - 11t$
10. (a) $y^2 + 5y$
    (b) $x^2 + 9x$

In Problems 11–26, solve each equation by completing the square. Give both the exact solutions and the two decimal approximations.

11. (a) $x^2 + 2x - 3 = 0$
    (b) $x^2 + 4x - 7 = 0$
12. (a) $x^2 - 12x - 7 = 0$
    (b) $x^2 + 6x - 11 = 0$
13. (a) $x^2 - 5x - 6 = 0$
    (b) $x^2 + 3x - 4 = 0$
14. (a) $x^2 - 7x = 8$
    (b) $x^2 + 11x = 12$
15. $3m^2 - 12m - 3 = 0$
16. $3u^2 - 18u - 7 = 0$
17. $4y^2 - 8y - 3 = 0$
18. $25y^2 - 25y - 14 = 0$
19. $5m^2 - 8m - 17 = 0$
20. $9x^2 - 27x - 14 = 0$
21. $4y^2 + 7y - 5 = 0$
22. $3t^2 - 8t - 4 = 0$
23. $7m^2 + 5m - 1 = 0$
24. $25x^2 - 50x - 21 = 0$
25. $16y^2 - 24y = 5$
26. $9t^2 - 30t = -21$

In Problems 27–30,

   (a) Find the output values for the given input values of the function whose graph is given.

   (b) Rewrite each equation in the equivalent form
$$f(x) = a(x - h)^2 + k$$
by completing the square.

   (c) Use the algebraic results from part (b) to find the coordinates of the lowest point on the graph.

   (d) Use the graph to visually indicate the range of the function.

27. $f(x) = x^2 - 2x - 5, f(0), f(1)$

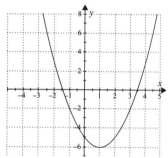

28. $f(x) = x^2 - 6x + 7, f(0), f(3)$

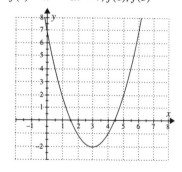

29. $f(x) = x^2 + 5x - 3, f(0), f\left(-\dfrac{5}{2}\right)$

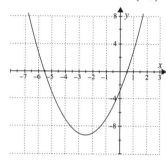

30. $f(x) = x^2 + 3x - 2, f(0), f\left(-\dfrac{3}{2}\right)$

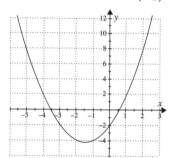

Applying the Concepts

31. **Suspension Bridge Cable:** The model
$$y = 0.0375x^2 - 1.5x + 30, 0 \le x \le 40$$
describes a parabolic cable for a suspension bridge, where y (in feet) is the height of the cable at a point above the road bed x (in feet) from the left support tower (Figure 4).

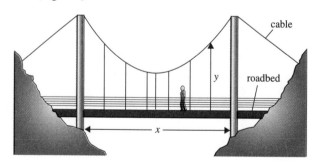

Figure 4

   (a) Use completing the square to rewrite this model in the form $y = a(x - h)^2 + k$.

   (b) Indicate where the lowest point on the cable occurs, and visually check on the diagram the smallest height of the cable from the road bed at the center of the bridge.

32. **Flight of a Baseball:** A ballplayer hits a high pop-up straight above home plate. If the bat meets the ball 4 feet above the ground and sends it up at a speed of 80 feet per second, then the height y (in feet) of the ball after t seconds is predicted by the model

$$y = -16t^2 + 80t + 4. \quad t \geq 0$$

   (a) Write the equation for this model in the form $y = a(t - h)^2 + k$ by completing the square.
   (b) Use part (a) to indicate the highest point reached by the ball (see Example 6 on page 279).

Developing and Extending the Concepts

33. Determine the value of k so that the trinomial is a perfect square.
   (a) $kx^2 - 10x + 1$
   (b) $9x^2 + kx + 25$
34. Solve the equation $x^2 - 2x - 15 = 0$
   (a) by factoring and
   (b) by completing the square.

In Problems 35–40, solve each formula for the indicated variable. Assume all symbols represent positive real numbers.

35. (a) $A = s^2$, for s
      (area of a square)
   (b) $S = 4\pi r^2$, for r
      (surface area of a sphere)

36. (a) $T = 2\pi\sqrt{l/g}$, for l
      (period of a pendulum)
   (b) $V = s^3$, for s
      (Volume of a cube)

37. (a) $A = (1 + r)^2$, for r
      (compound interest)
   (b) $h = \dfrac{1}{2}gt^2$, for t
      (free falling object)

38. (a) $F = \dfrac{m_1 m_2}{r^2}$, for r
      (Law of gravity)
   (b) $a^2 + b^2 = c^2$, for a
      (Pythagorean Theorem)
39. $-gt^2 + vt = s$, for t
      (falling object)
40. $S = 2\pi r^2 + 2\pi rh$, for r
      (surface area of a cylinder)

In Problems 41–46, use completing the square to obtain the complex solutions of each equation.

41. $3x^2 - 6x + 5 = 0$
42. $3x^2 + 12x + 31 = 0$
43. $9x^2 + 30x + 26 = 0$
44. $7x^2 - 2x + 5 = 0$
45. $4x^2 + 8x + 7 = 0$
46. $10x^2 + 5x + 2 = 0$
47. Let $f(x) = x^2 + 6x - 3$; find and simplify.
   (a) $f(-3 + 2\sqrt{3})$ (b) $f(-3 - 2\sqrt{3})$
48. Let $f(x) = 2x^2 + 3x + 2$; find and simplify.
   (a) $f\left(\dfrac{-3 + \sqrt{7}i}{4}\right)$ (b) $f\left(\dfrac{-3 - \sqrt{7}i}{4}\right)$

In Problems 49 and 50, one solution of each quadratic equation is given.

   (a) Find a value of a or b.
   (b) Use completing the square to determine the other solution.

49. $ax^2 + 7x - 2 = 0; x = 2$
50. $9x^2 + bx - 11 = 0; x = 1$

In Problems 51–54, match the function with its graph. Then use completing the square to rewrite each function in the form

$$f(x) = a(x - h)^2 + k.$$

Indicate the coordinates of the lowest point or the highest point on the graph of f.

51. $f(x) = x^2 - 6x + 8$
52. $f(x) = x^2 - 12x + 12$
53. $f(x) = 4x^2 - 20x + 20$
54. $f(x) = -4x^2 + 12x - 3$

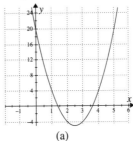

(a)

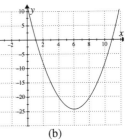

(b)

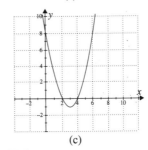

(c)

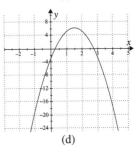

(d)

G In Problems 55–58, use a grapher to graph each function.
   (a) Estimate the x intercepts of each graph.
   (b) Use completing the square to solve each associated equation.
   (c) Compare the results of (a) and (b).

55. $f(x) = x^2 - 2x - 6$
      $x^2 - 2x - 6 = 0$
56. $f(x) = x^2 + 6x + 8$
      $x^2 + 6x + 8 = 0$
57. $f(x) = 6x^2 + 13x + 6$
      $6x^2 + 13x + 6 = 0$
58. $f(x) = 2x^2 + 11x + 15$
      $2x^2 + 11x + 15 = 0$

OBJECTIVES

1. Derive the Quadratic Formula
2. Introduce the Discriminant
3. Solve Applied Problems
4. Use Technology Exploration

6.3 The Quadratic Formula

Deriving the Quadratic Formula

Every quadratic equation can be written in the form

$$ax^2 + bx + c = 0, a \neq 0$$

This *general form* can be solved using the method of completing the square. The result is a formula that can be used to solve any quadratic equation.

$$ax^2 + bx + c = 0 \qquad \text{General form of quadratic equation}$$

$$ax^2 + bx = -c \qquad \text{Subtract } c \text{ from each side}$$

$$x^2 + \frac{b}{a}x = \frac{-c}{a} \qquad \text{Divide each side by } a$$

$$x^2 + \frac{b}{a}x + \frac{b^2}{4a^2} = \frac{-c}{a} + \frac{b^2}{4a^2} \qquad \text{Add } \left[\frac{1}{2}\left(\frac{b}{a}\right)\right]^2 = \frac{b^2}{4a^2} \text{ to each side}$$

$$\left(x + \frac{b}{2a}\right)^2 = \frac{b^2}{4a^2} - \frac{c}{a} \qquad \text{Express the left side as the square of a binomial}$$

$$\left(x + \frac{b}{2a}\right)^2 = \frac{b^2}{4a^2} - \frac{4ac}{4a^2} \qquad \text{Rewrite with a common denominator}$$

$$\left(x + \frac{b}{2a}\right)^2 = \frac{b^2 - 4ac}{4a^2} \qquad \text{Add fractions on right side}$$

$$\sqrt{[x + (b/2a)]^2} = \sqrt{(b^2 - 4ac)/(4a^2)} \qquad \text{Take the square root of both sides}$$

$$\left| x + \frac{b}{2a} \right| = \sqrt{(b^2 - 4ac)/(4a^2)}$$

$$x + \frac{b}{2a} = \pm\sqrt{(b^2 - 4ac)/(4a^2)}$$

$$x = \frac{-b}{2a} \pm \frac{\sqrt{b^2 - 4ac}}{2a} \qquad \text{Subtract } \frac{b}{2a} \text{ from each side}$$

$$x = \frac{-b \pm \sqrt{b^2 - 4ac}}{2a} \qquad \text{Combine the fractions}$$

It is important to note that the quadratic formula can always be used to solve quadratic equations and will produce the same results as any other method.

THE QUADRATIC FORMULA

A quadratic equation must be written in standard form before using the quadratic formula.

> If $ax^2 + bx + c = 0$ and $a \neq 0$, then the solutions of the equation are given by the **Quadratic Formula**
>
> $$x = \frac{-b \pm \sqrt{b^2 - 4ac}}{2a}$$

EXAMPLE 1   Using the Quadratic Formula

Solve the equation $2x^2 - 6x + 3 = 0$ by the quadratic formula.

Solution   Since the equation is written in standard form, we begin by identifying a, b, and c. Substituting $a = 2$, $b = -6$, and $c = 3$ in the formula, we have

$$x = \frac{-b \pm \sqrt{b^2 - 4ac}}{2a}$$

$$x = \frac{-(-6) \pm \sqrt{(-6)^2 - 4(2)(3)}}{2(2)}$$

$$x = \frac{6 \pm \sqrt{36 - 24}}{4}$$

$$x = \frac{6 \pm \sqrt{12}}{4} = \frac{6 \pm 2\sqrt{3}}{4} = \frac{3 \pm \sqrt{3}}{2}$$

The exact solutions of the equation are $(3 + \sqrt{3})/2$ and $(3 - \sqrt{3})/2$. The two decimal approximations of these solutions are 2.37 and 0.63 respectively. ◈

EXAMPLE 2   Solving Equations Using the Quadratic Formula

Solve each equation using the quadratic formula. Write the complex solutions in the form $a + bi$.

(a) $3x^2 + 4x - 4 = 0$ (b) $x^2 - 2x = -1$ (c) $(x - 1)(x + 3) = -5$

Solution   (a) Since the equation $3x^2 + 4x - 4 = 0$ is in standard form, $a = 3$, $b = 4$ and $c = -4$. Hence,

$$x = \frac{-b \pm \sqrt{b^2 - 4ac}}{2a}$$

$$x = \frac{-4 \pm \sqrt{(4)^2 - 4(3)(-4)}}{2(3)}$$

$$x = \frac{-4 \pm \sqrt{16 + 48}}{6}$$

$$x = \frac{-4 \pm 8}{6}$$

Therefore, the solutions are $(-4 + 8)/6 = 2/3$ and $(-4 - 8)/6 = -2$.

(b) The standard form of the equation $x^2 - 2x = -1$ is

$$x^2 - 2x + 1 = 0$$

Substituting $a = 1$, $b = -2$ and $c = 1$ in the quadratic formula, we have

$$x = \frac{-b \pm \sqrt{b^2 - 4ac}}{2a}$$

$$x = \frac{-(-2) \pm \sqrt{(-2)^2 - 4(1)(1)}}{2(1)}$$

$$x = \frac{2 \pm \sqrt{4 - 4}}{2}$$

$$x = \frac{2 \pm 0}{2} = 1$$

Therefore, there is only one solution, $x = 1$.

(c) The standard form of the equation $(x - 1)(x + 3) = -5$ is

$$x^2 + 2x + 2 = 0$$

where $a = 1$, $b = 2$ and $c = 2$. Substituting for a, b, and c in the quadratic formula, we have,

$$x = \frac{-b \pm \sqrt{b^2 - 4ac}}{2a}$$

$$x = \frac{-2 \pm \sqrt{2^2 - 4(1)(2)}}{2(1)}$$

$$x = \frac{-2 \pm \sqrt{4 - 8}}{2}$$

$$x = \frac{-2 \pm \sqrt{-4}}{2} = \frac{-2 \pm 2i}{2}$$

$$x = -1 \pm i$$

Therefore, the two complex solutions are $-1 + i$ and $-1 - i$. ◈

If the discriminant is negative and a, b and c are real numbers, the two solutions will be complex conjugates of each other. If $b^2 - 4ac$ is a perfect square, then the polynomial is factorable.

Introducing the Discriminant

The expression $b^2 - 4ac$ that occurs under the radical symbol in the quadratic formula is called the **discriminant**. It provides interesting information about the number and types of solutions of a quadratic equation. In Example 2 we solved three quadratic equations. Our results are summarized in the following table:

Equation in standard form	Discriminant $b^2 - 4ac$	Sign of Discriminant	Type of Solutions
$3x^2 + 4x - 4 = 0$	64	positive	two real solutions
$x^2 - 2x + 1 = 0$	0	none	one real solution
$x^2 + 2x + 2 = 0$	-4	negative	two complex solutions

The use of the discriminant can be generalized in the following test:

Discriminant Test for $ax^2 + bx + c = 0$, $a \neq 0$

Discriminant $b^2 - 4ac$	Type of Solutions
positive	two real solutions
zero	one real solution
negative	two complex solutions

EXAMPLE 3 Applying the Discriminant Test

Use the discriminant test to determine the number and type of solutions of each equation.

(a) $4x^2 - 5x + 1 = 0$ (b) $4x^2 - 12x + 9 = 0$ (c) $2x^2 + 3x + 2 = 0$

Solution (a) The discriminant $b^2 - 4ac = (-5)^2 - 4(4)(1) = 9$. Thus, the equation has two real solutions.

(b) The discriminant $b^2 - 4ac = (-12)^2 - 4(4)(9) = 0$. Thus, the equation has one real solution.

(c) The discriminant $b^2 - 4ac = (3)^2 - 4(2)(2) = -7$. Thus, the equation has two complex solutions.

Use the quadratic formula to check these results.   ◈

Solving Applied Problems

Although the quadratic formula was known to the Babylonians around 2000 B.C., the idea of the discriminant was not discovered until the 17th century by Sir Isaac Newton. The discussion of the type of solutions of a quadratic equation came after the invention of the Cartesian coordinate system. In the next example we will use the quadratic formula and the discriminant to help solve applied problems.

EXAMPLE 4 Solving a Diving Problem

An Olympic competitor stands on a 10 meter diving platform and jumps from the platform at an initial upward speed of 2.2 meters per second. The competitor's height h (in meters) above the surface of the water after t seconds is given by the model

$$h(t) = -4.9t^2 + 2.2t + 10, \ t \geq 0$$

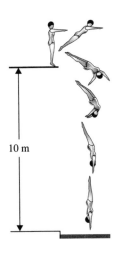

10 m

Use the quadratic formula to determine how many seconds it takes for the competitor to reach the following heights:
(a) the surface of the water.
(b) ten meters above the surface of the water.
(c) twelve meters above the surface of the water. Is this possible?
Round off each answer to two decimal places.

Solution   (a) The diver reaches the surface of the water when $h(t) = 0$, so the equation is
$$0 = -4.9t^2 + 2.2t + 10$$
Using the quadratic formula with $a = -4.9$, $b = 2.2$, and $c = 10$, we have
$$t = \frac{-b \pm \sqrt{b^2 - 4ac}}{2a}$$
$$t = \frac{-2.2 \pm \sqrt{(2.2)^2 - 4(-4.9)(10)}}{2(-4.9)}$$
$$t = 1.67 \text{ or } -1.22$$
The solutions are 1.67 and -1.22. We reject -1.22 since it does not make sense in this problem. This means the diver hits the surface of the water after 1.67 seconds.

(b) The diver reaches 10 meters above the surface of the water when $h(t) = 10$, so the equation is
$$10 = -4.9t^2 + 2.2t + 10 \text{ or } 4.9t^2 - 2.2t + 0 = 0$$
Using the quadratic formula with $a = 4.9$, $b = -2.2$, and $c = 0$, we have
$$t = \frac{-b \pm \sqrt{b^2 - 4ac}}{2a}$$
$$t = \frac{-(-2.2) \pm \sqrt{(-2.2)^2 - 4(4.9)(0)}}{2(4.9)}$$
$$t = 0.45 \text{ or } 0$$
So the solutions are 0.45 and 0. This means the diver was initially at a height of 10 meters at $t = 0$ and again at $t = 0.45$ on the way down to the water. This equation can also be solved by the factoring method.

(c) For $h = 12$, we have
$$12 = -4.9t^2 + 2.2t + 10 \text{ or } 4.9t^2 - 2.2t + 2 = 0$$
Using the quadratic formula with $a = 4.9$, $b = -2.2$, and $c = 2$, we have
$$t = \frac{-b \pm \sqrt{b^2 - 4ac}}{2a}$$
$$t = \frac{-(-2.2) \pm \sqrt{(-2.2)^2 - 4(4.9)(2)}}{2(4.9)}$$
$$t = 0.22 \pm 0.60i$$
The solutions are $0.22 \pm 0.60i$. The solutions are complex because the discriminant is negative. This means the diver will never reach a height of 12 meters. Figure 1 shows the relationship between time and the heights indicated in parts (a) and (b).   ◈

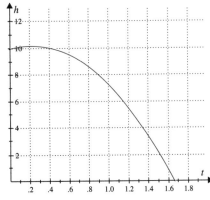

Figure 1

Using Technology Exploration

We can use a grapher to predict the number and type of solutions we get when we solve quadratic equations of the form $ax^2 + bx + c = 0$. The graph of the function $f(x) = ax^2 + bx + c$ can be used to support the prediction indicated by the discriminant. The graph of f may have two x intercepts when the discriminant is positive, one x intercept if the discriminant is zero, and no x intercepts if the discriminant is negative.

G EXAMPLE 5   Using a Grapher to Determine the Type of Solutions
Use a grapher to predict the number and type of solutions of the quadratic equations in Example 3. The associated functions are:
(a) $f(x) = 4x^2 - 5x + 1$
(b) $f(x) = 4x^2 - 12x + 9$
(c) $f(x) = 2x^2 + 3x + 2$

Solution   (a) Figure 2a shows the graph of f which crosses the x axis at two different points and the equation has two real unequal solutions.

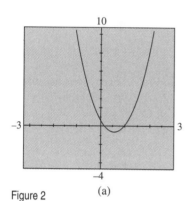

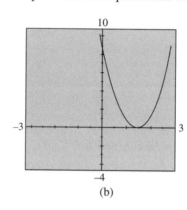

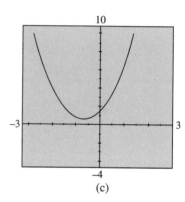

Figure 2   (a)   (b)   (c)

(b) Figure 2b shows the graph of f which crosses the x axis at one point and the equation has one real solution.

(c) Figure 2c shows the graph of f which does not cross the x axis and the equation has no real solutions. The solutions are complex conjugates.   ◈

◈ **PROBLEM SET 6.3**

Mastering the Concepts

In Problems 1–20, solve each equation by the quadratic formula. Approximate irrational solutions to two decimal places. Write complex solutions in the form $a + bi$.

1. (a) $x^2 - 5x + 4 = 0$
   (b) $x^2 - 2x - 3 = 0$
2. (a) $x^2 - 2x - 24 = 0$
   (b) $x^2 + 10x - 24 = 0$
3. (a) $3x^2 + 8x + 4 = 0$
   (b) $3x^2 - 19x + 20 = 0$
4. (a) $6x^2 - 13x - 8 = 0$
   (b) $6x^2 + 11x + 4 = 0$
5. $4t^2 - 8t + 3 = 0$
6. $2u^2 - 5u - 3 = 0$
7. $6y^2 - 7y - 5 = 0$
8. $6p^2 + 3p - 5 = 0$
9. $6z^2 - 7z - 2 = 0$
10. $2x^2 - 5x + 1 = 0$
11. $5y^2 - 2y + 7 = 0$
12. $9m^2 - 2m + 11 = 0$
13. $12p^2 - 4p + 3 = 0$
14. $15y^2 + 2y - 8 = 0$
15. $3x^2 - x + 7 = 0$
16. $6x^2 + 17x - 14 = 0$
17. $t(t + 8) + 8(t - 8) = 0$
18. $(t - 2)(t - 3) = 5$
19. $(y - 1)(y - 5) = 9$
20. $(t + 1)(2t - 1) = 4$

In Problems 21–28, use the discriminant to determine the number and type of solutions to each equation.

21. $x^2 + 6x - 7 = 0$
22. $x^2 - 9x + 20 = 0$
23. $t^2 + 3t + 5 = 0$
24. $x^2 + 15x + 37 = 0$
25. $9z^2 + 30z + 25 = 0$
26. $2x^2 + x + 5 = 0$
27. $2x^2 + x - 5 = 0$
28. $20x^2 - 11x + 3 = 0$

Applying the Concepts

29. **Free Falling Object:** An object is tossed vertically upward from a platform 4 feet above the ground. The height h (in feet) above the ground of the object after t seconds is given by the model

$$h(t) = -16t^2 + 192t + 4, 0 \le t \le 12$$

Use the quadratic formula to determine how many seconds it takes the object to reach the following heights:
(a) 500 feet   (b) 580 feet   (c) 600 feet

30. **Insect Population:** A biologist determines that the population P of insects in an experiment after t weeks is given by the model

$$P(t) = 100 + 24t + 2t^2, 0 \le t \le 40$$

When will the number of insects reach 3000?

31. **Height of a Diver:** During the 1992 Olympic Summer Games in Barcelona, Spain, Mark Lenzi of the United States won a gold medal in the springboard diving competition. If Mark springs into the air at an initial speed of 6.1 meters per second from a diving board 10 meters above the surface of the water, his height h (in meters) after t seconds is given by the model

$$h(t) = -4.88t^2 + 6.1t + 10, t \ge 0$$

After how many seconds is Mark 11 meters above the surface of the water?

32. **Water in a Pond:** The volume of water (in cubic meters) in an irrigation pond after t hours of pumping from a well is approximated by the model

$$V(t) = 3t^2 - 14t, t \ge 5$$

For how long must the water be pumped so that the pond will contain 185 cubic meters of water?

33. **Revenue Function:** Sometimes businesses can increase their revenues by lowering their prices. As they lower their prices, the number of units sold increases. Assume that a company's sales of its cellular telephones generate daily revenue R in dollars according to the function

$$R(x) = (128 - 0.1x) x, x \ge 0$$

where x is the quantity sold. Use the quadratic formula to find the quantity x the company must sell per day to have a daily revenue of $40,960.

34. **Profit Function:** Cellular phone sales (see Problem 33) generate a daily profit P in dollars according to the function

$$P(x) = (128 - 0.1x)x - (36x + 1000), x \ge 0$$

(a) Find the break-even point; that is, the level of sales that generates no profit and no loss. (Hint: Solve for x when $P(x) = 0$)

(b) Use the quadratic formula to find the sales x that will yield a daily profit of $20,160.

Developing and Extending the Concepts

In Problems 35 and 36, the solution of a quadratic equation is given.

(a) Compare the values to the symbols a, b, and c used in the quadratic formula

$$x = \frac{(-b \pm \sqrt{b^2 - 4ac})}{2a}$$

and identify a, b, and c.

(b) Evaluate the expression without the use of a calculator.

(c) Substitute a, b, and c in the equation $ax^2 + bx + c = 0$ to find the original equation.

35. (a) $x = \dfrac{9 \pm \sqrt{9^2 - 4(1)(8)}}{2(1)}$

(b) $x = \dfrac{-(-4) \pm \sqrt{(-4)^2 - 4(1)(3)}}{2(1)}$

36. (a) $x = \dfrac{-(-5) \pm \sqrt{(-5)^2 - 4(3)(2)}}{2(3)}$

(b) $x = \dfrac{-(-7) \pm \sqrt{(-7)^2 - 4(3)(-6)}}{2(3)}$

In Problems 37–40, use the quadratic formula to solve each equation. Round off the answers to two decimal places.

37. $0.17u^2 - 0.55u - 3.87 = 0$

38. $1.47y^2 - 3.82y - 5.71 = 0$

39. $1.32x^2 + 2.78x - 9.321 = 0$

40. $8.84x^2 - 71.41x - 94.03 = 0$

In Problems 41–44, use the quadratic formula to solve each equation for the indicated unknown.

41. $nx^2 + mnx - m^2 = 0$, for x.

42. $2ay^2 - 7ay - 4ab^2 = 0$, for y.

43. $LI^2 + RI + \dfrac{1}{C} = 0$, for I

44. $Mx^2 + 2Rx + K = 0$, for x.

In Problems 45–48, find the value of k so that the quadratic equation will have:

(a) one real solution,

(b) two real solutions, or

(c) no real solutions.

45. $-2x^2 - 3x + k = 0$ 46. $-4x^2 + k = 0$

47. $kx^2 - 4x + 1 = 0$ 48. $kx^2 - 15x + 5 = 0$

In Problems 49 and 50, use the fact that

$$x_1 = \frac{-b + \sqrt{b^2 - 4ac}}{2a} \text{ and } x_2 = \frac{-b - \sqrt{b^2 - 4ac}}{2a}$$

are solutions to the quadratic equation $ax^2 + bx + c = 0$, $a \ne 0$, to answer parts (a) and (b).

(a) Find the sum of the solutions, x_1 and x_2, and show that this sum is equal to

$$-\frac{b}{a}.$$

(b) Find the product of the solutions, x_1 and x_2, and show that this product is equal to

$$\frac{c}{a}.$$

49. $3x^2 - 5x - 2 = 0$; $x_1 = -1/3$ and $x_2 = 2$.

50. $2x^2 + 9x - 5 = 0$; $x_1 = 1/2$ and $x_2 = -5$.

In Problems 51–54, write a quadratic equation with integral values for a, b, and c, having the given solutions. Check your answers by using the results in parts (a) and (b) of the instructions for Problems 49 and 50.

51. (a) -6 and 3
    (b) -4 and -1

52. (a) 2 is the only solution.
    (b) -3 is the only solution.

53. (a) $-\sqrt{5}$ and $\sqrt{5}$
    (b) $2i$ and $-2i$

54. (a) $3 - 2i$ and $3 + 2i$
    (b) $3 - \sqrt{11}$ and $3 + \sqrt{11}$

In Problems 55–58, match the function with its graph. Also,

(a) find the x intercepts of each quadratic function.
(b) Find the coordinates of the lowest or highest points on the graph by writing each function in the form $f(x) = a(x - h)^2 + k$.

55. $f(x) = x^2 - 6x - 8$ 56. $f(x) = -x^2 + 6x + 8$
57. $f(x) = 2x^2 - 5x - 4$ 58. $f(x) = -x^2 + 5x + 4$

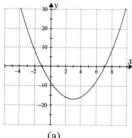

(a)

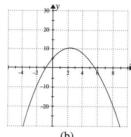

(b)

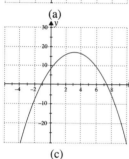

(c)

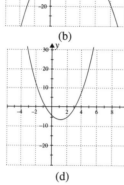

(d)

G In Problems 59–62, use a grapher to graph the functions f and g in the same viewing window. Then use the ZOOM and TRACE features to estimate where the graphs of f and g intersect. Confirm your results by setting $f(x) = g(x)$ and solve the resulting equation by the quadratic formula.

59. (a) $f(x) = 2x^2 - 8x + 4$
    (b) $g(x) = -2x^2 + 8x - 4$
60. (a) $f(x) = -5x^2 + 6x - 1$
    (b) $g(x) = 5x^2 - 6x + 1$
61. (a) $f(x) = 3x^2 - x - 1$
    (b) $g(x) = -3x^2 + x + 1$
62. (a) $f(x) = 2x^2 - x - 4$
    (b) $g(x) = -2x^2 + x + 4$

OBJECTIVES

1. Solve Equations Involving Rational Expressions
2. Solve Equations Involving Radicals
3. Solve Equations Reducible to Quadratic Form
4. Solve Applied Problems
5. Use Technology Exploration

6.4 Equations Reducible to Quadratic Form

In this section, we discuss various types of equations that can be solved by using methods for solving quadratic equations. These equations are: *rational equations, radical equations,* and *equations quadratic in form.*

Solving Equations Involving Rational Expressions

Recall from section 2.4 that the simplest way to solve most equations involving rational expressions is to multiply each side of the equation by the least common denominator (LCD) thereby eliminating the fractions from the equation. Remember, *whenever a variable appears in the denominator of any rational expression, we must check the proposed answer in the original equation for excluded values that cause division by zero.* In the following example, we solve an equation involving rational expressions that simplifies to a quadratic equation.

EXAMPLE 1   Solving an Equation with Rational Expressions
Solve the equation.

$$\frac{6}{x - 1} + \frac{6}{x} = 5$$

Solution   First we multiply each side of the equation by the LCD $= x(x - 1)$. Remember, any value of the variable that makes the denominator zero is not allowed. Thus, $x = 0$ and $x = 1$ cannot be solutions of the original equation.

$$\frac{6}{x - 1} + \frac{6}{x} = 5 \qquad \text{Original equation}$$

$$x(x - 1)\left[\frac{6}{x - 1} + \frac{6}{x}\right] = 5x(x - 1) \qquad \text{Multiply each side by the } LCD$$

$$6x + 6(x - 1) = 5x(x - 1) \qquad \text{Use the distributive property}$$

$$6x + 6x - 6 = 5x^2 - 5x \qquad \text{Simplify}$$

$$5x^2 - 17x + 6 = 0 \qquad \text{Write in standard form}$$

$$(5x - 2)(x - 3) = 0 \qquad \text{Factor the left side}$$

$$5x - 2 = 0 \quad \text{or} \quad x - 3 = 0 \qquad \text{Apply the zero factor property}$$

$$5x = 2 \quad \bigg| \quad x - 3 = 0$$

$$x = \frac{2}{5} \quad \bigg| \qquad x = 3$$

Since neither result causes division by zero in the original equation, the solutions are 2/5 and 3. These solutions should be checked in the original equation. ◈

Solving Equations Involving Radicals

In section 5.6, we solved simple equations involving radicals. Here, we solve radical equations that lead to quadratic equations. Recall that the strategy for solving equations with radicals includes isolating the radicals and then raising both sides of the equation to a power that will clear the radicals. *For even roots, we must check for extraneous solutions.*

In the next example, we will square both sides of the equation twice in order to eliminate the radical expressions.

EXAMPLE 2   Solving an Equation Involving Radicals
Solve the equation $\sqrt{3x - 2} - \sqrt{x} = 2$

Solution   The equation will be easier to solve with the radicals on opposite sides of the equation.

$$\sqrt{3x - 2} - \sqrt{x} = 2 \qquad \text{Original equation}$$

$$\sqrt{3x - 2} = \sqrt{x} + 2 \qquad \text{Isolate one radical}$$

$$(\sqrt{3x - 2})^2 = (\sqrt{x} + 2)^2 \qquad \text{Square each side of the equation}$$

$$3x - 2 = x + 4\sqrt{x} + 4 \qquad \text{Simplify}$$

$$2x - 6 = 4\sqrt{x} \qquad \text{Isolate the radical}$$

$$x - 3 = 2\sqrt{x} \qquad \text{Divide each side by 2}$$

$$(x - 3)^2 = (2\sqrt{x})^2 \qquad \text{Square each side of the equation}$$

$$x^2 - 6x + 9 = 4x \qquad \text{Simplify}$$

$$x^2 - 10x + 9 = 0 \qquad \text{Write in standard form}$$

$$(x - 9)(x - 1) = 0 \qquad \text{Factor the left side}$$

$$x - 9 = 0 \text{ or } x - 1 = 0 \qquad \text{Apply the zero factor property}$$

$$x = 9 \qquad x = 1$$

Check:

If $x = 9$, then	If $x = 1$, then
$\sqrt{3(9) - 2} - \sqrt{9}$	$\sqrt{3(1) - 2} - \sqrt{1}$
$\quad = \sqrt{25} - 3$	$\quad = \sqrt{1} - 1$
$\quad = 5 - 3 = 2$	$\quad = 1 - 1 = 0 \neq 2$

Therefore, 9 is a solution and 1 is an extraneous solution. ◈

Solving Equations Reducible to Quadratic Form

Some equations are not quadratic but can be treated as quadratic equations provided that a suitable substitution is made. After making the appropriate substitution, we can use methods for solving quadratic equations to solve the resulting equations. The following table shows some examples of equations that can be written in quadratic form.

Original Equation	Suggested Substitution	Equivalent Quadratic Equation
$x^4 - 5x^2 + 4 = 0$	$u = x^2, u^2 = x^4$	$u^2 - 5u + 4 = 0$
$(x - 2)^2 - 3(x - 2) - 4 = 0$	$u = x - 2, u^2 = (x - 2)^2$	$u^2 - 3u - 4 = 0$
$1/t^2 - 5/t + 4 = 0$	$u = 1/t, u^2 = 1/t^2$	$u^2 - 5u + 4 = 0$

EXAMPLE 3 Solving Equations Quadratic in Form

Solve each equation.

(a) $x^4 - 5x^2 + 4 = 0$

(b) $(x - 2)^2 - 3(x - 2) - 4 = 0$

Solution (a) Let $u = x^2$, $u^2 = x^4$, then substitute in the original equation.

$$x^4 - 5x^2 + 4 = 0 \qquad \text{Original equation}$$
$$u^2 - 5u + 4 = 0 \qquad \text{Substitute } u \text{ for } x^2$$
$$(u - 1)(u - 4) = 0 \qquad \text{Factor left side}$$
$$u - 1 = 0 \text{ or } u - 4 = 0 \qquad \text{Apply zero property}$$
$$u = 1 \qquad u = 4$$
$$x^2 = 1 \qquad x^2 = 4 \qquad \text{Replace } u \text{ by } x^2$$
$$x = \pm 1 \qquad x = \pm 2$$

The solutions are

$$-2, -1, 1, \text{ and } 2.$$

A common error is to solve for u and forget to solve for x.

(b) Let $u = x - 2$ and $u^2 = (x - 2)^2$, then substitute

$$(x - 2)^2 - 3(x - 2) - 4 = 0 \qquad \text{Original equation}$$
$$u^2 - 3u - 4 = 0 \qquad \text{Substitute } u \text{ for } x - 2$$
$$(u + 1)(u - 4) = 0 \qquad \text{Factor the left side}$$
$$u + 1 = 0 \qquad u - 4 = 0 \qquad \text{Apply the zero property}$$
$$u = -1 \qquad u = 4$$
$$x - 2 = -1 \qquad x - 2 = 4 \qquad \text{Replace } u \text{ by } x - 2$$
$$x = 1 \qquad x = 6$$

The solutions are

$$1 \text{ and } 6.$$

◈

Quite often when we make a substitution, we place a restriction on the new variable. This way we avoid possible extraneous solutions. The following table lists some restrictions imposed on the substitution.

Original Equation	Substitution	Restriction	New Equation
$x - 15\sqrt{x} = 100$	$u = \sqrt{x}$	$u \geq 0$	$u^2 - 15u = 100$
$28/x^2 + 1/x - 2 = 0$	$u = 1/x$	$u \neq 0$	$28u^2 + u - 2 = 0$
$1/x - 1/\sqrt{x} - 12 = 0$	$u = 1/\sqrt{x}$	$u > 0$	$u^2 - u - 12 = 0$

EXAMPLE 4   Solving an Equation with an Extraneous Solution

Solve the equation $x - 15\sqrt{x} = 100$

Solution   Let $u = \sqrt{x}$ and $u^2 = x$. Then substitute in the original equation.

$x - 15\sqrt{x} = 100$ Original equation

$u^2 - 15u - 100 = 0$ Substitute u for $\sqrt{x}$
$(u - 20)(u + 5) = 0$ Factor left side
$u - 20 = 0$ or $u + 5 = 0$ Apply zero factor property
$\qquad u = 20 \qquad\qquad u = -5$

$\qquad \sqrt{x} = 20 \qquad\qquad \sqrt{x} = -5$ Replace u by $\sqrt{x}$

$\qquad x = 400 \qquad\qquad x = 25$

A check shows that 25 is not a solution.
The restriction on u, $u \geq 0$, prohibits using $u = -5$ but allows $u = 20$.
Therefore, there is only one solution, namely $x = 400$.

Solving Other Equations Reducible to Quadratic Form

The following example is an equation involving rational exponents that can be reduced to equations which are quadratic in form.

EXAMPLE 5   Solving Equations with Rational Exponents

Solve the equation

$$z^{\frac{2}{3}} - z^{\frac{1}{3}} - 12 = 0.$$

Solution   Let

$$u = z^{\frac{1}{3}} \text{ and } u^2 = \left(z^{\frac{1}{3}}\right)^2 = z^{\frac{2}{3}}.$$

Since this substitution involves cube roots there are no restrictions.

$z^{2/3} - z^{1/3} - 12 = 0$ Original equation
$u^2 - u - 12 = 0$ Substitute u for $z^{1/3}$
$(u + 3)(u - 4) = 0$ Factor the left side
$u + 3 = 0$ or $u - 4 = 0$ Apply the zero factor property
$\qquad u = -3 \qquad\qquad u = 4$
$\qquad z^{1/3} = -3 \qquad\qquad z^{1/3} = 4$ Replace u by $z^{1/3}$
$\qquad z = (-3)^3 \qquad\qquad z = 4^3$
$\qquad z = -27 \qquad\qquad z = 64$

Therefore, the solutions are

$$-27 \text{ and } 64.$$

Solving Applied Problems

Now we examine applications and models that arise from equations that are quadratic in form.

EXAMPLE 6   Solving an Earth Science Application

In Earth Science it is determined that at an elevation h (in meters) above sea level, the temperature T (in Celsius) at which water boils is given by the model

$$h(T) = 580 (100 - T)^2 + 1000 (100 - T), \quad 95° \leq T \leq 100°$$

(a) Using this model, determine at what temperature will water boil at sea level.

(b) At what temperature will water boil at the top of a mountain 4320 meters above sea level?

Solution   (a) At sea level, $h = 0$, we are to solve the equation

$$0 = 580(100 - T)^2 + 1000(100 - T)$$
$$0 = 580u^2 + 1000u \qquad \text{Let } u = 100 - T$$
$$58u^2 + 100u = 0 \qquad \text{Divide each side by 10}$$
$$u(58u + 100) = 0 \qquad \text{Factor the left side}$$
$$u = 0 \quad \text{or} \quad 58u + 100 = 0$$
$$u = 0 \qquad\qquad u = -1.72$$
$$100 - T = 0 \quad \text{or} \quad 100 - T = -1.72$$
$$T = 100 \qquad\qquad T = 101.72$$

We reject $T = 101.72$ since it is not within the restriction on the model. The temperature at which water boils at sea level is 100°C.

(b) At the top of a mountain 4320 meters high, we have

$$4320 = 580(100 - T)^2 + 1000(100 - T)$$
$$4320 = 580u^2 + 1000u \qquad \text{Let } u = 100 - T$$
$$29u^2 + 50u - 216 = 0 \qquad \text{Divide each side by 20}$$
$$u = \frac{-b \pm \sqrt{b^2 - 4ac}}{2a}$$
$$u = \frac{-50 \pm \sqrt{(50)^2 - 4(29)(-216)}}{2(29)}$$
$$u = -3.72 \text{ or } 2$$
$$100 - T = -3.72 \qquad 100 - T = 2$$
$$T = 103.72 \qquad\qquad T = 98$$

We reject 103.72 since it is not in the restriction of the model. Therefore, the temperature at which water boils on top of the mountain is 98°C. ◇

Using Technology Exploration

In example 3(a), we solved the equation

$$x^4 - 5x^2 + 4 = 0$$

and obtained its four solutions algebraically. Now we use a grapher to examine the graph of the associated function

$$f(x) = x^4 - 5x^2 + 4$$

and approximate its x intercepts.

G EXAMPLE 7   Using a Grapher

Using a grapher, produce the graph of the function

$$f(x) = x^4 - 5x^2 + 4$$

on a given viewing window. Use the solutions of the equation $x^4 - 5x^2 + 4 = 0$ in example 3(a) to label the x intercepts of the graph of f.

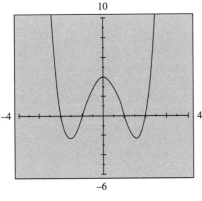

Figure 1

Solution   The x intercepts are found by setting $f(x) = 0$ and solving the resulting equation $x^4 - 5x^2 + 4 = 0$. The solutions of this equation are
$$-2, -1, 1 \text{ and } 2$$
(see example 3(a)). These solutions are the x intercepts of the graph in Figure 1.   ◇

The advantage in using a grapher to solve equations involving radicals is that there are no extraneous solutions on the graph. For example, let us solve Example 2 by graphing the associated function.

G **EXAMPLE 8**   Using a Grapher
Use a grapher to solve the equation $\sqrt{3x - 2} - \sqrt{x} = 2$ from Example 2.

Solution   The associated function is
$$f(x) = \sqrt{3x - 2} - \sqrt{x} - 2$$
and the x intercept of the graph of this function is the solution of the original equation (Figure 2). Using the ZOOM and TRACE features, we see that the graph crosses the x axis only at 9 which is the solution of the equation. Notice that the algebraic process in Example 2 shows
$$x = 1 \text{ and } x = 9$$
as solutions (1 is an extraneous solution); this graph only shows 9 as the solution of the equation.   ◇

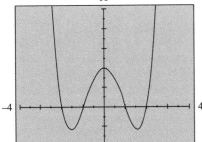

Figure 2

◇ PROBLEM SET 6.4

Mastering the Concepts

In Problems 1–6, solve each equation. Check your results.

1. $\dfrac{12}{x} - 7 = \dfrac{12}{1 - x}$

2. $\dfrac{1}{1 - x} + \dfrac{1}{2} = \dfrac{6}{x^2 - 1}$

3. $\dfrac{3 - 2x}{4x} - 4 = \dfrac{3}{4x - 3}$

4. $\dfrac{5}{4x + 16} - 1 = \dfrac{3}{4x - 8}$

5. $\dfrac{5}{x + 4} - \dfrac{3}{x - 2} = 4$

6. $\dfrac{2x - 5}{2x + 1} = \dfrac{7}{4} - \dfrac{6}{2x - 3}$

In Problems 7–14, solve each equation. Check your results.

7. $\sqrt{x} = \sqrt{x + 16} - 2$
8. $\sqrt{x + 2} = 5 - \sqrt{x - 3}$
9. $\sqrt{x + 12} = 2 + \sqrt{x}$
10. $\sqrt{6x + 7} = 1 + \sqrt{3x + 3}$
11. $\sqrt{y + 5} = 1 + \sqrt{y}$
12. $\sqrt{7 - 4t} = 1 + \sqrt{3 - 2t}$
13. $\sqrt{1 + 5x} = 1 + \sqrt{3x}$
14. $\sqrt{m + 4} + 1 = \sqrt{m + 11}$

In Problems 15–28, reduce each equation to quadratic form by making a suitable substitution. Check for extraneous solutions.

15. (a) $x^4 - 7x^2 + 12 = 0$
    (b) $y^{-2} - 7y^{-1} + 12 = 0$
    (c) $t - 7\sqrt{t} + 12 = 0$

16. (a) $x^4 - 13x^2 + 36 = 0$
    (b) $y^{-2} - 13y^{-1} + 36 = 0$
    (c) $z - 13\sqrt{z} + 36 = 0$

17. (a) $x^4 - 17x^2 + 16 = 0$
    (b) $x^{2/3} - 17x^{1/3} + 16 = 0$
    (c) $z - 17\sqrt{z} + 16 = 0$

18. (a) $y^4 - 10y^2 + 9 = 0$
    (b) $x^{2/3} - 10x^{1/3} + 9 = 0$
    (c) $z - 10\sqrt{z} + 9 = 0$

19. (a) $x^4 - 2x^2 + 1 = 0$
    (b) $y^6 - 2y^3 + 1 = 0$
    (c) $(3 - \sqrt{x})^2 - 2(3 - \sqrt{x}) + 1 = 0$

20. (a) $y^4 - 16y^2 + 64 = 0$
    (b) $y^6 - 16y^3 + 64 = 0$
    (c) $(t^2 - 8)^2 - 16(t^2 - 8) + 64 = 0$

21. $(x + 2)^2 = 9(x + 2) - 20$
22. $(3x - 1)^2 - 9(3x - 1) = 10$
23. $(y^2 + 2y)^2 - 2(y^2 + 2y) = 3$
24. $(y^2 + 2y)^2 - 14(y^2 + 2y) = 15$
25. $(3 - \sqrt{x})^2 + 2(3 - \sqrt{x}) = 3$
26. $(4 + \sqrt{x})^2 - 2(4 + \sqrt{x}) = 3$

27. $\left(3x - \dfrac{2}{x}\right)^2 + 6\left(3x - \dfrac{2}{x}\right) + 5 = 0$

28. $\left(y - \dfrac{5}{y}\right)^2 - 2\left(y - \dfrac{5}{y}\right) = 8$

Applying the Concepts

29. **Population:** The population P (in thousands) of a small town is expected to increase according to the mathematical model

$$P = 3t + 4 + \sqrt{3t + 4}, \, t \geq 0$$

where t is time in years. When will the population be 56,000?

30. **Travel Time:** A motorboat leaves a dock traveling at a constant rate making a 48 kilometer trip to an island downstream and then returns to the dock. Suppose that the current rate is 4 kilometers per hour. The time T (in hours) that it took for the entire trip is given by the equation

$$T = \dfrac{48}{r - 4} + \dfrac{48}{r + 4}$$

where r (in kilometers per hour) is the speed of the boat in still water. Find r when T is 5 hours.

Developing and Extending the Concepts

In Problems 31 and 32, solve each equation involving negative rational exponents.

31. $x^{1/2} - 1 - 12x^{-1/2} = 0$ 32. $x^{1/3} - 1 - 12x^{-1/3} = 0$

In Problems 33–40, find a substitution that will create a quadratic equation, and then solve the equation.

33. $x^2 + 6x - 6(x^2 + 6x - 2)^{1/2} = -3$

34. $2x^2 + x - 4(2x^2 + x + 4)^{1/2} = 1$

35. $\dfrac{m}{m - 1} - 4 = \dfrac{5m - 5}{m}$

36. $\dfrac{p^2}{p + 2} + \dfrac{2p + 4}{p^2} = 3$

37. $\dfrac{m + 1}{m} + 2 = \dfrac{3m}{m + 1}$

38. $\dfrac{q^2 + 2q + 1}{q^2} - \dfrac{2q}{q + 1} = \dfrac{6q}{q + 1}$

39. $\dfrac{x^2 + 1}{x} + \dfrac{4x}{x^2 + 1} - 4 = 0$

40. $\left(\dfrac{2x^2 + 1}{x}\right)^2 + 5\left(\dfrac{2x^2 + 1}{x}\right) + 6 = 0$

In Problems 41–48, find the x intercepts of the graph of each function. That is, find x when $f(x) = 0$.

41. $f(x) = x - 2\sqrt{x} - 8$
42. $f(x) = x - 10\sqrt{x} + 9$
43. $f(x) = x^4 - 29x^2 + 100$
44. $f(x) = x^4 - 17x^2 + 16$
45. $f(x) = (2x + 3)^2 - 18(2x + 3) + 65$
46. $f(x) = (5x - 1)^2 - 13(5x - 1) + 30$
47. $f(x) = x^{2/3} - 8x^{1/3} + 15$
48. $f(x) = x^{2/3} - 2x^{1/3} - 8$

49. Let $f(x) = \sqrt{2x - 1} + \sqrt{x + 3}$, find x such that $f(x) = 3$.

50. Let $f(x) = \sqrt{5x - 1} + \sqrt{x + 3}$, find x such that $f(x) = 4$.

G In Problems 51–56,
(a) find the x intercepts of the graph of each function algebraically. That is, find x when $f(x) = 0$.
(b) Label the x intercepts found in part (a) on the given graph of each function.
(c) Use a grapher with the given viewing window and the **ZOOM** and **TRACE** features to estimate the x intercepts of each graph. Do these answers agree with the answers in part (a)?

51. $f(x) = x^4 - 18x^2 + 81$ 52. $f(x) = -x^4 + 5x^2 - 4$

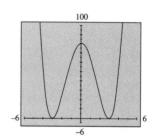

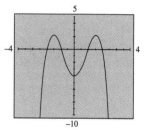

53. $f(x) = -x^4 + 8x^2 + 9$ 54. $f(x) = x^4 - 12x^2 + 27$

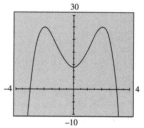

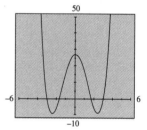

55. $f(x) = \sqrt{5x + 6} + \sqrt{3x - 2} - 6$

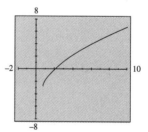

56. $f(x) = \sqrt{3x + 7} + \sqrt{x + 2} - 1$

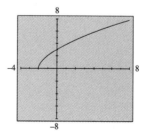

OBJECTIVES
1. Solve a Variety of Applied Problems
2. Solve Models by Using the Pythagorean Theorem
3. Solve Models Involving Rational Expressions

6.5 Quadratic Applications and Models

So far, we have solved many applied problems that were modeled by a given quadratic equation. In this section, we examine additional applied problems where the models have to be created, and the techniques to solve them that were developed in section 2.3. Recall the strategies developed by G. Polya for using mathematical models to solve applied problems.

> 1. Make a plan
> 2. Write the solution
> 3. Look back

Solving a Variety of Applied Problems

In working these applied problems, the emphasis should be on writing the mathematical models.

EXAMPLE 1   Solving a Number Problem

Find two consecutive odd, positive integers whose product is 195.

Solution   Make a Plan

The goal is to find two positive odd integers which differ by 2 and have a product of 195. Any negative number is not part of the domain of the application. Let x = the first odd integer, then
$x + 2$ = the next odd integer.
The following diagram is a visual description of the given.

First odd positive integer		Second odd positive integer		Product of the two integers
	•		=	
x	•	$x + 2$	=	195

Translating the problem into an equation, we have
$$x(x + 2) = 195$$

Write the Solution

Now we solve the equation.

$x(x + 2) = 195$	Original equation
$x^2 + 2x = 195$	Use the distributive property
$x^2 + 2x - 195 = 0$	Write in standard form
$(x + 15)(x - 13) = 0$	
$x + 15 = 0$ or $x - 13 = 0$	Apply the zero property
$x = -15 \qquad x = 13$	

We reject the solution -15 because it is not a positive integer. Since $x = 13$ and $x + 2 = 15$, the two consecutive odd integers are 13 and 15.

Look Back

The numbers, 13 and 15, are two consecutive positive odd integers. Their product, $(13)(15)$, is 195.   ◈

EXAMPLE 2   Solving a Geometric Model

An artist is planning an acrylic painting in the form of a rectangle whose dimensions are 2 feet by 3 feet, surrounded by a stark border of uniform width. To achieve vitality, the artist wants the total area of the rectangle including the border to be 12 square feet. Find the width of the border that will accomplish this.

Solution   Make a Plan

Let x (in feet) represent the width of the border. Figure 1 shows the painting and the border. Recall that the area, A, of a rectangle is

$$A = lw$$

here $2 + 2x = w$, the width and $3 + 2x = l$, the length. The following diagram is a visual description of the given.

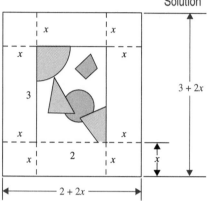

Figure 1

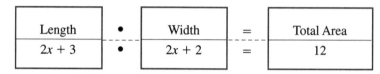

Length	•	Width	=	Total Area
$2x + 3$	•	$2x + 2$	=	12

Translating the problem into an equation, we have

$$(2x + 3)(2x + 2) = 12$$

Write the Solution

$(2x + 3)(2x + 2) = 12$	Original equation
$4x^2 + 10x + 6 = 12$	Simplify
$4x^2 + 10x - 6 = 0$	Write in standard form
$2x^2 + 5x - 3 = 0$	Divide each side by 2
$(2x - 1)(x + 3) = 0$	Factor
$2x - 1 = 0 \quad \text{or} \quad x + 3 = 0$	Apply the zero property
$x = 1/2 \quad \text{or} \quad x = -3$	

Since the width of the border cannot be negative, the width of the border must be 1/2 foot.

Look Back

The total area of the painting and the border is given by

$$(2(\tfrac{1}{2}) + 3)(2(\tfrac{1}{2}) + 2) = (4)(3) = 12$$

Solving Models by Using the Pythagorean Theorem

Recall the Pythagorean theorem. In any right triangle, the sum of the squares of the legs is equal to the square of the hypotenuse. Using the right triangle in Figure 2, we have,

$$a^2 + b^2 = c^2$$

Figure 2

EXAMPLE 3   Finding the Lengths of the Rafters of a Solar Collector

In order to support a solar panel collector at an angle that will make it most efficient, the roof trusses of a house are designed as right triangles. The rafters of the truss are the legs of the right triangle, and the base of the truss is the hypotenuse (Figure 3a). A solar collector is mounted on the same side as the shorter rafter. If that rafter is 10 feet shorter than the other rafter, and the base of each truss is 50 feet long, how long are the rafters?

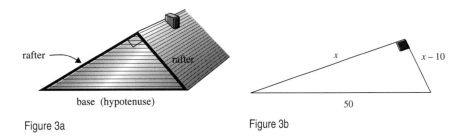

rafter

rafter

base (hypotenuse)

x

$x - 10$

50

Figure 3a

Figure 3b

Solution   Make a Plan

First, we must make sure that we understand the problem. The rafters are connected to each other at a right angle, making the base of the triangular truss the hypotenuse of a right triangle. The relationship between these sides is described by the Pythagorean theorem. The rafters are not the same size, one rafter being 10 feet shorter than the other.

Let x = the length of the long rafter and $x - 10$ = the length of the short rafter. Using the Pythagorean theorem with $a = x$, $b = x - 10$, and $c = 50$ (Figure 3b), we form the following equation:

$$a^2 + b^2 = c^2$$
$$x^2 + (x - 10)^2 = 50^2$$

Write the Solution

$$
\begin{aligned}
x^2 + (x - 10)^2 &= 50^2 && \text{Original equation} \\
x^2 + x^2 - 20x + 100 &= 2500 && \text{Simplify} \\
2x^2 - 20x - 2400 &= 0 && \text{Write in standard form} \\
x^2 - 10x - 1200 &= 0 && \text{Divide each side by 2} \\
(x + 30)(x - 40) &= 0 && \text{Factor}
\end{aligned}
$$

$$x + 30 = 0 \quad \text{or} \quad x - 40 = 0$$
$$x = -30 \quad \text{or} \quad x = 40$$

Because the length must be positive, the length of the long rafter is 40 feet and the length of the short rafter is $x - 10 = 40 - 10 = 30$ feet.

Look Back

Notice that $30^2 + 40^2 = 900 + 1600 = 2500 = 50^2$ and 50 feet is the length of the base. ◈

Solving Models Involving Rational Expressions

The next two examples are concerned with interest and distance. They are appropriately placed in this section because the rational expressions involved yield quadratic equations. They are analogous since the formula for interest, $I = Prt$, and the formula for distance, $d = rt$ involve the same mathematical operations.

EXAMPLE 4   Solving an Investment Problem

In planning a $10,000 investment portfolio, an investor decides to invest in two one-year ventures. In one year, the first earned $300 in interest and the second earned $500 in interest. If the interest rate on the second venture is 1% higher than that of the first, what is each interest rate?

Solution   Make a Plan

The simple interest formula is given by

$$I = Prt, \text{ if } t = 1, \text{ then } I = Pr \text{ and } P = \frac{I}{r}.$$

The following table summarizes the given information for $t = 1$ year:
    Let $r =$ the interest rate for the first investment and
$r + 0.01 =$ the interest rate for the second investment.

Investment	Interest (dollars)	Rate (percent)	Principal (dollars)
First	$300	r	$300/r$
Second	$500	$r + 0.01$	$500/(r + 0.01)$

The total investment of $10,000 is the sum of the two principals. The following diagram is a visual description of the given.

Principal in first venture	+	Principal in second venture	=	Total Principal
$\dfrac{300}{r}$	+	$\dfrac{500}{r + 0.01}$	=	$10,000

Translating the problem into an equation, we have

$$\frac{300}{r} + \frac{500}{r + 0.01} = 10,000$$

Write the Solution

$$\frac{300}{r} + \frac{500}{r + 0.01} = 10,000 \qquad \text{Original equation}$$

$$r(r + 0.01)\left[\frac{300}{r} + \frac{500}{r + 0.01}\right] = 10,000r(r + 0.01) \qquad \begin{array}{l}\text{Multiply by the}\\ \text{LCD, } r(r + 0.01)\end{array}$$

$$300(r + 0.01) + 500r = 10,000r(r + 0.01) \qquad \text{Simplify}$$
$$300r + 3 + 500r = 10,000r^2 + 100r \qquad \begin{array}{l}\text{Use the distributive}\\ \text{property}\end{array}$$

$$10,000r^2 - 700r - 3 = 0 \qquad \text{Write in standard form}$$

$$r = \frac{-b \pm \sqrt{b^2 - 4ac}}{2a}$$

$$r = \frac{-(-700) \pm \sqrt{(-700)^2 - 4(10000)(-3)}}{2(10000)}$$

$$r \approx -0.004 \text{ or } r \approx 0.074$$

Therefore, the interest rate for the first investment is $0.074 = 7.4\%$ and the rate for the second is $0.074 + 0.01 = 0.084 = 8.4\%$. The negative value for r is not appropriate for this situation.

Look Back

To check, $300/0.074 + 500/(0.074 + 0.01) \approx 10{,}006 \approx 10{,}000$

EXAMPLE 5   Solving a Rate Problem

A bicyclist trained for a 100 kilometer race every weekend. Her average speed for the second weekend was 5 kilometers per hour faster than her average speed for the first. By doing this, she cut her riding time by 40 minutes. What was her average speed for the first weekend?

Solution   Make a Plan

Solving the distance formula $d = rt$, for t yields $t = d/r$.

Let r = rate in kilometers per hour for the first weekend, then $r + 5$ = rate in kilometers per hour for the second weekend.

The following table summarizes the given information.

	Distance (kilometers)	rate (kilometers/hour)	time (hours)
First weekend	100	r	$100/r$
Second weekend	100	$r + 5$	$100/(r + 5)$

We form an equation by noting the time saved is 40 minutes or 2/3 hour.

$$\frac{100}{r} - \frac{100}{r + 5} = \frac{2}{3}$$

Write the Solution

$$\frac{100}{r} - \frac{100}{r + 5} = \frac{2}{3} \qquad \text{Original equation}$$

$$3r(r + 5)\left[\frac{100}{r} - \frac{100}{r + 5}\right] = \frac{2}{3}3r(r + 5) \qquad \text{Multiply by the } LCD, \; 3r(r + 5)$$

$$300(r + 5) - 300r = 2r(r + 5) \qquad \text{Use the distributive property}$$

$$300r + 1500 - 300r = 2r^2 + 10r \qquad \text{Simplify}$$

$$2r^2 + 10r - 1500 = 0 \qquad \text{Write in standard form}$$

$$r^2 + 5r - 750 = 0 \qquad \text{Divide each side by 2}$$

$$(r + 30)(r - 25) = 0$$

$$r + 30 = 0 \quad \text{or} \quad r - 25 = 0 \qquad \text{Apply the zero property}$$

$$r = -30 \qquad \qquad r = 25$$

Therefore, the bicyclist's average speed was 25 kilometers per hour for the first week.

Look Back

To check,

$$\frac{100}{25} - \frac{100}{(25 + 5)} = 4 - \frac{100}{30}$$

$$= \frac{12}{3} - \frac{10}{3} = \frac{2}{3}.$$

Notice how we can use the mathematical techniques we learned earlier to help us solve equations applicable to new models. Here, we multiplied each side of the equation by the LCD to clear the fractions, and then we solved the resulting quadratic equation by factoring.

PROBLEM SET 6.5

Mastering the Concepts

1. **Number Problem:** (a) Find two consecutive positive integers whose product is 30. (b) Find two consecutive negative integers whose product is 56.

2. **Number Problem:** (a) Find two consecutive odd integers whose product is 63. (b) Find two consecutive even integers such that the sum of their squares is 100.

3. **Window Design:** An artist is designing a rectangular stained glass window in such a way that the length of the window is 3 inches less than twice its width. What are the dimensions of the window if its area is 740 square inches?

4. **Gardening:** A homeowner has 60 feet of fencing to enclose a rectangular vegetable garden next to her house, with the house forming one side of the garden. What length and width are needed to enclose an area of 418 square feet and use all the fencing?

5. **Border of a Pool:** A rectangular swimming pool 20 feet wide and 60 feet long has a concrete walkway of uniform width as a border. If the total area of the walkway is 516 square feet, how wide is the walkway (Figure 4)?

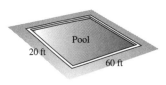

Figure 4

6. **Advertising:** A newspaper advertisement has the shape of a rectangle whose length is 2 centimeters more than its width. The newspaper charges $9 for each square centimeter. What are the dimensions of the ad if the total cost is $275?

7. **Television Antenna:** A rural homeowner has his television antenna held in place by three guy wires (Figure 5). Suppose that the distances to each of the stakes from the base of the antenna are the same. Also the distance from the base of the antenna to one stake is 7 feet shorter than the height of the antenna and the length of a guy wire is 13 feet. How long is the distance from the base of the antenna to a stake?

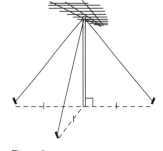

Figure 5

8. **Television Screen:** When we speak of a 20 inch T.V. set, we refer to the diagonal of the screen measuring 20 inches. Suppose the width and height of the screen (in inches) are w and h respectively (Figure 6).

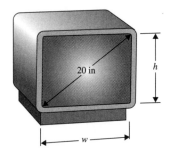

Figure 6

   (a) What is the width and height of a 20 inch T.V. set if its width is 4 inches more than its height?

   (b) What are the dimensions of a 25 inch T.V. set if its height is 5 inches less than its width?

9. **Camping Tent:** The entrance of a tent has the shape of an isosceles triangle. Suppose that the base of the entrance is 2 feet less than its height, and the area of the entrance is 24 square feet. Find the height of the triangular entrance.

10. **Geometry:** The sum of the lengths of one side of a right triangle and the hypotenuse is 32 centimeters. The other leg is 1 centimeter shorter than the hypotenuse. Find the length of each side of the triangle.

11. **Investment Portfolio:** An inheritance of $15,000 was invested in two one-year investments. In one year the first investment earned $540 in interest and the second earned $480 in interest. If the interest rate on the second investment is 2% higher than that of the first, what is each interest rate?

12. **Investment Portfolio:** An investor decided to invest $12,000 into two one-year investments, one a time deposit and the other in bonds. Suppose that the interest earned in one year on the time deposit is $560, and the interest earned in one year on the bonds is $450. If the interest rate on the bonds is 1% higher than that of the time deposit, what is each interest rate?

13. **Commuting:** A commuter takes a train to college 10 miles away from home. The return train travels along the same route at a rate 10 miles per hour faster. If the commuter spends a total of 50 minutes per day on the train, what is the rate of each train?

14. **Car Trips:** On a 50 mile trip, a motorist traveled 10 miles in heavy traffic and then 40 miles in less congested traffic. The average speed of the motorist in heavy traffic was 20 miles per hour less than the average speed in light traffic. What was each rate of speed if the total time for the trip is 1.5 hours?

Developing and Extending the Concepts

15. **Age Difference:** Two brothers were born in consecutive years. The product of their present ages is 156. How old are they now?

16. **Ecology:** In a predator-prey model, such as foxes and rabbits, the number of the prey (rabbits) is 10 times the number of predators (foxes). If the product of their populations is 36,000, what is the predator population?

17. **Gambling:** In a casino, two dice were thrown by a gambler. If the difference between the face values of the two dice is 2 and the product of their values is 24, what numbers appeared on the face of the two dice?

18. **Designing a Book:** A book designer determines that 88 square inches of print is needed on a page. In order to accomplish this, the height of the page should be two-thirds of the width. In addition, the top and bottom margins are to be 1 inch each, and the side margins are to be 2 inches each. What are the dimensions of the page?

19. **Drainage Pipe:** The cross-sectional area of a circular drainage pipe under a road is 260 square inches.
    (a) What is the diameter of the pipe in inches?
    (b) What is the diameter of the pipe in feet?

20. **Geometry:** The surface area A of a box with a square base whose width is w and height l is given by the formula
$$A = 2w^2 + 4lw$$
Find the width w of the base if $l = 5$ inches and $A = 69$ square inches (Figure 7).

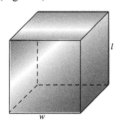

Figure 7

21. **Geometry:** The lengths of the shorter side, the longer side, and the hypotenuse of a right triangle are represented by consecutive even integers. Find the length of each side of the triangle.

22. **Geometry:** If increasing the lengths of the sides of a square by 8 feet results in a square whose area is 25 times the area of the original square, what was the length of a side of the original square?

23. **Microwave Oven:** Suppose that the width and height of the interior of a microwave oven are each 1 foot less than its length. What is the length of the interior of the oven if its interior capacity is 2 cubic feet?

24. **Rink Area:** A rectangular skating rink was originally 100 feet long and 70 feet wide. Due to increased demand, the owners decided to enlarge the rink to 13,000 square feet. If they add rectangular strips of equal width to one side and one end, maintaining the rectangular shape of the rink, how wide should these strips be?

25. **Golden Rectangle and Golden Ratio:** A **golden rectangle** is a rectangle possessing proportions in a certain ratio w/l, called the **golden ratio**, which has been of interest to artists and architects. The ancient Greeks used this ratio in their architecture even before the building of the Parthenon in Greece in the fifth century B.C. (Figure 8a).

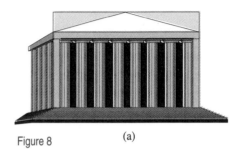

Figure 8   (a)

For instance, in a golden rectangle of width w and length l, the condition under which w/l is the golden ratio is that
$$\frac{w}{l} = \frac{l - w}{w}$$

(a) The ratio w/l in the rectangle in Figure 8(b) is $1/x$. For the small rectangle, the ratio of w/l is $(x - 1)/1$. Use this relation to find the exact golden ratio.

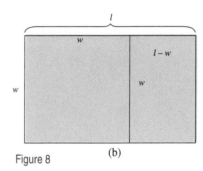

Figure 8   (b)

(b) Find the length l that yields a golden rectangle with a width of 5 inches.

26. **Geometry:** A diagonal of a polygon is a line segment between two nonadjacent vertices. Figure 9 shows all the diagonals of a polygon with eight vertices. A polygon with n vertices has a total of $[n(n - 3)]/2$ diagonals for $n \geq 3$.

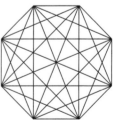

Figure 9

(a) How many vertices are on a polygon with 14 diagonals?
(b) Is there a polygon that has 17 diagonals?

27. **Compound Interest:** A firm invests $10,000 in a savings account for 2 years at a rate of interest r compounded annually. If a total of $12,544 is in the account at the end of the second year, what is the annual interest rate r? (See problem 42 on page 281.)

28. **Surface Area of an Oil Drum:** A standard oil drum is a cylinder of height 34 inches with a closed top and bottom. Find the radius r of such a drum if its total surface area is 990π square inches.

29. **Transmission Distance:** Two campus security officers with two-way radios leave the same point and walk at right angles to each other. One walks at 4 miles per hour; the other at 3 miles per hour. How long can they communicate with one another if each radio has a maximum range of 2 miles?

30. **Bicyclist:** Two bicyclists, each moving at a constant speed, leave the same point and travel at right angles to each other. One travels 5 kilometers per hour faster than the other. In $3\frac{1}{2}$ hours the bicyclists are 98 kilometers apart. How fast does each travel?

31. **Air Travel:** Two planes leave the same point and fly at the same altitude, one due north and the other due east. After 2 hours they are 1000 miles apart. If one plane averages 100 miles per hour more than the other, how fast are the planes flying?

32. **Car Travel:** A sales representative travels 280 miles at a certain speed. Traveling the same distance going home, the sales representative goes 10 miles per hour faster, and takes one hour less to travel the same distance. What is the salesperson's speed?

33. **Marathon Training:** In training for a marathon, two friends run 13.5 miles every Saturday. They can cut their training time by 18 minutes if they increase their average speed by 1.5 miles per hour. What is their current average speed for the 13.5 mile run?

34. **Truck Speed:** While moving, you drive a loaded rental truck 225 miles. After emptying the truck, you drive it another 150 miles to the nearest drop-off point at a speed of 5 miles per hour faster than before. If you drive a total of 8 hours, what was your speed with the empty truck?

35. **Rowing Rate:** A rowing crew takes $1\frac{1}{2}$ hours to complete a round trip, rowing 10 kilometers with the current and 10 kilometers against the current. If the rate of the current is 5 kilometers per hour, find the rate at which the crew can row in still water.

36. **Navigation:** The current in a shipping channel of the Detroit River flows at a rate of about 4 miles per hour. If it takes a tanker 5 hours to travel 24 miles up the river and back, how fast does the tanker travel in still water?

37. **Work:** Two conveyor belts can unload a shipment of red peppers in 4 hours. Working alone, the slower belt takes 6 hours longer than the faster belt to unload the peppers. How many hours would each belt take to complete the job by itself?

38. **Work:** A carpenter working alone can finish a job in 2 hours less time than an apprentice. Together they can complete the job in 7 hours. Find the time that each requires to complete the job alone.

39. **Volume of a Sandbox:** A child's sandbox will be made by cutting a 10-inch square from each corner of a square sheet of metal x by x and turning up the sides (Figure 10). The sandbox will hold 20 cubic feet of sand. What size piece of metal should be used? (Hint: 1 cubic foot $= 12^3$ cubic inches)

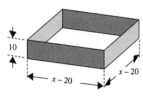

Figure 10

40. **Volume of Rain:** A 40-foot-long rectangular sheet of aluminum, 12 inches wide, is to be made into a rain gutter by turning up two sides so that they are perpendicular to the bottom (Figure 11). How much should be turned up in order to give the gutter a capacity of 37 gallons? (Hint: 1 gallon $= 231$ cubic inches).

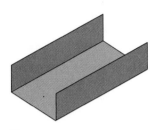

Figure 11

41. **Filling a Tank:** Two pipes connected to the same tank can fill the tank in 3 hours when used together. The larger pipe can fill the tank 2 hours faster than the small one. How long will it take the small pipe to fill the tank alone?

42. **Filling a Pool:** Two hoses are connected to a swimming pool. The pool can be filled in 4 hours if both hoses are used. How many hours are required for each hose to fill the pool alone, if the smaller hose requires 3 hours more than the larger hose?

43. **Marketing:** The marketing manager of a department store found that, on the average, 200 tennis rackets were sold monthly at the unit price of $80. However, for each $5 reduction in price, an extra 30 rackets were sold. How many rackets were sold at the reduced price if the total revenue last month was $19,250? (Hint: Revenue $=$ (price) (quantity).)

44. **Surface Area of a Silo:** A silo is a structure shaped like a right circular cylinder of radius r and height h topped by a half sphere with radius r (Figure 12).

    (a) Show that the surface area S of the silo is given by the formula.
    $$S = 2\pi rh + 2\pi r^2$$
    (Hint: use problem 35(b) on page 289.)

    (b) Find the base radius of the silo if $h = 30$ feet and $S = 800\pi$ square feet.

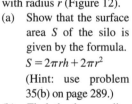

Figure 12

45. **Janitorial Services:** Two janitors working together can clean a fitness center in 5 hours. One experienced janitor can clean the center in 5 hours less than the new person. How long would it take each janitor working alone to clean the fitness center?

46. **Physics:** During a grand remodeling project, you decide to rewire your kitchen. You want one parallel circuit to serve the toaster and a series circuit to serve several lights. The maximum total resistance allowed by the circuit breaker is $R = 8$ ohms, where the total resistance R is given by the formula

$$\frac{1}{R} = \frac{1}{R_1} + \frac{1}{R_2}$$

If the light's total resistance R_2 is 5 ohms greater than the toaster's resistance R_1, what is the toaster's resistance?

47. **Economics:** A proprietor of a service center determines the demand for a popular brand of gasoline in a given week is given by the model $q_1 = 1200/p$, and the supply for the week is given by the model $q_2 = p - 110$, where q_1 and q_2 denote the number in thousands of gallons demanded and supplied each week respectively, at the price of p cents per gallon. Find the price per gallon at which the supply is equal to the demand.

48. **Landscaping:** A landscape designer is planning to plant a rectangular tulip bed on a circular plot of land so that the diagonals of the rectangle intersect at the center of the circle (Figure 13). If the length of the rectangular bed is 4 meters more than its width, and if the radius of the circular plot is 7 meters, what is the length of the rectangular bed?

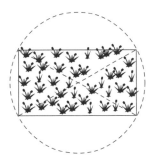

Figure 13

OBJECTIVES

1. Solve Quadratic Inequalities Algebraically
2. Solve Inequalities Graphically
3. Solve Rational Inequalities
4. Solve Applied Problems
5. Use Technology Exploration

6.6 Nonlinear Inequalities

In this section, we solve quadratic and rational inequalities algebraically and justify some solutions graphically.

Solving Quadratic Inequalities Algebraically

An inequality that can be written so that one side is a quadratic expression and the other side is zero is called a **quadratic inequality** in **standard form**. Examples are:

$$x^2 + 3x - 10 < 0 \qquad 6x^2 - 5x + 1 > 0$$
$$3x^2 + 2x - 6 \leq 0 \qquad x^2 - 8x + 7 \geq 0$$

A **solution** of a quadratic inequality in one variable is the set of values of the variable that makes the inequality a true statement. Let us consider the values of the quadratic expression

$$x^2 + 3x - 10$$

as we substitute test numbers for x. By doing so, we can draw a conclusion about whether the values of the expression are positive, negative or zero. To solve the inequality

$$x^2 + 3x - 10 < 0$$

we look for the values of x for which $x^2 + 3x - 10$ is *negative*. These values can be found by examining the graph of the quadratic function

$$f(x) = x^2 + 3x - 10$$

in Figure 1. The graph shows the x values for which $y = f(x)$ is positive are separated from the x values for which $y = f(x)$ is negative by values of x for which

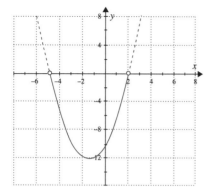

Figure 1

$y = f(x)$ is zero. That is, the x intercepts are the points where the y values change sign. These intercepts are obtained by solving the equation

$$x^2 + 3x - 10 = 0$$
$$(x + 5)(x - 2) = 0$$
$$x = -5 \text{ or } x = 2$$

It is not always necessary to graph the function $f(x) = x^2 + 3x - 10$ to solve this inequality. Instead, the two solutions -5 and 2 divide the number line into three open intervals $(-\infty, -5)$, $(-5, 2)$ and $(2, \infty)$ which we designate by A, B, and C respectively (Figure 2a).

It is important to note that if $x^2 + 3x - 10$ is negative for one value of x in a given interval, then the expression is negative for *any* value of x in that interval. The same holds true if the value of $x^2 + 3x - 10$ is positive for a particular value of x in an interval. To determine if $x^2 + 3x - 10 < 0$ in each interval, we choose a test number from each interval and substitute its value for x in the inequality $x^2 + 3x - 10 < 0$. If the resulting inequality is true, the interval containing the test point is a solution interval. The following table lists the results.

Interval	Test number	Test Value of $x^2 + 3x - 10$	Sign of $x^2 + 3x - 10$	Solution of inequality
A	-6	$(-6)^2 + 3(-6) - 10 = 8$	positive	False
B	0	$(0)^2 + 3(0) - 10 = -10$	negative	True
C	3	$(3)^2 + 3(3) - 10 = 8$	positive	False

The values of x in the interval $(-5, 2)$ satisfy the inequality. Therefore, the solution set is $(-5, 2)$ (Figure 2b).

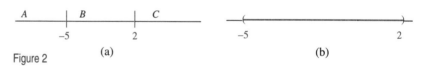

Figure 2   (a)   (b)

The method illustrated above can also be used to solve any inequality whose left side is a polynomial. The values of x for which $f(x) = 0$ are the x intercepts of the graph of f and are called the **cut-points**.

STRATEGY FOR SOLVING A POLYNOMIAL INEQUALITY—THE CUT-POINT METHOD

Step 1. Write the inequality in standard form.

Step 2. Set the polynomial equal to zero, and solve the resulting equation to determine the cut points.

Step 3. Arrange the solutions obtained from Step 2 in increasing order on a number line. These solutions will divide the number line into intervals.

Step 4. Choose a test number in each interval and determine if its value satisfies the original inequality.

Step 5. Draw a number line showing the information in Step 4. Then read the solution set for the original inequality by identifying all the intervals where the inequality is true.

EXAMPLE 1   Solving a Quadratic Inequality

(a) Solve the inequality $x^2 + x \geq 12$.
(b) Use part (a) to find the domain of the function $f(x) = \sqrt{x^2 + x - 12}$.

Solution   (a) Step 1. $x^2 + x - 12 \geq 0$ Write in standard form

Step 2. $x^2 + x - 12 = 0$ Solve related equation

$(x + 4)(x - 3) = 0$

$x = -4$ or $x = 3$

Step 3. The numbers -4 and 3 separate the number line into three intervals which we denote by A, B, and C (Figure 3a).

Step 4. Select a test number in each interval and determine if it satisfies the original inequality. These values are listed in the following table.

Interval	Test Number	Test Value of $x^2 + x - 12$	Sign of $x^2 + x - 12$	Solution of Inequality
A	-5	$(-5)^2 + (-5) - 12 = 8$	positive	True
B	0	$(0)^2 + (0) - 12 = -12$	negative	False
C	4	$(4)^2 + (4) - 12 = 8$	positive	True

Step 5. Figure 3b illustrates the information obtained in Step 4. The solution set is $x \leq -4$ or $x \geq 3$. In interval notation, the solution set is $(-\infty, -4]$ or $[3, \infty)$.

Figure 3   (a)   (b)

(b) Since the square root of a nonnegative expression is a real number, the domain of the function is the set of all values of x which make $x^2 + x - 12 \geq 0$. Using part (a), the solution set of this inequality is $(-\infty, -4]$ or $[3, \infty)$. Therefore, the domain of the function is $(-\infty, -4]$ or $[3, \infty)$.   ◈

Solving Inequalities Graphically

The process used in Example 1 suggests the following alternative method for solving inequalities graphically.

SOLVING INEQUALITIES GRAPHICALLY

Step 1. Write the inequality in standard form.
Step 2. Write the associated function, f, for the given inequality and sketch its graph.
Step 3. Identify points on the graph that correspond to the solution of the inequality. They are organized as follows:
   (a) $f(x) > 0$, points above the x axis.
   (b) $f(x) \geq 0$, points on or above the x axis.
   (c) $f(x) < 0$, points below the x axis.
   (d) $f(x) \leq 0$, points on or below the x axis.
Step 4. Find the x values that correspond to these points. These values of x form the solution set of the inequality.

EXAMPLE 2   Solving an Inequality Graphically

Solve the inequality $2x^2 + 3x - 2 \geq x^2 + 2x$ graphically.

Solution   Writing the inequality in standard form, we have

$$2x^2 + 3x - 2 \geq x^2 + 2x$$
$$x^2 + x - 2 \geq 0$$

The associated function is given by

$$f(x) = x^2 + x - 2$$

Figure 4 shows the graph of f whose x intercepts are the solutions of

$$x^2 + x - 2 = 0$$
$$(x + 2)(x - 1) = 0$$
$$x = -2, 1$$

The solutions of the inequality are those values of x for which the graph of f is on or above the x axis. It follows that the solution set of

$$f(x) \geq 0$$

is

$$x \leq -2 \text{ or } x \geq 1$$

or

the intervals

$$(-\infty, -2] \text{ or } [1, \infty).$$

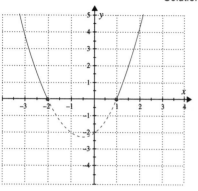

Figure 4

EXAMPLE 3   Solving a Polynomial Inequality Graphically

Figure 5 shows the graph of
$f(x) = x^3 + x^2 - 12x.$
Use the graph to solve the inequality $x^3 + x^2 - 12x \geq 0$.

Solution   The x intercepts of the graph are the solutions of the equation

$$x^3 + x^2 - 12x = 0$$
$$x(x + 4)(x - 3) = 0$$
$$x = -4, 0, 3.$$

The solutions of the inequality are those values of x for which the graph of f is on or above the x axis. It follows that the solution set of

$$f(x) \geq 0 \text{ is}$$
$$-4 \leq x \leq 0 \text{ or } x \geq 3$$

In interval form the solution is $[-4, 0]$ or $[3, \infty)$.

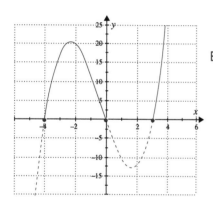

Figure 5

Solving Rational Inequalities

Inequalities such as

$$\frac{x - 3}{x + 2} < 0$$

or

$$\frac{x + 3}{5x - 7} \geq 0$$

where the left side contains rational expressions are examples of **rational inequalities in standard form**. These inequalities are solved using the same procedure as quadratic inequalities with one simple modification.

EXAMPLE 4   Solving a Rational Inequality

Solve the inequality

$$\frac{(x-4)}{(x+2)} \leq 0$$

Solution   Notice that a rational expression can change its algebraic sign only if the numerator or denominator changes algebraic sign. Hence, intervals where a rational expression is positive are separated from intervals where that expression is negative by values of x for which the numerator or denominator is zero. The numerator of

$$\frac{(x-4)}{(x+2)}$$

is zero when $x = 4$; its denominator is zero when $x = -2$. The numbers -2 and 4 separate the number line into three intervals which we denote by A, B, and C (Figure 6a).

Next we select a test number in each interval and determine if that value satisfies the original inequality. Those values are listed in the following table.

Interval	Test Number	Test Value of $\dfrac{(x-4)}{(x+2)}$	Sign of $\dfrac{(x-4)}{(x+2)}$	Solution of Inequality
A	-3	7	positive	False
B	0	-2	negative	True
C	5	$1/7$	positive	False

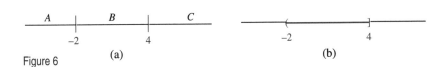

Figure 6   (a)   (b)

Figure 6b illustrates the information obtained from the above table. The solution set is

$$-2 < x \leq 4 \text{ or } (-2,4].$$

Observe that 4 belongs to the solution set because in this example the quotient

$$\frac{(x-4)}{(x+2)}$$

is equal to zero when $x = 4$. The value -2 must be excluded from the solution set because the denominator is zero when $x = -2$ and this causes the fraction to be undefined.   ◇

Solving Applied Problems

The next example shows how to use inequalities to solve applied problems and models.

EXAMPLE 5   Solving a Fireworks Problem

A firework shell is launched from a mortar on the ground with an initial speed of 64 feet per second. The height h (in feet) of the shell after t seconds from being launched is given by the model

$$h(t) = 64t - 16t^2$$

During what time interval will the shell be at least 48 feet above the ground?

Solution   When the shell is at least 48 feet above the ground, the time t will satisfy the inequality

$$h(t) = 64t - 16t^2 \geq 48$$

We solve the inequality by first solving the associated equation.

$$64t - 16t^2 = 48 \qquad \text{Associated equation}$$
$$16t^2 - 64t + 48 = 0 \qquad \text{Standard form}$$
$$16(t - 3)(t - 1) = 0$$
$$t = 3, 1$$

The numbers 1 and 3 separate the number line into three intervals, A, B, and C (Figure 7).

The solution set is [1,3], which means that the shell containing the firework is at least 48 feet high when the time is between and including 1 and 3 seconds.

If we graph the original model, $h(t) = 64t - 16t^2$, the h coordinate of each point on the graph represents the height of the shell for each value of time. To solve the inequality, we are looking for heights of at least 48 feet; that is, points on the graph whose second coordinate is greater than or equal to 48. Figure 8 shows the graph of $h(t) = 64t - 16t^2$. The points we are interested in are on or above the line $y = 48$. The solution is values of t belonging to the interval [1,3]. ◈

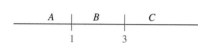

Figure 7

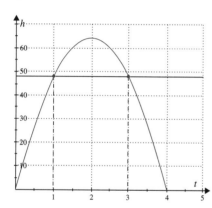

Figure 8

Using Technology Exploration

We can use a grapher to solve quadratic and rational inequalities as the next example shows.

G EXAMPLE 6   Using a Grapher to Solve an Inequality
Use a grapher to solve the inequality

$$2x^2 - 5x < 3$$

Solution   First we write the inequality in standard form:

$$2x^2 - 5x - 3 < 0$$

Next we graph the associated function:

$$f(x) = 2x^2 - 5x - 3$$

Figure 9 shows the graph of $y = 2x^2 - 5x - 3$ in the given viewing window. Using ZOOM and TRACE we see that the x intercepts of the graph are -0.5 and 3. For $y < 0$, we look for that portion of the graph that is below the x axis. From the graph we see that points on the graph with $y < 0$ have x coordinates in the interval between the two x intercepts. That is, all values of x such that $-0.5 < x < 3$ or x belongs to the interval $(-0.5,3)$. The solution set is $(-0.5,3)$. ◈

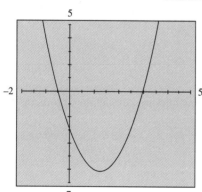

Figure 9

PROBLEM SET 6.6

Mastering the Concepts

In Problems 1–16, solve each inequality using the cut-point strategy. Write the solution set in interval notation and show the solution set on a number line.

1. $(x - 4)(x + 3) < 0$ 2. $(x - 2)(x + 3) < 0$
3. $(x + 5)(x - 2) \leq 0$ 4. $(x - 4)(x + 1) \leq 0$
5. $(x - 1)(x + 3) \geq 0$ 6. $(x - 3)(x + 1) \geq 0$
7. $x^2 - x - 2 < 0$ 8. $x^2 - x - 6 < 0$
9. $x^2 - 3x + 2 \geq 0$ 10. $x^2 - 5x - 6 \geq 0$
11. $2x^2 - 5 \geq -x^2$ 12. $3x^2 - 5 \leq -x^2 - 27x - 12$
13. $5x^2 + 6x + 1 \geq -4x^2$ 14. $12x^2 + 7x + 1 \geq -4x^2 - x$
15. $16x^2 - 24x \leq -9$ 16. $49x^2 + 1 \leq -14x$

In Problems 17–22, find the domain of each function.

17. $f(x) = \sqrt{x^2 + 2x - 3}$ (see Problem 5)

18. $g(x) = \sqrt{x^2 - 2x - 3}$ (see Problem 6)

19. $f(x) = \dfrac{1}{\sqrt{2 + x - x^2}}$ (see Problem 7)

20. $h(x) = \dfrac{1}{\sqrt{6 + x - x^2}}$ (see Problem 8)

21. $f(x) = \sqrt{x^2 - 3x + 2}$ (see Problem 9)

22. $g(x) = \sqrt{x^2 - 5x - 6}$ (see Problem 10)

In Problems 23–32, solve each inequality using the cut-point strategy. Write the solution set in interval notation and show the solution set on a number line.

23. $\dfrac{1}{x - 3} > 0$ 24. $\dfrac{1}{x + 2} > 0$

25. $\dfrac{x - 1}{x + 4} \leq 0$ 26. $\dfrac{x + 3}{x - 3} \leq 0$

27. $\dfrac{x + 1}{2 - x} \geq 0$ 28. $\dfrac{2 - x}{x + 3} \geq 0$

29. $\dfrac{2x - 1}{x + 2} > 0$ 30. $\dfrac{x}{x - 5} > 0$

31. $\dfrac{2x}{x + 4} \geq 0$ 32. $\dfrac{-x}{3 - x} \leq 0$

In Problems 33–38, find the domain of each function.

33. $f(x) = \dfrac{1}{\sqrt{x - 3}}$ (see Problem 23)

34. $g(x) = \dfrac{1}{\sqrt{x + 2}}$ (see Problem 24)

35. $f(x) = \sqrt{\dfrac{x - 1}{x + 4}}$ (see Problem 25)

36. $g(x) = \sqrt{\dfrac{x + 3}{x - 3}}$ (see Problem 26)

37. $f(x) = \sqrt{\dfrac{x + 1}{2 - x}}$ (see Problem 27)

38. $g(x) = \sqrt{\dfrac{2 - x}{x + 3}}$ (see Problem 28)

In Problems 39–42, solve each inequality by examining the graph of the corresponding function. Write the solution in interval notation and show the solution on a number line.

39. $f(x) = x(x - 3)(x + 3)$;
    $f(x) < 0$

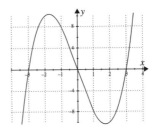

40. $f(x) = (x - 1)(x + 2)(x - 3)$;
    $f(x) > 0$.

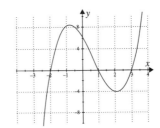

41. $f(x) = x^3 - 9x$;
    $f(x) \geq 0$

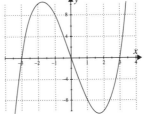

42. $f(x) = 4x - x^3$;
    $f(x) \leq 0$

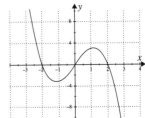

Applying the Concepts

43. **Total Profit:** A company determines that its total profit P (in dollars) is determined by the function

    $$P(x) = 360x - 2x^2$$

    where x is the number of items produced and sold.
    (a) Find all values of x for which the company makes a profit.
    (b) Find all values of x for which the company takes a loss.
    (c) Find all values of x for which the company makes a profit of at least \$16,000.

44. **Height of a Golf Ball:** The height h (in feet) of a golf ball above a fairway depends on the time t (in seconds) it has been in flight. A golfer hits a tee shot that has an approximate height after t seconds that is modeled by the function

$$h(t) = 80t - 16t^2$$

   (a) For what time intervals will the height of the golf ball be more than 64 feet above the ground?

   (b) For what time intervals will the height of the golf ball be less than 20 feet above the ground?

45. **Height of a Thrown Ball:** A ball is thrown vertically upward from the roof of an 80-foot building with an initial speed of 32 feet per second. Its height h (in feet) above the ground after t seconds is modeled by the function

$$h(t) = -16t^2 + 32t + 80$$

   (a) For what time intervals will the height exceed 87 feet?

   (b) For what time intervals will the height be less than 47 feet?

46. **Storage:** The formula for the surface area A of a cylinder is given by

$$A = \pi r^2 + 2\pi rh$$

where r is the radius and h is the height. Suppose a tomato soup can has a height of 6 inches.

   (a) For what measures of the radius will the surface area exceed 200 square inches?

   (b) For what measures of the radius will the surface area be less than 400 square inches?

Developing and Extending the Concepts

In Problems 47–58, solve each inequality algebraically. Write the solution in interval notation and show the solution on a number line.

47. $2x^3 - 8x \le 0$
48. $(x + 2)(x^2 - 5x + 4) < 0$
49. $(x + 3)(x^2 - x - 2) > 0$
50. $(x + 4)(x^2 - 2x - 8) \ge 0$
51. $\dfrac{(x - 1)(x - 3)}{x + 4} \le 0$
52. $\dfrac{(x + 2)(x - 1)}{x - 5} > 0$
53. $\dfrac{x + 1}{(x - 2)(x + 3)} > 0$
54. $\dfrac{x - 2}{(x + 2)(x + 1)} \le 0$
55. $\dfrac{x + 1}{x - 3} \le 1$
56. $\dfrac{x + 2}{x - 1} \ge 2$
57. $\dfrac{1}{x} \le 3$
58. $\dfrac{1}{x} \ge 2$

G In Problems 59–62, use a grapher to solve each inequality.

59. $(x + 3)(x^2 - x - 2) > 0$
60. $(x + 4)(x^2 - 2x - 8) \ge 0$
61. $\dfrac{(x - 1)(x - 3)}{x + 4} \le 0$
62. $\dfrac{(x + 2)(x - 1)}{x - 5} > 0$

OBJECTIVES

1. Graph Functions of the Form $f(x) = ax^2$
2. Graph Functions of the Form $f(x) = a(x - h)^2 + k$
3. Graph a Parabola by Completing the Square
4. Solve Applied Problems—Maximum and Minimum
5. Use Technology Exploration

6.7 Quadratic Functions

So far in this chapter, we have introduced quadratic functions of the form

$$f(x) = ax^2 + bx + c, \ a \ne 0$$

and indicated that their graphs are smooth curves. In this section, we develop techniques for graphing these functions in more detail. We consider the role played by the constant numbers a, b, and c in determining the shape and the position of their graphs. Familiarity with these constants will help describe models defined by certain functions. For instance, people often wonder why a golf player requires so many golf clubs. The reason is that the length of the club, its weight, and the angle of the striking face are designed to produce the right *parabolic* path of the golf ball at each point in the game.

Graphing Functions of the Form $f(x) = ax^2$

The simplest quadratic function is of the form

$$f(x) = ax^2$$

and the curve is called a **parabola**. In fact, the graph of any quadratic function is a *parabola*. The most basic quadratic functions are of the form

$$f(x) = x^2 \text{ and } g(x) = -x^2$$

These functions can be graphed by choosing some input values of x and computing the corresponding output values. The resulting ordered pairs are plotted and connected with smooth curves.

EXAMPLE 1   Graphing Quadratic Functions by Point Plotting

Produce the graphs of the functions

$$f(x) = x^2 \text{ and } g(x) = -x^2$$

by selecting input values and obtaining the corresponding output values, and then plot these points.

Solution   First we create a table showing the input and output values for the functions.

x	$f(x) = x^2$	$(x, f(x))$	$g(x) = -x^2$	$(x, g(x))$
-3	$f(-3) = (-3)^2 = 9$	$(-3, 9)$	$g(-3) = -(-3)^2 = -9$	$(-3, -9)$
-2	$f(-2) = (-2)^2 = 4$	$(-2, 4)$	$g(-2) = -(-2)^2 = -4$	$(-2, -4)$
-1	$f(-1) = (-1)^2 = 1$	$(-1, 1)$	$g(-1) = -(-1)^2 = -1$	$(-1, -1)$
0	$f(0) = (0)^2 = 0$	$(0, 0)$	$g(0) = -(0)^2 = 0$	$(0, 0)$
1	$f(1) = (1)^2 = 1$	$(1, 1)$	$g(1) = -(1)^2 = -1$	$(1, -1)$
2	$f(2) = (2)^2 = 4$	$(2, 4)$	$g(2) = -(2)^2 = -4$	$(2, -4)$
3	$f(3) = (3)^2 = 9$	$(3, 9)$	$g(3) = -(3)^2 = -9$	$(3, -9)$

Figures 1(a) and 1(b) show the smooth curves obtained from plotting the points in the above table.

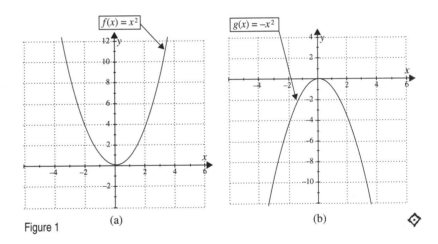

Figure 1   (a)   (b)

All quadratic functions have graphs which look like the ones in Figure 1. To examine the effect of the coefficient a when graphing

$$f(x) = ax^2,$$

we consider the graphs of

$$f(x) = 2x^2, \ g(x) = x^2, \text{ and } h(x) = \frac{1}{2}x^2$$

Once again, we construct a table of input-output values, plot the resulting points and connect them with smooth curves.

Notice the effect that the value of a has on the shape of each graph. The graph of $h(x) = (1/2)x^2$ is *flatter* than the graph of $g(x) = x^2$, while the graph of $f(x) = 2x^2$ is *thinner* than $g(x) = x^2$ (Figure 2).

x	$f(x) = 2x^2$	$g(x) = x^2$	$h(x) = (1/2)x^2$
-2	8	4	2
-1	2	1	1/2
0	0	0	0
1	2	1	1/2
2	8	4	2

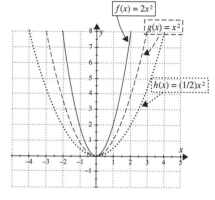

Figure 2

In general, all graphs of parabolas in Figure 3 are of the form

$$f(x) = ax^2,$$

and have the same basic shape. They have the following geometric characteristics shown in Table 1. The **vertex** is the highest or lowest point on the curve. The **axis of symmetry** is the line through the vertex which divides the parabola into two equal pieces.

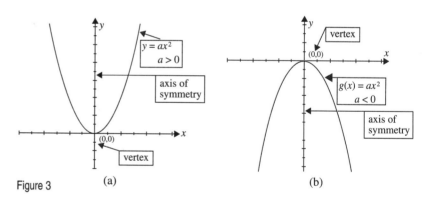

Figure 3   (a)   (b)

TABLE 1

Function	Graph Opens	Vertex	Axis of Symmetry	Domain	Range
$y = ax^2, a > 0$	Upward Figure 3a	(0,0) lowest point	y axis	$\mathfrak{R}$	$[0,\infty)$
$y = ax^2, a < 0$	Downward Figure 3b	(0,0) highest point	y axis	$\mathfrak{R}$	$(-\infty,0]$

Graphing Functions of the Form $f(x) = a(x - h)^2 + k$

Consider the graphs of the functions

$$y = x^2, f(x) = (x + 2)^2, g(x) = x^2 + 2, \text{ and } h(x) = (x + 2)^2 + 2$$

We begin by constructing input values and determine the corresponding output values in the following table:

x	$y = x^2$	$f(x) = (x + 2)^2$	$g(x) = x^2 + 2$	$h(x) = (x + 2)^2 + 2$
-2	4	0	6	2
-1	1	1	3	3
0	0	4	2	6
1	1	9	3	11
2	4	16	6	18

Figure 4 shows the graphs of these parabolas. Notice that these graphs have the same shape but their position is different. Table 2 shows the description of each graph.

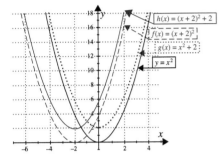

Figure 4

TABLE 2

Function	Vertex	Axis of Symmetry	Geometric Description of Graphs
$y = x^2$	(0,0)	$x = 0$	Basic graph of a parabola.
$f(x) = (x + 2)^2$	$(-2,0)$	$x = -2$	Looks like $y = x^2$, shifted 2 units to the left
$g(x) = x^2 + 2$	(0,2)	$x = 0$	Looks like $y = x^2$, shifted 2 units upward.
$h(x) = (x + 2)^2 + 2$	$(-2,2)$	$x = -2$	Looks like $y = x^2$, shifted 2 units to the left and 2 units up.

We generalize the above results by comparing the graph of $f(x) = a(x - h)^2 + k$ with the graph of $y = ax^2$ and examine the effects of h and k.

Graphing $f(x) = a(x - h)^2 + k$

Function	Vertex	Axis of Symmetry	Geometric Description of Graphs
$m(x) = a(x - h)^2$	(h,0)	$x = h$	Looks like $y = ax^2$, shifted horizontally h units to the right if $h > 0$ and $|h|$ units to the left if $h < 0$.
$g(x) = ax^2 + k$	(0,k)	$x = 0$	Looks like $y = ax^2$, shifted vertically k units upward if $k > 0$ and $|k|$ units downward if $k < 0$.
$f(x) = a(x - h)^2 + k$	(h,k)	$x = h$	Looks like $y = ax^2$, shifted h units horizontally to the right and k units vertically upward when $h > 0$ and $k > 0$.

EXAMPLE 2   Graphing $f(x) = a(x - h)^2 + k$

Sketch the graphs of $y = 2x^2, f(x) = 2(x + 2)^2 + 1$, and $g(x) = -2(x + 2)^2 + 1$ on the same coordinate system. Find the vertex of each parabola and its axis of symmetry. Also find the range of each function.

Solution   We select input x values and determine the corresponding output y values as shown in the accompanying table. Then we plot these points and connect them with a smooth curve (Figure 5).

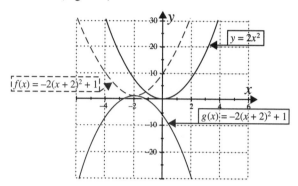

Figure 5

x	$y = 2x^2$	$f(x) = 2(x + 2)^2 + 1$	$g(x) = -2(x + 2)^2 + 1$
-2	8	1	1
-1	2	3	-1
0	0	9	-7
1	2	19	-17
2	8	33	-31

Notice that the graph of $f(x) = 2(x + 2)^2 + 1$ looks like $y = 2x^2$ but shifted horizontally 2 units to the left and vertically 1 unit upward. The graph of $g(x) = -2(x + 2)^2 + 1$ looks like $y = -2x^2$ but shifted horizontally 2 units to the left and vertically 1 unit upward.

The graphs of f and g (Figure 5) suggest that the range of f consists of all real numbers greater than or equal to 1. That is $[1, \infty)$; whereas, the range of g is the set $(-\infty, 1]$. ◈

Graphing a Parabola by Completing the Square

A quadratic function in the form

$$f(x) = ax^2 + bx + c, a \neq 0$$

can be converted to the form

$$f(x) = a(x - h)^2 + k$$

by using the techniques of completing the square as the next example shows.

EXAMPLE 3   Graphing a Function by Completing the Square

Rewrite $f(x) = 2x^2 - 12x + 10$ in the form $f(x) = a(x - h)^2 + k$ by completing the square. Find the vertex and the axis of symmetry. Also determine the intercepts and sketch the graph.

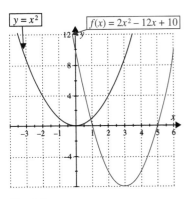

Figure 6

Solution   We complete the square to find the values of a, h, and k.

$$f(x) = 2x^2 - 12x + 10 \qquad \text{The given function}$$
$$= 2(x^2 - 6x) + 10 \qquad \text{Factor 2 from the first two terms.}$$
$$= 2(x^2 - 6x + 9) - 2(9) + 10 \qquad \text{Add and subtract } 2[\tfrac{1}{2}(6)]^2 = 2(9).$$
$$= 2(x - 3)^2 - 18 + 10 \qquad \text{Express the trinomial as a perfect square}$$
$$= 2(x - 3)^2 - 8$$

By comparing the latter form to the form $f(x) = a(x - h)^2 + k$, we have $a = 2$, $h = 3$, and $k = -8$. The graph of f looks like the graph of $y = x^2$ but it is thinner than this graph since $a = 2$. The graph of $f(x)$ is obtained by shifting the graph of $y = 2x^2$, 3 units to the right and 8 units downward. The vertex is $(3, -8)$ and the axis of symmetry is $x = 3$ (Figure 6).

To find the y intercept, we set $x = 0$ and solve for y. That is,
$$y = f(0) = 2(0)^2 - 12(0) + 10 = 10.$$
So the y intercept is 10. To find the x intercepts, we set $y = f(x) = 0$ and solve the original equation for x. That is, solve the equation $2x^2 - 12x + 10 = 0$ for x.

$$2x^2 - 12x + 10 = 0 \qquad \text{Original equation}$$
$$x^2 - 6x + 5 = 0 \qquad \text{Divide by 2}$$
$$(x - 1)(x - 5) = 0 \qquad \text{Factor the left side}$$
$$x - 1 = 0, x - 5 = 0$$
$$x = 1, 5 \qquad\qquad\qquad\qquad\qquad\qquad\qquad \text{◈}$$

So the x intercepts are 1 and 5 as in Figure 6.

The method used in example 3 can be generalized to find a formula for locating the vertex of a parabola. We can complete the square to rewrite $f(x) = ax^2 + bx + c$ in the form

$$f(x) = a\left[x - \frac{-b}{2a}\right]^2 + \frac{4ac - b^2}{4a}$$

If we compare the latter form to the form $f(x) = a(x - h)^2 + k$, we observe that $h = -b/(2a)$ and $k = (4ac - b^2)/(4a)$. That is, the vertex of the parabola $f(x) = ax^2 + bx + c$ is given by

$$(h,k) = \left(-\frac{b}{2a}, \frac{4ac - b^2}{4a}\right)$$

In practice, it is usually easier to compute the x coordinate of the vertex, $h = -b/(2a)$, and then evaluate $f(h)$ to find the y coordinate k. That is, the **coordinates of the vertex** may be written as

$$(h,k) = \left(-\frac{b}{2a}, f\left(-\frac{b}{2a}\right)\right)$$

We could have found the vertex of the parabola directly in example 3 by using the above formula.

We see that $a = 2$ and $b = -12$, so that

$$h = -\frac{b}{2a} = -\frac{-12}{2(2)} = 3$$

Substituting 3 for x into $f(x) = 2x^2 - 12x + 10$, we obtain k as follows:

$$k = f\left(\frac{-b}{2a}\right) = f(3) = 2(3)^2 - 12(3) + 10 = -8$$

Therefore, the vertex of the parabola is given by

$$(h,k) = (3, -8)$$

Solving Applied Problems—Maximum and Minimum

We indicated that if $a > 0$, the parabola $f(x) = ax^2 + bx + c$ opens upward. In this case, we say the y coordinate of the vertex is the **minimum value of f** (Figure 7a). Similarly, if $a < 0$, the parabola opens downward, and we say that the y coordinate of the vertex is the **maximum value of f** (Figure 7b).

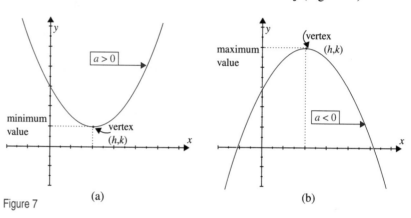

$\boxed{a > 0}$

minimum value

vertex (h,k)

(a)

(vertex (h,k)

maximum value

$\boxed{a < 0}$

(b)

Figure 7

Applied problems and models in which a quantity has to be *maximized* or *minimized* can be solved by finding the coordinates of the vertex, as the next example shows.

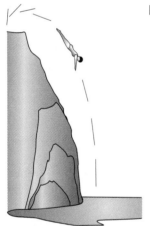

EXAMPLE 4 Finding the Maximum Height

A cliff diver's path from the top of a cliff 110 feet above the surface of the water after t seconds is approximated by the model

$$h = -t^2 + 2t + 110, \, t \geq 0$$

where h is the height in feet above the water.
(a) Find the maximum height of the diver.
(b) Find the time it takes for the diver to reach the maximum height.

Solution (a) To find the maximum height of the diver, we use

$$h = -t^2 + 2t + 110$$

and complete the square to find the vertex.

$h = -t^2 + 2t + 110$	Original equation
$= -(t^2 - 2t) + 110$	Factor -1
$= -(t^2 - 2t + 1) + 1 + 110$	Complete the square
$= -(t - 1)^2 + 111$	Perfect square trinomial

The vertex of the parabola is (1,111). Since $a = -1$ is a negative number, the y coordinate of the vertex is the maximum height of the diver, 111 feet.
(b) The maximum height occurs when $t = 1$. That is, the maximum height occurs after 1 second. ◈

Using Technology Exploration

We can use a grapher to determine the maximum or minimum of a quadratic function.

EXAMPLE 5   Designing a Gazebo

An architect is designing a rectangular gazebo edged with 144 feet of fancy railing.

(a) Create a model that expresses the area A in terms of the length x (in feet).

G (b) Use a grapher to graph A. From the graph, find the maximum value of A and the value of x at which A is a maximum.

(c) Find the dimensions of the gazebo that will yield the largest area.

(d) Complete the following table and plot the ordered pairs (length,area) $= (x, A)$. Determine whether these points lie on the graph. Use the data to estimate the dimensions of the gazebo.

Length, x	42	40	38	36	32	28	24
Area, A							

Solution   We use the formula for area A of a rectangle

$$A = xy$$

where x is the length and y is the width, both in feet.

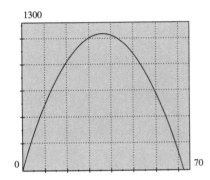

Figure 8

(a) We eliminate y, expressing the area A strictly as a function of the length x, by using the fact that there is 144 feet of railing material.

$$2x + 2y = 144 \qquad \text{Perimeter must be 144}$$
$$x + y = 72 \qquad \text{Divide by 2}$$
$$y = 72 - x \quad \text{Solve for } y$$

Substituting $72 - x$ for y in the area formula, we obtain,

$$A = xy = x(72 - x)$$
$$A(x) = 72x - x^2$$

(b) Figure 8 shows the graph of this function. By using the **ZOOM** and **TRACE** features, we see that the coordinate of the maximum point is $(36,1296)$.

(c) By completing the square, we write $A(x)$ as

$$A(x) = -x^2 + 72x$$
$$= -(x^2 - 72x + 36^2) + 36^2$$
$$= -(x - 36)^2 + 1296$$

so that the coordinates of the vertex are $(36,1296)$ and the x coordinate of the vertex is 36. Solving for y we have,

$$y = 72 - 36 = 36$$

Therefore, the dimensions of the gazebo are 36 feet by 36 feet.

(d) In order to better understand the problem, we look at some dimensions. Using the following table, several possible values for x and y are chosen so that a rectangular base can be enclosed with 144 feet.

Length, x	42	40	38	36	32	28	24
Width, y	30	32	34	36	40	44	48
Area, A	1260	1280	1292	1296	1280	1232	1152

By plotting the data points (x, A) we observe all these points fit on the graph in Figure 8. The table also shows that 1296 is greater than any other value for A. Therefore, the corresponding dimensions for the gazebo are 36 feet by 36 feet.   ◇

Some quadratic functions also fit descriptive data as the next example shows.

EXAMPLE 6 Analyzing a Cost Function

Number of Radios manufactured, x	Cost per radio, C (in dollars)
3	23
6	14
9	23

Suppose that the manufacturing cost C (in dollars) per radio and the number x of radios manufactured, are related by the data points in the following table.
(a) Construct a scattergram that exhibits the data.
(b) Assuming that the relationship is approximately quadratic, find a quadratic function $C(x) = ax^2 + bx + c$ that fits this data.
G (c) Use a grapher to graph the function in part (b).
(d) Assuming the function in part (b) provides an accurate relationship between C and x for this experiment, find the value of x that predicts the minimum cost C. What is the minimum cost for this value of x?

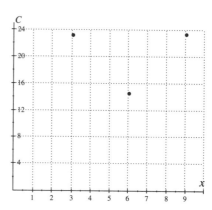

Figure 9

Solution (a) We plot the data points (x,C) as $(3,23)$, $(6,14)$, and $(9,23)$ and observe that they do not fit a linear function (Figure 9).
(b) The statement of the problem leads us to look for a quadratic function that fits these data points.

$$C(x) = ax^2 + bx + c$$

We substitute the given values of x and C into the above equation

$$C(x) = ax^2 + bx + c$$
$$23 = a(3)^2 + b(3) + c$$
$$14 = a(6)^2 + b(6) + c$$
$$23 = a(9)^2 + b(9) + c$$

By simplifying, we see that we need to solve the system

$$\begin{cases} 23 = 9a + 3b + c \\ 14 = 36a + 6b + c \\ 23 = 81a + 9b + c \end{cases}$$

Solving this system for a, b, and c, we have

$$a = 1, b = -12, \text{ and } c = 50$$

Thus the function

$$C(x) = x^2 - 12x + 50$$

fits the given data.
(c) Figure 10 shows the graph of C generated by a grapher.
(d) By plotting the points $(3,23)$, $(6,14)$, and $(9,23)$, we confirm that these points fit on the graph. It appears from the graph that the value of x that provides the minimum cost is 6 and the minimum cost is 14. Therefore, the number of radios that should be manufactured is 6 at a minimum cost of $14 each. ◈

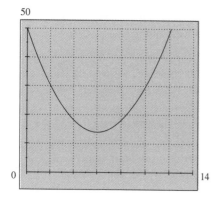

Figure 10

◈ PROBLEM SET 6.7

Mastering the Concepts

In Problems 1–4, sketch the graphs of each set of functions on the same coordinate system.

1. (a) $y = x^2$ (b) $y = 2x^2$ (c) $y = x^2/2$
2. (a) $y = x^2$ (b) $y = 4x^2$ (c) $y = x^2/4$
3. (a) $y = -x^2$ (b) $y = -3x^2$ (c) $y = -x^2/3$
4. (a) $y = -x^2$ (b) $y = -4x^2$ (c) $y = -x^2/4$

In Problems 5–8, sketch the graphs of each set of functions on the same coordinate system.

5. (a) $y = x^2$ (b) $y = x^2 + 2$ (c) $y = x^2 - 3$
6. (a) $y = -x^2$ (b) $y = -x^2 + 1$ (c) $y = -x^2 + 4$
7. (a) $y = -x^2$ (b) $y = -(x + 2)^2$ (c) $y = -(x - 1)^2$
8. (a) $y = x^2$ (b) $y = (x + 3)^2$ (c) $y = (x - 2)^2$

In Problems 9–20, sketch the graph of each quadratic function. Identify the vertex, the axis of symmetry, and find the domain and range of f.

9. $f(x) = (x - 2)^2 + 1$
10. $f(x) = (x + 2)^2 - 3$
11. $f(x) = -(x - 4)^2 + 2$
12. $f(x) = (x + 4)^2 - 1$
13. $f(x) = 1 - (x + 2)^2$
14. $f(x) = -1 - (x - 2)^2$
15. $f(x) = 2(x + 3)^2 + 2$
16. $f(x) = -2(x + 3)^2 - 3$
17. $f(x) = -3(x - 1)^2 + 1$
18. $f(x) = 3(x - 1)^2 + 5$
19. $f(x) = -\frac{1}{2}(x + 3)^2 + 1$
20. $f(x) = -\frac{1}{2}(x - 3)^2 + 1$

In Problems 21–28, rewrite $f(x)$ in the form
$$f(x) = a(x - h)^2 + k$$
by completing the square. Sketch the graph and find the vertex, the axis of symmetry, and the range of f.

21. (a) $f(x) = x^2 - 2x$ (b) $f(x) = -x^2 + 2x$
22. (a) $f(x) = x^2 + 4x$ (b) $f(x) = -x^2 - 4x$
23. $f(x) = x^2 - 6x + 10$
24. $f(x) = x^2 + 6x + 10$
25. $f(x) = -2x^2 - 4x - 4$
26. $f(x) = 2x^2 + 4x + 2$
27. $f(x) = 2x^2 + 6x + 9$
28. $f(x) = 2x^2 - 6x - 9$

In Problems 29–34, find the x and y intercepts, the vertex, and the axis of symmetry. Sketch the graph of f and find the range.

29. (a) $f(x) = 4 - x^2$ (b) $f(x) = x^2 - 4$
30. (a) $f(x) = 1 - x^2$ (b) $f(x) = x^2 - 1$
31. $f(x) = x^2 - 10x + 25$
32. $f(x) = 10x - x^2 - 25$
33. $f(x) = x^2 - 2x - 3$
34. $f(x) = -x^2 + 2x + 3$

In Problems 35–40, without graphing, find the vertex, the maximum value, or the minimum value.

35. $f(x) = 3(x - 7)^2 + 5$
36. $f(x) = 5(x + 2)^2 + 3$
37. $f(x) = -2(x + 6)^2 + 1$
38. $f(x) = 4[x - (1/2)]^2 + 2$
39. $f(x) = \frac{2}{3}(x - 1)^2 + 2$
40. $f(x) = -\frac{2}{5}(x + 5)^2 + 1$

Applying the Concepts

41. **Height of a Rocket:** A model rocket is launched and then it accelerates until the propellant burns out, after which it coasts upward to its highest point. The height d (in feet) of the rocket above the ground t seconds after the burn out, is modeled by the function

$$d(t) = -16t^2 + 192t, \, t \geq 0$$

(a) Calculate and explain the meaning of each output value, $d(5)$, $d(10)$ and $d(15)$.
(b) Write $d(t)$ in the form
$$d(t) = a(t - h)^2 + k$$
(c) Find the vertex of $d(t)$, and then use the d coordinate of the vertex to determine the maximum height attained by the rocket.
(d) How many seconds after the engines shut down does it take the rocket to reach the maximum height?
(e) Find the time for the rocket to hit the ground.
(f) Sketch the graph of d that describes the situation.

42. **Minimum Cost:** The cost C (in dollars) to produce x graphing calculators is modeled by the function

$$C(x) = 2x^2 - 800x + 92,000, \, 0 \leq x \leq 210$$

(a) Calculate and explain the meaning of each output value $C(0), C(10), C(100), C(200)$.
(b) Write $C(x)$ in the form
$$C(x) = a(x - h)^2 + k.$$
(c) Find the vertex, and then use the x coordinate of the vertex to determine the number of graphing calculators that must be produced to minimize cost.
(d) Determine the minimum cost.
(e) Sketch the graph of the function that describes the situation.

43. **Maximum Revenue:** A vending machine company that operates juice machines determines that the daily revenue R (in dollars) from selling juices at a price p (in dollars) per bottle is modeled by the function

$$R(p) = 270p - 150p^2, \, 0 \leq p \leq \$1.10$$

(a) Calculate the following values and explain the meaning of each: $R(0), R(0.45), R(0.75), R(0.85)$
(b) What is the price of a bottle of juice if the daily revenue is $120?
(c) Write $R(p)$ in the form $R(p) = a(p - h)^2 + k$.
(d) Find the vertex, then use the p coordinate of the vertex to determine the price per bottle to attain the maximum revenue.

44. **Volcanic Eruption:** In the 1968 volcanic eruption in Costa Rica, large blocks, thrown out by volcanic explosions, showered the surrounding area. The trajectories of their flight indicate that the blocks were propelled upward like mortar shells. The height d (in meters) of the blocks is approximated by the model

$$d(t) = -7.8t^2 + 358t, \; t \geq 0$$

where t is in seconds (Source: William Melson, Smithsonian Institution).

(a) Calculate and explain the meaning of each output value: $d(10)$, $d(20)$, $d(30)$, and $d(40)$.
(b) Write $d(t)$ in the form $d(t) = a(t - h)^2 + k$.
(c) Find the vertex of $d(t)$, and then use the second coordinate of the vertex to determine the maximum height attained by the blocks.
(d) How many seconds does it take a block to reach the maximum height?
(e) Find the time for a block to land on the ground.
(f) Sketch the graph of d that describes the situation.

45. **Maximum Area:** A homeowner wishes to build a rectangular observation deck whose perimeter is 40 meters.
(a) Let x (in meters) be the length of one side of the observation deck. Explain why the formula

$$A(x) = (20 - x)x$$

describes the area of the deck.
(b) What are the restrictions on the model of part (a)?
(c) Sketch the graph of this function and indicate the vertex.
(d) Use the information from part (c) to find the dimensions of the largest observation deck.
(e) What is the maximum area of the deck?

46. **Maximum Area:** A rancher has 40 meters of fencing to enclose a rectangular pen next to a barn, with the barn wall forming one side of the fence.
(a) Let x (in meters) be the length of the side of the pen that lies along the barn. Explain why the formula

$$A(x) = x\left[20 - \left(\frac{x}{2}\right)\right]$$

describes the area of the pen.
(b) What is the domain of A? Why?
(c) Sketch the graph of this function and indicate the vertex.
(d) Use the information from part (c) to find the dimensions of the pen that produces a maximum enclosed area.
(e) What is the maximum area of the pen?

47. **Maximum Area:** A farmer has 600 feet of fencing to enclose two rectangular holding pens for cattle (Figure 11).

Figure 11

(a) Let x (in feet) be the width of each pen. Explain why the formula

$$A(x) = x(300 - 1.5x)$$

describes the total area A (in square feet) for the pens.
(b) What is the domain of A? Why?
(c) Sketch the graph of A and indicate the vertex.
(d) What is the maximum total area for the holding pens?
(e) What are the dimensions of the largest total area for the pens?

48. **Maximum Area:** A total of 800 meters of fence are to be used to fence a rectangular plot of land divided into three equal portions by using parallel sides (Figure 12).

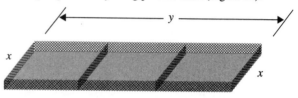

Figure 12

(a) Let x (in meters) be the length of each of the parallel sides and y is the length. Explain why the formula

$$A(x) = x(400 - 2x)$$

describes the entire rectangular area.
(b) What is the domain of A?
(c) Sketch the graph of A and indicate the vertex.
(d) What are the dimensions of the largest area?
(e) What is the maximum area?

Developing and Extending the Concepts

In Problems 49–54, match each condition A, B, or C with the graphs.

A. $b^2 - 4ac < 0$ B. $b^2 - 4ac > 0$ C. $b^2 - 4ac = 0$

49.
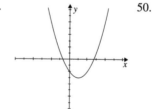

50.
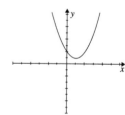

51.
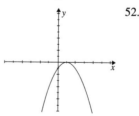

52.
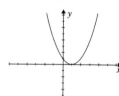

53.
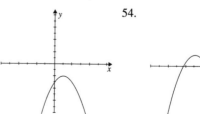

54.

In Problems 55 and 56, write an equation of the given parabola that has the shape of $f(x) = a(x - h)^2 + k$.

55.

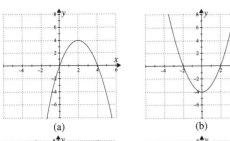

| (a) | (b) |

56.

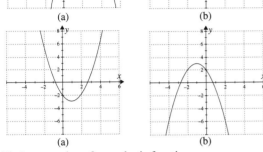

| (a) | (b) |

57. Find an equation of a quadratic function
$$F(x) = a(x - h)^2 + k$$
if the graph of F has the exact shape as the graph of
$$f(x) = 2(x + 1)^2 + 5$$
and $F(x)$ has a minimum value at the same point where
$$g(x) = -2(x - 3)^2 + 2$$
has a maximum value.

G 58. Use a grapher to graph the following functions on the same coordinate system:
$$y = x^2 + x, y = x^2 + 4x, \text{ and } y = x^2 - 4x$$
Find the vertex of each parabola.

59. **Data Analysis:** Find a quadratic function that best fits the data $(0,20)$, $(1,84)$, and $(5,20)$. Sketch the graph.

60. **Maximum Volume:** Consider a sandbox being built by cutting corners from a 12 foot by 12 foot sheet of metal (Figure 13).
   (a) Complete the following table.

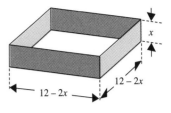

Figure 13

Height x	1	2	3	4	5	6
Length $12 - 2x$						
Width $12 - 2x$						
Volume $V = x(12 - 2x)^2$						

   (b) Use the table to find x that maximizes the volume.
   G (c) Use a grapher to graph V and find x that maximizes the volume.

61. (a) Find two numbers whose sum is 20 and whose product is a maximum.
   (b) Find two numbers that differ by 6 and whose product is a minimum.

OBJECTIVES

1. Graph Horizontal Parabolas
2. Graph Circles
3. Solve Applied Problems
4. Use Technology Exploration

6.8 Horizontal Parabolas and Circles

All the curves we considered so far are the graphs of functions. In this section, we consider curves whose graphs are not functions. These curves are circles and other forms of parabolas.

Graphing Horizontal Parabolas

In section 6.7 we indicated that the graph of quadratic functions of the form $f(x)$ or $y = a(x - h)^2 + k$ are parabolas. These *parabolas* open *upward* if $a > 0$ or open *downward* if $a < 0$. Parabolas can also open to the *left* or to the *right*. Although graphs of quadratic functions are parabolas, not all parabolas are quadratic functions since parabolas opening any way other than upward or downward fail the vertical line test.

Suppose that we interchange the role of x and y in the equation $y = ax^2 + bx + c$. We obtain a "new" equation $x = ay^2 + by + c$ which can be written in the form

$$x = a(y - k)^2 + h$$

by completing the square. The latter equation is referred to as a **standard equation** of a parabola that opens to the right if $a > 0$ or to the left if $a < 0$. *The parabola has a vertex at* (h,k) *and its axis of symmetry is* $y = k$. Figure 1 shows parabolas in standard form.

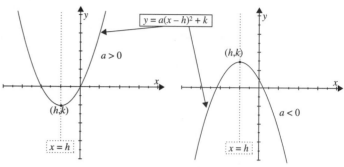

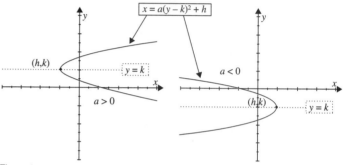

Figure 1

EXAMPLE 1   Graphing Parabolas

Graph each parabola. Indicate the vertex and axis of symmetry.

(a) $x = 3y^2$ (b) $x = -3y^2$

Solution   (a) The equation is written in standard form as

$$x = 3y^2 \text{ or } x = 3(y - 0)^2 + 0$$

with $a = 3, h = 0$, and $k = 0$. Its graph is a parabola that opens to the right since $a > 0$. Its vertex is at $(0,0)$ and its axis of symmetry is the line $y = 0$ (Figure 2a).

(b) The standard form of this equation is

$$x = -3y^2 \text{ or } x = -3(y - 0)^2 + 0$$

with $a = -3, h = 0$, and $k = 0$. Its graph is a parabola that opens to the left since $a < 0$. Its vertex is at $(0,0)$ and its axis of symmetry is the line $y = 0$ (Figure 2b).

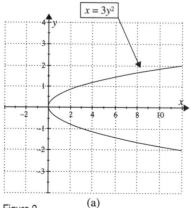

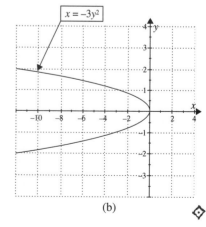

Figure 2   (a)   (b)

EXAMPLE 2   Graphing a Parabola

Sketch the graph of each parabola. Indicate the vertex and the axis of symmetry.
(a) $x = 2(y - 1)^2 + 3$ (b) $x = -y^2 - 2y + 15$

Solution   (a) The equation $x = 2(y - 1)^2 + 3$ is in the form $x = a(y - k)^2 + h$, with $a = 2$, $h = 3$, and $k = 1$. Since $a > 0$, the parabola opens to the right. Its vertex is $(3,1)$ and its axis of symmetry is the line $y = 1$ (Figure 3a).

(b) To write the equation $x = -y^2 - 2y + 15$ in standard form, we complete the square in y.

$x = -y^2 - 2y + 15$	Original equation
$x = -(y^2 + 2y) + 15$	Factor -1
$x = -(y^2 + 2y + 1) + 16$	Complete the square
$x = -(y + 1)^2 + 16$	

The equation is in standard form $x = a(y - k)^2 + h$, with $a = -1$, $h = 16$, and $k = -1$. The parabola opens to the left since $a < 0$, its vertex is at $(16, -1)$ and its axis of symmetry is the line $y = -1$ (Figure 3b) .

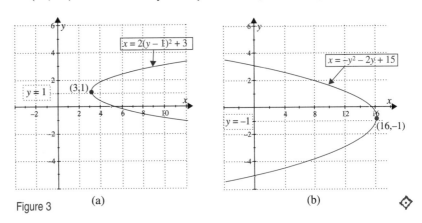

Figure 3   (a)   (b)

Graphing Circles

It is possible to recognize and graph equations of circles with relative ease. We know from geometry that a circle consists of all points that are a fixed distance r, its *radius*, from a fixed point C, its *center* (Figure 4a). Suppose that a circle is located in a Cartesian plane so that its center C is (h,k) and its radius is r. If a point $P = (x,y)$ is on the circle, its distance from $C = (h,k)$ has to be r units (Figure 4b). By the distance formula, x and y satisfy the equation

$$\sqrt{(x - h)^2 + (y - k)^2} = r$$

or

$$(x - h)^2 + (y - k)^2 = r^2$$

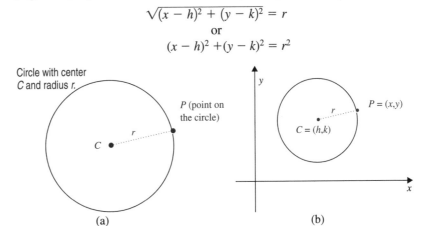

Circle with center
C and radius r.

P (point on the circle)

$P = (x,y)$

$C = (h,k)$

(a)   (b)

Figure 4

Conversely, any ordered pair (x,y) that satisfies this last equation defines a point $P = (x,y)$ that lies on the circle with center $C = (h,k)$ and radius r. Thus, we have the following general result:

STANDARD EQUATION OF A CIRCLE

> Any point (x,y) on the circle with center (h,k) and radius r must satisfy the equation
> $$(x - h)^2 + (y - k)^2 = r^2$$
> This equation is called the **standard equation of the circle**.

EXAMPLE 3   Graphing a Circle

Graph each circle. Indicate the center and radius.

(a) $x^2 + y^2 = 9$ (b) $(x - 1)^2 + (y - 2)^2 = 9$

Solution   (a) The equation $x^2 + y^2 = 9$ is written in standard form as
$$(x - 0)^2 + (y - 0)^2 = 9$$
Its center is at $(0,0)$ and its radius is 3 (Figure 5a).

(b) The equation $(x - 1)^2 + (y - 2)^2 = 9$ is in standard form with the center at $(1,2)$ and the radius 3 (Figure 5b).

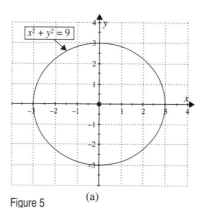
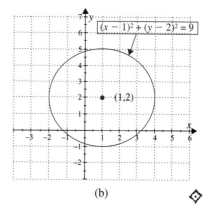

Figure 5   (a)   (b)

EXAMPLE 4   Finding the Standard Equation of a Circle

Find the standard equation for the circle of radius 4 with center at the point $(4,-1)$.

Solution   Substituting 4 for r and $(4,-1)$ for (h,k) in the equation
$$(x - h)^2 + (y - k)^2 = r^2$$
we get $(x - 4)^2 + (y - (-1))^2 = 4^2$
The standard equation is $(x - 4)^2 + (y + 1)^2 = 16$

At times, an alternative algebraic form of an equation of a circle is given as the next example shows.

EXAMPLE 5   Finding the Center and Radius of a Circle

Show that the graph of the equation
$$x^2 + y^2 + 6x - 2y - 15 = 0$$
is a circle by converting it to its standard equation. Determine the center, the radius, and sketch the circle.

Solution We will create a perfect square trinomial in both x and y as follows:

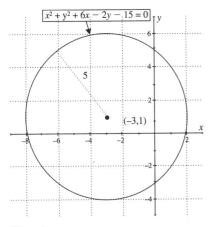

$$(x^2 + 6x) + (y^2 - 2y) = 15$$ Isolate variable terms on one side by grouping the x terms and the y terms

$$(x^2 + 6x + 9) + (y^2 - 2y + 1) = 15 + 9 + 1$$ Complete the square in each pair of parentheses

$$(x + 3)^2 + (y - 1)^2 = 25$$ Factor and simplify

We recognize the form of this last equation. Its graph is a circle of radius $r = 5$ and center $(h,k) = (-3,1)$ (Figure 6).

Figure 6

Solving Applied Problems

In the next example, we show the use of a circle in architecture.

EXAMPLE 6 Solving an Architecture Problem

The Norman style architecture of a window is characterized by massive construction, carved decorations and rectangles surmounted by semicircular arches (Figure 7). Suppose that the rectangular part is 4 feet wide and 4.5 feet in length.

(a) Find an equation of the semicircle using the coordinate system with the origin at the midpoint of the width of the rectangle.

(b) What is the height of the arch 1.5 feet to the right of the origin of the coordinate system?

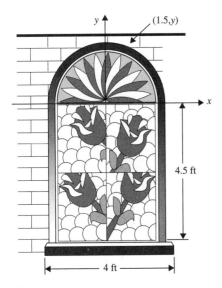

Figure 7

Solution (a) Since the center of the circle is at the origin of the coordinate system, $(h,k) = (0,0)$ and the radius is $4/2 = 2$. The standard equation of this circle is

$$x^2 + y^2 = 4$$

(b) The height of the arch 1.5 feet to the right of the origin is the value of the y coordinate of the point on the circle with an x coordinate of 1.5. This y coordinate is the solution of the equation $(1.5)^2 + y^2 = 4$ so we have

$$y^2 = 4 - 2.25$$
$$y = \sqrt{1.75} \approx 1.3$$

The clearance is approximately 1.3 feet.

Using Technology Exploration

Most graphers are only capable of graphing curves that are functions. Some graphers are able to graph a circle of the form

$$x^2 + y^2 = 4.$$

However graphs of circles and horizontal parabolas can be graphed by solving for y first, and then graphing both the upper half and the lower half of the curve.

EXAMPLE 7   Using a Grapher to Graph a Circle

G Use a grapher to graph the circle with equation
$$(x - 1)^2 + (y + 2)^2 = 9$$

Solution   In this case, we solve first for y in terms of x.

$$(x - 1)^2 + (y + 2)^2 = 9$$
$$(y + 2)^2 = 9 - (x - 1)^2$$
$$y + 2 = \pm\sqrt{9 - (x - 1)^2}$$
$$y = -2 \pm \sqrt{9 - (x - 1)^2}$$

Now we use a grapher with the given viewing window to graph

$$y = -2 + \sqrt{9 - (x - 1)^2} \text{ and } y = -2 - \sqrt{9 - (x - 1)^2}$$

The display screen will show an upper half and a lower half of the circle (Figure 8).   ◈

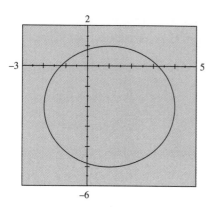

Figure 8

◈ PROBLEM SET 6.8

Mastering the Concepts

In Problems 1–12, graph each parabola. Indicate the vertex and the axis of symmetry of the parabola.

1. (a) $x = 2y^2$ (b) $x = -2y^2$

2. (a) $x = -\dfrac{1}{2}y^2$ (b) $x = \dfrac{1}{2}y^2$

3. (a) $x = \dfrac{4}{5}y^2$ (b) $x = -\dfrac{4}{5}y^2$

4. (a) $4x = -3y^2$ (b) $4x = 3y^2$
5. (a) $2x = -7y^2$ (b) $2x = 7y^2$
6. (a) $5x = -8y^2$ (b) $5x = 8y^2$
7. $x = 4(y - 1)^2 + 3$ 8. $x = -4(y + 1)^2 + 3$
9. $x = -12(y - 4)^2 - 7$ 10. $x = 8(y + 2)^2 + 7$
11. $x = 2(y - 4)^2 - 7$ 12. $x = -3(y + 2)^2 + 7$

In Problems 13–22, write each equation in standard form by completing the square. Indicate the vertex of the parabola and its axis of symmetry.

13. $x = y^2 + 4y - 3$ 14. $x = y^2 - 4y + 4$
15. $x = -2y^2 + 10y - 7$ 16. $x = -2y^2 + 4y - 3$
17. $x = -y^2 + 4y - 5$ 18. $x = y^2 - 4y + 1$
19. $x = y^2 + 2y + 3$ 20. $x = -y^2 + 2y + 4$
21. $x = 2y^2 + 4y - 3$ 22. $x = -2y^2 - 4y + 3$

In Problems 23–34, graph each circle. Indicate the center and radius.

23. (a) $x^2 + y^2 = 1$ (b) $x^2 + y^2 = 9$
24. (a) $x^2 + y^2 = 25$ (b) $x^2 + y^2 = 16$
25. $(x - 1)^2 + (y + 2)^2 = 4$
26. $(x - 4)^2 + (y + 1)^2 = 4$
27. $(x + 1)^2 + (y - 2)^2 = 4$
28. $(x + 4)^2 + (y - 2)^2 = 16$
29. $(x - 2)^2 + (y + 4)^2 = 16$
30. $(x + 2)^2 + (y - 4)^2 = 16$
31. $(x - 3)^2 + (y + 4)^2 = 25$
32. $(x + 3)^2 + (y - 1)^2 = 25$
33. $[x + (1/2)]^2 + [y - (1/3)]^2 = 1/4$
34. $[x - (1/4)]^2 + [y + (1/2)]^2 = 1/4$

In Problems 35–38, find an equation of a circle satisfying the given conditions.

35. (a) Center (0,0), radius 4   (b) Center (0,0), radius 3
36. (a) Center (3,4), radius 7   (b) Center (−5,2), radius 7
37. (a) Center (−4,3), radius $4\sqrt{3}$
     (b) Center (−5,−2), radius $2\sqrt{5}$
38. (a) Center (−3,−1), radius $3\sqrt{2}$
     (b) Center (1,−4), radius $2\sqrt{3}$

In Problems 39–44, write each equation in standard form by completing the square. Find the center and radius.

39. $x^2 + y^2 - 4x - 6y + 4 = 0$
40. $x^2 + y^2 + 8x - 6y - 15 = 0$
41. $x^2 + y^2 - 6x + 4y + 4 = 0$
42. $x^2 + y^2 - 8x - 10y + 5 = 0$
43. $x^2 + y^2 - 6x + 8y + 9 = 0$
44. $x^2 + y^2 + 4x + 4y - 8 = 0$

Applying the Concepts

45. Gear Design: Figure 9 shows two gears meshed together in such a way that the larger gear is represented by the circle with equation $x^2 + y^2 = 64$. The smaller gear is tangent to the larger one and is centered at the point $(-5,12)$. Find an equation of the smaller gear. Assume the point of tangency is on the line through the centers of the gears.

Figure 9

46. Watering Trough Construction: A cylindrical barrel is cut to make a watering trough in such a way that the trough is 8 inches high and 30 inches wide (Figure 10). What is the radius of the original barrel?

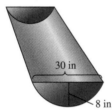

30 in

8 in

Figure 10

47. Bridge Design: An arch in the shape of a circle with a diameter measuring 50 meters is to support a bridge over a river (Figure 11). The center of the arch is to be 25 meters above the surface of the river.

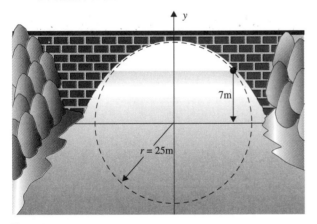

7m

$r = 25$m

Figure 11

(a) Find an equation of this semicircular arch.
(b) Find the point(s) on the arch if the water level rises 7 meters from the original surface of the river.

48. Dish Antenna Design: Suppose that an engineer has designed a dish antenna whose cross section is described by the model

$$y = 10x^2.$$

For the revision of the design, the engineer decides to rotate the dish antenna so that it opens to the right with its vertex on the x axis and 4 units to the right of the origin. Assuming the shape of the antenna is the same, what is the new equation of the dish antenna?

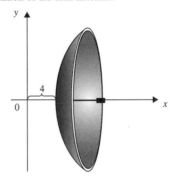

Developing and Extending the Concepts

In Problems 49 and 50, write each equation in standard form, and then find the vertex and axis of symmetry.

49. $2y^2 + 6y + 2x + 9 = 0$

50. $3x^2 + 12x + 6y + 13 = 0$

51. Find an equation of a circle so that the points $A = (-2,-3)$ and $B = (6,-3)$ are the ends of its diameter.

52. Find an equation of a circle whose center is at $(3,5)$ and is tangent to the x axis.

53. Without graphing, what can you say about the graph of
$$y = 3 - \sqrt{9 - (x - 2)^2}$$

54. Find an equation of a circle whose center is at $(-2,5)$ and whose circumference is 10π units.

G 55. Use a grapher with the given viewing window to graph each curve.
(a) $(x + 2)^2 + y^2 = 9$
   xMin $= -7$, xMax $= 5$, xScl $= 1$
   yMin $= -5$ yMax $= 5$, yScl $= 1$
(b) $x = -2y^2 + 4y - 3$
   xMin $= -6$, xMax $= 2$, xScl $= 1$
   yMin $= -5$, yMax $= 5$, yScl $= 1$
Note: If the graph seems distorted, use the ZOOMSQR feature of the grapher.

56. Find the equation for the given graphs.

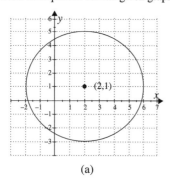

(a)

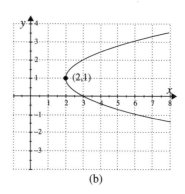

(b)

CHAPTER 6 REVIEW PROBLEM SET

In Problems 1 and 2, solve each equation by factoring.

1. (a) $x^2 - 14x - 15 = 0$
   (b) $6x^2 - 15x = 0$
   (c) $5x^2 = 2x + 24$

2. (a) $y^2 - 5y - 84 = 0$
   (b) $5t^2 - 10t = 0$
   (c) $4y(y + 1) = 3$

In Problems 3 and 4, obtain complex solutions of each equation. Check the solutions.

3. (a) $(x - 1)^2 + 9 = 0$
   (b) $(3m - 1)^2 + 25 = 0$

4. (a) $\left(x - \dfrac{3}{5}\right)^2 + \dfrac{16}{25} = 0$
   (b) $(2t + 3)^2 + 49 = 0$

In Problems 5 and 6, what term should be added to the given expression so it becomes a perfect square trinomial?

5. (a) $u^2 - 8u$
   (b) $x^2 + 11x$

6. (a) $x^2 - 14x$
   (b) $y^2 + 17y$

In Problems 7 and 8, solve each equation by completing the square. Give both the exact solutions and the two decimal approximations.

7. (a) $x^2 - 6x + 7 = 0$
   (b) $4x^2 + 13 = 12x$

8. (a) $y^2 + 4y + 21 = 0$
   (b) $2t^2 + 3 = 8t$

In Problems 9 and 10, solve each equation by the quadratic formula. Approximate irrational solutions to two decimal places. Write complex solutions in the form $a + bi$.

9. (a) $x^2 + 10x - 13 = 0$
   (b) $(2x - 1)(x + 2) = 5$

10. (a) $3 - x - x^2 = 0$
    (b) $4x^2 - x - 1 = 0$

In Problems 11 and 12, write a quadratic equation with integral values for a, b and c having the given solutions.

11. (a) $-4, -3$
    (b) $5 - \sqrt{7}, 5 + \sqrt{7}$
    (c) $3 - 4i, 3 + 4i$

12. (a) $-5, -5$
    (b) $-\sqrt{11}, \sqrt{11}$
    (c) $3 + \sqrt{5}i, 3 - \sqrt{5}i$

In Problems 13–18, solve each equation and check the results.

13. $\dfrac{3 - 2x}{4x} - 4 = \dfrac{3}{4x - 3}$

14. $\dfrac{2x - 5}{2x + 1} = \dfrac{7}{4} - \dfrac{6}{2x - 3}$

15. $\sqrt{5x + 5} + \sqrt{x - 4} = 5$

16. $\sqrt{5x + 1} = 1 + \sqrt{3x}$

17. $\sqrt{y - 3} + \sqrt{y} = 3$

18. $\sqrt{2x + 1} + \sqrt{8 - x} = 5$

In Problems 19–30, reduce each equation to a quadratic form by making a suitable substitution. Check for extraneous solutions.

19. (a) $x^4 - 5x^2 + 4 = 0$
    (b) $y^{-2} - 5y^{-1} + 4 = 0$

20. (a) $x^4 - 8x^2 + 16 = 0$
    (b) $y^{-2} - 8y^{-1} + 16 = 0$

21. (a) $1 - 2(x + 3)^2 - 3(x + 3)^4 = 0$
    (b) $1 - 2t^{-2} - 3t^{-4} = 0$

22. (a) $16(x + 1)^4 - 17(x + 1)^2 + 1 = 0$
    (b) $16t^{-4} - 17t^{-2} + 1 = 0$

23. $(x^2 + x)^2 - 8(x^2 + x) + 12 = 0$

24. $(y^2 + 4y)^2 - 17(y^2 + 4y) = 60$

25. $t^2 + 2t + 3(t^2 + 2t)^{-1} = 4$

26. $2t^2 - t - 6(2t^2 - t)^{-1} + 1 = 0$

27. $t^{2/3} - 5t^{1/3} + 6 = 0$

28. $t^{2/3} - 2t^{1/3} = 8$

29. $3x + x^{1/2} - 4 = 0$

30. $x^{2/5} + 8x^{1/5} + 16 = 0$

In Problems 31–46, find all x intercepts of the graph of the function. Round off to two decimal places when necessary.

31. (a) $f(x) = 3x^2 - 2 + x$
    (b) $f(x) = x^2 - 8x - 9$

32. (a) $f(x) = 10x^2 + 19x - 15$
    (b) $f(x) = -8x^2 + 26x - 15$

33. (a) $f(x) = 16x^3 - 34x^2 - 15x$
    (b) $f(x) = 3x^3 + 2x^2 - 27x - 18$

34. (a) $f(x) = 6x^3 - 43x^2 + 20x$
    (b) $f(x) = 8x^3 - 34x^2 + 30x$

35. $f(x) = 2x^2 + 18x + 4$

36. $f(x) = -x^2 + 5x + 8$

37. $f(x) = 0.1x^2 + 0.6x - 1.2$

38. $f(x) = 1.74x^2 - 2.04x + 6.2$

39. $f(x) = 2x^{-2} + x^{-1} - 1$

40. $f(x) = 6x^{-2} - x^{-1} - 12$

41. $f(x) = x - 13\sqrt{x} + 40$

42. $f(x) = x^{2/3} + 11x^{1/3} + 28$

43. $f(x) = x^4 - 7x^2 + 12$

44. $f(x) = -3x^4 - 14x^2 + 5$

45. $f(x) = \dfrac{1}{x-2} - 6\sqrt{\dfrac{1}{x-2}} - 40$

46. $f(x) = \left(\dfrac{x-1}{x}\right)^2 - 4\left(\dfrac{x-1}{x}\right) - 12$

In Problems 47–50, solve each inequality. Write the solution set in interval notation and illustrate the solution on a number line.

47. (a) $x^2 + 2x - 15 < 0$
    (b) $(x-1)(x+1)(x-2) > 0$

48. (a) $4x^2 - 3x \geq 10 + 3x^2$
    (b) $(x+1)(x-2)(x-4) > 0$

49. (a) $\dfrac{3x-1}{5x-7} \leq 0$ (b) $\dfrac{(x+1)(x-2)}{x-5} \leq 0$

50. (a) $\dfrac{7x-3}{9x+1} \geq 0$ (b) $\dfrac{x+2}{(x-2)(x+7)} > 0$

In Problems 51 and 52,

  (i) write $f(x)$ in the form $f(x) = a(x-h)^2 + k$.
  (ii) Find the vertex.
  (iii) Find the intercepts.
  (iv) Find the range.
  (v) Find the maximum or minimum value.
  (vi) Sketch the graph of f.

51. (a) $f(x) = -3x^2 - 12x - 8$
    (b) $f(x) = x^2 + 2x + 5$ (c) $f(x) = -x^2 + 4x - 8$

52. (a) $f(x) = -2x^2 - 20x + 47$
    (b) $f(x) = 2x^2 + 16x + 33$ (c) $f(x) = -2x^2 - 4x + 1$

In Problems 53 and 54, write each equation in standard form. Find the center and radius of each circle.

53. (a) $x^2 + y^2 + 8x - 6y + 21 = 0$
    (b) $x^2 + y^2 - 6x - 4y + 4 = 0$

54. (a) $x^2 + y^2 + 2x - 8y + 16 = 0$
    (b) $x^2 + y^2 + 6y = 0$

In Problems 55 and 56, write each equation in standard form. Find the vertex and sketch the graph of the parabola.

55. (a) $x = y^2 + y - 12$ (b) $x = -y^2 - y + 2$

56. (a) $x = y^2 + 3y - 10$ (b) $x = -y^2 - y + 6$

In Problems 57 and 58, solve each equation for the indicated variable.

57. (a) $s = vt + \dfrac{gt^2}{2}$, for t (b) $s = \pi r^2 + 2\pi rh$, for r

58. (a) $s = \pi r^2 + 2\pi r$, for r (b) $2N = k^2 - 3k$, for k

59. **Number Problem:** Find a number x such that the reciprocal of the expression $x - \sqrt{3}$ is equal to the expression $x + \sqrt{3}$.

60. **Falling Objects:** An object is thrown vertically upward from the ground with an initial speed of 60 feet per second. Its height h (in feet) t seconds after it is released is given by the model

$$h = 60t - 16t^2, \; t \geq 0$$

  (a) When is the object 50 feet high?
  (b) When will the object hit the ground?

61. **Camping Tent:** An entrance of a tent is in the shape of an isosceles triangle whose height is 1 foot less than twice its base. Find the height and the base of the triangular entrance if its area is 28 square feet. Round off to two decimal places.

62. **Car Travel:** A sales representative travels 280 miles at a certain speed. In the return trip, the car travels 5 miles per hour faster and as a result the trip takes 1 hour less. How fast is the sales representative traveling in each part of the trip?

63. **Law Enforcement:** A law enforcement officer estimates that the braking distance d (in feet) for a particular make of car is given by the model

$$d = 0.1v^2 + v, \; v \geq 0$$

where v is the speed of the car in miles per hour. What is the speed of the car at the time of an accident if it leaves skid marks measuring 120 feet?

64. **Jogging:** Two joggers leave the same point and travel at right angles to each other, each moving at a uniform speed. One jogs 1 mile per hour faster than the other. After 3 hours, they are 15 miles apart. Find the speed of each jogger.

65. **Investment Portfolio:** Suppose P_1 dollars are invested at an annual interest rate r (percent). After one year, an additional P_2 is added to the account at the same rate. At the end of the second year, the balance A in the account is given by the model

$$A = P_1(1 + r)^2 + P_2(1 + r)$$

An investor invests \$1000 in a savings account for 2 years. At the beginning of the second year, an additional \$2000 is invested. If a total of \$3368 is in the account at the end of the second year, what is the annual interest rate r?

66. **Boating:** A boat can travel 18 miles per hour in still water. It can travel 32 miles upstream and downstream in a total of 4 hours. What is the speed of the current?

67. **Advertising:** A printed poster is to have 3 inch margins at the top and bottom and 2 inch margins on the sides. The area of the printed portion is 48 square inches and the total area of the poster is 160 square inches. What is the possible length and width of the poster?

68. **Ticket Pricing:** Tickets to a concert cost \$30 for each customer. At this price, 800 people attend the concert. However, it is projected that for \$1 increase in a ticket price, the average attendance will decrease by 10. At what price will the receipts be \$30,250?

69. **Launched Object:** A projectile is shot straight upward from the ground with an initial speed of 64 feet per second. Its height (in feet) t seconds after it is shot is given by the model

$$h = 64t - 16t^2$$

  (a) Find the maximum height attained by the projectile.
  (b) Find the time it takes for the projectile to return to earth.

70. **Fencing a Pen:** A farmer wishes to fence a rectangular pen with 200 feet of fencing. Find the dimensions of the pen with maximum area.

71. **Ballistics:** A ball is thrown vertically upward from the ground with an initial speed of 64 feet per second. Its height h (in feet) after t seconds is given by the model

$$h = 64t - 16t^2$$

Find the time interval when the ball is at least 48 feet above the ground.

72. **Number Problem:** Find a number such that the sum of the number and its reciprocal is at most $5/2$.

 In Problems 73–76, use a grapher to graph each function. Use the x intercepts of the graph to solve the associated equation and inequality. Round off to two decimal places if necessary.

73. $f(x) = x^2 - 0.75x - 0.25$
$$x^2 - 0.75x - 0.25 = 0$$
$$x^2 - 0.75x - 0.25 > 0$$

74. $f(x) = 1.41x^2 + 5x + 1.41$
$$1.41x^2 + 5x + 1.41 = 0$$
$$1.41x^2 + 5x + 1.41 \le 0$$

75. $f(x) = 5x^2 - 13x - 8$
$$5x^2 - 13x - 8 = 0$$
$$5x^2 - 13x - 8 \le 0$$

76. $f(x) = 10x^2 - 13x - 3$
$$10x^2 - 13x - 3 = 0$$
$$10x^2 - 13x - 3 > 0$$

◆ CHAPTER 6 PRACTICE TEST

Solve each equation. Round off answers to two decimal places when necessary.

1. (a) $x^2 - 2x - 15 = 0$
   (b) $6x(x - 1) = 12$ (c) $2x^3 + 13x^2 + 21x = 0$
2. (a) $(x + 1)^2 + 4 = 0$
   (b) $x^2 - 4x + 2 = 0$ (c) $3x^2 - 18x - 5 = 0$
3. (a) $3x^2 - 5x + 4 = 0$
   (b) $9x^2 - 12x + 4 = 0$ (c) $3x^2 + 2x - 3 = 0$
4. $\sqrt{2x + 1} + \sqrt{x} = 1$
5. $x^{2/3} - 17x^{1/3} + 16 = 0$
6. Write a quadratic equation with integral coefficients a, b and c having -3 and 6 as solutions.

In Problems 7 and 8, solve the inequality and write the solution in interval notation. Show the solution on a number line. Use this solution to determine the domain of function f.

7. $2x^2 + x + 1 > 0$ $f(x) = \dfrac{1}{\sqrt{2x^2 + x + 1}}$

8. $\dfrac{2x + 3}{x - 5} \ge 0$ $f(x) = \sqrt{\dfrac{2x + 3}{x - 5}}$

9. Sketch the graph of the function
$$f(x) = -2(x - 1)^2 + 3.$$
Identify the vertex, the axis of symmetry, the domain and range of f.

10. Rewrite $f(x) = 2x^2 - 8x - 3$ in the form
$f(x) = a(x - h)^2 + k$.
Sketch the graph and find the range of f.

11. (a) Find the radius and center of the circle
$$x^2 + y^2 + 6x - 2y - 15 = 0$$
    (b) Sketch the graph of the parabola $x = 3(y + 1)^2 + 2$.

12. A homeowner is designing a rectangular garden in such a way that its length is 3 feet less than twice its width. What is the length and the width of the garden if its area is 104 square feet?

13. A cliff diver's path from the top of a cliff 120 feet above the surface of the water after t seconds is modeled by the function

$$h = -t^2 + 4t + 120, \; t \ge 0$$

where h is the height in feet above the water.
(a) Find the maximum height of the diver.
(b) Find the time it takes for the diver to reach the maximum height.

Chapter 7

EXPONENTIAL AND LOGARITHMIC FUNCTIONS

7.1 Exponential Functions

7.2 Logarithmic Functions

7.3 Evaluating Logarithms

7.4 Graphs of Logarithmic Functions

7.5 Properties of Logarithms

7.6 Exponential and Logarithmic Equations

You are probably familiar with bank accounts in which the interest is said to be *compounded*. A compounded interest model is devised to determine the interest on a savings account. The essential variable in the model is the number of times a year the interest is compounded. The interest may be compounded quarterly, monthly, weekly, or daily. As you will see in this chapter, it is also possible for interest to be compounded continuously. A continuous compounding function can be used as a powerful modeling tool. As an illustration, in 1626 Peter Minuit of the Dutch West India company purchased Manhattan Island from the Indians for $24. Assuming continuous compounding at a rate of 4% per year, how much will Manhattan be worth after 374 years? The answer to this question is provided in Example 5 on page 344.

In this chapter we introduce two closely related functions—*exponential* and *logarithmic functions*. These functions are widely used to develop mathematical models from real-life situations in fields such as business, seismology, engineering, psychology, and economics.

OBJECTIVES

1. Evaluate Exponential Functions
2. Graph Exponential Functions
3. Transform Graphs of Exponential Functions
4. Solve Applied Problems
5. Use Technology Exploration

7.1 Exponential Functions

When a *disease epidemic* like AIDS starts, the number of people with the disease is often expressed as a function of time since the start of the epidemic.

When a population grows over time, the *population increase* can be expressed as a function of time since the start of the population. When a person deposits money in an interest-bearing account, the *bank balance* increases as a function of time since the initial deposit. Each of these situations can be modeled by an *exponential function*.

Evaluating Exponential Functions

In section 5.1 we introduced the idea of exponential functions and we used the graphs of the functions to solve simple exponential equations. Recall that a function of the form

$$f(x) = b^x \quad \text{or} \quad f(x) = a \cdot b^{cx}$$

where b is a positive constant, $b \neq 1$, is called an **exponential function** with base b.

The restriction $b \neq 1$ is made to exclude the constant function $f(x) = 1^x = 1$

Additional examples of exponential functions are:

$$f(x) = 3^x, \ g(x) = \left(\frac{1}{2}\right)^x, \ \text{and} \ h(x) = 500(1.07)^{-2x}$$

Note that in exponential functions the variables are in the exponents.

We have seen how to use rational exponents for exponential functions. Using a calculator we can interpret values of exponential functions with irrational exponents, as the next example shows. As long as b is positive, b^x has a meaning for *any* input real number x. Also the properties of exponents still hold when exponents are real numbers.

EXAMPLE 1   Evaluating Exponential Functions

Consider the functions $g(x) = 5^{-x}$ and $g(x) = 1000(1.03)^x$. Find the output values for the given input values for each function at: $x = 1.2$, $x = \sqrt{3}$ and $x = -\sqrt{7}$. Round off each result to three decimal places.

Solution   Using a calculator and rounding off to 3 decimal places, we have

(a) $f(x) = 5^{-x}$ (b) $g(x) = 1000(1.03)^x$

$f(1.2) = 5^{-1.2} = 0.145$ $g(1.2) = 1000(1.03)^{1.2} = 1{,}036.107$

$f(\sqrt{3}) = 5^{-\sqrt{3}} = 0.062$ $g(\sqrt{3}) = 1000(1.03)^{\sqrt{3}} = 1{,}052.531$

$f(-\sqrt{7}) = 5^{-(-\sqrt{7})} = 70.681$ $g(-\sqrt{7}) = 1000(1.03)^{-\sqrt{7}} = 924.775$ ◈

Graphing Exponential Functions

In section 5.1 we graphed exponential functions. As the following examples show, the graphs of exponential functions have several distinctive features.

EXAMPLE 2   Graphing Exponential Functions

Sketch the graph of each exponential function.

(a) $f(x) = 3^x$ (b) $g(x) = \left(\dfrac{1}{3}\right)^x$

Use the graphs to indicate the domain and range of each function.

Solution   We compute output values for different input values for each of the two functions. Then we plot these points and connect them with a smooth curve since these functions are defined for all real input values.

x	-3	-2	-1	0	1	2	3
$f(x) = 3^x = y$	1/27	1/9	1/3	1	3	9	27
$g(x) = \left(\frac{1}{3}\right)^x = y$	27	9	3	1	1/3	1/9	1/27

A number of features should be noted about the graphs of these functions (Figure 1(a) and 1(b)). First, the y intercept of each graph is 1, but neither has an x intercept. The graphs suggest that the domain of each function consists of all real numbers $\mathcal{R}$, and the range of each consists of positive real numbers or $(0,\infty)$. We can also observe from Figure 1(a) that as the x values are increasing, the y values are also increasing; while Figure 1(b) shows that as the x values are increasing, the y values are decreasing.

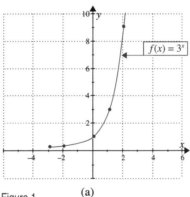   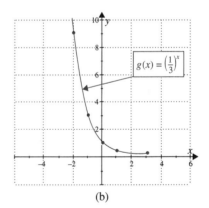

Figure 1   (a)   (b)

Table 1 summarizes these features.

TABLE 1

$f(x) = 3^x$	Domain $\mathcal{R}$	Range $(0,\infty)$	increasing
$g(x) = \left(\frac{1}{3}\right)^x$	Domain $\mathcal{R}$	Range $(0,\infty)$	decreasing

As illustrated in Figure 2a, the graphs of the functions $f(x) = b^x$, where $b > 1$, resemble the graph of $y = 2^x$. The larger the value of b, the faster the graph rises to the right. To see this, compare the graphs of $y = 2^x$ and $y = 3^x$. Similarly, if $0 < b < 1$, the graphs of $f(x) = b^x$ have the same general shape and characteristics as the graph of $y = (1/2)^x$. The smaller the value of b, the slower the graph falls to the right. Compare the graphs of $y = (1/2)^x$ and $y = (1/3)^x$ (Figure 2b).

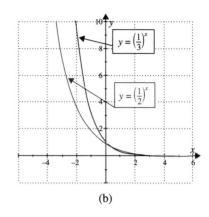

(a)

(b)

Figure 2

We have already seen how the value of the base b influences the graph of an exponential function. Now we choose an exponential function with an important base and analyze its graph.

One important base that is widely used in the study of exponential functions is the number e. The value of e can be obtained by approximating the expression $[1 + (1/n)]^n$ for different values of n. Table 2 shows output values of the expression rounded to six decimal places.

TABLE 2

n	1	2	4	12	365	8760
$[1 + (1/n)]^n$	2.000000	2.250000	2.441406	2.613035	2.714567	2.718127

The output values in the table suggest that as n increases, the expression $[1 + (1/n)]^n$ keeps getting closer and closer to a very important fixed number, denoted by e, in honor of the Swiss mathematician Leonhard Euler (1707–1783) (pronounced oiler). By using advanced methods and high speed computers, the numerical value of e has been calculated to thousands of decimal places. Rounded to six decimal places

$$e = 2.718282$$

EXAMPLE 3   Evaluating Exponential Functions

Let $f(x) = e^x$. Find each output value rounded to 4 decimal places. Also sketch the graph.

(a) $f(-3)$ (b) $f(-2)$ (c) $f(-1)$ (d) $f(0)$ (e) $f(1.2)$ (f) $f(\sqrt{3})$

Solution   Since $e > 1$, the graph of $f(x) = e^x$ has the same shape as the graph of $y = b^x$ for $b > 1$ (Figure 3). The output values are obtained by using a calculator.

(a) $f(-3) = e^{-3} = 0.0498$
(b) $f(-2) = e^{-2} = 0.1353$
(c) $f(-1) = e^{-1} = 0.3679$
(d) $f(0) = e^0 = 1.0000$
(e) $f(1.2) = e^{1.2} = 3.3201$
(f) $f(\sqrt{3}) = e^{\sqrt{3}} = 5.6522$

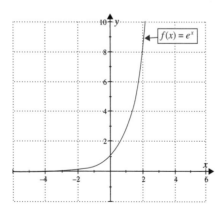

Figure 3

By plotting and connecting these points, we observe that each will lie on the graph of f. The graph suggests the domain of f consists of real numbers $\mathfrak{R}$ and the range consists of positive real numbers or $(0,\infty)$.

Transforming Graphs of Exponential Functions

As with the graphs of other functions, the graph of an exponential function can be reflected, stretched vertically or shifted horizontally and vertically to obtain graphs of other exponential functions.

EXAMPLE 4   Transforming Graphs

Sketch the graphs of $f(x) = e^x$, $g(x) = e^{x-2}$, and $h(x) = e^x - 2$ on the same coordinate system. Compare the graphs of f, g, and h.

Solution   First we create a table for input values from -2 to 2 for each of these functions and round the output values to 2 decimal places.

x	$f(x) = e^x$	$g(x) = e^{x-2}$	$h(x) = e^x - 2$
-2	0.14	0.02	-1.86
-1	0.37	0.05	-1.63
0	1.00	0.14	-1.00
1	2.72	0.37	0.72
2	7.39	1.00	5.39

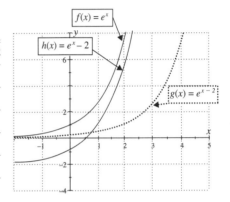

Figure 4

We then plot these points from the above table for each function and connect them with smooth graphs on the same coordinate system (Figure 4). Notice that the three graphs are identical in shape, but $g(x) = e^{x-2}$ is shifted 2 units to the right of $f(x) = e^x$. The graph of $h(x) = e^x - 2$ is obtained by shifting the graph of $f(x) = e^x$, 2 units downward. ◈

Solving Applied Problems

A large number of mathematical models using exponential functions are derived from the annual *compound interest* model

$$A = P(1 + r)^t$$

The above model can be generalized to the **compound interest** formula that was covered in section 5.1. Recall that

$$A = P\left(1 + \frac{r}{n}\right)^{nt}$$

where A is the *future value*, P is the *present value* (also known as the *principal*), r is the *annual interest rate* as a decimal, and t is the *number of years*. When interest is compounded n times per year for larger and larger values of n, we say that the interest is **compounded continuously**. In this situation, we use the formula

$$A = Pe^{rt}$$

for continuously compounded interest, where A, P, r, and t are defined as in the above formula.

EXAMPLE 5   Purchasing Manhattan Island

In 1626 Peter Minuit purchased Manhattan Island for $24 worth of trinkets and beads. Suppose that the $24 had been deposited in a forgotten savings account that paid 4% annual interest. Find the amount of this investment after 374 years if the interest is compounded (a) quarterly and (b) continuously.

Solution   (a) Using the formula

$$A = P\left(1 + \frac{r}{n}\right)^{nt}$$

with $P = 24$, $r = 0.04$, $n = 4$, and $t = 374$, we have

$$A = 24[1 + (0.04/4)]^{4(374)} \approx 69{,}982{,}011.38$$

Therefore, the $24 would have increased to $69,982,011.38.

(b) For continuously compounded interest, we use the formula

$$A = Pe^{rt}$$

with $P = 24$, $r = 0.04$, $t = 374$. So that

$$A = 24 \, e^{0.04(374)} \approx 75{,}380{,}096.83$$

In this case, the $24 would have increased to $75,380,096.83.   ◈

Exponential functions represented by the functions

$$A(t) = Pb^t \text{ and } A(t) = Pb^{-t}$$

in which either the quantity A *increases* or *grows exponentially* as a function of t, or the quantity A *decreases* or *decays exponentially* as a function of t, serve as models for diverse phenomena called **exponential growth and decay**. The simplest growth model predicts a reasonably accurate description of growth of population such as bacteria, zebra mussels, and cell cultures.

EXAMPLE 6   Modeling Spread of AIDS

In 1995 the World Health Organization estimated the number N (in millions) of people infected by HIV, the virus that leads to AIDS, was about 18 million people, and the number was growing after t years according to the model

$$N = 18(1.087)^t$$

Assuming no new medical breakthrough, we can use this model to predict the number of people, to the nearest million, infected by the year (a) 2003 and (b) 2010.

Solution   (a) $t = 0$ corresponds to the year 1995. The year 2003 corresponds to $t = 8$, so that

$$N = 18(1.087)^8 = 35.08$$

Therefore, the number of people infected in the year 2003 would be about 35 million people.

(b) $t = 15$ corresponds to the year 2010, so that

$$N = 18(1.087)^{15} = 62.91$$

Therefore, the number of people infected in the year 2010 would be about 63 million.   ◈

Using Technology Exploration

The use of a grapher can help find output values of a function for given input values as the next example shows.

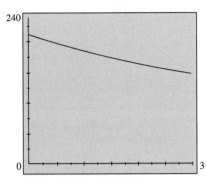

Figure 5

G EXAMPLE 7   Solving a Restaurant Management Problem

At a famous restaurant, the chef determines that the soup should be served to customers at a temperature of no less than 160°F. It has been determined that the cooling rate for this soup is 0.21°F per minute. According to a law from physics, known as *Newton's Law of Cooling*, the temperature T of the soup t minutes after it is removed from a boiling pot is given by the model

$$T = 68 + 144e^{-0.21t}, t \geq 0$$

(a) Use a grapher to graph the function T.
(b) Use the TRACE feature to predict the temperature after 1, 2, and 2.2 minutes.

Solution   (a) Figure 5 shows the graph of the function T with the given viewing window.

(b) Using the TRACE of the graph, we predict the temperature T after 1, 2, and 2.2 minutes to be 185°F, 163°F, and 159°F respectively. ✧

◈ PROBLEM SET 7.1

Mastering the Concepts

In Problems 1–8, use a calculator to approximate the output values for each function. Round off to 3 decimal places.

1. $f(x) = 3^x$
   (a) $f(1.52)$ (b) $f(-\sqrt{5})$
   (c) $f(3 - \sqrt{2})$ (d) $f(\pi)$

2. $g(x) = \left(\dfrac{2}{3}\right)^x$
   (a) $g(0.43)$ (b) $g(-1.71)$
   (c) $g(\sqrt{3})$ (d) $g(1 + \pi)$

3. $h(t) = 100(1.015)^t$
   (a) $h(0.21)$ (b) $h(3.2)$
   (c) $h(3)$ (d) $h(20)$

4. $F(x) = 0.75(0.94)^{x/2}$
   (a) $F(3)$ (b) $F(2.5)$
   (c) $F(-0.05)$ (d) $F(1.012)$

5. $f(x) = e^{2x}$
   (a) $f(1.3)$ (b) $f(-2.1)$
   (c) $f(2.4)$ (d) $f(\sqrt{2})$

6. $f(x) = 36e^{-0.2x}$
   (a) $f(-1.4)$ (b) $f(2.7)$
   (c) $f(-4)$ (d) $f(5.3)$

7. $A(x) = 2000\left(1 + \dfrac{0.06}{4}\right)^{2x}$
   (a) $A(4)$ (b) $A(7)$
   (c) $A(12)$ (d) $A(20)$

8. $P(x) = 40(1 + e^{-0.04x})^{-1}$
   (a) $P(-1)$ (b) $P(0.41)$
   (c) $P(-0.61)$ (d) $P(1.7)$

In Problems 9–12, each graph is a transformation of the graph of $y = 2^x$. Match the function with its graph.

9. $f(x) = 2^{x+1}$ 10. $f(x) = 2^x + 1$
11. $f(x) = -2^x$ 12. $f(x) = 2^x - 1$

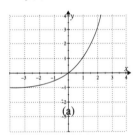

(a)

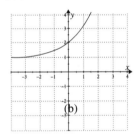

(b)

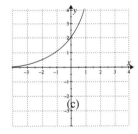

(c)

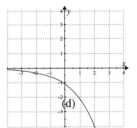

(d)

In Problems 13–16, each graph is a transformation of the graph of $y = e^{-x}$. Match each function with its graph.

13. $f(x) = e^{-x} - 1$ 　　 14. $f(x) = e^{-(x-1)}$
15. $f(x) = -e^{-x}$ 　　 16. $f(x) = e^{-x} + 1$

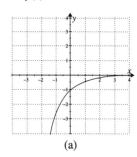

(a)

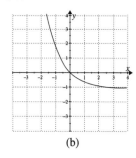

(b)

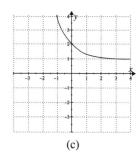

(c)

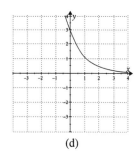

(d)

In Problems 17–22, sketch the graph of each function. Indicate the domain and range.

17. $f(x) = 5^x$ 　　 18. $g(x) = \left(\dfrac{1}{5}\right)^x$ 　　 19. $f(x) = 4^{-x}$

20. $g(x) = 6^{-x}$ 　　 21. $f(x) = e^{-x}$ 　　 22. $g(x) = -\left(\dfrac{1}{2}\right)^x$

In Problems 23–32, sketch each pair of functions on the same coordinate system. Compare the graphs of f and g.

23. (a) $f(x) = 3^x$ 　　 (b) $g(x) = 3^{-x}$
24. (a) $f(x) = 4^x$ 　　 (b) $g(x) = 4^x - 3$
25. (a) $f(x) = 2^x$ 　　 (b) $g(x) = 2^x + 3$
26. (a) $f(x) = -2^x$ 　　 (b) $g(x) = -2^{x+1}$
27. (a) $f(x) = 4^x$ 　　 (b) $g(x) = -4^{-x}$
28. (a) $f(x) = \left(\dfrac{1}{2}\right)^x$ 　　 (b) $g(x) = -\left(\dfrac{1}{2}\right)^x$
29. (a) $f(x) = 2^x$ 　　 (b) $g(x) = 5^x$
30. (a) $f(x) = 3^x$ 　　 (b) $g(x) = 7^x$
31. (a) $f(x) = e^x$ 　　 (b) $g(x) = e^x + 2$
32. (a) $f(x) = e^x$ 　　 (b) $g(x) = e^x - 4$

Applying the Concepts

33. **Compound Interest:** Find the amount of money owed at the end of 4 years for a $12,000 business loan at an annual interest rate of 8% compounded (a) quarterly (b) monthly (c) continuously.

34. **Compound Interest:** An investor has $2000 to deposit for 3 years. Shopping around, the investor discovers that it is being calculated in four different ways: 4% quarterly, 3.95% monthly, 3.9% daily, and 3.9% continuously. Calculate the amount of money the investor would have with each method after 3 years. Which method would yield the most return?

35. **Population Growth:** In 1994 the population of Mexico City was 15.5 million and growing at an annual rate of approximately 0.7% or 0.007. The number of people, N, t years after 1994 is given by the model

$$N = 15.5(1 + r)^t, \ t \geq 0$$

Using this model, predict the population of the city in (a) the year 2005 (b) the year 2010.

36. **Population Growth:** In 1798 the *English economist Thomas Malthus* invented a model for predicting the population growth P after t years (assuming that the birth rate b and the death rate d are fixed). The model is

$$P = P_o e^{kt}$$

where P_o is the initial population and k is the annual growth (or decay), $k = b - d$ and t is the number of years since the initial population. This model is known as the *Malthusian Model.* According to the U.S. Census Bureau, the population of the United States in 1990 was 244 million, the birth rate was 1.6% or 0.016 and the death rate was 0.86% or 0.0086. Suppose that the population grew according to the Malthusian Model. Predict the population in the year 2005 and in the year 2010.

37. **Salvage Value:** A sport utility vehicle was purchased for $31,000. Its value is expected to depreciate by 18% of the previous year's value so that its value V (in dollars) after t years is given by the model

$$V = 31,000(1 + r)^{-t}, \ t \geq 0$$

where r is the annual rate of depreciation. Find the value of this vehicle after (a) 2 years (b) 4 years and (c) 5 years.

38. **Caffeine Consumption:** Caffeine is a chemical stimulant that is found in coffee, teas, and some soft drinks. An average person eliminates 10% of this compound each hour after ingestion. Suppose an office worker drinks a cup of coffee that contains 50 milligrams of caffeine. The number of milligrams N of caffeine that remains in the worker's system after t hours is given by the model

$$N = 50(1 + r)^{-t}, \ t \geq 0$$

where r is the hourly rate for eliminating the compound after ingestion. How much caffeine remains in the worker's system after (a) 2 hours (b) 2.5 hours?

39. **Housing Values:** State and local agencies frequently use the formula
$$S = C(1 + r)^t$$
to assess the value S (in dollars) of a property t years after it was purchased at a price of C dollars, assuming an annual rate of inflation r. During the 1970s and 1980s, California real estate prices soared dramatically, with some areas seeing an average annual inflation rate as high as 14%. If a ranch house sold for $250,000 in March 1976, what was the price of the house in March of 1980, assuming 14% annual inflation?

40. **Cost of Living:** In 1990 the cost of tuition at a certain college was $10,000. The cost (in dollars) after t years is modeled by the function
$$C(t) = 10000e^{0.05t}$$
Use this model to predict the cost of tuition in the year (a) 2003 and (b) 2010.

41. **Environmental Science**: According to the *Bouguer–Lambert Law*, the percentage P of light that penetrates ordinary sea water to a depth of d feet is given by the function
$$P = e^{-0.044d}$$
What is the percentage of light penetration in sea water to a depth of (a) 4 feet? (b) 6 feet?

42. **Recycling:** In the *State of Michigan*, stores charge an additional $0.10 deposit for each bottle of beverage sold. Customers return these bottles for refunds, and then the stores return them to the beverage companies for recycling. Suppose that 65% of all bottles distributed will be recycled every year. If a company distributed 2.5 million bottles in 1 year, the number N (in millions) of recycled containers still in use after t years is given by the model
$$N = 2.5(0.65)^t, \ t \geq 0$$
How many recycled bottles (in millions) are still in use after (a) 1 year? (b) 3 years? (c) 5 years? Round off the answers to two decimal places.

43. **Radioactive Decay:** Certain radioactive elements decay exponentially. The decay model for a specific radioactive element is
$$A = A_o e^{-0.04463t}, \ t \geq 0$$
where A is the amount present after t days and A_o is the amount present initially. Assume there is a block of 50 grams of the element at the start. How much of the element remains after (a) 10 days? (b) 15 days? (c) 30 days?

44. **Baking a Cake:** Suppose that a baked cake is removed from an oven and cooled to a room temperature of 70°F. The temperature T of the cake after t minutes is given by the model
$$T = 70 + 280 \ e^{-0.14t}, \ t \geq 0$$
What is the temperature of the cake after (a) 5 minutes? (b) 10 minutes? and (c) 20 minutes?

Developing and Extending the Concepts

In Problems 45 and 46, state if the inequality is true or false.

45. (a) $e^{-1} < \dfrac{1}{3}$ (b) $\dfrac{1}{3} < e^{-1} < \dfrac{1}{2}$
    (c) $e^{-1} > \dfrac{1}{2}$

46. (a) $\sqrt{e} < 1$ (b) $1 < \sqrt{e} < 1.6$
    (c) $1.5 < \sqrt{e} < 2$

G 47. Use a grapher to graph the following functions.
$y = 2^x, y = 3^x, y = 4^x$, and $y = 5^x$ on the same viewing window $x\text{Min} = -2$, $x\text{Max} = 2$, $x\text{Scl} = 1$, $y\text{Min} = 0$, $y\text{Max} = 6$, $y\text{Scl} = 1$
   (a) What one point lies on all four graphs?
   (b) Compare the steepness of the graphs.

G 48. Use a grapher to graph the following functions.
$y = \left(\dfrac{1}{2}\right)^x$, $y = \left(\dfrac{1}{3}\right)^x$, $y = \left(\dfrac{1}{4}\right)^x$, and $y = \left(\dfrac{1}{5}\right)^x$ on the viewing window, $x\text{Min} = -2$, $x\text{Max} = 2$, $x\text{Scl} = 1$, $y\text{Min} = 0$, $y\text{Max} = 6$, $y\text{Scl} = 1$
   (a) What one point lies on all four graphs?
   (b) Compare the steepness of the graphs.

G 49. Let $f(x) = (2^x + 2^{-x})^2 - (2^x - 2^{-x})^2$ and $g(x) = 4$. Use a grapher to graph the functions f and g on the same viewing window. Compare the two graphs.

G 50. Use a grapher to graph each function $f(x) = e^x + e^{-x}$ and $g(x) = e^x - e^{-x}$. Use the graph to indicate the domain and range of each function.

51. **Harvesting Eucalyptus:** Some farmers grow eucalyptus for floral arrangements. The more they cut the branches, the fewer branches there will be for the next cutting. The number of bundles N harvested in one year is approximated by the model
$$N = 4000(1 + e^{-0.06(t - 2)}), \ 1 \leq t \leq 3$$
where t is the number of times the stems have been harvested within the year. How many bundles will be harvested when t equals (a) 1 (b) 2 and (c) 3? Round to the nearest bundle.

52. **Future Value:** The formula
$$F = m\left[\frac{(1 + i)^{kn} - 1}{i}\right]$$
represents the *future value F* (in dollars) of an annuity with deposits of m dollars made regularly k times each year for n years, with interest compounded k times per year at an annual rate r, where i is the periodic rate, $i = r/k$. How much is the account worth if $600 were deposited each month at an annual interest rate of 6% compounded monthly for a period of 2 years?

53. **Present Value:** The formula for the present value P (in dollars) of an annuity is given by

$$P = m\left[\frac{[(1 + i)^{kn} - 1]}{i(1 + i)^{kn}}\right]$$

where m (in dollars) represents the regular payments made k times per year for n years with interest compounded k times per year at an annual rate r for each of the n periods. Find the present value of a state lottery that pays 4 annual payments of $250,000 at an annual interest rate of 5.5%, where $i = r/k$.

54. **Retirement Installment:** A common method for financing retirement is to invest a large amount of money at a given rate r compounded annually and then withdraw a fixed amount of money from this investment each year. The amount of possible withdrawal W (in dollars) if an investment P (in dollars) is expected to last for t years is given by the model

$$W = \frac{Pr(1 + r)^t}{(1 + r)^t - 1}$$

where r is the annual interest rate. If $400,000 is invested in an account at a rate of 7% with the intention of making withdrawals for 12 years, how much can be withdrawn each year?

55. **Data Analysis:** In 1991 the population P (in millions) of India was about 866 million and growing after t years according to the table:

Year	1991	1992	1993	1994	1995	1996	1997
Population	866	883	894	917	934	952	970

(a) Construct a scattergram that exhibits the data.
(b) Assuming that this relationship is approximately exponential, determine how closely the function

$$P = 866e^{0.019t}$$

fits the data ($t = 0$ represents the year 1991).

G (c) Use a grapher to graph the function in part (b). Use the window: xMin = 0, xMax = 7, xScl = 1, yMin = 800, yMax = 1000, yScl = 100.

56. **Data Analysis:** According to the Television Bureau of Advertising, Inc. the percent p of TV households with VCR's is patterned by the following table.

Year t	80	88	89	90	91	92	93	94	95	96	97
Percent	1.1	58.0	64.6	68.6	71.9	75.0	77.1	77.2	77.5	77.6	77.7

(a) Construct a scattergram that exhibits the data.
(b) Assuming this relationship is exponential, determine how closely the function

$$P = 77.8\,[1 + 63.4e^{-0.638t}]^{-1}$$

fits the data ($t = 0$ represents the year 1980).

G (c) Use a grapher to graph the function in part (b). Use the window: xMin = 0, xMax = 17, xScl = 2, yMin = 0, yMax = 100, yScl = 10. This model is known as the *logistic model*.

OBJECTIVES

1. Describe the Inverse Function Geometrically
2. Find the Inverse of an Exponential Function—Logarithmic Function
3. Solve Logarithmic Equations of the Form $\log_b u = k$.
4. Solve Applied Problems
5. Use Technology Exploration

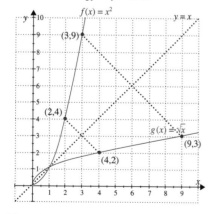

Figure 1

7.2 Logarithmic Functions

In this section, we introduce inverse exponential functions as a means of solving equations involving logarithms.

Describing the Inverse Function Geometrically

Many calculators, whether they are scientific or graphic, have both a squaring key (marked x^2) and a square root key (marked $\sqrt{x}$). For nonnegative numbers x, the functions represented by these keys "undo" each other in the sense that

$$x \xrightarrow{\text{squaring function}} x^2 \xrightarrow{\text{square-root function}} \sqrt{x^2} = x$$

$$x \xrightarrow{\text{square-root function}} \sqrt{x} \xrightarrow{\text{squaring function}} (\sqrt{x})^2 = x.$$

Two functions related in such a way that each "undoes" what the other "does" are said to be **inverses** of each other. Geometrically, we say that these *functions are inverses of each other if and only if the graph of g is the mirror image of the graph of f across the line $y = x$.* Using Figure 1, we observe that the graphs of $f(x) = x^2$, $x \geq 0$, and $g(x) = \sqrt{x}$ are the mirror images of each other with respect to the line whose equation is $y = x$. In this case, we say the function $g(x) = \sqrt{x}$ reverses the coordinates of the function $f(x) = x^2$, $x \geq 0$. That is, if the point (2,4) belongs to the graph of f, then the point (4,2) belongs to the graph of g. Similarly, if (3,9) belongs to the graph of f, then (9,3) belongs to the graph of g.

Note that f^{-1} is not the reciprocal of the function f. The reciprocal of f, $\frac{1}{f(x)}$ is denoted by $[f(x)]^{-1}$

Notice that the line segment connecting (2,4) and (4,2) is perpendicular to and cut in half by the graph of $y = x$. For this reason, *we say that the points (2,4) and (4,2) are symmetric with respect to the line $y = x$.* This notion can be generalized as follows: *the graph of $g(x) = \sqrt{x}$ is a reflection of the graph of $f(x) = x^2$, $x \geq 0$ with respect to the line $y = x$.* We call g *the inverse of f and denote g by f^{-1}.* Thus,

$$\text{if } f(x) = x^2, x \geq 0, \text{ then } f^{-1}(x) = \sqrt{x}.$$

EXAMPLE 1   Graphing a Function and Its Inverse
(a) Sketch the graph of the function $f(x) = 3^x$ and its inverse on the same coordinate system.
(b) Write the corresponding ordered pairs for f^{-1} whose coordinates are highlighted in the graph of f.

Solution   (a) The graph of the inverse of the function $f(x) = 3^x$ is obtained by reflecting the graph of f about the line $y = x$ (Figure 2).
(b) The ordered pairs corresponding to $(-2,1/9)$, (1,3), and (2,9) on the graph of f are $(1/9,-2)$, (3,1) and (9,2) on the graph of f^{-1}. ◈

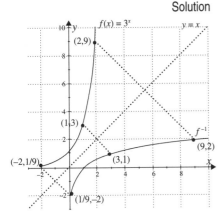

Figure 2

We can distinguish whether or not a function f has an inverse by drawing a horizontal line through the graph of f and determining if *this horizontal line intersects the graph of f more than once.* If this is the case, we say that *the function f has no inverse.* Using the horizontal line test, we see that the function f in Figure 3a has an inverse while the function g in Figure 3b does not. This is because any horizontal line in Figure 3a that intersects the graph of f (on its domain) crosses the graph exactly once, while every horizontal line in Figure 3b (in the domain of g) crosses the graph twice.

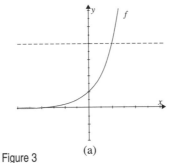

Figure 3   (a)   (b)

Finding the Inverse of an Exponential Function— The Logarithmic Function

We now turn our attention to answering the following question: If a function f has an inverse, how do we find the inverse function f^{-1}? Earlier, we observed that whenever (a,b) is on the graph of f, then (b,a) is on the graph of f^{-1}. This means that f^{-1} is the function that can be formed by interchanging the roles of the domain and range values of f. For instance, suppose that

$$f(x) = 3^x$$

First, we observe that any horizontal line that intersects the graph of f crosses this graph only once, and so f has an inverse f^{-1}. To write the equation for f^{-1}, we follow these steps:

Step 1. We start with $f(x) = 3^x$. Replace $f(x)$ by y, so that $y = 3^x$.
Step 2. Interchange x and y, to get $x = 3^y$.

Step 3. Solve for y. To solve for y, we introduce a new notation which we call the *logarithmic notation*. The symbol

$$\log_3 x$$

means "the power to which 3 is raised in order to produce a result of x". That is, $\log_3 x = y$ means $3^y = x$.

We say "the logarithm of x to the base 3" or "the log of x to the base 3". Thus, if $f(x) = 3^x$, then $f^{-1}(x) = \log_3 x$.

Also, if $f(x) = (1/2)^x$, then $f^{-1}(x) = \log_{1/2} x$.

More formally, we have the following definition.

DEFINITION—LOGARITHMIC FUNCTION

The *inverse of the exponential function $f(x) = b^x$*, where $b > 0$, $b \neq 1$, is the **logarithmic function**

$$g(x) = \log_b x.$$

That is, $y = \log_b x$ means the same as $x = b^y$.

The domain of the logarithm function g is the set of positive real numbers and its range is the set of real numbers, $\mathcal{R}$.

We know that $5^2 = 25$ and we may refer to 2 as the logarithm of 25 to the base 5. We write

$$\log_5 25 = 2$$

The value of this logarithm, 2, is the exponent to which 5 is raised to get 25. We call $5^2 = 25$ the *exponential form* and $\log_5 25 = 2$ the *logarithmic form*. These two equations are equivalent and as such may be used interchangeably as the next example shows.

EXAMPLE 2   Converting Exponential to Logarithm Form and Vice Versa

Determine the missing entries in the following table by changing logarithmic equations to exponential equations and vice versa.

Logarithmic Equation	Exponential Equation
$\log_2 16 = 4$	
$\log_9 3 = 1/2$	
	$5^{-2} = 1/25$
	$10^1 = 10$
	$3^y = 2x$

Solution

Logarithmic Equation	Exponential Equation
$\log_2 16 = 4$	$2^4 = 16$
$\log_9 3 = 1/2$	$9^{1/2} = 3$
$\log_5 (1/25) = -2$	$5^{-2} = 1/25$
$\log_{10} 10 = 1$	$10^1 = 10$
$\log_3 2x = y$	$3^y = 2x$

Solving Logarithmic Equations of the Form $\log_b u = k$

We can solve equations involving logarithms of the form

$$\log_b u = k$$

for any of the three variables by first converting the equation to exponential form. The solutions of these equations relies on the exponential property:

> For any real number b, $b > 0$ and $b \neq 1$, $b^u = b^v$ if and only if $u = v$

We considered this property in section 5.1 to solve simple exponential equations.

EXAMPLE 3   Solving Logarithmic Equations

Solve each equation.

(a) $\log_2 8 = x$ (b) $\log_5 y = 2$ (c) $\log_x 36 = 2$

Solution   We convert each equation to exponential form and solve.

(a) $\log_2 8 = x$ if and only if $2^x = 8$

$\quad\quad 2^x = 2^3$ Write 8 as a power of 2

$\quad\quad x = 3$ Equating exponents

(b) $\log_5 y = 2$ if and only if $y = 5^2$

$\quad\quad y = 25$

(c) $\log_x 36 = 2$ if and only if $x^2 = 36$

$\quad\quad x = \pm\sqrt{36} = \pm 6$

Since the base must be positive, we discard -6 and only use $x = 6$.   ◈

Since $b^x = y$ is equivalent to $x = \log_b y$, then evaluating logarithms to different bases often depends upon solving an equivalent exponential equation by inspection.

EXAMPLE 4   Determining Values of Logarithms

Find the exact value of each logarithm.

(a) $\log_3 81$ (b) $\log_2 4\sqrt{2}$ (c) $\log_{13} 1$ (d) $\log_7 7$

Solution   (a) Let $x = \log_3 81$. The plan calls for writing $x = \log_b y$ in its equivalent exponential form $b^x = y$ and solving this equation by inspection. The solution x to this exponential equation is the value of $\log_b y$. Similar ideas are used for parts (b), (c), and (d). The following table shows these methods.

	$x = \log_b y$	$b^x = y$	Value of $\log_b y$
(a)	$x = \log_3 81$	$3^x = 81$	Since $3^4 = 81$, then $\log_3 81 = 4$
(b)	$x = \log_2 4\sqrt{2}$	$2^x = 4\sqrt{2}$	Since $4\sqrt{2} = 2^2 2^{1/2} = 2^{5/2}$, then $\log_2 4\sqrt{2} = 5/2$
(c)	$x = \log_{13} 1$	$13^x = 1$	Since $13^0 = 1$, then $\log_{13} 1 = 0$
(d)	$x = \log_7 7$	$7^x = 7$	Since $7^1 = 7$, then $\log_7 7 = 1$ ◈

Examples 4 c and d can be generalized as follows:

> If $b > 0$ and $b \neq 1$, then
> (i) $\log_b b = 1$ and (ii) $\log_b 1 = 0$

Solving Applied Problems

There are many applications and models that are described by logarithmic functions as the next example shows.

EXAMPLE 5   Solving an Advertising Model

A company's sales S (in thousands of dollars) are related to its advertising expenditures x (in thousands of dollars) by the model

$$S = 100 + 81\log_3 (2x + 1), x \geq 0$$

Calculate the sales associated with advertising expenditures of (a) one thousand dollars (b) four thousand dollars.

Solution   (a) Let $x = 1$ and find S
$S = 100 + 81 \log_3 [2(1) + 1]$
$S = 100 + 81 \log_3 3$ Because $\log_3 3 = 1$
$S = 100 + 81 = 181$
Therefore, sales were 181 thousand dollars.

(b) Let $x = 4$ and find S
$S = 100 + 81 \log_3 (2(4) + 1)$
$S = 100 + 81 \log_3 9$ Because $\log_3 9 = 2$
$S = 100 + 81(2) = 262$
Therefore, sales were 262 thousand dollars.

Using Technology Exploration

A grapher can be used to visualize the results of the graphs of a function and its inverse. To see this, consider the following example.

G   EXAMPLE 6   Using a Grapher to Graph f^{-1}
Use a grapher to graph the functions

$$f(x) = 10^x \text{ and } f^{-1}(x) = \log_{10} x$$

on the same viewing window. (Note that $\log_{10} x$ appears as **LOG** on a calculator.

Solution   Figure 4 shows the graphs of f and f^{-1} on the same viewing window. Visually, it appears that the graphs are reflections of each other across the line $y = x$.

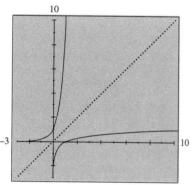

Figure 4

PROBLEM SET 7.2

Mastering the Concepts

In Problems 1–6, assume that the function f whose graph is given has an inverse.

(a) Sketch the graph of f^{-1} by reflecting the graph of f about the line $y = x$.

(b) Locate the corresponding ordered pairs for f^{-1} and name them.

1.

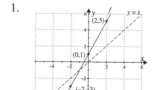

2.

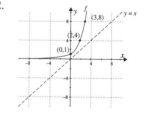

3.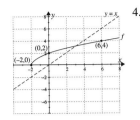

4.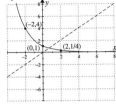

5.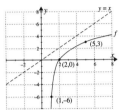

6.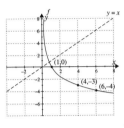

In Problems 7 and 8, determine whether or not the graph of each function has an inverse.

7. (a) 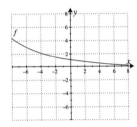   (b)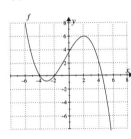

8. (a) 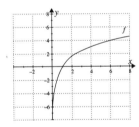   (b)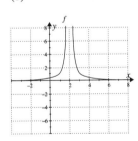

In Problems 9 and 10, assume f^{-1} exists.

9. (a) If $f(3) = 2$, find $f^{-1}(2)$.
   (b) If $f(5) = 11$, find $f^{-1}(11)$.
10. (a) If $f(4) = 60$, find $f^{-1}(60)$.
    (b) If $f(-7) = 4$, find $f^{-1}(4)$.

In Problems 11–16, sketch the graph of each exponential function f and then reflect the graph of the function about the line $y = x$ to obtain the graph of f^{-1}.

11. $f(x) = 4^x$
12. $f(x) = (1/4)^x$
13. $f(x) = (1/2)^x$
14. $f(x) = (2/3)^x$
15. $f(x) = -5^x$
16. $f(x) = -6^{-x}$

In Problems 17–20, write each equation as an equivalent logarithmic equation.

17. (a) $5^3 = 125$ (b) $4^{-2} = 1/16$
18. (a) $10^3 = 1000$ (b) $2^{-3} = 0.125$
19. (a) $\sqrt{9} = 3$ (b) $\left(\dfrac{1}{8}\right)^{-2/3} = 4$
20. (a) $\sqrt[5]{32} = 2$ (b) $100^{-3/2} = 0.001$

In Problems 21–24, write each equation as an equivalent exponential equation.

21. (a) $\log_9 81 = 2$ (b) $\log_6 36 = 2$
22. (a) $\log_3 9 = 2$ (b) $\log_6 216 = 3$
23. (a) $\log_{1/3} 9 = -2$ (b) $\log_{10} \dfrac{1}{10} = -1$
24. (a) $\log_{4/9} \dfrac{27}{8} = \dfrac{-3}{2}$ (b) $\log_4 2 = \dfrac{1}{2}$

In Problems 25–30, solve each equation for x.

25. (a) $\log_2 x = 2$ (b) $\log_3 x = 3$
26. (a) $\log_x 16 = 2$ (b) $\log_x 8 = 3$
27. (a) $\log_x 81 = 4$ (b) $\log_x 125 = 3$
28. (a) $\log_4 x = 1/2$ (b) $\log_{25} x = 1/2$
29. (a) $\log_3 9 = x$ (b) $\log_3 27 = x$
30. (a) $\log_3 1 = x$ (b) $\log_2 \dfrac{1}{16} = x$

In Problems 31–36, find the exact value of each logarithm.

31. (a) $\log_2 4$ (b) $\log_4 16$
32. (a) $\log_5 25$ (b) $\log_7 49$
33. (a) $\log_5 5$ (b) $\log_9 1$
34. (a) $\log_9 81$ (b) $\log_9 \dfrac{1}{3}$
35. (a) $\log_9 1/9$ (b) $\log_3 9\sqrt{3}$
36. (a) $\log_5 \dfrac{1}{125}$ (b) $\log_3 243$

Applying the Concepts

37. **Business:** A company's sales S (in thousands of units) of a product is given by the model
$$S = 80 + 40 \log_{10} (3t + 2)$$
where t is the number of years after the product is introduced. What were the sales after (a) 2 years (b) 10 years?

38. **Pollution:** A lake polluted by coliform bacteria is treated with bactericidal agents. Environmentalists estimate that t days after the treatment, the number N of viable bacteria per milliliters is given by the model
$$N = 10t - 100 \log (t/27), \ 1 \le t \le 10$$
Find the number of viable bacteria per milliliters after (a) 1 day (b) 3 days (c) 9 days.

Developing and Extending the Concepts

39. If $\log_2 x = 3$, find $\log_2 \dfrac{1}{x}$.

40. Let $f(x) = 1^x$, (a) find the domain and range of f.
    (b) Does $f^{-1}(x)$ exist?

41. If $f(2) = 6$, find x such that $7 + f^{-1}(x - 1) = 9$.

42. If $f(5) = 13$, find x such that $1 + f^{-1}(2x + 3) = 6$.

G In Problems 43–48, use a grapher to graph each function. Use the graph of each function to determine whether or not the given function has an inverse.

43. $f(x) = e^{-x^2}$

44. $f(x) = e^{x^2} + 1$

45. $f(x) = 3 + 2^{-x}$

46. $f(x) = 16 - 2^x$

47. $f(x) = -e^{-x} + 1$

48. $f(x) = 2 - e^x$

G In Problems 49–52, graph each logarithmic function by using a grapher with the capability of graphing the inverse of a function.

49. $f(x) = \log_5 x$, graph the inverse of $y = 5^x$.

50. $g(x) = \log_{1/5} x$, graph the inverse of $y = (1/5)^x$.

51. $h(x) = \log_{1/3} x$, graph the inverse of $y = (1/3)^x$.

52. $F(x) = \log_3 x$, graph the inverse of $y = 3^x$.

OBJECTIVES

1. Evaluate Logarithms
2. Evaluate the Inverse of a Logarithm
3. Evaluate Logarithmic Functions to Any Base
4. Solving Applied Problems
5. Use Technology Exploration

7.3 Evaluating Logarithms

Any positive real number b ($b \neq 1$) can serve as the base for a logarithm. In practice, however, the two bases that are most widely used are 10 and e.

Evaluating Logarithms

In this section we look closely at the logarithmic bases 10 and e. A logarithm with base 10 is called a **common logarithm**. Its value at a positive number x is denoted by **log x**. Thus,

$$\log x = \log_{10} x$$

Base 10 is important because of its pivotal role as the base of the decimal system. The other important base is e; the base of the natural exponential function. A logarithm with base e is called a **natural logarithm**, and its value at a positive number x is denoted by **ln x**. Thus,

$$\ln x = \log_e x$$

Since $\log_b b = 1$, it follows that

$$\log 10 = 1 \text{ and } \ln e = 1$$

A function f of the form

$$f(x) = \log x$$

is called a **common logarithm function**, whereas a function g of the form

$$g(x) = \ln x$$

is called a **natural logarithm function**.

To find the exact value of some common logarithms, the use of a calculator is not needed, as the next example shows.

EXAMPLE 1   Finding Exact Values of Common Logarithms

Let $f(x) = \log x$. Find each *exact* output value of the function.

(a) $f(10)$ (b) $f(100)$ (c) $f\left(\dfrac{1}{10}\right)$ (d) $f(\sqrt{10})$

Solution   Using the definition of logarithm, we find the exact output values for the given input values of this function as follows:

(a) $f(10) = \log 10 = 1$

(b) $f(100) = \log 100 = \log 10^2 = 2$

(c) $f\left(\dfrac{1}{10}\right) = \log \dfrac{1}{10} = \log 10^{-1} = -1$

(d) $f(\sqrt{10}) = \log \sqrt{10} = \log 10^{1/2} = \dfrac{1}{2}$

To evaluate a common logarithm of any positive real number that is not a power of 10, we use either a scientific calculator or a graphing calculator.

EXAMPLE 2   Evaluating a Common Logarithm Function.

Let $f(x) = \log x$, use a calculator to find each output value. Round off each answer to 4 decimal places.

(a) $f(364)$ (b) $f(4780)$ (c) $f(86.2)$ (d) $f(0.568)$ (e) $f(0.0841)$

Solution   Using a calculator and rounding off to 4 decimal places, the output values of this function are:

(a) $f(364) = \log 364 = 2.5611$

(b) $f(4780) = \log 4780 = 3.6794$

(c) $f(86.2) = \log 86.2 = 1.9355$

(d) $f(0.568) = \log 0.568 = -0.2457$

(e) $f(0.0841) = \log 0.0841 = -1.0752$

Notice that the results of parts (d) and (e) of the above example are negative because $10^{-0.2457} = 0.568$ and $10^{-1.0752} = 0.0841$. Using this pattern, we see that *a common logarithm of any number between 0 and 1 is always negative*. The reason is that the logarithm is an exponent of 10 that produces the number.

To find the exact value of a natural logarithm of powers of e, we rely on the definition of logarithm.

EXAMPLE 3   Finding the Exact Values of Natural Logarithms

Let $f(x) = \ln x$, find the exact output value.

(a) $f(e^3)$ (b) $f(\sqrt[5]{e})$ (c) $f(e^{-2})$

Solution   As in example 1, we use the definition of logarithm to obtain each output value.

(a) $f(e^3) = \ln e^3 = 3$

(b) $f(\sqrt[5]{e}) = \ln \sqrt[5]{e} = \ln e^{1/5} = \dfrac{1}{5}$

(c) $f(e^{-2}) = \ln e^{-2} = -2$

Natural logarithms occur in many applications and models from different fields. We can use a calculator to find approximate output values for given input values.

EXAMPLE 4   Evaluating a Natural Logarithm Function

Let $g(x) = \ln x$. Use a calculator to find each output value. Round off each answer to 4 decimal places.

(a) $g(8132)$ (b) $g(743)$ (c) $g(0.0436)$

Solution   Using a calculator and rounding off to 4 decimal places, the output values of this function are obtained as follows:

(a) $g(8132) = \ln 8132 = 9.0036$
(b) $g(743) = \ln 743 = 6.6107$
(c) $g(0.0436) = \ln 0.0436 = -3.1327$

Evaluating the Inverse of a Logarithm

Recall that the *inverse of a logarithmic function* is an exponential function and this inverse is called an **antilogarithm**. To find the inverse logarithm, we find the power of the base. For instance, recall that

$$\text{if } f(t) = \log t, \text{ then } f^{-1}(t) = 10^t$$
$$\text{if } f(t) = \ln t, \text{ then } f^{-1}(t) = e^t$$

To calculate the inverse of a common logarithm of a number x, we use the 10^x key which can be found on most calculators. For the inverse of a natural logarithm of x, we use e^x key.

EXAMPLE 5   Evaluating Inverse Logarithm Function

Find each value of x. Round off each answer to 2 decimal places.

(a) $\log x = 1.7782$ (b) $\log x = -1.8742$ (c) $\ln x = -0.6713$

Solution   (a) $\log x = 1.7782$ if and only if $x = 10^{1.7782} = 60.01$
(b) $\log x = -1.8742$ if and only if $x = 10^{-1.8742} = 0.01$
(c) $\ln x = -0.6713$ if and only if $x = e^{-0.6713} = 0.51$

Evaluating Logarithmic Functions to Any Base

Suppose that $f(x) = \log_3 x$ and we wish to find $f(7) = \log_3 7$. There is no logarithm base 3 on a calculator and 7 is not an obvious power of 3. So we must find another way. It is often useful to rewrite $\log_3 7$ in terms of the *common or natural* logarithm which can be directly evaluated by using a calculator. This is accomplished by using the following *change of base formula*.

CHANGE OF BASE FORMULA

If a, b and x are positive real numbers, and a and b are not equal to 1, then

$$\log_b x = \frac{\log_a x}{\log_a b}$$

Using the change of base formula, we can write $\log_b x$ as

$$\log_b x = \frac{\log x}{\log b} \qquad \text{or} \qquad \log_b x = \frac{\ln x}{\ln b}$$

EXAMPLE 6   Evaluating a Logarithm to Any Base

Let $f(x) = \log_3 x$. Find (a) $f(7)$ and (b) $f(2.5)$. Round off each answer to 4 decimal places.

Solution   We use the change of base formulas.

$$f(x) = \log_3 x = \frac{\log x}{\log 3} \text{ or } f(x) = \frac{\ln x}{\ln 3}.$$

We could choose either base and the value of $\log_3 x$ would be the same.

(a) $f(7) = \log_3 7 = \dfrac{\log 7}{\log 3} = 1.7712$

(b) $f(2.5) = \log_3 2.5 = \dfrac{\log 2.5}{\log 3} = 0.8340$ ◈

Solving Applied Problems

Applications such as measuring the strength of an earthquake and determining the acidity of a liquid are among several models that make use of logarithmic functions.

Seismologists measure the magnitude M of an earthquake using a logarithmic scale called the **Richter Scale**, in honor of the American seismologist Charles F. Richter (1900–1984). The energy E (in ergs) released by an earthquake of magnitude M can be approximated by the model

$$\log E = 11.4 + 1.5M$$

EXAMPLE 7   Finding Magnitudes of Earthquakes

Find the magnitude of each earthquake.
(a) The 1990 earthquake in Iran with $E \approx 8.9 \times 10^{22}$ ergs.
(b) The 1992 earthquake in southern California with $E \approx 3.2 \times 10^{22}$ ergs.

Solution   We substitute each value of E into the model

$$\log E = 11.4 + 1.5M$$

and solve for M.
(a) Replacing E by 8.9×10^{22} and using a calculator, we have

$$\log (8.9 \times 10^{22}) = 11.4 + 1.5M$$
$$22.9494 = 11.4 + 1.5M$$
$$1.5M = 11.5494$$
$$M = 7.7$$

Therefore, the magnitude of the earthquake in Iran was 7.7 on the Richter Scale.
(b) Replacing E by 3.2×10^{22}, we have

$$\log (3.2 \times 10^{22}) = 11.4 + 1.5M$$
$$22.5051 = 11.4 + 1.5M$$
$$1.5M = 11.1051$$
$$M = 7.4$$

Therefore, the magnitude of the earthquake in California was 7.4.   ◈

Using Technology Exploration

We have seen in section 7.2 that the relationship between logarithmic and exponential functions enabled us to find exact values of certain logarithmic functions such as $\log_4 16 = 2$ or $\log_5 125 = 3$. Graphers can also produce the graph of f by rewriting f in the form

$$f(x) = \log_4 x = \frac{\log x}{\log 4}$$

 **EXAMPLE 8** Use a Grapher to Graph a Logarithmic Function

Use a grapher to graph $f(x) = \log_3 x = (\log x)/(\log 3)$ on the given viewing window. Then use the **ZOOM** and **TRACE** features to find the following approximate output values (round off to 4 decimal places):

(a) $f(2.7)$ (b) $f(3.4)$ (c) $f(5.82)$ (d) $f(7.35)$

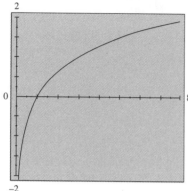

Solution   Figure 1 shows the graph of
$$f(x) = (\log x)/(\log 3).$$
The output values are:
(a) $f(2.7) = 0.9041$
(b) $f(3.4) = 1.1139$
(c) $f(5.82) = 1.6032$
(d) $f(7.35) = 1.8157$

Figure 1

 ## PROBLEM SET 7.3

Mastering the Concepts

In Problems 1 and 2, let $f(x) = \log x$, find the *exact* output values for the given input values.

1. (a) $f(1000)$ (b) $f(1/100)$

   (c) $f(\sqrt[3]{10})$ (d) $f\left(\dfrac{1}{\sqrt{10}}\right)$

2. (a) $f(10,000)$ (b) $f(0.01)$

   (c) $f(\sqrt[4]{10})$ (d) $f\left(\dfrac{1}{\sqrt[3]{10}}\right)$

In Problems 3–8, let $f(x) = \log x$, find each output value for the given input values. Round off to 4 decimal places.

3. (a) $f(317)$ (b) $f(3914)$
   (c) $f(39,102)$ (d) $f(235,567)$
4. (a) $f(731)$ (b) $f(9341)$
   (c) $f(42,651)$ (d) $f(141,329)$
5. (a) $f(5)$ (b) $f(0.5)$
   (c) $f(0.05)$ (d) $f(0.005)$
6. (a) $f(3.4)$ (b) $f(0.34)$
   (c) $f(0.034)$ (d) $f(0.0034)$
7. (a) $f(8.04)$ (b) $f(80.4)$
   (c) $f(0.804)$ (d) $f(0.0804)$
8. (a) $f(7.86)$ (b) $f(78.6)$
   (c) $f(0.786)$ (d) $f(0.0786)$

In Problems 9 and 10, let $g(x) = \ln x$, find the *exact* output values for the given input values.

9. (a) $g(e^2)$ (b) $g(e^{-1})$

   (c) $g(\sqrt{e})$ (d) $g\left(\dfrac{1}{\sqrt{e}}\right)$

10. (a) $g(e^4)$ (b) $g(e^{-3})$

   (c) $g(\sqrt[3]{e})$ (d) $g\left(\dfrac{1}{\sqrt[3]{e}}\right)$

In Problems 11–16, $g(x) = \ln x$, find each output value for the given input value. Round off each answer to 4 decimal places.

11. (a) $g(54)$ (b) $g(154)$
   (c) $g(1545)$ (d) $g(15,453)$
12. (a) $g(71)$ (b) $g(712)$
   (c) $g(7123)$ (d) $g(71,235)$
13. (a) $g(1.7)$ (b) $g(0.17)$
   (c) $g(0.017)$ (d) $g(0.0017)$
14. (a) $g(5)$ (b) $g(0.5)$
   (c) $g(0.05)$ (d) $g(0.005)$
15. (a) $g(0.712)$ (b) $g(0.759)$
   (c) $g(0.315)$ (d) $g(0.847)$
16. (a) $g(0.00512)$ (b) $g(0.00812)$
   (c) $g(0.00137)$ (d) $g(0.00413)$

In Problems 17–20, find each value of x. Round off each answer to 2 decimal places.

17. (a) $\log x = 0.4133$ (b) $\log x = 0.4871$
   (c) $\log x = 1.2945$
18. (a) $\ln x = 1.725$ (b) $\ln x = 0.3517$
   (c) $\ln x = 0.8991$
19. (a) $\ln x = 2.9781$ (b) $\ln x = -1.6289$
   (c) $\ln x = -1.8082$
20. (a) $\log x = -1.8971$ (b) $\log x = 2.4881$
   (c) $\log x = 0.2415$

In Problems 21–24, find each output value for the given input value. Use the change of base formula and round off each value to 4 decimal places.

21. $f(x) = \log_5 x$
   (a) $f(7)$ (b) $f(11)$
22. $g(x) = \log_6 x$
   (a) $g(11)$ (b) $g(8)$
23. $F(x) = \log_4 x$
   (a) $F(5)$ (b) $F(17)$
24. $h(x) = \log_7 x$
   (a) $h(13)$ (b) $h(4.3)$

Applying the Concepts

25. **Seismology:** In 1964 a severe earthquake in Alaska killed 131 people, releasing 10^{24} ergs. What was the magnitude of this earthquake? Round off the answer to one decimal place (see Example 7).

26. **Seismology:** Nowadays, seismologists are using a different version of Richter's ideas in which the magnitude M of an earthquake is computed as a function of the energy intensity I. One such model adopted by scientists is

    $$M = \log I$$

    Calculate the magnitude of the indicated earthquakes:
    (a) Mexico City, 1985, where $I = 125,890,000$,
    (b) Armenia, 1988, where $I = 6,310,000$ and
    (c) Guam, 1993, where $I = 100,000,000$.

27. **Audiology:** Audiologists generally agree that continuous exposure to a sound level in excess of 90 decibels for more than 5 hours daily may cause long-term hearing problems. Suppose that the **sound level** S (measured in decibels) is given by the *Fechner-Weber Law* in the form of the model

    $$S = 10 \log \frac{P}{P_0}$$

    where P is the power (in watts) of the sound being measured, and P_0 is the power of sound of the lowest threshold of human hearing. Find the loudness in decibels for the indicated sounds:

    (a) a supersonic jet where $\dfrac{P}{P_0} = 10^{11}$

    (b) a machine in a factory where $\dfrac{P}{P_0} = 11,000,000$

    (c) a rock concert where $\dfrac{P}{P_0} = 263,000,000$.

28. **Chemistry:** In chemistry the **pH of a substance** is defined by the model

    $$pH = -\log [H^+]$$

    where $[H^+]$ is the concentration form of hydrogen ions in the substance measured in moles per liter. The pH of distilled water is 7. A substance with a pH of less than 7 is called an acid, whereas a substance with a pH of more than 7 is called a base. Find the pH of each substance for the given concentration of hydrogen ions:
    (a) acid rain whose $[H^+]$ is 3.16×10^{-4}
    (b) a vinegar whose $[H^+]$ is 7.94×10^{-4}
    (c) a lemon juice whose $[H^+]$ is 5.01×10^{-3}
    (d) a tomato whose $[H^+]$ is 6.30×10^{-5}

29. **Chemistry:** Find the hydrogen ion concentration [H+] in moles per liter of each substance (see problem 28). Round off each answer to 2 decimal places.
    (a) vinegar; pH = 3.1   (b) beer; pH = 4.3
    (c) lemon juice; pH = 2.3   (d) bile; pH = 7.9

30. **Medicine:** Find the hydrogen ion concentration [H+] of a patient's blood (see problem 28) (a) when the patient's pH = 7.3, and (b) when the patient's pH = 7.6. Round each answer to two decimal places.

31. **Spread of Flu:** A person with a flu virus visited a school campus. The number t (in days) it takes the virus to infect n people is given by the model

    $$t = -16.3\ln\left(\frac{5450}{n} - \frac{4}{11}\right), \quad 4000 \leq n \leq 14,000$$

    How many days will it take to infect (a) 6000 people? (b) 10,000 people? (c) 13,000 people?

32. **Advertising:** A bicycle store determines that the number N of bicycles sold is related to the number x (in dollars) spent on advertising is given by the model

    $$N = 51 + 100 \ln\left(\frac{x}{100} + 2\right)$$

    How many bikes will be sold if (a) nothing is spent on advertising? (b) $1,000 is spent? (c) $5,000 is spent?

33. **Doubling Investment:** The number of years, n, required to double your investment when it is invested at an annual interest rate r (in decimal form) compounded annually is given by the model

    $$n = \frac{\ln 2}{\ln (1 + r)}$$

    Find the number of years it takes to double your money at each annual interest rate (a) 4% (b) 6% (c) 8%.

34. **Rule of 72:** Round off the answers in problem 33 to the nearest year and compare them with these numbers.

    $$\frac{72}{4} = 18, \frac{72}{6} = 12, \frac{72}{8} = 9$$

    Use this evidence to state a "rule of thumb" for approximating **doubling time** without using the model in problem 33. This rule of thumb, which has been long used by bankers, is called the **rule of 72**.

Developing and Extending the Concepts

In Problems 35–38, evaluate each logarithmic expression by using a calculator. Round off the answer to 4 decimal places.

35. (a) $\ln 4 + \ln \sqrt[3]{4}$ (b) $\ln (4 + \sqrt[3]{4})$

36. (a) $\ln\left(\dfrac{\sqrt{3}}{2}\right)$ (b) $\dfrac{\ln \sqrt{3}}{\ln 2}$

37. (a) $\ln (\sqrt{7} - \sqrt{3})$ (b) $\ln \sqrt{7} - \ln \sqrt{3}$

38. (a) $\ln \sqrt{11}$ (b) $\sqrt{\ln 11}$

39. By using a calculator and rounding off to 4 decimal places, we find that $e^1 = 2.7183$ and $e^2 = 7.3891$. Is it possible to find two positive consecutive integers a and b (without using a calculator) such that $a < \ln 4 < b$? If so, find them.

40. Let n be the number of letters in your last name.
    (a) Find $\ln n$ rounded to 4 decimal places.
    (b) Raise e to the power indicated by the number you found above. What kind of conclusion do you draw?

41. (a) Is $\log (\ln 1)$ defined as a real number? Explain.
    (b) Is $\ln (\log 0.5)$ defined as a real number? Explain.

42. Find a formula for converting:
    (a) a common logarithm to a natural logarithm.
    (b) a natural logarithm to a common logarithm.

43. Suppose that $f(x) = 2A \log x + 3B$, where A and B are constant real numbers, and $f(1) = 6$ and $f(10) = 16$. Find A and B.

44. Suppose that $g(x) = A \ln x + B$, where A and B are constant real numbers, and $g(e) = 5$ and $g(e^2) = 8$. Find A and B.
45. Suppose that $x = (\log_{49} 7)^{\log_7 49}$. What is the value of $\log_2 x$?
46. Which of the following expressions is not equal to 1?
    (a) $\ln e$ (b) $(\log e)^0$ (c) $\log 10$ (d) $\log 1$

$\boxed{\text{G}}$ In Problems 47 and 48, use a grapher to graph each function, and then use the $\boxed{\text{ZOOM}}$ and $\boxed{\text{TRACE}}$ features to approximate each value. Round off to 4 decimal places.
    (a) $f(4)$ (b) $f(3.84)$ (c) $f(5.75)$

47. $f(x) = \log_7 x = \dfrac{\log x}{\log 7}$ 48. $f(x) = \log_7 x = \dfrac{\ln x}{\ln 7}$

OBJECTIVES

1. Graph Logarithmic Functions to Any Base
2. Graph Common and Natural Logarithms
3. Solve Applied Problems
4. Use Technology Exploration

7.4 Graphs of Logarithmic Functions

In this section, we investigate graphs of logarithmic functions, which are the inverses of exponential functions.

Graphing Logarithmic Functions to Any Base

There are different ways for graphing logarithmic functions. One way for graphing $y = \log_b x$ is to first convert its equation to the equivalent exponential form $x = b^y$. Then use the point plotting method and connect the points with a smooth curve to sketch the graph of $x = b^y$.

Another way for graphing
$$y = \log_b x \text{ is to graph } y = b^x,$$
and then reflect the graph across the line $y = x$. Recall that the inverse of
$$f(x) = b^x \quad \text{is} \quad f^{-1}(x) = \log_b x.$$
The following diagram shows the graphs of
$$f(x) = b^x \quad \text{and} \quad f^{-1}(x) = \log_b x.$$

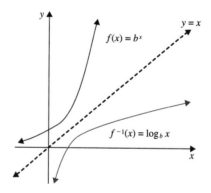

Notice that the graph of $f^{-1}(x) = \log_b x$ is obtained by reflecting the graph of $f(x) = b^x$ about the line $y = x$.

The following table lists some exponential functions and their inverses:

Exponential Functions	Inverse Function
$f(x) = 4^x$	$f^{-1}(x) = \log_4 x$
$f(x) = (1/3)^x$	$f^{-1}(x) = \log_{1/3} x$
$f(x) = 10^x$	$f^{-1}(x) = \log x$
$f(x) = e^x$	$f^{-1}(x) = \ln x$

EXAMPLE 1   Graphing a Logarithmic Function

Sketch the graph of

$$f(x) = \log_2 x$$

(a) by interchanging the roles of x and y and graphing

$$x = 2^y.$$

(b) by graphing

$$y = 2^x$$

and reflecting the graph across the line $y = x$.

Find the domain and range of the function f.

Solution   (a) The equation

$y = \log_2 x$ is equivalent to the equation $x = 2^y$.

If we graph the equation $x = 2^y$, then we have accomplished our objective. To do so, we select values for y and then compute values for x as the following table shows.

y	$x = 2^y$	Ordered pairs (x,y)
-2	$x = 2^{-2} = 1/2^2 = 1/4$	$(1/4, -2)$
-1	$x = 2^{-1} = 1/2^1 = 1/2$	$(1/2, -1)$
0	$x = 2^0 = 1$	$(1, 0)$
1	$x = 2^1 = 2$	$(2, 1)$
2	$x = 2^2 = 4$	$(4, 2)$

Now we plot these points and connect with a smooth curve (Figure 1a).

(b) First we graph

$$y = 2^x,$$

and reflect this graph across the line $y = x$. Notice that the curve in part (a),

$$x = 2^y,\ \text{looks like the graph of } y = 2^x,$$

except that it is reflected across the line $y = x$ (Figure 1b).

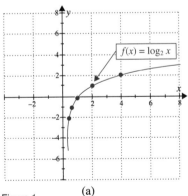
(a)

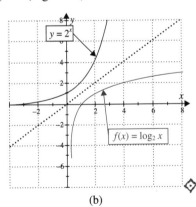
(b)

Figure 1

Each graph suggests that the domain of the function consists of all positive real numbers, $(0,\infty)$, and the range consists of all real numbers $\Re$.

EXAMPLE 2 Graphing a Logarithmic Function

Sketch the graph of each function

(a) $g(x) = \log_{1/2} x$ (b) $h(x) = -\log_{1/2} x$.

Indicate the domain and range of each function.

Solution First we make a table of ordered pairs for each function. Then sketch the graphs by connecting these points with a smooth curve.

(a) The graph of g in Figure 2a is a reflection of the graph of $y = (1/2)^x$ across the line $y = x$. The domain of g is $(0,\infty)$ and its range consists of all real numbers $\mathcal{R}$.

(b) Notice that the graph of $h(x) = -\log_{1/2} x$ in Figure 2b is a reflection of the graph of g across the x axis. The domain of h is $(0,\infty)$ and its range is $\mathcal{R}$.

x	$f(x) = \log_{1/2} x$	$g(x) = -\log_{1/2} x$
$\tfrac{1}{2}$	1	-1
1	0	0
2	-1	1
4	-2	2
8	-3	3

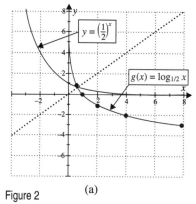

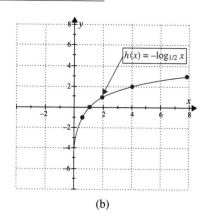

Figure 2   (a)   (b)

A number of features should be noted about the graphs of the functions $f(x) = \log_2 x$ in Figure 1a and $g(x) = \log_{1/2} x$ in Figure 2a. First, the x intercept of each graph is 1, but neither has a y intercept. The graphs suggest that the domain of each function consists of positive real numbers, $(0,\infty)$, and the range of each function consists of all real numbers $\mathcal{R}$. Figure 1a shows that as x values increase, the y values also increase, while Figure 2a shows as x values increase, the y values decrease. Table 1 summarizes these features.

TABLE 1

$f(x) = \log_2 x$	Domain $(0,\infty)$	Range $\mathcal{R}$	increasing
$g(x) = \log_{1/2} x$	Domain $(0,\infty)$	Range $\mathcal{R}$	decreasing

As illustrated in Figure 3a, the graphs of the functions of the form $f(x) = \log_b x$, where $b > 1$, resemble the graph of $y = \log_2 x$. The larger the value of b, the slower the graph rises to the right. Observe this in Figure 3a with the graphs of

$$y = \log_4 x, \, y = \log_3 x, \text{ and } y = \log_2 x.$$

Similarly, if $0 < b < 1$, the graphs of the form $f(x) = \log_b x$ have the same general shape and characteristics as the graph of $y = \log_{1/2} x$. The larger the value of b, the faster the graph falls to the right. Compare the graphs of

$$y = \log_{1/2} x, \, y = \log_{1/3} x,$$

and $y = \log_{1/10} x$ in Figure 3b.

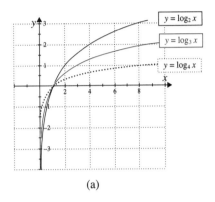

(a)

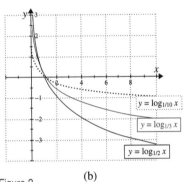

(b)

Figure 3

These graphs illustrate the following important features about the function

$$f(x) = \log_b x$$

1. The domain of f is the set of positive real numbers and the range is the set of real numbers.
2. The graph of f approaches but never touches the negative y axis.

Graph Common and Natural Logarithms

Two bases, 10 and e, are used in a variety of applications so that these functions require special attention.

EXAMPLE 3   Graphing Natural Logarithm Functions

Sketch the graph of each function.

(a) $f(x) = \ln x$ (b) $g(x) = \ln (x + 2)$

Compare the graphs of the functions in parts (a) and (b).

Solution   We make a table of input values and find the corresponding output values.

x	$f(x) = \ln x$	$g(x) = \ln (x + 2)$
-2	-2 is not in the domain of f	-2 is not in the domain of g
-1	-1 is not in the domain of f	$g(-1) = \ln (-1 + 2) = 0$
0	0 is not in the domain of f	$g(0) = \ln (0 + 2) = 0.69$
1	$f(1) = 0$	$g(1) = \ln (1 + 2) = 1.10$
2	$f(2) = 0.69$	$g(2) = \ln (2 + 2) = 1.39$
3	$f(3) = 1.10$	$g(3) = \ln (3 + 2) = 1.61$

We make a table of ordered pairs for each function, and then sketch the graphs that contain the points in the above table.

(a) Once again, keep in mind that the graph of $f(x) = \ln x$ in Figure 4a is a re-flection of the graph of $y = e^x$ across the line $y = x$. The domain of f is $(0,\infty)$ and its range consists of all real numbers $\mathcal{R}$.

(b) The graph of

$$g(x) = \ln (x + 2)$$

is obtained by shifting the graph of

$$f(x) = \ln x$$

horizontally 2 units to the left (Figure 4b). The domain of g is $(-2,\infty)$ and its range consists of all real numbers $\mathcal{R}$.

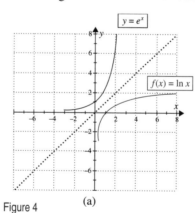
(a)

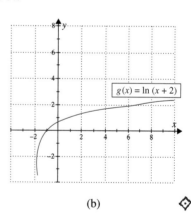
(b)

Figure 4

EXAMPLE 4   Graphing Common Logarithm Functions

Sketch the graph of each pair of functions on the same coordinate system.

(a) $y = \log x$ (b) $y = \log x$

and   and

$f(x) = -\log x$ $g(x) = \log (-x)$

Compare the graphs of each pair of functions.

Solution (a) The graph of $f(x) = -\log x$ is obtained by reflecting the graph of $y = \log x$ across the x axis (Figure 5a).

(b) The graph of $g(x) = \log(-x)$ is obtained by reflecting the graph of $y = \log x$ across the y axis (Figure 5b).

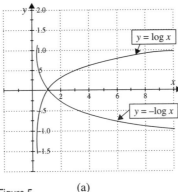

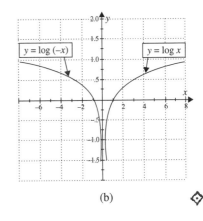

Figure 5   (a)   (b)

Solving Applied Problems

Logarithmic functions are used to model phenomena such as advertising, manufacturing, growth and decay of populations, and other sciences, as the next example shows.

G EXAMPLE 5   Finding the Time of a Skydiver
The time t (in seconds) required for a skydiver in free fall (before the parachute opens) to reach a velocity v (in feet per second) is given by the function

$$t = -5.47 \ln(1 - 0.006v), \ 0 \le t \le 18$$

Find the time required for the free fall velocity of the diver to be 100 feet per second, 130 feet per second and 160 feet per second. Round off to two decimal places.

Solution   We begin by substituting 100 for v in the equation

$$t = -5.47 \ln(1 - 0.006v)$$
$$\text{when } v = 100, t = -5.47 \ln(1 - 0.006(100)) = 5.01$$
$$\text{when } v = 130, t = -5.47 \ln(1 - 0.006(130)) = 8.28, \text{ and}$$
$$\text{when } v = 160, t = -5.47 \ln(1 - 0.006(160)) = 17.61.$$

Therefore, the times corresponding to the velocities in feet per second of 100, 130, and 160 are 5.01 seconds, 8.28 seconds, and 17.61 seconds respectively.

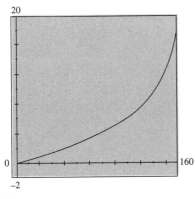

Figure 6

Using Technology Exploration

Figure 6 shows the graph of the model in Example 5 produced by a grapher on the given viewing window. Notice that according to this model, as the velocity of the skydiver increases, the time also increases.

The next example considers graphs of logarithmic functions with bases different from 10 and e.

Recall that since we can evaluate $\log_b x$ on a calculator by using the change of base formula, we can sketch the graph of such a logarithmic function on a grapher by using the change of base formula

$$h(x) = \log_b x = \frac{\log x}{\log b}$$

G EXAMPLE 6   Using a Grapher

Use a grapher to graph the functions

(a) $f(x) = \log_4 x$ and (b) $g(x) = \log_4 (x - 1)$ on the same viewing windows. Compare the two graphs, then find the domain and range of each function.

Solution   (a) We use the change of base formula to rewrite the function f in the form

$$f(x) = \frac{\log x}{\log 4}$$

Figure 7a shows the graph with the given viewing window. The graph suggests the domain of f consists of all positive real numbers, $(0, \infty)$ and the range consists of all real numbers, $\mathcal{R}$.

(b) Once again, we use the change of base formula to rewrite g as

$$g(x) = \frac{\log (x - 1)}{\log 4}$$

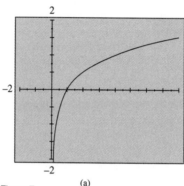
(a)

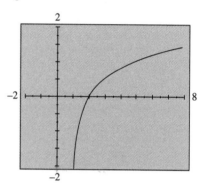
(b)

Figure 7

Figure 7b shows the graph of g produced by a grapher. The graph suggests the domain of g consists of all the real numbers greater than 1, that is, the interval $(1, \infty)$. The range of g consists of all the real numbers, $\mathcal{R}$.

Comparing the graphs of f and g, we see that the shapes are the same but g is horizontally shifted one unit to the right. ◈

◈ PROBLEM SET 7.4

Mastering the Concepts

1. Sketch the graph of $y = \log_3 x$:
   (a) by graphing $x = 3^y$,
   (b) by graphing $y = 3^x$ and reflecting the graph across the line $y = x$.
   (c) Find the domain and range of the function.
2. Sketch the graph of $y = \log_{1/3} x$:
   (a) by graphing $x = (1/3)^y = 3^{-y}$,
   (b) by graphing $y = (1/3)^x = 3^{-x}$ and reflecting the graph across the line $y = x$.
   (c) Find the domain and range of the function.

In Problems 3–6, each of the curves is the graph of a function of the form $y = \log_b x$. Find the base b if its graph contains the given point.

3.

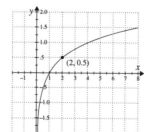

4.

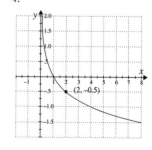

5.

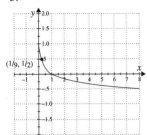

6.

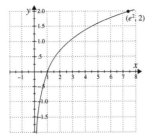

In Problems 7 and 8, use the graph of $H(x) = \log_3 x$ in Figure 3a on page 363 and shifting or reflecting to write a rule for the logarithmic function whose graph is shown.

7.

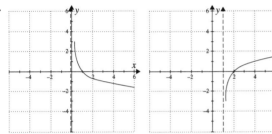

             (a)   (b)

8.

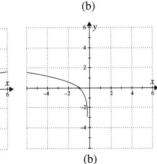

             (a)   (b)

In Problems 9–12, match the function with its graph.

9. $f(x) = \log_5(-x)$ 10. $g(x) = \log_5(x + 2)$

11. $h(x) = \log_5(x - 1)$ 12. $F(x) = -\log_5(x + 1)$

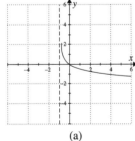

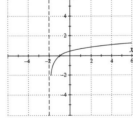

            (a)   (b)

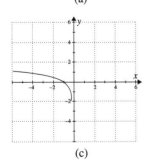

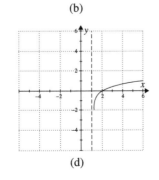

            (c)   (d)

In Problems 13–20, (a) sketch the graph of each pair of functions on the same coordinate system and (b) compare the graphs of each pair of functions.

13. $y = \log_5 x$ and $f(x) = \log_5(x + 1)$

14. $y = \log_{1/5} x$ and $f(x) = \log_{1/5}(x + 1)$

15. $y = \ln x$ and $f(x) = \ln(x + 4)$

16. $y = \log x$ and $f(x) = \log(x - 2)$

17. $y = \log x$ and $f(x) = -\log(x + 1)$

18. $y = \log_2 x$ and $f(x) = \log_2(1 - x)$

19. $y = \ln x$ and $f(x) = -\ln(-x)$

20. $y = \log x$ and $f(x) = -\log(x + 2)$

G   In Problems 21–30, use the change of base formula to determine the function which must be entered on a grapher in order to produce the graph of each function. Use the grapher to graph each function.

21. $f(x) = \log_3 x$ 22. $f(x) = \log_5 x$

23. $g(x) = \log_{1/3} x$ 24. $g(x) = \log_{3/7} x$

25. $h(x) = -2 \log_7 x$ 26. $h(x) = -3 \log_4 x$

27. $F(x) = -3 \log_{2/3} x$ 28. $F(x) = -\log_{2/3} x$

29. $g(x) = 2 + \log_3 x$ 30. $g(x) = 2 - \log_3 x$

Applying the Concepts

31.   **Advertising:** A company has determined that the number N of units of its product sold is related to the amount x (in thousand dollar increments) spent on advertising by the model

$$N = 3000 + 300 \log x, \; x \geq 1$$

    (a)   How many units are sold after spending \$10,000 on advertising?

    G (b)   Use a grapher with the given viewing window to graph the function N. xMin = 0, xMax = 100, xScl = 10, yMin = 2000, yMax = 4000, yScl = 100.

    (c)   Use the TRACE feature of the graph to determine the number of units sold after spending \$1000 on advertising.

    (d)   Use the TRACE feature to estimate how much should be spent on advertising if the number of units sold is 3300.

32.   **Manufacturing:** A manufacturer of auto parts has determined that the weekly profit P (in dollars) is given by the model

$$P = 1830 + 900 \log x, \; x > 0$$

where x is the number of parts produced each minute.

    (a)   What is the weekly profit (to the nearest dollar) if the number of parts produced each minute is 1000?

    G (b)   Use a grapher to graph the function P on the given viewing window. xMin = 0, xMax = 1000, xScl = 100, yMin = 1000, yMax = 5000, yScl = 500.

    (c)   Use the TRACE feature of the grapher to estimate the number of parts per minute (round off to the nearest whole number) that must be produced in order to have a weekly profit of approximately \$3630.

33. **Tornadoes:** The speed S of the wind (in miles per hour) near the center of a tornado is given by the model

$$S = 45 + 55 \log x, \ 0 < x \le 20$$

where x is the distance (in miles) the tornado travels. On July 2, 1997 a tornado struck portions of the Detroit metropolitan area covering a distance of approximately 17 miles.

(a) Approximate the speed of the wind near the center of the tornado.

G (b) Use a grapher to graph the function S in the given viewing window. xMin = 0, xMax = 20, xScl = 1, yMin = 0, yMax = 120, yScl = 25.

(c) Use the TRACE feature of the grapher to estimate the distance covered by a tornado whose speed is 100 miles per hour.

34. **Travel Industry**: A travel agency determines if x is the total number of hours worked per week by its employees, its revenue R (in dollars) is given by the model

$$R(x) = 1550 \ln (x + 1), \ 10 \le x \le 250$$

(a) What is the agency's weekly revenue if its employees worked 200 hours per week?

G (b) Use a grapher to graph the function R on the given viewing window xMin = 0, xMax = 250, xScl = 25, yMin = 0, yMax = 10000, yScl = 1000.

(c) Use the TRACE feature to estimate how many hours the employees worked per week, if the weekly revenue is $8,435?

35. **Forgetting Curve:** A group of students had an average score of 70 on an examination. As part of an experiment, the students were tested on the same material at monthly intervals thereafter without further study. The researcher determined that the average score A (as a percent), after t months was approximated by the function

$$A(t) = 70 - 6 \ln (t + 1), \ t \ge 0$$

(a) What was the average score after 4 months?

G (b) Use a grapher with the given viewing window to graph the function A. xMin = 0, xMax = 10, xScl = 1, yMin = 0, yMax = 100, yScl = 10.

(c) Use the TRACE feature of the grapher to determine after what time t was the average score 56.

36. **Pollution:** A lake polluted by bacteria is treated with a bactericidal agent. Environmentalists estimate that t days after the treatment, the number N of viable bacteria per milliliter is given by the model

$$N(t) = 10t - 30 - 100 \ln (0.1t), \ 1 \le t \le 12$$

(a) Find the number of bacteria per milliliter after 8 days.

G (b) Use a grapher to graph the function N on the given viewing window. xMin = 0, xMax = 13, xScl = 1, yMin = 0, yMax = 200, yScl = 25.

(c) Use the TRACE feature to estimate how many days after the treatment the number of bacteria per milliliter is 100.

Developing and Extending the Concepts

37. For what values of b will the graph of $f(x) = \log_b x$ (a) rise from left to right? (b) fall from left to right?

38. Suppose that $f(x) = \log_b x$. Determine the x intercept of the graph of f if   (a) $b > 1$ (b) $0 < b < 1$.

In Problems 39–42, determine f^{-1} for each of the given functions. Sketch the graphs of f and f^{-1} on the same coordinate system.

39. $f(x) = \ln x$ 40. $f(x) = \log_2 x$
41. $f(x) = \log_{1/3} x$ 42. $f(x) = \log x$

G In Problems 43–52, determine the domain of each function. You may use a grapher to help you determine the domain.

43. $f(x) = \ln (x - 2)$ 44. $g(x) = \log (3x + 2)$
45. $h(x) = \log (x^2 - 4)$ 46. $f(x) = \ln (9 - x^2)$
47. $f(x) = \log_3 |x + 2|$ 48. $g(x) = \log |2 - x|$
49. $f(x) = \ln \sqrt{x - 2}$ 50. $g(x) = \ln \dfrac{3}{\sqrt{x - 2}}$
51. $f(x) = \log (1 - x)$ 52. $g(x) = -\log (3 + x)$

G In Problems 53–58, use a grapher to graph each function. Find the domain of each function.

53. $f(x) = \ln|x|$ 54. $g(x) = |\ln x|$
55. $f(x) = \log \dfrac{1}{x}$ 56. $g(x) = \dfrac{1}{\log x}$
57. $f(x) = \ln x^3$ 58. $g(x) = \log \sqrt{x}$

OBJECTIVES

1. Use the Properties of Logarithms
2. Simplify Exponential and Logarithmic Expressions
3. Solve Applied Problems
4. Use Technology Exploration

7.5 Properties of Logarithms

Historically, the significance of logarithms was in simplifying numerical computations. Today, with the advances and easy access of technology, this use becomes trivial. However, the three properties that supported this historical use are still helpful in simplifying logarithmic expressions.

Using the Properties of Logarithms

Recall from section 7.3, that for $x > 0$, the equation

$$y = \log_b x \text{ is equivalent to the equation } b^y = x.$$

Substituting $\log_b x$ for y in the equation $b^y = x$, we get the property

$$b^{\log_b x} = x$$

For base e and base 10, the property takes the form

$$e^{\ln x} = x \text{ and } 10^{\log x} = x$$

For instance, $e^{\ln 7} = 7$ and $e^{\ln (x^2)} = x^2$.

Since logarithms are exponents, we can use the properties of exponents to derive some basic properties of logarithms.

We verify these properties as follows:

(i) Since $M = b^{\log_b M}$ and $N = b^{\log_b N}$,

$$MN = b^{\log_b M} \cdot b^{\log_b N} = b^{\log_b M + \log_b N}$$

By converting the equation to logarithmic form, we get

$$\log_b MN = \log_b M + \log_b N$$

(ii)   Since $\dfrac{M}{N} \cdot N = M$, we have

$$\log_b \left[\left(\frac{M}{N} N \right) \right] = \log_b M$$

$$\log_b \frac{M}{N} + \log_b N = \log_b M \qquad \text{Property (i)}$$

$$\log_b \frac{M}{N} = \log_b M - \log_b N \qquad \text{Subtract } \log_b N \text{ from each side}$$

(iii) Since $N = b^{\log_b N}$, then

$$N^r = (b^{\log_b N})^r = b^{r \log_b N} \qquad \text{Power rule for exponents}$$

After converting to logarithmic form, we get $\log_b N^r = r \log_b N$

Unfortunately, there is no simple property for $\log_b (M + N)$ or for $\log_b (M - N)$. In general,

(i) $\log_b (M + N) \neq \log_b M + \log_b N$ (ii) $\log_b (M - N) \neq \log_b M - \log_b N$

(iii) $\log_b MN \neq (\log_b M)(\log_b N)$ (iv) $\log_b (M/N) \neq (\log_b M)/(\log_b N)$

The basic properties of logarithms for the special cases of the common and natural logarithms are stated in Table 1.

TABLE 1 Basic Properties of Common and Natural Logarithms

Common Logarithms Base 10	Natural Logarithms Base e
(i) $\log MN = \log M + \log N$	(i) $\ln MN = \ln M + \ln N$
(ii) $\log (M/N) = \log M - \log N$	(ii) $\ln (M/N) = \ln M - \ln N$
(iii) $\log N^r = r \log N$	(iii) $\ln N^r = r \ln N$

In Property (iii), if the value of N is the same as the base, then we get $\log_b b^r = r \log_b b = r \cdot 1 = r$, so

$$\log_b b^r = r$$

Specifically, for $b = 10$ or $b = e$, we have

$$\log 10^r = r \text{ and } \ln e^r = r$$

For instance, $\ln e^{x^2 - x} = x^2 - x$.

The next two examples illustrate how the properties of logarithms are used to manipulate logarithmic expressions algebraically.

EXAMPLE 1   Using Properties of Logarithms

Write each expression as a sum or difference of multiples of logarithms.

(a) $\log_x 5y$ (b) $\ln\left(\dfrac{17}{5}\right)$ (c) $\log_5\left(\dfrac{x^2 y}{2}\right)$ (d) $\log (mn)^{3/5}$

Solution   We assume that all quantities whose logarithms are taken are positive.

(a) $\log_x 5y = \log_x 5 + \log_x y$ The Product Property

(b) $\ln\left(\dfrac{17}{5}\right) = \ln 17 - \ln 5$ The Quotient Property

(c) $\log_5\left(\dfrac{x^2 y}{2}\right) = \log_5 (x^2 y) - \log_5 2$ The Quotient Property

           $= \log_5 x^2 + \log_5 y - \log_5 2$ The Product Property

           $= 2 \log_5 x + \log_5 y - \log_5 2$ The Power Property

(d) $\log (mn)^{3/5} = (3/5) \log (mn)$ The Power Property

           $= (3/5)[\log m + \log n]$ The Product Property

           $= (3/5) \log m + (3/5) \log n$ ◈

EXAMPLE 2   Combining Logarithmic Expressions

Write each expression as a single logarithm.

(a) $\log_3 15 + \log_3 13$ (b) $\log 216 - \log 54$

(c) $3\log_2 x - \log_2 5x$ (d) $\ln (t - 2) + \ln (t + 2)$

Solution   We assume that all quantities whose logarithms are taken are positive.

(a) $\log_3 15 + \log_3 13 = \log_3(15)(13)$ The Product Property

                $= \log_3 195$

(b) $\log 216 - \log 54 = \log \dfrac{216}{54}$ The Quotient Property

                $= \log 4$

(c) $3\log_2 x - \log_2 5x = \log_2 x^3 - \log_2 5x$ The Power Property

                $= \log_2 \dfrac{x^3}{5x} = \log_2 \dfrac{x^2}{5}$ The Quotient Property

(d) $\ln (t - 2) + \ln (t + 2) = \ln (t - 2)(t + 2)$ The Product Property

                $= \ln (t^2 - 4)$ ◈

Simplifying Exponential and Logarithmic Expressions

At times we encounter expressions involving exponential and logarithmic functions with base e that have to be simplified by using the properties of logarithms.

EXAMPLE 3   Simplifying Expressions Involving Base e

Simplify each expression: (a) $e^{3\ln 2}$ and (b) $e^{4 - 2\ln 3}$

Solution   (a) $e^{3\ln 2} = e^{\ln 2^3} = 2^3 = 8$

(b) $e^{4-2\ln 3} = e^4 \cdot e^{-2\ln 3} = e^4 \cdot e^{\ln 3^{-2}} = e^4 \cdot 3^{-2}$

$$= \frac{e^4}{3^2} = \frac{e^4}{9}$$

◈

Solving Applied Problems

The properties of logarithms are used to rewrite formulas used in a variety of applications.

EXAMPLE 4   Determining Atmospheric Pressure at Different Altitudes

The amount of atmospheric pressure P (in pounds per square inch) at a height h (in miles) is given by the model

$$h = -\frac{1}{0.21}(\ln P - \ln 14.7)$$

(a) Rewrite the equation for this model for P in terms of h.

(b) Determine the atmospheric pressure 40 miles above sea level. Round off to four decimal places.

Solution   (a)

$-\dfrac{1}{0.21}(\ln P - \ln 14.7) = h$	Original equation
$\ln P - \ln 14.7 = -0.21h$	Multiply each side by -0.21
$\ln \dfrac{P}{14.7} = -0.21h$	The Quotient Property
$\dfrac{P}{14.7} = e^{-0.21h}$	Convert to exponential form
$P = 14.7e^{-0.21h}$	

(b) If $h = 40$ miles, then the above equation becomes

$P = 14.7e^{-0.21\,(40)}$

$P = 0.0033$

Thus, at a height of 40 miles the atmospheric pressure is 0.0033 pounds per square inch.   ◈

Using Technology Exploration

We can use a grapher to explore certain mathematical models as the next example shows.

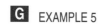 EXAMPLE 5 Using a Grapher

Physiologists determined that the average height h (in inches) of a young girl can be approximated by the model

$$h = 11.3 + 19.42 \ln A, 6 < A < 15$$

where A is the age of the girl (in years).
(a) Rewrite the equation for this model for A in terms of h.
(b) Use a grapher to graph the resulting equation in the given viewing window.
(c) Use the TRACE feature to determine the age of a young girl whose height is 48 inches, 55 inches, and 60 inches. Round to one decimal place.

Solution (a)

$$11.3 + 19.42 \ln A = h \qquad \text{Original equation}$$
$$19.42 \ln A = h - 11.3 \qquad \text{Subtract 11.3 from each side}$$
$$\ln A = \frac{h - 11.3}{19.42} \qquad \text{Divide each side by 19.42}$$
$$A = e^{(h-11.3)/19.42} \qquad \text{Convert to exponential form}$$

(b) Figure 1 shows the graph of this model in the given viewing window.
(c) Using the TRACE feature, this model predicts that a girl 48 inches high is 6.6 years old, 55 inches is 9.5 years old, and 60 inches tall is 12.3 years old.

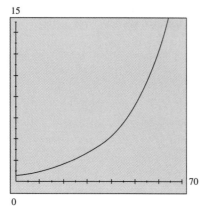

Figure 1

PROBLEM SET 7.5

Mastering the Concepts

In Problems 1–14, rewrite each expression as a sum or difference of multiples of logarithms. Assume all variables represent positive real numbers.

1. (a) $\log_3 5y$ (b) $\log_2 7x$
2. (a) $\log_4 uv$ (b) $\log_7 cd$
3. (a) $\log_5 \dfrac{x}{3}$ (b) $\log_3 \dfrac{t}{11}$
4. (a) $\log_3 \dfrac{x}{15}$ (b) $\log_4 \dfrac{p}{q}$
5. (a) $\log_4 x^{12}$ (b) $\log_2 p^{2/3}$
6. (a) $\log_5 w^{13}$ (b) $\log_2 u^{3/5}$
7. (a) $\ln \sqrt{Q}$ (b) $\ln \sqrt[3]{P}$
8. (a) $\log \sqrt[5]{x}$ (b) $\log \sqrt[4]{x}$
9. (a) $\log_3 \left(\dfrac{xy^2}{4}\right)$ (b) $\log_3 \left(\dfrac{x^3 y}{4}\right)$
10. (a) $\ln(xy)^{-3/5}$ (b) $\ln(x^2 y)^{4/3}$
11. (a) $\log (x^4 y)$ (b) $\log \sqrt[5]{xy}$
12. (a) $\ln \sqrt{xy^3}$ (b) $\ln \sqrt[7]{x^2 y}$
13. (a) $\log_5 \left(\dfrac{x^2}{y^4}\right)$ (b) $\log_3 \left(\dfrac{x^7}{y^2}\right)$
14. (a) $\ln \dfrac{\sqrt{x}}{y^2}$ (b) $\ln \dfrac{x^2}{\sqrt[3]{y}}$

In Problems 15–26, rewrite each expression as a single logarithm. Assume all variables represent positive real numbers.

15. (a) $\log_3 x + \log_3 4$ (b) $\log_5 3 + \log_5 y$
16. (a) $\log_4 y + \log_4 7$ (b) $\log u + \log v$
17. (a) $\ln 4 - \ln x$ (b) $\log x - \log 3$
18. (a) $\log_4 4 - \log_4 v$ (b) $\log w - \log u$
19. (a) $\log x^3 - \log x^2$ (b) $\log_3 y^2 - \log_3 y$
20. (a) $\ln x^2 - \ln y$ (b) $\log_3 u^2 + \log_3 v^3$
21. (a) $3 \log_2 x + 2 \log_2 x$ (b) $2 \log_3 y + \log_3 y^3$
22. (a) $2 \ln x + 3 \ln x^2$ (b) $4 \log y + \log y^2$
23. (a) $3 \log z - 2 \log y$ (b) $3 \log_2 4 - 2 \log_2 v$
24. (a) $3 \log u - 2 \log v$ (b) $5 \log u - \log v$
25. $\log (x - 3) + \log (x + 3)$ 26. $\ln (x + 5) + \ln (x - 5)$

In Problems 27–32, simplify each expression.

27. (a) $e^{2\ln 3}$ (b) $e^{-2\ln y}$
28. (a) $e^{-3\ln 2}$ (b) $e^{4\ln x}$
29. (a) $\ln e^{4x}$ (b) $\ln e^{-2y}$
30. (a) $\ln e^{\sqrt{x}}$ (b) $\ln e^{-4u}$
31. (a) $e^{1 - \ln x}$ (b) $e^{2 - 3\ln x}$
32. (a) $e^{4 + 3\ln x}$ (b) $e^{x + 2\ln x}$

33. **Chemistry:** The pH of a solution is given by

$$pH = \log \left[\frac{1}{H^+}\right]$$

when H^+ is the concentration of the hydrogen ion. Use the properties of logarithms to express pH in terms of $\log H^+$.

34. **Earthquakes:** The seismologist Charles Richter established the measure of the magnitude of an earthquake, in terms of its intensity, I, by means of the equation

$$M = \log I - \log I_0$$

where I_0 is a certain minimum intensity. Use the properties of logarithms to write the right side of this equation as a single logarithm.

Developing and Extending the Concepts

In Problems 35–38, rewrite each expression in terms of log 2 and log 5.

35. (a) $\log 10$ (b) $\log 0.4$
36. (a) $\log 2.5$ (b) $\log \sqrt{20}$
37. (a) $\log 0.1$ (b) $\log 0.16$
38. (a) $\log 250$ (b) $\log 0.005$
39. Suppose that $\log_b 2 = 0.69$, $\log_b 3 = 1.10$ and $\log_b 5 = 1.61$. Find the value of each expression.
     (a) $\log_b 7.5$ (b) $\log_b \sqrt{0.9}$ (c) $\log_b 1.5$
40. Simplify each expression.
     (a) $10^{\log 3x + \log 2x}$
     (b) $e^{\ln 6x^4 - \ln 2x^2}$

In Problems 41–46, rewrite each expression as a sum or difference of multiples of logarithms. Assume all variables represent positive numbers.

41. (a) $\log_4 \sqrt[7]{x^4 y^5}$ (b) $\log_5 \sqrt[5]{x\, y^4}$
42. (a) $\log_7 \left(\dfrac{x^3 \sqrt[4]{y}}{z^3} \right)$ (b) $\log_4 \left(\dfrac{u^4 v^5}{\sqrt[4]{z^3}} \right)$
43. $\log \left(\dfrac{c^7 \sqrt[9]{d^2}}{5\sqrt{f}} \right)$ 44. $\ln \sqrt{\dfrac{y}{y+3}}$
45. $\ln \left(4 \sqrt[3]{x} \sqrt{y} \right)$ 46. $\log \left(\dfrac{7}{\sqrt[3]{x^2 y^4}} \right)$

In Problems 47–52, rewrite each expression as a single logarithm.

47. (a) $\log_a \dfrac{x}{y} + \log_a \dfrac{y^2}{3x}$
     (b) $\log_2 \dfrac{x^2}{y} - \log_2 \dfrac{x^4}{y^2}$
48. (a) $2 \log_3 a - 3 \log_3 b - 4 \log_3 x$
     (b) $5 \log x - 5 \log y - 7 \log z$
49. $\ln \dfrac{a}{a-1} + \ln \dfrac{a^2 - 1}{a}$
50. $\log \dfrac{x+y}{z} - \log \dfrac{1}{x+y}$
51. $\log_3 (x^2 - 25) - \log_3 (x^2 - 4x - 5)$
52. $\log (3x^2 - 5x - 2) - \log (x - 2)$

OBJECTIVES

1. Solve Special Exponential Equations
2. Solve Exponential Equations with Different Bases
3. Solve Logarithmic Equations
4. Solve Applied Problems
5. Use Technology Exploration

7.6 Exponential and Logarithmic Equations

Many applications and models make use of exponential and logarithmic functions. In this section we develop methods to solve equations involving these functions.

Solving Special Exponential Equations

Recall from section 7.2 that we solved exponential equations using the following exponential equality.

EXPONENTIAL EQUALITY

Let b be a real number, $b > 0$ and $b \neq 1$.
$$b^u = b^v \text{ if and only if } u = v.$$

For example, to solve the equation
$$3^{2x+4} = 27$$
We write
$$27 = 3^3$$
so that
$$3^{2x+4} = 3^3$$
This means
$$2x + 4 = 3$$
$$2x = 3 - 4$$
$$2x = -1$$
$$x = -\dfrac{1}{2}$$

EXAMPLE 1   Solving Exponential Equations with Special Bases

Solve each equation.

(a) $9^t = 27^{4t-1}$
(b) $\left(\dfrac{1}{5}\right)^y = 25^{3y-1}$

Solution   (a) Because $9 = 3^2$ and $27 = 3^3$, we can rewrite the given equation as:

$$(3^2)^t = (3^3)^{4t-1}$$

or

$$3^{2t} = 3^{3(4t-1)}$$

We set the exponents equal to one another because the base is the same.

$$2t = 3(4t - 1) \qquad \text{Equate exponents}$$
$$2t = 12t - 3$$
$$-10t = -3 \qquad\qquad \text{Solve for } t$$
$$t = \frac{3}{10}$$

Therefore, the solution is $\dfrac{3}{10}$.

(b) Because $1/5 = 5^{-1}$ and $25 = 5^2$, we can rewrite the given equation as:

$$(5^{-1})^y = (5^2)^{3y-1}$$

or

$$5^{-y} = 5^{2(3y-1)}$$

It follows that

$$-y = 2(3y - 1) \qquad \text{Equate Exponents}$$
$$-y = 6y - 2$$
$$-7y = -2 \qquad\qquad \text{Solve for } y$$
$$y = \frac{2}{7}$$

Therefore, the solution is 2/7.

Solving Exponential Equations with Different Bases

Suppose that we were to solve the exponential equation

$$3^x = 7$$

Is it possible to write each side of $3^x = 7$ as a power of the same base? To do so, we make use of the following logarithmic equality.

LOGARITHMIC EQUALITY

> Let u, v, and b be positive numbers, with $b \neq 1$.
> $$\log_b u = \log_b v \text{ if and only if } u = v.$$

To solve an equation such as

$$3^x = 7,$$

we usually take the common or natural logarithm on both sides of the equation.

EXAMPLE 2   Solving Exponential Equations with Different Bases

Solve each equation.

(a) $5^x = 7$
(b) $e^{3x+7} = 8$

Round off each answer to two decimal places.

Solution   (a) We take the common logarithm of both sides of the equation:

$$5^x = 7 \qquad \text{Original equation}$$
$$\log 5^x = \log 7 \qquad \text{Take log of each side}$$
$$x \log 5 = \log 7 \qquad \text{Use the power property}$$
$$x = \frac{\log 7}{\log 5} \approx 1.21 \qquad \text{Simplify}$$

Therefore, the solution is approximately 1.21.

(b) Taking the natural logarithm of both sides of the equation, we have

$$e^{3x+7} = 8 \qquad \text{Original equation}$$
$$\ln e^{3x+7} = \ln 8 \qquad \text{Take ln of each side}$$
$$(3x + 7)\ln e = \ln 8 \qquad \text{Use the power property}$$
$$3x + 7 = \ln 8 \qquad \text{Note } \ln e = 1$$
$$x = \frac{\ln 8 - 7}{3} \approx -1.64 \qquad \text{Simplify}$$

Therefore, the solution is approximately -1.64.

Solving Logarithmic Equations

To solve equations involving logarithmic expressions such as

$$\log_2 (5x - 3) = 5$$

we first convert the equation to exponential form and then solve the equation. The next example details this process.

EXAMPLE 3   Solving a Logarithmic Equation

Solve each equation. Round off to two decimal places.

(a) $\log_2 (5x - 3) = 5$ (b) $\log_3 (2x - 1) = 2.43$ (c) $\log_2 (x^2 + 15x) = 4$

Solution   (a) We write the equation in the equivalent exponential form.

$$2^5 = 5x - 3 \qquad \text{Exponential equation}$$
$$32 = 5x - 3$$
$$35 = 5x$$
$$7 = x$$

Therefore, the solution is 7.

(b) We write the equation in the exponential form as follows:

$$3^{2.43} = 2x - 1 \qquad \text{Exponential equation}$$
$$2x - 1 = 14.4346 \qquad \text{Use a calculator}$$
$$2x = 15.4346$$
$$x \approx 7.72$$

Therefore, 7.72 is the approximate solution.

(c) The exponential form of the equation is given by

$$2^4 = x^2 + 15x \qquad \text{Exponential equation}$$
$$x^2 + 15x = 16$$
$$x^2 + 15x - 16 = 0 \qquad \text{Standard form}$$
$$(x + 16)(x - 1) = 0 \qquad \text{Factor}$$
$$x + 16 = 0 \qquad x - 1 = 0$$
$$x = -16 \qquad x = 1$$

The solutions are -16 and 1. These solutions check in the original equation.◆

Quite often we need to solve equations of the form

$$\log x + \log (x + 1) = 2$$

In this case, the properties of logarithms are needed to combine the terms of an equation before we convert it to its equivalent exponential form.

EXAMPLE 4   Solving Logarithmic Equations

Solve each equation.

(a) $\log x + \log (x + 3) = 1$ (b) $\log_4 (x + 3) - \log_4 x = 1$

Solution   (a) Remember the base of each logarithm here is 10.

$$\begin{aligned}
\log x + \log (x + 3) &= 1 & &\text{Original equation} \\
\log x(x + 3) &= 1 & &\text{Use the product property} \\
x(x + 3) &= 10^1 & &\text{Convert to exponential form} \\
x^2 + 3x - 10 &= 0 & &\text{Write in standard form} \\
(x + 5)(x - 2) &= 0 & &\text{Factor}
\end{aligned}$$

$$x + 5 = 0 \qquad x - 2 = 0$$
$$x = -5 \qquad x = 2$$

The solutions should be checked to avoid extraneous solutions.

Check: When $x = -5$, we have

$$\log (-5) + \log (-5 + 3) = \log (-5) + \log (-2)$$

Because the number -5 is not in the domain of the logarithmic function, it is not a solution.

When $x = 2$, we have

$$\log (2) + \log (2 + 3) = \log (2) + \log (5)$$
$$= \log (2)(5) = \log 10 = 1$$

So, the number 2 is an acceptable solution.

(b) $\log_4 (x + 3) - \log_4 x = 1$

$$\begin{aligned}
\log_4 \frac{x + 3}{x} &= 1 & &\text{Use the quotient property} \\
\frac{x + 3}{x} &= 4^1 & &\text{Convert to exponential form} \\
x + 3 &= 4x & &\text{Multiply each side by the LCD} \\
3 &= 3x \\
1 &= x
\end{aligned}$$

Check: When $x = 1$, we have

$$\log_4 (1 + 3) - \log_4 1 = \log_4 4 - \log_4 1 = 1 - 0 = 1$$

Therefore, the solution is 1.

Solving Applied Problems

Recall that the functions

$$P(t) = P_0 e^{kt} \text{ and } P(t) = P_0 e^{-kt} \text{ for } k > 0$$

describe *exponential growth* and *exponential decay* models respectively. The **doubling time** for the model $P(t) = P_0 e^{kt}$ is the amount of time necessary for the population to double its size. The rate of decay k for the model $P(t) = P_0 e^{-kt}$ is measured in terms of its **half life**, which is the period of time required for half of the quantity to decay.

EXAMPLE 5   Doubling Time of an Investment

An investment of $5000 offers an annual interest rate of 6% compounded quarterly. In t years it will grow to the amount A given by the function

$$A = 5000(1.015)^{4t}$$

(a) How long will it take to accumulate $6,734 in the account?

(b) How long will it take for the amount of this investment to double in value?

Solution (a) We replace A by 6,734 and solve the equation.

$$6{,}734 = 5000(1.015)^{4t} \qquad \text{Original equation}$$

$$(1.015)^{4t} = \frac{6734}{5000} \qquad \text{Divide each side by 5000}$$

$$\log (1.015)^{4t} = \log \frac{6734}{5000} \qquad \text{Take logarithm of each side}$$

$$4t\log (1.015) = \log \frac{6734}{5000} \qquad \text{Use the power property}$$

$$t = \frac{\log (6734/5000)}{4 \log (1.015)} \approx 5$$

Therefore, it takes approximately 5 years to accumulate $6,734.

(b) Replacing A by $2(5000) = 10{,}000$, we have

$$10{,}000 = 5000(1.015)^{4t} \qquad \text{Original equation}$$

$$(1.015)^{4t} = 2 \qquad \text{Divide each side by 5000}$$

$$\log(1.015)^{4t} = \log 2 \qquad \text{Take logarithm of each side}$$

$$4t\log(1.015) = \log 2 \qquad \text{Use the power property}$$

$$t = \frac{\log 2}{4 \log (1.015)} \approx 11.6$$

Therefore, it takes approximately 11.6 years to double the amount.   ◇

EXAMPLE 6   Finding the Half-life of a Radioactive Material

Potassium 42 is a radioactive isotope that is often used as a tracer by cardiologists. It decays at the rate of 5.4% per hour. The amount A remaining from an initial amount I, at t hours is given by the function

$$A = Ie^{-0.054t}$$

(a) Find the half-life of potassium 42. That is, find the time it takes for one-half of a sample of the isotope to decay. Round off to one decimal place.

(b) How long will it take for 80% of the sample to decay? Round off to one decimal place.

Solution (a) We must solve the equation

$$A = Ie^{-0.054t}$$

for t when $A = 0.5I$.

$$0.5I = Ie^{-0.054t}$$

$$e^{-0.054t} = 0.5 \qquad \text{Divide both sides by } I$$

$$-0.054t = \ln 0.5 \qquad \text{Take natural logarithm of both sides}$$

$$t = \frac{\ln 0.5}{-0.054} \approx 12.8$$

Thus, the half-life of the sample is approximately 12.8 hours.

(b) Now solve the equation for t when $A = 0.2I$.

$$0.2I = Ie^{-0.054t}$$

$$e^{-0.054t} = 0.2 \qquad \text{Divide both sides by } I$$

$$-0.054t = \ln 0.2 \qquad \text{Take natural logarithm of both sides}$$

$$t = \frac{\ln 0.2}{-0.054} \approx 29.8$$

Therefore, it takes approximately 29.8 hours for 80% of the sample to decay.◇

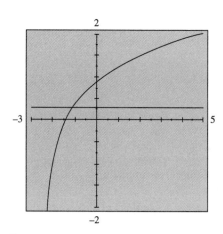

Figure 1

Using Technology Exploration

Exponential and logarithmic equations can also be solved by a grapher. The solutions are obtained by either approximating the x intercepts of the associated function or by graphing each side of the equation and estimating points of intersection.

G **EXAMPLE 7**   Solving an Equation by a Grapher

Use a grapher to solve the equation

$$5 \ln (x + 2.4) = 1.5$$

Solution   First we divide each side by 5 to obtain $\ln (x + 2.4) = 0.3$. Figure 1 shows the graphs of $y = \ln (x + 2.4)$ and $y = 0.3$ on a given viewing window. Using the **TRACE** feature of the grapher, we see that the x coordinate of the point of intersection is approximately -1.05. A check verifies that this is the solution of the equation.   ◈

◈ ## PROBLEM SET 7.6

Mastering the Concepts

In Problems 1–6, find the *exact* solution of each equation.

1. (a) $2^{-4x+1} = 64$ (b) $3^{x-5} = 27$
2. (a) $5^{2x-1} = 125$ (b) $3^{4x-5} = 81$
3. (a) $4^{3x} = 8^{x-1}$ (b) $5^{2x} = 25^{3x}$
4. (a) $25^{3x} = 5^{2x-3}$ (b) $\left(\dfrac{1}{3}\right)^{2x+7} = 81$
5. $3^{x^2+2x} = 27$ 6. $2^{3x^2-2x} = 32$

In Problems 7–20, solve each equation. Round off each answer to two decimal places.

7. $2^x = 7.3$ 8. $3^{2x} = 5.70$
9. $4^{-x} = 3.4$ 10. $5^{-x} = 8.9$
11. $7^{3x-2} = 48$ 12. $e^{2x} = 3.7$
13. $e^{-4x} = 3.7$ 14. $e^{2x+1} = 6.2$
15. $e^{2x+5} = 3.1$ 16. $5e^{-0.14x} = 23$
17. $2e^{0.23x} = 43.8$ 18. $4^{y+1} = 6.4$
19. $4^{8x-1} = 5^{1-3x}$ 20. $e^{3-7x} = 4^{1-5x}$

In Problems 21–28, find the *exact* solution of each equation.

21. $\log_5 (2x - 7) = 2$ 22. $\log_3 (2x + 3) = 2$
23. $\log_3 (x^2 + 2x) = 1$ 24. $\log_2 (3x^2 - 2x) = 5$
25. $\log_2 (x^2 - 6x) = \log_2 \dfrac{1}{8}$
26. $\log_5 x = 5\log_5 2 - 2\log_5 4$
27. $\log_b x = 3\log_b 2 + \dfrac{1}{2} \log_b 25 - \log_b 20$
28. $\log_b x = \log_b 6 + \dfrac{1}{2} \log_b 16 - \log_b 216$

In Problems 29–32, solve each equation. Round off each answer to two decimal places.

29. $\log_4 (3x + 1) = 0.6531$ 30. $\log_2 (5x + 3) = 1.7435$
31. $\log (7x - 1) = 1.0374$ 32. $\ln (8x - 3) = 0.7364$

In Problems 33–42, use the properties of logarithms to find the *exact* solution of each equation.

33. (a) $\log_4 x - \log_4 21 = 1/2$ (b) $\log_2 5x + \log_2 3 = 3$
34. (a) $\log_2 x + 2\log_2 3 = 3$ (b) $\log_2 x - (1/2)\log_2 16 = 64$
35. $\log x + \log (x + 3) = 1$
36. $\log_{12} x + \log_{12} (x + 1) = 1$
37. $\log (3x - 2) - \log (2x + 1) = 0$
38. $\ln (3x + 4) - \ln (2x + 8) = 0$
39. $\log_5 x + \log_5 (x - 4) = 1$
40. $\log (x^2 - 9) - \log (x + 3) = 2$
41. $\log (x^2 - 4) - \log (x - 2) = 2$
42. $\log_4 x + \log_4 (6x + 10) = 1$

Applying the Concepts

43. **Investment Portfolio:** Suppose that $7000 is invested in a savings account at an annual interest rate of 5% compounded quarterly. Use the model

$$A = 7000(1.0125)^{4t}$$

to determine the number of years for this investment to double. (Round off the answer to the nearest year.)

44. **Investment Portfolio:** Suppose that $5000 is invested in a mutual fund that earns an annual rate of return of 7% compounded continuously according to the model

$$A = 5000e^{0.07t}$$

When will this portfolio triple itself?

45. **Population Growth:** In 1995 the population of Brazil was about 158 million people, and the population grows approximately according to the model

$$P = 158e^{0.025t}$$

where t is the number of years after 1995, with $t = 0$ representing 1995 and P is in millions of people.
   (a) How many years after 1995 will the population of Brazil be 172 million? Round off to one decimal place.
   (b) How many years after 1995 will the population double? Round off to one decimal place.

46. **Population Growth:** In 1990 the population of Albuquerque, New Mexico, was about 385,000, and the population grows approximately according to the model

$$P = 385(1.049)^t$$

where t is the number of years after 1990, with $t = 0$ representing 1990 and P is in thousands.
   (a) In what year will the population be 1.3 million? Round off to the nearest year.
   (b) In what year will the population double? Round off to the nearest year.

47. **Salmon Fishing:** In 1991 the total revenue for a West Coast state from salmon fishing was $18.8 million, and the yearly economic impact ever since is given by the model

$$R = 18.8(0.78)^t, \ 0 \le t \le 15$$

where t is the number of years after 1991 and revenue R is in millions of dollars. Use this model to predict when the revenue will be 2.01 million dollars. Round off to the nearest year.

48. **Casino Gambling:** Suppose that a "big time" gambler visited Las Vegas and decided to spend his $200 gambling allotment on a 25-cent slot machine that pays back 93%. His strategy is to set aside all his winnings each day for playing the next day. Suppose that his gambling strategy follows the model

$$A = 200(0.93)^t$$

where A is the number of dollars after t days of playing. How many days does he expect to play and end with a balance of $10? Round off to the nearest day.

49. **Water Consumption:** Suppose that the water consumption in a large city increases at the rate of 8% each year. After t years, the water consumption is given by the model

$$C = C_0(1.08)^t$$

where C (in billions of gallons) is the water consumption after t years and C_0 is the water consumption when $t = 0$. In how many years will the water consumption triple? Round off the answer to the nearest year.

50. **Oceanography:** In oceanography, the *Beer-Lambert* Law states that the intensity, P, of light (as a decimal) that penetrates the surface of the ocean to a depth of x feet is approximated by the model

$$P = e^{-0.015x}$$

At what depth under the surface of the ocean will the intensity of light be 0.45% of the surface light intensity. Round off to the nearest foot.

51. **Ecology:** In 1986 the estimated number n of an endangered species of African Antelope was 7500 and was decreasing after t years according to the model

$$n = 7500e^{-0.062t}$$

In how many years will the population of antelope decline to 1500? Round off to the nearest tenth of a year.

52. **Archeology:** A wooden death mask was discovered in an ancient burial site in New Mexico. It was determined by archeologists that the amount A of carbon-14 left after t years is approximated by the model

$$A = Ie^{-0.000121t}$$

where I is the original amount of carbon-14. If 55% of the original carbon-14 of the wooden mask still remains, what is the approximate age of the mask? Round off to the nearest tenth of a year.

53. **Radioactive Decay:** The percentage P (in decimals) of radioactive carbon-14 found in a specimen t years after it begins to decay is modeled by the function

$$P = e^{-0.000121t}$$

   (a) After how many years of decay will 60% of it still be present?
   (b) What is the half-life of this element?

54. **Radioactive Decay:** Certain radioactive elements decay exponentially. The decay model of Cesium-137 after t years is given by the model

$$P = P_0e^{-0.023t}$$

Find the half-life of this element.

55. **Recycling:** A beverage company estimates that about 60% of all aluminum cans distributed are recycled each year. If the company distributes 2.5 million cans, the number N of cans still in use after t years is given by the model

$$N = 2.5(0.60)^t$$

How many years will it take for the number of cans in use to reach one-half of a million? Round off the answer to the nearest year.

56. **Present Value:** Suppose that $7225 is invested in a certificate of deposit (CD) and it will pay $10,000 at maturity. If the interest rate is 6.5% compounded continuously, how long will it take to reach maturity?

57. **Earth Science:** The atmospheric pressure P (in pounds per square inch) at an altitude of h feet above sea level is given by the model

$$\ln P = \ln 14.7 - 0.00005h$$

At what altitude will the atmospheric pressure be 11.4 pounds per square inch? Round off to the nearest ten feet.

58. **Business:** A computer software company discovers that its annual revenue R (in thousands of dollars) is related to its research and development expenditures x (in tens of thousands of dollars) by the model

$$R = 3750 + 475 \ln (2x + 2)$$

How much money should be spent on research and development for the company's revenue to be 5.5 million?

Developing and Extending the Concepts

In Problems 59–68, find the *exact* solution of each equation.

59. $3 + 4^{2x+1} = 35$

60. $\left(\frac{3}{5}\right)^{3-x} = \left(\frac{25}{9}\right)^{2x}$

61. $e^{x^2-1} = 1$

62. $(\sqrt{2})^{3-2x} = (\sqrt[3]{2})^{12x}$

63. $\frac{2^{x+1}}{8^x} = \frac{1}{4^x}$

64. $\left(\frac{1}{5}\right)^{x^2} = (125)^{(2x/3)-1}$

65. $\log |3x + 1| = 0$

66. $\log (4x - 5) - \log (x - 2) = \log (2x + 1)$

67. $\ln x^2 = (\ln x)^2$

68. $(\ln x)^2 + \ln x^3 - 4 = 0$

G In Problems 69–72, solve each equation (a) algebraically (b) graphically by using a grapher. Round off to two decimal places.

69. $3 + 2(1 + e^{x/2}) = 64$

70. $26 - 50(1 + 1.01e^{-0.05t}) = 12.3$

71. $3 + 4 \log\frac{x}{2} = 5.4$

72. $1 + \frac{2}{3} \ln (2x + 1) = -1$

73. **Skydiving:** After a skydiver jumps out of an airplane, the diver's speed S (in meters per second) is estimated by the model

$$S(t) = 20[1 - (0.6)^t], \quad 0 \le t \le 15$$

where t represents the time of free fall in seconds.

G (a) Use a grapher to graph the function.
(b) Use the graph to estimate the speed of the skydiver after 6 seconds.
(c) Use the graph to estimate the time when the speed of the skydiver is 19.66 meters per second.

74. **Radioactive Decay:** The number of years t it would take 200 milligrams of radium to decompose so that only x milligrams remain is given by the model

$$t = 2350[\ln 200 - \ln x]$$

G (a) Use a grapher to graph the function t.
(b) Use the graph to estimate the number of years it would take to have 18 milligrams remain.

75. **Advertising:** A company has determined that if its monthly advertising expenditure is x (in thousands of dollars), then the total monthly sales S (in thousands of dollars) is given by the model

$$S = 870 \ln (x + 5)$$

(a) What are the expected monthly sales totals (to the nearest thousand dollars) when the monthly expenditure for advertising is $10,000? $50,000? and $100,000?

G (b) Use a grapher to graph the function S. Then use the TRACE feature to determine the amount (to the nearest thousand dollars) that should be spent on advertising if the monthly sales totals are $2 million, $3 million, and $4 million.

76. **Spread of AIDS:** In 1985 the National Institute of Allergy and Infectious Diseases estimated the number of cases of AIDS reported was 8250. The following table relates the time t (in years after 1985) to the number of cases reported:

Time t (in years)	Number N of cases
0	8250
1	12,835
2	19,970
3	31,070
4	48,338
5	75,205
6	117,004

(a) Determine which of the following models fit this data:

$$N = 8250e^{0.84t}$$

$$N = 8250e^{0.442t}$$

$$N = 8250(2^{1.33t})$$

G (b) Use a grapher to graph the model that fits the data in part (a).
(c) Use the graph to predict after how many years will 800,000 cases occur.

◆ CHAPTER 7 REVIEW PROBLEM SET

In Problems 1 and 2, find

    (a) $f(-3)$ (b) $f(-2)$ (c) $f(0)$
    (d) $f(2.3)$ (e) $f(\sqrt{2})$
    Round off each answer to two decimal places.

1. $f(x) = 7^x$ 2. $f(x) = e^{-2x}$

In Problems 3 and 4, sketch the graph of each function. Indicate the domain and range.

3. (a) $f(x) = 7^x$ (b) $g(x) = e^{-x}$
4. (a) $f(x) = (0.3)^x$ (b) $g(x) = e^{3x}$

In Problems 5–8, sketch the graph of each pair of functions on the same coordinate system. Compare the graphs of f and g.

5. (a) $f(x) = 7^x$ (b) $g(x) = 7^x + 2$
6. (a) $f(x) = e^{-x}$ (b) $g(x) = -e^{-x}$
7. (a) $f(x) = 2^x$ (b) $g(x) = 2^{x-3}$
8. (a) $f(x) = 5^{-x}$ (b) $g(x) = 5^{-x} - 3$

In Problems 9 and 10, find each product. Round off each answer to two decimal places.

9. (a) $3^{\sqrt{2}} \cdot 3^{\sqrt{5}}$ (b) $(7^{\sqrt{2}})^{\sqrt{3}}$

10. (a) $7^{\sqrt{3}} \cdot 7^{-\sqrt{5}}$ (b) $(0.3^{\sqrt{7}})^{\sqrt{3}}$

In Problems 11 and 12, sketch the graph of the function f, and then reflect the graph of f about the line $y = x$ to obtain the graph of f^{-1}.

11. (a) $f(x) = 7^x$ (b) $f(x) = \left(\dfrac{1}{7}\right)^x$

12. (a) $f(x) = (0.3)^x$ (b) $f(x) = e^{2x}$

In Problems 13 and 14, find the value of each expression. Round off each answer to two decimal places.

13. $\dfrac{(46{,}000)(0.105)}{12[1 - (1 + \frac{0.105}{12})^{-12(30)}]}$

14. $\dfrac{1{,}200{,}000}{1 + (1200 - 1)e^{-0.4(3)}}$

In Problems 15 and 16, write each equation in logarithmic form.

15. (a) $4^{-2} = \dfrac{1}{16}$ (b) $5 = x^y$

16. (a) $\left(\dfrac{1}{3}\right)^{-2} = 9$ (b) $3 = m^{-n}$

In Problems 17 and 18, write each equation in exponential form.

17. (a) $\log_\pi \pi = 1$ (b) $-2 = \log_7 \dfrac{1}{49}$

18. (a) $\log_{1/3} 1 = 0$ (b) $2 = \log_4 x$

In Problems 19 and 20, find the exact value of each logarithm.

19. (a) $\log_2 8$ (b) $\log_9 3$
    (c) $\log_4 64$ (d) $\log_6 216$
20. (a) $\log_5 125$ (b) $\log_{125} 5$
    (c) $\log_9 9$ (d) $\log_{100} 10{,}000$

In Problems 21–24, find each value. Round off each answer to 4 decimal places.

21. (a) $\log 7806$ (b) $\ln 7.601$
    (c) $\log 0.0012$ (d) $\ln 0.0462$
22. (a) $\log 312.5$ (b) $\ln 276$
    (c) $\log 0.0072$ (d) $\ln 0.0043$
23. (a) $\log (\ln 71)$ (b) $\ln (\log 13)$
    (c) $\log (\ln 0.065)$ (d) $\ln (\log 0.065)$
24. (a) $\log (\ln 13)$ (b) $\ln (\log 27)$
    (c) $\log (\ln 0.5)$ (d) $\ln(\log 0.05)$

In Problems 25 and 26, find x. Round off each answer to two decimal places.

25. (a) $\log x = 0.4133$ (b) $\ln x = 0.4871$
    (c) $\log x = -1.7382$
26. (a) $\log x = 3.5412$ (b) $\ln x = -2.8913$
    (c) $\ln x = 0.7813$

In Problems 27 and 28, find the exact value of x.

27. (a) $\log_4 16 = x$ (b) $\log_3 9\sqrt{3} = x$
    (c) $\ln (17x - 33) = 0$
28. (a) $\log_{36} x = -1/2$ (b) $\log_{1/2} 16 = x$
    (c) $\ln (3x - 11) = 0$

In Problems 29 and 30, rewrite each expression as a sum or difference of multiples of logarithms. Assume all variables are positive real numbers.

29. (a) $\log_6 7x$ (b) $\ln 4x^3$
    (c) $\log 4xy^5$ (d) $\ln \sqrt[4]{x^2y^3}$
30. (a) $\ln 5x$ (b) $\log_3 5xy^2$
    (c) $\log_3 \sqrt[3]{xy}$ (d) $\ln \dfrac{\sqrt[3]{x}}{\sqrt[3]{54}}$

In Problems 31 and 32, rewrite each expression as a logarithm of one quantity. Assume all variables are positive real numbers.

31. (a) $\log \dfrac{3}{7} + \log \dfrac{14}{27}$ (b) $\ln (x + 1) - \ln x$
    (c) $\ln (xy + y^2) - \ln (xz + yz)$

32. (a) $\log_3 \dfrac{5}{12} + \log_3 \dfrac{4}{15}$ (b) $5\ln x - \ln y$
    (c) $\ln\left(\dfrac{x}{u} + x\right) - \ln\left(\dfrac{y}{u} + y\right)$

In Problems 33 and 34, simplify each expression. Assume all variables are positive real numbers.

33. (a) $e^{\ln x}$ (b) $e^{-2 - \ln 3}$ (c) $\ln e^{x^2 - 4}$
34. (a) $e^{-3\ln 2}$ (b) $e^{3 + 2\ln 2}$ (c) $\ln e^{x - x^2}$

In Problems 35–38, sketch the graphs of the following functions on the same coordinate system. Find the domain of each function.

35. (a) $f(x) = \log_3 x$ (b) $g(x) = -\log_3 x$
    (c) $h(x) = \log_3 (-x)$
36. (a) $f(x) = \log_4 x$ (b) $g(x) = -\log_4 x$
    (c) $h(x) = \log_4 (x + 1)$
37. (a) $f(x) = \ln x$ (b) $g(x) = \ln (x - 1)$
    (c) $h(x) = \ln |x + 1|$
38. (a) $f(x) = -\ln x$ (b) $g(x) = \ln (-x)$
    (c) $h(x) = \ln |x|$

In Problems 39–44, solve each equation for x. Round off when necessary to two decimal places.

39. (a) $2^{3x} = 8$ (b) $3^{x^2 - 2x} = 9$ (c) $8^{x^2} = 16^x$
40. (a) $3^{7x} = 81$ (b) $5^{x^2 - 3x} = 625$ (c) $3^{x^2 + 4x} = 9^{-2}$
41. (a) $2^x = 5$ (b) $7^{x+1} = 11$ (c) $e^x = 5^x$
42. (a) $4^x = 3$ (b) $5^{x+1} = 4$ (c) $8^{x^2} = 5^x$
43. (a) $\log x + \log(x - 48) = 2$
    (b) $\ln (3x + 5) - \ln (2x + 6) = 0$
44. (a) $\log (4x + 1) - \log (2x + 9) = 0$
    (b) $2 \log (x + 2) + \log 12 = \log (x + 2)$

45. **Compound Interest:** An initial deposit of $700 earns 4.3% interest, compounded quarterly. How much money will be in the account after 4 years?

46. **Compound Interest:** A company invests $210,000 for its pension plan at 9.5% compounded monthly.
    (a) What is the value of the plan after 5 years?
    (b) If they decide to change plans after 3 years, what is the value of the plan after 3 years of investment? Round to the nearest dollar.

47. **Population Growth:** Suppose that at the start of an experiment, 3000 bacteria are present in a culture. After t hours, the number N in the culture is growing according to the model

$$N(t) = 3000e^{0.06t}$$

   (a) How many bacteria are present 8 hours later? Round off to a whole number.
   (b) After how many hours will the number of bacteria be 50,000? Round to the nearest whole number.

48. **Cellular Phones:** A telephone company estimates that the number N (in thousands) of cellular phones in use by its customers is given by the model

$$N(t) = 0.4(1.74)^t$$

   where t is the number of years after 1990. Estimate the number of cellular phones which will be in use in the year 2004 and then in the year 2008. Round to the nearest whole number.

49. **Real Estate Appreciation:** Suppose that the value of real estate increases at a rate of 5% per year. The value V (in dollars) of a home purchased for P dollars after t years is modeled by the function

$$V(t) = P(1.05)^t, \quad t \geq 0$$

   If this trend continues, find the value of a home purchased for $75,000 and sold 10 years later. Round off to the nearest hundred dollars.

50. **Oceanography:** The intensity I of light (in lumens) for a certain location at a distance x meters below the surface of an ocean is modeled by the function

$$I = I_0(0.4)^x$$

   where I_0 is the intensity at the surface of the ocean. Find the intensity of light at a depth of 3 meters where I_0 is 8. Round off to two decimal places.

51. **Tornadoes:** The speed S of the wind (in miles per hour) near the center of a tornado is modeled by the equation

$$S(x) = 50 + 55 \log x, \quad 0 < x \leq 20$$

   where x is the distance (in miles) the tornado travels. On May 31, 1998, a tornado struck and destroyed the entire city of Spencer, South Dakota, covering a distance of approximately 10 miles. Approximate the speed of the tornado. Round to the nearest mile per hour.

52. **Manufacturing:** A manufacturer of computer monitors has determined that the weekly profit P (in dollars) is given by the model

$$P(x) = 2840 + 950 \ln x, \quad x > 0$$

   where x is the number of monitors produced each hour. What is the weekly profit (to the nearest dollar) if the number of monitors produced each hour is 500?

G In Problems 53–58, use a grapher to graph each function. Use the graph as a guideline to state the domain and range of each function.

53. $f(x) = e^x + e^{-x}$
54. $f(x) = e^{-x^2}$
55. $f(x) = -2 \log_{2/3} x$
56. $f(x) = 4 + \log_3 x$
57. $f(x) = \ln x^2$
58. $f(x) = \ln \sqrt{x}$

◆ CHAPTER 7 PRACTICE TEST

1. For each function below, find $f(1.73)$, $f(-\sqrt{3})$ and $f(2 - \sqrt{2})$. Round off each answer to two decimal places.
   (a) $f(x) = 2^x$ (b) $f(x) = (0.7)^x$
   (c) $f(x) = 4(1 + e^{-x})$

2. Sketch the graphs of the following functions on the same coordinate system. Describe the relationships between these functions.
   (a) $f(x) = 2^x$ (b) $g(x) = -2^x$ (c) $h(x) = 2^x + 3$

3. Sketch the graph of $f(x) = 3^x$, and then reflect the graph of f about the line $y = x$ to obtain the graph of f^{-1}.

4. (a) Write the equation $3^{-2} = \frac{1}{9}$ in logarithmic form.
   (b) Write the equation $\log_{1/5} 625 = -4$ in exponential form.

5. Find the *exact* value of each logarithm.
   (a) $\log_3 27$ (b) $\log_{10} 0.001$
   (c) $\log_2 8\sqrt{2}$ (d) $\log_5 0.04$

6. Solve each equation for x.
   (a) $\log_3 x = 2$ (b) $\log_3 9\sqrt{3} = x$
   (c) $\left(\frac{1}{3}\right)^x = 9^{3x-1}$ (d) $5^x = 11$
   (e) $\log_3(x + 2) - \log_3 x = 2$ (f) $e^{3x} = 4$

7. (a) Find each value of the function $f(x) = \log x$, (i) $f(735)$, (ii) $f(17.6)$, (iii) $f(0.0892)$. Round off each answer to 4 decimal places.
   (b) Find x, round off each answer to two decimal places. (i) $\log x = 1.4023$ (ii) $\ln x = 2.4163$ (iii) $\log x = -1.3271$ (iv) $\ln x = -2.3$

8. Sketch the graphs of the following functions on the same coordinate system. Find the domain of each function.
   (a) $f(x) = \log x$ (b) $g(x) = \log(-x)$
   (c) $h(x) = \log(x + 1)$

9. Rewrite each expression as a sum or difference of multiples of logarithms. Assume all variables are positive real numbers.

   (a) $\log_3 (8x)$

   (b) $\ln \dfrac{x}{7}$

   (c) $\ln (x^3 y^4)$

   (d) $\log \sqrt[3]{x^2 y}$

10. Rewrite each expression as a single logarithm. Assume all variables are positive real numbers.

    (a) $\log \dfrac{3}{8} + \log \dfrac{16}{15}$

    (b) $3 \ln x - 2 \ln y$

    (c) $\log_2 (x + 2) + 3 \log_2 x$

11. Find the value of each expression. Round off the answers to two decimal places.

    (a) $10^{-2.8192}$

    (b) $\log_7 23$

    (c) $e^{-0.72}$

12. Suppose that $5000 is invested in a high risk mutual fund that earns an annual rate of 17% compounded quarterly.

    (a) How much money is accumulated after 3 years?

    (b) When will this investment double itself?

13. Find the present value of a state lottery that in one year makes 2 payments of $150,000 at an interest rate of 4% compounded twice a year (see problem 53 section 7.1). Round to the nearest dollar.

Chapter 8

TOPICS IN ALGEBRA

8.1 Solutions of Linear Systems Using Matrices

8.2 Solutions of Linear Systems Using Determinants and Cramer's Rule

8.3 Linear Programming

8.4 Conic Sections: The Ellipse and Hyperbola

8.5 Systems of Nonlinear Equations and Inequalities

Collecting, storing, and analyzing data are routine activities in the world of business. For instance, a company running stores in different locations may want to compare sales generated by the stores from month to month. The sales figures can be represented by data arranged in a rectangular array of rows and columns called a *matrix*. Systems of linear equations can be solved using matrices. Example 6 on page 392 shows how matrices are used to solve a linear system of equations.

In this chapter we develop alternate methods for solving systems of linear equations using matrix operations and determinants. In addition, we consider linear programming—a topic that is of major importance in business applications. The chapter also includes some graphs of conic sections. These conics are used in presenting graphical solutions of nonlinear systems of equations.

OBJECTIVES

1. Write Linear Systems in Matrix Notation
2. Solve Linear Systems Using Row Equivalent Operations
3. Solve Applied Problems
4. Use Technology Exploration

8.1 Solutions of Linear Systems Using Matrices

Recall from sections 4.2 and 4.3 that we solved linear systems by using elimination. In this section we shall introduce some basic terminology and discuss new procedures for solving such systems.

Writing Linear Systems in Matrix Notation

Consider the system of two linear equations in two unknowns.

$$\begin{cases} 3x - y = 7 \\ 4x + 7y = 15 \end{cases}$$

Now we introduce a scheme for keeping track of the coefficients of the variables in the system in such a way that we do not have to write the x's and the y's. Referring to the system, we first check that the variables appear in the same order in each equation and that terms not involving variables are to the right of the equal signs. We then list the numbers that are involved in the equations as follows:

$$\begin{bmatrix} 3 & -1 & \vdots & 7 \\ 4 & 7 & \vdots & 15 \end{bmatrix}$$

This array of numbers is called the **augmented matrix** for the system. The term **matrix** is used to denote a rectangular array of numbers. The individual numbers are called the **elements** or **entries** of the matrix. We ordinarily enclose the elements of a matrix in *brackets*.

In the matrix below, the **rows** of a matrix are denoted by the horizontal entries while the **columns** are denoted by the vertical ones.

$$\begin{bmatrix} -1 & 2 & 5 \\ 2 & 0 & 1 \\ 3 & 4 & 6 \end{bmatrix} \begin{matrix} \leftarrow \text{Row 1} \\ \leftarrow \text{Row 2} \\ \leftarrow \text{Row 3} \end{matrix}$$

Column 1 Column 2 Column 3

When we use matrices to solve a system of linear equations, it is customary to draw a dashed vertical line to separate the last column (the column that corresponds to the constants).

We usually refer to a matrix by its *size* or *dimensions*. In this case we specify the number of rows and the number of columns in that order. For example, the above matrix is called a **3 by 3 (also written as 3 × 3)** matrix, with elements or entries $-1, 2, 5, 2, 0, 1, 3, 4$ and 6. The linear system in terms of x and y corresponding to the following augmented matrix

$$\begin{bmatrix} 7 & -2 & \vdots & 1 \\ 3 & 4 & \vdots & -5 \end{bmatrix}$$

is given by $\begin{cases} 7x - 2y = 1 \\ 3x + 4y = -5 \end{cases}$

The first column of this matrix lists the coefficients of the x variable, while the second column lists the coefficients of the y variable. The last column lists the constant numbers on the right side of the equations. If any variable does not appear in an equation, a "0" is inserted in the appropriate position in the array of coefficients and constants.

Solving Linear Systems Using Row Equivalent Operations

The basic method for solving this system is to replace the given system by a new one that has the same solution but which is easier to solve. The new system is generally obtained by a sequence of steps involving the following operations. These steps are identical to the steps used in the elimination method in chapter 4.
1. Multiply an equation by a nonzero constant.
2. Interchange two equations.
3. Add a multiple of one equation to another.

Since the rows of an augmented matrix correspond to the equations in the associated system, these three operations correspond to the following operations on the rows of the augmented matrix listed in Table 1.

TABLE 1 Row Operations

Row Operations Performed on a Matrix	Explanation of Operation	Symbolic Representation of Operation
1. Interchange two rows	Interchange row i and row j	$R_i \leftrightarrow R_j$
2. Multiply a row by a nonzero constant	Multiply each entry in row i by a constant k to get new row i, where $k \neq 0$	$kR_i \rightarrow R_i$
3. Add a constant multiple of a row to another row	Multiply each entry of row i by the constant k and add the resulting entries to each entry in row j to get a new row j	$kR_i + R_j \rightarrow R_j$

EXAMPLE 1   Solving a System by Matrices

Use matrices to solve the system.
$$\begin{cases} 3x - y = 15 \\ x + 2y = -2 \end{cases}$$

Solution   In the following presentation we solve the system of linear equations in two ways. On the left, we solve the system by operating on the equations in the system. On the right, we solve the same system by operating on the rows of the augmented matrix.

System of Equations	Augmented Matrix of System	Strategy
(A) $\begin{cases} 3x - y = 15 \\ x + 2y = -2 \end{cases}$	(A) $\begin{bmatrix} 3 & -1 & \vdots & 15 \\ 1 & 2 & \vdots & -2 \end{bmatrix}$	

First, interchange the two equations.

First, interchange the two rows.

| (B) $\begin{cases} x + 2y = -2 \\ 3x - y = 15 \end{cases}$ | (B) $\begin{bmatrix} 1 & 2 & \vdots & -2 \\ 3 & -1 & \vdots & 15 \end{bmatrix}$ | We want the first equation to have a leading coefficient of 1. |

Next, replace the second equation with the sum of -3 times the first equation and the second equation.

Next, replace the second row with the sum of -3 times the first row and the second row.

We want to eliminate the x term in the second equation.

(C) $\begin{cases} x + 2y = -2 \\ 0 - 7y = 21 \end{cases}$ (C) $\begin{bmatrix} 1 & 2 & | & -2 \\ 0 & -7 & | & 21 \end{bmatrix}$

Multiply the second equation by $-\frac{1}{7}$

Multiply the second row by $-\frac{1}{7}$

We want the second equation to have a leading coefficient of 1.

(D) $\begin{cases} x + 2y = -2 \\ \quad\quad y = -3 \end{cases}$ (D) $\begin{bmatrix} 1 & 2 & | & -2 \\ 0 & 1 & | & -3 \end{bmatrix}$

The system is said to be in triangular form.

The matrix is said to be in triangular form.

Thus, the last matrix (D) is an abbreviated form of the last system form (D). Clearly, $y = -3$. To get x, we substitute -3 for y in the first equation to obtain

$$x + 2(-3) = -2, \text{ so } x = 4.$$

Therefore, the solution is $(4, -3)$.

EXAMPLE 2 Solving a System of Two Equations Using Matrices

Write the associated augmented matrix of the system and solve it using row operations.

$$\begin{cases} 4x - 3y = 15 \\ x + 2y = 1 \end{cases}$$

Solution We first write the augmented matrix of the system:

$$\begin{bmatrix} 4 & -3 & | & 15 \\ 1 & 2 & | & 1 \end{bmatrix}$$

Next, we convert it into another matrix in triangular form. The easiest way to get a 1 in the upper left corner is to interchange rows 1 and 2:

$$\begin{bmatrix} 4 & -3 & | & 15 \\ 1 & 2 & | & 1 \end{bmatrix} R_1 \leftrightarrow R_2 \begin{bmatrix} 1 & 2 & | & 1 \\ 4 & -3 & | & 15 \end{bmatrix}$$

We want a 0 to be the first entry in row 2, so we multiply row 1 by -4 and then add the result to row 2:

$$\begin{bmatrix} 1 & 2 & | & 1 \\ 4 & -3 & | & 15 \end{bmatrix} -4R_1 + R_2 \to R_2 \begin{bmatrix} 1 & 2 & | & 1 \\ 0 & -11 & | & 11 \end{bmatrix}$$

Finally, to get 1 as the first nonzero entry in row 2, we multiply row 2 by $-1/11$:

$$\begin{bmatrix} 1 & 2 & | & 1 \\ 0 & -11 & | & 11 \end{bmatrix} -\frac{1}{11}R_2 \to R_2 \begin{bmatrix} 1 & 2 & | & 1 \\ 0 & 1 & | & -1 \end{bmatrix}$$

The last matrix is in the desired triangular form. By writing the associated system, we have

$$\begin{cases} x + 2y = 1 \\ \quad\quad y = -1. \end{cases}$$

Now, we replace y by -1 in the first equation to obtain $x + 2(-1) = 1$ or $x - 2 = 1$, so $x = 3$. Therefore, the solution is $(3, -1)$.

Let us review the strategy used to solve the system in example 2:

Step 1. Form the augmented matrix of the system.
Step 2. Use the elementary row operations to reduce the augmented matrix to triangular form.
Step 3. Write the corresponding system in triangular form.
Step 4. Use back-substitution to solve the resulting system.

We can avoid the necessity of back-substitution in Example 2 by reducing the last matrix even further. To do this, multiply row 2 by -2 and add the result to row 1 to obtain:

$$\begin{bmatrix} 1 & 2 & | & 1 \\ 0 & 1 & | & -1 \end{bmatrix} \quad -2R_2 + R_1 \rightarrow R_1 \quad \begin{bmatrix} 1 & 0 & | & 3 \\ 0 & 1 & | & -1 \end{bmatrix}$$

This final matrix is an especially simple one to interpret because the solution can be read directly from the entries in the last column. This matrix is said to be in *row-reduced echelon form.*

ROW-REDUCED ECHELON FORM
OF A MATRIX

A matrix is said to be in **row-reduced echelon form** if it satisfies the following conditions.
1. The first nonzero entry or leading term in each row is 1.
2. Any rows that consist entirely of 0's are below all rows that do not consist entirely of 0's.
3. The first nonzero entry in each row is to the right, one place or more, of the first nonzero entry in the preceding row.
4. In each column that contains a leading 1 of some row, all other entries are 0.

EXAMPLE 3   Determining Whether a Matrix Is in Row-Reduced Echelon Form

Which of the following matrices is a row-reduced echelon form matrix?

$$A = \begin{bmatrix} 1 & 4 & 0 \\ 0 & 0 & 1 \end{bmatrix} \qquad B = \begin{bmatrix} 1 & -2 & 0 & 5 \\ 0 & 0 & 1 & 7 \end{bmatrix} \qquad C = \begin{bmatrix} 1 & -2 & 0 & 5 & 7 \\ 0 & 0 & 1 & 3 & -4 \\ 0 & 0 & 0 & 0 & 0 \end{bmatrix}$$

$$D = \begin{bmatrix} 0 & 1 \\ 0 & -2 \end{bmatrix} \qquad E = \begin{bmatrix} 0 & 0 & 0 \\ 1 & 0 & 0 \end{bmatrix} \qquad F = \begin{bmatrix} 1 & 1 & 0 \\ 0 & 1 & 0 \\ 0 & 0 & 1 \end{bmatrix}$$

Solution   The matrices A, B, and C are in row-reduced echelon form because each satisfies the conditions of the above definition. However, the matrices D, E, and F are not in row-reduced echelon form. In matrix D, condition 1 is violated because the leading term in the second row is not 1. In matrix E condition 2 is violated. In matrix F, condition 4 is violated because the entry in row 1, column 2 is a 1 rather than 0.   ◈

EXAMPLE 4   Interpreting Row-Reduced Echelon Matrices

Each of the following augmented matrices is the row-reduced echelon matrix form corresponding to a linear system of equations. Solve each system.

(a) $\begin{bmatrix} 1 & 0 & | & 2 \\ 0 & 1 & | & 3 \end{bmatrix}$ (b) $\begin{bmatrix} 1 & 0 & | & 5 \\ 0 & 0 & | & 1 \end{bmatrix}$ (c) $\begin{bmatrix} 1 & 3 & | & 6 \\ 0 & 0 & | & 0 \end{bmatrix}$

Solution   (a) The matrix

$$\begin{bmatrix} 1 & 0 & | & 2 \\ 0 & 1 & | & 3 \end{bmatrix}$$

corresponds to the system

$$\begin{cases} x = 2 \\ y = 3 \end{cases}$$

so the system has one solution, (2,3).

(b) The matrix

$$\begin{bmatrix} 1 & 0 & | & 5 \\ 0 & 0 & | & 1 \end{bmatrix}$$

corresponds to the system

$$\begin{cases} x = 5 \\ 0 = 1 \end{cases}$$

Because of the contradiction in the second equation, we conclude that the system has no solution.

(c) The matrix

$$\begin{bmatrix} 1 & 3 & | & 6 \\ 0 & 0 & | & 0 \end{bmatrix}$$

corresponds to the system

$$x + 3y = 6$$

The last row of the matrix corresponds to $0 = 0$, which is true for all values of x and y. The system is consistent and the equations are dependent and the system has an infinite number of solutions.   ◈

We now establish a generalized method for solving linear systems based on the following steps.

SOLVING A SYSTEM BY MATRICES

Step 1. Represent the linear system with an augmented matrix.

Step 2. Use elementary row operations to transform the original matrix into a row-reduced echelon matrix.

Step 3. Change the reduced matrix from step 2 to the equation form of the linear system.

Step 4. Interpret the result in step 3 to obtain the solution of the system.

If a system has three equations and three variables, we can still perform a sequence of row operations. However, the corresponding augmented matrix has three rows and four columns.

EXAMPLE 5   Solving a System of Three Equations Using Matrices

Solve the system.

$$\begin{cases} x + 2y - z = 30 \\ 4x + 3y - 4z = 70 \\ 2x \quad\quad + 3z = -5 \end{cases}$$

Solution   We begin with writing the augmented matrix of the system.

$$\begin{bmatrix} 1 & 2 & -1 & | & 30 \\ 4 & 3 & -4 & | & 70 \\ 2 & 0 & 3 & | & -5 \end{bmatrix}$$

We then apply the elementary row operations to obtain another matrix of an equivalent system of equations.

$$\begin{bmatrix} 1 & 2 & -1 & | & 30 \\ 0 & -5 & 0 & | & -50 \\ 2 & 0 & 3 & | & -5 \end{bmatrix}$$ Add -4 times the first row to the second row. In symbols,
$$-4R_1 + R_2 \rightarrow R_2$$

$$\begin{bmatrix} 1 & 2 & -1 & | & 30 \\ 0 & -5 & 0 & | & -50 \\ 0 & -4 & 5 & | & -65 \end{bmatrix}$$ Add -2 times the first row to the third row.
$$-2R_1 + R_3 \rightarrow R_3$$

$$\begin{bmatrix} 1 & 2 & -1 & | & 30 \\ 0 & 1 & 0 & | & 10 \\ 0 & -4 & 5 & | & -65 \end{bmatrix}$$ Multiply the second row by $-\frac{1}{5}$.
$$-\frac{1}{5}R_2 \rightarrow R_2$$

$$\begin{bmatrix} 1 & 0 & -1 & | & 10 \\ 0 & 1 & 0 & | & 10 \\ 0 & -4 & 5 & | & -65 \end{bmatrix}$$ Add -2 times the second row to the first row.
$$-2R_2 + R_1 \rightarrow R_1$$

$$\begin{bmatrix} 1 & 0 & -1 & | & 10 \\ 0 & 1 & 0 & | & 10 \\ 0 & 0 & 5 & | & -25 \end{bmatrix}$$ Add 4 times the second row to the third row.
$$4R_2 + R_3 \rightarrow R_3$$

$$\begin{bmatrix} 1 & 0 & -1 & | & 10 \\ 0 & 1 & 0 & | & 10 \\ 0 & 0 & 1 & | & -5 \end{bmatrix}$$ Multiply the third row by $\frac{1}{5}$.
$$\frac{1}{5}R_3 \rightarrow R_3$$

$$\begin{bmatrix} 1 & 0 & 0 & | & 5 \\ 0 & 1 & 0 & | & 10 \\ 0 & 0 & 1 & | & -5 \end{bmatrix}$$ Add the third row to the first row.
$$R_3 + R_1 \rightarrow R_1$$

The last matrix is the row-reduced form and it corresponds to the system

$$\begin{cases} x = 5 \\ y = 10 \\ z = -5 \end{cases}$$

The solution of the system is $(5, 10, -5)$.

Solving Applied Problems

Many applied problems produce linear systems that can be solved by matrices as the next example shows.

EXAMPLE 6   Solving a Nutrition Problem

A cup of low fat yogurt contains a total of 32 grams of carbohydrates, protein, and fat. There are 12 more grams of carbohydrates than of fat. Protein and fat weigh as much as the carbohydrates. How many grams of protein, fat, and carbohydrates are there?

Solution   Make a Plan

We have a cup of low fat yogurt that contains 32 grams of carbohydrates, protein, and fat. If we let

x = the number of grams of protein
y = the number of grams of fat
z = the number of grams of carbohydrates

The equations are $x + y + z = 32$, $z = y + 12$, and $x + y = z$. The system of equations is

$$\begin{cases} x + y + z = 32 \\ y - z = -12 \\ x + y - z = 0 \end{cases}$$

Write the Solution

The augmented matrix that corresponds to this system is

$$\left[\begin{array}{ccc:c} 1 & 1 & 1 & 32 \\ 0 & 1 & -1 & -12 \\ 1 & 1 & -1 & 0 \end{array}\right]$$

Using the row operations, the row-reduced echelon matrix is

$$\left[\begin{array}{ccc:c} 1 & 0 & 0 & 12 \\ 0 & 1 & 0 & 4 \\ 0 & 0 & 1 & 16 \end{array}\right]$$

The associated system for the transformed matrix is

$$\begin{cases} x = 12 \\ y = 4 \\ z = 16 \end{cases}$$

So there are 12 grams of protein, 4 grams of fat, and 16 grams of carbohydrates.

Look Back

If we substitute these numbers in the system, they check.   ◇

Using Technology Exploration

Graphers can use matrix operations to reduce a matrix to row-reduced echelon form and solve linear systems of equations.

G EXAMPLE 7   Solving a Linear System by a Grapher
Use a grapher to solve the system.

$$\begin{cases} x + y + 2z + 3w = 5 \\ x - 2y + z + w = 5 \\ 3x + y + z - w = 2 \\ 2x - y + z + 4w = 8 \end{cases}$$

Solution   The system can be solved easily on a grapher by writing the associated augmented matrix and reducing it to row-reduced echelon form. Consult with your grapher's manual.

$$\begin{bmatrix} 1 & 1 & 2 & 3 & | & 5 \\ 1 & -2 & 1 & 1 & | & 5 \\ 3 & 1 & 1 & -1 & | & 2 \\ 2 & -1 & 1 & 4 & | & 8 \end{bmatrix} \rightarrow \begin{bmatrix} 1 & 0 & 0 & 0 & | & 1 \\ 0 & 1 & 0 & 0 & | & -1 \\ 0 & 0 & 1 & 0 & | & 1 \\ 0 & 0 & 0 & 1 & | & 1 \end{bmatrix}$$

This means that $x = 1$, $y = -1$, $z = 1$ and $w = 1$. The solution of the system is $(1, -1, 1, 1)$. ◇

PROBLEM SET 8.1

Mastering the Concepts

In Problems 1–4, find the augmented matrix for each system of linear equations.

1. $\begin{cases} 2x + 3y = 2 \\ x - 2y = 8 \end{cases}$

2. $\begin{cases} 7x - 8y = 9 \\ 4x + 3y = -10 \end{cases}$

3. $\begin{cases} x - 2y + 3z = 4 \\ 2x + y - 4z = 3 \\ -3x + 4y - z = -2 \end{cases}$

4. $\begin{cases} 4x - y + 3z = 6 \\ -8x + 3y - 5z = -6 \\ 5x - 4y = -9 \end{cases}$

In Problem 5–8, find a system of linear equations corresponding to each augmented matrix.

5. $\begin{bmatrix} 2 & 8 & | & 7 \\ 3 & -5 & | & 4 \end{bmatrix}$

6. $\begin{bmatrix} 5 & -6 & | & 4 \\ 3 & 7 & | & 8 \end{bmatrix}$

7. $\begin{bmatrix} 2 & 6 & -4 & | & 1 \\ 1 & 3 & -2 & | & 4 \\ 2 & 1 & -3 & | & 7 \end{bmatrix}$

8. $\begin{bmatrix} 1 & 3 & -1 & | & -3 \\ 3 & -1 & 2 & | & 1 \\ 2 & -1 & 1 & | & -1 \end{bmatrix}$

In Problems 9–12, use matrices to solve each linear system in two ways as in example 1 (a) on the left, use elementary operations to convert each system to an equivalent *triangular* form (b) on the right, show the augmented matrix corresponding to each equivalent linear system, and describe the corresponding elementary row operations.

9. $\begin{cases} x - 2y = -4 \\ 3x + 4y = 3 \end{cases}$

10. $\begin{cases} x - 2y = 8 \\ 2x + 3y = 2 \end{cases}$

11. $\begin{cases} x + 2y = 4 \\ 2x - y = 3 \end{cases}$

12. $\begin{cases} x + 3y = 7 \\ 3x - y = 1 \end{cases}$

In Problems 13–16, determine which matrix is in row-reduced echelon form.

13. (a) $\begin{bmatrix} 1 & -3 & 0 \\ 0 & 0 & 1 \end{bmatrix}$ (b) $\begin{bmatrix} 1 & 5 & 0 \\ 0 & 1 & -4 \end{bmatrix}$

14. (a) $\begin{bmatrix} 1 & 0 & 0 & 4 \\ 0 & 1 & 0 & -3 \end{bmatrix}$ (b) $\begin{bmatrix} 1 & 0 & 0 & 4 & -1 \\ 0 & 1 & 0 & 2 & 3 \end{bmatrix}$

15. (a) $\begin{bmatrix} 0 & 0 & 0 \\ 0 & 0 & 0 \end{bmatrix}$ (b) $\begin{bmatrix} 0 & 0 & 0 & 1 \\ 0 & 0 & 0 & 0 \end{bmatrix}$

16. (a) $\begin{bmatrix} 0 & 1 & 0 & 4 & -1 \\ 1 & 0 & 0 & 3 & 2 \end{bmatrix}$ (b) $\begin{bmatrix} 0 & 1 & 2 & 0 \\ 1 & 0 & 0 & 3 \\ 0 & 0 & 0 & 0 \end{bmatrix}$

In Problems 17–22, each row-reduced echelon matrix is the augmented matrix form of a corresponding system of linear equation. Solve each system.

17. $\begin{bmatrix} 1 & 0 & | & 2 \\ 0 & 1 & | & -3 \end{bmatrix}$

18. $\begin{bmatrix} 1 & 0 & | & -3/7 \\ 0 & 1 & | & 2/7 \end{bmatrix}$

19. $\begin{bmatrix} 1 & 0 & 0 & | & -3 \\ 0 & 1 & 0 & | & 2 \\ 0 & 0 & 1 & | & 4 \end{bmatrix}$

20. $\begin{bmatrix} 1 & 0 & 0 & | & 1/2 \\ 0 & 1 & 0 & | & 2/3 \\ 0 & 0 & 1 & | & -1/4 \end{bmatrix}$

21. $\begin{bmatrix} 1 & 0 & 0 & | & 2 \\ 0 & 1 & 0 & | & -3 \end{bmatrix}$

22. $\begin{bmatrix} 1 & 0 & 0 & | & -4 \\ 0 & 1 & 0 & | & 2 \end{bmatrix}$

In Problems 23–34, write the augmented matrix of each system, and then use row operations to rewrite the matrix in a row-reduced echelon form and solve the system.

23. $\begin{cases} x - y = 2 \\ x + y = 4 \end{cases}$

24. $\begin{cases} x + y = -3 \\ x - y = 21 \end{cases}$

25. $\begin{cases} x + 2y = 4 \\ 2x + y = 3 \end{cases}$

26. $\begin{cases} x + y = 0 \\ 2x - y = -2 \end{cases}$

27. $\begin{cases} 2x + 3y = 1 \\ 4x - 3y = -7 \end{cases}$

28. $\begin{cases} 2x - y = 3 \\ -2x + 5y = 7 \end{cases}$

29. $\begin{cases} 2x + 3y = -26 \\ 4x - 3y = 2 \end{cases}$

30. $\begin{cases} 3x + 2y = 7 \\ 6x + 3y = 12 \end{cases}$

31. $\begin{cases} x - 2y - 3z = -1 \\ x + 3y - 2z = 3 \\ 2x + y + z = 6 \end{cases}$

32. $\begin{cases} x + 3y - z = -3 \\ 3x - y + 2z = 1 \\ 2x - y + z = -1 \end{cases}$

33. $\begin{cases} -x + 3y + z = 1 \\ 2x + 5y = 3 \\ 3x + y - 2z = -2 \end{cases}$

34. $\begin{cases} x - 2y + 2z = 0 \\ 3y + z = 17 \\ 5x + 2y - z = -7 \end{cases}$

Applying the Concepts

In Problems 35–38, set up a linear system of equations, and then use the elementary row operations on the augmented matrix to solve the system.

35. **Coin Value:** A collection of 45 coins consists of dimes and quarters. The total value is $8.70. How many dimes and how many quarters are there?

36. **Age Problem:** The sum of the ages of two sisters is 15. The older sister is four times as old as the younger one. How old is each?

37. **Geometry:** The sum of the interior angles of a triangle is 180°. The measure of the smallest angle is one-third of the measure of the largest one. The measure of the middle-sized angle is 30° less than the measure of the largest angle. Find the measure of each angle.

38. **Investment Portfolio:** A retiree receives $345 in one year in simple interest from three investments totaling $5000. Part of the money is invested in bonds at 8% annual simple interest, part in certificates that earn 7% annual simple interest, and the rest in a mutual fund that earns 6% annual simple interest. There is $1500 more invested in certificates than in bonds. Find the amount invested in each?

Developing and Extending the Concepts

In Problems 39–42, each row-reduced echelon matrix is the augmented matrix form of a corresponding linear system of equations. Solve each system (if possible).

39. $\left[\begin{array}{ccc|c} 1 & 0 & -1 & 3 \\ 0 & 1 & -3 & 5 \end{array}\right]$

40. $\left[\begin{array}{ccc|c} 1 & 0 & 2 & -7 \\ 0 & 1 & 5 & 10 \end{array}\right]$

41. $\left[\begin{array}{ccc|c} 1 & 0 & 0 & -4 \\ 0 & 1 & 0 & 5 \\ 0 & 0 & 0 & 6 \end{array}\right]$

42. $\left[\begin{array}{ccc|c} 1 & 0 & 0 & 3 \\ 0 & 1 & 0 & 5 \\ 0 & 0 & 0 & -7 \end{array}\right]$

In Problems 43–50, write the augmented matrix of each system, and then use row operations to rewrite the matrix in row-reduced echelon form and solve the system. **G** Use a grapher in problems 49 and 50.

43. $\begin{cases} 2x - y + 4z = 8 \\ -3x + y - 2z = 5 \end{cases}$

44. $\begin{cases} 5x + 2y - z = 10 \\ y + z = -3 \end{cases}$

45. $\begin{cases} 2x - y + z = 0 \\ x - y - 2z = 0 \\ 2x - 3y - z = 0 \end{cases}$

46. $\begin{cases} x - 2y - 2z = 0 \\ x + y + z = 0 \\ 2x + y + z = 0 \end{cases}$

47. $\begin{cases} -x + y = -1 \\ 4x - 3y = 1 \\ 2x + y = -7 \end{cases}$

48. $\begin{cases} x + y = 1 \\ x - 2y = 13 \\ 2x + 3y = -2 \end{cases}$

49. $\begin{cases} x + 2y - z - 3w = 2 \\ x + y + 3z - 2w = -3 \\ -2x + y - 2z + w = -9 \\ 3x + y - 2z - w = 6 \end{cases}$

50. $\begin{cases} w + x - 2y - 5z = -1 \\ z + w - y + 2x = 1 \\ y + x + 3z - 2w = 2 \\ 3x - 2y - 4z - 2w = 1 \end{cases}$

OBJECTIVES

1. Find the Value of a Determinant
2. Apply Cramer's Rule for Systems of Two Linear Equations
3. Examine the Cases when $D = 0$ in Cramer's Rule
4. Apply Cramer's Rule to Systems of Three Linear Equations
5. Solve Applied Problems
6. Use Technology Explorations

8.2 Solutions of Linear Systems Using Determinants and Cramer's Rule

So far, we have solved linear systems of equations by substitution, elimination, and matrices. In this section we consider an alternative method called *Cramer's Rule* for solving these systems.

Finding the Value of a Determinant

Cramer's rule is based on the idea of a *determinant*. If a, b, c and d are any real numbers, the symbol

$$\begin{vmatrix} a & b \\ c & d \end{vmatrix} \text{ or } \det \begin{bmatrix} a & b \\ c & d \end{bmatrix}$$

is called a **2 by 2** (*also written as* **2 × 2**) **determinant** with *entries* or *elements* a, b, c, and d. Its value is defined by the number $ad - cb$; that is

$$\begin{vmatrix} a & b \\ c & d \end{vmatrix} = ad - cb$$

EXAMPLE 1   Finding the Value of a Determinant

Find the value of each determinant.

(a) $\begin{vmatrix} 3 & 4 \\ 2 & 7 \end{vmatrix}$ (b) $\begin{vmatrix} 5 & 4 \\ 6 & -7 \end{vmatrix}$ (c) $\begin{vmatrix} 3 & 5 \\ 2 & 6 \end{vmatrix}$

Solution   (a) $\begin{vmatrix} 3 & 4 \\ 2 & 7 \end{vmatrix} = 3(7) - 2(4) = 21 - 8 = 13$

(b) $\begin{vmatrix} 5 & 4 \\ 6 & -7 \end{vmatrix} = 5(-7) - 6(4) = -35 - 24 = -59$

(c) $\begin{vmatrix} 3 & 5 \\ 2 & 6 \end{vmatrix} = 3(6) - 2(5) = 18 - 10 = 8$ ◈

Applying Cramer's Rule for Systems of Two Linear Equations

In 1750 Gabriel Cramer (1704–1752), a professor of mathematics in Geneva, Switzerland, proved the general rule for solving systems of linear equations which now bears his name. Today we write Cramer's rule as follows.

CRAMER'S RULE FOR SYSTEMS OF TWO LINEAR EQUATIONS

Consider the system of two linear equations:

$$\begin{cases} ax + by = h \\ cx + dy = k \end{cases}$$

The solution of this system is given by

$$x = \frac{D_x}{D} \quad \text{and} \quad y = \frac{D_y}{D}, D \neq 0$$

where

$$D = \begin{vmatrix} a & b \\ c & d \end{vmatrix}, D_x = \begin{vmatrix} h & b \\ k & d \end{vmatrix} \text{ and } D_y = \begin{vmatrix} a & h \\ c & k \end{vmatrix}$$

In Cramer's rule, D is called the **coefficient determinant** because its entries are the coefficients of the unknowns in the system.

Notice that D_x is obtained by replacing the *first* column of D (the coefficients of x) by the constant column on the right side of the system, and that D_y is obtained by replacing the *second* column of D (the coefficients of y) by the constant column on the right.

EXAMPLE 2   Applying Cramer's Rule
Solve the system.

$$\begin{cases} 2x - y = 7 \\ x + 3y = 14 \end{cases}$$

Solution   In this system,

$$D = \begin{vmatrix} 2 & -1 \\ 1 & 3 \end{vmatrix} = 2(3) - 1(-1) = 7,$$

$$D_x = \begin{vmatrix} 7 & -1 \\ 14 & 3 \end{vmatrix} = 7(3) - 14(-1) = 35,$$

$$D_y = \begin{vmatrix} 2 & 7 \\ 1 & 14 \end{vmatrix} = 2(14) - 1(7) = 21$$

Hence, by Cramer's rule,

$$x = \frac{D_x}{D} = \frac{35}{7} = 5$$

and

$$y = \frac{D_y}{D} = \frac{21}{7} = 3$$

The solution is (5,3).

Examining the Cases when $D = 0$ in Cramer's Rule

Next we examine the two cases for which
$$D = 0.$$
These cases indicate geometrically that the system either involves two parallel lines or two coinciding lines.

EXAMPLE 3   Examining Systems with $D = 0$
Use Cramer's rule to solve the following systems:

(a) $\begin{cases} x - 2y = -6 \\ x - 2y = 2 \end{cases}$ (b) $\begin{cases} 3x - 4y = 20 \\ 6x = 8y + 40 \end{cases}$

Solution   (a) In this system,

$$D = \begin{vmatrix} 1 & -2 \\ 1 & -2 \end{vmatrix} = 1(-2) - 1(-2) = 0,$$

$$D_x = \begin{vmatrix} -6 & -2 \\ 2 & -2 \end{vmatrix} = -6(-2) - 2(-2) = 16, \text{ and}$$

$$D_y = \begin{vmatrix} 1 & -6 \\ 1 & 2 \end{vmatrix} = 1(2) - 1(-6) = 8$$

Hence,

$$x = \frac{D_x}{D} = \frac{16}{0}$$

and

$$y = \frac{D_y}{D} = \frac{8}{0},$$

and so the system has no solution. Actually, the system is inconsistent as can be seen from the graphs which show two parallel lines in Figure 1a.

(b) First, we rewrite the system in the form

$$\begin{cases} 3x - 4y = 20 \\ 6x - 8y = 40 \end{cases}$$

and so,

$$D = \begin{vmatrix} 3 & -4 \\ 6 & -8 \end{vmatrix} = -24 - (-24) = 0,$$

$$D_x = \begin{vmatrix} 20 & -4 \\ 40 & -8 \end{vmatrix} = -160 - (-160) = 0, \text{ and}$$

$$D_y = \begin{vmatrix} 3 & 20 \\ 6 & 40 \end{vmatrix} = 120 - 120 = 0$$

Hence,

$$x = \frac{D_x}{D} = \frac{0}{0}$$

and

$$y = \frac{D_y}{D} = \frac{0}{0},$$

In this case, the system has an infinite number of solutions and the system is consistent and the equations are dependent. Figure 1(b) shows two lines which coincide.

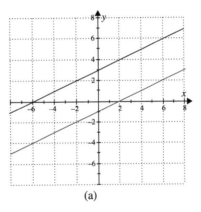

(a)

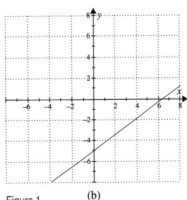

Figure 1   (b)

Keep in mind that Cramer's rule makes sense only when $D \neq 0$. If $D = 0$, the system is either inconsistent, or consistent and dependent.

Applying Cramer's Rule to Linear Systems of Three Equations

To extend Cramer's rule to linear systems of three equations in three variables, we begin by extending the definition of a determinant.

DEFINITION OF 3 × 3 DETERMINANT

The **value** of a 3 × 3 determinant is defined as follows:

$$\begin{vmatrix} a_1 & a_2 & a_3 \\ b_1 & b_2 & b_3 \\ c_1 & c_2 & c_3 \end{vmatrix} = a_1 \begin{vmatrix} b_2 & b_3 \\ c_2 & c_3 \end{vmatrix} - a_2 \begin{vmatrix} b_1 & b_3 \\ c_1 & c_3 \end{vmatrix} + a_3 \begin{vmatrix} b_1 & b_2 \\ c_1 & c_2 \end{vmatrix}$$

We refer to this definition as the **expansion formula** for a 3 × 3 determinant.

In the expansion formula, notice that each entry in the first row is multiplied by the 2×2 determinant that remains when the row and column containing the multiplier are (mentally) crossed out. Thus,

$$a_1 \text{ is multiplied by } \begin{vmatrix} a_1 & a_2 & a_3 \\ b_1 & b_2 & b_3 \\ c_1 & c_2 & c_3 \end{vmatrix} = \begin{vmatrix} b_2 & b_3 \\ c_2 & c_3 \end{vmatrix}$$

$$a_2 \text{ is multiplied by } \begin{vmatrix} a_1 & a_2 & a_3 \\ b_1 & b_2 & b_3 \\ c_1 & c_2 & c_3 \end{vmatrix} = \begin{vmatrix} b_1 & b_3 \\ c_1 & c_3 \end{vmatrix}$$

and

$$a_3 \text{ is multiplied by } \begin{vmatrix} a_1 & a_2 & a_3 \\ b_1 & b_2 & b_3 \\ c_1 & c_2 & c_3 \end{vmatrix} = \begin{vmatrix} b_1 & b_2 \\ c_1 & c_2 \end{vmatrix}$$

To **expand** a 3 × 3 determinant means to find its value by using the expansion formula. We emphasize: when expanding a 3 × 3 determinant, *don't forget the negative sign in front of the middle term.*

EXAMPLE 4   Evaluating a 3 × 3 Determinant

Use the expansion formula to evaluate the determinant.

$$\begin{vmatrix} 3 & 1 & 2 \\ -4 & 2 & 4 \\ 1 & 0 & 5 \end{vmatrix}$$

Solution

$$\begin{vmatrix} 3 & 1 & 2 \\ -4 & 2 & 4 \\ 1 & 0 & 5 \end{vmatrix} = 3 \begin{vmatrix} 2 & 4 \\ 0 & 5 \end{vmatrix} - 1 \begin{vmatrix} -4 & 4 \\ 1 & 5 \end{vmatrix} + 2 \begin{vmatrix} -4 & 2 \\ 1 & 0 \end{vmatrix}$$

$$= 3[2(5) - 0(4)] - 1[-4(5) - 1(4)] + 2[-4(0) - 1(2)]$$

$$= 3(10) - (-24) + 2(-2)$$

$$= 50$$

Now we can state Cramer's rule for solving a linear system of three equations in three variables.

CRAMER'S RULE FOR SOLVING 3 × 3
LINEAR SYSTEMS

The solution of the system

$$\begin{cases} a_1x + a_2y + a_3z = k_1 \\ b_1x + b_2y + b_3z = k_2 \\ c_1x + c_2y + c_3z = k_3 \end{cases}$$

is given by

$$x = \frac{D_x}{D}, y = \frac{D_y}{D}, \text{ and } z = \frac{D_z}{D}, D \neq 0$$

where

$$D = \begin{vmatrix} a_1 & a_2 & a_3 \\ b_1 & b_2 & b_3 \\ c_1 & c_2 & c_3 \end{vmatrix}, D_x = \begin{vmatrix} k_1 & a_2 & a_3 \\ k_2 & b_2 & b_3 \\ k_3 & c_2 & c_3 \end{vmatrix},$$

$$D_y = \begin{vmatrix} a_1 & k_1 & a_3 \\ b_1 & k_2 & b_3 \\ c_1 & k_3 & c_3 \end{vmatrix}, D_z = \begin{vmatrix} a_1 & a_2 & k_1 \\ b_1 & b_2 & k_2 \\ c_1 & c_2 & k_3 \end{vmatrix}$$

EXAMPLE 5 Using Cramer's Rule to Solve a System of Three Linear Equations

Solve the system by Cramer's rule.

$$\begin{cases} x + 2y + 3x = 7 \\ 4x - y - 3z = 7 \\ 5x + 2y - z = 11 \end{cases}$$

Solution Using Cramer's rule, we have

$$D = \begin{vmatrix} 1 & 2 & 3 \\ 4 & -1 & -3 \\ 5 & 2 & -1 \end{vmatrix} = 1 \begin{vmatrix} -1 & -3 \\ 2 & -1 \end{vmatrix} - 2 \begin{vmatrix} 4 & -3 \\ 5 & -1 \end{vmatrix} + 3 \begin{vmatrix} 4 & -1 \\ 5 & 2 \end{vmatrix}$$

$$= 1(7) - 2(11) + 3(13) = 24$$

$$D_x = \begin{vmatrix} 7 & 2 & 3 \\ 7 & -1 & -3 \\ 11 & 2 & -1 \end{vmatrix} = 7 \begin{vmatrix} -1 & -3 \\ 2 & -1 \end{vmatrix} - 2 \begin{vmatrix} 7 & -3 \\ 11 & -1 \end{vmatrix} + 3 \begin{vmatrix} 7 & -1 \\ 11 & 2 \end{vmatrix}$$

$$= 7(7) - 2(26) + 3(25) = 72$$

$$D_y = \begin{vmatrix} 1 & 7 & 3 \\ 4 & 7 & -3 \\ 5 & 11 & -1 \end{vmatrix} = 1 \begin{vmatrix} 7 & -3 \\ 11 & -1 \end{vmatrix} - 7 \begin{vmatrix} 4 & -3 \\ 5 & -1 \end{vmatrix} + 3 \begin{vmatrix} 4 & 7 \\ 5 & 11 \end{vmatrix}$$

$$= 1(26) - 7(11) + 3(9) = -24$$

$$D_z = \begin{vmatrix} 1 & 2 & 7 \\ 4 & -1 & 7 \\ 5 & 2 & 11 \end{vmatrix} = 1 \begin{vmatrix} -1 & 7 \\ 2 & 11 \end{vmatrix} - 2 \begin{vmatrix} 4 & 7 \\ 5 & 11 \end{vmatrix} + 7 \begin{vmatrix} 4 & -1 \\ 5 & 2 \end{vmatrix}$$

$$= 1(-25) - 2(9) + 7(13) = 48$$

so $x = \dfrac{D_x}{D} = \dfrac{72}{24} = 3, y = \dfrac{D_y}{D} = \dfrac{-24}{24} = -1, \text{ and } z = \dfrac{D_z}{D} = \dfrac{48}{24} = 2$

Thus, the solution is $(3, -1, 2)$.

Solving Applied Problems

Determinants can be used to solve applied problems from different fields.

EXAMPLE 6   Solving a Puzzle Problem

Suppose that you hire 4 people to do a small job that pays a sum of $125. What is the share of each individual if when the first person's share is increased by 4, the second's is decreased by 4, the third's is multiplied by 4, and the fourth's is divided by 4, all these results are equal?

Solution   Make a Plan

The four people agree to do the job for a total of $125. The conditions are as follows: When the first person's share is increased by 4, the second's is decreased by 4, the third's is multiplied by 4 and the fourth's is divided by 4, all the results are the same.

Suppose that

$$x = \text{the first person's share in dollars}$$
$$y = \text{the second person's share in dollars}$$
$$z = \text{the third person's share in dollars, and}$$
$$(125 - x - y - z) = \text{the fourth person's share in dollars.}$$

Then, we form the system as follows:

$$\begin{cases} x + 4 = y - 4 \\ y - 4 = 4z \\ 4z = \dfrac{125 - x - y - z}{4} \end{cases}$$

Write the Solution

First, we rewrite the system as follows:

$$\begin{cases} x - y = -8 \\ y - 4z = 4 \\ x + y + 17z = 125 \end{cases}$$

Then, we use Cramer's rule to solve the system.

$$D = \begin{vmatrix} 1 & -1 & 0 \\ 0 & 1 & -4 \\ 1 & 1 & 17 \end{vmatrix} = 25, \quad D_x = \begin{vmatrix} -8 & -1 & 0 \\ 4 & 1 & -4 \\ 125 & 1 & 17 \end{vmatrix} = 400$$

$$D_y = \begin{vmatrix} 1 & -8 & 0 \\ 0 & 4 & -4 \\ 1 & 125 & 17 \end{vmatrix} = 600 \quad D_z = \begin{vmatrix} 1 & -1 & -8 \\ 0 & 1 & 4 \\ 1 & 1 & 125 \end{vmatrix} = 125$$

so that,

$$x = \frac{D_x}{D} = \frac{400}{25} = 16, \ y = \frac{D_y}{D} = \frac{600}{25} = 24, \text{ and } z = \frac{D_z}{D} = \frac{125}{25} = 5$$

Therefore, the shares are as follows:

The first = $16, the second = $24, the third = $5, and
the fourth = $125 - 16 - 24 - 5 = 80.

Look Back

Notice that $16 + 4 = 24 - 4 = 4(5) = 80/4$

Using Technology Exploration

Graphers have a command that calculates the *determinant* of an nxn matrix.

G EXAMPLE 7   Using a Grapher

Use a grapher to find the value of the determinant

$$\begin{vmatrix} -2 & 3 & 4 & 0 \\ 1 & 0 & -2 & 2 \\ 0 & 1 & 1 & -1 \\ 3 & 1 & -2 & -1 \end{vmatrix}$$

Solution   First we enter the square matrix on a grapher.

$$\begin{bmatrix} -2 & 3 & 4 & 0 \\ 1 & 0 & -2 & 2 \\ 0 & 1 & 1 & -1 \\ 3 & 1 & -2 & -1 \end{bmatrix}$$

Then, depending upon your grapher, the command to evaluate the determinant of a matrix might be:

$$\det \begin{bmatrix} -2 & 3 & 4 & 0 \\ 1 & 0 & -2 & 2 \\ 0 & 1 & 1 & -1 \\ 3 & 1 & -2 & -1 \end{bmatrix}$$

This determinant has a value of -3.

PROBLEM SET 8.2

Mastering the Concepts

In Problems 1–8, evaluate each determinant.

1. (a) $\begin{vmatrix} 1 & -1 \\ 2 & 1 \end{vmatrix}$ (b) $\begin{vmatrix} 2 & 3 \\ 5 & 1 \end{vmatrix}$

2. (a) $\begin{vmatrix} 3 & -2 \\ 8 & -1 \end{vmatrix}$ (b) $\begin{vmatrix} 1 & 3 \\ 6 & -1 \end{vmatrix}$

3. (a) $\begin{vmatrix} 5 & -2 \\ 1 & 4 \end{vmatrix}$ (b) $\begin{vmatrix} 5 & 1 \\ 2 & -3 \end{vmatrix}$

4. (a) $\begin{vmatrix} -10 & 1 \\ 3 & 2 \end{vmatrix}$ (b) $\begin{vmatrix} 7 & 1 \\ -2 & 1 \end{vmatrix}$

5. $\begin{vmatrix} 1 & 2 & 4 \\ 2 & -3 & 1 \\ 3 & -1 & -2 \end{vmatrix}$ 6. $\begin{vmatrix} 1 & 1 & 1 \\ 2 & -1 & -1 \\ 1 & -1 & 2 \end{vmatrix}$

7. $\begin{vmatrix} 1 & 1 & 1 \\ 2 & 3 & -1 \\ 3 & 5 & 1 \end{vmatrix}$ 8. $\begin{vmatrix} 1 & 2 & 5 \\ 4 & 1 & 3 \\ 6 & 9 & -1 \end{vmatrix}$

In Problems 9–28, use Cramer's rule to solve each system.

9. $\begin{cases} 2x - y = 0 \\ x + y = 1 \end{cases}$ 10. $\begin{cases} -3x + y = 3 \\ -2x - y = -5 \end{cases}$

11. $\begin{cases} 3x + y = 1 \\ 9x + 3y = -4 \end{cases}$ 12. $\begin{cases} x + y = 30 \\ 2x - 2y = 25 \end{cases}$

13. $\begin{cases} 7x - 9y = 13 \\ 5x + 2y = 10 \end{cases}$ 14. $\begin{cases} 8x - 2y = 52 \\ 3x - 5y = 45 \end{cases}$

15. $\begin{cases} 7x + 4y = 1 \\ 9x + 4y = 3 \end{cases}$ 16. $\begin{cases} 3x + 7y = 16 \\ 2x + 5y = 13 \end{cases}$

17. $\begin{cases} -3x + y = 3 \\ -2x - y = -5 \end{cases}$ 18. $\begin{cases} 3x + 2y = 7 \\ -2x + 7y = 12 \end{cases}$

19. $\begin{cases} 2x - y + z = 3 \\ -x + 2y - z = 1 \\ 3x + y + 2z = -1 \end{cases}$ 20. $\begin{cases} x + y + z = 6 \\ 3x - y + 2z = 7 \\ 2x + 3y - z = 5 \end{cases}$

21. $\begin{cases} 2x - 3y = 4 \\ x + y - 2z = 1 \\ x - y - z = 5 \end{cases}$ 22. $\begin{cases} 3x + 2y + 2z = 6 \\ x - 5y + 6z = 2 \\ 6x - 8y = 12 \end{cases}$

23. $\begin{cases} x - y + z = 3 \\ 2x + 3y - 2z = 5 \\ 3x + y - 4z = 12 \end{cases}$ 24. $\begin{cases} x + y + z = 4 \\ x - y + 2z = 8 \\ 2x + y - z = 3 \end{cases}$

25. $\begin{cases} x + y + z = 2 \\ 2x + 3y - z = 3 \\ 3x + 5y + z = 8 \end{cases}$ 26. $\begin{cases} x + 2y + 5z = 4 \\ 4x + y + 3z = 9 \\ 6x + y + z = 21 \end{cases}$

27. $\begin{cases} x + y + z = 6 \\ x - y + 2z = 12 \\ 2x + y + z = 1 \end{cases}$ 28. $\begin{cases} x + 3y - z = 4 \\ 3x - 2y + 4z = 11 \\ 2x + y + 3z = 13 \end{cases}$

Applying the Concepts

In Problems 29–38, write a linear system of equations. Then use Cramer's rule to solve the system.

29. **Geometry:** Find the measure of each of two complementary angles if the measure of one angle is twice the measure of the other.

30. **Marketing:** A packing house dealer paid a total of $27,700 for two kinds of avocados—Hass and Bacon. The Hass were sold at a profit of 30% on the dealer's cost, but the Bacon started to spoil, resulting in a selling price of 3% loss on the dealer's cost. If a profit of $5965 was made on the total transaction, how much did the dealer pay for each kind of avocado?

31. **Student Loans:** A student's loans totaled $10,500. Part was a bank loan at 8.5% annual simple interest, and the rest was a Federal Education loan at 8% annual simple interest. After one year, the interest accumulated from both loans was $860. What was the original amount of each loan?

32. **Boating:** It took a motorboat 3 hours to make a trip downstream when the current is 6 miles per hour. It took 5 hours for the return trip against the current. What is the speed of the boat in still water?

33. **Nutrition:** A nutritionist wants to create a special diet out of three substances: A, B, and C. The requirements in the diet are 480 units of calcium, 190 units of iron, and 360 units of vitamins. The calcium, iron, and vitamin contents in substances A, B, and C are listed in Table 1. How many ounces of each substance are needed?

TABLE 1

Substances	Calcium Units per ounce	Iron Units per ounce	Vitamin Units per ounce
A	30	10	15
B	20	10	10
C	10	5	20

34. **Production:** Three assembly lines, A, B, and C, can produce 8400 TV dinners per day. Together, lines A and B can produce 4900 TV dinners, while lines B and C together can produce 5600 TV dinners. Find the number of TV dinners each assembly line can produce alone.

35. **Horticulture:** A horticulturist wishes to mix three types of fertilizer which contain 25%, 35%, and 40% nitrogen, respectively, in order to get a mixture of 4000 pounds of $35\tfrac{5}{8}\%$ nitrogen. The final mixture should contain three times as much of the 40% type as the 25% type. How much of each type is in the final mixture?

36. **Fast-Food Chain:** A fast-food chain sells three types of franchises: A, B, and C. Franchise A sells for $20,000, franchise B for $25,000, and franchise C for $30,000. In one year the company sold 18 franchises for a total of $430,000. If the number of franchise A sold was twice the number sold of franchise C, how many of each type did the company sell that year?

37. **Resource Management:** A natural resource department supplies three types of food to support species A, B, and C of fish in a lake environment. Each week 25,000 units of Food I, 20,000 units of Food II and 41,000 units of Food III are supplied to the lake. Suppose that a, b, and c denote the number of fish in the lake of species A, B, and C respectively. Assume that all food is consumed and the units of Food I, Food II, and Food III eaten by the fish are listed in Table 2. How many fish of each species can coexist in the lake?

TABLE 2

Species	Units of Food I	Units of Food II	Units of Food III
A	1	1	3
B	3	4	5
C	2	1	2

38. **Drinking Water:** A town uses three pumps to provide its drinking water. When pumps A, B, and C are all working, they can pump 74,000 gallons of water into a tank in 2 hours. If pump A runs for 4 hours and pump B runs for 2 hours, together they can pump 64,000 gallons of water. If pump B runs for 5 hours and pump C for 4 hours, together they can pump 120,000 gallons of water. How many gallons per hour does each pump provide?

Developing and Extending the Concepts

39. Find the value of $\begin{vmatrix} \sqrt{6} & -3\sqrt{5} \\ 2\sqrt{5} & 7\sqrt{6} \end{vmatrix}$.

40. Find the value of $\begin{vmatrix} \sqrt{7} - \sqrt{3} & 5 + \sqrt{2} \\ 5 - \sqrt{2} & \sqrt{7} + \sqrt{3} \end{vmatrix}$.

In Problems 41–46, solve each equation for x.

41. $\begin{vmatrix} x & x \\ 5 & 3 \end{vmatrix} = 2$

42. $\begin{vmatrix} x & -x \\ 5 & 3 \end{vmatrix} = 2$

43. $\begin{vmatrix} x+1 & x \\ x & x-2 \end{vmatrix} = -6$

44. $\begin{vmatrix} x-2 & 1 \\ x+1 & -2 \end{vmatrix} = 35$

45. $\begin{vmatrix} x & 4 & 5 \\ 0 & 1 & x \\ 5 & 2 & 0 \end{vmatrix} = 7$

46. $\begin{vmatrix} 5x & 0 & 1 \\ 2x & 1 & 2 \\ 3x & 2 & 3 \end{vmatrix} = 0$

47. (a) Find the value of $\begin{vmatrix} 1 & 0 & 2 \\ 4 & 6 & -1 \\ -1 & 0 & -1 \end{vmatrix}$.

(b) Interchange the first and the second rows, and then evaluate the "new" determinant.

(c) Compare the answers of (a) and (b).

48. (a) Find the value of $\begin{vmatrix} 5 & 8 & 3 \\ 0 & 2 & 10 \\ 0 & 0 & 7 \end{vmatrix}$.

(b) Find the product of the three numbers on the diagonal from the upper left to the bottom right.

(c) Compare the answers of (a) and (b).

49. (a) Find the value of the determinant $\begin{vmatrix} 1 & 0 & 2 \\ 1 & 1 & 1 \\ 2 & 1 & -1 \end{vmatrix}$.

(b) Multiply the second row of part (a) by 10 and find the value of the determinant.

(c) Compare the answers of (a) and (b).

50. (a) Find the value of the determinant $\begin{vmatrix} 2 & 1 & -1 \\ 1 & 0 & 0 \\ 3 & 1 & 0 \end{vmatrix}$.

(b) Interchange the first and third rows of part (a) and find the value of the determinant.

(c) Compare the answers of (a) and (b).

G In Problems 51 and 52, use a grapher to find the value of each determinant.

51. $\begin{vmatrix} 1 & 0 & -2 & 0 \\ 0 & 3 & 0 & 2 \\ 1 & 0 & 3 & 1 \\ 0 & 4 & 0 & 5 \end{vmatrix}$

52. $\begin{vmatrix} 1 & 0 & 0 & -3 \\ 1 & 4 & 3 & -2 \\ 2 & 5 & 0 & 1 \\ 2 & -2 & 5 & 1 \end{vmatrix}$

OBJECTIVES

1. Graph Solution Regions
2. Linear Programming Model
3. Solve Applied Problems
4. Use Technology Exploration

8.3 Linear Programming

In this section we discuss an important mathematical technique pertaining to the allocation of limited resources and scheduling that is very common in many practical business problems. The method is known as *linear programming*.

Graphing Solution Regions

In section 4.5 we introduced the solution set of linear inequalities. This solution set is also referred to as the **feasible solution** or the **feasible region** for the system, because coordinates of any point that lie within this region satisfy all the inequalities of the system simultaneously. The corners formed by the intersections of boundary lines of a feasible region are called **vertices** of the region.

EXAMPLE 1   Graphing a Feasible Region

Graph the feasible region and identify each vertex of the system.

$$\begin{cases} x \geq 0 \\ y \geq 0 \\ x + 2y \leq 8 \\ x + y \leq 5 \end{cases}$$

Solution   We begin by graphing the system of inequalities (Figure 1). By solving the pair of equations $x + 2y = 8$ and $x + y = 5$ we obtain the ordered pair $(2, 3)$. This point is the intersection of the two graphs and represents one vertex of the feasible region. Figure 1 shows the other three vertices $(0,0)$, $(0,4)$ and $(5,0)$.   ◈

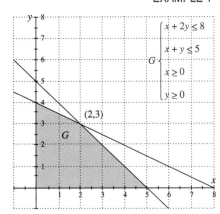

$$G\begin{cases} x + 2y \leq 8 \\ x + y \leq 5 \\ x \geq 0 \\ y \geq 0 \end{cases}$$

Figure 1

Linear Programming Model

Many problems in business, economics, and the sciences require that a function be maximized or minimized subject to limits imposed by available resources. These limits are referred to as **problem constraints**, and the function is called the **objective function**. When the objective function is a linear (first-degree) polynomial in two or more variables and the constraints are expressed as a system of nonstrict linear inequalities, we have a **linear programming problem**.

Here we consider only linear programming problems in which the objective function F depends on two variables x and y, and the graph of the system of linear inequalities that express the constraints is a *bounded* (enclosed) region in the xy plane. Thus, F has the form

$$F = ax + by + c,$$

where a, b, and c are constants and the constraint is a region bounded by a finite number of line segments meeting at a finite number of vertices. Under these circumstances we introduce the following strategy to solve linear programming problems.

STRATEGY FOR SOLVING A LINEAR PROGRAMMING PROBLEM

Step 1. Write a linear objective function.
Step 2. Write problem constraints using linear inequalities or linear equations.
Step 3. Graph the feasible region and find the vertices (corner points).
Step 4. Find the value of the objective function at each vertex. The largest and the smallest of these values are the maximum and minimum of the function respectively.

EXAMPLE 2 Finding the Maximum and Minimum Values of a Function

Find the maximum and minimum values of the objective function

$$F = 3x + 4y + 1,$$

subject to the constraints

$$\begin{cases} x \geq 0, \, y \geq 0 \\ x + 2y \leq 8 \\ x + y \leq 5 \end{cases}$$

Solution We begin by using the method described to sketch the graph of the system representing the constraints (Figure 1). The coordinates of the vertices of the feasible region are listed in Table 1 along with the corresponding values of $F = 3x + 4y + 1$ at each vertex. From Table 1, the minimum value of F, which is 1, occurs at (0,0); and the maximum value of F, which is 19, occurs at (2, 3). ◈

TABLE 1

Vertex	Value of $F = 3x + 4y + 1$
(0,0)	1 minimum
(5,0)	16
(2,3)	19 maximum
(0,4)	17

Solving Applied Problems

In the next example we use linear programming techniques to find the minimum value of an objective function subject to given constraints.

EXAMPLE 3 Find the Minimum of an Objective Function

A large school system wants to design a lunch menu containing two food items X and Y. Each ounce of X supplies 1 unit of protein, 2 units of carbohydrates, and 1 unit of fat. Each ounce of Y supplies 1 unit of protein, 1 unit of carbohydrates, and 1 unit of fat. The two items together must provide at least 7 units of protein, at least 10 units of carbohydrates, and no more than 8 units of fat per serving. If each ounce of X costs 12 cents and each ounce of Y costs 8 cents, how many ounces of each item should each serving contain to meet the dietary requirements at the lowest cost?

Solution   Make a Plan

Let x and y denote the number of ounces per serving of X and Y, respectively. Since each ounce of X costs 12 cents and each ounce of Y costs 8 cents, the cost (in cents) per serving is

$$F = 12x + 8y$$

Since each ounce of food X or of Y supplies 1 unit of protein, each serving will supply $x + y$ units of protein. To meet the protein requirement, we must have

$$x + y \geq 7.$$

Similarly, to meet the carbohydrate requirements, we must have

$$2x + y \geq 10,$$

and to meet the fat requirement,

$$x + y \leq 8,$$

Because the number of ounces per serving cannot be negative, we also have the conditions $x \geq 0$ and $y \geq 0$. Therefore, the problem is to minimize the objective function

$$F = 12x + 8y,$$

subject to the constraints

$$\begin{cases} x + y \geq 7 \\ 2x + y \geq 10 \\ x + y \leq 8 \\ x \geq 0, y \geq 0 \end{cases}$$

Write the Solution

The graph of the system representing these constraints is sketched in Figure 2, from which we find the vertices listed in Table 2. Thus, if each serving contains 3 ounces of item X and 4 ounces of item Y, all dietary requirements will be met at the minimum cost of 68 cents per serving.

TABLE 2

Vertex	Value of $F = 12x + 8y$
(7,0)	84
(8,0)	96
(2,6)	72
(3,4)	68 minimum

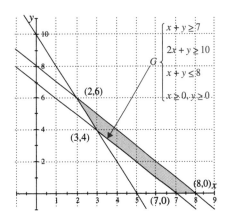

Figure 2

Look Back

The protein supplied by 3 ounces of X and 4 ounces of Y is
$$3(1) + 4(1) = 7$$
which is at least 7 units.
The carbohydrates supplied by X and Y are:
$$3(2) + 4(1) = 6 + 4 = 10$$
which is at least 10 units.
The fat supplied by foods X and Y is:
$$3(1) + 4(1) = 7$$
which is not more than 8.

◈

Using Technology Exploration

We can use a grapher to identify the feasible region, the vertices of this region, and maximize or minimize the objective function as the following example shows:

G EXAMPLE 4   Finding the Maximum of an Objective Function

A company manufactures two types of pens, standard and fine. They make a profit of $630 per gross (144 pens in a gross) on the standard pen and $220 per gross on the fine pen. The following table shows the material and hours needed to produce a gross of pens and the maximum resources available:

	Hours of Production	Gallons of Ink	Pounds of Plastic
Standard	3.7	1.3	4.5
Fine	2.1	0.3	1.7
Resources at most available	80	18	71

Determine how to allocate the resources so that profit is maximized.

Solution   The objective function for maximizing profit is
$$\text{Profit} = 630s + 220f$$
where s is the number of gross of standard pens manufactured and f is the number of gross of fine pens manufactured.
The constrains are defined by the inequalities
$$\begin{cases} 1.3s + 0.3f \le 18 \\ 4.5s + 1.7f \le 71 \\ 3.7s + 2.1f \le 80 \end{cases}$$
We must graph the equations to find the feasible region. Let s be the horizontal axis and f the vertical axis.
$$\begin{cases} f = \dfrac{-1.3s + 18}{0.3} \\ f = \dfrac{-4.5s + 71}{1.7} \\ f = \dfrac{-3.7s + 80}{2.1} \end{cases}$$

Then, we use a grapher to graph these equations (Figure 3).

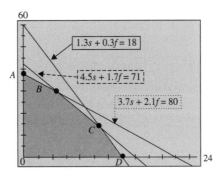

Figure 3

Using the ZOOM and TRACE features, we find the coordinates of A, B, C and D. So that, the vertices are at:

A. $(0, 38.1)$ where $3.7s + 2.1f = 80$ crosses the vertical axis.

B. $(4.1, 30.8)$ where $4.5s + 1.7f = 71$ and $3.7s + 2.1f = 80$ intersect.

C. $(10.8, 13.1)$ where $1.3s + 0.3f = 18$ and $4.5s + 1.7f = 71$ intersect.

D. $(13.8, 0)$ where $1.3s + 0.3f = 18$ crosses the horizontal axis.

The profit function can be evaluated on the grapher with the following results.

Vertex (s, f)	Profit function $P = 630s + 220f$
$(0, 38.1)$	$8382
$(4.1, 30.8)$	$9359
$(10.8, 13.1)$	$9686
$(13.8, 0)$	$8694

Therefore, to maximize profit, the company should produce 10.8 gross of standard pens and 13.1 gross of fine pens for a profit of $9686. ◈

◈ PROBLEM SET 8.3

Mastering the Concepts

In Problems 1–8, graph the constraints and identify each vertex of the feasible region.

1. $\begin{cases} x \geq 0, y \geq 0 \\ x + y \leq 2 \\ -x + y \leq 1 \end{cases}$

2. $\begin{cases} x \geq 0, y \geq 0 \\ x - y \leq 5 \\ x + y \geq 0 \end{cases}$

3. $\begin{cases} x \geq 0, y \geq 0 \\ -x + y \leq 1 \\ 3x + 4y \geq 32 \end{cases}$

4. $\begin{cases} x \geq 0, y \geq 0 \\ 2x - y \leq 7 \\ x + y \geq 2 \end{cases}$

5. $\begin{cases} x \geq 0, y \geq 0 \\ 2x + 3y \geq 12 \\ 2x + 5y \geq 16 \end{cases}$

6. $\begin{cases} x \geq 0, y \geq 0 \\ x + 3y \geq 9 \\ 2x + 3y \geq 12 \end{cases}$

7. $\begin{cases} x \geq 0, y \geq 0 \\ x - 2y \geq -10 \\ 2x + y \leq 10 \end{cases}$

8. $\begin{cases} x \geq 0, y \geq 0 \\ x + y \leq 4 \\ x - y \geq 4 \end{cases}$

In Problems 9–18, find the maximum and minimum values of the objective function F subject to the given constraints.

9. $F = 3x + 6y$

$$\begin{cases} x \geq 0, y \geq 0 \\ 2x + 3y \geq 12 \\ 2x + 5y \geq 16 \end{cases}$$

10. $F = 3x + 2y - 1$

$$\begin{cases} x \geq 0, y \geq 0 \\ x + 2y \leq 7 \\ 5x - 8y \leq -3 \end{cases}$$

11. $F = 5x + 4y$

$$\begin{cases} x \geq 0, y \geq 0 \\ x + 2y \leq 5 \\ x + 2y \geq 3 \end{cases}$$

12. $F = 4x - y + 7$

$$\begin{cases} x \geq 0, y \geq 0 \\ 3x + y \leq 36 \\ 3x + 8y \leq 120 \end{cases}$$

13. $F = 3x - 5y + 2$

$$\begin{cases} x \geq 0, y \geq 0 \\ x + y \leq 10 \\ x - 3y \geq -18 \end{cases}$$

14. $F = -10x + 5y + 3$

$$\begin{cases} x \geq 0, y \geq 0 \\ x + \dfrac{3}{2}y \geq -60 \\ x + y \geq -50 \end{cases}$$

15. $F = 2x + y$

$$\begin{cases} x \geq 0, y \geq 0 \\ 4x + y \leq 36 \\ 4x + 3y \leq 60 \end{cases}$$

16. $F = 2x - y$

$$\begin{cases} x \leq 3, y \leq 3 \\ x + y \geq 3 \end{cases}$$

17. $F = 5x + 2y$

$$\begin{cases} x \geq 0, y \geq 0 \\ x + 3y \leq 15 \\ 2x + y \leq 10 \end{cases}$$

18. $F = x + 2y$

$$\begin{cases} x \geq 0, y \geq 0 \\ 2x + y \leq 14 \\ 2x - 3y \geq 6 \end{cases}$$

Applying the Concepts

G In Problems 19–28, use a grapher if necessary.

19. **Stock Investment:** A total of at most $20,800 is invested in two companies: a high-risk electronics firm and an established automotive manufacturer. Stock in the electronics firm sells for $6 per share and stock in the automotive manufacturer sells for $86 per share. Suppose that an investor wishes to buy at least three times as many shares of the electronics stock as of the automotive stock. Write a system of inequalities in x and y describing these constraints and sketch a graph of the system.

20. **Spray Cans Display:** A hardware store has display space for at most 40 spray cans of rustproofing paint, x cans of red, and y cans of gray. If there are to be at least 10 cans of each type in the display, (a) sketch a graph showing the possible number of red and gray cans in the display, and (b) find all vertices of the graph of the feasible region.

21. **Manufacturing:** An electronics company manufactures two models of household smoke detectors. Model A requires 1 unit of labor and 4 units of parts; model B requires 1 unit of labor and 3 units of parts. If 90 units of labor and 320 units of parts are available, and if the company makes a profit of $5 on each model A detector and $4 on each model B detector, how many of each model should it manufacture to maximize its profit?

22. **Meat Mixtures:** A supplier has 105 pounds of leftover beef which must be sold before it spoils. Of this beef, 15 pounds is prime grade, 40 pounds is grade A, and the rest is utility grade. A local restaurant will buy ground beef consisting of 20% prime grade, 40% grade A, and 40% utility grade for 75 cents per pound. A hamburger stand will buy ground beef consisting of 40% grade A and 60% utility grade for 55 cents per pound. How much ground beef should the supplier sell to the restaurant and how much to the hamburger stand in order to maximize its revenue?

23. **Recycling:** A town operates two recycling centers. Each day that center I is open 300 pounds of glass, 200 pounds of paper, and 100 pounds of aluminum are deposited there. Each day that center II is open, 100 pounds of glass, 600 pounds of paper, and 100 pounds of aluminum are deposited there. The town has contracted to supply at least 1200 pounds of glass, 2400 pounds of paper, and 800 pounds of aluminum per week to a salvage company. The cost of supervision and maintenance at center I is $40.00 each day it is open, and center II costs $50.00 each day it is open. How many days a week should each center remain open to minimize the total weekly cost for supervision and maintenance, yet allow the town to fulfill its contract with the salvage company?

24. **Farming:** A family owns and operates a 312-acre farm on which it grows cotton and peanuts. The tasks necessary for growing and harvesting cotton require at most 27 hours of labor per acre and 45 hours for peanuts. The family is able to devote at most 9500 hours to these activities. If the profit for each acre of cotton grown is $152 and the profit for each acre of peanuts grown is $173, how many acres should be planted in cotton and how many in peanuts to maximize the family's profit?

25. **Corsages:** During prom season a florist sells two types of corsages: orchids and rosebuds. In the past, the total number of corsages sold has been between 30 and 40 inclusive, but never more than 30 orchids nor 20 rosebud corsages. If the florist makes $15.50 profit on each orchid corsage and $12 profit on each rosebud corsage, how many corsages of each type would yield a maximum profit?

26. **Hotel Booking:** For a resort hotel, a travel agent makes a profit of $8 for booking a double room over a weekend and $15 for booking a suite. The number of double rooms available per weekend is between 30 and 80 inclusive, and the number of suites is between 10 and 30 inclusive. The total number of rooms cannot exceed 80 per weekend. How many of each type should be booked to maximize profit?

27. **Budgeting:** A grower has 70 acres of planting fields in which to grow flowers and eucalyptus. It costs the grower $60 per acre to grow flowers and $30 per acre to grow eucalyptus. The budget for planting is $1800. It takes 3 days to plant an acre of flowers and 4 days to plant an acre of eucalyptus. The grower has a maximum of 120 days to plant the flowers. The flowers will bring a profit of $180 per acre, and the eucalyptus will bring a profit of $100 per acre. How many acres of each crop should be planted to maximize profit?

28. **Floral Arrangements:** A florist has two types of spring flowers for centerpieces: jonquils and tulips. The florist uses more than 13 flowers, but no more than 9 jonquils per arrangement. The florist finds the arrangements sell better if the ratio of tulips to jonquils is no more than 5 to 2. If tulips cost $8.00 per dozen and jonquils cost $6.50 per dozen, how many of each type of flower should the florist use per centerpiece to minimize the cost?

Developing and Extending the Concepts

In Problems 29 and 30, write a system of inequalities that describe the graph of the constraints and identify each vertex of the feasible region.

29.

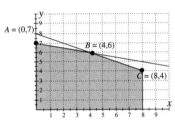

30.

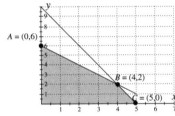

In Problems 31–34, find the maximum and minimum values of the objective function F subject to the given constraints.

31. $F = \dfrac{1}{2}x + \dfrac{3}{4}y$

$$\begin{cases} x \geq 0, y \geq 0 \\ x + \dfrac{1}{2}y \leq 40 \\ x + y \geq 10 \\ \dfrac{1}{3}x + y \leq 30 \end{cases}$$

32. $F = 0.15x + 0.1y$

$$\begin{cases} x \geq 0, y \geq 0 \\ 4x + 5y \leq 2000 \\ x + y \geq 100 \\ 12x + 5y \leq 3000 \end{cases}$$

33. $F = 7x + 14y$

$$\begin{cases} x \geq 0, \ y \geq 0 \\ x + y \leq 2 \\ 4x - y \leq 12 \\ x + 2y \leq 16 \end{cases}$$

34. $F = 12x + 8y$

$$\begin{cases} x \geq 0, \ y \geq 0 \\ x + y \geq 7 \\ x + y \leq 8 \\ 2x + y \geq 10 \end{cases}$$

35. For what positive values of the constants a and b will the objective function $F = ax + by$, subject to the constraints $x \geq 0$, $y \geq 0$, $x - y \leq 2$, and $2x + 3y \leq 9$, take on the same maximum value at the vertex $(0,3)$ and the vertex $(3, 1)$?

OBJECTIVES

1. Graph an Ellipse with Center at (0,0)
2. Graph a Hyperbola with Center at (0,0)
3. Solve Applied Problems
4. Use Technology Exploration

8.4 Conic Sections: The Ellipse and Hyperbola

The term *Conic Section* or *Conics* refer to various figures or sections formed by intersecting a right circular cone with a plane. If a plane is perpendicular to the axis of the cone, a *circle* is formed (Figure 1a). When the plane is not perpendicular to the axis of the cone, its intersection forms a *parabola* (Figure 1b), an *ellipse* (Figure 1c), or a *hyperbola* (Figure 1d).

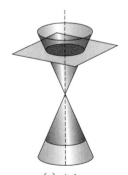

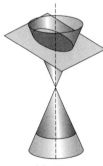

Figure 1   (a) circle   (b) parabola   (c) ellipse   (d) hyperbola

In sections 6.7 and 6.8, we discussed the circle and the parabola. In this section, we introduce the *ellipse* and the *hyperbola*.

Graphing an Ellipse with Center at (0, 0)

Ellipses are useful in providing mathematical models of a variety of physical phenomena ranging from art to astronomy. The geometric definition of an ellipse follows:

ELLIPSE

An **ellipse** is the set of all points P in a plane such that the sum of the distances from P to two fixed points is a constant. The fixed points are called the **foci** (plural of **focus**) of the ellipse.

We can use this definition to draw an ellipse by taking a string of length k and fastening its ends at two fixed points F_1 and F_2. Figure 2a shows two fixed points F_1 and F_2 and a string of length k stretched tightly about them to the point P. Hence, as P moves about, $\overline{|PF_1|} + \overline{|PF_2|}$ always has a constant value k. That is, as Figure 2b shows,

$$d_1 + m_1 = d_2 + m_2 = k \text{ , where } k \text{ is a constant.}$$

Thus, if a pencil point P is inserted into the string and moved so as to keep the string tight, it traces out an ellipse.

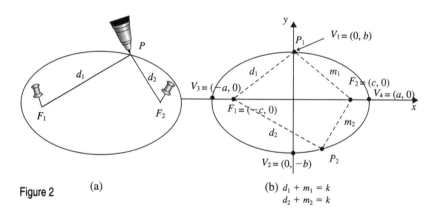

Figure 2   (a)

(b) $d_1 + m_1 = k$
$d_2 + m_2 = k$

The ellipse in Figure 2b has foci $F_1 = (-c,0)$ and $F_2 = (c,0)$ and y intercepts at the points $V_1 = (0,b)$ and $V_2 = (0, -b)$, where $c > 0$ and $b > 0$. The line segment $\overline{V_3V_4}$ where $V_3 = (-a, 0)$ and $V_4 = (a, 0)$ is called the **major axis**, and the line segment $\overline{V_1V_2}$ is called the **minor axis**. The foci, F_1 and F_2, lie on the major axis. The **center** of an ellipse is the midpoint of the major (or minor) axis. The **vertices** of an ellipse are the endpoints of the major axis.

As with the parabola, we can use the definition of the ellipse to derive its standard equation. We choose the x axis as the line containing the foci $F_1 = (-c,0)$ and $F_2 = (c,0)$, where $c > 0$. The origin is the midpoint of the line segment $\overline{F_1F_2}$. Also, we assume that the constant sum of the distances between

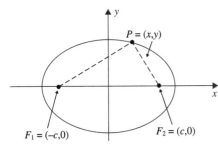

Figure 3

EQUATIONS OF AN ELLIPSE
WITH CENTER AT (0, 0)

any point P on the ellipse and the foci is $2a$. (This constant is written in the form $2a$ so that the equation of the ellipse will have a simple form.) In triangle F_1PF_2 (Figure 3), we know from geometry that the sum of the lengths of the two sides of the triangle, $(|\overline{PF_1}| + |\overline{PF_2}| = 2a)$, is greater than the length of the third side, $(|\overline{F_1F_2}| = 2c)$.

Thus, $2a > 2c$ or $a > c$. If we let $P = (x,y)$, we get

$$d(P,F_1) + d(P,F_2) = 2a$$

Using the distance formula and simplifying we get the following equations of the ellipse.

1. The **equation of an ellipse** centered at the origin with foci at $F_1 = (-c,0)$ and $F_2 = (c,0)$ and y intercepts $-b$ and b is given by

$$\frac{x^2}{a^2} + \frac{y^2}{b^2} = 1,$$

where $a^2 = b^2 + c^2$ (Figure 4a).

2. The **equation of an ellipse** centered at the origin with foci at $F_1 = (0,-c)$ and $F_2 = (0,c)$ and y intercepts $-a$ and a is given by

$$\frac{x^2}{b^2} + \frac{y^2}{a^2} = 1,$$

where $a^2 = b^2 + c^2$ (Figure 4b).

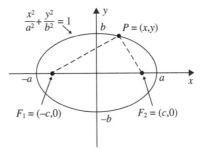

(a)   Horizontal major axis

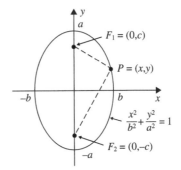

Figure 4   (b)   Vertical major axis

EXAMPLE 1   Graphing an Ellipse

Identify the intercepts, the vertices, the foci, and sketch the graph of each ellipse.

(a) $4x^2 + 9y^2 = 36$ (b) $25x^2 + 9y^2 = 225$

Solution   (a) To find the x intercepts, let $y = 0$ and solve for x. So,

$$x^2 = 9 \text{ or } x = \pm 3$$

To find the y intercepts, let $x = 0$ and solve for y, so

$$y^2 = 4 \text{ or } y = \pm 2$$

We write the equation in standard form by dividing each side by 36.

$$\frac{x^2}{9} + \frac{y^2}{4} = 1$$

Therefore, the vertices are

$$(-3,0) \text{ and } (3,0)$$

Here $a = 3$ and $b = 2$, so that $c^2 = a^2 - b^2 = 9 - 4 = 5$ so, $c = \pm\sqrt{5}$. Hence, the foci are $(-\sqrt{5},0)$ and $(\sqrt{5},0)$. Figure 5a shows the graph of the ellipse.

(b) To find the x intercepts, let $y = 0$

$$x^2 = 9 \text{ or } x = \pm 3$$

To find the y intercepts, let $x = 0$

$$y^2 = 25 \text{ or } y = \pm 5$$

We write the equation in standard form by dividing each side by 225.

$$\frac{x^2}{9} + \frac{y^2}{25} = 1$$

Therefore, the vertices are, $(0,-5)$, and $(0,5)$. Here, $a = 5$ and $b = 3$, so $c^2 = a^2 - b^2 = 25 - 9 = 16$ or $c = 4$. Hence, the foci are $(0,-4)$ and $(0,4)$. Figure 5b shows the graph of the ellipse.

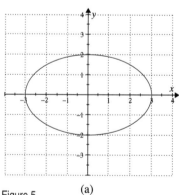   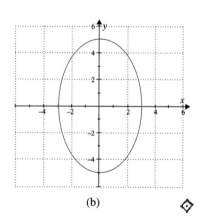

Figure 5   (a)   (b)

EXAMPLE 2   Finding an Equation of an Ellipse

Find an equation of the ellipse with foci

$$F_1 = (0,-4) \text{ and } F_2 = (0,4),$$

and the vertices

$$V_1 = (0,-6) \text{ and } V_2 = (0,6).$$

Sketch the graph.

Solution   The equation of the ellipse we are to determine is of the form

$$\frac{x^2}{b^2} + \frac{y^2}{a^2} = 1$$

So we must find a^2 and b^2. Because the foci are on the y axis, the vertical axis is the major axis of the ellipse. Since the vertices are $(0,-6)$ and $(0,6)$, the value of a is 6. Also with the foci at $(0,-4)$ and $(0,4)$, this means that $c = 4$. We compute b^2 as follows:

$$b^2 = a^2 - c^2 = 36 - 16 = 20$$

Hence, an equation of the ellipse is

$$\frac{x^2}{20} + \frac{y^2}{36} = 1$$

The graph is in Figure 6.

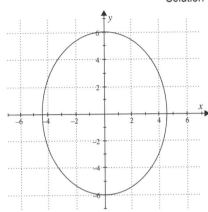

Figure 6

Graphing a Hyperbola with Center of (0, 0)

As with the ellipse, we give a geometric definition of a hyperbola.

HYPERBOLA

> A **hyperbola** is the set of all points P in the plane such that the absolute value of the difference of the distances from P to two fixed points is a constant positive number. The two fixed points are called the **focal points**, or **foci** of the hyperbola. The point halfway between the foci is the **center** of the hyperbola.

Figure 7 shows a geometric description of a hyperbola with foci F_1 and F_2 and with center C. The points P_1 and P_2 are on the hyperbola, and

$$\left| d_1 - m_1 \right| = \left| d_2 - m_2 \right| = k$$

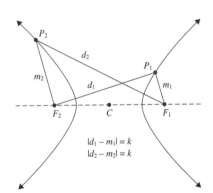

Figure 7

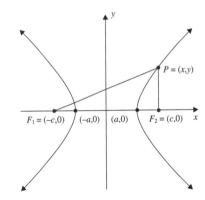

Figure 8

Suppose that a hyperbola is positioned in a Cartesian coordinate system in such a way that the center is at the origin and two foci $F_1 = (-c,0)$ and $F_2 = (c,0)$ lie on the x axis, with $c > 0$. To derive the standard equation of the hyperbola analytically, we let $P = (x,y)$ be a point on the hyperbola, and assume that the constant difference k equals $2a$ (Figure 8). Since $P = (x, y)$, we have

$$\left| d(P,F_1) - d(P,F_2) \right| = 2a$$

Equations in standard form for hyperbolas may be derived from the geometric definition. As we shall see the hyperbola may intercept the x or the y axis depending on its standard equation.

STANDARD EQUATIONS OF A HYPERBOLA

1. The **standard equation for a hyperbola** centered at the origin with vertices at $V_1 = (-a,0)$ and $V_2 = (a,0)$ and foci at $F_1 = (-c,0)$ and $F_2 = (c,0)$ is given by

$$\frac{x^2}{a^2} + \frac{y^2}{b^2} = 1,$$

where $a^2 + b^2 = c^2$ (Figure 9a).

2. The **standard equation for a hyperbola** centered at the origin with vertices at $V_1 = (0,-b)$ and $V_2 = (0,b)$ and foci at $F_1 = (0,-c)$ and $F_2 = (0,c)$ is given by

$$\frac{y^2}{b^2} + \frac{x^2}{a^2} = 1,$$

where $a^2 + b^2 = c^2$ (Figure 9b).

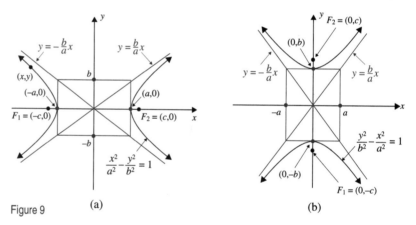

Figure 9   (a)   (b)

Some distinguishing features of a hyperbola with foci on the x axis are as follows:

1. The line segment that has the vertices as its end points is called the **transverse axis**.
2. The line segment that contains the center of the hyperbola and is perpendicular to the transverse axis is called the **conjugate axis**. Although the points $(0,-b)$ and $(0,b)$ are not on the hyperbola whose equation is $(x^2/a^2) - (y^2/b^2) = 1$, they are useful for describing its graph.
3. The line determined by the transverse axis is called the **axis of symmetry** of the hyperbola. For the hyperbola $(x^2/a^2) - (y^2/b^2) = 1$, the foci lie on the axis of symmetry, c units from the center, at points $(-c,0)$ and $(c,0)$.
4. The lines

$$y = -\frac{b}{a}x \text{ and } y = \frac{b}{a}x$$

are called the **asymptotes** of a hyperbola of either form

$$\frac{x^2}{a^2} - \frac{y^2}{b^2} = 1 \qquad \text{or} \qquad \frac{y^2}{b^2} - \frac{x^2}{a^2} = 1$$

EXAMPLE 3   Graphing a Hyperbola

Identify the intercepts, the vertices, and the foci. Find the equations of the asymptotes and sketch the graph.

(a) $4x^2 - 9y^2 = 36$ (b) $4y^2 - 25x^2 = 100$

Solution   (a) The x intercepts are found by setting $y = 0$ and solving for x.

$$4x^2 = 36 \text{ or } x^2 = 9, x = \pm 3$$

The x intercepts are -3 and 3 and the vertices of the hyperbola are $(-3,0)$ and $(3,0)$. The transverse axis of the hyperbola is horizontal, and there are no y intercepts. We write the equation in standard form.

$4x^2 - 9y^2 = 36$ Original equation

$\dfrac{x^2}{9} - \dfrac{y^2}{4} = 1$ Divide each side by 36

Here, $a = 3$, $b = 2$, so $c^2 = a^2 + b^2 = 9 + 4 = 13$ and $c = \sqrt{13}$. So the foci are $(-\sqrt{13},0)$ and $(\sqrt{13},0)$. Figure 10a shows the graph of the hyperbola. The equations of the asymptotes are

$$y = \frac{b}{a}x = \frac{2}{3}x \qquad \text{and} \qquad y = -\frac{b}{a}x = -\frac{2}{3}x$$

(b) The y intercepts are found by setting $x = 0$ and solving for y.

$$4y^2 = 100, y^2 = 25, y = \pm 5$$

The y intercepts are -5 and 5 and the vertices of the hyperbola are $(0,-5)$ and $(0,5)$. The transverse axis of the hyperbola is vertical, and there are no x intercepts. We write the equation in standard form.

$4y^2 - 25x^2 = 100$ Original equation

$\dfrac{y^2}{25} - \dfrac{x^2}{4} = 1$ Divide each side by 100

Here, $a = 2$, $b = 5$, so $c^2 = a^2 + b^2 = 4 + 25 = 29$ and $c = \sqrt{29}$. So the foci are $(0,-\sqrt{29})$ and $(0,\sqrt{29})$. Figure 10b shows the graph of the hyperbola. The equations of the asymptotes are

$$y = \frac{b}{a}x = \frac{5}{2}x \quad \text{and} \quad y = -\frac{b}{a}x = -\frac{5}{2}x$$

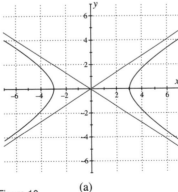

(a)

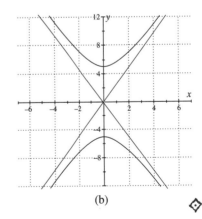

(b)

Figure 10

EXAMPLE 4   Finding an Equation of a Hyperbola

Write an equation of a hyperbola centered at the origin, where vertices are $(-4,0)$ and $(4,0)$, with no y intercept, and whose foci are $(-5,0)$ and $(5,0)$. Sketch the graph and find the equations of the asymptotes.

Solution   Here $a = 4$ and $c = 5$, so that

$c^2 = a^2 + b^2$ or $b^2 = c^2 - a^2$ and $b = \sqrt{25 - 16} = 3$. Therefore, an equation of the hyperbola is of the form

$$\frac{x^2}{a^2} - \frac{y^2}{b^2} = 1$$

so that

$$\frac{x^2}{16} - \frac{y^2}{9} = 1 \text{ (Figure 11)}$$

The equations of the asymptotes are given by

$$y = \frac{b}{a}x = \frac{3}{4}x \text{ and } y = -\frac{b}{a}x = -\frac{3}{4}x$$

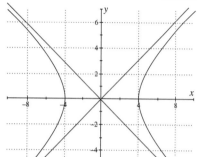

Figure 11

Solving Applied Problems

Ellipses and hyperbolas are of practical importance in many applications of mathematics as the next example shows.

EXAMPLE 5   Solving an Ellipse Problem

The park located between the White House and the Washington Monument in Washington, D.C. is elliptical in shape. The major axis is 458 meters long and its minor axis is 390 meters. What is the distance between the foci of this ellipse?

Solution   Let the center of this ellipse be at the origin. We know that $2a = 458$ and $2b = 390$ so that

$$a = 229 \text{ and } b = 195$$

An equation of this ellipse in standard form is

$$\frac{x^2}{(229)^2} + \frac{y^2}{(195)^2} = 1$$

$c^2 = a^2 - b^2 = (229)^2 - (195)^2 = 14{,}416$, so that $c \approx 120$. Therefore, the foci of this ellipse are approximately at $(-120,0)$ and $(120,0)$. Hence, the distance between the two foci is $120 - (-120) = 240$ meters.

Using Technology Exploration

We can use a grapher to graph equations of hyperbolas and ellipses as the next example shows.

G EXAMPLE 6   Using a Grapher to Graph an Ellipse

Use a grapher to graph the equation of the ellipse

$$\frac{x^2}{16} + \frac{y^2}{9} = 1$$

Solution   We can graph this equation by first solving for y in terms of x and then graphing each of the two halves of the ellipse on the same viewing window.

$$\frac{y^2}{9} = 1 - \frac{x^2}{16}$$

$$y^2 = 9\left(1 - \frac{x^2}{16}\right)$$

$$y = \pm 3\sqrt{1 - \frac{x^2}{16}}$$

So that

$$y = +3\sqrt{1 - \frac{x^2}{16}}$$

(top half) and

$$y = -3\sqrt{1 - \frac{x^2}{16}}$$

(bottom half). By graphing the top and bottom halves of the ellipse, we produce the graph of

$$\frac{x^2}{16} + \frac{y^2}{9} = 1,$$

(Figure 12).

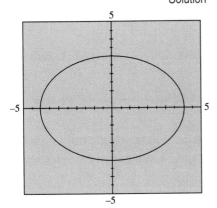

Figure 12

PROBLEM SET 8.4

Mastering the Concepts

In Problems 1–4, match each equation with its graph.

1. $4x^2 + 9y^2 = 36$
2. $x^2 - y^2 = 4$
3. $9x^2 + 4y^2 = 36$
4. $y^2 - x^2 = 4$

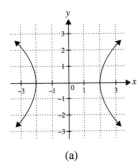

(a)

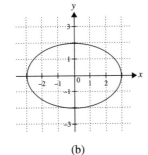

(b)

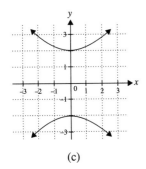

(c)

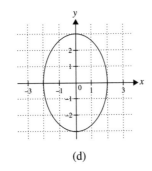

(d)

In Problems 5–12, identify the intercepts, the vertices and the foci. Sketch the graph of each ellipse.

5. $9x^2 + 16y^2 = 144$
6. $16y^2 + 25x^2 = 400$
7. $4y^2 + 16x^2 = 64$
8. $4x^2 + 9y^2 = 36$
9. $4x^2 + 25y^2 = 100$
10. $25x^2 + 10y^2 = 250$
11. $15x^2 + 3y^2 = 45$
12. $25x^2 + 15y^2 = 150$

In Problems 13–22, identify the intercepts, the vertices and the foci of each hyperbola. Also sketch the graph and find the equations of the asymptotes.

13. $25x^2 - 9y^2 = 225$
14. $25y^2 - 9x^2 = 225$
15. $9x^2 - 4y^2 = 36$
16. $9y^2 - 4x^2 = 36$
17. $x^2 - 9y^2 = 9$
18. $9x^2 - y^2 = 1$
19. $16y^2 - 4x^2 = 48$
20. $36x^2 - 9y^2 = 1$
21. $4y^2 - 5x^2 = 20$
22. $3y^2 - x^2 = 1$

In Problems 23–26, find the standard equation of the conic that satisfies the given conditions. Sketch the graph.

23. An ellipse with center at the origin, foci at $F_1 = (-\sqrt{2},0)$, $F_2 = (\sqrt{2},0)$ and vertices at $V_1 = (-2,0)$ and $V_2 = (2,0)$.
24. An ellipse with center at the origin, foci at $F_1 = (0,-4)$, $F_2 = (0,4)$ and vertices at $V_1 = (0,-5)$ and $V_2 = (0,5)$.
25. A hyperbola with center at the origin, foci at $F_1 = (0,-\sqrt{2})$, $F_2 = (0,\sqrt{2})$ and asymptotes at $y = -x$ and $y = x$.
26. A hyperbola with center at the origin, vertices at $V_1 = (-3,0)$, $V_2 = (3,0)$ and asymptotes at

$$y = -\frac{4}{3}x \text{ and } y = \frac{4}{3}x.$$

Applying the Concepts

27. **Dimension of an Arch:** An arch in the shape of the upper half of an ellipse is to support a bridge over a roadway 100 feet wide. The center of the arch is 30 feet above the center of the roadway, and the paved portion of the

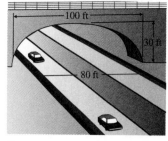

Figure 13

roadway is within a strip that is 80 feet wide (Figure 13).
   (a) Sketch the ellipse in an xy coordinate system, with horizontal major axis along the roadway and center at the origin.
   (b) Find the standard equation of the ellipse in part (a).
   (c) Confirm that the foci of the ellipse are located at the edges of the paved portion of the roadway.

28. **Communications Satellite:** A communications satellite travels in an elliptical orbit around the Earth. The radius of the Earth is about 6400 kilometers, and its center is located at one focus of the orbit. Suppose that the satellite moves in such a way that when it is closest to the center of the Earth, it is 2000 kilometers from the surface, and when it is farthest from the center, it is 10,000 kilometers from the surface.
   (a) Sketch the orbit in an xy coordinate system with the major axis lying on the x axis and with center at the origin.
   (b) Determine the standard equation of the ellipse in part (a).
   (c) What is the height of the satellite above the surface of the Earth at a point corresponding to the focus (at the center of the earth) of the elliptical orbit?

29. **Locating Explosions:** Two microphones are located at two detection stations, A and B, 1150 meters apart on an east-west line (Figure 14). An explosion occurs at an unknown location P. The sound of the explosion is detected by the microphone at station A exactly 2 seconds before it is detected by the microphone at station B. Assume that sound travels at a uniform speed of 130 meters per second.
   (a) Sketch the curve that describes the possible location of P in an xy coordinate system so that the origin is at the middle of the horizontal axis between A and B.
   (b) Find the standard equation of the curve in part (a).
   (c) Can the explosion at P be pinpointed at an exact location? Explain.

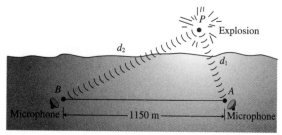

Figure 14

30. **Airplane Flight:** An airplane starts at a point directly north of the origin and flies northeasterly in a hyperbolic path. Transmitters located at the origin and 150 miles directly north of the origin send out synchronized radio signals. Instruments in the airplane measure the difference between the arrival times of the signals. By knowing the speed of the radio signals, it is determined that the distance from the plane to the transmitter at the origin is 50 miles farther than the distance to the other transmitter.
   (a) Sketch the path of the airplane in a coordinate system that satisfies the given conditions.
   (b) Find the standard equation of the curve in part (a).
   (c) Find the location of the airplane when it is 40 miles directly east of the line between the two transmitters.

Developing and Extending the Concepts

31. (a) Sketch the graph of the equation of the ellipse
$$36x^2 + 100y^2 = 3600.$$
   (b) Find the foci.
   (c) Find the length of the major axis.
   (d) Show that the points $(0, -6)$ and $(10, 0)$ are on the graph.
   (e) Verify that the sum of the distances from any of the above two points to each of the foci is equal to the length of the major axis.

32. (a) Sketch the graph of the equation of the hyperbola
$$4x^2 - 16y^2 = 144.$$
   (b) Find the foci.
   (c) Find the length of the transverse axis.
   (d) Show that the points $(10, -4)$ and $(6\sqrt{2}, 3)$ lie on the graph.
   (e) Verify that the difference of the distances from each of the above two points to each of the foci is equal to the length of the transverse axis.

In Problems 33–40, equations for ellipses and hyperbolas are given. Find an equation for the new ellipses and hyperbolas by shifting each center as specified.

33. $\dfrac{x^2}{9} + \dfrac{y^2}{4} = 1$ 2 units to the left and 1 unit down

34. $\dfrac{x^2}{4} + \dfrac{y^2}{9} = 1$ 3 units to the right and 2 units upward

35. $\dfrac{x^2}{9} + \dfrac{y^2}{16} = 1$ 3 units to the right and 4 units upward

36. $\dfrac{x^2}{9} + \dfrac{y^2}{1} = 1$ 2 units to the left and 3 units downward

37. $\dfrac{x^2}{4} - \dfrac{y^2}{1} = 1$ 2 units to the left and 1 unit upward

38. $\dfrac{x^2}{25} - \dfrac{y^2}{16} = 1$ 4 units to the right and 1 unit upward

39. $\dfrac{y^2}{16} - \dfrac{x^2}{4} = 1$ 3 units to the right and 3 units downward

40. $\dfrac{y^2}{4} - \dfrac{x^2}{1} = 1$ 3 units to the left and 2 units downward

G In Problems 41 and 42, use a grapher to graph the equations.
41. $9x^2 + 4y^2 = 36$ 42. $4x^2 - 9y^2 = 36$

OBJECTIVES

1. Solve Systems of Nonlinear Equations
2. Solve Systems of Nonlinear Inequalities
3. Solve Applied Problems
4. Use Technology Exploration

8.5 Systems of Nonlinear Equations and Inequalities

So far, the systems of equations and inequalities we have solved have been linear. In this section, we consider the solution of systems in which at least one of these equations or inequalities is nonlinear. Some examples are shown below:

$$\begin{cases} x^2 + y = 6 \\ 2x - 5y = -6 \end{cases}$$

and

$$\begin{cases} x + y \le 16 \\ x^2 + y \ge 4 \end{cases}$$

We use substitution and elimination to solve these systems algebraically.

Solving Systems of Nonlinear Equations

At times, we are able to solve nonlinear systems of equations by using various algebraic techniques such as substitution and elimination, as we did when solving linear systems.

EXAMPLE 1   Solving a Nonlinear System by Substitution

Solve the system by using substitution.

$$\begin{cases} x^2 - 2y = 0 \\ x + 2y = 6 \end{cases}$$

Solution   We solve the first equation for y to get

$$y = \frac{x^2}{2},$$

then we substitute the expression $x^2/2$ for y in the second equation to get

$$x + 2\left(\frac{x^2}{2}\right) = 6$$

$$x^2 + x - 6 = 0 \qquad \text{Write the equation in standard form}$$

$$(x + 3)(x - 2) = 0 \qquad \text{Factor}$$

$$x + 3 = 0 \qquad x - 2 = 0$$

$$x = -3 \qquad x = 2$$

Next, we find y by substituting these numbers for x in the first equation.

$$x = -3, \qquad y = \frac{x^2}{2}$$

$$= \frac{(-3)^2}{2} = \frac{9}{2}$$

$$x = 2, \qquad y = \frac{x^2}{2}$$

$$= \frac{(2)^2}{2} = 2$$

Thus, the two solutions are

$$\left(-3, \frac{9}{2}\right) \text{ and } (2,2).$$

The graph in Figure 1 shows the two points of intersection whose coordinates are

$$\left(-3, \frac{9}{2}\right), \text{ and } (2, 2). \qquad \diamond$$

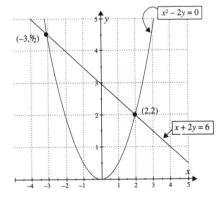

Figure 1

EXAMPLE 2   Solving a Nonlinear System by Elimination

Solve the system by elimination.

$$\begin{cases} x^2 + y^2 = 25 \\ x^2 + y = 13 \end{cases}$$

Solution   Using the elimination method and subtracting the second equation from the first, we get

$$y^2 - y = 12$$

Solving this equation for y by factoring, we have

$y^2 - y - 12 = 0$	Standard form
$(y + 3)(y - 4) = 0$	Factor
$y + 3 = 0 \qquad y - 4 = 0$	
$y = -3 \qquad \quad y = 4$	

Substituting for y in the second equation, we have

$$y = -3, \quad x^2 + y = 13$$
$$x^2 - 3 = 13$$
$$x^2 = 16$$
$$x = \pm 4$$

$$y = 4, \quad x^2 + y = 13$$
$$x^2 + 4 = 13$$
$$x^2 = 9$$
$$x = \pm 3$$

Therefore, the solutions are

$$(-4, -3),\ (4, -3),\ (-3, 4),\ \text{and}\ (3, 4).$$

The graph in Figure 2 shows the four points of intersection whose coordinates are

$$(-4, -3),$$
$$(4, -3),\ (-3, 4),\ \text{and}\ (3, 4). \qquad \diamondsuit$$

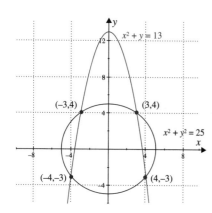

Figure 2

When the elimination method can be used, as in example 2, it is usually the most efficient way to solve the system. Otherwise, try the substitution method.

Solving System of Nonlinear Inequalities

In section 4.5, we solved linear systems of inequalities with two variables by first graphing each of the inequalities on the same coordinate system. The points common to all graphs describe the solution set of the system. The same approach is used to solve nonlinear systems of inequalities as the next example shows.

EXAMPLE 3   Graphing a System of Nonlinear Inequalities

Graph the solution of the system of inequalities.

$$\begin{cases} y > \dfrac{x^2}{4} \\ x + y \leq 4 \end{cases}$$

Solution   We begin by graphing each inequality separately and then determining where the regions overlap. Figure 3a shows the graph of $y > x^2/4$, as the shaded region above the curve $y = x^2/4$. Figure 3b shows the graph of the inequality $x + y \leq 4$ as the shaded region on and below the line $x + y = 4$. Thus, the graph of the solution set of the system of inequalities consists of points common to both regions; that is, the shaded region in Figure 3c.

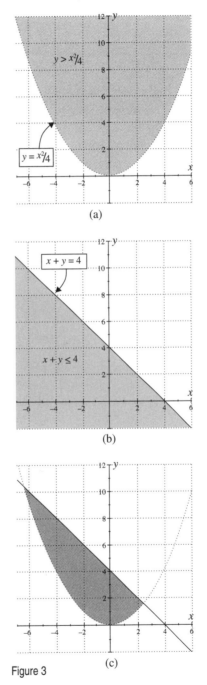

Figure 3

Solving Applied Problems

At times, we use nonlinear systems to solve applied problems as the next example shows.

EXAMPLE 4   Finding the Dimensions of a Television Screen

A 25-inch diagonal of a television is advertised as having a viewing area of 300 square inches. Find the width and height of the screen.

Solution   **Make A Plan**

Figure 4 shows the television in question. Since the viewing screen is almost rectangular, its area A is given by the formula

$$A = \text{length} \cdot \text{width}$$

Let $x =$ the width (in inches) of the screen, and $y =$ the height (in inches) of the screen. Using the Pythagorean theorem, we write

$$x^2 + y^2 = 25^2$$

The system that describes this situation is given by

$$\begin{cases} x^2 + y^2 = 25^2 \\ \quad\ xy = 300 \end{cases}$$

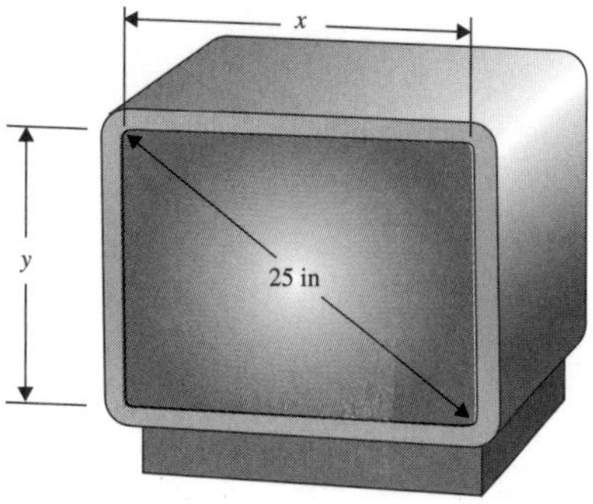

Figure 4

Write the Solution

Write $xy = 300$ as $y = \frac{300}{x}$ and replace y in the first equation by $\frac{300}{x}$.

$$x^2 + \left(\frac{300}{x}\right)^2 = 625 \qquad\qquad \text{Original equation}$$

$$x^2 + \frac{90{,}000}{x^2} = 625$$

$$x^4 + 90{,}000 = 625x^2 \qquad\qquad \text{Multiply each side by } x^2$$

$$x^4 - 625x^2 + 90{,}000 = 0 \qquad\qquad \text{Write in standard form}$$

$$(x^2 - 400)(x^2 - 225) = 0 \qquad\qquad \text{Factor}$$

$$x^2 - 225 = 0 \qquad x^2 - 400 = 0$$

$$x^2 = 225 \qquad\quad\ x^2 = 400$$

$$x = \pm 15 \qquad\qquad x = \pm 20$$

x cannot be negative in this situation, so x is either 15 or 20. If $x = 15$, then $y = 20$, and if $x = 20$, then $y = 15$. Because the width of a television screen is usually larger than its height, we conclude that the screen is 20 inches wide and 15 inches high.

Look Back

Although we have four solutions for each x and y, we use only the positive values. Here

$$x^2 + y^2 = 20^2 + 15^2 = 400 + 225 = 625, \text{ and}$$
$$xy = (20)(15) = 300 \qquad\qquad\qquad\qquad\qquad\qquad\ \diamondsuit$$

Using Technology Exploration

As with linear systems of equations, we can solve nonlinear systems of equations by using a grapher as the next example shows.

G EXAMPLE 5   Using a Grapher to Solve a Nonlinear System

Use a grapher to solve the system

$$\begin{cases} x^2 + y^2 = 4 \\ x^2 + y\ \ = 2 \end{cases}$$

by graphing both equations on the same viewing window.

Solution   The graph of $x^2 + y^2 = 4$ is a circle whose radius is 2 and whose center is at the origin. The graph of $y = -x^2 + 2$ is a parabola opening downward with vertex at $(0, 2)$. Both graphs are shown in Figure 5. Using the TRACE feature, we see that the points of intersection are approximately $(-1.73, -1)$, $(1.73, -1)$ and $(0, 2)$. The solutions can be confirmed by using substitution.

From the second equation $x^2 = 2 - y$

$$x^2 + y^2 = 4 \qquad \text{First equation}$$
$$2 - y + y^2 = 4 \qquad \text{Substitute } 2 - y \text{ for } x^2$$
$$y^2 - y - 2 = 0 \qquad \text{Write in standard equation}$$
$$(y - 2)(y + 1) = 0 \qquad \text{Factor}$$
$$y - 2 = 0, \quad y + 1 = 0$$
$$y = 2, \qquad y = -1$$

$$x^2 = 2 - y \qquad\qquad x^2 = 2 - y$$
$$x^2 = 2 - 2 = 0 \qquad\quad x^2 = 2 - (-1) = 3$$
$$x = 0 \qquad\qquad\qquad x = \pm\sqrt{3}$$

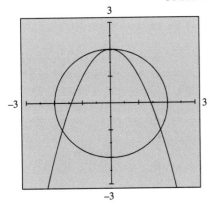

Figure 5

Therefore, the exact solutions are $(-\sqrt{3}, -1)$, $(\sqrt{3}, -1)$, and $(0, 2)$. Approximately, these solutions rounded to two decimal points are given by $(-1.73, -1)$, $(1.73, 1)$, and $(0, 2)$.   ◇

◆ PROBLEM SET 8.5

Mastering the Concepts

In Problems 1–6, solve each system by substitution.

1. $\begin{cases} x - y = 1 \\ x^2 + y = 5 \end{cases}$

2. $\begin{cases} x - 2y = 3 \\ x^2 + y^2 = 24 \end{cases}$

3. $\begin{cases} 3x - y = 2 \\ x^2 + y^2 = 20 \end{cases}$

4. $\begin{cases} x + y = 6 \\ x^2 + y^2 = 20 \end{cases}$

5. $\begin{cases} 3x + 2y = 1 \\ 3x^2 - y^2 = 4 \end{cases}$

6. $\begin{cases} x - y^2 = 0 \\ x^2 + 2y^2 = 24 \end{cases}$

In Problems 7–12, solve each system by elimination.

7. $\begin{cases} x^2 - 2y^2 = 1 \\ x^2 + 4y^2 = 25 \end{cases}$

8. $\begin{cases} x^2 - 4y^2 = -15 \\ -x^2 + 3y^2 = 11 \end{cases}$

9. $\begin{cases} x^2 + 9y^2 = 33 \\ x^2 + y^2 = 25 \end{cases}$

10. $\begin{cases} x^2 + 5y^2 = 70 \\ 3x^2 - 5y^2 = 30 \end{cases}$

11. $\begin{cases} x^2 + 2y^2 = 20 \\ 2x^2 - y^2 = 20 \end{cases}$

12. $\begin{cases} 3x^2 - y^2 = 3 \\ 4x^2 + 3y^2 = 43 \end{cases}$

In Problems 13–20, match each system with its graph, then use an appropriate method to solve the system.

13. $\begin{cases} x^2 - 25y^2 = 20 \\ 2x^2 + 25y^2 = 88 \end{cases}$

14. $\begin{cases} 3x^2 - 8y^2 = 40 \\ 5x^2 + y^2 = 81 \end{cases}$

15. $\begin{cases} 2x^2 - 3y^2 = 6 \\ 3x^2 + 2y^2 = 35 \end{cases}$

16. $\begin{cases} x^2 - y^2 = 7 \\ x^2 + y^2 = 25 \end{cases}$

17. $\begin{cases} 6x + 2y = 1 \\ 3x^2 - y^2 = -4 \end{cases}$

18. $\begin{cases} 5x - 3y = 10 \\ x^2 - y^2 = 4 \end{cases}$

19. $\begin{cases} 2x + 3y = 7 \\ x^2 + y^2 - 4y = 8 \end{cases}$

20. $\begin{cases} x - y = -1 \\ x^2 + 3y^2 = 12 \end{cases}$

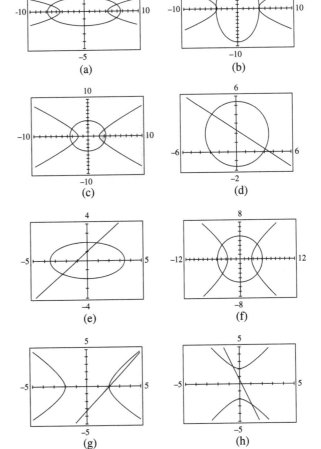

In Problems 21–28, graph the solution of each system of inequalities.

21. $\begin{cases} -x^2 + y \geq 0 \\ x + y < 1 \end{cases}$

22. $\begin{cases} \sqrt{x} - y > 0 \\ x - 9y \leq 0 \end{cases}$

23. $\begin{cases} x^2 + y \leq 0 \\ x + y > -2 \end{cases}$

24. $\begin{cases} y > x - 3 \\ y \leq \sqrt{x - 1} \\ y \geq 0 \end{cases}$

25. $\begin{cases} x \geq 0, y \geq 0 \\ y \leq 4 - x^2 \end{cases}$

26. $\begin{cases} x^2 + y^2 \leq 1 \\ x + y > 1 \end{cases}$

27. $\begin{cases} 4x^2 - y^2 \geq 0 \\ x + y < 9 \end{cases}$

28. $\begin{cases} y \geq (x - 1)^2 + 2 \\ 2x - 3y \leq -9 \end{cases}$

Applying the Concepts

29. **Computer Monitor Design:** An electronics engineer designs a rectangular computer monitor screen with a 10 inch diagonal and a viewing area of 48 square inches. Find the dimensions of the screen (Figure 6).

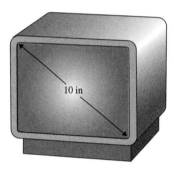
Figure 6

10 in

30. **Salt Storage:** A certain northern state stores road salt (used to melt ice and snow in winter), in buildings made of galvanized sheet iron in the shape of right circular cones (Figure 7). Each building has a height h equal to 1.5 times the radius r of its base, and each building has a volume of 12,000 cubic feet. Find h and r rounded off to two decimal places.

Figure 7

31. **Construction:** Figure 8 shows a child's sandbox made of galvanized sheet iron with a square base of side x and of height y. If the volume of the sandbox is 50 cubic feet, and the combined area of the sides and the bottom is 65 square feet, find the dimensions x and y of the box. Round off to one decimal place.

Figure 8

32. **Investment Portfolio:** The annual simple interest earned on an investment is $170. If the interest had been 1% higher, the interest earned from this investment would have been $238. What was the amount of the investment and what was the lower interest rate?

33. **Engineering Design:** An engineer designs a rectangular solar collector with a surface area of 750 square feet. If the length of the collector is 30 times its width, find its length and width.

34. **Template Design:** An artist designs a plastic template in the shape of a right triangle with one base length of 60 centimeters and an area of 1500 square centimeters. Find the lengths of the sides of the triangular template. Round off to one decimal place.

35. **Fencing:** A farmer uses 30 meters of fence to enclose a rectangular pen next to a barn, with the barn wall forming one side of the pen (Figure 9). What are the dimensions of the pen if it encloses an area of 108 square meters?

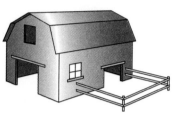
Figure 9

36. **Fencing:** A rancher wishes to fence two adjacent rectangular corrals on all sides with a total of 100 meters of fence. What are the dimensions of the smaller corral, if one corral is a square region and is 144 square meters more than the other? Round off to one decimal place.

37. **Puzzle:** Find two positive numbers whose sum is 15 and the difference of their squares is also 15.

38. **Number Problem:** The sum of the squares of the digits of a certain two-digit number is 61. The product of the number and the number with the digits reversed is 3640. What is the number?

Developing and Extending the Concepts

In Problems 39–42, solve each system algebraically.

39. $\begin{cases} -4x^2 + 3xy = 11 \\ 2x^2 - xy = 11 \end{cases}$

40. $\begin{cases} -2uv + 3v^2 = 8 \\ uv + v^2 = 6 \end{cases}$

41. $\begin{cases} (p - 2)^2 + (q - 2)^2 = 2 \\ pq = 1 \end{cases}$

42. $\begin{cases} (u - 4)^2 + (v + 4)^2 = 25 \\ u + v = 7 \end{cases}$

G In Problems 43–46, use a grapher to approximate the solutions of each system. Round off each answer to two decimal places.

43. $\begin{cases} x^2 - 3y = 1 \\ x - y = 1 \end{cases}$

44. $\begin{cases} \sqrt{x} - y = 0 \\ x^2 + y^2 = 6 \end{cases}$

45. $\begin{cases} x - y = 21 \\ \sqrt{x} + \sqrt{y} = 7 \end{cases}$

46. $\begin{cases} \dfrac{3}{x^2} - \dfrac{2}{y^2} = 1 \\ \dfrac{7}{x^2} - \dfrac{6}{y^2} = 2 \end{cases}$

In Problems 47–50, solve each system of inequalities.

47. $\begin{cases} -x^2 + y^2 \le 64 \\ x^2 - y^2 \ge 4 \end{cases}$

48. $\begin{cases} \dfrac{x^2}{36} + \dfrac{y^2}{16} \ge 1 \\ \dfrac{x^2}{16} + \dfrac{y^2}{36} \le 1 \end{cases}$

49. $\begin{cases} x^2 + y^2 \ge 4 \\ \dfrac{x^2}{4} + \dfrac{y^2}{25} \le 1 \end{cases}$

50. $\begin{cases} x^2 - y^2 \ge 9 \\ x^2 + y^2 \le 25 \end{cases}$

◈ CHAPTER 8 REVIEW PROBLEM SET

In Problems 1 and 2, write the augmented matrix for each system.

1. (a) $\begin{cases} x + 4y = 5 \\ 3x - 11y = 8 \end{cases}$
   (b) $\begin{cases} 2x + 3y + 4z = -1 \\ -x + 4y - z = -15 \\ x - y - 5z = 10 \end{cases}$

2. (a) $\begin{cases} x - 3y = -1 \\ 2x + y = 5 \end{cases}$
   (b) $\begin{cases} x + 2y - z = 4 \\ -x + 3y - 2z = 5 \\ 2x - y + z = 0 \end{cases}$

In Problems 3 and 4, write the system corresponding to the augmented matrix.

3. (a) $\begin{bmatrix} 4 & 1 & | & 2 \\ -2 & 3 & | & -1 \end{bmatrix}$
   (b) $\begin{bmatrix} 1 & 1 & | & 1 & 0 \\ 3 & 1 & | & 1 & 2 \\ 2 & -1 & | & 2 & 3 \end{bmatrix}$

4. (a) $\begin{bmatrix} 1 & 7 & | & 1 \\ 3 & -2 & | & 4 \end{bmatrix}$
   (b) $\begin{bmatrix} 1 & -2 & | & 3 & 3 \\ 4 & -3 & | & 1 & -5 \\ 5 & -5 & | & 4 & -2 \end{bmatrix}$

In Problems 5 and 6, each row-reduced matrix is the augmented matrix form of a corresponding linear system of equations. Solve each system.

5. (a) $\begin{bmatrix} 1 & 0 & | & 5 \\ 0 & 1 & | & -3 \end{bmatrix}$
   (b) $\begin{bmatrix} 1 & 0 & | & 0 & -4 \\ 0 & 1 & | & 0 & -6 \\ 0 & 0 & | & 1 & 3 \end{bmatrix}$

6. (a) $\begin{bmatrix} 1 & 0 & | & 6 \\ 0 & 1 & | & -\frac{1}{2} \end{bmatrix}$
   (b) $\begin{bmatrix} 1 & 0 & | & 0 & 4 \\ 0 & 1 & | & 0 & -3 \\ 0 & 0 & | & 1 & 2 \end{bmatrix}$

In Problems 7–10, write the augmented matrix for each system, and then use the row operations to solve the system.

7. $\begin{cases} 2x + 3y = 4 \\ 3x - y = -5 \end{cases}$

8. $\begin{cases} x - 3y = 7 \\ 4x + y = 5 \end{cases}$

9. $\begin{cases} -x + 2y + 8z = 1 \\ 2x + y - z = 2 \\ x + 4y + 6z = 4 \end{cases}$

10. $\begin{cases} x - 4y - 2z = -10 \\ x + 2y - z = 23 \\ 2x + 5y + z = 43 \end{cases}$

In Problems 11–16, find the value of each determinant.

11. (a) $\begin{vmatrix} 5 & 2 \\ 2 & -3 \end{vmatrix}$
    (b) $\begin{vmatrix} 2 & 8 \\ 3 & 7 \end{vmatrix}$

12. (a) $\begin{vmatrix} 2 & -5 \\ 1 & 6 \end{vmatrix}$
    (b) $\begin{vmatrix} 3 & 1 \\ 2 & -4 \end{vmatrix}$

13. $\begin{vmatrix} 1 & 1 & 2 \\ 2 & -3 & -1 \\ 3 & 7 & -2 \end{vmatrix}$

14. $\begin{vmatrix} 2 & 1 & 2 \\ 1 & 3 & -1 \\ 3 & -1 & -2 \end{vmatrix}$

15. $\begin{vmatrix} 3 & 2 & -1 \\ 1 & -1 & 2 \\ 5 & 3 & -4 \end{vmatrix}$

16. $\begin{vmatrix} 1 & -1 & 2 \\ 3 & 1 & 1 \\ 2 & -1 & 5 \end{vmatrix}$

In Problems 17 and 18, solve each equation for x.

17. $\begin{vmatrix} x + 1 & x \\ x & x - 2 \end{vmatrix} = -6$

18. $\begin{vmatrix} x & 4 & 5 \\ 0 & 1 & x \\ 5 & 2 & 0 \end{vmatrix} = 7$

In Problems 19–22, use Cramer's rule to solve each system.

19. $\begin{cases} x - y = 3 \\ 2x + y = 3 \end{cases}$

20. $\begin{cases} 5x + 2y = 3 \\ 2x + 3y = -1 \end{cases}$

21. $\begin{cases} x - y + 2z = 0 \\ 3x + y + z = 2 \\ 2x - y + 5z = 5 \end{cases}$

22. $\begin{cases} 3x + 2y - z = -4 \\ x - y + 2z = 13 \\ 5x + 3y - 4z = -15 \end{cases}$

In Problems 23–26, (a) graph the constraints and identify each vertex of the feasible region. (b) Find the maximum and minimum value of the objective function subject to the constraints in part (a).

23. $F = 2x + y + 3$
    $\begin{cases} x \ge 0, y \ge 0 \\ x + y \le 6 \\ x + y \ge 2 \end{cases}$

24. $F = 2x + 3y + 1$
    $\begin{cases} x \ge 0, y \ge 0 \\ 3x + y \le 15 \\ 2x + 4y \le 40 \end{cases}$

25. $F = 3x - y$
    $\begin{cases} x \ge 4, y \ge 3 \\ x + y \ge 8 \end{cases}$

26. $F = 2x + 7y$
    $\begin{cases} x \ge 0, y \ge 0 \\ x + 2y \le 20 \\ x + 7y \le 35 \end{cases}$

In Problems 27–34, identify the intercepts, the foci and sketch the graph of each conic.

27. $4x^2 + 9y^2 = 36$
28. $25x^2 + 4y^2 = 100$
29. $7x^2 + 8y^2 = 56$
30. $11x^2 + 5y^2 = 55$
31. $7x^2 - 16y^2 = 112$
32. $16y^2 - 4x^2 = 64$
33. $9x^2 - y^2 = 9$
34. $25x^2 - 64y^2 = 1600$

In Problems 35–38, find the standard equation of each conic with the given information. Also sketch the graph.

35. The hyperbola with vertices at $(-3,0)$ and $(3,0)$, center at $(0,0)$ and one focus at $(5,0)$.

36. The ellipse with vertices at $(-5,0)$, $(5,0)$ and y intercepts 3 and -3.

37. The ellipse with a vertex at $(0,7)$, center at $(0,0)$, and a focus at $(0,4)$.

38. The hyperbola with center $(0,0)$, focus at $(4,0)$ and a vertex at $(2,0)$.

In Problems 39–42, solve each system algebraically. Check the solutions by graphing each equation and determining the points of intersection.

39. $\begin{cases} 3x - 4y = 25 \\ x^2 + y^2 = 25 \end{cases}$

40. $\begin{cases} 2x - y = 2 \\ x^2 + 2y^2 = 12 \end{cases}$

41. $\begin{cases} x^2 - y^2 = 21 \\ x^2 + y^2 = 29 \end{cases}$

42. $\begin{cases} 3x^2 - 2y^2 = 27 \\ 7x^2 + 5y^2 = 63 \end{cases}$

43. **Health Food:** A health food store has two different kinds of granola-cashew nut selling at $4.20 per pound, and golden granola selling at $3.40 per pound. How much of each kind should be mixed to produce a 200 pound mixture selling at $3.72 per pound?

44. **Geometry:** The perimeter of a rectangular waterbed is 270 inches. If 5 times its width equals 4 times its length, what is the length and width of the waterbed?

45. **Loan:** Suppose that at the end of 8 months, the balance on a business loan including simple annual interest is $2563, and at the end of 18 months, the balance is $2754. How much money was borrowed and what is the annual rate of interest?

46. **Trucking:** An owner of a 5000 gallon fuel truck loads his truck with gasoline and kerosene. For a shipment of fuel, the owner charges $0.15 for delivering each gallon of gasoline and $0.18 for delivering each gallon of kerosene. How many gallons of each fuel should the trucker load in order to charge $780 for the load of fuel?

47. **Investment:** A businesswoman invested a total of $62,000 in three mutual funds. At the end of one year, one fund had returned 15%, the second returned 8% and the third returned -3% respectively on their parts of the investment. If the total return on the entire investment for a year was $6230, and if the businesswoman invested $8000 more in the 15% fund than she did in the 8% fund, how much did she invest in each fund? Round to the nearest dollar.

48. **Farming:** A farmer owns and operates a 62-acre grove on which he grows oranges and avocados. On the average, the task of picking oranges requires at most 6 hours of labor per acre, and avocados require at most 8 hours of labor per acre. Suppose that up to 640 hours of labor are reserved for harvesting. If the profit for each acre of oranges grown is $1,730 and the profit for each acre of avocados grown is $1,520, how many acres should be planted in oranges and how many in avocados to maximize the farmer's profit?

49. **Management:** A manufacturer makes two products, A and B. The cost per unit for A is $40 and for B is $25. The firm must make at least 75 units of A and at least 100 units of B. If each unit of A sells for $75 and each unit of B for $30, and the revenue from the sale of A and B must be at least $2250, find the number of units of each product it must produce in order to minimize cost.

50. **Carpeting:** A homeowner purchases carpets for two square rooms with a combined area of 52 square yards. The price of the first carpet is $13 per square yard, and the price of the second is $18 per square yard. If the total cost of the carpets is $856, find the dimensions of each square room.

51. **Investment:** One investor receives $500 annual income from an investment. His brother invested $4000 more, but at an annual interest rate of 1% less, and receives an annual income of $560. Find the amount and the interest rate of each investment.

52. **Travel Time and Rate:** Linda drove her car for a trip of 350 miles. Her boyfriend made the same trip at a speed 5 miles per hour faster than Linda and required 20 minutes less time. Find the time and the speed for Linda to make the trip.

G Some graphers can display the reduced form of a matrix, the value of a determinant and the graph of a conic. In Problems 53–56, explore the capabilities of your grapher to produce each of the following:

53. Find the row reduced echelon form of the augmented matrix of

$$\begin{cases} x_1 + 2x_2 - 3x_3 = -4 \\ -x_1 - x_2 + x_3 + 4x_4 = 4 \\ 3x_1 + x_2 - x_3 - 6x_4 = 0 \\ 2x_1 - x_2 + x_3 - 2x_4 = 7 \end{cases}$$

54. Use Cramer's rule to solve the system.

$$\begin{cases} x_1 + x_2 + x_3 = 1 \\ x_1 + 2x_3 + 5x_4 = -3 \\ 2x_1 - x_2 + 4x_4 = 2 \\ 2x_1 + x_3 + 9x_4 = 5 \end{cases}$$

55. Display the graph of $x^2 + y^2 + 8x - 6y + 1 = 0$.

56. Display the graph of $x^2 - y^2 - 2x + 4y - 12 = 0$.

CHAPTER 8 PRACTICE TEST

1. Write the augmented matrix for each system.

   (a) $\begin{cases} 3x + 7y = 16 \\ 2x + 5y = 13 \end{cases}$

   (b) $\begin{cases} x + y + z = 6 \\ 3x - y + 2z = 7 \\ 2x + 3y - z = 5 \end{cases}$

2. Write the system corresponding to the augmented matrix.

   (a) $\begin{bmatrix} 3 & -4 & | & 1 \\ -2 & 3 & | & 5 \end{bmatrix}$

   (b) $\begin{bmatrix} 1 & 0 & 1 & | & -2 \\ 2 & 3 & 0 & | & 1 \\ -2 & 1 & -3 & | & 5 \end{bmatrix}$

3. Each row-reduced matrix is the augmented matrix form of a corresponding linear system of equations. Solve each system.

   (a) $\begin{bmatrix} 1 & 0 & | & -3 \\ 0 & 1 & | & 7 \end{bmatrix}$

   (b) $\begin{bmatrix} 1 & 0 & 0 & | & -2 \\ 0 & 1 & 1 & | & 2 \\ 0 & 0 & 0 & | & 0 \end{bmatrix}$

4. Write the augmented matrix for each system, and then use the row operations to solve the system.

   (a) $\begin{cases} 7x + 4y = 1 \\ 9x + 54y = 3 \end{cases}$

   (b) $\begin{cases} x + y + z = 9 \\ 27x + 9y + 3z = 93 \\ 8x + 4y + 2z = 36 \end{cases}$

5. Evaluate each determinant.

   (a) $\begin{vmatrix} -4 & 3 \\ 3 & -2 \end{vmatrix}$

   (b) $\begin{vmatrix} 3 & 0 & -1 \\ 3 & 1 & -1 \\ 1 & 1 & 2 \end{vmatrix}$

6. Use Cramer's rule to solve each system.

   (a) $\begin{cases} 3x + 4y = 7 \\ -2x + 3y = 1 \end{cases}$

   (b) $\begin{cases} 3x + y + z = 5 \\ 5x - y + z = 5 \\ 3x + 2y - z = 4 \end{cases}$

7. (a) Graph the constraints and identify each vertex of the feasible region

   $$\begin{cases} x \geq 0, y \geq 0 \\ x + y \leq 4 \\ 2x + y \leq 6 \end{cases}$$

   (b) Find the maximum value of the objective function

   $$F = 3x + 6y$$

   subject to the constraints in part (a).

8. Identify the intercepts, the vertices, and the foci; sketch the graph for each conic.

   (a) $4x^2 + y^2 = 4$

   (b) $4x^2 - 9y^2 = 36$

9. Find the standard equation of the hyperbola whose vertices are $(-5,0)$ and $(5,0)$, center at $(0,0)$ and one focus at $(7,0)$.

10. Solve the system $\begin{cases} 2x + y = 10 \\ x^2 + y^2 = 25 \end{cases}$

11. A gold membership for a fitness center includes an initial fee for joining as well as monthly dues. The total cost after 9 months' membership is $385 and after 14 months' membership is $560. Find both the initial fee and the monthly dues.

Chapter 9

SEQUENCES, SERIES, AND THE BINOMIAL THEOREM

9.1 Sequences
9.2 Series and Summation
9.3 Arithmetic Sequences and Series
9.4 Geometric Sequences and Series
9.5 The Binomial Theorem

Botanists have examined plants from all over the world and studied the arrangement of their flowers, leaves, stems, and seeds. They discovered that the arrangements fall into a limited number of patterns called *sequences*. The numbers that occur most often in the description of plant element arrangements form a sequence of numbers called the *Fibonacci Sequence*. This sequence is arranged as follows:

$$1, 1, 2, 3, 5, 8, 13, 21, 34, 55, 89, 144, 233, \ldots$$

It is used in a wide variety of applications in fields such as botany, physics, and ecology. Example 5 on page 433 shows its use in describing the production of pairs of rabbits that grow to maturity in a cycle of one month.

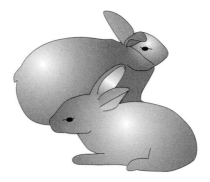

In this chapter we introduce three topics that are very important for advanced courses. The two main topics are *sequences* and *series* and they are the center of our attention in this chapter. With the growing availability of computers, these topics are playing increasingly important roles in real world situations. The third topic is the *binomial theorem*, which provides a valuable technique for expanding binomials to any positive integer power.

OBJECTIVES

1. Describe Sequences
2. Define a Sequence Recursively
3. Find the General Term of a Sequence
4. Graph Sequences
5. Solve Applied Problems
6. Use Technology Exploration

9.1 Sequences

In this section we give a mathematical definition of a *sequence*. Later we concentrate on finding the sum of the terms of a sequence called a *series*.

Describing Sequences

Intuitively, a **sequence** is a collection or list of numbers arranged in a definite order. The individual numbers of a sequence are called **terms**. Examples of sequences are:

$$2, 4, 6, 8, 10 \qquad 2, 4, 6, 8, 10, \ldots$$

The sequence in the first example contains five terms, a finite number; it is called a **finite sequence**. The second sequence contains infinitely many terms (indicated by the three periods called an *ellipsis*), and is called an **infinite sequence** that lists the positive multiples of 2. The definition of a sequence can be clearly stated by using the terminology of functions. A **sequence** is a function with the positive integers as its set of inputs (domain). When the domain is apparent, we refer to either a finite or an infinite sequence as a sequence. Following tradition, we write the function values in the form

$$a_1, a_2, a_3, \ldots, a_n, \ldots$$

These elements in the range of a sequence are called its **terms**. The terms are named with a letter, such as a and a subscript corresponding to the position of the term. In the above sequence, a_1 is referred to as the **first term,** a_2 is the **second term,** and a_n is the **nth term** or the **general term**.

The notation $\{a_n\}$, in which the general term is enclosed in braces, is also used to denote a sequence. To specify a particular sequence, we give a rule by which the nth term, a_n, is determined. This is often done by means of a formula such as

$$a_n = (n + 1)^2$$

In that case, we generate the terms of a sequence by substituting values for n in the formula, where

$$n = 1, 2, 3, \ldots$$

EXAMPLE 1   Determining the Term of a Sequence
Find the first five terms of each sequence.

(a) $a_n = (-1)^n$

(b) $a_n = 3 - \dfrac{1}{n}$

Solution   To find the first five terms of each sequence we substitute the positive integers 1, 2, 3, 4, and 5, in turn, for n in the formula for the general term.

(a) We have

$$a_1 = (-1)^1 = -1, \qquad a_2 = (-1)^2 = 1, \qquad a_3 = (-1)^3 = -1$$
$$a_4 = (-1)^4 = 1, \qquad\qquad \text{and} \qquad\qquad a_5 = (-1)^5 = -1$$

Thus, the first five terms of the sequence $\{a_n\}$ are

$$-1, 1, -1, 1, -1.$$

(b) We have

$$a_1 = 3 - \frac{1}{1} = 2, \qquad a_2 = 3 - \frac{1}{2} = \frac{5}{2}, \qquad a_3 = 3 - \frac{1}{3} = \frac{8}{3}$$

$$a_4 = 3 - \frac{1}{4} = \frac{11}{4}, \qquad \text{and} \qquad a_5 = 3 - \frac{1}{5} = \frac{14}{5}$$

Therefore, the first five terms of the sequence $\{a_n\}$ are

$$2, \frac{5}{2}, \frac{8}{3}, \frac{11}{4}, \frac{14}{5}$$

Defining a Sequence Recursively

So far, the formulas used for the nth term, a_n, of a sequence have expressed the nth term as a function of n. In another approach, a formula that relates the general term a_n of a sequence to one or more of the terms that come before it is called a **recursion formula**. A sequence that is specified by giving the first term (or the first few terms) together with a recursion formula is said to be defined **recursively**.

EXAMPLE 2   Using a Recursion Formula

Find the first five terms of the sequence defined recursively by

$$a_1 = 1 \text{ and } a_n = na_{n-1}, \ n \geq 2$$

Solution   We are given $a_1 = 1$. By the recursion formula $a_n = na_{n-1}$, we have

$$a_2 = 2a_{2-1} = 2a_1$$
$$= 2(1) = 2$$

Now, using the fact that $a_2 = 2$, we find that

$$a_3 = 3a_{3-1} = 3a_2$$
$$= 3(2) = 6$$

Continuing in this way, we have

$$a_4 = 4a_{4-1} = 4a_3$$
$$= 4(6) = 24$$
$$\text{and} \quad a_5 = 5a_{5-1} = 5a_4$$
$$= 5(24) = 120$$

Finding the General Term of a Sequence

Now we look at the problem in reverse. We often have the terms of a sequence and want to write a formula that defines the sequence. That is, if the first few terms of a sequence are known and the sequence continues in some indicated pattern, then we can predict a formula for a_n in terms of n.

EXAMPLE 3   Finding the General Term of a Sequence

Predict the general term of each sequence.

(a) $4, 7, 10, 13, \ldots$ (b) $1, -2, 4, -8, 16, \ldots$

Solution   (a) Notice that
$$7 - 4 = 3, \ 10 - 7 = 3, \ 13 - 10 = 3$$
and so on. Since the domain of a sequence consists of the positive integers, then the expression $3n + 1$ defines the general term. That is, the general term
$$a_n = 3n + 1, \text{ where } n = 1, 2, 3, 4, \ldots$$

(b) These terms are the powers of 2 with alternating signs, so the general term a_n is given by
$$a_n = (-1)^{n+1} 2^{n-1}, \ n = 1, 2, 3, 4, \ldots$$

Graphing Sequences

In graphing a sequence, we designate the sequence numbers a_n as outputs and positive integers n as inputs. As usual, we use the horizontal axis for the inputs n and the vertical axis for the outputs
$$a_n = f(n).$$
Then we plot

(n, a_n) as discrete points in a coordinate system.

EXAMPLE 4   Graphing a Sequence

(a) Graph the first five terms of the sequence whose general term is
$$a_n = 2n - 1 \text{ for } 1 \le n \le 5$$

(b) Graph the function
$$f(x) = 2x - 1 \text{ for } x \ge 1.$$

Solution   (a) To find the first five terms of the sequence, we substitute $n = 1, 2, 3, 4, 5$ into the general term $a_n = 2n - 1$ to get
$$a_1 = 2(1) - 1 = 1, \qquad a_2 = 2(2) - 1 = 3, \qquad a_3 = 2(3) - 1 = 5,$$
$$a_4 = 2(4) - 1 = 7, \qquad a_5 = 2(5) - 1 = 9$$

The graph of this sequence consists of the following discrete points

$(1,1), (2,3), (3,5), (4,7)$ and $(5,9)$ (Figure 1a)

(b) The graph of
$$f(x) = 2x - 1 \text{ for } x \ge 1$$
is a straight line (Figure 1b).

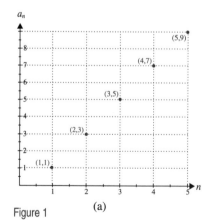

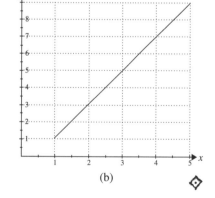

Figure 1   (a)   (b)

Solving Applied Problems

We use a recursion formula to investigate one of the most famous sequences, which was introduced by the Italian mathematician Leonardo Fibonacci in 1202 AD and is defined as follows:

$$a_1 = 1, a_2 = 1 \text{ and } a_n = a_{n-1} + a_{n-2}, \text{ for } n \geq 3.$$

EXAMPLE 5   Using a Fibonacci Sequence

Suppose that we have a pair of rabbits that grow to maturity in one month. How many pairs of rabbits can be produced from the single pair in a year if every month each pair begets a new pair that from the second month onward becomes productive?

Solution   Assume that the rabbits in the initial pair are newborn, so that they produce the first new pair after two months. The first month we have one pair of rabbits, the second month we still have one pair of rabbits, but they are mature enough to mate. By the third month, we have our first new pair of rabbits as well as the original pair. The number of pairs present after 12 months is given by:

$$a_1 = 1 \qquad\qquad\qquad a_2 = 1$$
$$a_3 = a_2 + a_1 = 1 + 1 = 2 \qquad a_4 = a_3 + a_2 = 2 + 1 = 3$$
$$a_5 = a_4 + a_3 = 3 + 2 = 5 \qquad a_6 = a_5 + a_4 = 5 + 3 = 8$$
$$a_7 = a_6 + a_5 = 8 + 5 = 13 \qquad a_8 = a_7 + a_6 = 13 + 8 = 21$$
$$a_9 = a_8 + a_7 = 21 + 13 = 34 \qquad a_{10} = a_9 + a_8 = 34 + 21 = 55$$
$$a_{11} = a_{10} + a_9 = 55 + 34 = 89 \qquad a_{12} = a_{11} + a_{10} = 89 + 55 = 144$$

Therefore, the first twelve terms of the sequence are:

$$1, 1, 2, 3, 5, 8, 13, 21, 34, 55, 89, 144$$

This pattern of numbers is called a Fibonacci sequence.   ◈

Using Technology Exploration

Graphers equipped with ⬭Tblset⬭ feature will create and graph sequences.

G   EXAMPLE 6   Using a Grapher

Use a grapher to graph the continuous function $f(x) = -2x + 3$ for $x \geq 1$ and the first five ordered pairs of the sequence $a_n = -2n + 3$.

Solution   Figure 2a shows the graph of $f(x) = -2x + 3$ is a line with slope -2 and y intercept 3 on the indicated viewing window. The first five ordered pairs of the sequence $a_n = -2n + 3$ are $(1,1)$, $(2,-1)$, $(3,-3)$, $(4,-5)$, $(5,-7)$. Figure 2b shows the graph of the sequence with the same size viewing window.

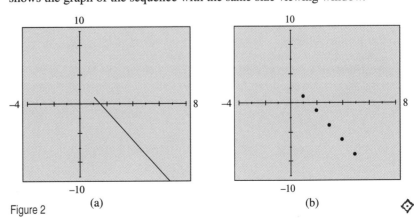

Figure 2   (a)   (b)   ◈

PROBLEM SET 9.1

Mastering the Concepts

In Problems 1–12, find the first five terms in each sequence.

1. $a_n = n - 4$
2. $a_n = \dfrac{n + 4}{n}$
3. $a_n = \dfrac{n(n + 2)}{2}$
4. $a_n = \dfrac{n + 1}{n + 2}$
5. $a_n = \dfrac{n(n - 3)}{2}$
6. $a_n = \dfrac{n(n + 3)}{2}$
7. $a_n = \dfrac{1}{n^2}$
8. $a_n = \dfrac{n^2 - 2}{2}$
9. $a_n = 2^{n - 3}$
10. $a_n = 3^{n - 4}$
11. $a_n = (-1)^n + 3$
12. $a_n = \dfrac{3}{n(n + 1)}$

In Problems 13–20, find the general term a_n for each sequence.

13. $2, 4, 6, 8, 10, \ldots$
14. $3, 6, 9, 12, 15, \ldots$
15. $1, 3, 5, 7, 9, \ldots$
16. $1, \dfrac{1}{8}, \dfrac{1}{27}, \dfrac{1}{64}, \dfrac{1}{125}, \ldots$
17. $1, 8, 27, 64, 125, \ldots$
18. $1, 16, 81, 256, 625, \ldots$
19. $-3, 6, -9, 12, -15, \ldots$
20. $2, -4, 6, -8, 10, \ldots$

In Problems 21–28, find the first five terms of each sequence defined recursively as indicated.

21. $a_1 = 1, a_{n+1} = \dfrac{1}{2}a_n$
22. $a_1 = 2, a_{n+1} = 2a_n$
23. $a_1 = 1, a_{n+1} = a_n + \dfrac{1}{2}$
24. $a_1 = 1, a_{n+1} = a_n - \dfrac{1}{2}$
25. $a_1 = -10, a_n = a_1 + (n - 1)(10)$
26. $a_1 = 10, a_n = a_1 + (n - 1)(-10)$
27. $a_1 = -2, a_n = 3a_{n-1}, n \geq 2$
28. $a_1 = 3, a_n = 4a_{n-1}, n \geq 2$

Applying the Concepts

29. **Pay Raise:** A person is offered a job with an annual salary of $20,000, and a raise of 4% at the end of each year. Write a sequence showing the salary at the end of 2 years, 3 years, 4 years and 5 years.

30. **Drug Measurement:** A patient takes 15 milligrams of a drug each morning. Suppose that 80% of the drug is eliminated each day. Write out the first six terms of the sequence where a_n is the amount of the drug present in the patient's body immediately after the nth dose.

31. **Investment:** One dollar is deposited in a savings account that pays an annual interest rate of 6%. If no money is withdrawn, what is the amount accrued in the account after the first, second, and third years?

32. **Bacteria Growth:** A colony of bacteria doubles every day. If the colony has 100 bacteria at the beginning of an experiment, find the number of bacteria after 6 days.

Developing and Extending the Concepts

In Problems 33–36, find the first five terms in each sequence.

33. $a_n = (-1)^n(n + 2)$
34. $a_n = \dfrac{(-1)^n}{2n + 1}$
35. $a_n = \sqrt{n + 1} - \sqrt{n}$
36. $a_n = n \log 10^{n - 1}$

In Problems 37–40, find the indicated term in each sequence.

37. The 10th term of $a_n = 3n + 2$
38. The 15th term of $a_n = 2n + 3$
39. The 20th term of $a_n = \dfrac{n - 2}{n + 2}$
40. The 7th term of $a_n = 5(2)^{3 - n}$
41. Write the first four terms of the sequence.
$$a_n = \sqrt{n} + \sqrt{n + 9 - 6\sqrt{n}}$$
Round off each answer to two decimal places.
42. Write the first four terms of the sequence
$a_n = \sqrt{n} + \sqrt{|\sqrt{n} - 3|}$.
Round off each answer to two decimal places.
43. Predict the general term of the sequence
$\sqrt{5}, \sqrt{8}, \sqrt{11}, \sqrt{14}, \ldots$
44. Predict the general term of the sequence
$-3, 3, -3, 3, -3, 3, \ldots$

In Problems 45–48, write the first five terms of each sequence defined recursively.

45. $a_1 = 1, a_{n+1} = 3 + 2a_n, n \geq 2$
46. $a_1 = 2, a_{n+1} = 2a_{n-1} - 3, n \geq 2$
47. $a_1 = 1, a_2 = 1, a_{n+2} = a_{n+1} + a_n, n \geq 3$
48. $a_1 = 5, a_2 = 7, a_{n+2} = a_{n+1} + a_n, n \geq 3$

G In Problems 49–52, use a grapher to graph the continuous function f and the first five ordered pairs of each sequence a_n in separate viewing windows.

49. $f(x) = 3x - 4$; $\quad a_n = 3n - 4$
50. $f(x) = 1 - 3x$; $\quad a_n = 1 - 3n$
51. $f(x) = 2^x$; $\quad a_n = 2^n$
52. $f(x) = \dfrac{2x + 1}{x}$; $\quad a_n = \dfrac{2n + 1}{n}$

OBJECTIVES

1. Describe a Series
2. Use Summation Notation
3. Use Properties of Summation
4. Solve Applied Problems
5. Use Technology Exploration

9.2 Series and Summation

In many applied problems, it is sometimes necessary to add the terms of a sequence. Special terminology, notation, and techniques have been developed for that purpose.

Describing a Series

The sum of the terms of a sequence is called a **series**. The sum of the first n terms of a sequence $\{a_n\}$ is written in the form

$$a_1 + a_2 + a_3 + \ldots + a_n$$

Because sequences can be finite or infinite, series can also be *finite* or *infinite*. A series of the form

$$a_1 + a_2 + a_3 + \ldots + a_n + \ldots$$

is called an **infinite series**. The term a_n in the above series is called the **general term**.

EXAMPLE 1   Find the Sum of a Series

Find the sum of the terms of the sequence.
5, 8, 11, 14, 17, 20, 23, 26, 29, 32, 35, 38

Solution   Of course, we can add these terms by using a calculator.
Without using a calculator, we pair these numbers as follows.

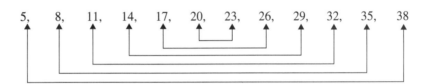

There are 12 numbers, and $\frac{12}{2} = 6$ pairs, each adding up to 43. The sum is $6(43) = 258$. ◈

Using Summation Notation

The sum of the first n terms of a sequence $a_1, a_2, a_3, \ldots, a_n$ may be written in the form

$$a_1 + a_2 + a_3 + \ldots + a_n$$

This sum can be written using a compact notation denoted by the Greek capital letter *sigma*, Σ as follows:

$$\sum_{k=1}^{n} a_k = a_1 + a_2 + a_3 + \ldots + a_n$$

The integer k is called the **index** of the summation. The notation $k = 1$ under Σ indicates that the summation starts at 1, which is called the **lower limit** of the summation. The value n above Σ indicates the **upper limit** of the summation. Letters such as i, j, k, and n are commonly used to represent indices of the summation. For instance, we may write

$$\sum_{k=1}^{n} 3^k = \sum_{i=1}^{n} 3^i = \sum_{j=1}^{n} 3^j = 3^1 + 3^2 + 3^3 + \ldots + 3^n$$

EXAMPLE 2   Using Summation Notation

Evaluate each sum.

(a) $\displaystyle\sum_{k=1}^{3} (4k^2 - 3k)$ (b) $\displaystyle\sum_{j=2}^{5} \frac{j-1}{j+1}$ (c) $\displaystyle\sum_{k=1}^{4} 5$

Solution   (a) Since $k = 1$ and $n = 3$, we substitute 1, 2, and 3 for k, and then add the resulting terms.

$$\sum_{k=1}^{3} (4k^2 - 3k) = [4(1)^2 - 3(1)] + [4(2)^2 - 3(2)] + [4(3)^2 - 3(3)]$$

$$= 1 + 10 + 27 = 38$$

(b) Since $j = 2$ and $n = 5$, we substitute 2, 3, 4, and 5 for j, and then add the resulting terms.

$$\sum_{j=2}^{5} \frac{j-1}{j+1} = \frac{2-1}{2+1} + \frac{3-1}{3+1} + \frac{4-1}{4+1} + \frac{5-1}{5+1}$$

$$= \frac{1}{3} + \frac{2}{4} + \frac{3}{5} + \frac{4}{6} = \frac{21}{10}$$

(c) This expression represents the sum of the first 4 terms of the sequence $a_k = 5$ specifically,

$$\sum_{k=1}^{4} 5 = 5 + 5 + 5 + 5 = 20$$

EXAMPLE 3   Writing a Series Using Summation Notation

Express each series using summation notation.

(a) $3 + 5 + 7 + 9 + 11$ (b) $\dfrac{1}{2} - \dfrac{1}{4} + \dfrac{1}{8} - \dfrac{1}{16} + \dfrac{1}{32} - \dfrac{1}{64}$

Solution   (a) The general term of this series is given by

$$a_k = 2k + 1, \text{ where } 1 \le k \le 5.$$

Thus, this sum is written in summation notation as

$$\sum_{k=1}^{5} (2k + 1)$$

(b) The sequence is of the form

$$\frac{1}{2}, \quad -\frac{1}{4}, \quad \frac{1}{8}, \quad -\frac{1}{16}, \quad \frac{1}{32}, \quad -\frac{1}{64}$$

Using the factor $(-1)^{k+1}$ to make the signs alternate between the positive and negative, we write the sum as

$$\sum_{k=1}^{6} (-1)^{k+1}\left(\frac{1}{2}\right)^k$$

There may be more than one way to express a sum in summation notation.

Using Properties of Summation

In calculations involving sums in sigma notation, the following properties may be used.

Let $\{a_n\}$ and $\{b_n\}$ be sequences. For every positive integer n and any constant c, we have

1. $\displaystyle\sum_{k=1}^{n} c = nc$
   Sum of a constant.

2. $\displaystyle\sum_{k=1}^{n} ca_k = c\sum_{k=1}^{n} a_k$
   Constant times a sum.

3. $\displaystyle\sum_{k=1}^{n} (a_k + b_k) = \sum_{k=1}^{n} a_k + \sum_{k=1}^{n} b_k$
   Sum of sums.

4. $\displaystyle\sum_{k=1}^{n} (a_k - b_k) = \sum_{k=1}^{n} a_k - \sum_{k=1}^{n} b_k$
   Difference of sums.

EXAMPLE 4   Using Properties of Summation

Suppose that $\displaystyle\sum_{k=1}^{10} a_k = 30$ and $\displaystyle\sum_{k=1}^{10} b_k = 45$, evaluate each summation.

(a) $\displaystyle\sum_{k=1}^{10} (2a_k + 3b_k)$

(b) $\displaystyle\sum_{k=1}^{10} (a_k + 5)$

Solution
(a) $\displaystyle\sum_{k=1}^{10} (2a_k + 3b_k) = \sum_{k=1}^{10} 2a_k + \sum_{k=1}^{10} 3b_k$ Use property 3.

$\displaystyle = 2\sum_{k=1}^{10} a_k + 3\sum_{k=1}^{10} b_k$ Use property 2

$= 2(30) + 3(45)$

$= 60 + 135 = 195$

(b) $\displaystyle\sum_{k=1}^{10} (a_k + 5) = \sum_{k=1}^{10} a_k + \sum_{k=1}^{10} 5$ Use property 3

$= 30 + 10(5)$

$= 30 + 50 = 80$

Solving Applied Problems

A story is told about the great German mathematician Carl Friedrich Gauss (1777–1858). When he was 10 years old, his teacher wanted to keep his young students busy for a considerable period. He asked the class to add the numbers from 1 to 100. Within only a few moments, young Gauss raised his hand and said that he had the answer. The teacher was amazed at the boy's speed. How did he find the sum so quickly? His technique was based on adding the first and last terms and multiplying by half the number of terms. That is,

Because $1 + 100 = 101$, $2 + 99 = 101$, $3 + 98 = 101$, and so on, all he needed to finish the problem is to find the number of such pairs, which is 50. The sum is $50(101) = 5050$. Gauss's pairing method was used in finding the sum of even terms of the sequence in example 1. His method also works with an odd number of terms of a sequence as the next example shows.

EXAMPLE 5   Finding a Sum of Salary

Suppose that a starting salary for an employee is $15,000 during the first year and the employee receives an $8,000 pay increase at the end of each subsequent year. What are the employee's earnings over the first 7 years?

Solution   The employee's salary over the first 7 years is given by the sum of the terms of the sequence.

$15,000, $23,000, $31,000, $39,000, $47,000, $55,000, $63,000

That is,

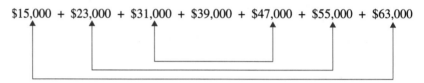

$15,000 + $23,000 + $31,000 + $39,000 + $47,000 + $55,000 + $63,000

Using Gauss's pairing method, there are $3\frac{1}{2}$ pairs each adding to $78,000. The sum is $3\frac{1}{2}(78,000) = $273,000.   ◇

Using Technology Exploration

Summation formulas frequently occur in calculus, where a polynomial is used to approximate given functions as the next example shows.

G EXAMPLE 6   Graphing a Function on a Grapher

A polynomial used to approximate the function $f(x) = \ln(1 + x)$ is given by

$$P_n(x) = \sum_{k=0}^{n} \frac{(-1)^k x^{k+1}}{k + 1}$$

(a) Let $n = 3$ and express $P_3(x)$ in expanded form.
(b) Use a grapher to graph $y = f(x)$ and $y = P_3(x)$ on the same viewing window.

Solution   (a) Letting $n = 3$ and expanding the summation, we have

$$P_3(x) = \sum_{k=0}^{3} \frac{(-1)^k x^{k+1}}{k + 1} = \frac{x}{1} - \frac{x^2}{2} + \frac{x^3}{3} - \frac{x^4}{4}$$

(b) Figure 1 shows the graphs of

$$y = \ln(1 + x) \text{ and } y = P_3(x)$$

on the same viewing window of a grapher. Notice that the graphs of f and P_3 are close.   ◇

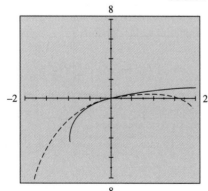

Figure 1

PROBLEM SET 9.2

Mastering the Concepts

In Problems 1 and 2, use Gauss's system of pairing numbers to find the sum of the terms of each sequence.

1.   4, 9, 14, 19, 24, 29, 34, 39, 44, 49, 54, 59, 64, 69
2.   11, 17, 23, 29, 35, 41, 47, 53, 59, 65, 71, 77, 83

In Problems 3–14 evaluate each sum.

3. $\displaystyle\sum_{k=1}^{5} k$

4. $\displaystyle\sum_{k=1}^{4} \frac{2^4}{k+1}$

5. $\displaystyle\sum_{i=1}^{10} 2i(i-1)$

6. $\displaystyle\sum_{k=0}^{4} 3^{2k}$

7. $\displaystyle\sum_{k=2}^{5} 2^{k-2}$

8. $\displaystyle\sum_{i=2}^{6} \frac{1}{i}$

9. $\displaystyle\sum_{k=1}^{3} (2k+1)$

10. $\displaystyle\sum_{k=1}^{5} (3k^2 - 5k + 1)$

11. $\displaystyle\sum_{i=2}^{4} \frac{i}{i+1}$

12. $\displaystyle\sum_{k=1}^{4} k^k$

13. $\displaystyle\sum_{k=1}^{100} 5$

14. $\displaystyle\sum_{i=3}^{7} (i+2)$

In Problems 15–22 express each series using summation notation.

15. $4 + 8 + 12 + 16 + 20 + 24$
16. $5 + 10 + 15 + 20 + 25 + 30$
17. $4 - 1 - 6 - 11 - 16 - 21$
18. $3 + 8 + 13 + 18 + 23 + 28$
19. $-1 - 4 - 7 - 10 - 13 - 16$
20. $-8 - 1 + 6 + 13 + 20 + 27$
21. $\dfrac{1}{1.2} + \dfrac{1}{2.3} + \dfrac{1}{3.4} + \dfrac{1}{4.5} + \dfrac{1}{5.6}$
22. $\dfrac{1}{1.2^3} + \dfrac{1}{2.3^3} + \dfrac{1}{3.4^3} + \dfrac{1}{4.5^3}$

In Problems 23–28, suppose $\sum_{k=1}^{30} a_k = 25$ and $\sum_{k=1}^{30} b_k = 15$. Use the properties of summation to evaluate each sum.

23. $\displaystyle\sum_{k=1}^{30} (4a_k + b_k)$

24. $\displaystyle\sum_{k=1}^{30} (a_k + 7b_k)$

25. $\displaystyle\sum_{k=1}^{30} (5a_k - 4)$

26. $\displaystyle\sum_{k=1}^{30} (4b_k + 5)$

27. $\displaystyle\sum_{k=1}^{30} (2a_k + 3b_k + 2)$

28. $\displaystyle\sum_{k=1}^{30} (3a_k - 4b_k + 8)$

In Problems 29–32, the formula

$$\bar{x} = \frac{x_1 + x_2 + x_3 + \dots + x_n}{n} = \frac{\sum_{k=1}^{n} x_k}{n}$$

determines the *arithmetic mean* of the sequence $x_1, x_2, x_3, \dots, x_n$. Find the arithmetic mean for the numbers in each sequence.

29. 1, 5, 9, 13, 17, 21, 25, 29
30. 3, 6, 9, 12, 15, 18, 21, 24
31. 1.3, 3.3, 5.3, 7.3, 9.3, 11.3
32. 1.1, 3.3, 5.5, 7.7, 9.9, 11.11

Applying the Concepts

33. **Investment:** Each year $1000 is deposited into an account for 5 years. The amount of money accumulated is approximated by the model $a_k = 1{,}000(1 + 0.04)^k$, where $k = 1, 2, 3, 4, 5$, and a_k is the amount accumulated during the kth year for each $1000 deposited. Use this model to determine the total amount of investment assuming no withdrawals.

34. **Distance Traveled:** A skier travels 4 feet down a mountain slope during the first second of descent. During the next second the skier travels 9 feet, and during the third second the skier travels 14 feet, and so on. During the kth second the skier's distance traveled is given by

$$a_k = 4 + 5(k - 1), k = 1, 2, 3, \dots, n$$

Find the total distance traveled by the skier during the first 10 seconds of descent.

Developing and Extending the Concepts

In Problems 35–38, compare the following sums. Round off to 4 decimal places.

35. (a) $\displaystyle\sum_{n=1}^{20} 1$ (b) $\displaystyle\sum_{n=1}^{20} 2$

36. (a) $\displaystyle\sum_{n=1}^{20} (n+1)$ (b) $\displaystyle\sum_{n=1}^{20} n + \sum_{n=1}^{20} 1$

37. (a) $\displaystyle\sum_{n=1}^{10} \frac{1}{n}$ (b) $\displaystyle\sum_{n=1}^{10} (-1)^n \frac{1}{n}$

38. (a) $\displaystyle\sum_{n=1}^{10} \frac{1}{n^2}$ (b) $\displaystyle\sum_{n=1}^{10} \frac{1}{n^3}$

In Problems 39–44, use the summation formulas

$$\sum_{k=1}^{n} k = \frac{n(n+1)}{2} \quad\text{and}\quad \sum_{k=1}^{n} k^2 = \frac{n(n+1)(2n+1)}{6}$$

and the properties of summation to evaluate each sum.

39. $\displaystyle\sum_{k=1}^{20} (3k + 4)$

40. $\displaystyle\sum_{k=1}^{20} (4k - 3)$

41. $\displaystyle\sum_{k=1}^{10} (2k^2 + 3k)$

42. $\displaystyle\sum_{k=1}^{10} (-3k^2 + 5k)$

43. $\displaystyle\sum_{k=1}^{15} \frac{k(k-1)}{2}$

44. $\displaystyle\sum_{k=1}^{15} \frac{k(2k+1)}{3}$

In Problems 45 and 46, solve each equation for x.

45. $\displaystyle\sum_{n=4}^{6} nx = 30$

46. $\displaystyle\sum_{n=3}^{5} (nx + 3) = 45$

G For Problems 47–50, consider the data $x_1, x_2, x_3, \ldots, x_n$. In statistics, the *standard deviation* of this data is given by

$$s = \sqrt{\frac{1}{n-1} \sum_{k=1}^{n} (x_k - \bar{x})^2}, \text{ where } \bar{x} = \frac{1}{n} \sum_{k=1}^{n} x_k$$

Use the formula and the STAT feature on a grapher to find the standard deviation of the given sequence.

47. 5, 9, 11, 2, 7, 1
48. 3, 5, 5, 7, 10
49. 1, 3, 4, 6, 7
50. 4, 2, 2, 8, 5, 3

G In Problems 51 and 52, the function

$$P_n(x) = \sum_{k=1}^{n} \frac{2}{2k-1} \left(\frac{x-1}{x+1} \right)^{2k-1}$$

is used as an approximation for $\ln x$.

G 51. (a) Express $P_2(x)$ in expanded form.
(b) Use a grapher to graph
$$y = \ln x \text{ and } y = P_2(x)$$
on the viewing window.
$x\text{Min} = -1, x\text{Max} = 12, x\text{Scl} = 1,$
$y\text{Min} = -4, y\text{Max} = 3, y\text{Scl} = 1$

G 52. (a) Express $P_3(x)$ in expanded form.
(b) Use a grapher to graph
$$y = \ln x \text{ and } y = P_3(x)$$
on the viewing window
$x\text{Min} = -1, x\text{Max} = 12, x\text{Scl} = 1$
$y\text{Min} = -4, y\text{Max} = 3, y\text{Scl} = 1$

OBJECTIVES

1. Define an Arithmetic Sequence
2. Find the Sum of an Arithmetic Sequence
3. Solve Applied Problems
4. Use Technology Exploration

9.3 Arithmetic Sequences and Series

Many sequences are generated by starting with an initial term, and then repeatedly adding a fixed number. Such a sequence is known as an *arithmetic sequence*. The general term of this type of a sequence corresponds to a linear function.

Defining an Arithmetic Sequence

Consider the following sequence

$$3, 7, 11, 15, 19, \ldots$$

Note that each term of this sequence is 4 more than the previous one, except the first one. That is, consecutive terms differ by 4 since

$$a_2 - a_1 = 7 - 3 = 4$$
$$a_3 - a_2 = 11 - 7 = 4$$
$$a_4 - a_3 = 15 - 11 = 4$$

and so on.

This sequence is an example of an arithmetic sequence, which has a feature that consecutive terms differ by a constant value. More precisely, we have the following definition:

ARITHMETIC SEQUENCES

A sequence of the form
$$a_1, a_2, a_3, \ldots, a_n, \ldots$$
is called an **arithmetic sequence** if there is a constant d such that any two consecutive terms differ by d. That is, for every positive integer $n \geq 1$.

$$a_{n+1} - a_n = d$$

The constant d is called the **common difference**.

In the above sequence, each term is created from the previous one (except for the first) by adding the common difference d. That is, every term of the sequence can be written in terms of the first term a_1 and common difference d as follows:

First term: $a_1 = a_1$
Second term: $a_2 = a_1 + d$
Third term: $a_3 = a_2 + d = (a_1 + d) + d = a_1 + 2d$
Fourth term: $a_4 = a_3 + d = (a_1 + 2d) + d = a_1 + 3d$

and so on. This pattern leads to a formula for the nth term which is given by

$$a_n = a_1 + (n - 1)d,\ n \geq 1$$

EXAMPLE 1   Finding the nth Term of a Sequence

Find the nth term of the arithmetic sequence 7, 10, 13, 16, ... then find a_{10} and a_{20}.

Solution   The first term is 7, and the common difference is

$$d = a_2 - a_1 = 10 - 7 = 3$$

Using the formula for the nth term a_n, we have

$$
\begin{aligned}
a_n &= a_1 + (n - 1)d \\
&= 7 + (n - 1)3 \qquad \text{Replace } a_1 \text{ by 7, } d \text{ by 3} \\
&= 7 + 3n - 3 \\
&= 4 + 3n \qquad\qquad \text{Simplify}
\end{aligned}
$$

Thus, the nth term is $a_n = 4 + 3n$, so that

$$
\begin{aligned}
a_{10} &= 4 + 3\,(10) \\
&= 4 + 30 = 34
\end{aligned}
$$

and

$$
\begin{aligned}
a_{20} &= 4 + 3\,(20) \\
&= 4 + 60 = 64
\end{aligned}
$$

EXAMPLE 2   Finding the Number of Terms of a Sequence

Find the number of terms of the finite arithmetic sequence

$$-9, -2, 5, 12, 19, \ldots, 96$$

Solution   Here $a_1 = -9$, $d = -2 - (-9) = 7$ and $a_n = 96$.
Substituting these values into the formula for a_n, we have

$$
\begin{aligned}
a_n &= a_1 + (n - 1)d \\
96 &= -9 + (n - 1)7 \\
96 &= -9 + 7n - 7 \\
96 &= -16 + 7n \\
112 &= 7n \\
16 &= n
\end{aligned}
$$

Therefore, there are 16 terms of this sequence.

EXAMPLE 3   Finding a Specific Term of a Sequence

Find the 50th term of the arithmetic sequence whose common difference is 4 and whose 10th term is 37.

Solution   Using $a_n = a_1 + (n - 1)d$, with $d = 4$, $a_{10} = 37$, and $n = 10$, we have

$$37 = a_1 + (10 - 1)4$$
$$37 = a_1 + 36$$
$$1 = a_1$$

Therefore,

$$a_{50} = a_1 + (n - 1)d$$
$$= 1 + (50 - 1)4$$
$$= 1 + 196 = 197$$

◈

Finding the Sum of an Arithmetic Sequence

Associated with any arithmetic sequence, with common difference d,

$$a_1, a_2, a_3, \ldots, a_n$$

is an **arithmetic series** of the form

$$S_n = \sum_{k=1}^{n} a_k = a_1 + a_2 + a_3 + \ldots + a_n$$

Suppose we are to find the value of the sum S_n of the first n terms of this sequence.

Using Gauss's pairing method in section 9.2 leads us to a general formula for the sum of the first n terms of an arithmetic sequence. This is accomplished by adding the first and last terms and multiplying by half the number of terms. That is, the sum S_n of the first n terms of an arithmetic sequence is given by

$$S_n = \sum_{k=1}^{n} a_k = \frac{n}{2}(a_1 + a_n)$$

Using the formula $a_n = a_1 + (n - 1)d$ for the nth term, we find the following alternative form for the sum S_n:

$$S_n = \frac{n}{2}[2a_1 + (n - 1)d]$$

EXAMPLE 4   Finding the Sum of an Arithmetic Sequence

Find the sum of the first 20 terms of the arithmetic sequence

$$2, 6, 10, 14, 18, \ldots$$

Solution   In this sequence, the first term is $a_1 = 2$ and the common difference is $d = 6 - 2 = 4$. Since we want to find the sum of the first 20 terms, we have $n = 20$. Substituting these values into the alternative form for the sum, we get

$$S_n = \frac{n}{2}[2a_1 + (n - 1)d]$$
$$= \frac{20}{2}[2(2) + (20 - 1)4]$$
$$= 10(4 + 76) = 800$$

◈

EXAMPLE 5 Finding the Sum of an Arithmetic Sequence

Find the sum of the terms of the finite arithmetic sequence

$$2, 5, 8, 11, 14, 17, 20, \ldots, 86$$

Solution The first term of this sequence is $a_1 = 2$, and the common difference is $d = 5 - 2 = 3$. We need to find the number of terms in the sequence such that $a_n = 86$. We substitute these values into the formula for a_n to get

$$a_n = a_1 + (n - 1)d$$
$$86 = 2 + (n - 1)3$$
$$84 = (n - 1)3$$
$$28 = n - 1$$
$$29 = n$$

Thus, the sequence contains 29 terms. Using the formula for the sum of an arithmetic sequence, we have

$$S_n = \frac{n}{2}(a_1 + a_n)$$
$$= \frac{29}{2}(2 + 86)$$
$$= 1{,}276$$

Thus, the sum of this sequence is 1,276.

Solving Applied Problems

Many applied problems have solutions leading to various arithmetic sequences and series as the next example shows.

EXAMPLE 6 Finding the Number of Seats in an Auditorium

A concert theater has 25 rows. The first row has 40 seats, the second row has 42 seats, the third row has 44 seats, and so on in an arithmetic sequence. Use a series to find the total number of seats in this auditorium.

Solution Here $a_1 = 40$, $d = 2$, and $n = 25$. Substituting these values in the formula for a_n we get

$$a_n = a_1 + (n - 1)d$$
$$a_{25} = 40 + (25 - 1)(2)$$
$$= 40 + 48 = 88$$

The total number of seats is obtained from the formula

$$S_n = \frac{n}{2}(a_1 + a_n)$$
$$= \frac{25}{2}(40 + 88) \qquad \text{Substitute 88 for } a_n$$
$$= 1600$$

Therefore, there are 1600 seats in this auditorium.

Using Technology Exploration

Recall that the graph of a sequence

$$a_1, a_2, a_3, \ldots, a_n$$

is the graph of the ordered pairs

$$(1, a_1), (2, a_2), (3, a_3), \ldots, (n, a_n)$$

The graph of an arithmetic sequence can give us additional information about its characteristics and properties. If we know the general term of an arithmetic sequence, most graphers will graph the sequence and give us any term we wish. Also many graphers use the summation notation to find the sum of an arithmetic sequence.

G EXAMPLE 7   Using a Grapher to Graph an Arithmetic Sequence
Use a grapher to graph the first five ordered pairs of the sequence

$$11, 8, 5, 2, -1, \ldots$$

Compare the graph of this sequence to the graph of a linear function.

Solution   The first term is $a_1 = 11$ and $d = 8 - 11 = -3$, so the nth term of this sequence is given by

$$a_n = a_1 + (n - 1)d$$
$$= 11 + (n - 1)(-3) = 11 - 3n + 3$$
$$= 14 - 3n$$

The first five ordered pairs are

$$(1, 11), (2, 8), (3, 5), (4, 2), \text{ and } (5, -1).$$

Figure 1a shows the graph of this sequence. The graph suggests that the graph of an arithmetic sequence is a set of collinear points which corresponds to the graph of the linear function.

$$f(x) = 14 - 3x \text{ (Figure 1b)}$$

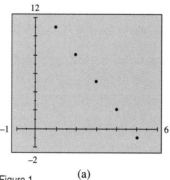

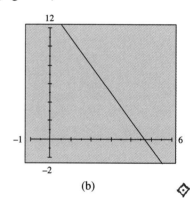

Figure 1   (a)   (b)

◈ **PROBLEM SET 9.3**

Mastering the Concepts

In Problems 1–10, find (a) a_n (b) a_{10} and (c) a_{20} in each arithmetic sequence.
1. $-2, 1, 4, 7, \ldots$
2. $-9, -5, -1, 3, \ldots$
3. $11, 8, 5, 2, \ldots$
4. $25, 16, 7, -2, \ldots$
5. $2, 5, 8, 11, \ldots$
6. $7, 12, 17, 22, \ldots$
7. $5, -1, -7, -13, \ldots$
8. $3, -1, -5, -9, \ldots$
9. $-9, -6, -3, 0, \ldots$
10. $1, -1, -3, -5, \ldots$

In Problems 11–16, find the number of terms of each finite sequence.

11. $5, 10, 15, 20, \ldots, 125$
12. $5, 2, -1, -4, \ldots, -55$
13. $-4, -1, 2, 5, 8, \ldots, 83$
14. $2, 6, 10, 14, \ldots, 118$
15. $-7, -2, 3, 8, 13, \ldots, 143$
16. $-6, -3, 0, 3, \ldots, 114$

In Problems 17–32, find the indicated term of each arithmetic sequence with the given information.

17. a_{10} if $a_4 = 8$ and $d = 7$
18. a_{30} if $a_3 = 15$ and $d = -2$
19. a_6 if $a_5 = 16$ and $d = 14$
20. a_{15} if $a_2 = 8$ and $d = -6$
21. a_{12} if $a_1 = 15$ and $d = -8$
22. a_{30} if $a_2 = 13$ and $d = 4$
23. a_{10} if $a_1 = 9$ and $d = -4$
24. a_{25} if $a_1 = 3$ and $d = -0.5$

25. a_1 and a_{20} if $a_5 = 8$ and $a_{10} = 23$
26. a_1 and a_{20} if $a_{10} = -26$ and $a_{14} = -7$
27. a_1 and a_{15} if $a_5 = 2$ and $a_{11} = -10$
28. a_1 and a_{20} if $a_{10} = 3$ and $a_{16} = 21$
29. a_1 and a_{20} if $a_{15} = 16$ and $a_{19} = -12$
30. a_1 and a_{50} if $a_{10} = -11$ and $a_{40} = -71$
31. a_1 and a_{40} if $a_5 = 3$ and $a_{50} = 30$
32. a_1 and a_{15} if $a_{10} = -37$ and $a_{12} = -45$
33. Find the sum of the first 10 terms of an arithmetic sequence whose first term is 1 and whose common difference is 3.
34. Find the sum of the first 15 terms of an arithmetic sequence whose first term is $\frac{1}{2}$ and whose common difference is $\frac{1}{2}$.
35. Find the sum of the first 8 terms of an arithmetic sequence whose first term is -5 and whose common difference is $\frac{3}{7}$.
36. Find the sum of the first 12 terms of an arithmetic sequence whose first term is 11 and whose common difference is -2.

In problems 37–44, find the sum of the terms of each finite arithmetic sequence.

37. $4, 10, 16, 22, \ldots, 184$
38. $7, 19, 31, 43, \ldots, 247$
39. $-12, -3, 6, 15, \ldots, 123$
40. $-15, -7, 1, 9, \ldots, 225$
41. $100, 95, 90, 85, \ldots, 10$
42. $4, 8, 12, 16, \ldots, 92$
43. $7, 11, 15, 19, \ldots, 127$
44. $120, 116, 112, 108, \ldots, 12$

Applying the Concepts

45. **Bricklayer's Pattern:** A bricklayer wishes to arrange 376 bricks in a pile. The bricks were stacked so that there will be 1 brick in the top row, 4 bricks in the second row, 7 in the third row, and so on in an arithmetic sequence. Can this be done? If so, (a) how many rows will it take? (b) and how many bricks will be in the bottom row?

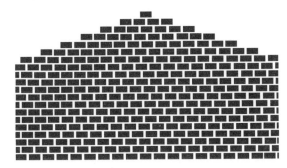

46. **Advertising:** During a special sales promotion, an automobile salesperson received a bonus of $100 for the first car sold, $105 for the second car sold, $110 for the third car sold, and so on in an arithmetic sequence. Find (a) the bonus received for the 30th car sold, and (b) the total bonus received for selling x cars.

47. **Accumulated Earnings:** A college professor started working in 1978 in a teaching job that started at an annual salary of $25,600 and each year received a $1400 pay raise. (a) How much would she be paid during the 1998 school year? (b) What would be the total salary paid from 1978 through 2002?

48. **Accumulated Savings:** If a newspaper carrier saved $1 on September 1, another $2 on September 2, another $3 on September 3, and so on, how much money was saved during the month of September (30 days)?

49. **Drilling a Well:** A well-drilling contractor charges $1.50 to drill the first foot, $1.65 for the second foot, $1.80 for the third foot, and so on in an arithmetic sequence. At this rate find (a) the cost to drill the last foot of a well 200 feet deep, and (b) the total cost to drill the well.

50. **Pollution Standards:** A company is informed that if it fails to meet EPA pollution standards by a fixed date, it is to be fined $2,000 the first day, and each day thereafter the fine would be increased by $400. What is the fine for a company that fails to meet the standards for 10 days after the deadline?

Developing and Extending the Concepts

51. Find the sixth term of the arithmetic sequence
$$a + 24b, 4a + 20b, 7a + 16b, \ldots$$

52. Find the sixteenth term of the arithmetic sequence
$$7a^2 - 4b, 2a^2 + 7b, -3a^2 + 18b, \ldots$$

53. Given an arithmetic sequence with $a_1 = -2$, $d = 7$ and $a_n = 138$, find n.

54. Given an arithmetic sequence with $a_3 = 25$, $d = -14$ and $a_n = -507$, find n.

G In Problems 55–58, use a grapher to graph the first five ordered pairs of each arithmetic sequence. Compare the graph of each sequence to a graph of a corresponding linear function.

55. $23, 17, 11, 5, -1, \ldots$ 56. $10, 13, 16, 19, 22, \ldots$
57. $-3, 0, 3, 6, 9, \ldots$ 58. $-10, -6, -2, 2, \ldots$

59. (a) Find the general terms a_n and b_n of each arithmetic sequence
$$2, 5, 8, 11, \ldots \text{ and } 7, 9, 11, 13, \ldots$$
    (b) What can you say about the sequence whose terms are the sum of the corresponding terms of the sequences in part (a)?
    (c) How about the sequence whose general term corresponds to the product of the corresponding terms of the sequences in part (a)?

60. Consider the sequence a_n defined by $a_n = n^2$. Suppose that b_n is another sequence defined by
$$b_n = a_n - a_{n-1}$$
show that $b_1, b_2, b_3, \ldots$ is an arithmetic sequence.

OBJECTIVES

1. Define a Geometric Sequence
2. Find a Sum of a Geometric Sequence
3. Evaluate Infinite Geometric Series
4. Solve Applied Problem
5. Use Technology Exploration

9.4 Geometric Sequences and Series

In this section we focus on another important sequence called a *geometric sequence*, where each term is determined by multiplying the previous term, except the first, by a fixed number.

Defining a Geometric Sequence

Consider the sequence

$$3, 6, 12, 24, 48, \ldots$$

Each term of this sequence, except the first, is 2 times the previous one. That is, the ratio of consecutive terms is 2.

$$\frac{a_2}{a_1} = \frac{6}{3} = 2$$

$$\frac{a_3}{a_2} = \frac{12}{6} = 2$$

$$\frac{a_4}{a_3} = \frac{24}{12} = 2$$

and so on.

This sequence is an example of a *geometric sequence*, which is characterized by the feature that the ratio of consecutive terms is constant.

More precisely, we have the following definition:

GEOMETRIC SEQUENCE

A sequence of the form

$$a_1, a_2, a_3, \ldots, a_n, \ldots$$

is called a **geometric sequence** if there is a constant r, called the **common ratio** such that

$$\frac{a_{n+1}}{a_n} = r,$$

for n, a positive integer.

EXAMPLE 1   Finding the Common Ratio of a Geometric Sequence

Find the common ratio of the sequence

$$-4, 16, -64, 256, \ldots$$

Solution   To find the common ratio r of this sequence, choose any two adjacent terms and divide the second by the first. Choosing the second and the third terms of this sequence, we have $r = (-64)/16 = -4$

Therefore, the common ratio is -4.   ◈

Each term of a geometric sequence can be expressed in terms of the first term a_1 and the common ratio r as follows

First term:	$a_1 = a_1$
Second term:	$a_2 = a_1 r$
Third term:	$a_3 = a_2 r = (a_1 r)r = a_1 r^2$
Fourth term:	$a_4 = a_3 r = (a_1 r^2)r = a_1 r^3$
Fifth term:	$a_5 = a_4 r = (a_1 r^3)r = a_1 r^4$
nth term:	$a_n = a_{n-1} r = (a_1 r^{n-2})r = a_1 r^{n-1}$

which suggests the following formula for the general term a_n:

$$a_n = a_1 r^{n-1}$$

EXAMPLE 2 Finding the nth Term of a Geometric Sequence

Find the nth term of each geometric sequence.

(a) $3, 6, 9, 12, \ldots$ (b) $-32, 16, -8, 4, \ldots$

Solution (a) The first term is $a_1 = 3$ and the common ratio $r = \frac{6}{3} = 2$. Using the formula for the nth term of a geometric sequence, we get

$a_n = a_1 r^{n-1}$

$a_n = 3(2)^{n-1}$ Replace a_1 by 3 and r by 2

(b) Here $a_1 = -32$ and $r = \frac{16}{-32} = -\frac{1}{2}$. Using the formula for the nth term, we have

$a_n = a_1 r^{n-1}$

$a_n = -32\left(-\frac{1}{2}\right)^{n-1}$ Replace a_1 by -32 and r by $-\frac{1}{2}$

EXAMPLE 3 Finding a Specific Term of a Geometric Sequence

Suppose that the first two terms of a geometric sequence are 2 and 4, respectively. Find (a) the term of the sequence that is equal to 512 (b) the fifteen term a_{15}.

Solution (a) Since the first two terms of the sequence are: $a_1 = 2$ and $a_2 = 4$, we conclude that $r = 2$.

If $a_n = 512,$ then $512 = 2\,(2^{n-1})$

or $2^9 = 2^{1+n-1} = 2^n$

so that $n = 9$

Therefore, the ninth term of the sequence is 512.

(b) The general term of this sequence is given by

$a_n = 2(2^{n-1}) = 2^n$ Replace n by 15

$a_{15} = 2^{15}$ $= 32,768$

Finding a Sum of a Geometric Series

Consider the geometric sequence

$$a_1, a_1 r, a_1 r^2, a_1 r^3, \ldots, a_1 r^{n-1}$$

with first term a_1 and common ratio r.
The associated sum

$$S_n = a_1 + a_1 r + a_1 r^2 + a_1 r^3 + \ldots + a_1 r^{n-1}$$

is called a **geometric series**. The value of S_n is the sum of n terms of the sequence.

For example, consider the sequence

$$5, 15, 45, 135, 405, \ldots, 5(3^{n-1})$$

The sum of the first five terms of this sequence is given by

$S_1 = 5$

$S_2 = 5 + 15 = 20$

$S_3 = 5 + 15 + 45 = 65$

$S_4 = 5 + 15 + 45 + 135 = 200$

$S_5 = 5 + 15 + 45 + 135 + 405 = 605$

Following this pattern, the sum S_n is given by

$$S_n = \frac{5}{2}(3^n - 1).$$

Notice that the sequence and its sum were based on powers of 3. This suggests that there is a formula for finding the sum of the terms of a geometric sequence based on the powers of r.

SUM OF GEOMETRIC SEQUENCE

> The **sum** S_n of the first n terms of a geometric sequence with first term a_1 and common ratio r is given by
> $$S_n = \frac{a_1(1 - r^n)}{1 - r}, r \neq 1$$

EXAMPLE 4 Finding the Sum of a Geometric Sequence

Find the sum of the first 10 terms of the geometric sequence whose first term is $\frac{1}{2}$ and whose common ratio is 2.

Solution   We substitute $n = 10$, $a_1 = \frac{1}{2}$, and $r = 2$ in the formula
$$S_n = \frac{a_1(1 - r^n)}{1 - r}, r \neq 1$$

and obtain
$$S_{10} = \frac{\frac{1}{2}(1 - 2^{10})}{1 - 2} = \frac{\frac{1}{2}(-1{,}023)}{-1} = 511.5$$

EXAMPLE 5 Finding the First Five Terms of a Geometric Series

If the sum of the first five terms of a geometric series is $\frac{61}{27}$, find the first five terms of this sequence that has a common ratio $r = -\frac{1}{3}$.

Solution   Substituting $S_n = \frac{61}{27}$ and $r = -\frac{1}{3}$ in the formula for the sum, we get

$$S_n = \frac{a_1(1 - r^n)}{1 - r}$$

$$\frac{61}{27} = \frac{a_1\left[1 - \left(-\frac{1}{3}\right)^5\right]}{1 - \left(-\frac{1}{3}\right)} \qquad \text{Replace } r \text{ by } -\frac{1}{3} \text{ and } n \text{ by } 5$$

$$\frac{61}{27} = \frac{\frac{244}{243}}{\frac{4}{3}} a_1$$

$$\frac{61}{27} = \frac{61}{81} a_1$$

$$a_1 = 3$$

The first five terms are $3, -1, \frac{1}{3}, -\frac{1}{9}, \frac{1}{27}$.

Evaluating an Infinite Geometric Series

Consider the geometric series
$$1 + \frac{1}{2} + \frac{1}{2^2} + \frac{1}{2^3} + \cdots + \frac{1}{2^{n+1}} + \cdots$$

Here $a_1 = 1$ and $r = \frac{1}{2}$, so the sum S_n of the first n terms is given by
$$S_n = \frac{a_1(1 - r^n)}{1 - r} = \frac{1(1 - (\frac{1}{2})^n)}{1 - \frac{1}{2}} = 2\left[1 - \left(\frac{1}{2}\right)^n\right]$$
$$= 2 - \frac{2}{2^n} = 2 - \frac{1}{2^{n-1}}$$

Now, as n gets larger and larger, the expression $(1/2^{n-1})$ gets closer and closer to zero. In fact, by taking n large enough, $(1/2^{n-1})$ can be made as close to zero as we please. This leads to the conclusion that

$$\frac{1 - (\frac{1}{2})^n}{1 - \frac{1}{2}} = 2 \text{ as } n \text{ gets large enough.}$$

Similar reasoning works for obtaining the sum of any infinite geometric series with common ratio r for which $|r| < 1$.

SUM OF AN INFINITE
GEOMETRIC SEQUENCE

If the common ratio r is between -1 and 1, the **sum** S of an infinite number of terms of a geometric sequence with first term a_1 is given by

$$S = \frac{a_1}{1 - r}$$

EXAMPLE 6   Finding the Sum of an Infinite Geometric Sequence
Find the indicated sum of infinite number of terms for each geometric sequence.

(a) $\dfrac{3}{10} + \dfrac{3}{100} + \dfrac{3}{1,000} + \dfrac{3}{10,000} + \ldots$ (b) $10 - 4 + 1.6 - 0.64 + \ldots$

Solution   (a) The common ratio

$$r = \frac{\frac{3}{100}}{\frac{3}{10}} = \frac{1}{10}$$

and

$$a_1 = \frac{3}{10}$$

so that

$$S = \frac{a_1}{1 - r} = \frac{\frac{3}{10}}{1 - \frac{1}{10}}$$

$$= \frac{\frac{3}{10}}{\frac{9}{10}} = \frac{3}{9}$$

$$= \frac{1}{3}$$

(b) Here

$$r = \tfrac{-4}{10} = \tfrac{-2}{5} \text{ and } a_1 = 10.$$

So

$$S = \frac{a_1}{1 - r} = \frac{10}{1 - (-\frac{2}{5})}$$

$$= \frac{10}{\frac{7}{5}} = \frac{50}{7}$$

Solving Applied Problems

Geometric sequences and series are often used to solve applied problems from different fields as the next example shows.

EXAMPLE 7   Finding the Distance of a Child Swing

A child swing starts with an initial distance of 18 feet and goes $\frac{7}{9}$ of the prior distance on each subsequent swing.
(a) How far does the swing travel through the nth swing?
(b) What is the total distance traveled through the sixth swing?
(c) How far does the swing travel before eventually coming to rest?

Solution   (a) The child swing travels 18 feet through the first swing, $18(\frac{7}{9})$ feet through the second swing, $18(\frac{7}{9})^2$ feet through the third swing, and so on. The distance the swing travels during the nth swing is given by

$$a_n = 18\left(\frac{7}{9}\right)^{n-1}$$

(b) The total distance traveled through the sixth swing is given by

$$S_n = \frac{a_1(1 - r^n)}{1 - r}$$

$$S_6 = \frac{18(1 - (\frac{7}{9})^6)}{1 - \frac{7}{9}} \qquad \text{Replace } n \text{ by 6}$$

$$= \frac{18\left[\frac{413,792}{531,441}\right]}{\frac{2}{9}}$$

$$= 63.07$$

(c) The total distance traveled before eventually coming to rest is given by

$$S = \frac{a_1}{1 - r}$$

$$= \frac{18}{1 - \frac{7}{9}} \qquad \text{Replace } a_1 \text{ by 18 and } r \text{ by } \frac{7}{9}$$

$$= \frac{18}{\frac{2}{9}}$$

$$= 81$$

The swing travels 81 feet before it comes to rest.

Using Technology Exploration

In section 9.3 we noted that an arithmetic sequence corresponds to a linear function. By graphing a geometric sequence on a grapher, we can see that geometric sequences correspond to exponential functions.

G EXAMPLE 8   Using a Grapher

Use a grapher to graph the first five ordered pairs and the corresponding exponential function on different viewing windows.

$$a_n = \left(\frac{2}{3}\right)^n \text{ and } f(x) = \left(\frac{2}{3}\right)^x, x \geq 1$$

Solution Substituting 1, 2, 3, 4 and 5 for n into $a_n = \left(\frac{2}{3}\right)^n$ we get

$$a_1 = \left(\frac{2}{3}\right)^1 = \frac{2}{3} \qquad a_2 = \left(\frac{2}{3}\right)^2 = \frac{4}{9} \qquad a_3 = \left(\frac{2}{3}\right)^3 = \frac{8}{27}$$

$$a_4 = \left(\frac{2}{3}\right)^4 = \frac{16}{81} \qquad a_5 = \left(\frac{2}{3}\right)^5 = \frac{32}{243}$$

The first five ordered pairs of this sequence are given by

$$\left(1, \frac{2}{3}\right), \left(2, \frac{4}{9}\right), \left(3, \frac{8}{27}\right), \left(4, \frac{16}{81}\right), \text{ and } \left(5, \frac{32}{243}\right).$$

Figure 1a shows the plot of these points and Figure 1b shows the graph of $f(x) = \left(\frac{2}{3}\right)^x$.

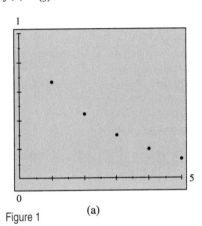

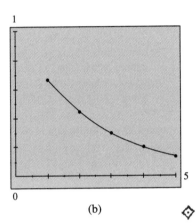

Figure 1
(a)
(b)

PROBLEM SET 9.4

Mastering the Concepts

In Problems 1–14, determine whether each sequence is arithmetic, geometric or neither. If it is geometric find the common ratio.

1. $2, 6, 18, 54, \ldots$
2. $1, -2, 4, -8, \ldots$
3. $81, 54, 36, 24, \ldots$
4. $2, 4, 8, 16, \ldots$
5. $9, -6, 4, -\frac{8}{3}, \ldots$
6. $3, 9, 27, 81, \ldots$
7. $64, -32, 16, -8, \ldots$
8. $18, -6, 2, \frac{2}{3}, \ldots$
9. $1, \frac{1}{5}, \frac{1}{25}, \frac{1}{125}, \ldots$
10. $\frac{4}{9}, \frac{1}{6}, \frac{1}{16}, \frac{1}{32}, \ldots$
11. $9, 3, -3, -9, \ldots$
12. $1, 4, 9, 16, \ldots$
13. $7, 19, 31, 43, \ldots$
14. $-15, -7, 1, 9, \ldots$

In Problems 15–26, find (a) the general term a_n, and (b) the eighth term a_8 for each geometric sequence.

15. $-4, 2, -1, \frac{1}{2}, \ldots$
16. $\frac{1}{8}, \frac{1}{4}, \frac{1}{2}, 1, \ldots$
17. $32, 16, 8, 4, \ldots$
18. $1, 1.03, (1.03)^2, (1.03)^3, \ldots$
19. $6, 12, 24, 48, \ldots$
20. $3, 6, 12, 24, \ldots$

21. $5, 10, 20, 40, \ldots$
22. $\frac{1}{4}, -8, 256, -8192, \ldots$
23. $\frac{1}{2}, -2, 8, -32, \ldots$
24. $\frac{1}{3}, -1, 3, -9, \ldots$
25. $8, 4, 2, 1, \ldots$
26. $-1, -\frac{3}{2}, -\frac{9}{4}, -\frac{27}{8}, \ldots$

27. If the first two terms of a geometric sequence are 2 and 1 respectively, which term of the sequence is 1/16?
28. If the first two terms of a geometric sequence are $3\sqrt{3}$ and 9, respectively, which term of the sequence is $243\sqrt{3}$?
29. If the third term of a geometric sequence is 9/4 and the sixth term is 243/256. What is its first term?
30. The sixth term of a geometric sequence is 27, and the common ratio is −1/3. Find the third term.
31. Find the sum of the first 6 terms of the geometric sequence whose first term is 3/2 and whose common ratio is 2.
32. Find the sum of the first 10 terms of the geometric sequence whose first term is 6 and whose common ratio is 1/2.
33. Find the sum of the first 12 terms of the geometric sequence whose first term is −4 and whose common ratio is −2.

34. Find the sum of the first 8 terms of the geometric sequence whose first term is 5 and whose common ratio is $-1/2$.

35. Find the sum of the first 8 terms of the geometric sequence
$$\frac{1}{2}, \frac{3}{2}, \frac{9}{2}, \frac{27}{2}, \cdots$$

36. Find the sum of the first 5 terms of the geometric sequence
$$-\frac{1}{24}, \frac{1}{12}, -\frac{1}{6}, \frac{1}{3}, \cdots$$

37. Find the sum of the first 7 terms of the geometric sequence
$$16, 4, 1, \frac{1}{4}, \cdots$$

38. Find the sum of the first 5 terms of the geometric sequence
$$3, 12, 48, 192, \ldots$$

39. The sum of the first 5 terms of a geometric sequence is 31 and the common ratio is 2. Find the first four terms of the sequence.

40. The sum of the first 4 terms of a geometric sequence is 20 and the common ratio is -2. Find the first five terms of the sequence.

In problems 41–48, find the sum of each infinite geometric series.

41. $1 + \frac{1}{2} + \frac{1}{4} + \frac{1}{8} + \ldots$ 42. $3 + 1 + \frac{1}{3} + \frac{1}{9} + \ldots$

43. $8 + 4 + 2 + 1 + \ldots$ 44. $1 + \frac{2}{3} + \frac{4}{9} + \frac{8}{27} + \ldots$

45. $54 - 18 + 6 - 2 + \ldots$ 46. $4 - 2 + 1 - \frac{1}{2} + \ldots$

47. $\frac{4}{5} + \frac{8}{15} + \frac{16}{45} + \frac{32}{135} + \ldots$

48. $\frac{4}{3} + \frac{2}{9} + \frac{1}{27} + \ldots$

Applying the Concepts

49. **Monthly Earning:** A firm offers an employee a starting salary of $24,000 with 5% annual increases for 6 years.
    (a) List the annual salaries during each of the 6 years as a sequence.
    (b) Does the annual salary during the fourth year of employment represent a 15% raise over the original salary? Explain.
    (c) Suppose that 5% annual raise continued for 10 years. Determine the annual salary during the tenth year of employment.

50. **Depreciation:** For tax purposes, a company depreciates the value of a company car by 20% every year. The car was originally purchased for $32,000.
    (a) List the depreciated values of the car during each of the first 4 years as a sequence.
    (b) Does the depreciated value of the car during the fourth year represent a 60% decrease? Explain.
    (c) Determine the depreciated value of the car during the seventh year.

51. **Rebound Distance:** A super ball dropped from the top of the Eiffel Tower (300 meters high) rebounds two thirds of the distance fallen. (Figure 2)

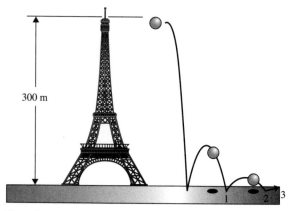

Figure 2

    (a) How far (up and down) will the ball have traveled when it hits the ground for the fifth time?
    (b) How far will it travel before it comes to rest?

52. **Investment:** At the beginning of each month $1,000 is deposited into an annuity for a college professor that pays 7% compounded monthly. Determine the balance of this investment after
    (a) 2 years   (b) 5 years   (c) 10 years.

53. **Orange Grower:** An orange grower receives $10,000 for the oranges he sold for the first year of his young trees. If each succeeding year his income increases 10%. What would be the grower's total income over the first 8 years?

54. **Pendulum Swings:** The tip of a pendulum swings through an arc of 8 inches and each arc thereafter is 11/13 of the length of the preceding one (Figure 3). How far does the tip move before the pendulum comes to rest?

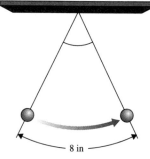

Figure 3

Developing and Extending the Concepts

55. If the first term of a geometric sequence is 1/16, the nth term is 32 and the common ratio is 2, find the number of terms n.

56. If the first term of a geometric sequence is 250, the nth term is 32 (2/5), and the common ratio is 3/5, find the sum of the first n terms.

57. If the sum of the first n terms of an infinite geometric sequence is 35 and the first term a_1 is 7, find the common ratio r.

58. If the sum of the first n terms S of an infinite geometric sequence is 20, and the common ratio r is $-1/2$, find the first term a_1.
59. The repeating decimal $0.\overline{7}$ can be written as the infinite geometric series $0.7 + 0.07 + 0.007 + 0.0007 + \ldots$, with $a_1 = 0.7$ and $r = 0.1$. Use the sum formula to show that the infinite sum is $7/9$. Note that this is another way to convert $0.\overline{7}$ to a fraction in the form a/b.
60. The repeating decimal $0.\overline{14}$ can be written as the infinite geometric series as $0.14 + 0.0014 + 0.000014 + 0.00000014 + \ldots$ with $a_1 = 0.14$ and $r = 0.01$. Use the formula for the sum of an infinite series to convert $0.\overline{14}$ to a fraction.

61. If $16, x, 9/4$ is a geometric sequence, how can we find the value of x?
62. Does there exist an infinite geometric series for which $a_1 = 2$ and its sum $S = 1$? Give examples to justify your answer.

G In Problems 63–66, use a grapher to graph the first five ordered pairs of the sequence and the corresponding exponential function on separate viewing windows, $x \geq 1$.

63. $a_n = 2^n, f(x) = 2^x$
64. $a_n = \dfrac{1}{2^n}, f(x) = \dfrac{1}{2^x}$
65. $a_n = \left(\dfrac{3}{2}\right)^n, f(x) = \left(\dfrac{3}{2}\right)^x$
66. $a_n = (1.1)^n, f(x) = (1.1)^x$

OBJECTIVES

1. Compute Binomial Coefficients by Pascal's Triangle
2. Compute Binomial Coefficients Using Factorials
3. Expand Binomials Using the Binomial Theorem
4. Find a Specific Term in a Binomial Expansion
5. Solve Applied Problems
6. Use Technology Exploration

9.5 The Binomial Theorem

In many applied problems in statistics, calculus, and other branches of mathematics, we are required to write expressions such as $(a + b)^n$ as a sum, which we call the **expanded form** of $(a + b)^n$. In this section we will develop a formula that enables us to accomplish this objective.

Computing Binomial Coefficients by Pascal's Triangle

In section 2.4 we considered the special products $(a + b)^2$ and $(a + b)^3$. We often work with expressions of the form $(a + b)^n$, where n is a positive integer. Since the expression $a + b$ is a binomial, the formula for expanding $(a + b)^n$ is called the **binomial theorem**.

We can expand $(a + b)^n$, for small values of n, by using direct calculations. For instance,

$$(a + b)^1 = a + b$$
$$(a + b)^2 = a^2 + 2ab + b^2$$
$$(a + b)^3 = a^3 + 3a^2b + 3ab^2 + b^3$$
$$(a + b)^4 = a^4 + 4a^3b + 6a^2b^2 + 4ab^3 + b^4$$
$$(a + b)^5 = a^5 + 5a^4b + 10a^3b^2 + 10a^2b^3 + 5ab^4 + b^5$$

The pattern holds for the expansion of $(a + b)^n$, where n is any positive integer. The following rules are used for this expansion:

1. There are $n + 1$ terms.
2. The first term is a^n and the last term is b^n.
3. The powers of a decrease by 1 and the powers of b increase by 1 for each term.
4. The sum of the exponents of a and b is n for each term.
5. Each term has a numerical coefficient that must be determined.

One way to display the coefficients in the expansion of $(a + b)^n$ for $n = 1, 2, 3, \ldots$ is the following array of numbers (Figure 1), known as **Pascal's triangle**, named in honor of the French mathematician Blaise Pascal (1623–1662):

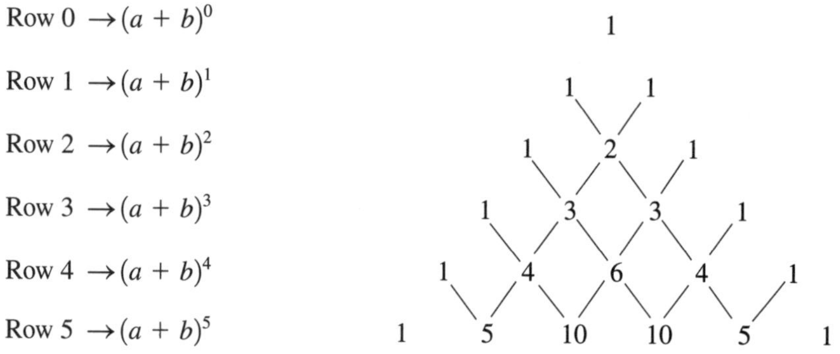

Row 0 $\rightarrow (a + b)^0$

Row 1 $\rightarrow (a + b)^1$

Row 2 $\rightarrow (a + b)^2$

Row 3 $\rightarrow (a + b)^3$

Row 4 $\rightarrow (a + b)^4$

Row 5 $\rightarrow (a + b)^5$

```
            1
          1   1
        1   2   1
      1   3   3   1
    1   4   6   4   1
  1   5   10   10   5   1
```

Figure 1

The first and last numbers in each row are always 1. The other numbers can be found by adding the pair of numbers from the preceding row as indicated by the V's. For example,

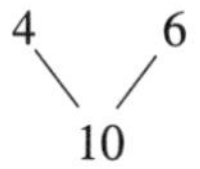

indicates that 10 was obtained by adding 4 and 6.

The advantage of Pascal's triangle is that it allows us to calculate the binomial coefficients using addition rather than multiplication. The disadvantage, however, is that it is impractical for large powers, since it requires us to calculate all rows preceding the one in which we are interested.

EXAMPLE 1   Using Pascal's Triangle to Expand a Binomial

Use Pascal's triangle to expand $(3x + 2y)^5$.

Solution   The coefficients are given in Row 5 of Figure 1, and the sum of the exponents of each term is 5. So that

$$(3x + 2y)^5 = (3x)^5 + 5(3x)^4(2y) + 10(3x)^3(2y)^2 + 10(3x)^2(2y)^3 + 5(3x)(2y)^4 + (2y)^5$$
$$= 243x^5 + 810x^4y + 1080x^3y^2 + 720x^2y^3 + 240xy^4 + 32y^5 \quad \diamondsuit$$

Computing Binomial Coefficient Using Factorials

We indicated that the disadvantage in using Pascal's triangle is that it becomes cumbersome when a binomial contains a larger power. However, by using the idea of *factorial notation*, we can determine the binomial coefficient more efficiently. The symbol $n!$ (read "n factorial") is defined as follows.

$$n! = 1 \cdot 2 \cdot 3 \cdot 4 \cdot \ldots \cdot (n - 1)n$$
$$\text{or}$$
$$n! = n(n - 1)(n - 2)\ldots \cdot 2 \cdot 1$$

Here are some examples of the value of factorials.

$$1! = 1$$
$$2! = 2 \cdot 1 = 2$$
$$3! = 3 \cdot 2 \cdot 1 = 6$$
$$4! = 4 \cdot 3 \cdot 2 \cdot 1 = 24$$
$$5! = 5 \cdot 4 \cdot 3 \cdot 2 \cdot 1 = 120$$

We also define

$$0! = 1$$

Notice that $(n - 1)! = (n - 1)(n - 2) \ldots \cdot 3 \cdot 2 \cdot 1$
Therefore,

$$n! = n(n - 1)!$$
n is a positive integer

The following example shows how to simplify expressions involving factorial notation:

EXAMPLE 2   Simplifying Expressions Involving Factorials

Find the value of each expression.

(a) $\dfrac{7!}{5!}$ (b) $\dfrac{8!}{3!\,5!}$ (c) $\dfrac{(n+1)!}{(n-1)!}$

Solution   (a) $\dfrac{7!}{5!} = \dfrac{7 \cdot 6 \cdot 5 \cdot 4 \cdot 3 \cdot 2 \cdot 1}{5 \cdot 4 \cdot 3 \cdot 2 \cdot 1} = 7 \cdot 6 = 42$

(b) $\dfrac{8!}{3! \cdot 5!} = \dfrac{8 \cdot 7 \cdot 6 \cdot 5 \cdot 4 \cdot 3 \cdot 2 \cdot 1}{(3 \cdot 2 \cdot 1)(5 \cdot 4 \cdot 3 \cdot 2 \cdot 1)} = 8 \cdot 7 = 56$

(c) $\dfrac{(n+1)!}{(n-1)!} = \dfrac{(n+1)(n)[(n-1)!]}{(n-1)!} = (n+1)n = n^2 + n$

Now, we can express the coefficients in a binomial expansion in terms of factorials. The following notation is used in its formulation.

THE BINOMIAL COEFFICIENT

The expression $\dbinom{n}{k}$ is called a **binomial coefficient**, where

$$\dbinom{n}{k} = \dfrac{n!}{k!\,(n-k)!}, \; n \geq k$$

EXAMPLE 3   Evaluating Binomial coefficients

Evaluate each binomial coefficient.

(a) $\dbinom{5}{3}$ (b) $\dbinom{2}{2}$ (c) $\dbinom{3}{0}$

Solution   (a) $\dbinom{5}{3} = \dfrac{5!}{3!\,(5-3)!} = \dfrac{5!}{3!\,2!} = \dfrac{3! \cdot 4 \cdot 5}{3!\,2!} = \dfrac{20}{2!} = 10$

(b) $\dbinom{2}{2} = \dfrac{2!}{2!\,(2-2)!} = \dfrac{2!}{2!\,0!} = \dfrac{1}{0!} = \dfrac{1}{1} = 1$

(c) $\dbinom{3}{0} = \dfrac{3!}{0!\,(3-0)!} = \dfrac{3!}{0!\,3!} = \dfrac{1}{0!} = \dfrac{1}{1} = 1$

Expanding Binomials Using the Binomial Theorem

If we use factorials and the above pattern for the coefficients, we can use a formula called the *Binomial Theorem* to generalize the expansion of a binomial for any positive integer power.

THE BINOMIAL THEOREM

Let a and b be real numbers or expressions and n any positive integer. Then the **binomial expansion** of $(a + b)^n$ is given by

$$(a+b)^n = \binom{n}{0}a^n + \binom{n}{1}a^{n-1}b + \binom{n}{2}a^{n-2}b^2 + \ldots + \binom{n}{n-1}ab^{n-1} + \binom{n}{n}b^n$$

EXAMPLE 4   Using the Binomial Theorem

Expand the binomial $(x + 2y)^4$.

Solution   Using the Binomial Theorem with $a = x$, $b = 2y$ and $n = 4$, we have,

$$(x + 2y)^4 = \binom{4}{0}x^4 + \binom{4}{1}x^3(2y) + \binom{4}{2}x^2(2y)^2 + \binom{4}{3}x(2y)^3 + \binom{4}{4}(2y)^4$$

$$= \frac{4!}{0!(4-0)!}x^4 + \frac{4!}{1!(4-1)!}x^3(2y) + \frac{4!}{2!(4-2)!}x^2(2y)^2$$

$$+ \frac{4!}{3!(4-3)!}x(2y)^3 + \frac{4!}{4!(4-4)!}(2y)^4$$

$$= x^4 + 8x^3y + 24x^2y^2 + 32xy^3 + 16y^4 \qquad \diamond$$

Finding a Specific Term in a Binomial Expansion

Suppose we want to find a specific term, say the $(k + 1)^{st}$ term, in the binomial expansion of $(a + b)^n$. If we examine each term in the expansion of $(a + b)^n$, we can write

$$\text{The first term as } \binom{n}{0}a^nb^0$$

$$\text{The second term as } \binom{n}{1}a^{n-1}b^1$$

$$\text{The third term as } \binom{n}{2}a^{n-2}b^2$$

and so on. This can be generalized by using a formula for finding a *specific term* without writing the entire expansion.

FINDING A SPECIFIC TERM

The $(k + 1)^{st}$ term of the binomial expansion of $(a + b)^n$ is

$$\binom{n}{k}a^{n-k}b^k$$

EXAMPLE 5   Finding a Specific Term of a Binomial

Find and simplify the indicated term in each expansion.

(a) the fifth term of $(x - y)^8$

(b) the term involving x^7 in the expansion of $(2 - x)^{12}$.

Solution   (a) First we note that $5 = 4 + 1$. Thus, $k = 4$, $a = x$, $b = -y$ and $n = 8$. So the fifth term of the expansion is:

$$\binom{8}{4}x^{8-4}(-y)^4 = \binom{8}{4}x^4y^4$$

$$= \frac{8\,!}{4\,!\,(8-4)\,!}x^4y^4$$

$$= 70x^4y^4$$

(b) For the term x^7, $k = 7$. If $n = 12$, $a = 2$, and $b = -x$, then

$$\binom{n}{k}a^{n-k}b^k = \binom{12}{7}2^{12-7}(-x)^7$$

$$= -\frac{12\,!}{7\,!\,(12-7)\,!}2^5x^7 = -25{,}344x^7$$

Solving Applied Problems

Many important applications from different fields involve activities that require the use of binomial expansion.

EXAMPLE 6   Solving Compound Interest Problem

Suppose that $1000 is invested in an account paying 4% annual interest compounded annually. Use the binomial expansion $1000(1 + 0.04)^6$ to compute the balance at the end of the sixth year.

Solution   Substituting $P = 1000$, $r = 4\% = 0.04$, $n = 1$ and $t = 6$ in the compound interest formula to get

$$S = P\left(1 + \frac{r}{n}\right)^{nt}$$

$$= 1000(1 + 0.04)^6$$

$$= 1000[1 + 6(0.04) + 15(0.04)^2 + 20(0.04)^3 +$$

$$+ 15(0.04)^4 + 6(0.04)^5 + (0.04)^6]$$

$$= \$1265.32$$

Notice that the direct calculation of the expression $1000(1.04)^6$ will also result in $1265.32.

Using Technology Exploration

Most graphers are equipped with a key that is capable of evaluating expressions of the form $\binom{n}{r}$. On a grapher, the notation $_nC_r$ is used to represent $\binom{n}{r}$.

EXAMPLE 7   Using a Grapher to Evaluate $_nC_r$

G Use a grapher to evaluate

(a) $\binom{5}{2}$ (b) $\binom{6}{4}$

Solution   On a grapher,

(a) $\binom{5}{2} = {}_5C_2 = 10$ (b) $\binom{6}{4} = {}_6C_4 = 15$

PROBLEM SET 9.5

Mastering the Concepts

In Problems 1–6, use Pascal's triangle to expand each binomial expression.

1. $(x + 2y)^4$
2. $(2x - y)^5$
3. $(y + 1)^6$
4. $(x^2 + 3y)^5$
5. $(y^2 - 2x)^4$
6. $(2 + x/y)^6$

In Problems 7–16, write each expression in expanded form and simplify.

7. $\dfrac{4!}{6!}$
8. $\dfrac{10!}{5!7!}$
9. $\dfrac{2!}{4! - 3!}$
10. $\dfrac{1}{4!} + \dfrac{1}{3!}$
11. $\dfrac{3!8!}{4!7!}$
12. $\dfrac{4!6!}{8! - 5!}$
13. $\dfrac{0!}{0!}$
14. $\dfrac{(n - 2)!}{(n - 1)!}$
15. $\dfrac{(n + 1)!}{(n - 3)!}$
16. $\dfrac{(n + k)!}{(n + k - 2)!}$

In Problems 17–24, evaluate each binomial coefficient.

17. $\binom{8}{3}$
18. $\binom{7}{2}$
19. $\binom{10}{4}$
20. $\binom{5}{4}$
21. $\binom{8}{8}$
22. $\binom{7}{6}$
23. $\binom{12}{10}$
24. $\binom{16}{14}$

In Problems 25–32, expand each expression by using the binomial theorem and simplify each term.

25. $(x + 2)^5$
26. $(a - 2b)^4$
27. $(x^2 + 4y^2)^3$
28. $(1 - a^2)^4$
29. $(a^3 - a^{-1})^6$
30. $\left(1 - \dfrac{x}{y^2}\right)^5$
31. $\left(2 + \dfrac{x}{y}\right)^5$
32. $\left(x + \dfrac{y}{2}\right)^6$

In Problems 33–36, find the first four terms of each expansion and simplify.

33. $(x^2 - 2a)^{10}$
34. $\left(2a - \dfrac{1}{b}\right)^6$
35. $(\sqrt{x/2} + 2y)^7$
36. $\left(\dfrac{x}{2} + \dfrac{1}{a}\right)^{11}$

In Problems 37–44, find the first five terms in each expansion and simplify.

37. $(x + y)^{16}$
38. $(a^2 + b^2)^{12}$
39. $(a - 2b^2)^{11}$
40. $(a + 2y^2)^8$
41. $(x - 2y)^7$
42. $\left(1 - \dfrac{x^2}{y}\right)^8$
43. $(a^3 - a^2)^9$
44. $\left(x + \dfrac{1}{2y}\right)^{15}$

In Problems 45–52, find the specified term of each binomial expansion.

45. $(x + y)^6$; fourth term
46. $(x + 2y)^8$; sixth term
47. $(t - 2)^{10}$; fifth term
48. $(2p + 3q)^4$; middle term
49. $(x - 3y)^{11}$; middle term
50. $(x + 5)^{36}$; fourth term
51. $(2x + 5y)^{19}$; term containing y^7
52. $(3p - 2q)^{21}$; term containing q^9

In Problems 53–56, find the indicated term of each expression.

53. $\left(2x^2 - \dfrac{a^2}{3}\right)^9$, seventh term
54. $(x + \sqrt{a})^{12}$, middle term
55. $\left(a + \dfrac{x^3}{3}\right)^9$, term containing x^{12}
56. $\left(2\sqrt{y} - \dfrac{x}{2}\right)^{10}$, term containing y^4

Applying the Concepts

57. **Depreciation:** A small business depreciates a company car that costs $25,000 at an annual rate of 15% for the first 5 years. Use the binomial expansion of $25{,}000(1 - 0.15)^5$ to determine the value of the car after 5 years.

58. **Population Growth:** The population of a small town grows at the rate of 3% per year, with an initial population of 20,000. Use the binomial expansion of $20{,}000(1 + 0.03)^4$ to determine the size of the city after 4 years. Round off to the nearest person.

Developing and Extending the Concepts

In Problems 59 and 60, find the binomial expansion of each expression.

59. $(1 + \sqrt{2})^5 + (1 - \sqrt{2})^5$
60. $(1 - \sqrt{3})^4 - (1 + \sqrt{3})^4$
61. Find a formula for $(a + b + c)^n$.
62. Find the expansion of $(x^2 + y^2 + z^2)^4$.
63. Expand and simplify the expression $\dfrac{(x + h)^7 - x^7}{h}$.
64. Solve each equation for x.
    (a) $\binom{4}{x} = \binom{4}{1}$
    (b) $\binom{6}{x} = \binom{6}{2}$

G In Problems 65 and 66, use a grapher to evaluate each expression.

65. (a) $\binom{4}{2}$
    (b) $\binom{5}{3}$
66. (a) $\binom{7}{2}$
    (b) $\binom{8}{7}$

CHAPTER 9 REVIEW PROBLEM SET

In Problems 1 and 2, find the first five terms of the sequence with the specified general term a_n.

1. (a) $a_n = (-1)^n n$ (b) $a_n = \dfrac{n}{2^{n+1}}$

    (c) $a_1 = 3, a_{n+1} = 2a_n, n \geq 2$

2. (a) $\begin{array}{l} a_2 = 1 \\ a_n = \dfrac{(-1)^n}{n^2} \end{array}$ (b) $a_n = \dfrac{1 + (-1)^n}{2 + 3n}$

    (c) $a_1 = -2, a_{n+1} = -a_n, n \geq 1$

In Problems 3 and 4, find the general term a_n for each sequence.

3. 3, 12, 27, 48, 75, . . .   4. 1, 8, 27, 64, 125, . . .

In Problems 5 and 6, write out and evaluate each sum.

5. (a) $\displaystyle\sum_{k=1}^{4} (3k + 2)$ (b) $\displaystyle\sum_{k=1}^{5} (-1)^k 3^k$

6. (a) $\displaystyle\sum_{k=0}^{4} 2^{-k}$ (b) $\displaystyle\sum_{k=1}^{4} (3 + (-1)^k)$

7. Express the series $2 + 4 + 6 + 8 + 10 + 12$ using summation notation.

8. Suppose that $\displaystyle\sum_{k=1}^{13} a_k = 4$ and $\displaystyle\sum_{k=1}^{13} b_k = -12$, evaluate

$\displaystyle\sum_{k=1}^{13} (3a_k - 2b_k)$.

In Problems 9–12, find the indicated term in each arithmetic sequence.

9. (a) a_{15} if 5, 9, 12, 17, . . .
    (b) a_{15} if 2, −2, −6, −10, . . .
10. (a) a_{30} if 27, 33, 39, 45, . . .
    (b) a_{20} if 5, 1, −3, −7, . . .
11. (a) a_{20} if $a_1 = 2$ and $d = 5$
    (b) a_{30} if $a_2 = -2$ and $a_{15} = -210$
12. (a) a_{40} if $a_1 = 26$ and $d = -10$
    (b) a_1 if $a_3 = -18$ and $a_{10} = -53$

In Problems 13–16, find the sum of the first n terms, S_n, for the given value of n of each arithmetic sequence.

13. (a) S_{15}; 4, 10, 16, 22, . . .
    (b) S_{10}; −15, −7, 1, 9, . . .
14. (a) S_{20}; −12, −3, 6, 15, . . .
    (b) S_{17}; 7, 19, 31, 43, . . .
15. S_{10} if $a_1 = 4, d = 3$
16. S_{11} if $a_1 = 5, d = 2$

In Problems 17 and 18, find the number of terms n of each finite arithmetic sequence.

17. $-2, 1, 4, 7, \ldots, 43$ 18. $51, 45, 39, 33, \ldots, -123$

In Problems 19–24, find the indicated term in each geometric sequence.

19. a_{10} if 1, 2, 4, 8, . . .   20. a_n if $3, 1, \dfrac{1}{3}, \dfrac{1}{9}, \ldots$

21. a_{20} if $4, 1, \dfrac{1}{4}, \dfrac{1}{16}, \ldots$ 22. a_{15} if $2, -\dfrac{2}{3}, \dfrac{2}{9}, \ldots$

23. a_1 if $a_2 = 15$ and $a_5 = -1875$

24. a_6 if $a_3 = -\dfrac{1}{4}$ and $r = \dfrac{1}{2}$

In Problems 25 and 26, find the sum of the first n terms S_n of a geometric sequence for a given value of n.

25. S_{10} if $a_1 = -\dfrac{1}{9}$ and $r = 3$

26. S_{10} if $a_1 = -4$ and $r = -2$

In Problems 27–32, find the sum of each infinite geometric series.

27. $10 + 4 + 1.6 + 0.64 + \ldots$
28. $10 + 6 + 3.6 + 2.16 + \ldots$
29. $5 + 5(0.3) + 5(0.3)^2 + \ldots$
30. $40 + 40\left(\dfrac{1}{4}\right) + 40\left(\dfrac{1}{4}\right)^2 + \ldots$
31. $3 + \dfrac{3}{5} + \dfrac{3}{5^2} + \dfrac{3}{5^3} + \ldots$
32. $0.25 + \dfrac{0.25}{4} + \dfrac{0.25}{4^2} + \dfrac{0.25}{4^3} + \ldots$

In Problems 33–36, find the value of each expression.

33. (a) $\dfrac{5! \cdot 2!}{3! \cdot 6!}$ (b) $\dfrac{(n-3)!}{(n-2)!}$

34. (a) $\dfrac{8!}{4! \cdot 6!}$ (b) $\dfrac{(n+1)!}{(n-2)!}$

35. (a) $\dbinom{8}{3}$ (b) $\dbinom{7}{2}$

36. (a) $\dbinom{15}{12}$ (b) $\dbinom{14}{11}$

In Problems 37–40, use Pascal's triangle to expand each binomial expression and simplify.

37. $(x + 3y)^4$ 38. $(2x - 3y)^5$
39. $(3x + 2)^5$ 40. $(3x - 4y)^4$

In Problems 41–44, use the binomial theorem to expand each binomial expression and simplify.

41. $(2x - y)^6$ 42. $\left(\dfrac{x}{2} + 3\right)^4$

43. $(p^2 + 1)^4$ 44. $(a + 2\sqrt{b})^4$

In Problems 45–48, find the specified term of each binomial expansion and simplify.

45. The fourth term of $(3x + 4y)^{11}$
46. The seventh term of $(2x + 5y)^{10}$
47. The term containing $x^4 y^8$ of $(x + y)^{12}$
48. The term containing $x^5 y^9$ of $(2x - y)^{14}$

49. **Investment:** Suppose that $2000 is deposited into an account that pays 4% annual interest compounded quarterly, the balance A_n in the account after n quarters is given by

$$A_n = 2000\left(1 + \frac{0.04}{4}\right)^n$$

Determine (a) the first five terms of this sequence.
(b) the balance in the account after 4 years.

50. **Depreciation:** A company depreciates a machine that costs $40,000 by $2500 each year. Determine
    (a) the value of the machine after 1 year, 2 years, 3 years, and 4 years.
    (b) the value of the machine after the nth year.

51. **Population Growth:** A colony of bacteria triples each hour. If there are 1000 bacteria initially, determine:
    (a) the number of bacteria after the first, second, and third hours.
    (b) the number of bacteria after the nth hour.

52. **Savings:** A father gives his son 5 cents on the first day of a year, 10 cents the next day, 15 cents the following day, and so on. If the pattern continues until the last day of the year, how much money will the son accumulate for one year? Assume that the year has 365 days.

53. **Depreciation:** A new car costs $24,000, and suppose that it depreciates by 15% of its value each year.
    (a) What is the value of the car at the beginning of its fifth year?
    (b) What is its total depreciation for 4 years?

54. **Puzzle Problem:** Find the sum of the numbers in each row of the first seven rows of Pascal's triangle. Can you see a pattern?

G 55. Use a grapher to graph the first five ordered pairs of each sequence. Compare the graph of each sequence to the graph of the corresponding linear function.
    (a) $f(x) = 3 - 2x, x \geq 1, a_n = 3 - 2n$
    (b) $f(x) = 4x + 1, x \geq 1, a_n = 4n + 1$

G 56. Use a grapher to graph the first five ordered pairs of each sequence and the corresponding exponential function.
    (a) $f(x) = 3^x, x \geq 1, a_n = 3^n$
    (b) $f(x) = 2^{-x}, x \geq 1, a_n = 2^{-n}$

◆ CHAPTER 9 PRACTICE TEST

1. Find the first five terms of each sequence.
   (a) $a_n = 2 + n^2$ (b) $a_n = 5^{n-1}$
   (c) $a_1 = 2, a_{n+1} = 3a_n, n \geq 1$

2. Find the general term a_n for each sequence.
   (a) $19, 13, 7, 1, -5, \ldots$ (b) $1, -\frac{1}{7}, \frac{1}{49}, -\frac{1}{343}, \ldots$

3. Suppose that $\sum_{k=1}^{20} a_k = 7$ and $\sum_{k=1}^{20} b_k = 3$, and evaluate each sum.
   (a) $\sum_{k=1}^{20} (4a_k + 3b_k)$ (b) $\sum_{k=1}^{20} (8a_k - 4b_k)$

4. Given the arithmetic sequence $5, -1, -7, -13, \ldots$, find
   (a) a_n (b) a_{10} (c) S_{20}

5. Given the geometric sequence $3, 6, 12, 24, 48, \ldots$ find
   (a) a_n (b) a_{15} (c) S_{10}

6. Find the sum of the infinite geometric series
   $8 + 4 + 2 + 1 + \ldots$.

7. If the first term of a geometric sequence is $1/8$, the nth term is 32 and the common ratio is 2, find the number of terms n.

8. Simplify each expression.
   (a) $\frac{4! \cdot 6!}{8! - 5!}$ (b) $\binom{7}{4}$

9. Find the first five terms of the expansion of $(2x + 3y)^{11}$ and simplify.

10. Find the fourth term of the binomial expansion of $(x + 2y)^{13}$.

11. A small company had sales of $200,000 during its first year of operation and sales increased by $30,000 per year during each successive year. Find
    (a) the company's sales during the eighth year.
    (b) the total sales during its first 8 years of operation.

ANSWERS TO ODD NUMBERED PROBLEMS AND REVIEWS
ALL ANSWERS TO PRACTICE TESTS

--CHAPTER R--

PROBLEM SET R1 (on page 5)

1a. 4 + 5 = 9
1b. 5 + 4 = 9
1c. 4 + 5 = 5 + 4
3.   x·5 > x/5

5a. The product of 5 and a number is 20.
5b. The product of 5 and a number less 4 is 20.

7a. The sum of a number and 2 is less than 10.
7b. 10 is greater than the sum of a number and 2.

9a. 6·6 = 36
9b. 3·3·3·3·3·3 = 729

11a. 2^2xy^3
11b. $2^23^2y^2z$

13a. 8
13b. 8

15a. 33   17. 65
15b. 33   19. 45

21a. 6   23. 42
21b. 3

25a. 60   27a. 9   29a. 5(3 + 4) = 35
25b. 60   27b. 9   29b. 5·3 + 5·4 = 35

31a. x = 10(20 + y)
31b. x = 20 + 10y

33a. Three times a number less one is not equal to 5.
33b. Three times the difference of a number and 1 is equal to 5.

35a. (4 + 5)·2 + 3 = 21
35b. (4 + 5)·(2 + 3) = 45
35c. 4 + 5·(2 + 3) = 29

37a. 4 + 3·(8 - 1) + 6 = 31
37b. 4 + 3·8 - 1 + 6 = 33
37c. (4 + 3)·(8 - (1 + 6)) = 7

39a. 0
39b. undefined

41. 173
43. 0

45a. 3, +, 4, /, 5
45b. (,3,+,4,),/,5
45c. 4,+,5,*,(,3,-,2,),/,10

47a. 2(95,000) - y
47b. $185,000, $183,000

49a. $\dfrac{w - 300}{5}$ 49b. $760

PROBLEM SET R2 (on page 10)

1a. ∅,{0},{1},{0,1}
1b. ∅, {a}, {b}, {c}, {a,b}, {a,c}, {b,c}, {a,b,c}

3a. {x|x ≤ 0}
3b. {x|x < 0}
3c. {x|x ≥ 0}

5a. {-5,-4,-3,-2,...,3,4,5}
5b. {9,10,11}
5c. {x| x ≠ 5}
5d. {x| x ε I and x ≠ 5}

7a.

7b.

9a. 0.9
9b. 0.09
9c. 0.009

11a. $0.\overline{2}$
11b. $0.\overline{5}$
11c. $0.\overline{7}$

13a. integer, rational
13b. natural number, integer, rational

15a. irrational
15b. natural, integer, rational

17a. 3.1416
17b. 3.1429
17c. 3.1411

19a. 2.9998
19b. 3.0000
19c. 3.0033

21a. -3/5

21b. 2/3

23a. distributive
23b. commutative
23c. commutative

25a. commutative
25b. associative
25c. commutative
27. symmetric
29. substitution

31.

33a. 19·7 + 82·7
33b. (82 + 19)·7
33c. 7(19 + 82)
35a. 25/5
35b. $-\sqrt{9}$,0, 25/5

35c. $-\sqrt{9}$,0, 25/5, 25/4
35d. $-\sqrt{5}$
35e. all are real

37a. $3.60
37b. $2.40 + $1.20
37c. a = 0.10, b = 0.05, c = 24

39. 8 ÷ 4 = 2 but 4 ÷ 4 = 0.5
41. 2^n

43a. distributive
43b. additive inverse
45a. associative

45b. multiplicative identity
45c. distributive
45d. additive inverse
45e. additive identity

PROBLEM SET R3 (on page 16)

1a. 25   3a. -2   5a. 5 - (-3)   7a. 10   9a. 2   11a. -18
1b. 25   3b. -3   5b. -3 - 1   7b. -10   9b. 2   11b. 4
1c. -25   3c. -4
13a. -20   15a. 5000   15c. -5000   17a. -225   19a. 22/15   21a. 4
13b. 20   15b. 5000   15d. -5000   17b. 4500   19b. -22/15   21b. 32
                                                        19c. -22/15

23a. 25   25. -15   29. undefined   31a. 25   33a. -64   35. -135
23b. -20   27. 0   31b. -25   33b. -64   37. -12
39a. -2   41a. 36   43a. -25/7   45a. |18° - (-2°)| 47a. 20 - 4 + 7 - 5
39b. -4   41b. 36   43b. 1   45b. 20°   47b. 18 yard line

PROBLEM SET R4 (on page 22)

1a. 1/3	3a. -13/19	5a. 2/3	7a. 80%	9a. 37.5%	11a. 7/100
1b. 9/14	3b. -5/17	5b. 3/5	7b. 35%	9b. 87.5%	11b. 1/4
13a. 1/200	15a. 7/200	17a. 30	19. 20%	21a. 3/5	23a. 5/8
13b. 1/160	15b. 53/500	17b. 8.4		21b. 2/3	23b. 2/9
25. 7/5	29. 1/50	33a. 3/5	35a. -8/13	37a. 11/16	39. 16/99
27. -1/3	31a. 7/4	33b. 1	35b. 2/25	37b. 68.75%	41. 1/5
	31b. 10/11			37c. 11/5	43. 46,000,000 acres

PROBLEM SET R5 (on page 26)

1a. 243	3a. 4^{12}	7a. 13^3	9a. $-x^5$	11a. 12^6	13a. y^{65}
1b. x^5	3b. x^{12}	7b. x^{30}	9b. $(-x)^5 = -x^5$	11b. x^{600}	13b. $-y^{65}$
	5. y^{12}				

15a. $-\dfrac{4x^2}{y^2}$ 17. $x^4 y^7$ 21a. 3.85×10^5 23a. 9.6×10^{14} 27. $\dfrac{a^4 + b^4}{a^2 b^2}$ 29. $-\dfrac{27a^9 b^{21}}{8}$

21b. 3.85×10^8 23b. $4.\overline{16} \times 10^5$

15b. $-\dfrac{16x^2}{y^2}$ 19. $\dfrac{y^4}{x^8}$ 21c. 3.8×10^4 25. 1.77×10^5

31. $\dfrac{a^3 b^4}{3}$ 33. $a^9 b^{20}$ 35a. x^2 37a. 3.584×10^8 km   39. closest star: 3.31×10^{23} m

35b. x 37b. 3.0×10^8 in   farthest star: 7.57×10^{23} m

37c. 3.5×10^{10} barrels   41. 9.0345×10^{24} atoms

PROBLEM SET R6 (on page 32)

1. 0.0588 m^2, $58,800$ m^2 5. $14,256$ ft   9. 112 in^2 13. 15 in^2 17. π cm^2

3. 0.01 m^2, $10,000$ mm^2 7. 24 cm^2 19. $V = 125$ in^3

11. $\dfrac{49\pi}{4}$ cm^2 15. $2 + \dfrac{\pi}{2}$ cm^2 $S = 150$ in^2

21. $V = 57.76$ m^3 23. $V = 16.49$ ft^3 25. $V = 1149.76$ yd^3 27. yes, $5^2 + 12^2 = 13^2$

     $S = 103.36$ m^2 $S = 39.25$ ft^2 $S = 530.66$ yd^2 29. no

31. yes, $12^2 + 35^2 = 37^2$

33. \$360   37. 2200 mi   41a. 77°F   43a. 62.6°F   45a. \$241.73

35. 4%   39. 285.7 ft/sec   41b. -3.9°C   43b. -8.3°C   45b. \$53.20

45c. \$294.93

47a. cube at 600 ft^2 49a. 57.6 acre   51. 0.35 hours   55a. \$1469.33

47b. cube at 1000 ft^3 49b. \$18,000/acre   53a. \$1400   55b. \$2938.66

53b. \$2800

CHAPTER R REVIEW PROBLEM SET (on page 34)

1a. $5 + (-5 + 3)$ 3a. $3 \cdot 3 \cdot 3 \cdot 3 \cdot 3$ 5a. associative   5c. commutative

1b. $(-2)(-6) - 3$ 3b. $3^3 \cdot x^4$ 5b. identity for   5d. additive

1c. $-30/5 - 2$ multiplication   inverse

5e. identity for addition   9a. 1

5f. distributive   7a. $0.\overline{3}$ 7d. $0.\overline{18}$ 9b. -3

5g. associative & commutative   7b. $0.2\overline{7}$ 7e. $-0.0\overline{70}$ 11a. 1/2

5h. distributive   7f. 0.3750, 0.2778, 0.2857,   11b. -1

7c. $0.\overline{538461}$ 0.1818, -0.0700   11c. 9/2

11d. 1/3   11g. 55/48   13a. 7.3×10^8

11e. 1/24   11h. 31/120   13b. 235,000   15a. 51.8°F

11f. 23/60   11i. 5/12   13c. 10^5 15b. $8.\overline{3}$ °C

11j. 3/40

17a. $2(100 + 70) - 123$ 19. \$225   23. 1/5   27. 152.91 units^2

17b. \$117   21. \$750   25. 32.7 ft

--CHAPTER 1 --

PROBLEM SET 1.1 (on page 41)

1a. degree 2, 3 of x^2, -7 of x, 6 constant   3a. 0   5a. 60

1b. degree 2, -7 of x^2, -2 of x, 4 constant   3b. 0   5b. -3

3c. 15   5c. -8

3d. 48   5d. -35

7. $7x^2 - x - 1$ 11. $4t^3 + 2t^2 - 11t - 6$ 13. $x^6 - x^5 - 2x^4 + 2x^3 - 3x^2 + 3x + 4$

9. $t^2 + 10t + 3$ 15. $5x^2 + 2x - 2$

17. $2t^3 + 2t^2 - t - 2$ 19. $w^2 + w + 9$ 21. $6x + 10$ 23. $10y^4 - 5y^3 + 3y^2 + 3y - 8$

25a. 64 ft   25c. 55 ft   27a. 10 + 4x   29a. $-6x^2 + 420x - 500$

25b. 60 ft   25d. 48 ft   27b. 26 ft   29b. \$6250, \$6850, \$5500

31. $2xy^2 - 3x^2 y - x^2 y^2 + xy + 5$ 39a. both = 0

33. $u^3 - u - 4$ 37. $-\dfrac{4}{3}x^3 - \dfrac{7}{2}x^2 + \dfrac{7}{3}x - \dfrac{7}{4}$ 39b. same value

35. $6x^3 y^2 - 9x^2 y - 6$ 41. $20x^2 - 39x + 15$

43a. any number without a   43b. degree 2   45. 0

      variable

PROBLEM SET 1.2 (on page 46)

1a. $6x^6$	3a. $-30y^6$	5a. $-2y^2 + 3y$	7a. $15y^6 + 35y^5$
1b. $-30t^7$	3b. $60p^8$	5b. $3t^3 + 2t^2$	7b. $-4x^4 - 6x^3 - 10x^2$

9. $9x^3 + 21x^2 - 2x - 8$ 11. $3y^3 - 10y^2 - 4y + 35$ 13. $4u^4 - 4u^3 + 2u^2 + 6u - 12$

15. $9x^3 + 21x^2 - 2x - 8$ 17. $m^5 + 2m^4 + 10m^2 - 9m + 12$ 19. $y^4 - 3y^3 - 4y^2 - 9y + 9$

21. $x^5 - x^4y - 7x^3y^2 + 4x^2y^3 + 2xy^4 - y^5$

23a. $x^2 + 3x + 2$ 25a. $x^2 + 2x - 15$

23b. $x^2 - 3x + 2$ 25b. $x^2 - 2x - 15$

27a. $5x^2 - x - 4$ 29. $10x^4 + 13x^2 - 3$ 31a. $w^2 - 49$ 33a. $x^2 + 2x + 1$

27b. $5x^2 + xy - 4y^2$ 31b. $49 - w^2$ 33b. $x^2 - 2x + 1$

35a. $16y^4 + 40y^2z + 25z^2$ 37a. $x^3 + y^3$ 39a. $x^2 - 5x$ 41a. $7200 - 180x - 2x^2$

35b. $16y^4 - 40y^2z + 25z^2$ 37b. $x^3 - y^3$ 39b. $24 \; ft^2$ 41b. 6250 boxes

43. $8x^3 - 8x^2 + 2x$ 47. $9x^3 - 12x^2 - 80x - 64$ 51. $81x^4 - 216x^3 + 216x^2 - 96x + 16$

45. $-2x^3 - 9x^2 + 17x + 84$ 49. $24x^4 - 36x^3 + 18x^2 - 3x$ 53. $x^5 + 2x^4 + 10x^2 - 9x + 12$

55. $x^4 - 2x^2y^2 + y^4$

PROBLEM SET 1.3 (on page 53)

1a. $x(x - 1)$ 3a. $ab(a - b)$ 5a. $12x^2y(x - 4y)$ 7a. $9mn(m + 2n - 3)$

1b. $2x(2x + 1)$ 3b. $17x^2y(xy - 2)$ 5b. $x(4x^2 - 2x + 1)$ 7b. $4xy^2(2 + 6xy + y)$

9a. $(2a + b)(3x + 5y)$ 11a. $(a + b)(x + y)$ 13a. $(a - c)(b^2 - d)$ 15a. $(a + b)(x + y + 1)$

9b. $(2m + 3)(x - 1)$ 11b. $(x^2 + 1)(a + b)$ 13b. $(2 - y)(x^2 + z^2)$ 15b. $(2x - 1)(a + b + c)$

17a. $(x + 4)(x + 12)$ 19a. $(x + 3)(x + 12)$ 21a. $(x - 4)(x + 6)$ 23a. $(x - 16)(x + 1)$

17b. $(x - 4)(x - 12)$ 19b. $(x + 18)(x + 2)$ 21b. $(x + 4)(x - 6)$ 23b. $(x - 8)(x + 2)$

25a. $(x - 9)(x + 2)$ 27a. $(2w + 1)(w + 3)$ 29a. $(2x + 3)(3x + 2)$ 31a. $-(w + 4)(2w - 3)$

25b. $(x - 15)(x - 2)$ 27b. $2(w^2 + 10w - 3)$ 29b. $(6y - 7)(y + 1)$ 31b. $(2x + 9)(9x + 10)$

33a. $(x + 10)(x - 10)$ 35a. $(4x + 7)(4x - 7)$ 37a. $(4u + 5v)(4u - 5v)$ 39a. $(x + 15)^2$

33b. $(10 + x)(10 - x)$ 35b. $(4x - 7y^2)(4x + 7y^2)$ 37b. $(5m + 7n)(5m - 7n)$ 39b. $(x - 15y)^2$

41a. $(x + 1)(x^2 - x + 1)$ 43a. $(w - 2yz)(w^2 + 2wyz + 4y^2z^2)$ 45a. $(3x - 1)(3x + 1)(9x^2 + 1)$

41b. $(y + 5)(y^2 - 5y + 25)$ 43b. $(4z + 3b)(16z^2 - 12bz + 9b^2)$ 45b. $(4x + y)(4x - y)(16x^2 + y^2)$

47a. $(x - 1)(x^2 + x + 1)(x^6 + x^3 + 1)$ 49a. $(x^2 + xy + y^2)(x^2 - xy + y^2)$

47b. $(x + 2)(x^2 - 2x + 4)(x^6 - 8x^3 + 64)$ 49b. $(2x^2 - 2xy + y^2)(2x^2 + 2xy + y^2)$

51a. $5x(x - 7)(x - 4)$ 53. $7x - 2$ 55a. $\pi(8x - 3)(2x + 1)$

51b. $yz^2(x - 4)(x + 3)$ 55b. $65\pi \; in^2$

57a. $3x(x - 2y)(x + 2y)(x^2 + 4y^2)$ 59a. $(2u - 3)(2u + 3)(4u^2 - 6u + 9)(4u^2 + 6u + 9)$

57b. $3t(t - 2)(t^2 + 2t + 4)$ 59b. $3t^2(t^2 + 3)(t^4 - 3t^2 + 9)$

61a. $(2x + y - 1)(x^2 + y)$ 63a. $b = 7, -7, 2, -2$

61b. $(a + b + c)(2x - 1)$ 63b. $b = \pm 25, \pm 14, \pm 11, \pm 10$

PROBLEM SET 1.4 (on page 58)

1a. 1   3a. 3   7a. $\dfrac{2y}{3x}$ 9a. y 11a. $\dfrac{x + 1}{x - 1}$

1b. -8   3b. -6   7b. $2x - 1$ 9b. x

1c. 2   5a. 6, -7   11b. $\dfrac{x - 3}{x}$

1d. 0   5b. -3, 2

13a. $\dfrac{y + 2}{4y}$ 15a. $\dfrac{x + 4}{x + 7}$ 17a. $\dfrac{x - y}{x + y}$ 19a. $\dfrac{x + 2y}{3x + 4y}$ 21a. $6xy$

21b. $39xy$

13b. $\dfrac{x + 2}{3x}$ 15b. $\dfrac{x + 4}{x + 2}$ 17b. $\dfrac{3x - 6y}{x + 2y}$ 19b. $\dfrac{x - y}{x + y}$ 23a. $20m^7y^4$

23b. $18x$

25a. $9a^3(a - b)^2$ 31a. yes

25b. $m^2n(m + n)$ 31b. yes   37a. $\dfrac{7}{x - y}$ 37b. $\dfrac{-4}{x - y}$ or $\dfrac{4}{y - x}$

33. no

27. $2x^2 - 9x + 9$ 35. yes

29. $2x^2 + x - 6$

39. $\dfrac{46}{3}$ ft   41. $-\dfrac{x + y}{2x + y}$ 43. $\dfrac{w + z}{y + z}$ 45. $x^n - 3$

PROBLEM SET 1.5 (on page 63)

1a. $\dfrac{8}{9y}$ 3. $\dfrac{2x}{3y^3}$ 7. $\dfrac{1}{2}$ 11a. $\dfrac{5}{2}$ 13. $75yz$

15. $\dfrac{1}{64b^2c}$

1b. $\dfrac{2y}{9x}$ 5. $-10x^2y^2z$ 9. $\dfrac{5}{3x}$ 11b. $\dfrac{15}{16}$

17. $-\dfrac{7z}{4y}$ 21. $\dfrac{7}{m - n}$ 25. $x^2 - 16x + 48$ 29. $1 - 3a$

27. $\dfrac{1}{(x - y)(2x + y)}$ 31. $\dfrac{(y - 1)^2}{(y + 1)^2}$

19. $\dfrac{x}{3}$ 23. $\dfrac{x^2}{5(x - 1)}$

33. $\dfrac{x-3}{3x+1}$　　35. $-\dfrac{t+2}{t+5}$　　37. $\dfrac{x-1}{x+2}$　　39. $\dfrac{a^2-b^2}{a^3}$　　41. $\dfrac{25y^2}{24x^2}$

43. $\dfrac{1}{(x+1)^2}$　　45. x^2+4x+3　　47. $\dfrac{(x-1)^2(x+1)}{x^2+1}$　　49. $x-2$　　51. $x-2+\dfrac{-8}{2x-1}$

53. $4x-1+\dfrac{2}{2x-3}$　　55. $x^2+5x+14+\dfrac{35}{x-2}$　　57. $x^3+x^2-4x+9+\dfrac{-8}{x+1}$　　59. x^2+3x+4

61. $x+2$　　63a. $\dfrac{x}{y}$　　　65. $(2m+n)(3m+5n)$　　67. $\dfrac{2(x+2)(x+3)}{5(x^2+3x+9)}$

63b. 20 mpg

69. $\dfrac{1-a}{b(a-3)(b+1)}$　　71. $\dfrac{2x}{x+1}$　　73a. 1　　73c. x^2　　73e. all values, $x\neq 0$

73b. x　　73d. 1　　73f. no values

PROBLEM SET 1.6 (on page 68)

1. $\dfrac{3}{4x}$　　3. $\dfrac{-8}{5x}$　　5. $\dfrac{8}{7x}$　　7. $\dfrac{1}{2t}$　　9. $\dfrac{t}{2k}$

11. $\dfrac{2a+3}{a+3}$　　13. $\dfrac{15}{y-5}$　　15. $\dfrac{2}{x+2}$　　17. $\dfrac{1}{4-x}$　　19. $\dfrac{6}{y+1}$

21. $\dfrac{21x+5y}{35y}$　　23. $\dfrac{y+12}{4}$　　25. $\dfrac{ab+a^3-b^2}{b^3}$　　27. $\dfrac{30+9t+4t^2}{6t^2}$　　29. $\dfrac{8x+7}{(x+2)(x-1)}$

31. $\dfrac{15y-57}{(y-5)(y-3)}$　　33. $\dfrac{x^2-x+2}{(3x+2)(x-4)}$　　35. $\dfrac{5x-3}{x(x^2-1)}$　　37. $\dfrac{2m^2+4}{m(m-6)(m+1)}$　　39. $\dfrac{9x-26}{(a-6)(a-2)(a+2)}$

41. $\dfrac{3-4c}{(c-3)(c-2)(c+3)}$　　43. $\dfrac{5x^2-12xy^2+21y}{12x^3y^3}$　　45. $\dfrac{2}{a+1}$　　47. $\dfrac{x^2-6x}{x^2-9}$　　49. $-\dfrac{17t^2+t-9}{3t^2-27}$

51a. $\dfrac{x}{60}+\dfrac{y}{60}=\dfrac{x+y}{60}$　　53. $\dfrac{3}{x(x+3)}$　　55a. $\dfrac{4x^3+79x^2+256x}{4x+16}$　　57a. $\dfrac{16}{15}$　　59. $\dfrac{x^4-y^4}{x^2y^2}$

51b. $28.\overline{3}$ hours　　　　55b. \$11,575.96　　57b. -6.9908　　　61. 2

63a.

$\dfrac{65}{24}=2.708\overline{3}$　　63b. $\dfrac{x^4+4x^3+12x^2+24x+24}{24}=\dfrac{65}{24}$　　65a. $\dfrac{y+x}{xy}$　　65b. 5/6

65c. 2/5, not equal

63c. $e\approx 2.71828182846$

PROBLEM SET 1.7 (on page 75)

1a. $\dfrac{4y}{3x^2}$　　3a. $-\dfrac{3}{10a}$　　5a. -xy　　9. $\dfrac{3x+1}{2x-3}$　　11a. $\dfrac{5}{6}$

5b. -6xy

1b. $\dfrac{x}{z}$　　3b. $-\dfrac{2x}{y}$　　7. $-\dfrac{1}{3a+3}$　　　11b. $\dfrac{3}{4}$

13a. $-\dfrac{y}{15ab^2}$　　15. $\dfrac{y(x^2-y)}{x^2(y-1)}$　　17. $\dfrac{1}{5x}$　　19. $\dfrac{2y+7x}{8y-6x}$　　21. $\dfrac{m}{m-2}$

13b. $\dfrac{1}{3b}$

23. $\dfrac{-c-d}{4}$　　25. $\dfrac{-1-2t}{3t+2}$　　27. $\dfrac{1+x}{1-x}$　　29. $\dfrac{1}{u-v}$　　31. $x-3$

33. $\dfrac{3t+2}{2t-1}$　　35. $\dfrac{m^2-mn-n}{m+mn-n^2}$　　37a. $\dfrac{1}{y^2}$　　39a. m^2　　41. $\dfrac{128}{3}$

37b. x^6　　39b. b^5

43. $\dfrac{1}{a^2b^2+ab^3}$　　45. $\dfrac{x+y}{xy}$　　47. $\dfrac{x+2}{3}$　　49. $\dfrac{-h-6}{9h^2+54h+81}$　　51a. 2 hrs

51b. 10 hrs

51c. 20 weeks

51d. 0.3 mpg

53a. \$411.98　　55. $\dfrac{-2x}{x^2+1}$　　59a. $\dfrac{7}{5}$　　59c. $\dfrac{41}{29}$

53b. \$3775.04

57. $\dfrac{y-1}{2y-1}$　　59b. $\dfrac{17}{12}$　　59d. $\dfrac{x^4+x^3+3x^2+2x+1}{x^4+3x^2+1}$

PROBLEM SET 1.8 (on page 80)

1.

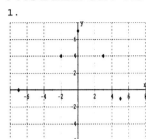

1. (3,4) in quadrant I
(-2,4) in quadrant II
(5,-1) in quadrant IV
and (0,-6) on y axis
(-7,0) on x axis
(0,7) on y axis

3. (-3,4)
5. (-8,0)
7. (-5,-5)

9. (0,2), (1,0)

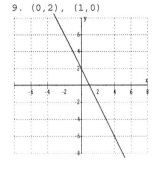

11. (0,3), (-6,0)

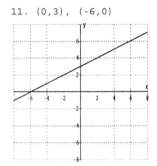

13. (0,2), (2/3,0)

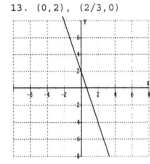

15. (0,-3), (2,0)

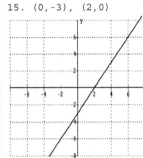

17. (0,3) no x intercept

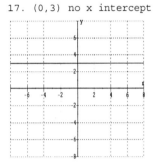

19. (0,-5), (5/3,0)

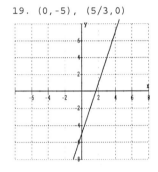

21. (0,4), (3,0)

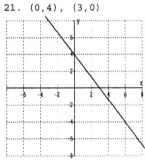

23. no y intercept (-2,0)

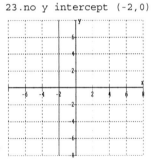

25. (0,0)

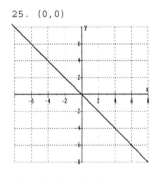

27. (0,4),(4,0)

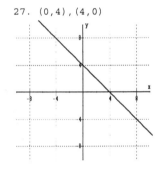

29. (0,-10), (20,0)

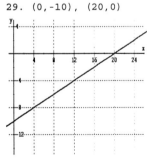

31. (0,6), (-6/5,0)

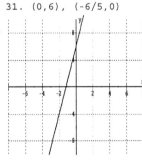

33. (0,2), (5,0)

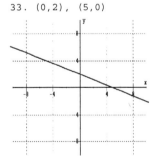

35. (0,15), (7.5,0)

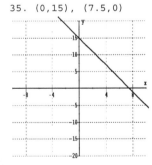

37. (0,0.5), (-6,0)

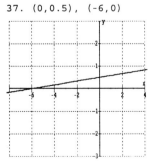

39. II
41. IV
43. I or III

45.

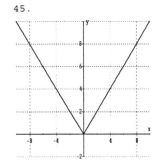

47.

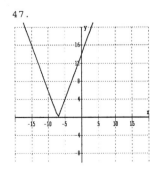

49. (0,5), (7,0)   51. (0,12), (27,0)   53. (0,-10), (20,0)

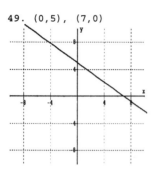   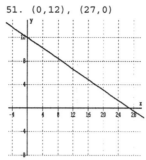   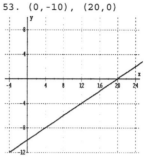

CHAPTER 1 REVIEW PROBLEM SET (on page 81)

1a. 4
1b. 5, 1, 2, -1, -13
1c. 81

3. $4x^3 - 2x^2 + 2x$
5. $2y^3 + y^2 + 8y - 9$
7. $a^3 - 3a^2b + 3ab^2 - b^3$

9. $12y^2 - 9xy - 81x^2$
11. $9x^2 + 42x + 49$
13. $7xy(x - 3y^2)$

15. $(m^2 + n)(5a - b)$
17. $9u(u - 3v)(u + 3v)$
19. $(m - 9)(m + 4)$

21. $2ab(a - 7b)(a + 2b)$
23. $5x - 1$
25. $x - 2$

27a. $\dfrac{1}{3m - 3n}$

27b. $\dfrac{2y + 3}{y}$

29. $\dfrac{x - 4}{x + 6}$

31. $\dfrac{x - 2}{x}$

33. $3a^2 + 3ab$
35a. 1

35b. $\dfrac{2t}{(2t - 1)(2t + 1)}$

37. $\dfrac{x^2 + 3x}{(2x - 3)(2x + 3)(x + 2)}$

39. $\dfrac{3m + 1}{(m - 1)(m + 1)(2m + 3)}$

41. $\dfrac{4(x - 4)^2}{x(x - 2)(x + 2)}$

43. 1

45. $\dfrac{m + 1}{m - 3}$

47. $\dfrac{(x - 2)(x + 2)}{(x - 4)(x + 1)}$

49. y

51. $\dfrac{-1}{x + y}$

53a. -$836,600
53b. $400
53c. -$6,600

55a. 3x + 6
55b. 129

57a. $\pi r^2 + 4r^2$
57b. $r^2(\pi + 4)$
57c. 28.6 m^2

59a. 474

59b. $\dfrac{2v_1v_2}{v_1 + v_2}$

61. 2 and -5
63. 2.5 and 5

CHAPTER 1 PRACTICE TEST (on page 83)

1a. degree of 2; 7, -3, 5, 39, 59
1b. degree of 3; 4, 0, 1, -1, -6

2a. $5x^2 + 2x + 3$
2b. $5x^2 - 8x + 7$
2c. $4x^3 + 4x^2 - 5x - 3$
2d. $2x^2 + 15x + 62$

3a. $49x^2 + 14x + 1$
3b. $1 - 10x + 25x^2$
3c. $9x^2 - 4$
3d. $x^3 - 8$

4a. $(3m - 10n)(3m + 10n)$
4b. $(a - b + c)(a + b - c)$
4c. $(x - 4)(y^2 + 2)$
4d. $(x + 17)(x - 1)$
8. (0,2), (-6,0)

5a. $\dfrac{3x + 1}{a - b}$

5b. $\dfrac{2x - 3y}{3x + 2y}$

6a. $\dfrac{2x + 4}{3}$

6b. 1

6c. $\dfrac{7x + 3}{4x^2 - 4}$

6d. $\dfrac{b^2 + 2ab - a^2}{a^2b - b^3}$

7a. $\dfrac{3x - 3}{3x - 7}$

7b. $\dfrac{y}{y + x}$

9a. (12 - 2x)(16 - 2x)

9b. $192 - 56x + 4x^2$
    A = 60 in^2

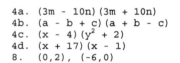
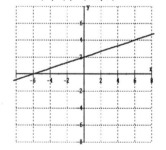

--CHAPTER 2--

PROBLEM SET 2.1 (on page 90)

1a. 8
1b. 0
21. no solution
47. 800
49. 0.25

3a. 10
3b. 22/3
23. 210
25. 2
51a. identity
51b. identity

5a. 13/4
5b. -2
27. 59/8
29. 1
53a. inconsistent
53b. identity

7a. -17/3
7b. 1/2
31. 1
33. -10/19

9. 1
11. 5
35. -63/10
37. -4/29
55. 86 nickels, 91 dimes, 43 quarters

13. -13/43
15. 8/3
39. 5
41. 2
57a. $135
57b. 600 mi

17. 3
19. -11/2
43. -5
45. 35/6
59a. $130,200
59b. $160,000

61. -1.5
63a. 2
65a. 2

63b.
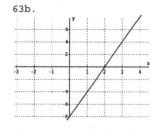

65b.
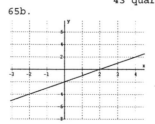

67. If a = b, then a + c = b + c

69. 9°

PROBLEM SET 2.2 (on page 96)

1. $5c$

3. $\dfrac{c - b}{a}$

5. $-\dfrac{b}{2}$

7. $-\dfrac{5a}{3}$

9. $-c/2$
11. $7d/2m$
13. $-7b$

15. $(12 - 4x)/3$

17. $(99 - 11x)/9$

19. $(63 - 27x)/14$

21. $100x, + 3$

23. $(4x - 90)/23$

25. $\dfrac{A}{rt + 1}$

27. $\dfrac{Ty}{x}$

29. $\dfrac{P}{0.7}$

31. $\dfrac{y - b}{m}$

33. $\dfrac{C}{2\pi}$

35. $\dfrac{P - 2w}{2}$

37a. $3V/h$
37b. $3V/B$
39a. $V + 32t$

39b. $\dfrac{V_o - V}{32}$

41a. $bx/(b - y)$
41b. $ay/(a - x)$

43. 240 ft
45. $\$22.52$

47a. 113.04 ft^3
47b. $V/\pi r^2$
47c. 6 ft
49a. $T/0.15 + 1000$
49b. $\$2680$
49d. $\$114$

49c.
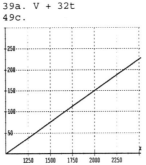

55a.
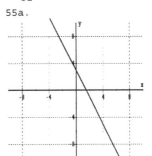

55b.
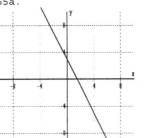

51a. $\dfrac{V}{\pi r^2} - \dfrac{4}{3}r$
51b. 15.61 m

53a. $C = 0.33 + 0.22(w - 1)$

53b. $w = \dfrac{C - 0.33}{0.22} + 1$

53c. 11 oz

55a. $y = 3 - 2x,\ 3/2$
55b. $y = 4x,\ 0$
57. $7/a$
59. $-17/a^2$

61a. $0.5N(N - 3)$
61b. 9 diagonals
63a. 1.60 in
63b. 3.60 in

PROBLEM SET 2.3 (on page 106)

1. $18, 20, 22$
3. $4, 6, 8$
5. $2, 4, 6$
7. 32 ft by 15 ft

9. 199m by 399m
11. 8 in by 10 in
13. 11 ft from one end
15. 27 ft, 29 ft, 31 ft

17. 32 cm, 96 cm, 52 cm
21. $\$45,000$
19. $\$20,000$ at 9%, $\$12,000$ at 8.5%
23. $\$1067.86$ @ 4.25%; $\$2382.14$ @ 7.75%
25. 25 gal   27. 300 lbs @ 30%, 100lbs @ 10% 29. 15 tons sand; 30 tons gravel
31. 350 mi @ 35 mph; 850 mi @ 85 mph   33. 20 mph   35. bus: 52.5 mph; train: 83.5 mph
37. 30 tens, 90 fives   39. son 11, father 36 41. $\$1029$ 43. $\$50$
45. $\$23.54$ 47. 135 49. 9% 51. 22 weeks
53. 2500 m   59a. $59, 60, 61$ 61. 14 m by 26 m   63. 128 in^3
55. 2.7 minutes   59b. $58, 60, 62$
57. $\$9.00$ 59c. not possible

PROBLEM SET 2.4 (on page 115)

1. $x \neq 0,\ x = -1/2$
3. $t \neq 0,\ t = 5$
5. $x \neq 0,\ x = 80$
7. $x \neq 0,\ x = 1$

9. $y \neq 0,\ y = 6/5$
11. $x \neq -4,\ x = 1$
13. $x \neq -5/4,\ x = 15/8$
15. $x \neq -7,\ x = -35/2$

17. $x \neq \pm 3,\ x = 15$
19. $y \neq 2,3,\ y = 4$
21. $t \neq 3,\ t = 3/4$
23. $y \neq -1,\ y = 1/4$

25. $x \neq -1,\ x = 4$
27. $x \neq 1$, no solution
29. $x \neq -2, -4,\ x = -1$
31. $x \neq 2$, extraneous solution

33. $x \neq 4$, extraneous solution

37. $a, b \neq 0,\ y = \dfrac{bc + d}{a}$

39. $x = \dfrac{6a - 7b}{4}$

41. $7, 28$
43. $\$1,110$

35. $y \neq 0$, no solution

45. 1875 women
47. 6 hrs
49. 15 hrs
51. $3\ 3/4$ hrs

53. 60 mph
55. 55 mph, 70 mph
57. 32.1 km/hr
59. $34, 51$ teeth
63. $\$265.50$

61a. $R = \dfrac{r_1 r_2 r_3}{r_1 r_2 + r_1 r_3 + r_2 r_3}$

61b. $E_1 = \dfrac{ER_1}{R + R_1}$

65a. s wave: 3 km/sec; p wave: 5 km/sec
65b. 18.75 km

PROBLEM SET 2.5 (on page 124)

1a. $(-1, 1]$

1b. $[-1, 1)$

3a. $[-1, \infty)$

3b. $(-\infty, -1)$

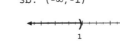

5a. $(-\infty, 2]$ or $[4, \infty)$

5b. $(-\infty, \infty)$

7a. $(-\infty, \infty)$

7b. $[-10, -10]$

9a. addition
9b. multiplication
9c. addition
9d. multiplication

11. $x \geq 1,\ [1, \infty)$

13. $x > -2,\ (-2, \infty)$

15. $x \geq 1,\ [1, \infty)$

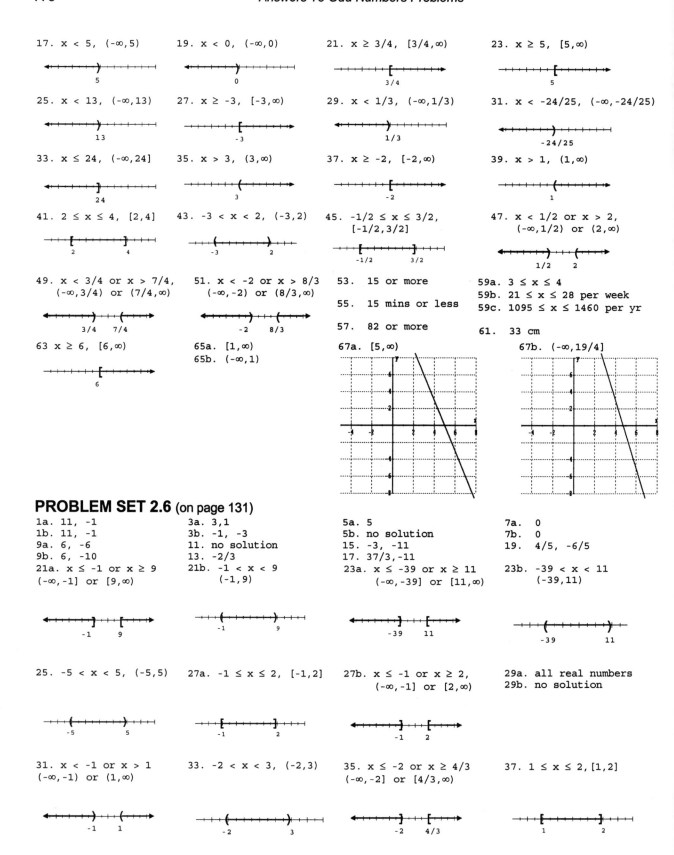

17. x < 5, (-∞,5)

19. x < 0, (-∞,0)

21. x ≥ 3/4, [3/4,∞)

23. x ≥ 5, [5,∞)

25. x < 13, (-∞,13)

27. x ≥ -3, [-3,∞)

29. x < 1/3, (-∞,1/3)

31. x < -24/25, (-∞,-24/25)

33. x ≤ 24, (-∞,24]

35. x > 3, (3,∞)

37. x ≥ -2, [-2,∞)

39. x > 1, (1,∞)

41. 2 ≤ x ≤ 4, [2,4]

43. -3 < x < 2, (-3,2)

45. -1/2 ≤ x ≤ 3/2, [-1/2,3/2]

47. x < 1/2 or x > 2, (-∞,1/2) or (2,∞)

49. x < 3/4 or x > 7/4, (-∞,3/4) or (7/4,∞)

51. x < -2 or x > 8/3, (-∞,-2) or (8/3,∞)

53. 15 or more

55. 15 mins or less

57. 82 or more

59a. 3 ≤ x ≤ 4
59b. 21 ≤ x ≤ 28 per week
59c. 1095 ≤ x ≤ 1460 per yr

61. 33 cm

63 x ≥ 6, [6,∞)

65a. [1,∞)
65b. (-∞,1)

67a. [5,∞)

67b. (-∞,19/4]

PROBLEM SET 2.6 (on page 131)

1a. 11, -1
1b. 11, -1
9a. 6, -6
9b. 6, -10
21a. x ≤ -1 or x ≥ 9
(-∞,-1] or [9,∞)

3a. 3,1
3b. -1, -3
11. no solution
13. -2/3
21b. -1 < x < 9
(-1,9)

5a. 5
5b. no solution
15. -3, -11
17. 37/3,-11
23a. x ≤ -39 or x ≥ 11
(-∞,-39] or [11,∞)

7a. 0
7b. 0
19. 4/5, -6/5
23b. -39 < x < 11
(-39,11)

25. -5 < x < 5, (-5,5)

27a. -1 ≤ x ≤ 2, [-1,2]

27b. x ≤ -1 or x ≥ 2, (-∞,-1] or [2,∞)

29a. all real numbers
29b. no solution

31. x < -1 or x > 1
(-∞,-1) or (1,∞)

33. -2 < x < 3, (-2,3)

35. x ≤ -2 or x ≥ 4/3
(-∞,-2] or [4/3,∞)

37. 1 ≤ x ≤ 2, [1,2]

39. x ≤ -1 or x ≥ 5
   (-∞,-1] or [5,∞)

41a. no solution
41b. -7

43a. x ≠ -6,
   (-∞,-6) or (-6,∞)

43b. (-∞,∞)
45. 96.3° ≤ T ≤ 100.9°
47. 195 v ≤ x ≤ 245 v
49. 3.5 m ≤ x ≤ 4.5 m

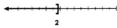

51. 2.3 cc ≤ x ≤ 2.7 cc
53a. |x - 3| = 8;
   x = 11, -5
53b. |x - 3| ≤ 8,
   -5 ≤ x ≤ 11

55a. |x| < 5
55b. |x| ≤ 4

57a. [-2,5]
57b. (-∞,-2) or (5,∞)

59a. [-9,6]
59b. (-∞,-8) or (0,∞)

CHAPTER 2 REVIEW PROBLEM SET (on page 132)

1. 3
3. 1/2
5. 100
7. -8
9. -15
11. 11/2
13. 62/11
15. 1, 2
17. 0, 1
19. 3

21. 3
23. $\dfrac{5 - y}{2}$
25. $\dfrac{2A - ah}{h}$
27. $\dfrac{S - 2\pi R^2}{2\pi R}$
29. $\dfrac{ac}{a - c}$

31. x ≤ 2, (-∞,2]
33. x > 1, (1,∞)
35. -3 ≤ x ≤ 3, [-3,3]
37. 2 < x < 3, (2,3)

39. -2 ≤ x ≤ 5,
   [-2,5]

41. -4 < x < 0,
   (-4,0)

43. x ≤ -12 or x ≥ 20,
   (-∞,-12] or [20,∞)

45. x < -14 or x > 10,
   (-∞,-14) or (10,∞)

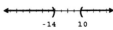

47. -11/3 ≤ x ≤ 1,
   [-11/3,1]

49a. -2/3
49b. x is any real
   number

51. 1 ≤ x ≤ 4, [1,4]

53. 35 and 50
55. 10 ft by 24 ft
57. 24 tons, 36 tons
59. 112, 224
61. 4 4/9
63. 8.25 minutes

65. x = -5
67. x = 0
69. x = 2

CHAPTER 2 PRACTICE TEST (on page 134)

1a. -85
1b. -1
1c. 12.$\overline{6}$

2a. 1
2b. no solution

3a. 8, -8
3b. 3, -5/3
3c. -5/3

4a. -3a/7
4b. $\dfrac{A}{RT + 1}$

5a. x ≤ -3

5b. -6 ≤ x ≤ 0

5c. x ≤ 5 or x > 8

6a. x ≤ -3, (-∞,-3]

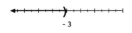

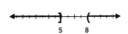

6b. x < -3, (-∞,-3)

6c. -4 < x ≤ 1,
   (-4,1]

6d. x ≥ 4/13, [4/13,∞)

6e. -3 ≤ x ≤ 3, [-3,3]

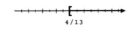

6f. x ≤ 3 or x ≥ 5,
   (-∞,3] or [5,∞)

7. $1,500

8. 4 inches

9. 3

10. x > -5

-- CHAPTER 3 --

PROBLEM SET 3.1 (on page 141)

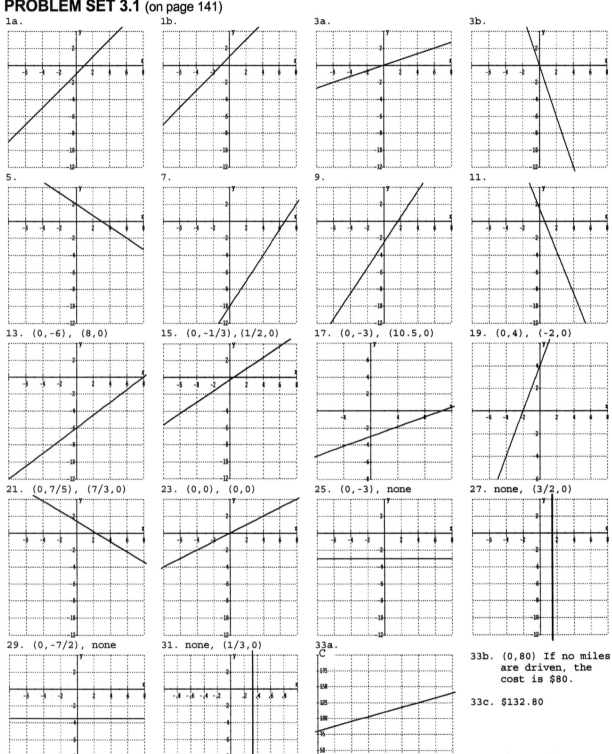

1a.   1b.   3a.   3b.

5.   7.   9.   11.

13. (0,-6), (8,0)   15. (0,-1/3),(1/2,0)   17. (0,-3), (10.5,0)   19. (0,4), (-2,0)

21. (0,7/5), (7/3,0)   23. (0,0), (0,0)   25. (0,-3), none   27. none, (3/2,0)

29. (0,-7/2), none   31. none, (1/3,0)   33a.

33b. (0,80) If no miles
      are driven, the
      cost is $80.

33c. $132.80

35a.

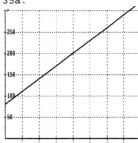

35b. (0,80) The fee to begin service is $80.

35c. 7 months

37.

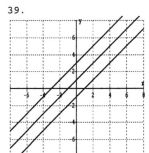

39.

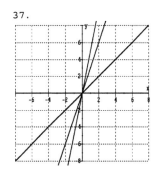

41a.
Lines are expected since the equations are of degree 1.

41b.

x	y = 3x + 4	y	x = 3y + 4
-4	(-4,-8)	-4	(-8,-4)
0	(0,4)	0	(4,0)
2	(2,10)	2	(10,2)

41c.

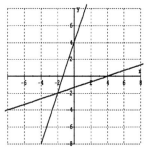

41d.

(-2,-2)

43. (0,3) (6,0)

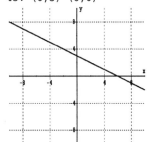

45. (0,-11) (-5.5,0)

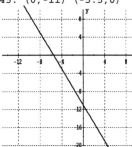

47a. 2y = 3x + 4

$$y = \frac{3x}{2} + 2$$

(0,2) and (-4/3,0)

47a.

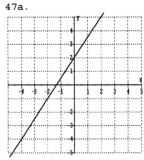

47b. 3y = 2x - 5

$$y = \frac{2x}{3} - \frac{5}{3}$$

(0,-5/3) and (5/2,0)

47b.

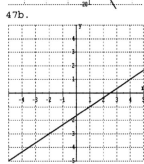

49a. $5,000
49b. $2000
49c. $500, $1000, $2000
49d. the value of the printer is diminishing

PROBLEM SET 3.2 (on page 152)

1. 3/2
3. 2/3

5. 5/2, rises

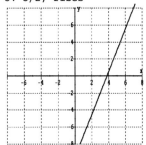

7. -1, falls

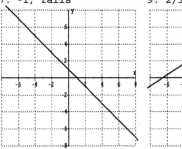

9. 2/3, rises

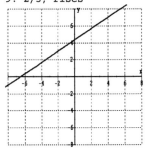

11. 3/4, rises

13. 0, horizontal

15. undefined, vertical

17. -1/2, falls

19. 20/11, rises

21. 2, rises

23.

25.

27.

29.

31a. y = -3x - 6
     m = -3, y int. -6

31b. y = x - 4
     m = 1, y int. -4

33a. y = -x
     m = -1, y int. 0

33b. y = x
     m = 1, y int. 0

35a. y = 2
     m = 0, y int. 2

35b. x = -2, m none
     y int. none

37. parallel lines
     with m = 4

39. m_1 = 4, m_2 = -1/4
     perpendicular

41. parallel lines with
     m = 3/2

43. m_1 = 2/3,
     m_2 = 3/2, neither

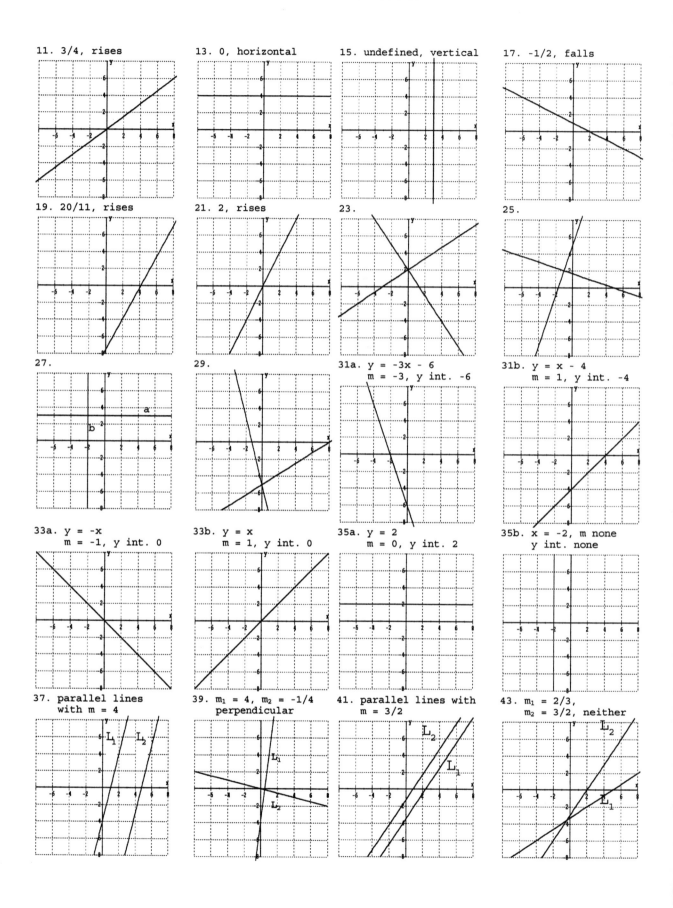

45a. m = 1/12 =
        0.0$\overline{83}$ = 8.$\overline{3}$%
45b. 31.2 ft.

47a. m = 3/40 = 7.5%
47b. 0.6 ft.

49a. 25/8
49b. 8.75°F

51a. M = (-1,7)
51b. M = (-7/2,7/2)

53. M = (6.02,6.92)

55a. y = 6
55b. slope

57a. W = 6w
57b. earth weight is 6
      times moon weight

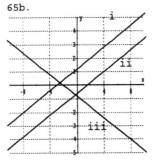

59a. k = 12.5
59b. k = m = 12.5
59c. 75 cm.
59d.

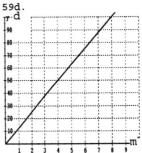

61a. 4, -2, 6, 0, 8, -10
61b. 2, 4, 6
61c. 5
61d. 3, 7
61e. 7

63a. It is a right
      triangle with
      angle B = 90°

63b. It is a right
      triangle with
      angle B = 90°

65a. m_i = 2/5,
      m_{ii} = 2/5,
      m_{iii} = -2/5
      lines i and ii
      are parallel

65b.

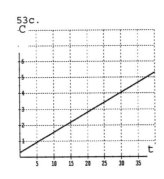

PROBLEM SET 3.3 (on page 161)

1. $y = \frac{4}{5}x + \frac{2}{5}$

3. y = -3x + 1

5. y = 1

7. $y = -\frac{1}{6}x + \frac{2}{3}$

9. y = 2x - 3

11. $y = -\frac{2}{3}x + 2$

13. y = 3x + 2

15. $y = -\frac{5}{8}x + 5$

17a. y - 2 = 5(x + 1)

17b. y = 5x + 7

19a. $y + 1 = -\frac{3}{7}(x - 5)$

19b. $y = -\frac{3}{7}x + \frac{8}{7}$

21a. $y = \frac{3}{8}x$

21b. $y = \frac{3}{8}x$

23. y = -4

25a. $y + 1 = \frac{5}{3}(x + 1)$

25b. $y = \frac{5}{3}x + \frac{2}{3}$

27a. y - 3 = -4(x - 2)
27b. y = -4x + 11
29. x = 4

31a. $y + 8 = -\frac{1}{4}(x + 5)$

31b. $y = -\frac{1}{4}x - \frac{37}{4}$

33. 2x - y = -7
35. 2x + 3y = 8
37. 4x - 5y = 32
39. 4x + 3y = -26
41. 8x + y = 37

43. 37x + 70y = -127
45. x + 0·y = -1

47a. $y = -\frac{1}{2}x + \frac{1}{2}$

47b. x + 2y = 1

49a. $y = \frac{3}{2}x + \frac{7}{2}$

49b. 3x - 2y = -7

51a. $y = \frac{1}{2}x + 3$

51b. x - 2y = -6

53a. (12,1.80)
      (20,2.80)

53b. $C = \frac{1}{8}t + 0.3$

53d. C = $5.30
53e. t = 46.8 min.

53c.

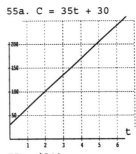

55a. C = 35t + 30

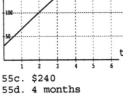

55c. $240
55d. 4 months

57a. V = -1280t + 7500
57b.

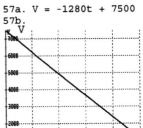

57c. $4940

59a. P = 132 - 0.6t
     P = 176 - 0.8t

59b. 108 ≤ P ≤ 144

59c. t = 50

61a. m = -b/a

$y = -\dfrac{b}{a}(x - a)$

ay = - bx + ab
bx + ay = ab

$\dfrac{x}{a} + \dfrac{y}{b} = 1$

61b. i. $\dfrac{x}{5} + \dfrac{y}{6} = 1$

61b. ii. $-\dfrac{x}{2} + \dfrac{y}{7} = 1$

63a. m = (2,2)

63b. $y - 2 = -\dfrac{3}{5}(x - 2)$

65a.

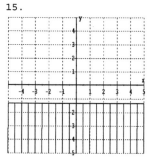

65c. m = $0.25, each
     bagel costs $0.25.
65d. $50 initial cost

67a.
     y = 0.125x + 1.675

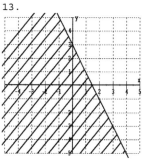

67b. points are not in
     line.

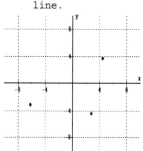

PROBLEM SET 3.4 (on page 169)

1a. iv
1b. i
1c. vi

3a. $y > \dfrac{2}{3}x + 2$

3b. $y \geq -x + 3$

3c. $y > \dfrac{1}{2}x + 1$

5a. line and below

5b. line and above

7.

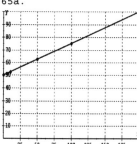

9.

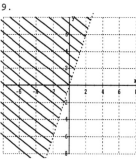

11.

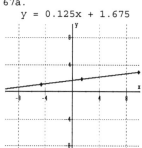

13.

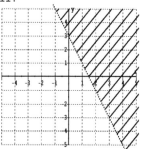

15.

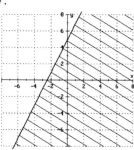

17.

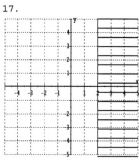

19.

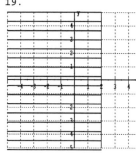

21a. y ≤ 3x + 3 and
     y ≥ 3x - 6

21b. y ≥ 1 or y < -2

21c. y > -2x - 4 and
     y ≤ -2x + 4

23.
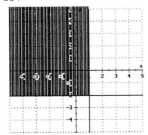

25.
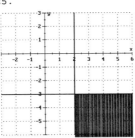

27.
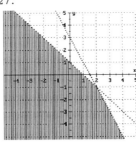

29.
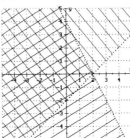

31.
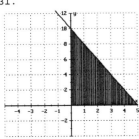

33. $\begin{cases} B + 2A \le 100 \\ A \ge 0, B \ge 0 \end{cases}$

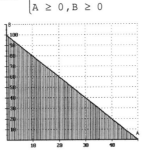

35a. x = # of $8 tickets
y = # of $5 tickets

$\begin{cases} 8x + 5y \ge 3000 \\ x \ge 0, y \ge 0 \end{cases}$

35b.
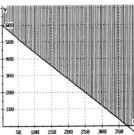

37a. B
37b. A or C

39. Not possible, no
graph

41.
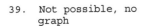

43.
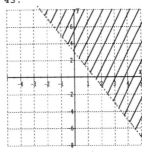

45a. x < 0 and y > 0

45b. x > 0 and y ≠ 0

47. x = # level I and
y = # level II
employees

$\begin{cases} 0 \le x \le 20 \\ 0 \le y \le 15 \\ 50x + 40y \le 1400 \end{cases}$

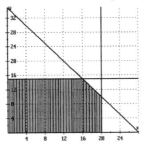

PROBLEM SET 3.5 (on page 178)

1a. is a function
Domain = {1,2,3,5}
Range = {5,7,9,13}

1b. is a function
Domain = {-4,-3,3}
Range = {16,9}

3a. yes
Domain = {each citizen}
Range = {all social
security numbers}

3b. yes
Domain = {person's
age}
Range (that person's
height}

5a. f(-2) = -6
5b. f(0) = 0
5c. f(2) = 6
13a. f(0) = -14
13b. f(7/6) = 0
13c. f(3) = 22

7a. f(-2) = -4
7b. f(0) = -2
7c. f(2) = 0
15a. f(t) = 7t + 3
15b. f(t + 2) = 7t + 17
15c. f(t) + 2 = 7t + 5
15d. f(t + 2) - f(2) = 14

9a. f(0) = 3
9b. f(3/4) = 0
9c. f(-1/2) = 5
17a. g(2t) = 5 - 8t
17b. g(t + h) = 5 - 4t - 4h
17c. g(t) + h = 5 - 4t + h
17d. g(t + h) - g(2t) = 4t - 4h

11a. h(0) = 1/5
11b. h(-1/3) = 0
11c. h(2) = 7/5
19. 3
21. -9
23 -5/2

25.

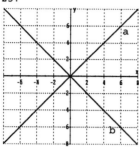

27.

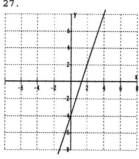

29.

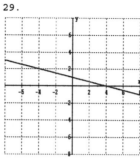

31.

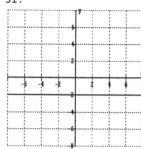

33a. yes
33b. [0,3.5]
33c. [0,210]
33d. 90 miles
33e. 3 hours

35a. yes
35b. no

37a. Domain = (-∞,2]
      Range = [0,∞)
37b. Domain = [-3,3]
      Range = [0,9]

39a. $C = \dfrac{1}{5}m + 24$

39b. $50
39c. 10 miles
39d. $84

41a. C = 350t

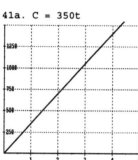

41c. 875 calories

43a. $20,500

43b. 2006

43c. no, since 1 ≤ t ≤ 10

45a. S = 3l - 22
45b. 9.5

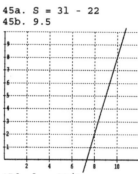

45d. l > 22/3

47a. 5x + 5
47b. x - 9

49a. $6x^2 + 17x - 14$
49b. $3x^2 - 2x$

51.   f(x) = 2x + 1

53.   any vertical line

55a.

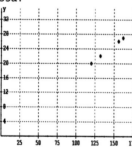

55b. $y = \dfrac{1}{6}x$

55c. ⅙ anything that
      weights 6 pounds on
      earth will weight 1
      pound on the moon.

55d. 192

57a. f(1/2) = 0;
      g(1/2) = 0;
      f(0) = 1;
      g(0) = 1

57b.

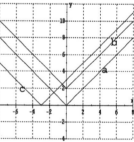

57c. Domain = (-∞, ∞)
      Range = [0,∞)

59a. [0,∞)
59b. [3,∞)
59c. [0,∞)

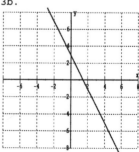

CHAPTER 3 REVIEW PROBLEM SET (on page 182)

1a.

1b.

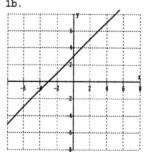

3a.

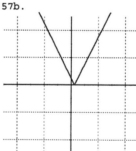

3b.

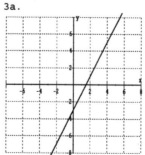

5.
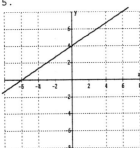

7. x int = 5, y int = -2
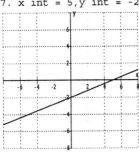

9. x int = -3, no y int
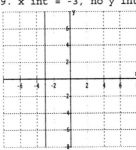

11. slope = 1, rises
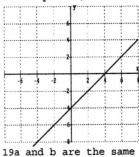

13. slope = -1/2,
    falls
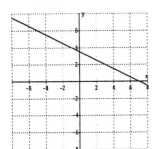

15. slope = 0,
    horizontal
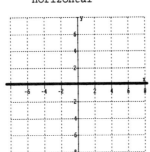
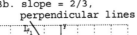

17.
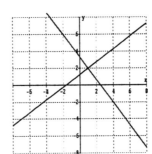

19a and b are the same
$$y = \frac{3}{4}x - \frac{9}{4}$$
slope = 3/4; y int -9/4
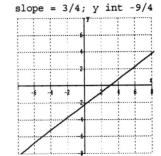

21. slope = 2,
    parallel
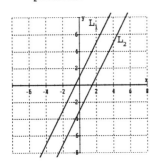

23a. slope = -3/2
23b. slope = 2/3,
     perpendicular lines
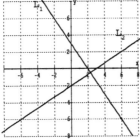

25a. y = 3x - 2
25b. y = -2x - 8
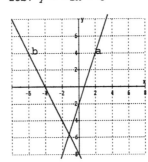

27a. y = 4
27b. x = 5
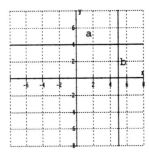

29. $y - 3 = \frac{3}{2}(x-2)$
    3x - 2y = 0
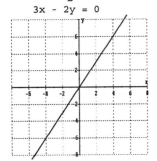

31. $y - 5 = \frac{2}{3}(x + 7)$
    2x - 3y + 29 = 0
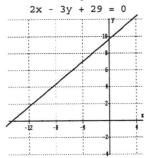

33a.
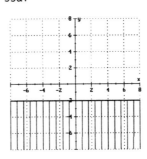

33b.
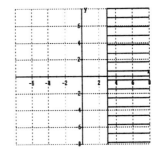

35.

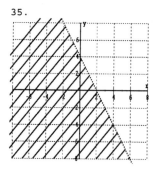

37.

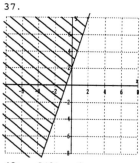

39a.

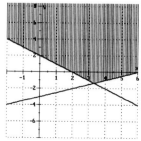

39b.

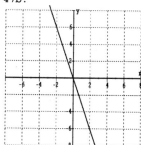

41a. f(3) = 9

41b. f(t + 1) = 2t + 5

41c. f(-2) = -1

41d. f(t + 1) - f(t) = 2

43a. f(0) = 2

43b. $\dfrac{1}{f(0)} = \dfrac{1}{2}$

43c. f(-2) = 0

43d. f(-1) = $\sqrt{3}$

45. -5

47a.

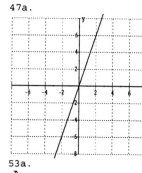

47b.

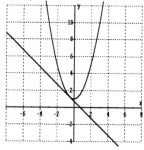

53b. A intercept = 21
means number of
gallons at the
start. The m
intercept is 350,
the number of
miles traveled
when the tank is
empty.
53c. 9 gallons
53d. 262.5 miles

55a. C = $\dfrac{1}{20}$ N + 5

55c. $42.50

49.

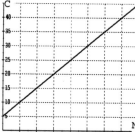

51a. yes
51a. Domain = [-4,4]
      Range = [-4,0]

51b. not a function

53a.

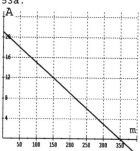

55b.

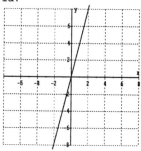

57a. t = 10N + 490

57b. slope = 10, each
month the number
of tons increases
by 10.

57c. 8 months

59. y = 2x

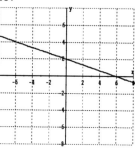

61. y = -x + 3/4

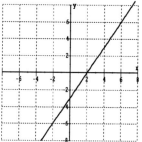

CHAPTER 3 PRACTICE TEST (on page 183)

1a.

1b.

1c.

2a. (2,0), (0,-3)

2b. (3,0), no y int

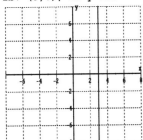

2c. no x int, (0,-1)

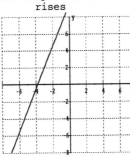

3a. slope = 5/2, rises

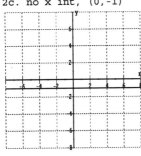

3b. slope = -2/3, falls

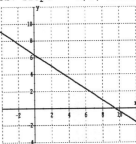

4. $y = \dfrac{5}{3}x - 5$,
   slope = 5/3, y int -5

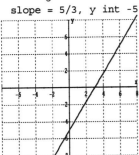

5. slope of L_1 = -3/4
   slope of L_2 = 4/3
   Lines are
   perpendicular

6a. y = -x + 5
    x + y - 5 = 0

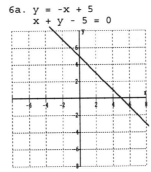

6b. $y = -\dfrac{4}{3}x - \dfrac{19}{3}$
    4x + 3y + 19 = 0

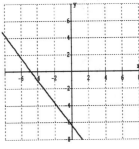

6c. $y = -\dfrac{3}{7}x - \dfrac{18}{7}$
    3x + 7y + 18 = 0

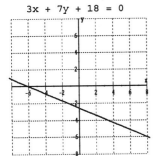

7a.

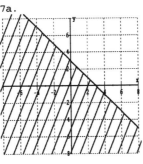

7b.

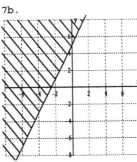

7c.

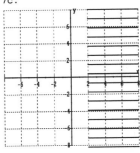

8.

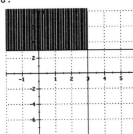

9a. g(0) = 4
9b. g(4) = 0
9c. $[g(2)]^2 = 12$
9d. $g(\sqrt{7}) = 3$

10. Domain = [-6,10]
    Range = [-6,2]

11a. y= 0.45x + 4500
11b. $4725
11c. 1900 cones

12a. V = -1000x + 6000
12b. $6000
12c. $1000

13a. N = -50P + 4000
13b. 3000 videos
13c. $10

14. they are the same

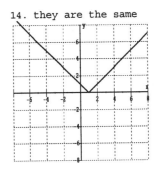

-- CHAPTER 4 --

PROBLEM SET 4.1 (on page 191)

1. (-1,9)

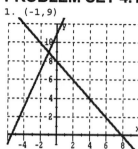

3. (1,-2)

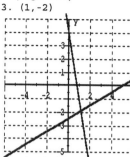

5. (1/3,-2)

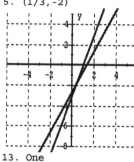

7. no solution

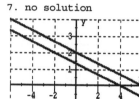

9. no solution

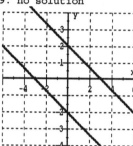

11. infinite number of solutions

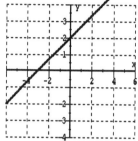

13. One

15. None

17. Infinite

19.
$$\begin{cases} y = -x + 6 \\ y = x - 4 \end{cases}$$

19. consistent and independent

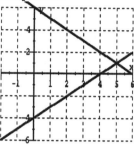

21.
$$\begin{cases} y = -\frac{2}{3}x + \frac{5}{3} \\ y = -\frac{2}{3}x + \frac{7}{6} \end{cases}$$

inconsistent

21.

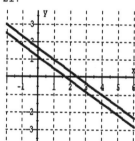

23.
$$\begin{cases} y = -\frac{2}{3}x + \frac{5}{3} \\ y = -\frac{2}{3}x + \frac{10}{3} \end{cases}$$

inconsistent

23.

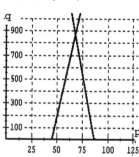

25.
$$\begin{cases} y = \frac{3}{4}x + \frac{1}{2} \\ y = \frac{3}{4}x - \frac{7}{4} \end{cases}$$

no solution

27.
$$\begin{cases} y = -x + 0.5 \\ y = x + 3 \end{cases}$$

one solution

29.
$$\begin{cases} y = x - 6 \\ y = x - 6 \end{cases}$$

infinite number of solutions

31b. (68,874)

33a. The lines have different slopes.

33b. The lines have the same slope but different x and y intercepts.

33c. The lines have the same slope and the same intercepts.

35a.
$$\begin{cases} y = mx + b_1 \\ y = mx + b_2 \\ \text{where } b_1 \neq b_2 \end{cases}$$

35b.
$$\begin{cases} y = ax + b \\ y = cx + d \\ \text{where } a = c \text{ and } b = d \end{cases}$$

37. k = -2

39. k = -2

41. $m_1 = m_2, \; b_1 \neq b_2$

43. $9

45. $90.33

47.
$$\begin{cases} x + y = -1 \\ x - y = 11 \end{cases} \quad (5,-6)$$

49a. $y = -70x + 901\frac{1}{3}$

49b. $(-70, \; 901\frac{1}{3})$

PROBLEM SET 4.2 (on page 198)

1. (4,3)
3. (7/3,-4/3)
5. (7/11,-10/11)
7. (1,5)
9. (159,-186)
11. consistent dependent
13. (3,1)
15. (7/3,-4/3)
17. (0,6)
19. (-6,9)
21. (-10,0)
23. (1,1)
25. (-5,3)
27. (-3/2,27/2)
29. (22/25, 21/25)
31. (3,2)
33. (2,-1)
35. (2/5,21/5)
37. (9,-5/3)
39. (10,24)
41. consistent & dependent
43. inconsistent
45. (12,15) consistent & independent
47a. 50

47b.

59a. a = 4/7,

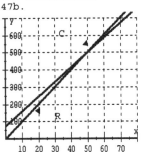

47c. A loss occurs when
x < 50.

51. (-3,4)

59a. b = 40/7

47d. A profit occurs when
x > 50

49. $6000

53. (10,13)

59b. c = 4

55. (1/13,1/17)

61. $y = \frac{3}{4}x + \frac{11}{4}$

57. (1/4,-1/5)

PROBLEM SET 4.3 (on page 204)

1. yes
3. (1,1,1)
5. (-5,5/3,28/3)
7. (2,2,1)
9. (5,2,1)
11. (94/35,1/7,1/5)
13. (1,-3,-2)
15. (3,1,0)
17. no solution,
inconsistent, the
three planes have no
common intersection.
19. infinite number of
solutions,
consistent and
dependent, the
planes intersect
along a line
21. no solution,
inconsistent, the
three planes have
no common
intersection.
23. infinite number
of solutions,
consistent and
dependent, there
is only one plane

25. (6,-3,12) consistent
and independent.
The three planes
intersect at a point
27. (2,0,2) same as #25
29. a = -2
b = 144
c = -80
31. a = 1/12 ≈ 0.08
b = -1.5
c = 23/3 ≈ 7.67

33. $\begin{cases} x + y - z = 6 \\ -x + 2y + z = 0 \\ -2x + z = -7 \end{cases}$

The number of systems is infinite.

35. (2,3,-1)
37. (4,3,2)
39. (2,-1,3)

41. (3,4,6)
43. a = 5, b = 3, c = 7
45a. (0.3589, -0.1204, 0.0595)
45b. (0.1, 0.3, 2)

PROBLEM SET 4.4 (on page 213)

1. l = 250 ft w = 200 ft,
Cost = $70,833.33
3. W = 18 in.
5. A = 6.5 oz, B = 4.5 oz
7. airliner = 502.23 mph,
wind = 33.48 mph

9. 22⅓ kph and 17⅓ kph
11. $25,000 @ 6% and
$15,000 @ 7%
13. $12,000 @ 10.5%,
$18,000 @ 6%
15. 4 lbs @ $9,
16 lbs @ $4

17. 30 oz @ 30%,
20 oz @ 90%
19. $6000 in CD, $9000
in stocks, $5000 in
municipal bonds
21. 3000 $20 tickets,
6000 $15 tickets,
11,000 $10 tickets
23. 15 model A,
10 model B,
20 model C

25. 33, 21
27. 17 dimes,
28 quarters
29. Tower is 437 m;
mast is 175 m
31. More than 58
visits

33. 185 pairs of tennis
shoes, 75 pairs
running shoes
35. Bride has $1000,
groom has $3000
37. $21,750 @ 6%,
$14,600 @ 8.5%
39. $0.90 for
juice; $0.80
for bagel;
$0.25 for
coffee

41. 13 dryers, 9 washers,
26 microwaves
43. 380 lbs A, 60 lbs B
160 lbs C.
45. 32 nickels, 8 quarters

PROBLEM SET 4.5 (on page 220)

1.

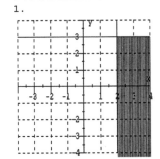

3.

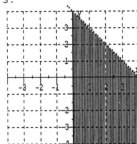

5.

7.

$\begin{cases} 0 \le x \le 4 \\ 0 \le y \le 2 \end{cases}$

9.

$\begin{cases} -x + y \le 2 \\ 4x - y \le 4 \end{cases}$

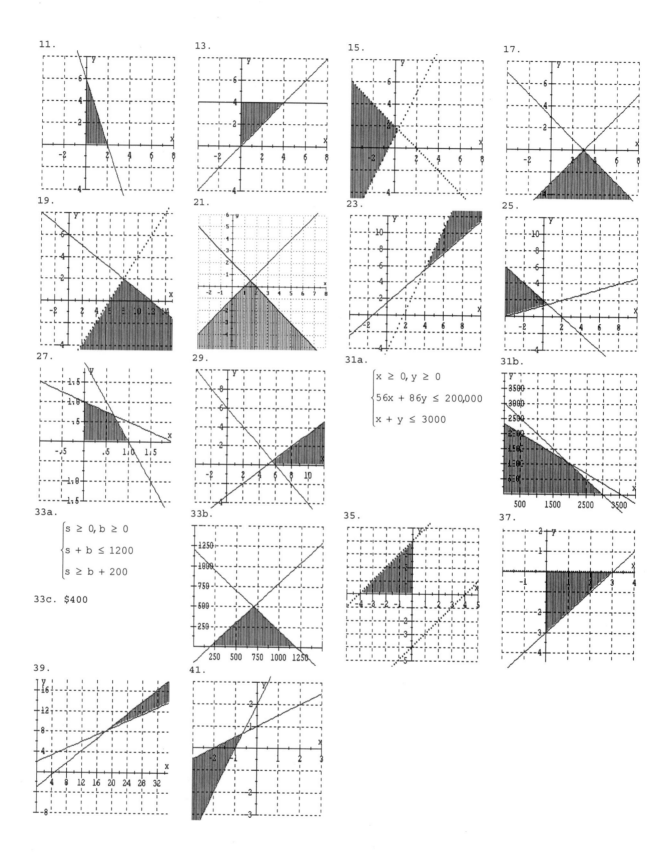

31a.
$$\begin{cases} x \geq 0, y \geq 0 \\ 56x + 86y \leq 200{,}000 \\ x + y \leq 3000 \end{cases}$$

33a.
$$\begin{cases} s \geq 0, b \geq 0 \\ s + b \leq 1200 \\ s \geq b + 200 \end{cases}$$

33c. $400

CHAPTER 4 REVIEW PROBLEM SET (on page 221)

1a. (3,2)
1b. y = -x + 5
    y = x - 1
1c. consistent and
    independent

3a. no intersection
3b. y = -2x + 6
    y = -2x + 4
3c. inconsistent
5. (7/3,1/3)

7. (12,6)
9. (2,0) consistent and
    independent
11. (-1,-2) consistent
    and independent
13. (-1,13/4)

15. (4,3,2)
17. (1,0,0)
19. no solution
21. infinite number of
    solutions
23. (1/2,1/2,1/2)

25.
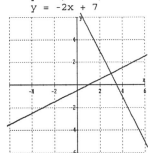

27.
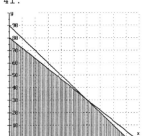

29.
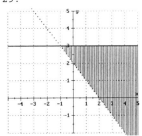

31.
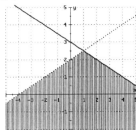

33. 106°, 74°

35. $570, $320

37. 4000, 1000

39. 440 mph, 560 mph

41.
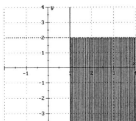

43. (-1.00,3.25)

45. (2,1)

CHAPTER 4 PRACTICE TEST (on page 222)

1a. (3,1)
    y = x/2 - 1/2
    y = -2x + 7

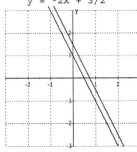

1b. no intersection
    y = -2x + 1
    y = -2x + 3/2

1c. identical lines
    y = x/2 + 1/2
    y = x/2 + 1/2

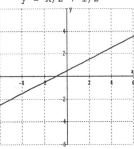

2a. consistent and
    independent

2b. inconsistent

2c. consistent and
    dependent

3a. (-1,-5), consistent

3b. (5,10), consistent

3c. no solution;
    inconsistent

4. (-2,5,3) is a
   solution

5a.

5b.
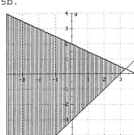

6. 30, 25

7. $8,460, $4,230

-- CHAPTER 5 --

PROBLEM SET 5.1 (on page 231)

1a. 1   3a. 1   5a. 1/25   7a. 9/16   9a. 25   11a. 9/8

1b. 1   3b. 1   5b. 1/16   7b. 9   9b. x^4

1c. 1   3c. 7   5c. 1/49   7c. x   9c. y^3 11b. $\dfrac{y^2}{x^4}$

13a. 49

13b. $\dfrac{1}{5^5} = \dfrac{1}{3125}$

11c. $\dfrac{b^5}{a^2}$

15a. $\dfrac{1}{x^5}$

15b. $\dfrac{1}{y^9}$

17a. $\dfrac{1}{256}$

17b. p^8

19a. $\dfrac{25}{p^8}$

19b. $\dfrac{1}{p^5 q^{10}}$

21a. $\dfrac{g^4}{p^6}$

21b. $\dfrac{x^2}{y^2}$

23. 1/9

25. $\dfrac{x^2}{y^{10}}$

27a. x = 2
27b. x = 3
29a. x = -1
29b. x = 1

31. x = -3, x = -2
33. V = 9.2×10^{-15} cm^3
35. 3×10^2 hrs
37. S = $10.08 per hr

39. S = $2,188.02
41. 1.77×10^{11}
43. 1.67×10^4
45. $1,268.24, $5,636.36

47a. 1/9
47b. 1
47c. 9
49a. 1/16
49b. 1
49c. 16
51. Domain = Reals
    Range = (0,∞)
    graph (c)
53. Domain = Reals
    Range = (0,∞)
    graph (a)

55a. x intercept = 2

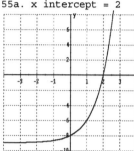

55a.

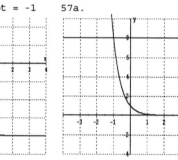

55b. x intercept = 3

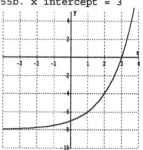

55b.

57a. x intercept = -1

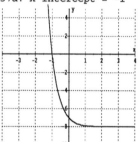

57a.

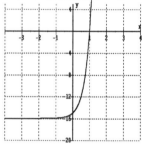

57b. x intercept = 1

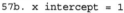

57b.

59a. a + b
59b. x^6

61a. $\dfrac{x^3 y^6}{27}$

61b. $\dfrac{u^{28} z^4}{v^{20}}$

63a. $x^{-2} y^{-4}$
63b. $5(2 + a)^{-5}$

PROBLEM SET 5.2 (on page 238)

1a. 5
1b. 4
3a. 4
3b. -2

5a. 0.2
5b. -0.6

7a. 3
7b. not real, the
    square of any real
    number is
    nonnegative.

9a. -2/3
9b. not real, the
    square of any real
    number is
    nonnegative.

11a. -1/2
11b. 3/4
13a. -11
13b. 11

15a. -3
15b. 3

17a. -2x
17b. 6y

19a. $2x^3 y^2$
19b. $2xy^2$

21a. 5x
21b. -3x

23a. y^7
23b. -3x

25a. -2x
25b. not real

27a. $2|x|$
27b. $-5|x|$

29a. $3|x|$
29b. y^2

31a. x
31b. $|x - 3|$

33a. 10
33b. $\sqrt{13} \approx 3.61$

35a. 5
35b. $\sqrt{241} \approx 15.52$

37a. $\sqrt{103.61} \approx 10.18$
37b. $\sqrt{165.64} \approx 12.87$

39a. 1
39b. 1.73
39c. 2.65

41a. -2.08
41b. -2
41c. -1.71

43a. -5.39
43b. -2
43c. -5.39

45. $|\overline{AB}| = 3\sqrt{13}$
    $|\overline{AC}| = 6$
    $|\overline{BC}| = 9$

45. yes, Δ
    ABC is a
    right
    triangle

47a. 3,000
47b. 4,300

49a. 1.7 sec
49b. 10.5 sec

51. 7.78 km/sec
53. A = 22.9 ft^2

55a. n odd
55b. n even

57a. $\sqrt{2}$
57b. 1
57c. 0

59a. 1
59b. 0
59c. not real

61a. $-\sqrt{2}$
61b. -1
61c. 0

63a. $x \geq 4$

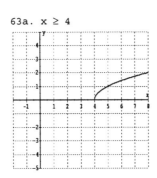

63b. $x \leq 4$

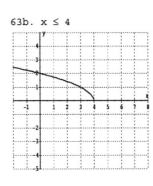

65a.

x	3	4	7	9
f(x)	0	1	2	2.45

65b. Domain = $[3, \infty)$
     Range = $[0, \infty)$

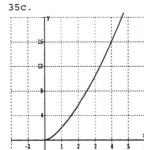

65d. $(7,2)$, $x = 7$ 67c. 64 times larger
67a. 8 times larger   67d. The number is increased by the cube of the factor.
67b. 27 times larger

PROBLEM SET 5.3 (on page 245)

1a. 6
1b. 1/7
1c. 3
13a. 1/16
13b. 1/9

3a. -3
3b. 3
3c. -2
15a. $\sqrt[7]{5}$
15b. $\dfrac{1}{\sqrt[4]{6}}$

5a. -3/2
5b. 5/9
5c. 3/4
17a. $\sqrt[7]{x^2}$
17b. $\sqrt[4]{y^3}$

7a. 4
7b. -16
19a. $\sqrt[3]{(8m)^2}$
19b. $\sqrt[5]{(4xy^2)^2}$

9a. -8
9b. 4
21a. $7^{1/2}$
21b. $3^{1/2}$
23a. $m^{1/3}$
23b. $x^{3/5}$

11a. not real
11b. not real
25a. $(7x^2)^{2/5}$
25b. $(3xy)^{2/4}$

27a. 2
27b. x

29a. 25
29b. $\dfrac{1}{x^{9/2}}$

31a. $16p^{12}$
31b. $\dfrac{y}{5}$

33a. x
33b. $x^{1/2}y^{1/3}$

35. $\dfrac{1}{xy^6}$

37. $\dfrac{2x^3}{5y^2}$

39a. $x^2 + 1$
39b. $x^{1/2} + 1$
53a. $x - 1/y$
53b. $x^3 - y^3$

41a. 16
41b. 0
41c. 9
55a. 4
55b. 3.31
55c. 0

43a. 1/4
43b. 1
43c. 1/9
55d. 1
55e. 4

45. $s \approx 11.5$
    $s \approx 21$

47a. \$3,663.65
47b. \$4,018.76

49. $x^{35/4}y^{17/2}$
51. $x^{2/3}y^{19/6}$

57a. Domain of g = $[0, \infty)$ 57a. Domain of h = $[0, \infty)$
     Range of g = $[0, \infty)$ Range of h = $[-8, \infty)$
                                    57b. $x = 4$

PROBLEM SET 5.4 (on page 250)

1a. $3\sqrt{3}$
1b. $-3\sqrt[3]{2}$

3a. $\dfrac{\sqrt{5}}{2}$
3b. $-\dfrac{1}{2}\sqrt[3]{5}$

5a. $4\sqrt{3x}$
5b. $2y\sqrt[3]{3}$

7a. $4c^2\sqrt{2c}$
7b. $-2p^2\sqrt[3]{4}$

9a. $\dfrac{\sqrt{7}}{2x}$
9b. $-2p\sqrt[5]{p}$

11a. $\dfrac{\sqrt{3}}{5x}$
11b. $\dfrac{\sqrt{5}}{3x^2}$

13a. 2a
13b. $x\sqrt{2}$

15a. $x^2\sqrt[3]{5}$
15b. x^2

17. $5ab^2c^3$
19. $3xy^3z^4$

21a. $2\sqrt[3]{4}$
21b. x

23a. $\sqrt[6]{432}$
23b. $\sqrt[6]{5488}$

25a. $\sqrt[6]{x^5}$
25b. $\sqrt[4]{x^3}$

27a. $\sqrt[6]{(5x)^5}$

27b. $2\sqrt[6]{2x^2y^3}$

29a. $2yz\sqrt[5]{2x^3y}$
29b. \$18,664,000

31a. 1.6266
31b. 1.6266
31c. $\sqrt[4]{7} = \sqrt{\sqrt{7}}$

33a. 1.732
     1.316
     1.147
     1.071
     1.035

33b. seems to
     approach 1

35a. $f(1) - 2$, $f(2) = 5.66$, $f(2.5) = 7.91$
     $g(1) = 2$, $g(2) = 5.66$, $g(2.5) = 7.91$
35b. $x \geq 0$
35d. Domain = Range = $[0, \infty)$
35e. yes

35c.

37. $x^2\sqrt[8]{x^3y}$

39. $\dfrac{54x^2\sqrt{x}}{5}$

41. $\dfrac{5}{2}\sqrt[3]{p^2q}$

PROBLEM SET 5.5 (on page 257)

1a. $8\sqrt{7}$ 3a. $8\sqrt{5}$ 5. $4\sqrt{x} + 5\sqrt{y}$ 9. $-10\sqrt{2}$ 13. $18\sqrt[3]{3}$

1b. $14\sqrt{11}$ 3b. $7\sqrt[5]{2}$ 7. $5\sqrt{2}$ 11. $18\sqrt{5}$ 15. $3\sqrt{3x} - \sqrt{3y}$

17a. $\sqrt{6} - 2$ 19a. $\sqrt{3x} - 3\sqrt{2}$ 21a. $2\sqrt{33} - 44$ 23a. $34 + 3\sqrt{6}$ 25a. $4x - 12\sqrt{x} + 9$
17b. $2\sqrt{3} - 15$ 19b. $4\sqrt{N} - 3N$ 21b. $7 + \sqrt{5}$ 23b. $7 - 4\sqrt{3}$ 25b. -4
                                                                                                                                    27a. 42
                                                                                                                                    27b. 9x - 121

29a. $\dfrac{2\sqrt{3}}{3}$ 31a. $\sqrt{2}$ 33a. $\dfrac{8\sqrt{11}}{77}$ 35. $\dfrac{5\sqrt[3]{3}}{3}$ 39. $7(\sqrt{10} - 3)$

29b. $\dfrac{3\sqrt{21}}{7}$ 31b. $\dfrac{\sqrt{10}}{4}$ 33b. $\dfrac{2\sqrt{5x}}{3x}$ 37. $3(\sqrt{5} + 2)$ 41. $\sqrt{5} - \sqrt{2}$

43. $\dfrac{8\left(2\sqrt{7} + \sqrt{5}\right)}{23}$ 45. $\dfrac{\sqrt{35} - 2\sqrt{7}}{7}$ 47. $\dfrac{\sqrt{21} + 3}{6}$ 49a. $\dfrac{4 + 3\sqrt{3}}{6}$ 49b. $\dfrac{-3 + 2\sqrt{11}}{5}$

51a. 54 in²   53c. $\left(\sqrt{63} + \sqrt{2}\right)\left(\sqrt{63} - 3\sqrt{2}\right)$ 55a. $\dfrac{17\sqrt{3}}{6} \approx 4.91$ 55c. $\dfrac{5\sqrt{15}}{3} \approx 6.45$
51b. 31.84 in²
53a. 34.55 $= 63 - 3\sqrt{126} + \sqrt{126} - 6$
53b. 34.55 $= 57 - 2\sqrt{126}$ 55b. $\dfrac{38\sqrt{7}}{7} \approx 14.36$

57a. 1.41, 2.73, 3.41, 4.45
57b. (b)   59a. 8.12, 3.73, 2.41, 1
                  59b. (a)
57c. Domain = [1,∞), Range = [$\sqrt{2}$,∞)   59c. Domain = (-∞,0], Range = [1,∞)

61a. $\sqrt{m}\,(m - 2m^3 + 3m^4)$ 63a. $\dfrac{5x\sqrt{2x}}{y}$ 63b. $28\sqrt[3]{x}$

61b. $2y^2\,\sqrt[5]{xy^3} - 8y\sqrt[5]{x^2y} + x^2\,\sqrt[5]{xy^3}$

65a.   65b. 1.41, 3.41, 4.45,   65c.
                                       5.99, 6.63
$\dfrac{2}{\sqrt{x+2} - \sqrt{x}} \cdot \dfrac{\sqrt{x+2} + \sqrt{x}}{\sqrt{x+2} + \sqrt{x}}$ 65d. Domain of f =   67. $\dfrac{3}{\sqrt{10} + 2}$
                                       Domain of g = [0,∞)
$= \dfrac{2\left(\sqrt{x+2} + \sqrt{x}\right)}{2}$ Range of f = Range   69.
                                       of g = [$\sqrt{2}$,∞). $15 + \sqrt{2} + \sqrt{3} +$
                                       f and g are
$= \sqrt{x+2} + \sqrt{x}$ algebraically $\sqrt{4} + \ldots + \sqrt{15}$
                                       equivalent ≈ 54.47 in

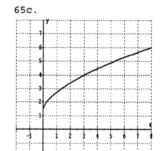

PROBLEM SET 5.6 (on page 263)

1a. 9   3a. 25   5. 25   9. 11/2   13. no solution
1b. 10   3b. -4   7. 3   11. 5   15. no solution
17. 7/2   21. -10   25. 1   29a. ± 8   31a. ± 3
19. 0   23. 7   27. 1   29b. ± 3   31b. ± 2
33. -7, 3   37. -1/4, 7/4   41. -243   45. ± 128   49. 2, 4
35. -7/2, 5/2   39. ± 8   43. ± 64   47. 1/81   51. -7, 9
53a. 301.72 in²   55a. 18.0 ft   57. g earth = 32.08 ft/sec²   59. 2145 lbs.
53b. 14,657,415 mi²   g moon = 5.35 ft/sec²   61. no solution
          55b. $\left(\sqrt[4]{x - 2}\right)^2 = 10^2 + 15^2$ 63. 8
                                                                                                   65. -2/3
                  105,627
67a. 16   69a. no solution
67b. -16   69b. 0
67c. graph b   69c. graph d

PROBLEM SET 5.7 (on page 270)

1a. 4i   3a. $-2i\sqrt{2}$ 5a. -3i/4   7a. 2   9. $8\sqrt{2}\,i$ 13. $-2(\sqrt{5} + \sqrt{6})$
1b. 9i   3b. $6i\sqrt{2}$ 5b. -5i/2   7b. 4   11. -10   15. x = 5, y = 4

17a. -1 + 9i   19a. 1 - i   21. 0 + 6i   25. 28 + 14i   29. 21 - i   33. 45 + 28i
17b. 13 - 17i   19b. 0 + 4i   23. 3 - 6i   27. 14 - 8i   31. 27 - 5i   35. 41 + 0i

37. 169 + 0i

39a. $\dfrac{4}{25} - \dfrac{3}{25}$ i

39b. -1 + 0i

41a. $\dfrac{35}{41} + \dfrac{28}{41}$ i

41b. $-\dfrac{3}{26} + \dfrac{37}{26}$ i

43a. $\dfrac{7}{13} - \dfrac{4}{13}$ i

43b. $\dfrac{66}{125} - \dfrac{12}{125}$ i

45a. i
45b. -i
45c. -1

47a. 1
47b. i
47c. i

49. 4 - 2i

51. 0 + 0i

53. -3 - 7i

55. 0 + 0i

57. $\dfrac{10}{841} + \dfrac{21}{1682}$ i

59a.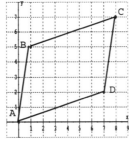

59b. $\overline{AC} = \sqrt{113}$
59c. 8 + 7i
59d. $\sqrt{113}$

CHAPTER 5 REVIEW PROBLEM SET (on page 271)

1a. 3
1b. 40/9
1c. 1/500
1d. $\dfrac{y^6}{x^4}$

3a. 81
3b. 1/8
3c. b^5

5a. $\dfrac{1}{x^{13/20}}$

5b. $\dfrac{1}{x^3}$

5c. $\dfrac{x^9 y^{12}}{8}$

7a. $\dfrac{y^5}{x}$

7b. $\dfrac{a^7}{b^6}$

7c. $\dfrac{4}{x^8 y^4}$

9a. -2
9b. 1

11a. 9
11b. -4

11c. not a real number, an even index and a negative radicand

13a. 1/8
13b. 2
13c. 1/7776

15a. $(-7)^{1/3}$

15b. $x^{6/5}$

17a. $\dfrac{1}{\sqrt[3]{7^2}}$

17b. $\sqrt[5]{(x+1)^2}$

19a. $5\sqrt{5}$
19b. -5
19c. $2t\sqrt[4]{2t}$

21a. 2/5

21b. $-\dfrac{\sqrt[3]{7}}{3}$

21c. $\dfrac{\sqrt[5]{3}}{2}$

23a. t/2
23b. 2x/3

23c. $\dfrac{\sqrt[3]{5}}{x^3}$

25a. $5x^3$
25b. $2x^2$
27a. $4x^2 y^3$
27b. 2x

29a. $\dfrac{y^6}{8x^3}$

29b. $\dfrac{8}{x^3 y^6}$

31a. $8\sqrt{2}$
31b. $10\sqrt{2}$

33a. $5\sqrt{7x}$
33b. $4\sqrt[3]{2y}$

35a. $12 + 4\sqrt{5}$
35b. $4x - 12\sqrt{xy} + 9y$

37a. 6
37b. $a - b^2$

39a. $2x - \sqrt{xy} - y$
39b. $3x + 5y\sqrt{x} - 2y^2$

41a. $\dfrac{2\sqrt{7}}{7}$

41b. $\dfrac{3\sqrt{2}}{10}$

43a. $2 - \sqrt{3}$
43b. $-\sqrt{14} - \sqrt{21}$

45. $\dfrac{x + 2\sqrt{x} + 1}{x - 1}$

47a. $-\dfrac{2}{\sqrt{5} - 3}$

47b. $\dfrac{4}{\sqrt{21} - 3}$

49a. $\sqrt{10} \approx 3.16$
49b. $\sqrt{181} \approx 13.45$

51a. 64
51b. 6

53a. 11
53b. 1, -3

55a. 0 + 2i
55b. -12 + 0i
55c. 2 + 0i

57a. 5 + 4i
57b. 35 + 5i
57c. 5 + 0i

59a. $\dfrac{14}{17} + \dfrac{5}{17}$ i

59b. $\dfrac{5}{13} + \dfrac{1}{13}$ i

61a. 4
61b. 1
61c. 0.12
61d. 5.28

63a. 2.65
63b. 1
63c. 2.86
63d. 0

65a. 1/4
65b. undefined
65c. 0.47
65d. 0.38

67. $1,235.02
69. 5.92 ft
71. Domain = [-1/3, ∞)
Range = [-4, ∞)
x intercept 5

71.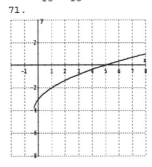

73a.
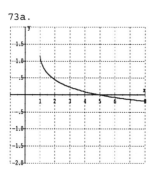

73b. Domain = [1,∞)
     Range = (-∞,1.236]
73c. x intercept 5

75a.
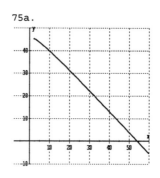

75b. Domain = [2,∞)
     Range = (-∞,45]
75c. x intercept 54.23

CHAPTER 5 PRACTICE TEST (on page 273)

1a. 1	2a. 4	4a. -3
1b. 49/9	2b. -1	4b. 5x
1c. 81	2c. 3/2	4c. 3x
		4d. not possible

1d. $\dfrac{1}{x^4}$

3a. 2
3b. not a real number
3c. -5
3d. -3

1e. $\dfrac{1}{xy^5}$

5. d = 10
6a. f(0) = -2
6b. f(-19) = -3

7a. 1/2
7b. 8
7c. 9
7d. 1/16

8a. $x^{2/7}$

8b. $\sqrt[11]{x^3}$

9a. 1/5
9b. 9
9c. 2

10a. f(-8) = 4
10b. f(0) = 0
10c. f(27) = 9

11a. $3\sqrt[3]{2}$

11b. $\dfrac{\sqrt{3}}{5x}$

11c. $-3x\sqrt[3]{3x}$

11d. $x\sqrt{15}$

12. $\sqrt[6]{432}$

13a. $6\sqrt{3}$
13b. $3\sqrt{5} + 3\sqrt{2}$
13c. $19 + 4\sqrt{21}$
13d. 5

14a. $\dfrac{\sqrt{21}}{3}$

14b. $\sqrt{3} - 1$

15a. 1
15b. $1 + \sqrt{2}$
15c. $2 + \sqrt{5}$

16a. ± 8
16b. 4
16c. 10

16d. $\pm \dfrac{5}{2}$

17a. 3 + 3i
17b. -5 + 2i

17c. 12 - i

17d. $\dfrac{1}{13} + \dfrac{5}{13}i$

17e. 1

18. $6.38

--CHAPTER 6--

PROBLEM SET 6.1 (on page 280)

1a. -1, -6	3a. 2, 4	5a. 3	7a. -1, 8	9a. -5,7
1b. 1, 11	3b. 1, 36	5b. 2/3	7b. -11, 6	9b. -7, 5
11a. -5/2, 2/3	13a. -5/2, 4/3	15a. 0, 7	17. -3/2, 11	21. -4/3, 3/2
11b. -2/3, 5/2	13b. -2/5, 7/2	15b. 0, 9/2	19. ± 3	23. -1, 1/3
25. -7/2, 1/9	29. 0, 2, -3, -7/2	35a. 32, 12, 0, 0	37a. 20,0,-7,-15,0	37c. -3, 5
27. 1, -2, 3/2	31. -2, ± 1	35c. 0, 4	37b. yes	37d. same
	33. -12, -1/2, 0	35d. same		
39. 4 sec	41a. t ≥ 0	43. 3/2, 3		51. c, 4, -1
	41b. 100 ft	45. no solution	49. 0, $\dfrac{1}{b-a}$	53. b, 5
	41c. 5 min	47. -3a, 5a		

55. The solutions of a quadratic
    equation are the same as the x
    intercepts of the graph of the
    associated function.

57. 5, -2

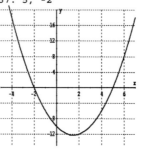

59. 8,-3

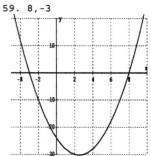

PROBLEM SET 6.2 (on page 287)

1a. ± 8i	3a. $\dfrac{-1 \pm 6i}{2}$	5. 2 ± 2i	7a. 9	9a. 49/4
1b. ± 3i	3b. -2 ± 5i		7b. 16	9b. 121/4

11a. -3, 1

11b. $-2 \pm \sqrt{11}$
    $\approx -5.32, 1.32$

13a. -1, 6
13b. -4, 1

15. $2 \pm \sqrt{5}$
   $\approx 4.24, -0.24$

17. $\dfrac{2 \pm \sqrt{7}}{2}$
   $\approx 2.32, -0.32$

19. $\dfrac{4 \pm \sqrt{101}}{5}$
   $\approx 2.81, 1.21$

21. $\dfrac{-7 \pm \sqrt{129}}{8}$

$\approx 0.54, -2.29$

23. $\dfrac{-5 \pm \sqrt{53}}{14}$

$\approx -0.88, 0.16$

25. $\dfrac{3 \pm \sqrt{14}}{4}$

$\approx -0.19, 1.69$

27a. -5, -6
27b. $(x - 1)^2 - 6$
27c. (1,-6) low point
27d. $[-6,\infty)$

29a. -3, 77/16
29b. $(x + 5/2)^2 - 37/4$
29c. (-5/2,-37/4) lowest
29d. $[-37/4,\infty)$

31a. $0.0375(x - 20)^2 + 15$
31b. (20,15) lowest

33a. 25
33b. ± 30

35a. $\sqrt{A}$

35b. $\dfrac{1}{2}\sqrt{\dfrac{S}{\pi}}$

37a. $\sqrt{A} - 1$

37b. $\sqrt{\dfrac{2h}{g}}$

39. $\dfrac{v \pm \sqrt{v^2 - 4gs}}{2g}$

41. $\dfrac{3 \pm \sqrt{6}i}{3}$

43. $\dfrac{-5 \pm i}{3}$

45. $\dfrac{-2 \pm \sqrt{3}i}{2}$

47a. 0
47b. 0

49a. -3
49b. 1/3

51. graph c
$(x - 3)^2 - 1$
lowest point (3,-1)

53. graph a
$4(x - 5/2)^2 - 5$
lowest point (5/2,-5)

55.
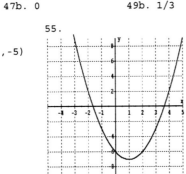

55a. -1.6, 3.6
55b. $1 \pm \sqrt{7}$
55c. same

57a. -1.5, -0.6
57b. -1.5, -2/3
57c. same

57.
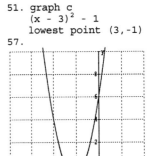

PROBLEM SET 6.3 (on page 294)

1a. 1,4
1b. -1,3

3a. -2, -2/3
3b. 4/3, 5

5. 1/2, 3/2
7. -1/2, 5/3

9. $\dfrac{7 \pm \sqrt{97}}{12}$

$\approx -0.24, 1.40$

11. $\dfrac{1 \pm \sqrt{34} i}{5}$

13. $\dfrac{1 \pm 2\sqrt{2} i}{6}$

15. $\dfrac{1 \pm \sqrt{83} i}{6}$

17. $-8 \pm 8\sqrt{2}$

$\approx -19.31, 3.31$

19. $3 \pm \sqrt{13}$

$\approx -0.61, 6.61$

21. D = 64, 2 unequal real solutions

23. D = -11, 2 complex solutions

25. D = 0, one real solution

27. D = 41,2 unequal real solutions

29a. 3.76 and 8.24 sec
29b. 6 sec

29c. $\dfrac{12 \pm \sqrt{5} i}{2}$, the object was never 600 ft above the ground

31. 0.19 and 1.06 sec
33. 640

35a. a = 1
b = -9
c = 8
x = 8, 1
$x^2 - 9x + 8 = 0$

35b. a = 1
b = -4
c = 3
x = 1, 3
$x^2 - 4x + 3 = 0$

37. -3.42, 6.66
39. -3.91, 1.81

41. $\dfrac{-mn \pm m\sqrt{n^2 + 4n}}{2n}$

43. $\dfrac{-CR \pm \sqrt{C^2R^2 - 4CL}}{2CL}$

45a. k = -9/8
45b. k > -9/8
45c. k < -9/8

47a. k = 4
47b. k < 4
47c. k > 4

49a. -(5/3) = -1/3 + 2
49b. -2/3 = 2·(-1/3)

51a. $x^2+ 3x - 18 = 0$
51b. $x^2+ 5x + 4 = 0$

53a. $x^2 - 5 = 0$
53b. $x^2 + 4 = 0$

55a. graph A; -1.12, 7.12
55b. low point (3,-17)

57a. graph D; -0.64, 3.14
57b. low point (5/4,-57/8)

59. (0.59,0), (3.41,0)

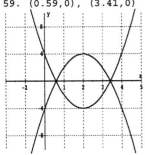

61. (-0.43,0), (0.77,0)

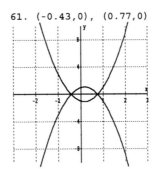

PROBLEM SET 6.4 (on page 301)

1. 3/7, 4
3. 1/4, 1/2
5. -5/2,1
7. 9
9. 4
11. 4
17a. ± 1, ± 4
19a. ± 1
21. 2. 3
13. 0, 3
15a. ± 2, ± √3
17b. 1, 4096
19b. 1
23. ± 1, -3
15b. 1/4, 1/3
17c. 256, 1
19c. 4
25. 4, 36
15c. 9, 16
27. 1/3 -2, 2/3, -1
29. 15 years
35. 1/2, 5/4
39. 1
43. ± 2, ± 5
31. 16
33. -9, 3, -3 ± 2√3
37. -1/4
41. 16
45. 1,5
47. 27, 125
55a. 2
49. 1
51a. ± 3
53a. ± 3
55b. (2,0)
51b. (-3,0), (3,0)
53b. (-3,0), (3,0)
55c. 2 yes
51c. -3,3 yes
53c. -3,3 yes

PROBLEM SET 6.5 (on page 308)

1a. 5,6
3. 20 in by 37 in
7. 5 ft
11. 6% and 8%
15. 12, 13 years
1b. -8,-7
5. 3 ft
9. 8 ft
13. 20, 30 mph
17. 4,6
19a. 18.19 in
27. 12%
33. 7.5 mph
19b. 1.52 ft
25a. $\dfrac{-1 + \sqrt{5}}{2}$
29. 24 min
35. 15 kph
21. 6,8,10
25b. $\dfrac{5 + 5\sqrt{5}}{2}$ in
31. 300, 400 mph
37. 6, 12 hrs
23. 2 ft
39. 78.8 in
41. 7.16 hrs
43. 350 or 330 rackets
45. 13.1, 8.1 hrs
47. 120 cents per gal

PROBLEM SET 6.6 (on page 317)

1. (-3,4)
3. [-5,2]
5. (-∞,-3] or [1,∞)
7. (-1,2)
9. (-∞,1] or [2,∞)

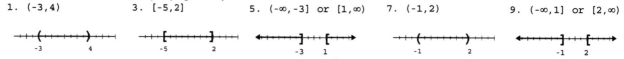

11. $\left(-\infty,-\dfrac{\sqrt{15}}{3}\right]$ or $\left[\dfrac{\sqrt{15}}{3}, \infty\right)$
13. (-∞,∞)
15. {3/4}
17. (-∞,-3] or [1,∞)

19. (-`1,2)

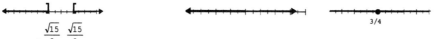

21. (-∞,1] or [2,∞)

23. (3,∞)
25. (-4,1)
27. [-1,2)
29. (-∞,-2) or (1/2,∞)

31. (-∞,-4) or [0,∞)
33. (3,∞)
39. (-∞,-3) or (0,3)
41. [-3,0] or [3,∞)

35. (-4,1)

37. [-1,2)

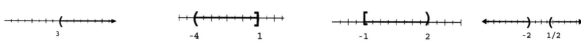

43a. (0,180)
43b. (180,∞)
43c. [80,100]
45a. 0.25 < t < 1.75 sec
45b. 2.75 < t < 3.45 sec
47. (-∞,-2] or [0,2]
49. (-3,-1) or (2,∞)
51. (-∞,-4) or [1,3]

53. (-3,-1) or (2,∞)
57. (-∞,0) or [1/3,∞)
59. (-3,-1) or (2,∞)
61. (-∞,-4) or [1,3]

55. (-∞,3)

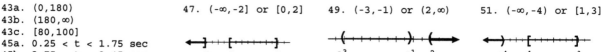

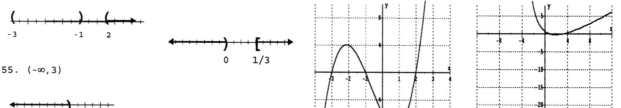

PROBLEM SET 6.7 (on page 327)

1.
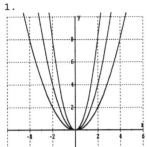

3.
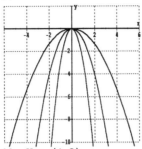

5.
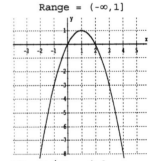

7.

9. V = (2,1)
    Axis x = 2
    Domain = Reals
    Range = [1,∞)

11. V = (4,2)
    Axis x = 4
    Domain = Reals
    Range = (-∞,2]

13. V = (-2,1)
    Axis x = -2
    Domain = Reals
    Range = (-∞,1]

15. V = (-3,2)
    Axis x = -3
    Domain = Reals
    Range = [2,∞)

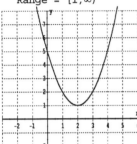

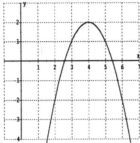

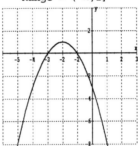

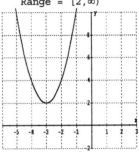

17. V = (1,1)
    Axis x = 1
    Domain = Reals
    Range = (-∞,1]

19. V = (-3,1)
    Axis x = -3
    Domain = Reals
    Range = (-∞,1]

21a. $(x - 1)^2 - 1$
     V = (1,-1)
     Axis x = 1
     Range = [-1,∞)

21b. $-(x - 1)^2 + 1$
     V = (1,1)
     Axis x = 1
     Range = (-∞,1]

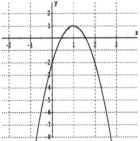

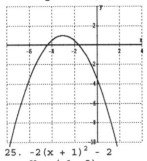

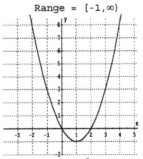

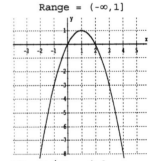

23. $(x - 3)^2 + 1$
    V = (3,1)
    Axis x = 3
    Range = [1,∞)

25. $-2(x + 1)^2 - 2$
    V = (-1,-2)
    Axis x = -1
    Range = (-∞,-2]

27. $2(x + 1.5)^2 + 4.5$
    V = (-3/2,9/2)
    Axis x = -3/2
    Range = [9/2,∞)

29a. x int = ± 2
     y int = 4
     V = (0,4)
     Axis x = 0
     Range = (-∞,4]

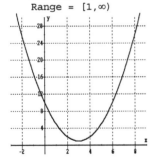

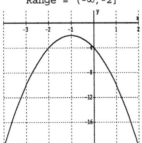

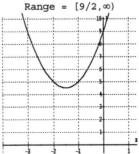

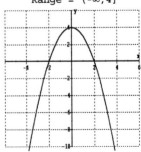

29b. x int = ± 2
     y int = -4
     V = (0,-4)
     Axis x = 0
     Range = [-4,∞)

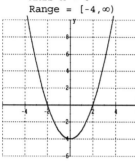

31. x int = 5
    y int = 25
    V = (5,0)
    Axis x = 5
    Range = [0,∞)

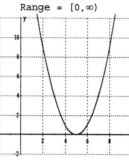

33. x int = -1,3
    y int = -3
    V = (1,-4)
    Axis x = 1
    Range = [-4,∞)

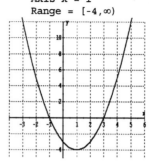

35. V = (7,5) min = 5
37. V = (-6,1) max = 1
39. V = (1,2) min = 2

41a. 560 ft, 320 ft,
     already on ground
41b. -16(t - 6)² + 576
41c. V = (6,576)
     max = 576 ft
41d. 6 sec
41e. 12 sec

41.

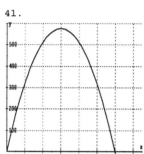

43a. 0, $91.13, $118.13, $121.13
43b. $0.80 or $1.00
43c. R(p) = -150(p - 0.9)² + 121.50
43d. V = (0.9,121.5) max $121.50

45a. length = 20 - x
45b. 0 < x < 20
45c. V = (10,100)
45d. 10 m by 10 m
45e. 100 m²

45c.

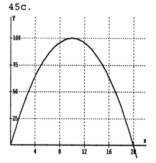

47a. length $\frac{600 - 3x}{2}$

47b. 0 < x < 200
47c. V = (100, 15,000)
47d. max area = 15,000 ft²
47e. w = 100 ft
     l = 150 ft

47c.

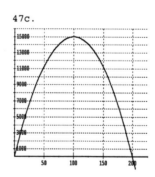

49. B

51. C

53. A

55a. -(x - 2)² - 4
55b. x² - 4

57. 2(x - 3)² + 2
59. -16x² + 80x + 20

59.

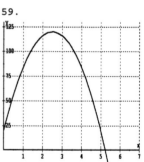

61a. 10 and 10

61b. -3 and 3

PROBLEM SET 6.8 (on page 334)

1a. V = (0,0)
    axis y = 0

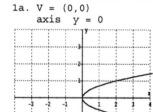

1b. V = (0,0)
    axis y = 0

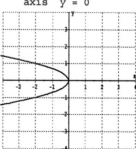

3a. V = (0,0)
    axis y = 0

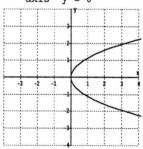

3b. V = (0,0)
    axis y = 0

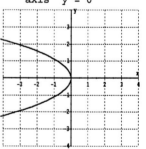

5a. V = (0,0)
    axis y = 0

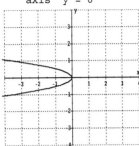

5b. V = (0,0)
    axis y = 0

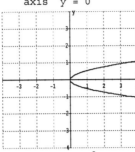

7. V = (3,1)
   axis y = 1

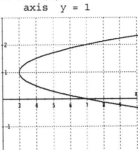

9. V = (-7,4)
   axis y = 4

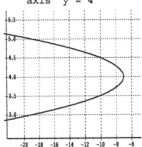

11. V = (-7,4)
    axis y = 4

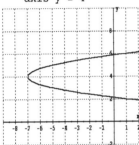

13. x = (y + 2)² - 7
    V = (-7,-2)
    axis y = -2
15. x = -2(y - 5/2)² + 11/2
    V = (11/2,5/2)
    axis y = 5/2
17. x = -(y - 2)² - 1
    V = (-1,2)
    axis y = 2
19. x = (y + 1)² + 2
    V = (2,-1)
    axis y = -1
21. x = 2(y + 1)² - 5
    V = (-5,-1)
    axis y = -1

23a. C = (0,0) r = 1

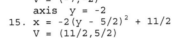
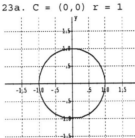

23b. C = (0,0) r = 3

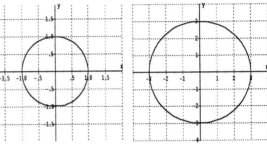

25. C = (1,-2), r = 2

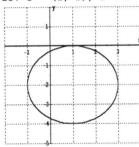

27. C = (-1,2) r = 2

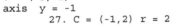
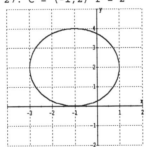

29. C = (2,-4) r = 4

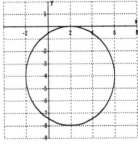

31. C = (3,-4) r = 5

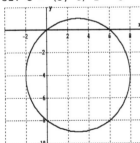

33. C = (-1/2,1/3) r = 1/2

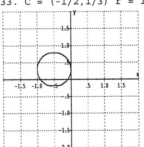

35a. x² + y² = 16
35b. x² + y² = 9
37a. (x + 4)² + (y - 3)² = 48
37b. (x + 5)² + (y + 2)² = 20
39. (x - 2)² + (y - 3)² = 9
     C = (2,3) r = 3
41. (x - 3)² + (y + 2)² = 9
     C = (3,-2) r = 3
43. (x - 3)² + (y + 4)² = 16
     C = (3,-4) r = 4
45. (x + 5)² + (y - 12)² = 25

47a. √(x² + y²) = 25
47b. 24 m

49. x = -(y + 3/2)² - 9/4
     V = (-9/4,-3/2)
     axis y = -3/2

51. (x - 2)² + (y + 3)² = 16

53. graph is a circle
    C = (2,3), r = 3

55a.

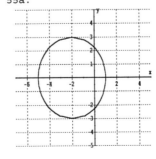

55b.

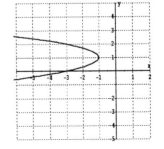

CHAPTER 6 REVIEW PROBLEM SET(on page 336)

1a. -1, 15
1b. 0, 5/2
1c. -2, 12/5

3a. 1 ± 3i

3b. $\dfrac{1 \pm 5i}{3}$

5a. 16
5b. 121/4

7a. 3 ± $\sqrt{2}$
    ≈ 1.59, 4.41
7b. 3/2 ± i

9a. -11.16, 1.16
9b. -2.77, 1.27
11a. $x^2 + 7x + 12 = 0$
11b. $x^2 - 10x + 18 = 0$
11c. $x^2 - 6x + 25 = 0$

13. 1/2, 1/4
15. 4
17. 4

19a. ± 1, ± 2
19b. 1, 1/4
21a. -3 ± i, -3 ± $\sqrt{3}$/3
21b. ± $\sqrt{3}$, ± i

23. 1, 2, -2, -3
25. 1, -3, -1 ± $\sqrt{2}$
27. 8, 27

29. 1
31a. -1, 2/3
31b. -1, 9
33a. -3/8, 0, 5/2
33b. -3, -2/3, 3

35. -8.77, -0.23
37. -7.58, 1.58
39. -1, 2
41. 25, 64
43. ± 1.73, ± 2
45. 2.01

47a. (-5,3)

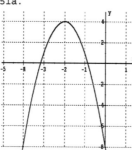

47b. (-1,1) or (2,∞)

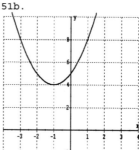

49a. [1/3,7/5)

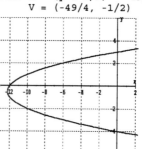

49b. (-∞,-1] or [2,5)

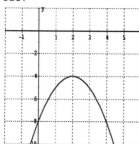

51a. -3(x + 2)² + 4
51a. (-2,4)
51a. x = -3.15, -0.84
    y int = -8
51a. (-∞,4]
51a. max 4

51a.

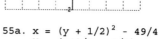

51b.

51b. (x + 1)² + 4
51b. (-1,4)
51b. x int = none
    y int = 5
51b. [4,∞)
51b. min 4

51c.

51c. -(x - 2)² - 4
51c. (2,-4)
51c. x int = none
    y int = -8
51c. (-∞,-4]
51c. min -4

53a. C = (-4,3), r = 2
53b. C = (3,2), r = 3

55a. x = (y + 1/2)² - 49/4
    V = (-49/4, -1/2)

55b. x = -(y + 1/2)² + 9/4
    V = (9/4, -1/2)

57a. $\dfrac{- V \pm \sqrt{V^2 + 2gs}}{g}$

57b. $\dfrac{- \pi h \pm \sqrt{\pi^2 h^2 + \pi s}}{\pi}$

59. ± 2

61. height = 10.1 ft
    base = 5.55 ft
63. 30 mph

65. 9%

67. 10 in by 16 in
or
    24 in by 6.67 in
69a. 64 ft
69b. 4 sec
71. 1 ≤ t ≤ 3

73. -0.25, 1
73. (-∞,-0.25) or (1,∞)

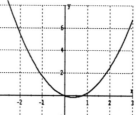

75. -0.51, 3.11
75. [-0.51,3.11]

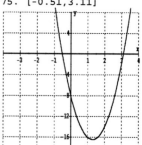

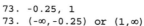

CHAPTER 6 PRACTICE TEST (on page 338)

1a. -3, 5
1b. -1,2
1c. -3, -7/2, 0

2a. -1 ± 2i
2b. 2 ± $\sqrt{2}$ ≈ 0.59, 3.41
2c. $\dfrac{9 \pm 4\sqrt{6}}{3}$
    ≈ 6.27, -0.27

3a. $\dfrac{5 \pm \sqrt{23}\,i}{6}$
3b. 2/3
3c. $\dfrac{-1 \pm \sqrt{10}}{3}$
    ≈ 0.72, -1.39

4. 0
5. 1, 4096
6. a(x^2 - 3x - 18) = 0
   a any integer

7. all reals

8. (-∞,-3/2] or (5,∞)

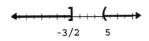

9. V = (1,3)
   axis x = 1
   Domain = Reals
   Range = (-∞,3]

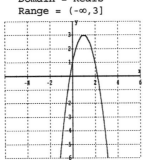

10. 2$(x - 2)^2$ - 11
    Range = [-11,∞)

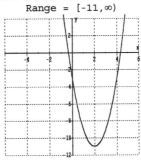

11a. C = (-3,1) r = 5
11b.

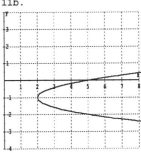

12. 8 ft by 13 ft

13a. 124 ft

13b. 2 sec

--CHAPTER 7--

PROBLEM SET 7.1 (on page 345)

1a. 5.312
1b. 0.086
1c. 5.710
1d. 31.544

3a. 100.313
3b. 104.880
3c. 104.568
3d. 134.686

5a. 13.464
5b. 0.015
5c. 121.510
5d. 16.919

7a. 2252.985
7b. 2463.511
7c. 2859.006
7d. 3628.037

9.. C
11. D
13. B
15. A

17. Domain = (-∞,∞)
    Range = (0,∞)

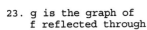

19. Domain = (-∞,∞)
    Range = (0,∞)

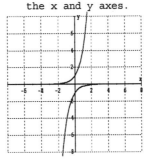

21. Domain = (-∞,∞)
    Range = (0,∞)

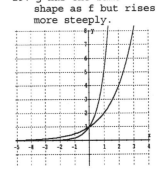

23. g is the graph of
    f reflected through
    the y axis

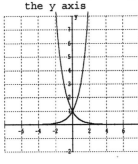

25. g is the graph of f
    moved up 3 units

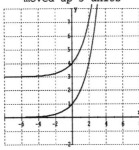

27. g is the graph of f
    reflected through
    the x and y axes.

29. g has the same
    shape as f but rises
    more steeply.

31. g is the graph of
    f move up 2 units

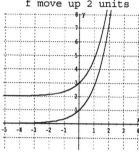

33a. $16,473.43
33b. $16,507.99
33c. $16,525.53
35a. 16.7 million
35b. 17.3 million
37a. $22,263.72
37b. $15,989.46
37c. $13,550.39
39. $422,240
41a. 0.84 ft
41b. 0.77 ft
43a. 32.0 gm
43b. 25.6 gm
43c. 13.1 gm

45a. F
45b. T
45c. F

47a. (0,1)
47b. as the base
     increases the
     steepness of the
     graph increases for
     x > 0

47.

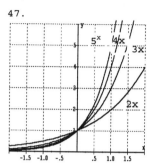

49. the graphs are the
    same

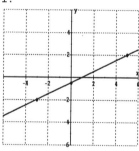

51a. 8247
51b. 8000
51c. 7767

53. $876,287.53

55a.

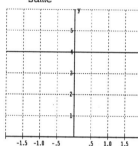

55b.

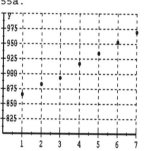

The graph is very close
to the data.

PROBLEM SET 7.2 (on page 352)

1.

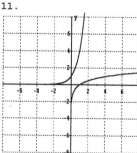

3.

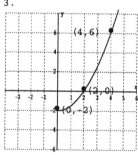

5.

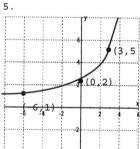

7a. yes
7b. no

9a. 3
9b. 5

11.

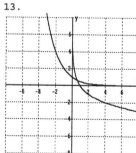

13.

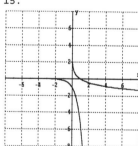

15.

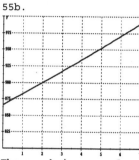

17a. log5 125 = 3
17b. log4 1/16 = -2
19a. log9 3 = 1/2
19b. log1/8 4 = -
2/3
21a. 92 = 81
25b. 62 = 36
23a. (1/3)-2 = 9
23b. 10-1 = 1/10
25a. 4
25b. 27
27a. 3
27b. 5
29a. 2
29b. 3

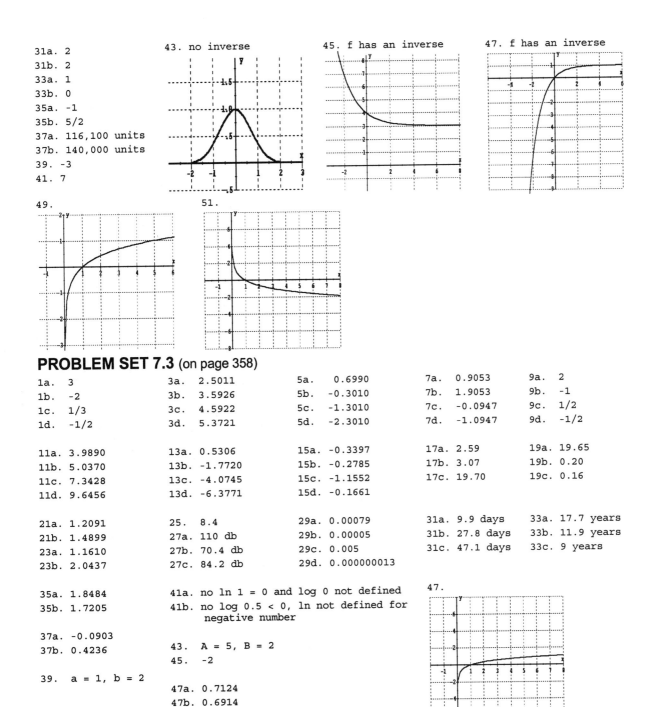

31a. 2
31b. 2
33a. 1
33b. 0
35a. -1
35b. 5/2
37a. 116,100 units
37b. 140,000 units
39. -3
41. 7

43. no inverse

45. f has an inverse

47. f has an inverse

49.

51.

PROBLEM SET 7.3 (on page 358)

1a. 3
1b. -2
1c. 1/3
1d. -1/2

3a. 2.5011
3b. 3.5926
3c. 4.5922
3d. 5.3721

5a. 0.6990
5b. -0.3010
5c. -1.3010
5d. -2.3010

7a. 0.9053
7b. 1.9053
7c. -0.0947
7d. -1.0947

9a. 2
9b. -1
9c. 1/2
9d. -1/2

11a. 3.9890
11b. 5.0370
11c. 7.3428
11d. 9.6456

13a. 0.5306
13b. -1.7720
13c. -4.0745
13d. -6.3771

15a. -0.3397
15b. -0.2785
15c. -1.1552
15d. -0.1661

17a. 2.59
17b. 3.07
17c. 19.70

19a. 19.65
19b. 0.20
19c. 0.16

21a. 1.2091
21b. 1.4899
23a. 1.1610
23b. 2.0437

25. 8.4
27a. 110 db
27b. 70.4 db
27c. 84.2 db

29a. 0.00079
29b. 0.00005
29c. 0.005
29d. 0.000000013

31a. 9.9 days
31b. 27.8 days
31c. 47.1 days

33a. 17.7 years
33b. 11.9 years
33c. 9 years

35a. 1.8484
35b. 1.7205

37a. -0.0903
37b. 0.4236

39. a = 1, b = 2

41a. no ln 1 = 0 and log 0 not defined
41b. no log 0.5 < 0, ln not defined for
     negative number

43. A = 5, B = 2
45. -2

47a. 0.7124
47b. 0.6914
47c. 0.8989

47.

PROBLEM SET 7.4 (on page 366)

1.

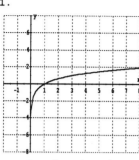

    Domain: $(0, \infty)$
    Range: $(-\infty, \infty)$

3. 4

5. 1/81

7a. $y = -\log_3 x$
7b. $y = \log_3 (x-1)$

9. c

11. d

13. This is the graph of $\log_5 x$ shifted one unit to the left.

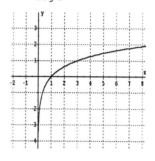

15. This is the same as the graph of ln x shifted 4 unit left.

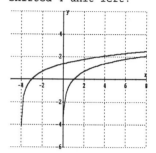

17. Reflect the graph of log x through the x axis and shift left one unit.

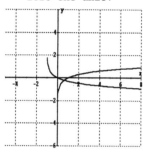

19. ln x is reflected through the x axis and then through the y axis.

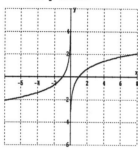

21. $\dfrac{\log x}{\log 3}$

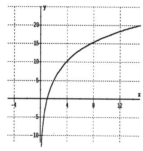

23. $\dfrac{\log x}{\log \frac{1}{3}}$

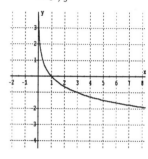

25. $-2\,\dfrac{\log x}{\log 7}$

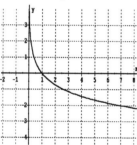

27. $-3\,\dfrac{\log x}{\log \frac{2}{3}}$

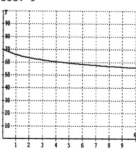

29. $2 + \dfrac{\log x}{\log 3}$

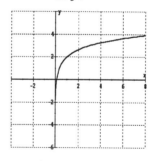

31a. 3300 units
31c. 3000 units
31d. $10,000

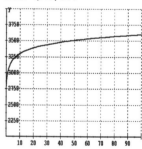

33a. 112.7 mph
33c. 10 mi

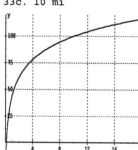

35a. 60
35c. 9

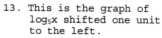

37a. b > 1
37b. 0 < b < 1

39.

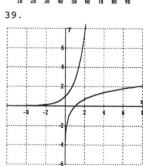

41.

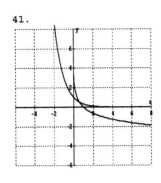

43. $(2,\infty)$
45. $(-\infty,-2)$, $(2,\infty)$

47. $x \neq -2$
49. $x > 2$

51. $x < 1$
53. $x \neq 0$

55. $(0, \infty)$
57. $x > 0$

PROBLEM SET 7.5 (on page 372)

1a. $\log_3 5 + \log_3 y$
1b. $\log_2 7 + \log_2 x$

3a. $\log_5 x - \log_5 3$
3b. $\log_3 t - \log_3 11$

5a. $12\log_4 x$

5b. $\frac{2}{3}\log_2 p$

7a. $\frac{1}{2}\ln Q$

7b. $\frac{1}{3}\ln p$

9a. $\log_3 x + 2\log_3 y - \log_3 4$
9b. $3\log_3 x + \log_3 y - \log_3 4$

11a. $4\log x + \log y$
11b. $\frac{1}{5}\log x + \frac{1}{5}\log y$

13a. $2\log_5 x - 4\log_5 y$
13b. $7\log_3 x - 2\log_3 y$

15a. $\log_3 4x$
15b. $\log_5 3y$

17a. $\ln \frac{4}{x}$

17b. $\log \frac{x}{3}$

19a. $\log x$
19b. $\log_3 y$

21a. $\log_2 x^5$
21b. $\log_3 y^5$

23a. $\log \frac{z^3}{y^2}$

23b. $\log_2 \frac{64}{v^2}$

25. $\log(x^2 - 9)$
27a. 9
27b. y^{-2}
29a. $4x$
29b. $-2y$

31a. $\frac{e}{x}$

31b. $\frac{e^2}{x^3}$

33. $-\log[H^+]$
35a. $\log 5 + \log 2$
35b. $\log 2 - \log 5$

37a. $-\log 2 - \log 5$
37b. $2\log 2 - 2\log 5$
39a. 2.02
39b. 0.05
39c. 0.41

41a. $\frac{4}{7}\log_4 x + \frac{5}{7}\log_4 y$

41b. $\frac{1}{5}\log_5 x + \frac{4}{5}\log_5 y$

43. $7\log c + \frac{2}{9}\log d - \log 5 - \frac{1}{2}\log f$

45. $\ln 4 + \frac{1}{3}\ln x + \frac{1}{6}\ln y$

47a. $\log_a \frac{y}{3}$

47b. $\log_2 \frac{y}{x^2}$

49. $\ln(a + 1)$

51. $\log_3 \frac{x + 5}{x + 1}$

PROBLEM SET 7.6 (on page 378)

1a. $-5/4$
1b. 8
17. 13.42
19. 0.19

3a. -1
3b. 0
21. 16
23. -3,1

5. -3,1
7. 2.87

25. $\frac{12 \pm \sqrt{146}}{4}$

9. -0.88
11. 1.33
27. 2
29. 0.49

13. -0.33
15. -1.93
31. 1.70
33a. 42
33b. 8/15

35. 2
37. 3
39. 5
55. 3 years
57. 5090 ft

41. 98
43. 13.9 years

69. 6.77

45a. 3.4 years
45b. 27.7 years
47. 2000

71. 7.96

49. 14 years
51. 26.0

53a. 4222 years
53b. 5728 years

73b. 19 m/s
73c. 8.04 sec

59. 3/4
61. ± 1

63. no solution
65. 0, -2/3

67. 1, e^2

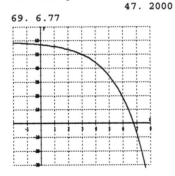

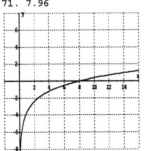

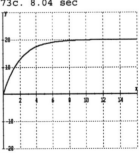

75a. $2,356,000, $3,486,000, $4,049,000

75b. $5,000, $26,000, $94,000

75.

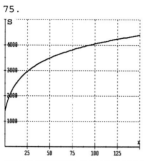

CHAPTER 7 REVIEW PROBLEM SET (on page 380)

1a. 0.00
1b. 0.02
1c. 1
1d. 87.85
1e. 15.67

3a. Domain = reals
    Range = $(0, \infty)$

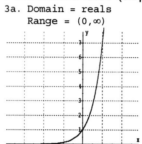

3b. Domain = reals
    Range = $(0, \infty)$

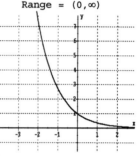

5.

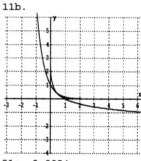

7.

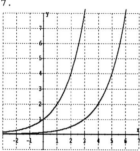

9a. 55.16
9b. 117.51

11a.

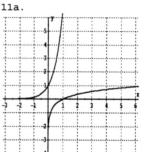

11b.

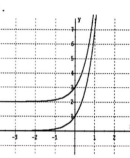

13. 420.78

15a. $\log_4 \frac{1}{16} = -2$

15b. $\log_x 5 = y$

17a. $\pi^1 = \pi$
17b. $1/49 = 7^{-2}$

19a. $\log_2 8 = 3$
19b. $\log_9 3 = 1/2$
19c. $\log_4 64 = 3$
19d. $\log_6 216 = 3$

21a. 3.8924
21b. 2.0283
21c. -2.1427
21d. -3.0748

23a. 0.6297
23b. 0.1079
23c. not real
23c. not real

25a. 2.59
25b. 1.63
25c. 0.02

27a. 2
27b. 2.5
27c. 2

29a. $\log_6 7 + \log_6 x$
29b. $\ln 4 + 3\ln x$
29c. $\log 4 + \log x + 5\log y$

31a. $\log \frac{2}{9}$

31b. $\ln \frac{x+1}{x}$

31c. $\ln \frac{y}{z}$

33a. x

33b. $\dfrac{1}{3e^2}$

33c. x^{2-4}

35. Domain of f = $(0, \infty)$
    Domain of g = $(0, \infty)$
    Domain of h = $(-\infty, 0)$

37. Domain of f = $(0, \infty)$
    Domain of g = $(1, \infty)$
    Domain of h = $x \neq -1$

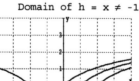

39a. 1
39b. 2.73, -0.73
39c. 0, 4/3

41a. 2.32
41b. 0.23
41c. 0, 0.77

43a. 50
43b. 1

45. $830.61

47a. 4,848 bacteria   49. $122,200   53. Domain = reals   55. Domain = $(0,\infty)$ 57. Domain = $x \neq 0$
47b. 47 hrs   51. 105 mph   Range = $[2,\infty)$ Range = Reals   Range = Reals

CHAPTER 7 PRACTICE TEST (on page 382)

1a. 3.32, 0.30, 1.50
1b. 0.54, 1.85, 0.81
1c. 4.71, 26.61, 6.23

2.

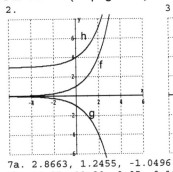

3.

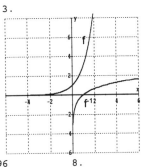

4a. $\log_3 1/9 = -2$

4b. $\left(\dfrac{1}{5}\right)^{-4} = 625$

5a. 3
5b. -3
5c. 3.5
5d. -2

6a. 9
6b. 2.5
6c. 2/7
6d. 1.49
6e. 1/4
6f. 0.46

7a. 2.8663, 1.2455, -1.0496
7b. 25.25, 11.20, 0.05, 0.10

8a. Domain = $(0,\infty)$
8b. Domain = $(-\infty,0)$
8c. Domain = $(-1,\infty)$

8.

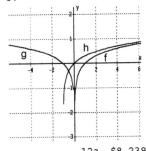

9a. $\log_3 8 + \log_3 x$
9b. $\ln x - \ln 7$
9c. $3\ln x + 4\ln y$
9d. $\frac{2}{3} \log x + \frac{1}{3} \log y$

10a. $\log 2/5$

10b. $\ln \dfrac{x^3}{y^2}$

10c. $\log[x^3(x + 2)]$

11a. 0.00
11b. 1.61
11c. 0.49

12a. $8,239.16
12b. 4.16 years
13. $291,234.00

--CHAPTER 8--

PROBLEM SET 8.1 (on page 393)

1.
$$\begin{bmatrix} 2 & 3 & \vdots & 2 \\ 1 & -2 & \vdots & 8 \end{bmatrix}$$

3.
$$\begin{bmatrix} 1 & -2 & 3 & \vdots & 4 \\ 2 & 1 & -4 & \vdots & 3 \\ -3 & 4 & -1 & \vdots & -2 \end{bmatrix}$$

5.
$$\begin{cases} 2x + 8y = 7 \\ 3x - 5y = 4 \end{cases}$$

7.
$$\begin{cases} 2x + 6y - 4z = 1 \\ x + 3y - 2z = 4 \\ 2x + y - 3z = 7 \end{cases}$$

9. $(-1, 3/2)$
11. $(2,1)$

13a. yes
13b. no, condition 4

15a. yes
15b. yes

17. $(2, -3)$
19. $(-3, 2, 4)$

21. $(2, -3, \text{any number})$

23. $(3,1)$
25. $(2/3, 5/3)$

27. $(-1,1)$
29. $(-4,-6)$

31. $(7/3, 2/3, 2/3)$
33. $(7/3, -1/3, 13/3)$

35. 17 dimes, 28 quarters
37. 30°, 60°, 90°

39. $x = z + 3$
    $y = 3z + 5$
    $z = \text{any number}$

41. no solution
43. $x = 2z - 13$
    $y = 8z - 34$
    $z = \text{any number}$

45. $(0,0,0)$
47. $(-2,-3)$

49. $\left(\dfrac{19}{12}, -\dfrac{145}{24}, -\dfrac{15}{8}, -\dfrac{85}{24}\right)$

PROBLEM SET 8.2 (on page 401)

1a. 3
1b. -13

3a. 22
3b. -17

5. 49
7. 4

9. $(1/3, 2/3)$
11. no solution, inconsistent system

13. $\left(\dfrac{116}{59}, \dfrac{5}{59}\right)$

15. $(1, -3/2)$
17. $(2/5, 21/5)$

19. $\left(\dfrac{19}{4}, -\dfrac{3}{4}, -\dfrac{29}{4}\right)$

21. $\left(-5, -\dfrac{14}{3}, -\dfrac{16}{3}\right)$

23. $(3, -1, -1)$
25. $(-1, 2, 1)$

27. $\left(-5, \dfrac{5}{3}, \dfrac{28}{3}\right)$

29. 30°, 60°
31. $4000, $6500

33. A: 10 oz
    B: 5 oz
    C: 8 oz

35. 25%: 500 lbs
    35%: 2000 lbs
    40%: 1500 lbs

37. $x_1 = 10,000$
    $x_2 = 1000$
    $x_3 = 6000$

39. 72
41. -1
43. 4

45. 2, 8
47a. 6
47b. -6
47c. opposites

49a. -4
49b. -40
49c. 10 times larger
51. 35

PROBLEM SET 8.3 (on page 407)

1. (0,1), (1/2,3/2),
   (0,0), (2,0)

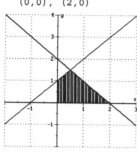

3. (4,5), (10.67,0)

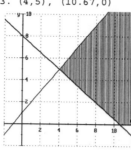

5. (0,4), (8,0), (3,2)

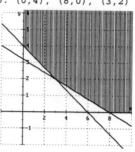

7. (0,5), (5,0),
   (2,6)

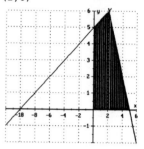

9. min: 21
   max: 24

11. min: 6
    max: 25

13. min: -28
    max: 32

15. min: 0
    max: 24

17. min: 0
    max: 25

19. $\begin{cases} x \geq 0, y \geq 0 \\ x \geq 3y \\ 6x + 86y \leq 20800 \end{cases}$

19.

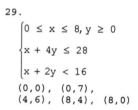

21. A: 50, B: 40

23. center I for 6 days
    center II for 2 days

25. 30 orchid
    10 rosebud

27. 24 acres of flowers
    12 acres of eucalyptus

29.

$\begin{cases} 0 \leq x \leq 8, y \geq 0 \\ x + 4y \leq 28 \\ x + 2y < 16 \end{cases}$

(0,0), (0,7),
(4,6), (8,4), (8,0)

31. min: 5
    max: 30
33. min: 0
    max: 28
35. 3a = 2b

PROBLEM SET 8.4 (on page 417)

1. B
3. D

5. x int ± 4, y int ± 3
   vertices (± 4,0)
   foci (± $\sqrt{7}$,0)

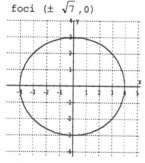

7. x int ± 2, y int ± 4
   vertices (0,± 4)
   foci (0,± 2$\sqrt{3}$)

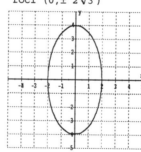

9. x int ± 5, y int ± 2
   vertices (± 5,0)
   foci (±$\sqrt{21}$,0)

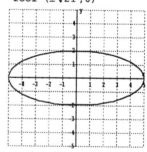

11. x int ± $\sqrt{3}$, y int ± $\sqrt{15}$
    vertices (0,± $\sqrt{15}$)
    foci (0,± 2$\sqrt{3}$)

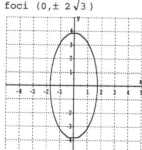

13. x int ± 3
    vertices (± 3,0)
    foci (± $\sqrt{34}$,0)
    asymptotes y = ± 5x/3

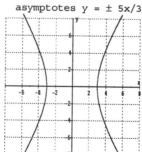

15. x int ±2
    vertices (± 2,0)
    foci (± $\sqrt{13}$,0)
    asymptotes y = ± 3x/2

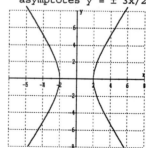

17. x int ± 3
    vertices (± 3,0)
    foci (± $\sqrt{10}$,0)
    asymptotes y = ± x/3

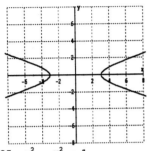

19. y int ± $\sqrt{3}$
    vertices (0,±$\sqrt{3}$)
    foci (0,± $\sqrt{15}$)
    asymptote y = ± x/2

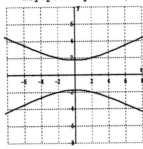

21. y int ± $\sqrt{5}$
    vertices (0,± $\sqrt{5}$)
    foci (0,± 3)
    asymptote y = ± $\sqrt{5}$ x/2

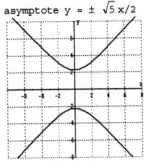

23. $\dfrac{x^2}{4} + \dfrac{y^2}{2} = 1$

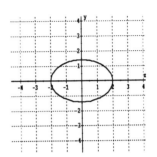

25. $y^2 - x^2 = 1$

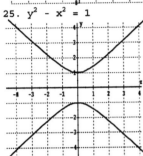

27a.

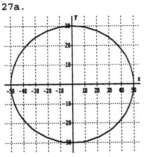

27b. $\dfrac{x^2}{2500} + \dfrac{y^2}{900} = 1$
    foci (± 40,0)

29b.
$\dfrac{x^2}{130^2} - \dfrac{y^2}{560.1^2} = 1$
    for x > 0, y ≥ 0

29c. although $d_2 - d_1$
    is 260 m, the
    exact distances
    cannot be found.

29a.

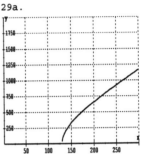

31a.

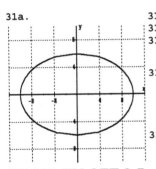

31b. foci (± 8,0)
31c. 20
31d. 36·0 + 100(-6)² = 3600,
    36(10)² + 100 (0) = 3600

31e. $d_1 = \sqrt{8^2 + 6^2} = 10$ and
    $d_2 = \sqrt{8^2 + 6^2} = 10$ so
    $d_1 + d_2 = 20$

33. $\dfrac{(x + 2)^2}{9} + \dfrac{(y + 1)^2}{4} = 1$

35.
$\dfrac{(x - 3)^2}{9} + \dfrac{(y - 4)^2}{16} = 1$

37.
$\dfrac{(x + 2)^2}{4} - \dfrac{(y - 1)^2}{1} = 1$

39.
$\dfrac{(y + 3)^2}{16} - \dfrac{(x - 3)^2}{4} = 1$

41.

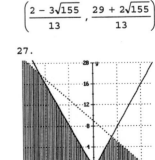

PROBLEM SET 8.5 (on page 423)

1. (-3,-4), (2,1)
3. (-4/5,-22/5), (2,4)

5. $\left(\dfrac{-3 + 2\sqrt{15}}{3}, 2 - \sqrt{15}\right), \left(\dfrac{-3 - 2\sqrt{15}}{3}, 2 + \sqrt{15}\right)$

7. (± 3,2), (± 3,-2)

9. (± 2$\sqrt{6}$,1), (± 2$\sqrt{6}$,-1),

11. (± 2$\sqrt{3}$,2), (± 2$\sqrt{3}$,-2)

13. A (± 6,4/5), (±6, 4/5)

15. C (± 3,2), (± 3,-2)

17. H
$\left(\dfrac{1 + \sqrt{11}}{4}, \dfrac{-1 - 3\sqrt{11}}{4}\right)$
$\left(\dfrac{1 - \sqrt{11}}{4}, \dfrac{-1 + 3\sqrt{11}}{4}\right)$

19. D
$\left(\dfrac{2 + 3\sqrt{155}}{13}, \dfrac{29 - 2\sqrt{155}}{13}\right)$
$\left(\dfrac{2 - 3\sqrt{155}}{13}, \dfrac{29 + 2\sqrt{155}}{13}\right)$

21.

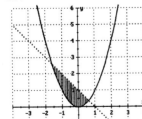

23.

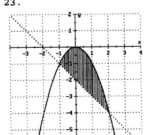

25.

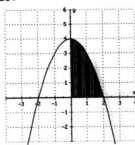

27.

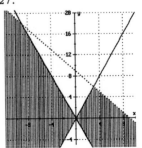

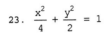
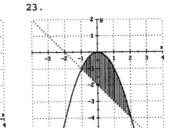
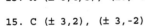

29. 6 in by 8 in

31. x = 5ft, y = 2 ft or
    x = 4.3ft, y = 2.7ft

33. 150 ft by 5 ft

35. 8 m by 12 m or
    6 m by 18 m

37. 8 and 7

39.

$$\left(\sqrt{22}, \frac{3}{2}\sqrt{22}\right)\!\left(-\sqrt{22},-\frac{3}{2}\sqrt{22}\right)$$

41. (1,1)

43. (1,0), (2,1)

45. (25,4)

47.

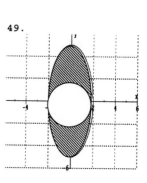

49.

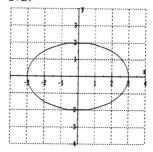

CHAPTER 8 REVIEW PROBLEM SET (on page 425)

1a.
$$\begin{bmatrix} 1 & 4 & \vdots & 5 \\ 3 & -11 & \vdots & 8 \end{bmatrix}$$

1b.
$$\begin{bmatrix} 2 & 3 & 4 & \vdots & -1 \\ -1 & 4 & -1 & \vdots & -15 \\ 1 & -1 & -5 & \vdots & 10 \end{bmatrix}$$

3a.
$$\begin{cases} 4x + y = 2 \\ -2x + 3y = -1 \end{cases}$$

3b.
$$\begin{cases} x + y + z = 0 \\ 3x + y + z = 2 \\ 2x - y + 2z = 3 \end{cases}$$

5a. (5,-3)
5b. (-4,-6,3)
7. (-1,2)

9.   (0.5,0.95,-0.05)
11a. -19
11b. -10

13. 60
15. 14

17. 4
19. (2,-1)
21. (-1,3,2)

23a.

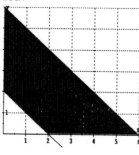

23b. min = 5
     max = 15

25b. min = 8
     max = 12

27b. x int ± 3
     y int ± 2
     foci (±$\sqrt{5}$,0)

25a.

27a.

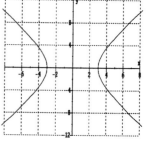

29. x int ± 2$\sqrt{2}$
    y int ± $\sqrt{7}$
    foci (± 1,0)

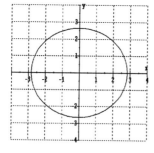

31. x int ± 4
    y int none
    foci (±$\sqrt{23}$,0)

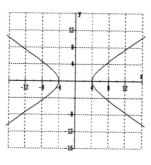

33. x int ± 1
    y int none
    foci (±$\sqrt{10}$,0)

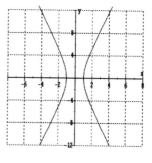

35. $\dfrac{x^2}{9} - \dfrac{y^2}{16} = 1$

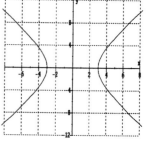

37. $\dfrac{x^2}{33} + \dfrac{y^2}{49} = 1$

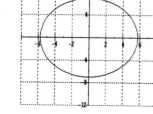

39. (3,-4)

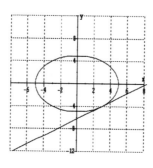

41. (± 5,2), (± 5,-2)

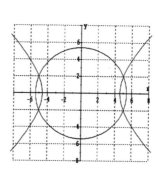

43. 80 lbs at $4.20,
    120 lbs at $3.40

45. $2410.20 at 9.5%

47. $30,931 at 15%
    $22,931 at 8%
    $ 8,138 at -3%

49. 75 A products
    100 B products

51. $10,000 at 5%
    $14,000 at 4%

53.

$$\begin{bmatrix} 1 & 0 & 0 & 0 & \vdots & 3 \\ 0 & 1 & 0 & 0 & \vdots & -2 \\ 0 & 0 & 1 & 0 & \vdots & 1 \\ 0 & 0 & 0 & 1 & \vdots & 1 \end{bmatrix}$$

55.

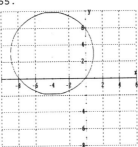

CHAPTER 8 PRACTICE TEST (on page 427)

1a.

$$\begin{bmatrix} 3 & 7 & \vdots & 16 \\ 2 & 5 & \vdots & 13 \end{bmatrix}$$

1b.

$$\begin{bmatrix} 1 & 1 & 1 & \vdots & 6 \\ 3 & -1 & 2 & \vdots & 7 \\ 2 & 3 & -1 & \vdots & 5 \end{bmatrix}$$

2a.
$$\begin{cases} 3x - 4y = 1 \\ -2x + 3y = 5 \end{cases}$$

2b.
$$\begin{cases} 2x \quad\;\; + z = -2 \\ 2x + 3y \quad\;\; = 1 \\ -2x + y - 3z = 5 \end{cases}$$

3a. (-3,7)
3b. infinite number of solutions, x = -2, y = 2 - z

4a.
$$\begin{bmatrix} 7 & 4 & \vdots & 1 \\ 9 & 54 & \vdots & 3 \end{bmatrix}$$
(7/57, 2/57)

4b.
$$\begin{bmatrix} 1 & 1 & 1 & \vdots & 9 \\ 27 & 9 & 3 & \vdots & 93 \\ 8 & 4 & 2 & \vdots & 36 \end{bmatrix}$$
(2,3,4)

5a. -1
5b. 7
6a. (1,1)
6b. (1,1,1)

7a. vertices at (0,0), (0,4), (2,2), (3,0)
7b. max 24

8a. x int ± 1, y int ± 2
     vertices (0,± 2)
     foci (0,± $\sqrt{3}$)

8b. x int ± 3, y int none
     vertices (± 3,0)
     foci (± $\sqrt{13}$,0)

9. $\dfrac{x^2}{25} - \dfrac{y^2}{24} = 1$

10. (5,0), (3,4)

11. initial fee $70
     monthly fee $35

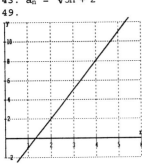

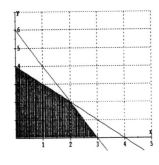

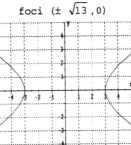

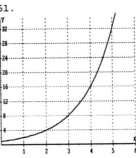

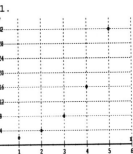

--CHAPTER 9--

PROBLEM SET 9.1 (on page 434)

1. -3, -2, -1, 0, 1
3. 3/2, 4, 15/2, 12, 35/2
5. -1, -1, 0, 2, 5
7. 1, 1/4, 1/9, 1/16, 1/25
9. 1/4, 1/2, 1, 2, 4
11. 2, 4, 2, 4, 2
13. $a_n = 2n$
15. $a_n = 2n - 1$
17. $a_n = n^3$
19. $a_n = (-1)^n 3n$
21. 1, 1/2, 1/4, 1/8, 1/16
23. 1, 3/2, 2, 5/2, 3
25. -10, 0, 10, 20, 30
27. -2, -6, -18, -54, -162
29. $20,800, $21,632, $22,497.28, $23,397.17

31. $1.06, $1.12, $1.19
33. -3, 4, -5, 6, -7
35. $\sqrt{2}$ - 1, $\sqrt{3} - \sqrt{2}$, 2 - $\sqrt{3}$, $\sqrt{5}$ - 2, $\sqrt{6} - \sqrt{5}$

37. 32
39. 9/11
41. 3, 3, 3, 3
43. $a_n = \sqrt{3n + 2}$
45. 1, 5, 13, 29, 61
47. 1, 1, 2, 3, 5, 8

49.

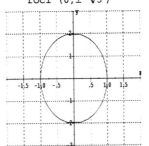

49.

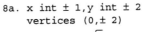

51.

51.

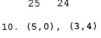

PROBLEM SET 9.2 (on page 439)

1. 511

3. 15

5. 660

7. 15

9. 15

11. 133/60
13. 500

15. $\displaystyle\sum_{k=1}^{6} 4k$

17. $\displaystyle\sum_{k=1}^{6} (9 - 5k)$

19. $\displaystyle\sum_{k=1}^{6} (2 - 3k)$

21. $\displaystyle\sum_{k=1}^{5} \frac{1}{1.1k + 0.1}$

23. 115

25. 5

27. 155

29. 15

31. 6.3

33. $5632.98

35a. 20
35b. 40, twice as large

37a. 2.9290
37b. -0.6456

39. 710
41. 935

43. 560
45. 2

47. 3.92
49. 2.39

51a. $\dfrac{2(x-1)}{x+1} + \dfrac{2}{3}\left(\dfrac{x-1}{x+1}\right)^{3}$

51b.

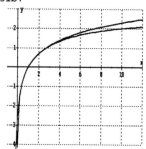

PROBLEM SET 9.3 (on page 444)

1a. $a_N = 3N - 5$
1b. $a_{10} = 25$
1c. $a_{20} = 55$

3a. $a_N = 14 - 3N$
3b. $a_{10} = -16$
3c. $a_{20} = -46$

5a. $a_N = 3N - 1$
5b. $a_{10} = 29$
5c. $a_{20} = 59$

7a. $a_N = 11 - 6N$
7b. $a_{10} = -49$
7c. $a_{20} = -109$

9a. $a_N = -12 + 3N$
9b. $a_{10} = 18$
9c. $a_{20} = 48$

11. N = 25

13. N = 30

15. N = 31

17. $a_{10} = 50$

19. $a_6 = 30$

21. $a_{12} = -73$
23. $a_{10} = -27$

25. $a_1 = -4$,
    $a_{20} = 53$

27. $a_1 = 10$
    $a_{15} = -18$

29. $a_1 = 114$
    $a_{20} = -19$

31. $a_1 = 3/5$
    $a_{40} = 24$

33. $S_{10} = 145$
43. $S_{31} = 2077$

35. $S_8 = -28$
45a. 16 rows
45b. 46 bricks

37. $S_{31} = 2914$
47a. $53,600
47b. $1,060,000

39. $S_{16} = 888$
49a. $31.35
49b. $3,285

41. $S_{19} = 1045$
51. 16a + 4b
53. 21

55.

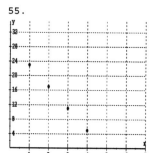

55.

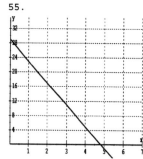

57.

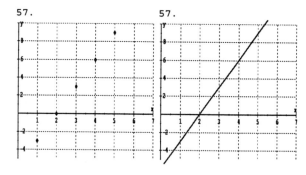

57.

59a. $a_N = 3N - 1$, $b_N = 2N + 5$

59b. $a_N = 5N + 4$

59c. $a_N = 6N^2 + 13N - 5$

PROBLEM SET 9.4 (on page 451)

1. geometric, r = 3

3. geometric, r = 2/3

5. geometric, r = -2/3

7. geometric, r = -1/2

9. geometric, r = 1/5
11. arithmetic, d = -6
13. arithmetic, d = 12

15a. $(-4)\left(-\dfrac{1}{2}\right)^{N-1}$
15b. 1/32

17a. $32\left(\dfrac{1}{2}\right)^{N-1}$
17b. 1/4

19a. $6(2)^{N-1}$
19b. 768

21a. $5(2)^{N-1}$
21b. 640

23a. $\dfrac{1}{2}(-4)^{N-1}$
23b. -8192

25a. $8\left(\dfrac{1}{2}\right)^{N-1}$
25b. 1/16

27. 6th
29. 4

31. 94.5
33. 5460

35. 1640
37. 21.3320

39. 1, 2, 4, 8
41. 2

43. 16
45. 81/2
47. 12/5

49a. $24,000, $25,200, $26,460, $27,783, $29,172.15, $30,630.76

49b. no, because the raises are geometric not arithmetic

49c. $37,231.88

51a. 1262.96 m
51b. 1500 m
53. $114,358.88
55. 10
57. 4/5

59. $0.7\left(\dfrac{1}{1-\dfrac{1}{10}}\right) = \dfrac{7}{9}$

61. $x = 16r$ and $9/4 = xr$, so $x = \pm 6$

63a. 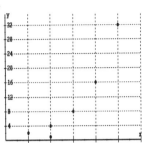 63b. 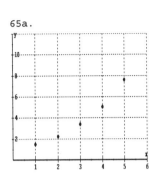 65a. 65b.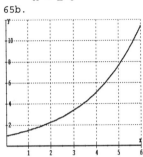

PROBLEMS SET 9.5 (on page 458)

1. $x^4 + 8x^3y + 24x^2y^2 + 32xy^3 + 16y^4$

3. $y^6 + 6y^5 + 15y^4 + 20y^3 + 15y^2 + 6y + 1$

5. $16x^4 - 32x^3y^2 + 24x^2y^4 - 8xy^6 + y^8$

7. 1/30
9. 1/9
11. 2
13. 1
15. $(N + 1)(N)(N - 1)(N - 2)$
17. 56
19. 210
21. 1
23. 66

25. $x^5 + 10x^4 + 40x^3 + 80x^2 + 80x + 32$

27. $x^6 + 12x^4y^2 + 48x^2y^4 + 64y^6$

29. $a^{18} - 6a^{14} + 15a^{10} - 20a^6 + 15a^2 - 6a^{-2} + a^{-6}$

31. $32 + 80\dfrac{x}{y} + 80\dfrac{x^2}{y^2} + 40\dfrac{x^3}{y^3} + 10\dfrac{x^4}{y^4} + \dfrac{x^5}{y^5}$

33. $x^{20} - 20ax^{18} + 180\,a^2x^{16} - 960a^3x^{14}$

35. $\dfrac{x^3}{8}\sqrt{\dfrac{x}{2}} + \dfrac{7x^3y}{4} + 21x^2y^2\sqrt{\dfrac{x}{2}} + 70x^2y^3$

37. $x^{16} + 16x^{15}y + 120x^{14}y^2 + 560x^{13}y^3 + 1820x^{12}y^4$

39. $a^{11} - 22a^{10}b + 220a^9b^2 - 1320a^8b^3 + 5280a^7b^4$

41. $x^7 - 14x^6y + 84x^5y^2 - 280x^4y^3 + 560x^3y^4$

43. $a^{27} - 9a^{26} + 36a^{25} - 84a^{24} + 126a^{23}$

45. $20x^3y^3$

47. $3360t^6$

49. 6th term: $-112266x^6y^5$
    7th term: $336798x^5y^6$

51. $1.6 \times 10^{13}x^{12}y^7$

53. $\dfrac{224}{243}x^6a^{12}$

55. $\dfrac{14}{9}a^5x^{12}$

57. $11,092.63

59. 82

61. $\binom{N}{0}(a + b)^N + \binom{N}{1}(a + b)^{N-1}c + \binom{N}{2}(a + b)^{N-2}c^2 + \ldots + \binom{N}{N-1}(a + b)^1c^{N-1} + \binom{N}{N}c^N$

65a. 6
65b. 10

63. $7x^6 + 21x^5h + 35x^4h^2 + 35x^3h^3 + 21x^2h^4 + 7xh^5 + h^6$

CHAPTER 9 REVIEW PROBLEM SET (on page 459)

1a. -1, 2, -3, 4, -5

1b. 1/4, 1/4, 3/16, 1/8, 5/64

1c. 3, 6, 12, 24, 48

3. $3N^2$

5a. 38

5b. -183

7. $\displaystyle\sum_{k=1}^{6} 2k$

9a. 61
9b. -54
11a. 97
11b. -450

13a. 690
13b. 210
15. 175
17. 16

19. 512

21. 1.4552×10^{-11}

23. -3

25. -29524/9

27. 50/3

29. 50/7

31. 15/4

33a. 1/18

33b. $\dfrac{1}{N - 2}$

35a. 56

35b. 21

37. $x^4 + 12x^3y + 54x^2y^2 + 108xy^3 + 81y^4$

39. $243x^5 + 810x^4 + 1080x^3 + 720x^2 + 240x + 32$

41. $64x^6 - 192x^5y + 240x^4y^2 - 160x^3y^3 + 60x^2y^4 - 12xy^5 + y^6$

43. $p^8 + 4p^6 + 6p^4 + 4p^2 + 1$

45. $69,284,160x^8y^3$

47. $495x^4y^8$

49a. $2020, $2040.20, $2060.60, $2081.21, $2102.02

49b. $2345.16

51a. 3000, 9000, 27 000

51b. $1000(3^N)$

53a. $12,528

53b. $11,472

55a. $y = 3 - 2x$

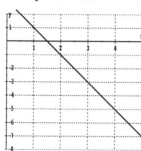

55a. $a_n = 3 - 2n$

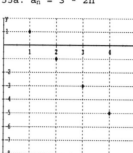

55b. $y = 4x + 1$

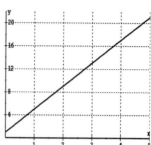

55b. $a_n = 4n + 1$

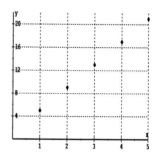

CHAPTER 9 PRACTICE TEST (on page 460)

1a. 3, 6, 11, 18, 27

1b. 1, 5, 25, 125, 625

1c. 2, 6, 18, 54, 162

2a. 25 - 6d

2b. $-1\left(-\dfrac{1}{7}\right)^N$

3a. 37

3b. 44

4a. 5 - 6(N - 1) or 11 - 6N

4b. -49

4c. -1040

5a. $3 \cdot 2^{N-1}$

5b. 49,152

5c. 3069

6. 16

7. 9

8a. 144/335

8b. 35

9. $2048x^{11} + 33,792x^{10}y + 253,440x^9y^2 + 1,140,480x^8y^3 + 3,421,440x^7y^4$

10. $2288x^{10}y^3$

11a. $410,000

11b. $2,440,000

Index

A

Abscissa, 77
Absolute value
  applied problems, 129–130
  of a complex number, 269
  cut point method solving inequalities, 128–129
  definition, 12, 126
  equations, 126–127
  grapher use in solving, 130–131
  inequalities, 127–129
Addition
  of complex numbers, 266, 270
  of integers, 13–14
  and order of operations, 4
  of polynomials, 39–40
  property of equations, 86
  of radical expressions, 252–253
  of rational expressions, 65–67
  of rational numbers, 19–20
Additive inverse, 9
Algebraic expressions
  definition, 4
  and English phrases, 100, 122
Analogous problems, 106
Antilogarithm, 356
Applied problems, 99–106
  analogous, 106
  arithmetic series, 443
  binomial expansion, 457
  circles, 333
  Cramer's rule, 400
  dimensions, 68, 102
  economics, 189–190
  ellipse, 416
  exponential functions, 229–231, 343–344
  factoring, 52
  formulas, 94–95, 371
  geometric series, 449–450
  geometry, 28–30, 45, 101, 206, 250, 256, 304
  grapher used in, 106, 140, 190, 198, 212–213, 324–326, 344–345, 406–407
  inequalities, 122–123, 129–130, 168, 218–219, 315–316
  investment and finance, 30–31, 103, 122, 210–211, 229, 244, 306, 344, 376–377, 457
  linear equations, 158–160
  linear functions, 176–177
  linear programming, 404–406
  logarithmic equations, 376–377
  logarithmic function(s), 352, 357, 365–366, 371
  manufacturing, 130, 140, 176, 198, 212
  matrices, 392
  mixture, 104, 208
  motion, 105, 114–115, 209, 307
  nonlinear systems, 422
  numbers, 100, 112, 269, 303
  polynomials, 40, 45
  projectile, 40, 279, 292–293, 316, 324
  quadratic equations, 299–300
  quadratic functions, 279, 286–287, 292–293, 324
  radicals, 236–237, 249–250, 256, 261–262
  rational exponents, 244
  rational expressions, 57, 63, 68, 74–75, 112–115
  science, 31–32, 123, 177, 228, 236–237, 279, 300, 357, 377
  scientific notation, 26, 228
  sequences, 433
  series, 437–438
  slope, 150, 158–159
  solution format, 99
  systems of equations, 189–190, 197–198, 203, 206–213
  systems of inequalities, 218–219
  work, 113
  See also Mathematical modeling
Area, of geometric figures, 28
Arithmetic operations, 2
Arithmetic sequence, 440–444
  definition, 440
  grapher use with, 443–444
  nth term, 441
  sum of terms, 442–443
Arithmetic series, 442
  applied problems, 443
Associative property for addition, 9
Associative property for multiplication, 9
Asymptotes, hyperbola, 414
Augmented matrix, 386, 390
Axis of symmetry
  hyperbola, 414
  parabola, 320, 330

B

Base
  exponential notation, 3
  logarithmic, 350, 354, 356

Binomial coefficients, 455
  grapher use with, 457
Binomial expansion, 456
  applied problems, 457
Binomial theorem, 453–457
Binomials
  definition, 38
  multiplication of, 43–44
Boundary line, 164
Bounded region, 404
Braces, and order of operations, 4
Brackets, and order of operations, 4
Break-even point, in business, 197–198

C

Calculator, graphing. *See* Grapher technology
Calculators, inverse keys, 10
Car payment, 74–75
Cardano, Geronimo, 264, 269
Cartesian coordinate system, 77
  and complex numbers, 269
Cartesian plane, 77
Center
  of an ellipse, 410
  of a circle, 331, 332–333
  of a hyperbola, 413
Change of base formula, 356
Circle(s)
  applied problems, 333
  area and circumference, 28
  conic section, 409
  finding center, 332–333
  grapher use with, 333–334
  graphing, 331–333
  standard equation, 332
Circular cylinder, volume and surface area, 29
Circumference, of a circle, 28
Coefficient, 38
  leading, 38
Coefficient determinant, 395
  when value is zero, 396–397
Columns of a matrix, 386
Common denominators
  involving dimensions, 68
  rational expressions, 65
Common difference, 440
Common logarithms, 354–355
  graphing functions of, 364–365
  properties of, 369
Common ratio, 446
Commutative property for addition, 9
Commutative property for multiplication, 9

Completing the square, 282–287
  graphing parabolas by, 322–323
  solving quadratic equations by, 284–285
Complex numbers, 264–270
  absolute value of, 269
  conjugates, 267
  defined, 265
  equal, 266
  operations with, 266–268, 270
  as solutions to quadratic equations, 282, 285
  standard form, 265
Complex plane, 269
Complex rational expressions, simplifying, 70–73
Compound inequalities, 121–122
Compound interest, 31, 229, 244
  continuous, 343
Computer, graphing software. *See* Grapher technology
Conditional equations, 89
Conic section, 409
Conjugate axis, hyperbola, 414
Conjugates, 255
  complex, 267
Consistent system, 187–189, 197, 202–203, 397
Constant, 38
Conversion factors, 28
Coordinate axes, 77
Coordinate axis (number line), 7
Coordinate(s), 7, 77
  vertex of a parabola, 323
Counting numbers, 8
Cramer, Gabriel, 395
Cramer's rule
  applied problems, 400
  coefficient determinant, 395
  system of three linear equations, 399
  system of two linear equations, 395
Cross products
  in a proportion, 19
  in rational expressions, 56
  in solving equations, 111
Cube, volume and surface area, 29
Cube root, 233
Cut-point method, 128
  polynomial inequalities, 312
Cut-points, 312

D
Decimal form of a rational number, 8–9
Degree of a nonzero term, 38
Degree of a polynomial, 38
Demand equation, 189
Dependent equations, 187–189, 197, 202–203, 397
Dependent variable, 173
Descartes, René, 264
Determinant(s)
  definition (2x2), 395
  definition (3x3), 398

evaluating, 395, 398
  expansion formula, 398
  grapher evaluating, 401
Difference of two cubes, 45
  factoring, 51
Difference of two numbers, 14
Difference of two squares, 45
  factoring, 51
Dimensional analysis, 27–28
Dimensions, 27
  and common denominators, 68
Discriminant, quadratic formula, 292
Distance between two points, 235–236
Distributive property, 9, 47
  combining polynomials, 39
  combining radicals, 252
  solving equations using, 88
Dividend, 61
Division
  of complex numbers, 267–268
  of integers, 15–16
  and order of operations, 4
  property of equations, 86
  of radical expressions, 254–255
  of rational expressions, 60–61
  of rational numbers, 20
Division of polynomials, 61–62
Divisor, 61
Domain of a function, 171, 175–176
Domain of a sequence, 430
Doubling time, 376

E
Echelon form, row-reduced, 389
Element of a set, 6
Elimination method
  solving a linear system by, 194–196, 200–201
  solving a nonlinear system by, 420
Ellipse
  applied problem, 416
  center, 410
  conic section, 409
  definition, 410
  determining an equation of, 412–413
  equation of, 411
  foci, 410
  grapher use with, 416–417
  graphing, 412
  major axis, 410
  minor axis, 410
  vertices, 410
Empty set, 7
Equality, properties of, 10
Equation(s)
  absolute value, 126–127, 130
  addition and subtraction property, 86
  circle, 332
  conditional, 89
  definition, 86
  dependent, 187–189, 197, 202–203
  ellipse, 411
  equivalent, 86
  exponential, 228–229

extracting square roots, 260–261
  first degree, 87
  hyperbola, 414
  identity, 89
  inconsistent, 89–90
  independent, 187–189, 202
  linear (*see* Linear equations)
  of lines, 155–159
  literal, 92–93
  multiplication and division property, 86
  with numerical fractions, 89
  power rule, 259
  quadratic in form, 298–300
  radical, 259–261, 297
  with rational exponents, 261, 299
  with rational expressions, 110–115, 296–297
  reducible to quadratic form, 296–301
  solution (root) of, 86
  symmetric property, 86
  systems (*see* Systems of equations)
Equivalent
  equations, 86
  inequalities, 119
  rational expressions, 55
  rational numbers, 18
  systems, 195, 200
Euler, Leonhard, 171, 264, 342
Expanded form of exponential notation, 3
Expanded form of the power of a binomial, 453
Expanding a determinant, 398
Exponential equations, 228–229
  with different bases, 374–375
  grapher use with, 231
  with special bases, 374
Exponential expressions, 226–227
Exponential form, 3, 350
Exponential functions, 229, 340–345
  applied problems, 230, 343–344
  with base e, 342
  comparing graphs, 341–343
  evaluating, 340, 342
  grapher use for, 231, 344–345
  graphing, 340–343
Exponential growth and decay, 344, 376
Exponential notation, 3
Exponential property of equality, 228, 351, 373
Exponents, 3
  negative, 72–73, 226
  positive integer, 23–26
  properties of, 23–25, 227
  rational, 241–245
  using, 3
  zero, 72, 226
Expression(s)
  algebraic, 4
  evaluating, 5
Extraneous solution, 111
  equations quadratic in form, 298–299
  radical equations, 259

F

Factor of an expression, 3
Factored form, 3
Factorial notation, 454–455
Factoring
    applied problems involving, 52
    by combining methods, 51–52
    by grouping, 48
    by trial and error, 50
    common factors, 47
    completely, 47
    definition, 47
    difference of two cubes, 51
    difference of two squares, 51
    perfect square trinomial, 51
    polynomials, 47–52
    sum of two cubes, 51
    trinomials, 48–50
Factors, 47
    prime, 47
Feasible region (solution), 403
Fibonacci, Leonardo, 433
Fibonacci sequence, 433
Finite
    sequence, 430
    series, 435
    set, 7
First degree equations, 87
Focal points (foci) of a hyperbola, 413
Foci (focus) of an ellipse, 410
FOIL method, 44
Formulas
    applications of, 27–32
    business, 30–31, 229
    containing rational expressions, 74
    definition, 27
    distance between two points, 236
    geometric (*see* Geometric formulas)
    as literal equations, 92
    mathematical modeling using,
        94–95
    science, 31–32
    solving, 94
    using in applied problems, 94–95
Fraction bars, and order of operations,
    4
Fractions
    decimal form, 8–9
    See also Rational expressions
Function(s), 171–178
    definition, 171
    evaluating, 172–173
    exponential, 229
    graphing, 174–175
    inverses of, 348–349
    logarithmic, 349–350, 354–357
    notation, 172
    objective, 404
    quadratic (*see* Quadratic functions)
    zeros of, 278
Fundamental principle of fractions,
    55
Fundamental principle of rational numbers,
    18, 21

G

Gauss, Carl Friedrich, 264, 269, 437
General term
    of a sequence, 430, 431–432
    of a series, 435
Geometric formulas
    area, 28
    perimeter, 28
    surface area, 29
    volume, 29
Geometric sequence
    common ratio, 446
    definition, 446
    finding terms of, 447
    grapher use with, 450
    sum of an infinite, 449
    sum of terms, 447–448
Geometric series, 447
    applied problems, 449–450
    infinite, 448–449
Grapher technology
    absolute value, 130–131
    absolute value function, 178
    applied problems, 106, 140, 190,
        197–198, 213, 324–326, 344–345,
        406–407
    arithmetic sequence, 443–444
    binomial coefficients, 457
    circles, 333–334
    determinants, 401
    ellipse, 416–417
    exponential functions, 231, 344–345
    geometric sequences, 450
    inequalities, 123–124, 168
    INTERSECTION feature, 262
    linear equations, 79–80, 90, 95–96,
        150–151
    linear programming, 406–407
    linear regression model, 159–160
    logarithms, 352, 357–358, 365–366,
        371–372, 378
    matrices, 392–393
    nonlinear systems of equations,
        422–423
    polynomials, 300–301
    quadratic equations, 279–280, 293–294
    quadratic functions, 287, 293–294,
        324–326
    radicals, 237–238, 250, 256, 262,
        300–301
    rational equations, 115
    rational exponents, 244–245
    REGRESSION feature, 160
    ROOT feature, 262
    sequences, 433
    SHADE feature, 168, 219
    SIMULT feature, 204, 213
    summation formulas, 438
    systems of equations, 190, 197–198,
        204, 213, 392–393
    systems of inequalities, 219–220
    TRACE feature, 79, 140, 178
    window settings, 79
    ZOOM feature, 79, 178

Graphing calculator. *See* Grapher technology
Graph(s), 77
    circles, 331–333
    exponential functions, 230–231,
        340–343
    functions, 174–175
    linear equations, 78–80, 136–140
    linear inequalities, 118–119
    logarithmic functions, 360–366
    number line, 7
    parabolas, 318–323, 329–331
    plotting points in plane, 77–78
    quadratic functions, 278–279, 318–323
    and slope, 142–149
    solving inequalities, 313–314
    square-root function, 237
    systems of equations, 186–190, 197
    systems of inequalities, 216–220
Greatest common factor (GCF), 47
Grouping symbols, and order of operations,
    4
Growth and decay, exponential, 344

H

Half life, 376–377
Half planes, 164
Horizontal change (run), 143
Horizontal lines
    equation of, 156
    graphing, 138–139
    slope of, 144–146
How to Solve It, 99
Huygens, Christian, 237
Hyperbola
    asymptotes, 414
    axis of symmetry, 414
    center, 413
    conic section, 409
    conjugate axis, 414
    definition, 413
    determining an equation of, 416
    focal points (foci), 413
    graphing, 415
    standard equation, 414
    transverse axis, 414
Hypotenuse, of a right triangle, 30

I

Identity
    for addition, 9
    for multiplication, 9
Identity equation, 89
Imaginary axis, 269
Imaginary numbers, pure, 265
Imaginary part of a complex number, 265
Imaginary unit, 264
Inconsistent equations, 89–90
Inconsistent system, 187–189, 196, 197,
    202, 396
Independent equations, 187–189, 202
Independent variable, 173
Index of a radical, 234
Index of a summation, 435

Inequalities
absolute value, 127–129
applied problems, 122–123, 129–130,
168, 218–219, 315–316
equivalent, 119
grapher use in solving, 316
graphical solution, 313–314
graphs of, 118–119
linear (*see* Linear inequalities)
nonlinear, 311–316
nonstrict, 118
polynomial, 312, 314
properties of, 119
quadratic, 311–314
rational, 314–315
satisfying, 118
sides (members) of, 118
solution set of, 119
strict, 118
symbols (signs) for, 117–118
Infinite
sequence, 430
series, 435
set, 7
Infinity, positive and negative notation, 118
Input values (domain), 171
Integer exponents, properties of, 227
Integers
addition of, 13–14
definition, 8
division of, 15–16
multiplication of, 14–15
subtraction of, 14
Intercepts, 78
using to graph equations, 137–138
Interest
compound, 31
computing, 30–31
simple, 103
Interval notation, 118
Inverse keys, calculators, 10
Inverses of functions, 348–349
Irrational numbers, 9

L
Leading coefficient, 38
Least common denominator (LCD), 21, 66
solving equations using, 89, 110–111
table for constructing, 66–67
Legs of a right triangle, 30
Leibniz, Gottfried, 171
Like radicals, 252
Like terms, 38
Linear equations, 78
examples of, 87
grapher used in, 79–80, 90, 95–96
graphing, 78–80, 136–140
in one variable, 86–90
slope-intercept form, 147–148
standard form, 136
strategy for solving, 87
systems of (*see* Linear systems)
in two variables, 136
using a grapher to solve, 90

Linear functions, 174
Linear inequalities, 117–124
applied problems, 122–123
compound, 121–122, 166–167
grapher use in solving, 123–124
graphing, 118–119, 164–168
solving, 119–124
systems of, 216–220
in two variables, 164–166
See also Inequalities
Linear programming, 403–407
applied problems, 404–406
grapher use in, 406–407
model for, 404
Linear regression models, 159
Linear system of equations in four variables,
solving using grapher, 392–393
Linear systems of equations in three variables, 200–204
applied problems, 210–212
equivalent, 200
grapher use in solving, 204
solving by Cramer's rule, 399
solving by elimination, 200–201
solving by matrices, 390–391
Linear systems of equations in two variables, 186
algebraic solution, 193–198
applied problems, 206–209
consistent, 187–189, 397
equivalent, 195
grapher use in solving, 190, 197–198,
213
graphical solution, 186–190
inconsistent, 187–189, 396
in matrix notation, 386
solving by Cramer's rule, 395–396
solving by elimination, 194–196
solving by matrices, 387–388, 390
solving by substitution, 193–194
triangular form, 388
Literal equations, 92–93
Logarithmic equations, 351, 375–376
applied problems, 376–377
Logarithmic form, 350
Logarithmic function(s)
applied problems, 352, 357, 365–366,
371
common, 354
definition, 350
evaluating, 354–358
grapher use with, 352, 357–358,
365–366, 371–372
graphing, 360–366
inverse of, 350, 356
natural, 354
to any base, 356–357
Logarithmic notation, 350
Logarithmic property of equality, 374
Logarithms
combining expressions, 370
common, 354
determining values of, 351
evaluating, 354–358

natural, 354
properties of, 368–372
Long division of polynomials, 61–62
Lower limit of a summation, 435
Lowest terms
of a rational expression, 55
of a rational number, 18

M
Major axis of an ellipse, 410
Market equilibrium point, 189
Mathematical modeling, 94
with formulas, 94–95
with the Pythagorean Theorem, 304–305
with rational expresssions, 305–307
with systems of equations, 206–213
See also Applied problems
Matrices (matrix)
applied problems, 392
augmented, 386, 390
columns, 386
elements (entries) of, 386
grapher use with, 392–393
notation, 386
row operations, 387
row-reduced echelon form, 389–390
rows, 386
size or dimensions of, 386
solving systems using, 387–388
triangular form, 388
writing linear systems using, 386
Maximum and minimum values
in linear programming, 404
parabolas, 324
using a grapher, 324–326
Minor axis of an ellipse, 410
Minuit, Peter, 344
Mixture problems, 104
system of equations, 208
Modeling. *See* Mathematical modeling
Monomials
definition, 38
multiplication of, 42
Motion problems, 105, 114–115
quadratic equations, 307
system of equations, 209
Multiplication
of binomials, 43–44
of complex numbers, 267
of integers, 14–15
of monomials, 42
and order of operations, 4
of polynomials, 42–45
property of equations, 86
of radical expressions, 253
of rational expressions, 59–61
of rational numbers, 19–20
square roots of negative numbers, 265
Multiplicative inverse, 9, 16

N
Natural logarithms, 354, 355–356
graphing functions of, 364
properties of, 369

Natural numbers, 8
Negative exponent, definition, 72–73
Negative of a number, 7–8
Negative numbers, 7
Negative slope, 145–146
Newton, Sir Isaac, 292
Newton's Law of Cooling, 345
Nonlinear inequalities, 311–316
Nonlinear systems
    of equations, 419–420, 422–423
    of inequalities, 420–421
nth root
    definition, 233
    finding, 233–235
    principal, 234
nth term of a sequence, 430
Null set, 7
Number line, 7
    addition and subtraction on, 13
Number problems, 100, 112, 269, 303

O
Objective function, 404
    finding maximum and minimum of,
        404–407
Opposite of a number, 7–8
Order of operations, 4
    and integers, 16
    and rational expressions, 61
Ordered pair, 77, 136, 164
Ordered triple, 200
Ordinate, 77
Origin, 7, 77
Output values (range), 171

P
Parabola(s), 318
    axis of symmetry, 320, 330
    completing the square, 322–323
    conic section, 409
    graphing, 318–323, 329–331
    horizontal, 329–331
    vertex of, 320, 323, 324, 330
Parallel lines
    equation of, 156–157
    and slope, 148–149, 156–157
Parentheses, and order of operations, 4
Pascal, Blaise, 454
Pascal's triangle, 454
Pendulum, length of, 237–238
Percent, definition, 18
Perfect cubes, 234
Perfect square trinomial
    creating, 283
    factoring, 51
Perfect squares, 234
Perimeter, of geometric figures, 28
Perpendicular lines
    equation of, 156–157
    and slope, 148–149, 156–157
Plotting points in the plane, 77–78
Point plotting method, 78, 136–137
    quadratic functions, 319
Point-slope equation, 155–156

Polya, George, 99
Polynomial, 312, 314
Polynomial equations, solving by factoring,
    278
Polynomial inequalities, 312, 314
Polynomials
    applied problems involving, 40–41, 45
    combining, 39–40
    definition, 38
    evaluating, 39
    factoring, 47–52
    graphing, 300–301
    long division of, 61–62
    multiplication of, 42–45
    special products, 44–45
Positive integers, 8
Positive numbers, 7
Positive slope, 145–146
Power rule for equations, 259
Power(s), 3
    and order of operations, 4
    properties for exponents, 23–24
Powers of i, 268–269
Prime factors, 47
Prime number, 21
Principal nth root, 234
Principal square root, 265
Problem constraints, 404
Problem solving process, 99
Product
    of integers, 14–15
    property of exponents, 23–24
    of rational numbers, 19–20
Projectile problems, 40, 279, 292–293, 316
Proportion, definition, 19
Pure imaginary numbers, 265
Pythagorean Theorem, 30
    applied problem, 262
    converse of, 30, 236
    distance formula, 235–236
    models using, 304–305

Q
Quadrants, 77
Quadratic equations, 276
    applied problems and modeling, 279,
        292–293, 299–300, 303–307
    complex solutions, 282, 285
    factoring method of solving, 276–277
    grapher use with, 279–280, 293–294
    solving by completing the square,
        284–285
    standard (general) form, 276, 290
Quadratic form, equations in, 298–300
Quadratic formula, 290
    applied problems, 292–293
    derivation, 290
    discriminant, 292
    solving equations using, 290–291
Quadratic functions, 276, 278, 318–326
    analyzing graphs of, 278–279, 286
    applied problems, 279, 286–287,
        292–293, 324
    grapher use with, 287, 293–294, 324–326

graphing, 318–323
    point plotting method, 319
    zeros, of, 278
Quadratic inequalities, 311–314
Quotient, 15, 61
    property for exponents, 25

R
Radical equations, 259–261
    grapher use with, 262, 301
    strategy for solving, 259
    that lead to quadratics, 297
Radicals, 233–238
    applied problems, 236–237, 249–250, 256
    converting to exponential form, 242, 249
    evaluating, 235
    grapher used with, 250, 256, 301
    like, 252
    nth root, 233
    operations with, 252–256
    properties of, 247
    simplest form, 248
    simplifying, 235, 247–250
Radicand, 234
Radius of a circle, 331, 332
    finding, 332–333
Range of a function, 171, 175–176
Ratio, definition, 18
Rational exponents, 241–245
    applied problems, 244
    converting to radical form, 242, 249
    definition, 241
    equations with, 261, 299
    evaluating expressions having, 242
    grapher use with, 244–245
    properties of, 242–243
    simplified form, 248
Rational expressions
    addition and subtraction of, 65–67
    applied problems, 57, 63, 68, 74–75
    building to higher terms, 56
    complex, 70–73
    definition, 54
    division of, 60–61
    equations involving, 110–115, 296–297
    equivalent, 55
    evaluating, 54
    grapher use with, 115
    models involving, 305–307
    multiplication of, 59–61
    reducing, 55
    signs of, 56–57
    undefined, 54–55
Rational inequalities, 314–315
Rational numbers
    addition of, 20–22
    decimal form, 8
    definition, 8
    division of, 20
    equivalent, 18
    multiplication of, 19–20
    reducing and simplifying, 17–18
    subtraction of, 20–22
Rationalizing the denominator, 254–255

Rationalizing factor, 255
Real axis, 269
Real numbers, 9
    properties of, 9–10
Real part of a complex number, 265
Reciprocals, 16
Rectangle, area and perimeter, 28
Rectangular coordinate system, 77
Rectangular solid, volume and surface area,
    29
Recursion formula, 431
Reducing
    rational expressions, 55
    rational numbers, 17–18
    to lowest terms, 55
Reflexive property of equality, 10
Regression model, 159–160
Remainder, 61
Repeating decimal, 8
Resistance of an electrical circuit, 74
Richter, Charles F., 357
Richter Scale, 357
Rise, 143
Root (solution) of an equation, 86
Root(s), finding the nth, 233–235
Row-reduced echelon form, matrix,
    389–390
Rows of a matrix, 386, 387
Run, 143

S
Salvage value, 158
Scale, 7
Scatter plot/diagram, 159–160
Scientific notation, 26
Sequence(s)
    arithmetic, 440–444
    defined recursively, 431
    definition, 430
    domain, 430
    Fibonacci, 433
    finite, 430
    general term (nth term), 430, 431–432
    geometric, 446–451
    grapher use with, 433
    graphing, 432
    infinite, 430
    terms of, 430
Series, 435
    applied problems, 437–438
    arithmetic, 442
    geometric, 447
    writing using summation notation, 436
Set notation, 6–7
Set-builder notation, 7
Sigma notation, 435
Similar terms, 38
Simplifying
    rational expressions, 55
    rational numbers, 17–18
Slope of a line, 142–151
    applied problems, 150
    definition, 143

Slope-intercept form, 147–148, 155
Solution set, 119, 164
Solution(s)
    of an equation, 86, 136
    extraneous, 111
    linear inequalities, 119, 164
    quadratic inequalities, 311
    systems of equations, 186, 200
Special products of polynomials, 44–45
Sphere, volume and surface area, 29
Square, area and perimeter, 28
Square root, 233
    of negative numbers, 265
    principal, 265
Square root function, 237
Square root property, solving equations us-
    ing, 260–261, 282
Squaring a binomial, 45
Straight line depreciation, 158
Subset, 7
Substitution method
    solving a linear system by, 193–194
    solving a nonlinear system by, 419
Substitution property of equality, 10
Subtraction
    of complex numbers, 266
    of integers, 14
    and order of operations, 4
    of polynomials, 39–40
    property of equations, 86
    of radical expressions, 252–253
    of rational expressions, 65–67
    of rational numbers, 20–22
Sum
    of an arithmetic sequence, 442–443
    of an infinite geometric sequence,
      448–449
    of a geometric sequence, 448–449
    of terms of a sequence, 435–436
    of two cubes, 45, 51
    See also Addition
Summation notation, 435–436
    grapher use with, 438
    properties, 437
    writing a series using, 436
Supply and demand equations, 189–190
Surface area, 29
    applied problems, 250
Symbols, translating into words, 2–3
Symmetric property of equality, 10, 86
Systems of equations
    consistent, 187–189, 196–197, 202–203
    equivalent, 195, 200
    inconsistent, 187–189, 196, 197, 202
    linear (*see* Linear systems of equations)
    modeling using, 206–213
    nonlinear, 419–420
Systems of linear inequalities, 216–220

T
Temperature conversion, 32, 95, 159
    using a grapher, 140
Terminating decimal, 8

Terms, definition, 4
Terms of a sequence, 430
Test point, inequalities, 164
Tolerances, 129–130
Transitive property of equality, 10
Translating words into symbols and vice
    versa, 2–3, 5, 100–101
Transverse axis, hyperbola, 414
Trial and error factoring, 50
Triangle
    area and perimeter, 28
    right, 30
Triangular form, 388
Trinomials
    definition, 38
    factoring, 48–50
    perfect square, 283

U
Undefined rational expressions, 54–55
Unit length, 7
Upper limit of a summation, 435

V
Value of an expression, 3
Variable, definition, 2
Vertex of a parabola, 320, 330
    coordinates of, 323
Vertical change (rise), 143
Vertical line test, 175
Vertical lines
    equation of, 156
    graphing, 138–139
    slope of, 144–146
Vertices of an ellipse, 410
Vertices of a feasible region, 403
Volume, of geometric figures, 29

W
Wessel, C., 269
Words, translating into symbols, 2–3, 5,
    100–101, 122
Work problems, 113

X
x axis, 77
x coordinate, 77
x intercept, 78, 137
xy plane, 77

Y
y axis, 77
y coordinate, 77
y intercept, 78, 137
Yûnis, Ibn, 237

Z
Zero, divison involving, 2
Zero exponent, definition, 72
Zero factor property, 276
Zero polynomial, 38
Zero slope, 145, 146
Zeros of a function, 278